T0211346

Alternative Reproductive Tactics
An Integrative Approach

Consistent variation in the reproductive behavior of males and females within a species is an evolutionary puzzle. How two forms of male can develop in one species, for example, and how such variation can be maintained in one population at the same time, offers a special opportunity to study the evolution and functional causes of phenotypic variation, which is a general problem in the field of evolutionary biology. By integrating both proximate (physiological) and ultimate (evolutionary) perspectives and by covering a great diversity of species, *Alternative Reproductive Tactics* addresses this exciting topic of longstanding interest, bringing together a multitude of information in an accessible form that is ideal for graduate students and researchers in evolutionary biology, behavior, and reproductive physiology.

RUI F. OLIVEIRA is Professor of Psychobiology, Chair of the Psychobiology Graduate Program, and Head of the Research Centre at the Instituto Superior de Psicologia Aplicada, Portugal. His research focuses on the neuroendocrine mechanisms of behavioral plasticity in vertebrates, particularly in fish. He is the current President of the Portuguese Ethological Society, Chief-Editor of the *Acta Ethologica* journal and Consulting Editor of the *Hormones and Behavior* journal. He also served on the Council of the Association for the Study of Animal Behaviour.

MICHAEL TABORSKY is Professor of Zoology at the University of Bern, Switzerland, and Head of the Department of Behavioural Ecology. His research focus is on evolutionary mechanisms of cooperative behavior and advanced sociality, and on the ultimate causes of alternative reproductive tactics. He is Editor-in-Chief of *Ethology* and has been President of the Ethologische Gesellschaft and Secretary General of the International Council of Ethologists.

H. JANE BROCKMANN is Professor of Zoology at the University of Florida. Her research focuses on animal behavior and sexual selection in insects and other invertabrates. She has been President of the Animal Behavior Society, Secretary General of the International Council of Ethologists, Editor of the journal *Ethology*, and she is presently Editor-in-Chief of *Advances in the Study of Behavior*.

Alternative Reproductive Tactics
An Integrative Approach

Edited by
Rui F. Oliveira
Istituto Superior de Psicologia Aplicada
Michael Taborsky
University of Bern
H. Jane Brockmann
University of Florida

CAMBRIDGE
UNIVERSITY PRESS

Shaftesbury Road, Cambridge CB2 8EA, United Kingdom

One Liberty Plaza, 20th Floor, New York, NY 10006, USA

477 Williamstown Road, Port Melbourne, VIC 3207, Australia

314–321, 3rd Floor, Plot 3, Splendor Forum, Jasola District Centre, New Delhi – 110025, India

103 Penang Road, #05–06/07, Visioncrest Commercial, Singapore 238467

Cambridge University Press is part of Cambridge University Press & Assessment,
a department of the University of Cambridge.

We share the University's mission to contribute to society through the pursuit of
education, learning and research at the highest international levels of excellence.

www.cambridge.org
Information on this title: www.cambridge.org/9780521540063

First published 2008

A catalogue record for this publication is available from the British Library

Library of Congress Cataloging-in-Publication data
Oliveira, Rui F.
    Alternative reproductive tactics : an integrative approach / Rui F. Oliveira, Michael Taborsky, H. Jane Brockmann.
        p.   cm.
    Includes bibliographical references.
    ISBN 978-0-521-83243-4 (hardback) – ISBN 978-0-521-54006-3 (pbk.)
    1.   Sexual behavior in animals. I.   Taborsky, Michael. II.   Brockmann, H. Jane. III.   Title.

    QL761.O45 2008
    591.56′2–dc22
    2008000262

ISBN   978-0-521-83243-4   Hardback
ISBN   978-0-521-54006-3   Paperback

Contents

vi   Contents

Contributors

VÍTOR C. ALMADA
Unidade de Investigação em Eco-Etologia
Instituto Superior de Psicologia Aplicada
Rua Jardim do Tabaco 34
1149–041 Lisboa
Portugal

SUZANNE H. ALONZO
427 Osborn Memorial Labs
Yale University
PO Box 208106
New Haven, CT 06520
USA

ANDREW H. BASS
Department of Neurobiology and Behavior
Cornell University
W233 Seeley G. Mudd Hall
Ithaca, NY 14853
USA

H. JANE BROCKMANN
Department of Zoology
University of Florida
Gainesville, FL 32611
USA

RYAN CALSBEEK
Center for Tropical Research
Institute of the Environment
University of California at Los Angeles
Los Angeles, CA 90095
USA

ADELINO V. M. CANÁRIO
Centro de Ciências do Mar
Universidade do Algarve
Campus de Gambelas
800–117 Faro
Portugal

LAUREN M. CHAN
Department of Ecology and Evolutionary Biology
Cornell University
A406a Corson Hall
Ithaca, NY 14853
USA

JANIS L. DICKINSON
Laboratory of Ornithology
Cornell University
159 Sapsucker Rd
Ithaca, NY 14850
USA

DOUGLAS J. EMLEN
Division of Biological Sciences
The University of Montana
Missoula, MT 59812
USA

PAUL M. FORLANO
Department of Neurobiology and Behavior
Cornell University
W233 Seeley G. Mudd Hall
Ithaca, NY 14853
USA

DAVID M. GONÇALVES
Unidade de Investigação em Eco-Etologia
Instituto Superior de Psicologia Aplicada
Rua Jardim do Tabaco 34
1149–041 Lisboa
Portugal

RICHARD D. HOWARD
Department of Biological Sciences
Purdue University
West Lafayette, IN 47907
USA

WALTER D. KOENIG
Department of Neurobiology and Behavior
Cornell University
W233 Seeley G. Mudd Hall
Ithaca, NY 14853
USA

OLIVER KRÜGER
Department of Zoology
University of Cambridge
Downing Street
Cambridge CB2 3EJ
United Kingdom

JEFFREY R. LUCAS
Department of Biological Sciences
Purdue University
West Lafayette, IN 47907
USA

PETER K. MCGREGOR
Department of Animal Behaviour
Zoological Institute
Copenhagen University
Tagensvej 16
DK 2200, Copenhagen N
Denmark

BRYAN NEFF
Department of Biology
University of Western Ontario
London, Ontario N6A 5B7
Canada

RUI F. OLIVEIRA
Unidade de Investigação em Eco-Etologia
Instituto Superior de Psicologia Aplicada
Rua Jardim do Tabaco 34
1149–041 Lisboa
Portugal

JOANA I. ROBALO
Unidade de Investigação em Eco-Etologia
Instituto Superior de Psicologia Aplicada
Rua Jardim do Tabaco 34
1149–041 Lisboa
Portugal

ALBERT F. H. ROS
Unidade de Investigação em Eco-Etologia
Instituto Superior de Psicologia Aplicada
Rua Jardim do Tabaco 34
1149–041 Lisboa
Portugal

JOANNA M. SETCHELL
School of Life Sciences
University of Surrey Roehampton
Whitelands College, West Hill
London SW15 3SN
United Kingdom

STEPHEN M. SHUSTER
Department of Biological Sciences
Northern Arizona University
Flagstaff, AZ 86011
USA

BARRY SINERVO
Earth and Marine Sciences A308
University of California
Santa Cruz, CA 95064
USA

MICHAEL TABORSKY
Division of Behavioural Ecology
Zoological Institute
University of Bern
Wohlenstrasse 50A
CH-3032 Hinterkappelen
Switzerland

JERRY O. WOLFF
Department of Biology
University of Memphis
Ellington Hall Room 103
Campus Box 526080
Memphis, TN 38152
USA

KELLY R. ZAMUDIO
Department of Ecology and Evolutionary Biology
Cornell University
E209 Corson Hall
Ithaca, NY 14853
USA

Preface

The study of alternative reproductive tactics (ARTs) is a hot topic in evolutionary and behavioral ecology. ART refers to consistent variation in the reproductive behavior (involving, e.g., mating, nesting, fighting) of males or females within one population. This variation offers a special opportunity to study the evolution and functional causes of phenotypic variation, a general problem in evolutionary biology. A large body of published data exists on ARTs, but there has been no conceptual unification of the available information, nor any strong effort to integrate it into a general framework. Apart from a recent book by S. M. Shuster and M. J. Wade (2003) (*Mating Systems and Strategies*, Princeton, NJ: Princeton University Press) that addresses the topic of ARTs within the larger scope of mating systems, there has been no major publication covering this topic. Moreover, the few reviews available in the literature are taxon specific and do not fully integrate the proximate and ultimate levels of analysis in understanding ARTs. Clearly, integration of data, concepts, and analysis levels is overdue. In trying to meet this challenge, Rui Oliveira joined forces with Michael Taborsky and Jane Brockmann, who were among the contributors to a symposium on ARTs at the 27th International Ethological Conference in Tübingen. The three of us have complementary connections to active researchers in the field and all three have been studying ARTs from very different perspectives. Following the Tübingen conference we started a fruitful discussion in order to establish the plan of a book, identifying the areas to be covered and who would be the most appropriate to write about each of quite a few essential topics. We decided to use an integrative approach to the field inviting people both from the area of ultimate causes (evolutionary and behavioral ecology) and from the field of proximate mechanisms (behavioral physiology). We asked authors to write a chapter on a specific issue selected by us and not to write short reviews about their own work, so that this book can be more than the sum of its parts. It is with great pleasure that we thank all the authors for investing a lot of time and effort in writing their chapters to meet these requirements, and for their generous patience with successive delays in finalizing the book.

Since ARTs can be viewed as a model system for studying the evolution of variation, which is one of the central questions in evolutionary biology, the potential audience for this book is very broad, including readers interested in animal behavior, life histories, phenotypic plasticity, biological game theory, evolutionary theory, and ecology in general. Thus, although this is not a textbook we hope it may become a reference for postgraduate courses in the above-mentioned areas.

We have organized the book in four parts, together with an opening general introduction and a final concluding chapter.

In the opening chapter we try to clarify concepts and address the levels at which questions about the evolution of ARTs should be asked. This is important because alternative hypotheses often turn out to be simply a matter of asking questions at different levels. Therefore, in this chapter we also attempt to provide a framework, language, and theoretical basis for studying ARTs. It is important to note, though, that we did not impose our framework on the other authors of this volume. In accordance with the topic, which deals with the intriguing wealth of biological variation between individuals of a population, we intended to allow different concepts and frameworks within the pages of this book. As with ARTs existing in a population, only time will show which concepts will finally persist.

Part I of the book summarizes the study of ultimate causes and origins of ARTs. It opens with a chapter on the evolution, life histories, and adaptiveness of ARTs viewed within the scope of alternative allocation phenotypes (Brockmann and Taborsky). It is followed by a chapter in which the use of comparative methods is proposed as a tool with a large potential to reveal phylogenetic patterns of ARTs in the study of their evolution (Almada and Robalo). This part ends with a chapter where dynamic game modeling is applied to the study of ARTs (Lucas and Howard).

Part II summarizes our current knowledge of the proximate mechanisms of ARTs. It starts with a chapter on the interaction between genetic and environmental factors on the development of ARTs (Emlen), which is followed by two chapters on neural (Bass and Forlano) and endocrine mechanisms underlying ARTs (Oliveira, Canário, Ros).

Part III is a compilation of taxonomic reviews with the goal to provide an overview of the occurrence of ARTs in the animal kingdom. This part covers most animal taxa for which ARTs have been described, namely insects (Brockmann), crustaceans (Shuster), fish (Taborsky), amphibians (Zamudio and Chan), reptiles (Calsbeek and Sinervo), birds (Krüger), nonprimate mammals (Wolff), and primates (Setchell). The few examples of ARTs in invertebrates not covered by these taxonomic reviews were addressed where appropriate in text boxes of other chapters.

Part IV is a compilation of chapters on emerging perspectives on ARTs, such as the role of animal communication in the evolution of ARTs (Gonçalves, Oliveira, and McGregor), the relationship between ARTs and mate choice for good genes or good care (Neff), a co-evolutionary approach to sexual conflict and ARTs within a sex (Alonzo), and the viewing of cooperative breeding as an ART (Koenig and Dickinson).

In the final chapter the editors reflect on what emerges from all the contributions to this book as the current status of the study of ARTs, and the prospects for future research in this field. It is a summarizing chapter attempting to pull all the topics together, pointing to major questions and suggesting the importance of studying the evolution of ARTs integratively.

All chapters were reviewed by the three editors and by external reviewers. We are very pleased that the authors made an extraordinary effort in considering the editors' and reviewers' comments in revising and amending their contributions. A large body of external reviewers made a significant contribution to the contents of this book and we would like to express our gratitude to their extensive criticisms and suggestions about each manuscript.

Apart from the dedicated work of authors and reviewers, this book would not have been possible in its current format without the contribution and support from various people and institutions. We are grateful to Desmond Morris who kindly permitted us to use one of his paintings for the cover of this book. Desmond Morris was one of the earliest ethologists to describe ARTs (in sticklebacks: Morris, D. 1954. The causation of pseudofemale and pseudomale behaviour: a further comment. *Behaviour* 7, 46–56) so we felt his artwork was particularly appropriate for this book. The Mayor Gallery (London) kindly provided technical support for the reproduction of this painting. The editors at Cambridge University Press, Martin Griffiths, Clare Georgy, and Shana Coates, were both patient and supportive and we thank them for their efforts.

Grace Kiltie worked directly with HJB on the formatting and copy-editing of all chapters. She also compiled the remissive index for the book and the figure and table lists. Her commitment and enthusiasm were decisive for the success of this enterprise. Joana Jordão also provided valuable help during the final phase of proofreading. RFO is supported by the Portuguese Foundation for Science and Technology (FCT; Pluriannual programme: R and D unit 331/2001) and the editing of this book was supported partially by a FCT research grant. ISPA and RFO generously hosted MT and HJB in Portugal during September 2004. MT was supported by SNF grant 3100A0-105626. HJB was supported by the National Science Foundation, Florida Foundation and Department of Zoology, University of Florida. Finally, Catarina and João Oliveira, Xana Lopes, Barbara Taborsky, and Tom Rider merit special thanks. They unconditionally supported us along the long and winded ways of compiling this book.

1 · The evolution of alternative reproductive tactics: concepts and questions

MICHAEL TABORSKY, RUI F. OLIVEIRA, AND H. JANE BROCKMANN

CHAPTER SUMMARY

Here we outline the meaning of the term alternative reproductive tactics, or ARTs, and discuss why the existence of ARTs is so widespread in animals. We ask what we need to know to understand the evolution of ARTs and the importance of general principles such as frequency dependence, density dependence, and condition dependence, and what we need to know about proximate mechanisms involved in the regulation of ARTs to comprehend evolutionary patterns. We discuss current issues in the study of ARTs and list 12 questions that we think need particular attention. Throughout we shall provide representative examples of ARTs in animals to illustrate the ubiquitous nature of this phenomenon.

1.1 WHAT IS THE MEANING OF *ALTERNATIVE REPRODUCTIVE TACTIC*?

1.1.1 Alternative

The concept of ARTs refers to *alternative* ways to obtain fertilizations in both males and females. In its most common use, this term refers to traits selected to maximize fitness in two or more alternative ways in the context of intraspecific and intrasexual reproductive competition. In general, alternative phenotypes are characterized by a discontinuous distribution of traits evolved towards the same functional end. Examples include size dimorphism, color polymorphism, dimorphic morphological structures involved in the monopolization of resources or mates, and various behavioral alternatives such as territoriality vs. floating, monopolization vs. scramble competition, or investment in primary access to a resource vs. social parasitism. Individuals allocate resources to either one or the other (mutually exclusive) way of achieving the same functional end using evolved decision-making rules (Brockmann 2001).

It is important to note here that in the study of allocation decisions in general, and ARTs in particular, any expression of *continuous* variation of traits is not regarded as *alternative* tactics. Discontinuity in morphological and physiological traits is often difficult to determine (Eberhard and Gutiérrez 1991, Emlen 1996, Kotiaho and Tomkins 2001). In behavioral traits, in contrast, discontinuities may seem easier to measure because of their visibility to observers. For example, there may be overlap between male types of dung beetles in their expression of horns and body sizes, but it is very clear-cut whether these male types fight for access to females or copulate without investing in primary access to mates (Kotiaho and Tomkins 2001; see also Hunt and Simmons 2000). However, subtle discontinuities might exist in any phenotype, including behavior (e.g., when the performance of alternative tactics depends on condition or situation: Brockmann and Penn 1992, Brockmann 2002). In a nutshell, in the context of ARTs, *alternative* refers to traits that show a *discontinuous* distribution.

1.1.2 Reproductive

We speak of alternative reproductive tactics when conspecific, intrasexual competitors find different solutions to *reproductive* competition. It is irrelevant whether the observed variation happens within or between individuals, but reproductive discontinuity within one population at the same time is of essence. In a general sense the concerned traits are alternative responses to competition from members of the same sex. Examples are males either courting females or forcing copulations, as in guppies and other poeciliid fishes (Bisazza 1993, Bisazza and Pilastro 1997), or females either digging burrows for their eggs or usurping those dug by others, as in digger wasps (Brockmann and Dawkins 1979, Brockmann *et al.* 1979). It is irrelevant whether adaptations to reproductive competition are mainly

Alternative Reproductive Tactics, ed. Rui F. Oliveira, Michael Taborsky, and H. Jane Brockmann. Published by Cambridge University Press.

Box 1.1   Examples of ARTs in animals

Reference to literature on ARTs in taxa mentioned here is given in the text of this chapter and in other chapters of this book.

Molluscs	Phallic and aphallic males
Horseshoe crabs	Males attached to females and satellites
Mites	Fighter and scrambler males
Crustacea	Mate guarding vs. searching in amphipods
	Three alternative male mating types in isopods
Insects	Calling and noncalling males in crickets
	Winged and wingless male morphs in bladder grasshoppers
	Single- and joint-nest foundresses in social wasps
	Color and horn polymorphisms in male damselflies and beetles
	Territorial vs. roaming males in dragonflies
Fishes	Bourgeois males and reproductive parasites in sunfish, salmonids, wrasses, cichlids, blennies, and gobies
	Bourgeois males and helpers or satellites in ocellated wrasse, cooperative cichlids, and anabantoids
	Courting and coercive males in poeciliids
Amphibians	Calling males and silent interlopers in frogs and toads
Reptiles	Differently colored males with different mating tactics in lizards
Birds	Courting males and satellites in lekking birds such as ruffs
	Pair and extra-pair matings in many monogamous species (e.g., red-winged blackbirds, blue tits)
	Single vs. joint courtship in manakins
	Nesting oneself or dumping eggs elsewhere (i.e., intraspecific brood parasitism) in many anatids
	Dominant breeders with helpers that share in reproduction in scrubwrens, *Campylorhynchus* wrens, and dunnocks
Mammals	Bourgeois males and satellites in ungulates such as waterbuck and kob
	Displaying/defending males and harassing interlopers in fallow deer
	Harem owners and opportunistic, submissive group males in many primates
	Flanged and unflanged males in orang-utans

What is *not* an ART?

Cooperative breeding, if helpers do not share in reproduction (*reproduction* is a necessary component of an *alternative reproductive tactic*)

Interspecific brood parasitism, as heterospecifics are not reproductive competitors

Sex change, even though in species with alternative tactics *within* one sex bourgeois and parasitic options in this phase may determine the threshold for the optimal timing of sex change (e.g., in wrasses with two or more male reproductive tactics: Munoz and Warner 2003)

Simultaneous hermaphroditism, as shedding sperm is not an *alternative* to shedding eggs among competitors for fertilizations

Infanticide, because it is not a *reproductive* tactic (i.e., to obtain fertilizations or produce offspring, even though it may indirectly contribute to this end)

Pure scramble competition for reproduction without discontinuous phenotypic variation

Alternative phenotypes in nonreproductive contexts (e.g., foraging or trophic polymorphisms such as left- and right-jawed fish, castes, and age polyethism in social insects when the different morphs do not engage in reproductive competition; polymorphisms that involve both males and females such as winged and wingless forms in some insects, alternative migratory patterns and diapause patterns; seasonal polyphenism that does not involve reproductive characters or individuals; and color polymorphisms caused by apostatic prey selection or other anti-predator strategies) or partly resulting from intrasexual, intersexual, or natural selection mechanisms. For example, the evolution of courting and sneaking tactics in a species may be subject to intrasexual rivalry, but it may also be influenced by mate choice (intersexual selection) and by the tactic-specific potential to evade predation (natural selection). Alternatively, there may be specialization of same-sex conspecifics in exploiting different reproductive niches. Irrespective of the underlying selection mechanisms, ultimately the existence of the two alternative tactics will be the expression of different solutions to reproductive competition. Interspecific brood parasitism, for example, is not an ART, because it is *not* the result of reproductive competition; neither are phenomena like infanticide, sex change, or age polyethism in social insects (see Box 1.1).

1.1.3 Tactic

In a general sense *tactic* refers to a trait or set of traits serving a particular function. In the context of ARTs, tactics usually involve behavioral traits, but the term is by no means restricted to behavioral phenotypes. For instance, various types of horns in a male population of horned beetles may be expressions of alternative reproductive tactics (Emlen 1997, Emlen and Nijhout 2000, Moczek and Emlen 2000); so are color morphs of some lizards (Sinervo and Lively 1996) and male genitalia in certain snails (Doums *et al.* 1998; see Box 1.1). Often, suites of behavioral, morphological, and physiological traits are associated in creating alternative phenotypes within a species (e.g., in plainfin midshipman fish: Bass and Andersen 1991, Bass 1992, 1996, Brantley *et al.* 1993, Brantley and Bass 1994).

We do not think that a distinction between "tactic" and "strategy" is useful here, because these two terms relate to the same issue, but at different levels. A distinction is often made in evolutionary game theory models (Maynard Smith 1982) where strategy relates to a particular life-history pattern or "genetically based program" (Gross 1996), and tactic classifies the application of rules that are part of a strategy (i.e., the phenotype: Shuster and Wade 2003). When analyzing empirical data, usually our potential for inference is limited to the level of phenotype, even if we are ultimately interested in the *evolution* of traits and hence in the effect on genotype frequencies. However, most often we lack information about underlying genotypes. For instance, we do not know whether different genotypes are involved at all or whether phenotypic traits are the expression of conditional variation produced by *exactly the same* genotype

(Shuster and Wade 2003). This may not be so bad in the end (see Grafen's [1991] discussion on "the phenotypic gambit"). The difference made between phenotypic traits produced by same or different genotypes has heuristic importance for (game theory) evolutionary models, but it ignores the fact that virtually all phenotypic traits are the product of genotypic *and* environmental influence (West-Eberhard 1989, 2003, Scheiner 1993). Hence, in reality the borders between the terms "strategy" and "tactic" are vague and flexible. The underlying mechanisms are usually unknown (i.e., to which extent patterns are genetically determined) at a point when we have not yet studied a phenomenon extensively but nonetheless wish to communicate about it. Therefore, we prefer an operational use of terms here instead of one encumbered with functional implications, just as in the sex-allocation literature (Charnov 1982; see Brockmann 2001). In short, we regard "tactic" and "strategy" as synonymous but prefer the use of "tactic" because we mainly deal with phenotypes and because of the connotations of the term strategy.

In essence, "alternative reproductive tactics" refers to discontinuous behavioral and other traits selected to maximize fitness in two or more alternative ways in the context of intraspecific and intrasexual reproductive competition. Individuals allocate resources to either one or the other (mutually exclusive) way of achieving the same functional end using evolved decision-making rules. This concept may apply to any major taxon, but we shall confine our discussion to the animal kingdom.

1.2 WHERE, WHEN, AND WHY DO WE EXPECT TO FIND ARTs?

We expect to find ARTs whenever there is fitness to be gained by pursuing different reproductive tactics and when intermediate expressions of a reproductive trait are either not possible (e.g., there is nothing in between nesting oneself and dumping eggs in conspecifics' nests: Yom-Tov 1980, 2001) or selected against by disruptive selection (e.g., benefits of large size for bourgeois tactics and of small size for parasitic[1] tactics: Taborsky 1999). Most often we find

---

[1] The term "bourgeois" tactic refers to individuals investing in privileged access to mates, by behavioral (e.g., defense, courtship), physiological (e.g., pheromones), or morphological means (e.g., secondary sexual characters). The "parasitic" tactic, in contrast, is performed by individuals exploiting the investment of bourgeois conspecifics. In general discussions of the

ARTs when there is investment to be exploited by same-sex competitors (Brockmann and Dawkins 1979, Wirtz 1982, Field 1992, 1994, Andersson 1994, Taborsky 1994, 1998, 2001, Villalobos and Shelly 1996, Hogg and Forbes 1997, Tallamy 2005). In principle, this is possible in both sexes, but because of the unavoidable higher investment of females (even parasitic females assume the costs of egg production), ARTs are expected to evolve more often in the male sex. It is worth emphasizing that anisogamy biases not only the intensity of sexual selection between the sexes, but consequently also the evolution of ARTs.

Investment in the privileged access to mates or fertilizable gametes bears costs (Taborsky et al. 1987, Simmons et al. 1992, Bailey et al. 1993, Lens et al. 1994, Prestwich 1994, Cordts and Partridge 1996, Grafe 1996, Hoback and Wagner 1997, Reinhold et al. 1998, Grafe and Thein 2001, Thomas 2002, Yoccoz et al. 2002, Basolo and Alcaraz 2003, Ward et al. 2003, Barboza et al. 2004, Pruden and Uetz 2004, Wagner 2005; but see Hack 1998, Kotiaho and Simmons 2003). It may involve (1) the production of conspicuous signals that may not only attract mates but also predators and competitors (Andersson 1994); (2) the construction of energetically demanding structures for mate attraction and brood care (Hansell 1984, 2005); or (3) parental investment to protect, provision, and raise offspring (Clutton-Brock 1991). Individuals using parasitic tactics may omit these costs and exploit their competitors' investment to gain access to mates or fertilizable gametes (Wirtz 1982, Miller 1984, Taborsky et al. 1987, Tomkins and Simmons 2000; see Taborsky 1994 for review). Often they use secretive "sneaking" tactics or fast "streaking" that cannot be easily overcome by the exploited bourgeois males (Warner and Hoffman 1980, Gross 1982, Westneat 1993, Kempenaers et al. 1995, 2001, Hall and Hanlon 2002, Correa et al. 2003, Sato et al. 2004; see Taborsky 1994, Westneat and Stewart 2003 for review). Alternatively, males using parasitic tactics may receive resources required for mating, or brood care for their offspring from bourgeois males also by force (van den Berghe 1988, Sinervo and Lively 1996, Mboko and Kohda 1999).

Cooperative behavior may be applied as an alternative to a purely parasitic tactic when individuals attempt to benefit from the effort of bourgeois competitors (Taborsky et al. 1987, Martin and Taborsky 1997, Dierkes et al. 1999, Taborsky 2001, Oliveira et al. 2002). Competing individuals may cooperate or "trade" with resource holders by paying for access to reproductive options by mutualism or reciprocity (Reyer 1984, 1986, Taborsky 1984, 1985, Lejeune 1985, Taborsky et al. 1987, Hatchwell and Davies 1992, Hartley et al. 1995, Davies et al. 1996, Magrath and Whittingham 1997, Martin and Taborsky 1997, Whittingham et al. 1997, Balshine-Earn et al. 1998, Johnstone and Cant 1999, Clutton-Brock et al. 2002, Oliveira et al. 2002, Richardson et al. 2002, Double and Cockburn 2003, Dickinson 2004, Huck et al. 2004, Webster et al. 2004, Bergmüller and Taborsky 2005). The relationships between such cooperating competitors are usually asymmetric, particularly in their resource-holding potential. The mechanisms regulating and stabilizing such cooperative relationships between reproductive competitors have been the target of much recent research (Vehrencamp 1983, Keller and Reeve 1994, Magrath and Whittingham 1997, Balshine-Earn et al. 1998, Johnstone and Cant 1999, Johnstone 2000, Kokko et al. 2002, Kokko 2003, Skubic et al. 2004, Bergmüller and Taborsky 2005, Bergmüller et al. 2005, Stiver et al. 2005), but there is still a great need for further integration of theory and empirical data.

Females may benefit from exploiting the nests built by other females (e.g., in digger wasps: Brockmann and Dawkins 1979, Brockmann et al. 1979, Field 1992, 1994) or by dumping eggs in another female's nest (or mouth) that will be cared for by the owner (i.e., intraspecific brood parasitism in insects: Eickwort 1975, Tallamy 1985, 2005, Müller et al. 1990, Brockmann 1993, Zink 2003; fish: Ribbink 1977, Yanagisawa 1985, Kellog et al. 1998; and birds: Yom-Tov 1980, 2001, Rohwer and Freeman 1989, Petrie and Møller 1991, Eadie and Fryxell 1992, Lyon 2003, Griffith et al. 2004). In this way, preparation of breeding sites and brood care can be spared by applying parasitic tactics (Sandell and Diemer 1999), or productivity can be increased (Tallamy and Horton 1990, Brown and Brown 1997, Ahlund and Andersson 2002, Zink 2003).

We may find ARTs also when animals use different niches for reproduction (such as temporally varying habitats). Selection may then favor multiple phenotypes that are specialized to exploit reproductive opportunities in each niche. Intermediate phenotypes will not be as effective as specialized ones when using the available options (Shuster

---

function of alternative tactics, these terms are preferable to the more descriptive terms often used in particular case studies (e.g., bourgeois males have been named guarders, territorials, primary males, parentals, nest males, type 1 males, or cuckolds, while parasitic males have been referred to as sneakers, streakers, satellites, hiders, pseudo-females, type 2 males, or cuckolders). For a discussion of reasons to use "bourgeois" and "parasitic" as collective, functional terms for ARTs see Taborsky (1997).

and Wade 2003). In this case, the frequency of morphs will depend on the reproductive potential in each niche (Zera and Rankin 1989, Mole and Zera 1993, Denno 1994, Langellotto et al. 2000, Langellotto and Denno 2001; see also Chapter 2 of this book).

1.3 WHICH EVOLUTIONARY PROCESSES ARE CAUSING THE PATTERNS WE FIND IN ARTs?

A major objective in evolutionary biology is to understand processes by which alternative phenotypes are created and maintained within populations (West-Eberhard 1986, Skúlason and Smith 1995, Smith and Skúlason 1996). This includes the question for the existence of two sexes (Parker et al. 1972), polymorphisms for the use of food and habitat (Sage and Selander 1975, Snorrason et al. 1994, Skúlason and Smith 1995, Robinson and Wilson 1996, Smith and Skúlason 1996), laterality (Hori 1993, McGrew and Marchant 1997, Nakajima et al. 2004), locomotion and migration patterns (Berthold and Querner 1982, Verspoor and Cole 1989, Berthold et al. 1990, Hindar and Jonsson 1993, Kaitala et al. 1993, Biro and Ridgway 1995, Smith and Skúlason 1996), predator evasion (Taborsky et al. 2003, Chipps et al. 2004), and the existence of reproductive "producers" and "scroungers" in same-sex conspecifics (Taborsky 1994, 2001, Gross 1996, Brockmann 2001). To understand the discontinuity of *reproductive* tactics, we should first look at the options of the involved players; that is, we should first know the patterns before disentangling the processes causing them. How do competitors achieve fertilizations? How divergent are the alternative tactics? Do individuals differ consistently in their tactics or are they choosing tactics according to circumstances? To identify underlying processes, we may analyze ARTs at three different levels of classification (Taborsky 1998).

1.3.1 Selection

Alternative tactics evolve when there is fitness to be gained by pursuing divergent allocation tactics. There are two principal conditions favoring the evolution of ARTs:

(1) Investment may be there to be exploited by conspecific, same-sex competitors, as we have outlined above. In the chosen sex, sexual selection leads to high investment in structures promoting mate acquisition. This includes secondary sexual signals that indicate quality (indirect benefits to mates) and supplying resources and brood care (direct benefits). Sexual selection has two major effects in this context; firstly, it causes variation in the success of the chosen sex (Darwin 1871). If some males are able to obtain several mates, others will end up without success (depending on the operational sex ratio: Shuster and Wade 2003), which selects for the pursuit of alternative tactics. Secondly, exploiting the investment of competitors without paying their costs may result in higher fitness (Fu et al. 2001). Both consequences of strong sexual selection set the stage for the evolution of ARTs. Indeed, a positive relationship between strong sexual selection and the evolution of ARTs has been observed (Gadgil 1972, Gross 1996, Sinervo 2001), although there may be negative feedback mechanisms involved as well (Jones et al. 2001a, b, Reichard et al. 2005).

(2) Different reproductive niches may exist for conspecific, same-sex competitors (see Chapter 2). This may occur when reproductive habitats differ discontinuously (Denno 1994, Langellotto and Denno 2001, Hiebeler 2004) or when competitors differ in some important feature as a result of natural selection (e.g., food niches or predation may select for body-size divergence: Pigeon et al. 1997, Lu and Bernatchez 1999, Jonsson and Jonsson 2001, Trudel et al. 2001, Kurdziel and Knowles 2002, Taborsky et al. 2003, Snorrason and Skúlason 2004). Little is known about the consequences of such polymorphisms on reproductive tactics (but see Kurdziel and Knowles [2002] for a notable exception) or about what is cause and what is effect (e.g., is a particular size dimorphism caused by natural selection favoring divergence, with respective consequence for reproduction, or does it result from ARTs caused by sexual selection as outlined above, with respective consequences regarding other aspects of life such as feeding and predator evasion? See Parker et al. 2001).

1.3.2 Flexibility

On the individual level, alternative tactics may be performed at the same time (simultaneous ARTs), in succession (sequential ARTs), or they may be fixed for life (fixed ARTs: Taborsky 1998) (Figure 1.1). This is a general feature of allocation patterns (Brockmann 2001), as found also in sex allocation (simultaneous and sequential hermaphroditism,

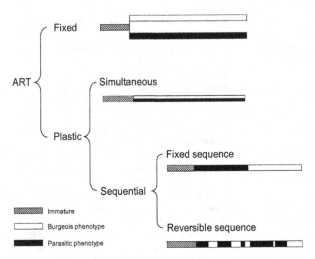

Figure 1.1 Alternative reproductive tactics can be fixed over a lifetime or plastic. In the latter case, they may be performed at the same time interval (simultaneous ARTs) or in a fixed or reversible sequence (sequential ARTs). See text for examples.

and gonochorism: Charnov 1982). Simultaneous and sequential ARTs are the product of a flexible or plastic response to conditions. A flexible response (i.e., phenotypic plasticity) may be beneficial if conditions vary either with regard to the physical or social environment of an animal, or its own physical condition (West-Eberhard 2003). If momentary conditions are highly unpredictable (e.g., number of potential partners, quality and number of current competitors in the neighborhood, tactic-dependent risk), there is selection for simultaneous ARTs as found, for example, in many fishes (Keenleyside 1972, Rowland 1979, Jennings and Philipp 1992; reviewed in Taborsky 1994), anurans (Perrill et al. 1982, Fukuyama 1991, Lucas et al. 1996, Byrne and Roberts 2004; reviewed in Halliday and Tejedo 1995), and birds (Westneat 1993, Kempenaers et al. 1995, 2001; reviewed in Westneat 2003). If conditions change with ontogeny, which applies in particular for organisms with indeterminate growth, sequential ARTs may be the optimal response (e.g., Warner et al. 1975, Magnhagen 1992, de Fraipont et al. 1993, Dierkes et al. 1999, Alonzo et al. 2000, Utami et al. 2002). If conditions either change rarely during the lifetime of an individual or change is unpredictable, fixed ARTs may be selected for (Shuster and Wade 2003). Additional factors influencing the existence (and coexistence) of fixed and flexible ARTs are differences in success between tactics and the costs of plasticity (Plaistow et al. 2004).

1.3.3 Origin of variation

Discontinuous phenotypic variation may originate from monomorphic or polymorphic genotypes (Austad 1984, Gross 1996, Shuster and Wade 2003). In genetically uniform individuals, the response to reproductive competition may be triggered by current conditions or by developmental switches; individual tactics differ due to diverging conditions, despite the same underlying genetic architecture. For example, individuals finding themselves in an unfavorable condition may do best by adopting an alternative tactic to the monopolization of mates, thereby doing "the best of a bad job" (Dawkins 1980). If resource availability varies strongly during development, the decision to adopt one or the other tactic may depend on the passing of a threshold; an individual passing a size threshold, for example, may do best by continuing to grow to adopt a bourgeois reproductive tactic later, while if this threshold is not passed, it may pay to reproduce early and in a parasitic role (note that in some salmonid fishes, it works the other way round; see below). Size thresholds may be important particularly for short-lived animals in seasonal habitats: early-born individuals have more time to grow in favorable conditions, so they will be larger at the start of reproduction. Such "birthdate effects" (Taborsky 1998) apparently influence the occurrence of ARTs in temperate fish (see Thorpe 1986). Thresholds in growth rates can also influence the choice of tactic (Hutchings and Myers 1994); fast-growing male salmon may start to reproduce earlier, while slow growers delay reproduction and end up in the bourgeois role as a consequence of prolonged growth (Thorpe and Morgan 1980, Thorpe 1986, Gross 1991). In anadromous salmonids this is linked to highly divergent feeding conditions between reproductive sites (oligotrophic rivers) and productive foraging areas (sea habitats: Healey *et al.* 2000, Vollestad *et al.* 2004).

Discontinuous alternative reproductive tactics may result also from polymorphic genotypes, regardless of whether variation is due to major gene effects or polygenic origin. Examples are known from a wide taxonomic range – from mites (Radwan 1995, 2003) and isopods (Shuster and

---

Box 1.2   The origin of male polymorphisms in acarid mites

"Fighter" and "scrambler" males occur in a number of acarid mites belonging to at least three genera (*Sancassania, Rhizoglyphus, Schwiebia*: Woodring 1969, Radwan 1995, 2001). Fighter males can kill competitors by puncturing their cuticle with a modified third pair of legs. Fighters may outcompete scramblers in low-density situations, but not at high densities, where they suffer from frequent and costly fights (Radwan 1993). Both a genetic polymorphism and a conditional expression of tactics with strong environmental influence during development have been found in different species of this group. In *Rhizoglyphus robini*, fighters sire higher pro-

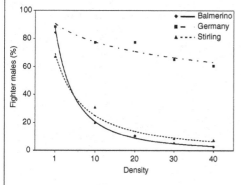

Figure 1.2 The proportion of fighter males emerging in laboratory populations of the acarid mite *Sancassania berlesei* depends on density. In an experiment, larvae originating from three different field populations were introduced into vials either alone or in groups of 10–40 individuals. While the majority of lone males turned into the fighter morph, the proportions of fighter males declined with increasing density, especially strongly in mites coming from two out of three natural populations. (After Tomkins *et al.* 2004.)

portions of fighters and the heritability of the male morphs is high; however the genetic mechanism underlying this polyphenism is not yet understood (Radwan 1995, 2003). Colony size and density have no effect on morph frequency, but diet provided during development does, with fewer fighters emerging under poor conditions (Radwan 1995). In this species, fighters survive longer, independently of colony density and morph ratio in the population (Radwan and Bogacz 2000, Radwan and Klimas 2001). Surprisingly, morph fitness was not found to be negatively frequency dependent, as would be expected if a genetic polymorphism is stabilized at an evolutionarily stable state (ESS) condition (Radwan and Klimas 2001).

In *R. echinopus*, no significant heritability of male morph was found, but the probability of males turning into fighters depended on chemical signals associated with colony density (Radwan 2001). In *Sancassania berlesei*, the decision by males to turn into fighters or scramblers strongly depends on social and food conditions during development (Timms *et al.* 1980, 1982, Radwan 1993, 1995, Radwan *et al.* 2002, Tomkins *et al.* 2004). In small or low density populations the proportion of fighter males is higher (Figure 1.2). Chemical (pheromonal) signals are used to determine tactic choice (Timms *et al.* 1980, Radwan *et al.* 2002), but the final-instar nymph weight is also important with heavier nymphs being more likely to become fighters, albeit at some costs: same-weight final-instar nymphs produced smaller fighters than scramblers (Radwan *et al.* 2002). Even though there is no indication of single-locus inheritance of morphs in this species, there is evidence for genetic covariance between sire status and offspring morph and considerable heritability of morph expression due to an adaptive response of the threshold reaction norm (Tomkins *et al.* 2004, Unrug *et al.* 2004). Data from this species are compatible with the status-dependent ESS model (Gross 1996), but a critical test showing that fitness functions of the alternative tactics cross at the phenotypic switch point is still missing (Tomkins *et al.* 2004).

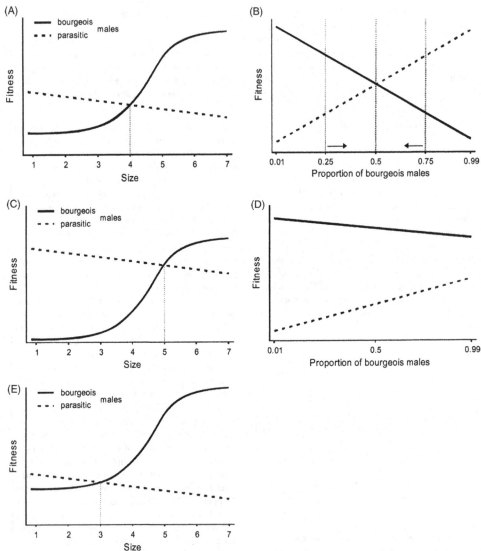

Figure 1.3 (A) When male competitors in a population show either bourgeois or parasitic reproductive behavior depending on condition, e.g., body size, and their fitness functions cross at a given size, males should switch from one to the other tactic at this intersection. (B) The fitness of males may depend also on the relative numbers of both types of males. If the fitness functions depending on relative frequencies of both male types cross, tactic frequencies in the population should converge towards the point of intersection. How does this relate to size-dependent tactic choice? We assume here that the fitness lines cross at a bourgeois male proportion of 0.5; if the population comprises 75% bourgeois males at some point, the average fitness of individuals performing the parasitic tactic would increase relative to that of bourgeois males. (C) The effect of this situation on optimal size-dependent tactic choice: while the fitness of bourgeois males drops due

Box 1.3   Do fitness curves always cross?

When condition-dependent fitness functions differ between bourgeois and parasitic males and the lines cross, tactic frequencies should depend on this point of intersection (Gross 1982, 1996) (Figure 1.3A). In addition, the fitness of each type of male may depend on the proportions of both male types in the population, resulting in frequency-dependent selection: the more parasitic males compete amongst each other, the less it may pay to choose this tactic (Figure 1.3B), which feeds back on condition-dependent tactic choice (Figure 1.3C and 1.3E). However, cases in which individuals differ in quality demonstrate that frequency dependence is not necessarily involved in the evolution of ARTs. Take a species with early- and late-born males in a seasonal environment that have very different lengths of growth periods before the first winter. In a short-lived species, reproduction may occur only within one reproductive season, i.e., after the first winter. Early- and late-born males will differ in size because they encountered different good growth conditions in their first year during time periods of different period lengths (e.g., Mediterranean wrasses: Alonzo et al. 2000). Large males

may do best by monopolizing resources and access to females; small males may do best by parasitizing the reproductive effort of large males because they are not able to compete with their larger conspecifics when performing a bourgeois tactic. The average reproductive success of the small males may never reach the same level as the average success of the higher-quality (large) males, even if they are rare in the population, because their small size may act as a constraint on getting access to fertilizable gametes. The result will be ARTs that are not stabilized by frequency-dependent selection (Figure 1.3D). Parasitic males will still persist in the population because males differ in quality due to differing growth conditions, as outlined above. Quality differences between individuals due to developmental constraints are very widespread (Schlichting and Pigliucci 1998), but hitherto, they have not been dealt with in this context in much detail. In theoretical models Mart Gross and Joe Repka (Repka and Gross 1995, Gross and Repka 1998a, b) showed that equilibria between alternative tactics causing unequal fitnesses may be evolutionarily stable; this approach has been criticized, however, because of unrealistic assumptions (Shuster and Wade 2003).

Wade 1991, Shuster 1992) to fish (Ryan et al. 1992), lizards (Sinervo and Lively 1996), and birds (Lank et al. 1995, 1999). In this case, genotype frequencies underlying ARTs are believed to be balanced by frequency-dependent selection, leading to equal lifetime fitness expectations of individuals using different tactics (Shuster and Wade 1991, Ryan et al. 1992, Repka and Gross 1995; but see Boxes 1.2 and 1.3; see also Chapter 2 of this book).

The relative importance of genetic monomorphism with conditional responses as opposed to genetic polymorphism for the evolution of ARTs has been extensively debated

(Pienaar and Greeff 2003; see Gross 1996, Shuster and Wade 2003 for review). The vast majority of described cases of ARTs involves some conditional responses of reproductive competitors (Gross 1996, Lank et al. 1999). Because a conditional choice of tactics has been associated with genetic monomorphism, it has been argued that genetic polymorphisms play only a minor role in the causation of ARTs (Gross 1996, Shuster and Repka 1998a). This view has been challenged (Shuster and Wade 2003). Why is this debate of general interest? To appreciate the importance of the issue, we need to consider the implications of these two

Caption for Fig. 1.3 (cont.)   to competition among males of this type, the fitness of parasitic males increases, which means that males should switch to the bourgeois tactic at a larger size (size "5" instead of "4" as depicted in (A)). If in contrast only 25% of the males in the population perform the bourgeois tactic, the fitness of bourgeois males will increase because of low competition while that of parasitic males will drop due to competition of these males when exploiting the relatively small number of bourgeois males in the population (see (E)). Males should switch earlier now from the parasitic to the bourgeois tactic. Note that for simplicity in this graphical model we assume that the relative frequency of both tactics (as shown in (B)) influences the pay-

off of males in a similar way over the whole range of sizes, i.e., the intercept of the fitness function changes, but not its shape or slope (cf. panels (C) and (E), and panel (A)). (D) It may be, however, that the fitness functions depending on relative frequencies of both male types do not cross, i.e., bourgeois males may always do better than parasitic males, regardless of the proportions of males in the population performing either tactic (see explanation in text). If this is the case, frequency-dependent selection will not determine tactic choice in the population. Tactic choice and hence tactic frequencies will then depend only on other factors (like male phenotypic quality such as size, which may be determined by developmental constraints).

potential mechanisms. When animals act according to conditions without any genetic component responsible for the type of response (i.e., tactic performance is not heritable), (a) the form and frequency of this response is not subject to selection (Shuster and Wade 2003), which precludes adaptive evolution, (b) different tactics may result in unequal fitness (Repka and Gross 1995, Gross and Repka 1998b), and (c) the frequencies of tactics may be independent of each other and of their relative success (see Box 1.3). When tactic choice is under genetic control and heritable, frequency-dependent selection will lead to (a) a fitness equilibrium associated with alternative tactics and (b) stable frequencies of ARTs in the population (Ryan *et al.* 1992), or (c) oscillations of tactics if no stable equilibrium can be reached (particularly if more than two ARTs exist in a population: Shuster 1989, Shuster and Wade 1991, Sinervo and Lively 1996). According to the "status-dependent selection model" (Gross 1996), the assumption of conditional tactics based on genetic monomorphism coincides with unequal fitnesses of players, except at the switch point where an individual is expected to change from one tactic to another. On the contrary, a genetic polymorphism can only persist if the lifetime fitnesses of players are equal or oscillating (Slatkin 1978, 1979, Shuster and Wade 1991, 2003).

It would be naïve to assume that ARTs will be either "genetically" or "environmentally" determined (Caro and Bateson 1986). In reality, many if not most dimorphic traits seem to be threshold traits (Roff 1996) influenced by quantitative trait loci: morph expression depends on whether a "liability" value is above or below a threshold (Falconer and Mackay 1996). In the context of ARTs this was shown for the expression of different male morphs in mites with the help of selection experiments, by which the threshold reaction norm was shifted (Unrug *et al.* 2004) (see Box 1.2). In this scenario, developmental pathways may change abruptly, e.g., at a particular size, producing different phenotypes on either side of the threshold (Emlen and Nijhout 2000, Nijhout 2003, Lee 2005). The operation of genetically based developmental thresholds means that trait expression is both conditional *and* heritable. It allows alternative phenotypes to evolve largely independently from each other, which greatly increases the scope for the evolution of alternative tactics (West-Eberhard 1989, 2003; see also Tomkins *et al.* 2005).

If adaptive evolution is not underlying conditional ARTs (as argued by Shuster and Wade 2003), why do they exist in the first place, why are conditional decisions

apparently the rule rather than the exception, and why do genetic polymorphisms associated with ARTs appear to be rare? One may ponder whether these concepts are sufficient to explain the evolution of ARTs. The problem is that in this discussion, conditional response and the genetic basis of tactics apparently have been separated from each other. More realistically, the thresholds or developmental switch points involved in tactic choice have a genetic basis and will therefore be subject to selection and adaptive evolution (Tomkins *et al.* 2004). In other words, phenotypic plasticity is heritable, and genetically based plastic traits vary among individuals of a population (see Chapter 5 of this book). Conditional responses may have a genetic basis but still lead to different lifetime reproductive successes of tactics (Hazel *et al.* 1990). This issue needs further theoretical treatment (see Shuster and Wade 2003).

1.4   INTEGRATING ACROSS LEVELS: PROXIMATE AND ULTIMATE CAUSES OF ARTs

How do the proximate mechanisms underlying the expression of ARTs relate to their evolution? An important aspect in our understanding of ARTs is the degree of divergence between tactics, which may functionally relate to the underlying mechanisms (e.g., pleiotropic effects if genetic determination is involved, or variance in ontogenetic conditions). In this context it is necessary to understand the proximate mechanisms involved to be able to interpret observed patterns. A distinction should be made, for instance, between alternative phenotypes that diverge only in behavioral traits or also in the expression of morphological and anatomical traits. Since behavior is often more labile than morphology and anatomy, the mechanisms underlying the expression of behavioral variation should be more flexible than those underlying morphological and anatomical variation.

Hormonal regulation is usually involved in the expression of alternative reproductive behavior (Brantley *et al.* 1993, Oliveira *et al.* 2005). Ketterson and Nolan (1999) proposed that one could distinguish between adaptations and exaptations (*sensu* Gould and Vrba 1982) in hormone-dependent traits by assessing whether these traits arose either in response to selection on circulating hormone levels or in response to variation in the responsiveness of the target tissues to invariant hormone levels (Figure 1.4). In the former case, selection probably did not act on all correlated

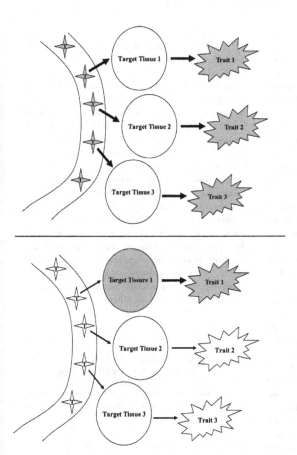

Figure 1.4 A model of two alternative evolutionary mechanisms underlying the hormonal regulation of alternative reproductive phenotype expression. Upper panel Alternative reproductive traits arise in response to selection on circulating hormone levels, whereby selection may not act on all correlated traits. Lower panel Alternative reproductive traits arise in response to selection on responsiveness of target tissues to (possibly invariant) hormone levels; here, traits result from the independent reaction of target tissue sensitivity to constant hormone levels.

traits and thus the ones that subsequently conferred an advantage to its carriers should be viewed as exaptations. In the latter case, selection probably acted independently on target tissue sensitivity to constant hormone levels, for example by varying density of receptors or the expression of enzymes for particular biosynthetic pathways. ARTs that involve the differential development of androgen-dependent traits within the same phenotype, such as the differentiation of larger testes in parasitic males without displaying secondary sex characters, suggest a compart- mentalization of androgen effects on different target tissues that can be achieved by varying the densities of androgen receptors in different targets. Therefore, ARTs that involve the compartmentalization of different endocrine-mediated traits probably evolved as adaptations, whereas ARTs in which there are no compartmentalization effects (e.g., conditional tactics, such as the facultative use of sneaking behavior by nest-holder males in sticklebacks: Morris 1952, Rico et al. 1992) rather represent exaptations. This approach stresses the importance that studies of proximate mechanisms may have to increasing our understanding of the evolution of alternative reproductive phenotypes.

The neural mechanisms behind alternative reproductive behavior patterns may involve the structural reorganization of neural circuits underlying the expression of reproductive behavior, or alternatively biochemical switching of existing circuits by neuromodulators (Zupanc and Lamprecht 2000). The former mechanisms may involve synaptogenesis, the regulation of apoptosis, and neurogenesis and thus should be associated with a slower, discontinuous but long-lasting expression of phenotypic plasticity. In contrast, the latter mechanisms should be associated with faster, gradual, and transient changes. These potential neural mechanisms underlying phenotypic plasticity may interact with hormonal mechanisms: structural (re)organization of neural circuits can be influenced by organizational effects of hormones during well-defined sensitive periods in the life of an individual, while biochemical switches can be driven by activational effects of hormones on central pathways underlying behavior (for a review on organizational vs. activational effects of hormones in vertebrates see Arnold and Breedlove 1985). Therefore, it is expected that simultaneous or reversible conditional tactics that may require rapid and transient changes in neural activity are mediated by biochemical switches influenced by hormones in an activational fashion (Zupanc and Lamprecht 2000, Hofmann 2003), whereas both fixed tactics involving the organization of the phenotype early in development and sequential tactics with a fixed sequence that involve a post-maturational reorganization of the phenotype are mediated by structural reorganization of neural networks. Concomitantly, the role of hormones in the expression of the different types of tactics would differ, with organizational (or reorganizational) effects predicted to be associated with fixed and fixed-sequence tactics, and activational effects expected in simultaneous or reversible conditional tactics (Moore 1991, Moore et al. 1998, Oliveira 2005).

Knowledge of the proximate mechanisms underlying the expression of ARTs may help to understand their evolution. Ketterson and co-workers (Ketterson et al. 1996, Ketterson and Nolan 1999) have proposed the use of phenotypic engineering to investigate the evolution of endocrine-mediated traits. This approach is based on the exogenous administration of hormones to study ecological consequences of the development of hormone-dependent traits. This approach can help to identify the costs and benefits associated with particular traits specific to each tactic as well as the evolutionary scenario in which ARTs evolved. A cost–benefit analysis of ARTs in teleosts, for instance, would help to identify costs associated with specific tactics imposed by their underlying physiological mechanisms, which may act as constraints for the evolution of ARTs. For example, bourgeois males usually display a set of androgen-dependent behavioral traits that help them to compete with other males for resources or females (e.g., through territoriality), which suggests that costs associated with maintaining high androgen levels should be associated with the bourgeois tactic (e.g., increased energy consumption, effects on immunocompetence, increased risk of predation, and a higher incidence of injuries from agonistic interactions: e.g., Wingfield et al. 1999, 2001, Ros et al. 2006). Therefore, knowledge of the physiological mechanisms underlying the expression of ARTs may shed light on the evolutionary landscapes in which they might have evolved by helping to identify proximate mechanisms that act as mediators of adaptive traits or as potential physiological constraints imposed by pleiotropic-like effects of hormones on the evolution of ARTs.

1.5 CURRENT ISSUES: WHAT ARE THE QUESTIONS WE NEED TO SOLVE?

Based on the above discussion and arguments we should like to emphasize 12 important questions regarding the evolution of ARTs.

(1) To what extent are thresholds and developmental switches responsible for the evolution of decision rules? In other words, is there genetic variance involved in the conditional response?

(2) If there is sufficient genetic variance among individuals of a population, to what degree are thresholds and developmental switch points subject to selection? An experimental approach would be desirable here.

(3) The occurrence of ARTs is apparently related to the intensity of sexual selection and to the existence of an opportunity to exploit the investment of same-sex conspecific competitors to acquire mates or fertilizations. These potential causes of the evolution of ARTs are not independent; however, they may independently influence the evolution of decision rules. Is one or the other of these factors more important (or of sole importance), or are additional factors involved?

(4) Is the observed intrasexual variation in reproductive phenotypes necessarily adaptive, or are there sometimes constraints (e.g., because a certain part of the population faces inferior conditions during ontogeny, causing significant intrasexual size variation; see Box 1.3) that may produce ARTs?

(5) Are there particular environmental circumstances (both physical and social) that favor either a combination between genetic monomorphism and conditional response or a genetic polymorphism underlying ARTs, either with or without conditional response components?

(6) The expression of ARTs may be fixed for an individual or flexible over a lifetime (Figure 1.1): on the proximate level, to what extent are they caused by structural (re)organization of neural networks, and what organizational and activational hormonal effects regulate fixed vs. plastic alternative phenotypes?

(7) What are the selective regimes favoring the evolution of fixed vs. plastic, simultaneous, or sequential ARTs? That is, which environmental conditions and intrinsic factors (i.e., constraints and life-history patterns) may take effect? Are fixed phenotypes associated with genetic polymorphisms and flexible ones with genetic monomorphism?

(8) How does discontinuous phenotypic variation among competitors that evolved in other functional contexts (e.g., by food niches or predation scenarios) affect the evolution of ARTs?

(9) What causes intermediate types to be less successful than "pure" alternatives? That is, why is selection disruptive?

(10) What controls tactic frequencies? Is frequency-dependent (Repka and Gross 1995) and density-dependent (Tomkins and Brown 2004) selection involved if tactics are purely conditional (which may cause unequal average fitnesses)? When do crossing fitness curves predict relative tactic frequencies (see Box 1.3)?

(11) What processes cause tactics to stabilize at an equilibrium frequency or to oscillate?

(12) Why do particular phenotypes take the form they do? Why are particular solutions so frequent across a wide range of taxa (e.g., female mimicry in males)?

Most of these questions have been asked before in various contexts and often with focus on certain examples, and some have been partially answered either on an empirical or theoretical basis. However, for most if not all of them, we lack enough crucial information to be able to give an answer at the level of specific examples and on a more general basis. This is not an exhaustive list. Of course there are other questions and details we need to consider (e.g., see Box 1.2 and other chapters of this book), but we believe that finding answers to these 12 questions will significantly advance our understanding of ARTs.

References

Ahlund, M. and Andersson, M. 2002. Female ducks can double their reproduction. *Nature* **414**, 600–601.

Alonzo, S. H., Taborsky, M., and Wirtz, P. 2000. Male alternative reproductive behaviors in a Mediterranean wrasse, *Symphodus ocellatus*: evidence from otoliths for multiple life-history pathways. *Evolutionary Ecology Research* **2**, 997–1007.

Andersson, M. 1994. *Sexual Selection*. Princeton, NJ: Princeton University Press.

Arnold, A. B. and Breedlove, S. M. 1985. Organizational and activational effects of sex steroids on brain and behavior: a reanalysis. *Hormones and Behavior* **19**, 469–498.

Austad, S. N. 1984. A classification of alternative reproductive behaviors and methods of field-testing ESS models. *American Zoologist* **24**, 309–319.

Bailey, W. J., Withers, P. C., Endersby, M., and Gaull, K. 1993. The energetic costs of calling in the bush-cricket *Requena verticalis* (Orthoptera, Tettigoniidae, Listroscelidinae). *Journal of Experimental Biology* **178**, 21–37.

Balshine-Earn, S., Neat, F. C., Reid, H., and Taborsky, M. 1998. Paying to stay or paying to breed? Field evidence for direct benefits of helping behavior in a cooperatively breeding fish. *Behavioral Ecology* **9**, 432–438.

Barboza, P. S., Hartbauer, D. W., Hauer, W. E., and Blake, J. E. 2004. Polygynous mating impairs body condition and homeostasis in male reindeer (*Rangifer tarandus tarandus*). *Journal of Comparative Physiology B* **174**, 309–317.

Basolo, A. L. and Alcaraz, G. 2003. The turn of the sword: length increases male swimming costs in swordtails. *Proceedings of the Royal Society of London B* **270**, 1631–1636.

Bass, A. H. 1992. Dimorphic male brains and alternative reproductive tactics in a vocalizing fish. *Trends in Neurosciences* **15**, 139–145.

Bass, A. H. 1996. Shaping brain sexuality. *American Scientist* **84**, 352–363.

Bass, A. H. and Andersen, K. 1991. Intra- and inter-sexual dimorphisms in the sound-generating motor system in a vocalizing fish: motor axon number and size. *Brain, Behavior and Evolution* **37**, 204–214.

Bergmüller, R. and Taborsky, M. 2005. Experimental manipulation of helping in a cooperative breeder: helpers "pay to stay" by pre-emptive appeasement. *Animal Behaviour* **69**, 19–28.

Bergmüller, R., Heg, D., and Taborsky, M. 2005. Helpers in a cooperatively breeding cichlid stay and pay or disperse and breed, depending on ecological constraints. *Proceedings of the Royal Society of London B* **272**, 325–331.

Berthold, P. and Querner, U. 1982. Partial migration in birds: experimental proof of polymorphism as a controlling system. *Experientia* **38**, 805–806.

Berthold, P., Mohr, G., and Querner, U. 1990. Control and evolutionary potential of obligate partial migration: results of a 2-way selective breeding experiment with the blackcap (*Sylvia atricapilla*). *Journal für Ornithologie* **131**, 33–45.

Biro, P. A. and Ridgway, M. S. 1995. Individual variation in foraging movements in a lake population of young-of-the-year brook charr (*Salvelinus fontinalis*). *Behaviour* **132**, 57–74.

Bisazza, A. 1993. Male competition, female mate choice and sexual size dimorphism in poeciliid fishes. *Marine Behavior and Physiology* **23**, 257–286.

Bisazza, A. and Pilastro, A. 1997. Small male mating advantage and reversed size dimorphism in poeciliid fishes. *Journal of Fish Biology* **50**, 397–406.

Brantley, R. K. and Bass, A. H. 1994. Alternative male spawning tactics and acoustic signals in the plainfin midshipman fish *Porichthys notatus* Girard (Teleostei, Batrachoididae). *Ethology* **96**, 213–232.

Brantley, R. K., Wingfield, J. C., and Bass, A. H. 1993. Sex steroid levels in *Porichthys notatus*, a fish with alternative reproductive tactics, and a review of the hormonal bases for male dimorphism among teleost fishes. *Hormones and Behavior* **27**, 332–347.

Brockmann, H. J. 1993. Parasitizing conspecifics: comparisons between Hymenoptera and birds. *Trends in Ecology and Evolution* **8**, 2–4.

Brockmann, H. J. 2001. The evolution of alternative strategies and tactics. *Advances in the Study of Behavior* **30**, 1–51.

Brockmann, H. J. 2002. An experimental approach to altering mating tactics in male horseshoe crabs (*Limulus polyphemus*). *Behavioral Ecology* **13**, 232–238.

Brockmann, H. J. and Dawkins, R. 1979. Joint nesting in a digger wasp as an evolutionarily stable preadaptation to social life. *Behaviour* **71**, 203–245.

Brockmann, H. J. and Penn, D. 1992. Male mating tactics in the horseshoe crab, *Limulus polyphemus*. *Animal Behaviour* **44**, 653–665.

Brockmann, H. J., Grafen, A., and Dawkins, R. 1979. Evolutionarily stable nesting strategy in a digger wasp. *Journal of Theoretical Biology* **77**, 473–496.

Brown, C. and Brown, M. 1997. Fitness components associated with alternative reproductive tactics in cliff swallows. *Behavioral Ecology* **9**, 158–171.

Byrne, P. G. and Roberts, J. D. 2004. Intrasexual selection and group spawning in quacking frogs (*Crinia georgiana*). *Behavioral Ecology* **15**, 872–882.

Cade, W. 1979. The evolution of alternative male reproductive strategies in field crickets. In M. S. Blum and N. A. Blum (eds.) *Sexual Selection and Reproductive Competition in Insects*, pp. 343–379. New York: Academic Press.

Caro, T. M. and Bateson, P. 1986. Organization and ontogeny of alternative tactics. *Animal Behaviour* **34**, 1483–1499.

Charnov, E. L. 1982. *The Theory of Sex Allocation*. Princeton, NJ: Princeton University Press.

Chipps, S. R., Dunbar, J. A., and Wahl, D. H. 2004. Phenotypic variation and vulnerability to predation in juvenile bluegill sunfish (*Lepomis macrochirus*). *Oecologia* **138**, 32–38.

Clutton-Brock, T. H. 1991. *The Evolution of Parental Care*. Princeton, NJ: Princeton University Press.

Clutton-Brock, T. H., Russell, A. F., Sharpe, L. L., et al. 2002. Evolution and development of sex differences in cooperative behavior in meerkats. *Science* **297**, 253–256.

Cordts, R. and Partridge, L. 1996. Courtship reduces longevity of male *Drosophila melanogaster*. *Animal Behaviour* **52**, 269–278.

Correa, C., Baeza, J. A., Hinojosa, I. A., and Thiel, M. 2003. Male dominance hierarchy and mating tactics in the rock shrimp *Rhynchocinetes typus* (Decapoda: Caridea). *Journal of Crustacean Biology* **23**, 33–45.

Darwin, C. 1871. *The Descent of Man, and Selection in Relation to Sex*. London: John Murray.

Davies, N. B., Hartley, I. R., Hatchwell, B. J., and Langmore, N. E. 1996. Female control of copulations to maximize male help: a comparison of polygynandrous alpine accentors, *Prunella collaris*, and dunnocks, *P. modularis*. *Animal Behaviour* **51**, 27–47.

Dawkins, R. 1980. Good strategy or evolutionarily stable strategy? In G. W. Barlow and J. Silverberg (eds.) *Sociobiology: Beyond Nature/Nurture?* pp. 331–367. Boulder, CO: Westview Press.

de Fraipont, M., FitzGerald, G. J., and Guderley, H. 1993. Age-related differences in reproductive tactics in the three-spined stickleback, *Gasterosteus aculeatus*. *Animal Behaviour* **46**, 961–968.

Denno, R. F. 1994. The evolution of dispersal polymorphisms in insects: the influence of habitats, host plants and mates. *Researches on Population Ecology* **36**, 127–135.

Dickinson, J. L. 2004. A test of the importance of direct and indirect fitness benefits for helping decisions in western bluebirds. *Behavioral Ecology* **15**, 233–238.

Dierkes, P., Taborsky, M., and Kohler, U. 1999. Reproductive parasitism of broodcare helpers in a cooperatively breeding fish. *Behavioral Ecology* **10**, 510–515.

Double, M. C. and Cockburn, A. 2003. Subordinate superb fairy-wrens (*Malurus cyaneus*) parasitize the reproductive success of attractive dominant males. *Proceedings of the Royal Society of London B* **270**, 379–384.

Doums, C., Viard, F., and Jarne, P. 1998. The evolution of phally polymorphism. *Biological Journal of the Linnean Society* **64**, 273–296.

Eadie, J. M. and Fryxell, J. M. 1992. Density dependence, frequency dependence, and alternative nesting strategies in goldeneyes. *American Naturalist* **140**, 621–641.

Eberhard, W. G. and Gutiérrez, E. E. 1991. Male dimorphisms in beetles and earwigs and the question of developmental constraints. *Evolution* **45**, 18–28.

Eickwort, G. 1975. Gregarious nesting of the mason bee *Hoplitis anthocopoides* and the evolution of parasitism and sociality among megachilid bees. *Evolution* **29**, 142–150.

Emlen, D. J. 1996. Artificial selection on horn length/body size allometry in the horned beetle *Onthophagus acuminatus* (Coleoptera: Scarabaeidae). *Evolution* **50**, 1219–1230.

Emlen, D. J. 1997. Alternative reproductive tactics and male-dimorphism in the horned beetle *Onthophagus acuminatus* (Coleoptera: Scarabaeidae). *Behavioral Ecology and Sociobiology* **41**, 335–341.

Emlen, D. J. and Nijhout, H. F. 2000. The development and evolution of exaggerated morphologies in insects. *Annual Review of Entomology* **45**, 661–708.

Falconer, D. S. and Mackay, T. F. C. 1996. *Introduction to Quantitative Genetics*, 4th edn. New York: Longman.

Field, J. 1992. Intraspecific parasitism as an alternative reproductive tactic in nest-building wasps and bees. *Biological Reviews* **67**, 79–126.

Field, J. 1994. Selection of host nests by intraspecific nest-parasitic digger wasps. *Animal Behaviour* **48**, 113–118.

Fu, P., Neff, B. D., and Gross, M. R. 2001. Tactic-specific success in sperm competition. *Proceedings of the Royal Society of London B* **268**, 1105–1112.

Fukuyama, K. 1991. Spawning behaviour and male mating tactics of a foam-nesting treefrog, *Rhacophorus schlegelii*. *Animal Behaviour* **42**, 193–199.

Gadgil, M. 1972. Male dimorphism as a consequence of sexual selection. *American Naturalist* **106**, 574–580.

Gould, S. J. and Vrba, E. S. 1982. Exaptation: a missing term in the science of form. *Paleobiology* **8**, 4–15.

Grafe, T. U. 1996. Energetics of vocalization in the African reed frog (*Hyperolius marmoratus*). *Comparative Biochemistry and Physiology A* **114**, 235–243.

Grafe, T. U. and Thein, J. 2001. Energetics of calling and metabolic substrate use during prolonged exercise in the European treefrog *Hyla arborea*. *Journal of Comparative Physiology B* **171**, 69–76.

Grafen, A. 1991. Modelling in behavioural ecology. In J. R. Krebs and N. B. Davies (eds.) *Behavioural Ecology*, pp. 5–31. Oxford, UK: Blackwell Scientific.

Griffith, S. C., Lyon, B. E., and Montgomerie, R. 2004. Quasi-parasitism in birds. *Behavioral Ecology and Sociobiology* **56**, 191–200.

Gross, M. R. 1982. Sneakers, satellites and parentals: polymorphic mating strategies in North-American sunfishes. *Zeitschrift für Tierpsychologie* **60**, 1–26.

Gross, M. R. 1991. Salmon breeding behavior and life history evolution in changing environments. *Ecology* **72**, 1180–1186.

Gross, M. R. 1996. Alternative reproductive strategies and tactics: diversity within sexes. *Trends in Ecology and Evolution* **11**, 92–98.

Gross, M. R. and Repka, J. 1998a. Inheritance in the conditional strategy. In L. A. Dugatkin and H. K. Reeve (eds.) *Game Theory and Animal Behavior*, pp. 168–187. Oxford, UK: Oxford University Press.

Gross, M. R. and Repka, J. 1998b. Stability with inheritance in the conditional strategy. *Journal of Theoretical Biology* **192**, 445–453.

Hack, M. A. 1998. The energetics of male mating strategies in field crickets (Orthoptera: Gryllinae: Gryllidae). *Journal of Insect Behavior* **11**, 853–867.

Hall, K. C. and Hanlon, R. T. 2002. Principal features of the mating system of a large spawning aggregation of the giant Australian cuttlefish *Sepia apama* (Mollusca: Cephalopoda). *Marine Biology* **140**, 533–545.

Halliday, T. M. and Tejedo, M. 1995. Intrasexual selection and alternative mating behaviour. In H. Heatwole and B. K. Sullivan (eds.) *Amphibian Biology*, vol. 2, pp. 469–517. Chipping Norton, NSW: Surrey Beatty and Sons.

Hansell, M. H. 1984. *Animal Architecture and Building Behaviour*. Harlow, UK: Longman.

Hansell, M. 2005. *Animal Architecture*. Oxford, UK: Oxford University Press.

Hartley, I. R., Davies, N. B., Hatchwell, B. J., *et al.* 1995. The polygynandrous mating system of the alpine accentor, *Prunella collaris*. 2. Multiple paternity and parental effort. *Animal Behaviour* 49, 789–803.

Hatchwell, B. J. and Davies, N. B. 1992. An experimental study of mating competition in monogamous and poyandrous dunnocks, *Prunella modularis*. 2. Influence of removal and replacement experiments on mating systems. *Animal Behaviour* 43, 611–622.

Hazel, W. N., Smock, R., and Johnson, M. D. 1990. A polygenic model for the evolution and maintenance of conditional strategies. *Proceedings of the Royal Society of London B* 242, 181–187.

Healey, M. C., Henderson, M. A., and Burgetz, I. 2000. Precocial maturation of male sockeye salmon in the Fraser River, British Columbia, and its relationship to growth and year-class strength. *Canadian Journal of Fisheries and Aquatic Sciences* 57, 2248–2257.

Heath, D. D., Rankin, L., Bryden, C. A., Heath, J. W., and Shrimpton, J. M. 2002. Heritability and Y-chromosome influence in the jack male life history of chinook salmon (*Oncorhynchus tshawytscha*). *Heredity* 89, 311–317.

Hiebeler, D. 2004. Competition between near and far dispersers in spatially structured habitats. *Theoretical Population Biology* 66, 205–218.

Hindar, K. and Jonsson, B. 1993. Ecological polymorphism in arctic charr. *Biological Journal of the Linnean Society* 48, 63–74.

Hoback, W. W. and Wagner, W. E. 1997. The energetic cost of calling in the variable field cricket, *Gryllus lineaticeps*. *Physiological Entomology* 22, 286–290.

Hofmann, H. A. 2003. Functional genomics of neural and behavioral plasticity. *Journal of Neurobiology* 54, 272–282.

Hogg, J. T. and Forbes, S. H. 1997. Mating in bighorn sheep: frequent male reproduction via a high-risk "unconventional" tactic. *Behavioral Ecology and Sociobiology* 41, 33–48.

Hori, M. 1993. Frequency-dependent natural selection in the handedness of scale-eating cichlid fish. *Science* 393, 216–219.

Huck, M., Lottker, P., and Heymann, E. W. 2004. The many faces of helping: possible costs and benefits of infant carrying and food transfer in wild moustached tamarins (*Saguinus mystax*). *Behaviour* 141, 915–934.

Hunt, J. and Simmons, L. W. 2000. Maternal and paternal effects on offspring phenotype in the dung beetle *Onthophagus taurus*. *Evolution* 54, 936–941.

Hutchings, J. A. and Myers, R. A. 1994. The evolution of alternative mating strategies in variable environments. *Evolutionary Ecology* 8, 256–268.

Jennings, M. J. and Philipp, D. P. 1992. Female choice and male competition in longear sunfish. *Behavioral Ecology* 3, 84–94.

Johnstone, R. A. 2000. Models of reproductive skew: a review and synthesis. *Ethology* 106, 5–26.

Johnstone, R. A. and Cant, M. A. 1999. Reproductive skew and the threat of eviction: a new perspective. *Proceedings of the Royal Society of London B* 266, 275–279.

Jones, A. G., Walker, D., Kvarnemo, C., Lindstrom, K., and Avise, J. C. 2001a. How cuckoldry can decrease the opportunity for sexual selection: data and theory from a genetic parentage analysis of the sand goby, *Pomatoschistus minutus*. *Proceedings of the National Academy of Sciences of the United States of America* 98, 9151–9156.

Jones, A. G., Walker, D., Lindstrom, K., Kvarnemo, C., and Avise, J. C. 2001b. Surprising similarity of sneaking rates and genetic mating patterns in two populations of sand goby experiencing disparate sexual selection regimes. *Molecular Ecology* 10, 461–469.

Jonsson, B. and Jonsson, N. 2001. Polymorphism and speciation in Arctic charr. *Journal of Fish Biology* 58, 605–638.

Kaitala, A., Kaitala, V., and Lundberg, P. 1993. A theory of partial migration. *American Naturalist* 142, 59–81.

Keenleyside, M. H. A. 1972. Intraspecific intrusions into nests of spawning longear sunfish (Pisces: Centrarchidae). *Copeia* 272–278.

Keller, L. and Reeve, H. K. 1994. Partitioning of reproduction in animal societies. *Trends in Ecology and Evolution* 9, 98–103.

Kellog, K. A., Markert, J. A., Stauffer, J. R., and Kocher, T. D. 1998. Intraspecific brood mixing and reduced polyandry in a maternal mouth-brooding cichlid. *Behavioral Ecology* 9, 309–312.

Kempenaers, B., Verheyen, G. R., and Dhondt, A. A. 1995. Mate guarding and copulation behavior in monogamous and polygynous blue tits: do males follow a best-of-a-bad-job strategy? *Behavioral Ecology and Sociobiology* 36, 33–42.

Kempenaers, B., Everding, S., Bishop, C., Boag, P., and Robertson, R. J. 2001. Extra-pair paternity and the reproductive role of male floaters in the tree swallow

(*Tachycineta bicolor*). *Behavioral Ecology and Sociobiology* 49, 251–259.

Ketterson, E. D. and Nolan Jr., V. 1999. Adaptation, exaptation, and constraint: a hormonal perspective. *American Naturalist* 154, S4–S25.

Ketterson, E. D., Nolan Jr., V., Cawthorn, M. J., Parker, P. G., and Ziegenfus, C. 1996. Phenotypic engineering: using hormones to explore the mechanistic and functional bases of phenotypic variation in nature. *Ibis* 138, 1–17.

Kokko, H. 2003. Are reproductive skew models evolutionary stable? *Proceedings of the Royal Society of London B* 270, 265–270.

Kokko, H., Johnstone, R. A., and Wright, J. 2002. The evolution of parental and alloparental effort in cooperatively breeding groups: when should helpers pay to stay? *Behavioral Ecology* 13, 291–300.

Kotiaho, J. S. and Simmons, L. W. 2003. Longevity cost of reproduction for males but no longevity cost of mating or courtship for females in the male-dimorphic dung beetle *Onthophagus binodis*. *Journal of Insect Physiology* 49, 817–822.

Kotiaho, J. S. and Tomkins, J. L. 2001. The discrimination of alternative male morphologies. *Behavioral Ecology* 12, 553–557.

Kurdziel, J. P. and Knowles, L. L. 2002. The mechanisms of morph determination in the amphipod *Jassa*: implications for the evolution of alternative male phenotypes. *Proceedings of the Royal Society of London B* 269, 1749–1754.

Langellotto, G. A. and Denno, R. F. 2001. Benefits of dispersal in patchy environments: mate location by males of a wing-dimorphic insect. *Ecology* 82, 1870–1878.

Langellotto, G. A., Denno, R. F., and Ott, J. R. 2000. A trade-off between flight capability and reproduction in males of a wing-dimorphic insect. *Ecology* 81, 865–875.

Lank, D. B., Smith, C. M., Hanotte, O., Burke, T., and Cooke, F. 1995. Genetic polymorphism for alternative mating behaviour in lekking male ruff *Philomachus pugnax*. *Nature* 378, 59–62.

Lank, D. B., Coupe, M., and Wynne-Edwards, K. E. 1999. Testosterone-induced male traits in female ruffs (*Philomachus pugnax*): autosomal inheritance and gender differentiation. *Proceedings of the Royal Society of London B* 266, 2323–2330.

Lee, J. S. F. 2005. Alternative reproductive tactics and status-dependent selection. *Behavioral Ecology* 16, 566–570.

Lejeune, P. 1985. Etude écoéthologique des comportements reproducteurs et sociaux des Labridae méditerranéens des genres *Symphodus* Rafinesque, 1810 et *Coris* Lacepède, 1802. *Cahiers d'Ethologie* 5, 1–208.

Lens, L., Wauters, L. A., and Dhondt, A. A. 1994. Nest-building by crested tit *Parus cristatus* males: an analysis of costs and benefits. *Behavioral Ecology and Sociobiology* 35, 431–436.

Lu, G. Q. and Bernatchez, L. 1999. Correlated trophic specialization and genetic divergence in sympatric lake whitefish ecotypes (*Coregonus clupeaformis*): support for the ecological speciation hypothesis. *Evolution* 53, 1491–1505.

Lucas, J. R., Howard, R. D., and Palmer, J. G. 1996. Callers and satellites: chorus behaviour in anurans as a stochastic dynamic game. *Animal Behaviour* 51, 501–518.

Lyon, B. E. 2003. Egg recognition and counting reduce costs of avian conspecific brood parasitism. *Nature* 422, 495–499.

Magnhagen, C. 1992. Alternative reproductive behaviour in the common goby, *Pomatoschistus microps*: an ontogenetic gradient? *Animal Behaviour* 44, 182–184.

Magrath, R. D. and Whittingham, L. A. 1997. Subordinate males are more likely to help if unrelated to the breeding female in cooperatively breeding white-browed scrubwrens. *Behavioral Ecology and Sociobiology* 41, 185–192.

Martin, E. and Taborsky, M. 1997. Alternative male mating tactics in a cichlid, *Pelvicachromis pulcher*: a comparison of reproductive effort and success. *Behavioral Ecology and Sociobiology* 41, 311–319.

Maynard Smith, J. 1982. *Evolution and the Theory of Games*. Cambridge, UK: Cambridge University Press.

Mboko, S. K. and Kohda, M. 1999. Piracy mating by large males in a monogamous substrate-breeding cichlid in Lake Tanganyika. *Journal of Ethology* 17, 51–55.

McGrew, W. C. and Marchant, L. F. 1997. On the other hand: current issues in and meta-analysis of the behavioral laterality of hand function in nonhuman primates. *Yearbook of Physical Anthropology* 40, 201–232.

Miller, P. L. 1984. Alternative reproductive routines in a small fly, *Puliciphora borinquenensis* (Diptera: Phoridae). *Ecological Entomology* 9, 293–302.

Moczek, A. P. and Emlen, D. J. 2000. Male horn dimorphism in the scarab beetle *Onthophagus taurus*: do alternative tactics favor alternative phenotypes? *Animal Behaviour* 59, 459–466.

Mole, S. and Zera, A. J. 1993. Differential allocation of resources underlies the dispersal-reproduction trade-off in the wing-dimorphic cricket, *Gryllus rubens*. *Oecologia* 93, 121–127.

Moore, M. C. 1991. Application of organization-activation theory to alternative male reproductive strategies: a review. *Hormones and Behavior* 25, 154–179.

Moore, M. C., Hews, D. K., and Knapp, R. 1998. Hormonal control and evolution of alternative male phenotypes: generalizations of models for sexual differentiation. *American Zoologist* 38, 133–151.

Morris, D. 1952. Homosexuality in the ten-spined stickleback (*Pygosteus pungitius* L.). *Behaviour* 4, 233–261.

Müller, J. K., Eggert, E. K., and Dressel, J. 1990. Intraspecific brood parasitism in the burying beetle, *Necrophorus vespilloides* (Coleoptera: Silphidae). *Animal Behaviour* 40, 491–499.

Munoz, R. C. and Warner, R. R. 2003. A new version of the size-advantage hypothesis for sex change: incorporating sperm competition and size–fecundity skew. *American Naturalist* 161, 749–761.

Nakajima, M., Matsuda, H., and Hori, M. 2004. Persistence and fluctuation of lateral dimorphism in fishes. *American Naturalist* 163, 692–698.

Nijhout, H. F. 2003. Development and evolution of adaptive polyphenisms. *Evolution and Development* 5, 9–18.

Oliveira, R. F. 2005. Neuroendocrine mechanisms of alternative reproductive tactics in fish. In K. A. Sloman, R. W. Wilson, and S. Balshine (eds.) *Fish Physiology*, vol. 24, *Behavior and Physiology of Fish*, pp. 297–357. New York: Elsevier.

Oliveira, R. F., Carvalho, N., Miranda, J., et al. 2002. The relationship between the presence of satellite males and nest-holders' mating success in the Azorean rock-pool blenny *Parablennius sanguinolentus parvicornis*. *Ethology* 108, 223–235.

Oliveira, R. F., Ros, A. F. H., and Gonçalves, D. M. 2005. Intra-sexual variation in male reproduction in teleost fish: a comparative approach. *Hormones and Behavior* 48, 430–439.

Parker, G. A., Baker, R. R., and Smith, V. G. F. 1972. The origin and evolution of gamete dimorphism and the male–female phenomenon. *Journal of Theoretical Biology* 36, 529–553.

Parker, H. H., Noonburg, E. G., and Nisbet, R. M. 2001. Models of alternative life-history strategies, population structure and potential speciation in salmonid fish stocks. *Journal of Animal Ecology* 70, 260–272.

Perrill, S. A., Gerhardt, H. C., and Daniel, R. E. 1982. Mating strategy shifts in male green treefrogs (*Hyla cinerea*): an experimental study. *Animal Behaviour* 30, 43–48.

Petrie, M. and Møller, A. P. 1991. Laying eggs in other's nests: intraspecific brood parasitism in birds. *Trends in Ecology and Evolution* 6, 315–320.

Pienaar, J. and Greeff, J. M. 2003. Different male morphs of *Otitesella pseudoserrata* fig wasps have equal fitness but are not determined by different alleles. *Ecology Letters* 6, 286–289.

Pigeon, D., Chouinard, A., and Bernatchez, L. 1997. Multiple modes of speciation involved in the parallel evolution of sympatric morphotypes of lake whitefish (*Coregonus clupeaformis*, Salmonidae). *Evolution* 51, 196–205.

Plaistow, S. J., Johnstone, R. A., Colegrave, N., and Spencer, M. 2004. Evolution of alternative mating tactics: conditional versus mixed strategies. *Behavioral Ecology* 15, 534–542.

Prestwich, K. N. 1994. The energetics of acoustic signaling in anurans and insects. *American Zoologist* 34, 625–643.

Pruden, A. J. and Uetz, G. W. 2004. Assessment of potential predation costs of male decoration and courtship display in wolf spiders using video digitization and playback. *Journal of Insect Behavior* 17, 67–80.

Radwan, J. 1993. The adaptive significance of male polymorphism in the acarid mite *Caloglyphus berlesei*. *Behavioral Ecology and Sociobiology* 33, 201–208.

Radwan, J. 1995. Male morph determination in two species of acarid mites. *Heredity* 74, 669–673.

Radwan, J. 2001. Male morph determination in *Rhizoglyphus echinopus* (Acaridae). *Experimental and Applied Acarology* 25, 143–149.

Radwan, J. 2003. Heritability of male morph in the bulb mite, *Rhizoglyphus robini* (Astigmata, Acaridae). *Experimental and Applied Acarology* 29, 109–114.

Radwan, J. and Bogacz, I. 2000. Comparison of life-history traits of the two male morphs of the bulb mite, *Rhizoglyphus robini*. *Experimental and Applied Acarology* 24, 115–121.

Radwan, J. and Klimas, M. 2001. Male dimorphism in the bulb mite, *Rhizoglyphus robini*: fighters survive better. *Ethology Ecology and Evolution* 13, 69–79.

Radwan, J., Unrug, J., and Tomkins, J. L. 2002. Status-dependence and morphological trade-offs in the expression of a sexually selected character in the mite, *Sancassania berlesei*. *Journal of Evolutionary Biology* 15, 744–752.

Reichard, M., Bryja, J., Ondrackova M., et al. 2005. Sexual selection for male dominance reduces opportunities for female mate choice in the European bitterling (*Rhodeus sericeus*). *Molecular Ecology* 14, 1533–1542.

Reinhold, K., Greenfield, M. D., Jang, Y. W., and Broce, A. 1998. Energetic cost of sexual attractiveness: ultrasonic advertisement in wax moths. *Animal Behaviour* 55, 905–913.

Repka, J. and Gross, M. R. 1995. The evolutionarily stable strategy under individual condition and tactic frequency. *Journal of Theoretical Biology* 176, 27–31.

Reyer, H. U. 1984. Investment and relatedness: a cost/benefit analysis of breeding and helping in the pied kingfisher (*Ceryle rudis*). *Animal Behaviour* 32, 1163–1178.

Reyer, H. U. 1986. Breeder–helper interactions in the pied kingfisher reflect the costs and benefits of cooperative breeding. *Behaviour* 96, 277–303.

Ribbink, A. J. 1977. Cuckoo among Lake Malawi cichlid fish. *Nature* 267, 243–244.

Richardson, D. S., Burke, T., and Komdeur, J. 2002. Direct benefits and the evolution of female-biased cooperative breeding in Seychelles warblers. *Evolution* 56, 2313–2321.

Rico, C., Kuhnlein, U., and FitzGerald, G. J. 1992. Male reproductive tactics in the threespine stickleback: an evaluation by DNA fingerprinting. *Molecular Ecology* 1, 79–87.

Robinson, B. W. and Wilson, D. S. 1996. Genetic variation and phenotypic plasticity in a trophically polymorphic population of pumpkinseed sunfish (*Lepomis gibbosus*). *Evolutionary Ecology* 10, 631–652.

Roff, D. A. 1996. The evolution of threshold traits in animals. *Quarterly Review of Biology* 71, 3–35.

Rohwer, F. C. and Freeman, S. 1989. The distribution of conspecific nest parasitism in birds. *Canadian Journal of Zoology* 67, 239–253.

Ros, A. F. H., Oliveira, R. F., Bouton, N., and Santos, R. S. 2006. Alternative male reproductive tactics and the immunocompetence handicap in the Azorean rock pool blenny, *Parablennius parvicornis*. *Proceedings of the Royal Society of London B* 273, 901–909.

Rowland, W. J. 1979. Stealing fertilizations in the fourspine stickleback, *Apeltes quadracus*. *American Naturalist* 114, 602–604.

Ryan, M. J., Pease, C. M., and Morris, M. R. 1992. A genetic polymorphism in the swordtail, *Xiphophorus nigrensis*: testing the prediction of equal fitnesses. *American Naturalist* 139, 21–31.

Sage, R. D. and Selander, R. K. 1975. Trophic radiation through polymorphism in cichlid fishes. *Proceedings of the National Academy of Sciences of the United States of America* 72, 4669–4673.

Sandell, M. I. and Diemer, M. 1999. Intraspecific brood parasitism: a strategy for floating females in the European starling. *Animal Behaviour* 57, 197–202.

Sato, T., Hirose, M., Taborsky, M., and Kimura, S. 2004. Size-dependent male alternative reproductive tactics in the shell-brooding cichlid fish *Lamprologus callipterus* in Lake Tanganyika. *Ethology* 110, 49–62.

Scheiner, S. M. 1993. Genetics and evolution of phenotypic plasticity. *Annual Review of Ecology and Systematics* 24, 35–68.

Schlichting, C. D. and Pigliucci, M. 1998. *Phenotypic Evolution: A Reaction Norm Perspective*. Sunderland, MA: Sinauer Associates.

Shuster, S. M. 1989. Male alternative reproductive strategies in a marine isopod crustacean (*Paracerceis sculpta*): the use of genetic markers to measure differences in fertilization success among a-, B-, and g-males. *Evolution* 43, 1683–1698.

Shuster, S. M. 1992. The reproductive behaviour of alpha, beta and gamma-male morphs in *Paracerceis sculpta*, a marine isopod crustacean. *Behaviour* 121, 231–258.

Shuster, S. M. and Wade, M. J. 1991. Equal mating success among male reproductive strategies in a marine isopod. *Nature* 350, 606–611.

Shuster, S. M. and Wade, M. J. 2003. *Mating Systems and Strategies*. Princeton, NJ: Princeton University Press.

Simmons, L. W., Teale, R. J., Maier, M., *et al.* 1992. Some costs of reproduction for male bush-crickets, *Requena verticalis* (Orthoptera, Tettigoniidae): allocating resources to mate attraction and nuptial feeding. *Behavioral Ecology and Sociobiology* 31, 57–62.

Sinervo, B. 2001. Runaway social games, genetic cycles driven by alternative male and female strategies, and the origin of morphs. *Genetica* 112, 417–434.

Sinervo, B. and Lively, C. M. 1996. The rock–paper–scissors game and the evolution of alternative male strategies. *Nature* 380, 240–243.

Skubic, E., Taborsky, M., McNamara, J. M., and Houston, A. I. 2004. When to parasitize? A dynamic optimization model of reproductive strategies in a cooperative breeder. *Journal of Theoretical Biology* 227, 487–501.

Skúlason, S. and Smith, T. B. 1995. Resource polymorphisms in vertebrates. *Trends in Ecology and Evolution* 10, 366–370.

Slatkin, M. 1978. Equilibration of fitnesses by natural selection. *American Naturalist* 112, 845–859.

Slatkin, M. 1979. Evolutionary response to frequency-dependent and density-dependent interactions. *American Naturalist* 114, 384–398.

Smith, T. B. and Skúlason, S. 1996. Evolutionary significance of resource polymorphisms in fishes, amphibians, and birds. *Annual Review of Ecology and Systematics* 27, 111–133.

Snorrason, S. S. and Skúlason, S. 2004. Adaptive speciation in northern freshwater fishes: patterns and processes. In U. Dieckmann, H. Metz, M. Doebeli and D. Tautz (eds.) *Adaptive Speciation*, pp. 210–228. Cambridge, UK: Cambridge University Press.

Snorrason, S. S., Skúlason, S., Jonsson, B., *et al.* 1994. Trophic specialization in arctic charr *Salvelinus alpinus* (Pisces, Salmonidae): morphological divergence and ontogenic niche shifts. *Biological Journal of the Linnean Society* 52, 1–18.

Stiver, K. A., Dierkes, P., Taborsky, M., Gibbs, H. L., and Balshine, S. 2005. Relatedness and helping in fish: examining the theoretical predictions. *Proceedings of the Royal Society of London B* 272, 1593–1599.

Taborsky, B., Dieckmann, U., and Heino, M. 2003. Unexpected discontinuities in life-history evolution under size-dependent mortality. *Proceedings of the Royal Society of London B* 270, 713–721.

Taborsky, M. 1984. Broodcare helpers in the cichlid fish *Lamprologus brichardi*: their costs and benefits. *Animal Behaviour* 32, 1236–1252.

Taborsky, M. 1985. Breeder–helper conflict in a cichlid fish with broodcare helpers: an experimental analysis. *Behaviour* 95, 45–75.

Taborsky, M. 1994. Sneakers, satellites, and helpers: parasitic and cooperative behavior in fish reproduction. *Advances in the Study of Behavior.* 23, 1–100.

Taborsky, M. 1997. Bourgeois and parasitic tactics: do we need collective, functional terms for alternative reproductive behaviours? *Behavioral Ecology and Sociobiology* 41, 361–362.

Taborsky, M. 1998. Sperm competition in fish: bourgeois males and parasitic spawning. *Trends in Ecology and Evolution* 13, 222–227.

Taborsky, M. 1999. Conflict or cooperation: what determines optimal solutions to competition in fish reproduction? In R. F. Oliveira, V. Almada, and E. Gonçalves (eds.) *Behaviour and Conservation of Littoral Fishes*, pp. 301–349. Lisbon: Instituto Superior de Psicologia Aplicada.

Taborsky, M. 2001. The evolution of parasitic and cooperative reproductive behaviors in fishes. *Journal of Heredity* 92, 100–110.

Taborsky, M., Hudde, B., and Wirtz, P. 1987. Reproductive behaviour and ecology of *Symphodus (Crenilabrus) ocellatus*, a European wrasse with four types of male behaviour. *Behaviour* 102, 82–118.

Tallamy, D. W. 1985. "Egg dumping" in lace bugs (*Gargaphia solani*, Hemiptera: Tingidae). *Behavioral Ecology and Sociobiology* 17, 357–362.

Tallamy, D. W. 2005. Egg dumping in insects. *Annual Review of Entomology* 50, 347–370.

Tallamy, D. W. and Horton, L. A. 1990. Costs and benefits of the eggdumping alternative in *Gargaphia* lace bugs (Hemiptera: Tingidae). *Animal Behaviour* 39, 352–359.

Thomas, R. J. 2002. The costs of singing in nightingales. *Animal Behaviour* 63, 959–966.

Thorpe, J. E. 1986. Age at first maturity in Atlantic salmon, *Salmo salar*: freshwater period influences and conflicts with smolting. In D. J. Meerburg (ed.) *Salmonid Age at Maturity*, pp. 7–14. Ottawa, ON: Department of Fisheries and Oceans.

Thorpe, J. E. and Morgan, R. I. G. 1980. Growth rate and smolting rate of progeny of male Atlantic salmon parr (*Salmo salar* L.). *Journal of Fisheries Biology* 17, 451–459.

Timms, S., Ferro, D. N., and Waller, J. B. 1980. Suppression of production of pleomorphic males in *Sancassania berlesei* (Michael) (Acari: Acaridae). *International Journal of Acarology* 6, 91–96.

Timms, S., Ferro, D. N., and Emberson, R. M. 1982. Andropolymorphism and its heritability in *Sancassania berlesei* (Michael) (Acari, Acaridae). *Acarologia* 22, 391–398.

Tomkins, J. L. and Brown, G. S. 2004. Population density drives the local evolution of a threshold dimorphism. *Nature* 431, 1099–1103.

Tomkins, J. L. and Simmons, L. W. 2000. Sperm competition games played by dimorphic male beetles: fertilization gains with equal mating access. *Proceedings of the Royal Society of London B* 267, 1547–1553.

Tomkins, J. L., LeBas, N. R., Unrug, J., and Radwan, J. 2004. Testing the status-dependent ESS model: population variation in fighter expression in the mite *Sancassania berlesei*. *Journal of Evolutionary Biology* 17, 1377–1388.

Tomkins, J. L., Kotiaho, J. S., and LeBas, N. R. 2005. Matters of scale: positive allometry and the evolution of male dimorphisms. *American Naturalist* 165, 389–402.

Trudel, M., Tremblay, A., Schetagne, R., and Rasmussen, J. B. 2001. Why are dwarf fish so small? An energetic analysis of polymorphism in lake whitefish (*Coregonus clupeaformis*). *Canadian Journal of Fisheries and Aquatic Sciences* 58, 394–405.

Unrug, J., Tomkins, J. L., and Radwan, J. 2004. Alternative phenotypes and sexual selection: can dichotomous handicaps honestly signal quality? *Proceedings of the Royal Society of London B* 271, 1401–1406.

Utami, S. S., Goossens, B., Bruford, M. W., de Ruiter, J. R. and van Hooff, J. A. R. A. 2002. Male bimaturism and reproductive success in Sumatran orang-utans. *Behavioral Ecology* 13, 643–652.

van den Berghe, E. P. 1988. Piracy: a new alternative male reproductive tactic. *Nature* 334, 697–698.

Vehrencamp, S. L. 1983. A model for the evolution of despotic versus egalitarian societies. *Animal Behaviour* 31, 667–682.

Verspoor, E. and Cole, L. J. 1989. Genetically distinct sympatric populations of resident and anadromous Atlantic salmon, *Salmo salar*. *Canadian Journal of Zoology/Revue Canadienne de Zoologie* 67, 1453–1461.

Villalobos, E. M. and Shelly, T. E. 1996. Intraspecific nest parasitism in the sand wasp *Stictia heros* (Fabr.) (Hymenoptera: Sphecidae). *Journal of Insect Behavior* 9, 105–119.

Vollestad, L. A., Peterson, J., and Quinn, T. P. 2004. Effects of freshwater and marine growth rates on early maturity in male coho and Chinook salmon. *Transactions of the American Fisheries Society* 133, 495–503.

Wagner, W. E. 2005. Male field crickets that provide reproductive benefits to females incur higher costs. *Ecological Entomology* 30, 350–357.

Ward, S., Speakman, J. R., and Slater, P. J. B. 2003. The energy cost of song in the canary, *Serinus canaria*. *Animal Behaviour* 66, 893–902.

Warner, R. R. and Hoffman, S. G. 1980. Population density and the economics of territorial defense in a coral reef fish. *Ecology* 61, 772–780.

Warner, R. R., Robertson, D. R., and Leigh, E. G. J. 1975. Sex change and sexual selection. *Science* 190, 633–638.

Webster, M. S., Tarvin, K. A., Tuttle, E. M., and Pruett-Jones, S. 2004. Reproductive promiscuity in the splendid fairy-wren: effects of group size and auxiliary reproduction. *Behavioral Ecology* 15, 907–915.

West-Eberhard, M. J. 1986. Alternative adaptations, speciation, and phylogeny. *Proceedings of the National Academy of Sciences of the United States of America* 83, 1388–1392.

West-Eberhard, M. J. 1989. Phenotypic plasticity and the origins of diversity. *Annual Review of Ecology and Systematics* 20, 249–278.

West-Eberhard, M. J. 2003. *Developmental Plasticity and Evolution*. New York: Oxford University Press.

Westneat, D. F. 1993. Temporal patterns of within-pair copulations, male mate guarding, and extra-pair events in eastern red-winged blackbirds (*Agelaius phoeniceus*). *Behaviour* 124, 267–290.

Westneat, D. F. and Stewart, I. R. K. 2003. Extra-pair paternity in birds: causes, correlates, and conflict. *Annual Review of Ecology Evolution and Systematics* 34, 365–396.

Whittingham, L. A., Dunn, P. O., and Magrath, R. D. 1997. Relatedness, polyandry and extra-group paternity in the cooperatively-breeding white-browed scrubwren (*Sericornis frontalis*). *Behavioral Ecology and Sociobiology* 40, 261–270.

Wingfield, J. C., Jacobs, J. D., Soma, K., *et al.* 1999. Testosterone, aggression, and communication: ecological bases of endocrine phenomena. In M. D. Hauser and M. Konishi (eds.) *The Design of Animal Communication*, pp. 257–283. Cambridge, MA: MIT Press.

Wingfield, J. C., Lynn, S. E., and Soma, K. K. 2001. Avoiding the "costs" of testosterone: ecological bases of hormone–behavior interactions. *Brain, Behavior, and Evolution* 57, 239–251.

Wirtz, P. 1982. Territory holders, satellite males and bachelor males in a high density population of waterbuck (*Kobus ellipsiprymnus*) and their associations with conspecifics. *Zeitschrift für Tierpsychologie* 58, 277–300.

Woodring, J. P. 1969. Observations on the biology of six species of acarid mites. *Annals of the Entomological Society of America* 62, 102–108.

Yanagisawa, Y. 1985. Parental strategy of the cichlid fish *Perissodus microlepis*, with particular reference to intraspecific brood "farming out." *Environmental Biology of Fish* 12, 241–249.

Yoccoz, N. G., Mysterud, A., Langvatn, R., and Stenseth, N. C. 2002. Age- and density-dependent reproductive effort in male red deer. *Proceedings of the Royal Society of London B* 269, 1523–1528.

Yom-Tov, Y. 1980. Intraspecific nest parasitism in birds. *Biological Reviews of the Cambridge Philosophical Society* 55, 93–108.

Yom-Tov, Y. 2001. An updated list and some comments on the occurrence of intraspecific nest parasitism in birds. *Ibis* 143, 133–143.

Zera, A. J. and Rankin, M. A. 1989. Wing dimorphism in *Gryllus rubens*: genetic basis of morph determination and fertility differences between morphs. *Oecologia* 80, 249–255.

Zink, A. G. 2003. Intraspecific brood parasitism as a conditional reproductive tactic in the treehopper *Publilia concava*. *Behavioral Ecology and Sociobiology* 54, 406–415.

Zupanc, G. K. H. and Lamprecht, J. 2000. Towards a cellular understanding of motivation: structural reorganization and biochemical switching as key mechanisms of behavioral plasticity. *Ethology* 106, 467–477.

Part I
Ultimate causes and origins of alternative reproductive tactics

2 · Alternative reproductive tactics and the evolution of alternative allocation phenotypes

H. JANE BROCKMANN AND MICHAEL TABORSKY

CHAPTER SUMMARY

Alternative reproductive tactics (ARTs) are part of a much larger class of alternative phenotypes that include sex allocation and alternative life histories. We examine the evolution of ARTs by drawing on the much larger base of theory from sex-allocation and life-history evolution. Insights into how alternative tactics evolve (their maintenance in populations, the evolution of their underlying mechanisms and flexibility, the evolution of morph differences and morph frequencies) are derived from principles developed for understanding the evolution of sex, sex determination, hermaphroditism, sexual dimorphism, and sex ratios.

2.1 INTRODUCTION

Darwin (1871) was fascinated by variation. In part this was because so many scholars at the time emphasized typological thinking and ignored the biological variation around them. But more importantly Darwin realized that heritable variation was at the heart of his theory. If variants showed differential survival and if those characteristics were passed on to their offspring, then evolution occurred. He understood that if one form were just a little more successful than the other, then the variant with the higher success would prevail. This understanding led him to worry about cases in which discrete variation was maintained at a stable frequency in one population. These worries included social insect castes, sexual dimorphism, and alternative forms of one sex (Shuster and Wade 2003).

Variation within one population is usually continuous but under some circumstances, discrete, discontinuous patterns of variation evolve and are maintained. Sexual dimorphism is the most obvious case. Sons and daughters are alternative, parental allocation tactics for achieving the same functional end, reproduction (Charnov 1982) (Box 2.1). Alternative, discrete forms can also be found in

life-history patterns where two forms differ in their schedules of age-specific maturation, dispersal, and reproduction. For example, male bluegill sunfish show two life-history trajectories that are maintained in populations over generations (Gross 1984, 1991a) (see Figure 2.2). Some males mature quickly and begin to breed at a young age and small size (parasitic tactic). Other males take longer to mature and begin breeding at a later age and larger size (bourgeois tactic). Once mature, the larger males invest in guarding nests, attracting females, and providing brood care, whereas the smaller males obtain fertilizations by joining spawning pairs. When quite small they release sperm by sneaking into nests (Philipp and Gross 1994) or, when older and similar in size to females, by mimicking female behavior (Dominey 1980). As adults the two male forms are ARTs, but during development they represent alternative life-history pathways.

Sex allocation, alternative life histories, and ARTs are part of a much larger set of alternative allocation phenotypes (Waltz and Wolf 1984, Lloyd 1987, Brockmann 2001). They include mimicry, color, and other protective polymorphisms (Turner 1977, Brönmark and Miner 1992, Sword 1999, Gonçalves et al. 2004), trophic polymorphism (Collins and Cheek 1983, Pfennig 1992, Skúlason and Smith 1995), partial migration (Kaitala et al. 1993), seasonal and phase polyphenism (Greene 1989, 1999, Moran 1992, Sword 1999, Sword et al. 2000), predator-induced reaction norms (Dodson 1989), alternative germination strategies in plants (Mathias and Kisdi 2002), caste polymorphism and polyethism in social insects (Wheeler 1991), and producer-scrounger systems (Barnard and Sibly 1981, Barnard 1984, Giraldeau and Livoreil 1998, Giraldeau and Caraco 2000). Although very diverse, alternative allocation phenotypes share important features. First, like all forms of phenotypic plasticity, alternative allocation phenotypes occur in one of three general patterns (Barnard and Sibly 1981, Gross 1984,

Alternative Reproductive Tactics, ed. Rui F. Oliveira, Michael Taborsky, and H. Jane Brockmann. Published by Cambridge University Press.

Box 2.1 Sex-allocation theory explained using red deer (*Cervus elephus*) as an example

Sex allocation results from maternal decisions (Figure 2.1A) about how to allocate resources between son and daughter production (resources a mother puts into a son that are not put into a daughter). Since every offspring in the next generation has exactly one mother and one father, son and daughter production are equally successful on average (although there is often higher variance in the fitness of sons). This means that the equilibrium – evolutionarily stable state (ESS) – is 50% female and 50% male offspring or equal investment in sons and daughters. If for some reason there were too

many sons in the population (above the 50% son ESS), then the success of son production would drop and selection would favor females that produce more daughters, thus returning the population to the ESS. So, the rarer sex is more successful, i.e., fitness is frequency dependent (Figure 2.1B).

Many species invest equally in a son as in a daughter but this is not always the case. When a female red deer bears a son, she tends not to have another offspring in the following year, whereas when she bears a daughter, she will probably reproduce again in the following year (Clutton-Brock *et al*. 1982). This means that females tend to invest more in sons than in daughters. This is because females that invest more have sons that are in better

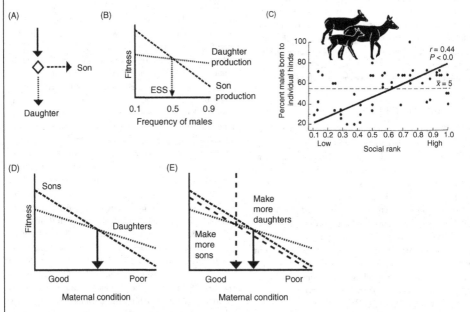

Figure 2.1 (A) Sex allocation is about the distribution of resources between son and daughter production. (B) The ESS occurs at the frequency where the production of sons and daughters is equally successful. The fitness of sons (or daughters) in the population is higher when they are rare, i.e., success is frequency dependent and tends to return the population to the ESS. (C) Birth sex ratios produced by individual female red deer differing in social rank over their lifespans. Measures of maternal rank were based on the ratio of animals that the subject threatened or displaced to animals that threatened or displaced it. (D) The result given in C means that sex allocation is in part status dependent, resulting from crossing fitness curves. Females should switch from daughter to son production at the age or condition that maximizes fitness. (E) If this status-dependent condition rule given in (D) results in too many sons (as might happen when conditions are good), then frequency-dependent selection will act to change the decision rule and return the population to the ESS.

physical shape at the time of weaning. Well-nourished sons are more likely to become dominant adults and thus increase their mother's fitness (Clutton-Brock et al. 1988). In species where one sex is more costly to produce than the other, as in red deer, selection favors an equilibrium sex ratio at fewer numbers of the more expensive sex. Selection favors equalizing maternal investment in the two sexes and not numbers of each sex.

In red deer, as in many other species, sex allocation decisions are both environment and status dependent even though the species has chromosomal sex determination. If a female is in good condition and of high rank, she is more likely to bear a son, and if she is in poor condition, she is more likely to have a daughter (Clutton-Brock et al. 1986) (Figure 2.1C). It is also known that more sons are born to dominant females when population density is low than when it is high (Kruuk et al. 1999).

Clearly, then, both condition dependence and frequency dependence are operating in sex-allocation decisions (Figure 2.1D). How do the two interact? If the condition- or status-dependent sex-allocation rule were to result in too many sons, i.e., above the ESS (as might occur when conditions are very good), then the average success of sons would drop. This would mean that females that had a slightly different decision rule that resulted in fewer sons would be favored. In this way the population would return to the ESS where the animals would be using a slightly different decision rule that would result in equal success through son and daughter production (Figure 2.1E).

Taborsky 1994, 2001, Schlichting and Pigliucci 1998, Moore et al. 1998, Brockmann 2001, Pigliucci 2001, Piersma and Drent 2003, Shuster and Wade 2003) (Figure 1.1). (a) In some cases the two phenotypes are inflexible alternatives (e.g., dioecious species, fixed tactics), but in other cases, (b) individuals may switch from one phenotype to the other during their adult lives (e.g., hermaphroditic species, conditional tactics). This switch may occur at a particular age or size or under certain social conditions (e.g., sequential hermaphroditism, phenotypic flexibility), or (c) some animals can flexibly change back and forth between phenotypes as adults (e.g., simultaneous hermaphroditism, life-cycle staging). Charnov (1982, 1986) showed that the basic principles of sex-allocation theory apply equally to dioecious and hermaphroditic patterns. Second, the mechanisms controlling alternative phenotypes include genetic differences (e.g., chromosomal sex determination), environmental or social differences (e.g., temperature- or behavior-dependent sex determination: Crews 1993; Karplus 2005), or combinations of these possibilities (Brockmann 2001). Third, alternative phenotypes occur when individuals make decisions that commit them to investing limited time or resources to a particular course of action, i.e., a "decision" is made. The time or resources that are invested in one option are then no longer available for investing in the other option, i.e., trade-offs exist. In fact, any mechanisms that result in the correlated expression of traits can cause trade-offs (Stearns 1992, Angilletta et al. 2003). Fourth, although various processes can maintain variation in populations (e.g., mutation, pleiotropy, drift, heterozygous advantage), alternative allocation phenotypes are usually found to be adaptations. Each route that an individual might follow has its own net benefit and cost for the individual; if differences in success are in part due to heritable differences, then the mechanisms that underlie decisions evolve through natural selection. Often one route invades the population rather than the other, and in such cases, the route that evolves is the one that maximizes long-term reproductive success (pure strategy). More interesting are cases in which two or more phenotypes (e.g., two sexes, two life-history patterns, or two mating tactics) are maintained in the population over generations.

If we are to understand the evolution of alternative allocation phenotypes, we must understand both how animals execute a particular decision and why that decision is favored by natural selection. What are the sources of information available to individuals (either during development or as adults); how is this information processed and translated into action; what are the mechanisms that allow individuals to switch into a particular morph or from one morph into another; and what is the nature of the trade-offs involved? These proximate topics are discussed in Chapters 5–7. Here we concentrate on ultimate aspects: why alternative phenotypes evolve. We divide this topic into five sections, recognizing that these are not mutually exclusive topics (Taborsky 1998, 1999, 2001, Brockmann 2001).

(1) Evolution of alternative phenotypes. What favors the evolution of two or more phenotypes in one population and how does the success of a phenotype depend on the relative frequencies of alternative phenotypes in

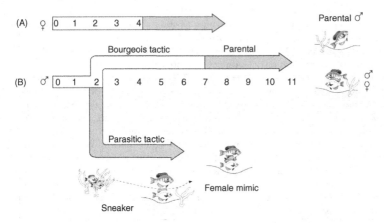

Figure 2.2 Alternative life-history and reproductive tactics in bluegill sunfish. (A) Females mature after 4 years whereas bourgeois males (B) mature at 7 years and as adults prepare and guard nests in which females spawn. Parasitic males mature at a much younger age and smaller size and do not build or guard nests but nonetheless fertilize eggs by sneaking into nests during spawning between a female and a parental male. (Redrawn from Gross 1984, 1991a.)

the population? When thinking about this question in sex allocation, we are asking about the evolution and maintenance of sex (why two sexes exist and why they are maintained in the population). When studying alternative life histories or alternative reproductive tactics, we are asking about the evolutionary processes that favor the initial appearance and coexistence of multiple tactics in one population.

(2) Evolution of underlying mechanisms. Why are alternative phenotypes sometimes controlled by a genetic polymorphism, sometimes by maternal factors, and sometimes by a condition-dependent switch in the individual? With sex allocation this is a question about the evolution of sex-determining mechanisms. With alternative life histories and mating tactics, this is a question about the evolution of gene–environment interactions, such as reaction norms and phenotypic plasticity.

(3) Evolution of phenotype flexibility and rigidity. Why are some phenotypes irreversible and others not? In sex allocation this is a problem in understanding why some species are dioecious whereas others are hermaphroditic, and among the hermaphroditic species, why some are sequential and others simultaneous hermaphrodites. With life histories and ARTs, we need to understand why it is that in some cases animals switch from one pattern to another, either depending on current conditions or in sequence over a lifetime, whereas in other cases, individuals permanently follow one tactic or the other throughout their reproductive lives.

(4) Evolution of alternative phenotypes or dimorphism. Why do alternative phenotypes differ and what explains the extent of the differences? In sex allocation this is a question about the evolution of sexual dimorphism; in alternative life-history or reproductive tactics, this is a question about the suite of characters that covary with each form and how these traits co-evolve.

(5) Evolution of phenotypic frequencies. What are the selective pressures that affect the relative frequencies of different phenotypes in a population? In sex allocation this is a question about the evolution of sex ratios, and in alternative life-history or reproductive tactics, this is a question about the frequency of the different tactics in the population.

Although these five questions intersect in various ways, we shall discuss each in turn as a way of organizing the multiple problems involved in understanding the evolution of alternative phenotypes. Because sex-ratio theory is so well developed and the closely related problem of alternative phenotype frequencies is so poorly developed in the literature, we place greater emphasis on this question than on the others. We will conclude by discussing some lessons for understanding ARTs that are derived from using this broader view of alternative allocation phenotypes.

2.2 EVOLUTION OF ALTERNATIVE PHENOTYPES: WHAT FAVORS MULTIPLE PHENOTYPES IN ONE POPULATION?

The evolutionary processes that maintain alternative phenotypes in populations are much the same despite the diversity of functional contexts, which include sex allocation, alternative life histories, and ARTs (Waltz and Wolf 1984, Brockmann 2001). In the most general case, the evolution of alternative phenotypes has four requirements: (a) a mechanism by which distinct (rather than continuous) phenotypes can develop (such as a developmental switch) (see Section 2.3); (b) heritable variation in the mechanisms controlling the expressions of the alternative phenotypes (Section 2.3; see also Chapter 5); (c) selection that favors multiple phenotypes rather than one optimal phenotype, i.e., disruptive selection (Section 2.2.2); and (d) crossing fitness curves (Section 2.2.1). Taken together these four processes are required for the maintenance of distinct alternative phenotypes in a population.

2.2.1 Crossing fitness curves and selection

Discrete phenotypic variants or morphs are maintained in populations when their fitness curves cross (Waltz 1982, Gross 1996, Taborsky 1999, Brockmann 2001) (Box 1.2). This occurs when individuals must choose (during development or as adults) between mutually exclusive alternatives and when those decisions present individuals with functional trade-offs (Halama and Reznick 2001). For example, a cichlid fish *Perissodus microlepis* from Lake Tanganyika sneaks up on other fish to eat their scales (Hori 1993) (Figure 2.3). It must move extremely quickly and so it has evolved an asymmetric mouth: left-jawed fish attack from the right rear and right-jawed from the left rear. Host fish learn to avoid attacks so if all parasites were right-jawed, then a rare left-jawed morph would have an advantage. The success of a morph depends on its frequency in the population, and the two morphs are maintained at a frequency such that they are equally successful (frequency-dependent selection). In other cases (conditional tactics), animals use decision rules or switch points to change tactics so that they will derive the highest possible fitness based on their circumstances (sometimes referred to as "best of a bad job": Lee 2005). The success of a tactic may depend on some external factor in the environment (environment-dependent tactics such as temperature or daylength or social conditions

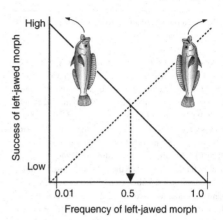

Figure 2.3 Frequency-dependent selection acting on two morphs of the scale-eating cichlid *Perissodus microlepis*. This fish from Lake Tanganyika eats the scales of other fish. It comes in two forms – a right-jawed morph that is most effective at removing scales from the left side of its host and a left-jawed morph that removes scales from the right side of its host. The figure shows the hypothesized model for the evolution of the two morphs. The success of the two morphs depends on their frequency in the population (each does well when rare). The frequency of the two morphs in one population over 9 years has cycled around 0.5 (arrow), the predicted evolutionarily stable state (ESS). (Redrawn from Hori 1993.)

such as density) and individuals should switch from one tactic to the other to maximize fitness depending on that factor (or a correlate of that factor) (Box 2.1). In still other cases, the tactic that maximizes fitness depends on some character in the individual such as its age or physical condition (Box 2.2) or its rate of growth (condition-dependent or status-dependent tactics) (Box 2.3). When decision-making switch points for conditional tactics are heritable (Emlen 1996, Tomkins 1999) (see Figure 5.7) and different decisions result in differential fitness, then the decision switch point evolves; the position of the switch point is the product of selection within that population (Hazel et al. 1990, Roff 1996, Moczek et al. 2002, Moczek 2003, Tomkins and Brown 2004, Tomkins et al. 2004). With environment-, condition-, or status- dependent tactics, the average fitness of the two forms is not necessarily equal (Repka and Gross 1995, Gross and Repka 1998a, b). The reason that we see both tactics in the population at the same time is that some individuals (or individuals at some points) can maximize fitness by following one decision path,

Box 2.2 Alternative mating tactics in horseshoe crabs
(*Limulus polyphemus*)

Male horseshoe crabs show two mating tactics: young males
arrive on the beach clasped to females with whom they
spawn as the female lays eggs in the beach sand; older males
arrive alone (Brockmann and Penn 1992, Penn and Brockmann 1995) (Figure 2.4A) and crowd around the nesting
couples as satellites. They engage in sperm competition with
the attached males and other satellites (Brockmann *et al.*
1994, 2000). They are quite successful, fertilizing on average
40% of the female's eggs when there are one or two satellites;
less if there are more satellites. Coming ashore as a satellite is
not simply a consequence of a male being unable to find or
hold onto a female. Rather, when males are experimentally
prevented from attaching, older and younger individuals
differ in their decisions to come ashore as satellites: older
males are more likely to come ashore and take up a satellite
position, whereas younger males are more likely to remain at
sea, presumably looking for females (Brockmann 2002). The

hypothesis to explain this difference is that the two tactics are
maintained by a condition-dependent divergence in optimal
behavior that depends mainly on age and that leads to
crossing fitness curves (Figure 2.4B): when males are young,
they have higher fitness by attaching and when older (or in
poorer condition), they have higher fitness by no longer
searching for unattached females at sea but by seeking out
nesting pairs on the beach (status- or condition-dependent
selection). Environmental conditions, female density, and
operational sex ratio likely to affect the lines and hence
the position of the switch point. Furthermore, the success of
the satellite tactic is likely to be affected by its frequency in
the population, since males that are spawning in groups of
three or more satellites have reduced fertilization success
(the first two satellites average 40% of the fertilizations each
if they are located over the female's incurrent canal, but
in larger groups the satellites reduce the success of other
satellites). This means that frequency-dependent effects
may be superimposed on condition-, status-, and environment-dependent effects on fitness.

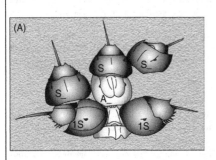

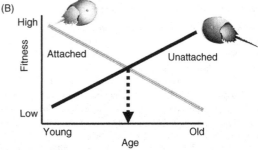

Figure 2.4 Alternative mating tatics in male horseshoe crabs
Limulus polyphemus. (A) Satellite male horseshoe crabs (indicated
as S) crowd around a nesting pair (attached male indicated as A).
Satellites that remain over the female's incurrent canal (indicated
as 1S) have higher fertilization success (average 40% each) than
those in other positions. (B) Model of spawning tactics that

captures the observation that young males are more likely to be
attached and presumably have higher fitness when attached than
older males. The arrow shows the age or condition at which a male
should change from coming ashore with a female (attached) and
coming ashore without a female (unattached) to engage in sperm
competition as a satellite.

whereas others (or at other points) can maximize fitness by
following a different path.

The decision rules that control alternative tactics are
often based on multiple factors that vary spatially and temporally. For example, the success of alternative tactics in
bluegill sunfish is affected by the amount of available cover,

population density, and predation pressure (Gross 1984). In
sex-allocation decisions of red deer (Box 2.1), environment-
dependent as well as condition-dependent effects influence
offspring sex, the population sex ratio, and the success
associated with son and daughter production for females
in different conditions. Similarly, in horseshoe crabs

Box 2.3   A model for the evolution of alternative tactics in male salmon (*Oncorhynchus kisutch*)

Males occur in two life-history phenotypes, hooknose and jack (Gross 1991b) (Figure 2.5A). The anadromous hooknose males swim to the ocean at the end of their first year where they mature, returning a year later to spawn with the females. Jack males mature precociously in their natal streams where they begin breeding in their first year. Females return to natal streams where they deposit

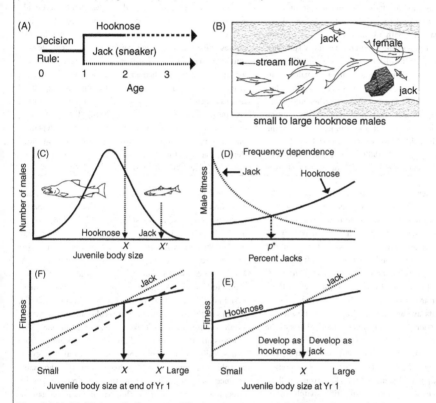

Figure 2.5   Model of the interaction between condition dependence and frequency dependence in the life-history decisions (and the ultimate reproductive tactics) of male coho salmon.
(A) Males have two life-history paths: some mature early and become a jack whereas others delay development and become a hooknose male. (B) The two kinds of males differ in their mating behavior. Hooknose males fight for position around females, whereas jacks slip into nests during spawning and fertilize eggs through sperm competition. (C) The decision to follow one life history rather than the other depends on the rate at which the juvenile is developing. Those that are developing the fastest (right end of distribution) become jacks, i.e., they mature at the end of their first year, whereas those that are growing more slowly become hooknose males, i.e., they go to sea for a year and return a year later. (D) The success of the jack tactic depends on its frequency in the population. The sneaker tactic does better when it is rare because the small, nonfighting males can get closer to the female; when sneakers are more common, then hooknose males become more vigilant. Also, since females will not spawn spontaneously with jacks, the success of the jacks depends on the frequency of jacks in the population. (E) Male proximity to spawning females by either fighting or sneaking (a measure of male fitness) is affected by body size. Fighting is most effectively done by large males and sneaking is most effectively done by small males, so fitness curves for the two tactics cross. (F) The condition-dependent switch, X, is shown by the solid lines and the arrow. However, if the condition-dependent developmental decision rule results in so many jacks that their average fitness drops (dashed line), then a new rule, a new switch point will be favored, X', at which the two tactics have equal success.

Box 2.3   (Cont.)

their eggs in nests (Figure 2.5B). Hooknose males fight viciously for access to females for spawning, and the largest males are most likely to fertilize the female's eggs. The small males, which are only 3.5% of the body mass of hooknose males, do not fight for position around the females but wait along the edges of the stream. When females are spawning with a hooknose male, a jack swims through the nest depositing sperm on the newly laid eggs, and in some cases his fertilization success is high (Foote *et al.* 1997). Females will not spawn with jack males spontaneously, but they will spawn if jacks are around (Thomaz *et al.* 1997). The decision about whether to develop into a hooknose or jack male is heritable and condition dependent (Figure 2.5C): if the animal is developing quickly, then he is likely to switch to the jack life history (Thorpe and Morgan 1978, 1980, Thorpe *et al.* 1983). The jack/hooknose tactics are frequency dependent (Figure

2.5D) since the success of a jack depends on the proportion of jacks in the population (they do well when rare: Hutchings and Myers 1988, Thomaz *et al.* 1997). Gross (1985) estimated the success of fighting and sneaking males and found that sneaking was more successful when the animals were small and fighting was more successful when the animals were large; males of intermediate size did poorly using either tactic – hence crossing fitness curves (Figure 2.5E). If this condition-dependent, developmental decision rule results in so many jacks that their average fitness drops, then a new rule, a new switch point (Figure 2.5F) should be favored that brings the frequency of jacks back to the point where their success is equal to that of the hooknose males (Gross 1991b). The result then is an evolutionary interplay between condition dependence, through developmental switch points, and frequency dependence in the evolution of morph frequencies (Hutchings and Myers 1994).

(Box 2.2), weather, tidal conditions, and female density differentially affect the relative success of the two tactics. This means that the best tactic for an individual to follow and the success associated with that tactic may vary, so when observations are made over a span of time or under different environmental conditions, we see different tactics and different rates of success (Waltz and Wolff 1988, Tomkins 1999, Lee 2005). Furthermore, in each of these examples, if for any reason one tactic is produced in such numbers that its average fitness drops, then frequency-dependent selection will act to reduce the numbers of the more common tactic. This means that frequency-dependent effects act along with condition-, status-, and environment-dependent effects on the fitness of alternative phenotypes (Brockmann 2001). The result will be different evolved switch points (or multiple switch points) for different populations (Moczek 2003, Tomkins and Brown 2004, Tomkins *et al.* 2004).

One characteristic of mating systems in which ARTs are common is intense sexual competition (Shuster and Wade 2003). This means that high variance in fitness exists among individuals so that many derive zero benefit despite high investment. These unsuccessful individuals must be included in calculations of the average fitness for each tactic. This means that even if some individuals derive low fitness from an alternative tactic, such as reduced investment in weapons, this benefit may still be greater than the *average* of the apparently "more successful" but higher variance tactic (Shuster and Wade 2003).

Density dependence often interacts with frequency and condition dependence in the evolution of alternative phenotypes including sex allocation (Cade 1981, Gross 1991a, Philipp and Gross 1994, Lucas *et al.* 1996, Kruuk *et al.* 1999, Sinervo *et al.* 2000). A common effect is that frequency dependence may be much stronger at high than at low densities (or may not operate at all at low densities: Eadie and Fryxell 1992). For example, in the male mimicry system of the damselfly *Ischnura ramburi* (Sirot *et al.* 2003), females of this species occur in two color morphs: some are brightly colored and look like males, called andromorphs, whereas others are brown and different in appearance from males, called gynomorphs (one locus, two alleles, gynomorphs dominant). The hypothesis to explain the maintenance of male mimicry in this and related species (Robertson 1985) is that females that mimic males gain an advantage by wasting less time in risky, expensive, and time-consuming copulation (and copulation attempts) when compared with gynomorphic females. The effectiveness of the male mimicry changes with the frequency of andromorphs in the population since males learn about the male-like females, which breaks their mimicry (Cordero and Egido 1998, Cordero *et al.* 1998, Sirot and Brockmann 2001, van Gossum *et al.* 2001). The hypothesis is that at high densities male mimicry increases andromorph success, whereas at low densities andromorphs gain little relative to gynomorphs. So population density and operational sex ratio may influence the response to frequency-dependent selection (Cordero 1992, Cordero *et al.* 1998, Andres *et al.* 2002).

2.2.2 Two types of disruptive selection: alternative phenotypes competing for the same or different resources

For multiple phenotypes to evolve from a continuously variable character, disruptive selection must act on the population (Danforth and Desjardins 1999). Disruptive selection may occur either when individuals of one population compete for access to the same resource, or when multiple resources (niches) exist that can be exploited better by different morphs. When selection acts against individuals of intermediate phenotype, a developmental switch mechanism that achieves two (or more) discrete phenotypes with few intermediates (either through genetic polymorphism or phenotypic plasticity or both) is strongly favored (Emlen and Nijhout 2000) (see Chapter 5). Once this bimodal expression of the trait has evolved, it allows the uncoupling and independent evolution of the two phenotypes toward increasing specialization (Danforth and Desjardins 1999, West-Eberhard 2003).

Disruptive selection is often caused by nonlinear costs and benefits (Gadgil 1972), which is illustrated in Figure 2.6. Costs (C) to the individual of investing in some trait such as horns or fighting are modeled as increasing linearly (Figure 2.6 upper graph). This means that individuals at the right-hand end of the distribution are paying a high cost for the exaggerated traits they bear, whereas those at the left without the trait are paying none. Benefits (B) are modeled as increasing at an accelerating rate so that exaggerated benefits accrue to those that invest the most (on the graph $B - C > 0$). If in addition some net benefit accrues to those that invest nothing ($B - C > 0$) and if individuals

making intermediate investments derive the least net benefit (they are paying the cost but not deriving the benefit, $B - C < 0$), then two fitness peaks result (Isvaran and St. Mary 2003). Condition or status dependence may influence which individuals end up developing particular alternative tactics if fitness and condition are correlated.

So far we have been viewing selection for alternative phenotypes as resulting from individuals taking different routes to acquire the same resource such as mates. A different perspective views the habitat as heterogeneous and sees alternative phenotypes within one population as adaptations to different niches (Waltz and Wolf 1984, Via 1994, Skúlason and Smith 1995, Smith and Skúlason 1996, Mathias and Kisdi 2002). For example, crickets and many other insects have two, discrete life histories: a dispersing, flight-capable morph (LW) with larger wing muscles, more fat, and higher metabolism but which mature later and with reduced fecundity, and a nondispersing, flightless form (SW) that matures earlier and has higher fecundity (Figure 2.7). Mole and Zera (1992, 1994) have shown that in crickets the two morphs do not differ in consumption or efficiency of food use, but the proportion of assimilated nutrients converted into biomass is higher for SW morphs than for LW. The differences between the morphs are not due to direct competition between developing structures but to the differential allocation of resources among different body functions, i.e., different allocation rules (Zhao and Zera 2002) that are heritable (Zera and Rankin 1989). Clearly, the two life-history patterns amount to alternative allocation rules that guide development and metamorphosis into two different adult phenotypes.

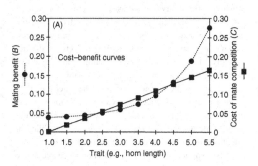

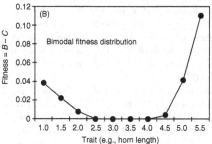

Figure 2.6 A model of disruptive selection on a sexually selected trait with nonlinear benefits. (A) Costs of a trait increase linearly whereas benefits show an increasing gain with increasing trait size. When the cost–benefit curves cross (when $B - C > 0$), selection results in (B) a bimodal distribution of the trait. Individuals with intermediate trait values pay the cost of the trait but do not derive the benefit (when $B - C < 0$), and thus there is selection against the intermediates. (Based on Gadgil 1972; Isvaran and St. Mary 2003.)

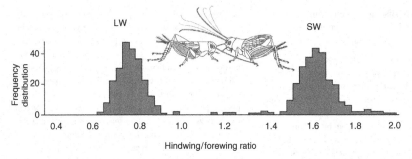

Figure 2.7 Long- and short-winged forms of the field cricket *Gryllus rubens* occur in one population with few intermediates. Although the two forms are known to be a dispersal polymorphism, long-winged (LW) and short-winged (SW) males differ in their ability to win fights and find females. The SW morph is larger (on right in picture) and wins when fighting males of the LW morph (on left). This means that both natural selection and sexual selection are involved in the maintenance of the wing dimorphism in the population. (Redrawn from Walker and Sivinski 1986.)

What favors selection for the two morphs? LW and SW are usually viewed as a dispersal polymorphism. For example, the planthopper *Prokelisia dolus* feeds on the sap of *Spartina*, a grass of intertidal marshes (Denno 1994, Denno *et al.* 1996) (Figure 2.8). When the planthoppers are living in permanent habitats with abundant food plant that is contiguous, SW females are favored, but when planthoppers are living in a sparse habitat where the food plant is less persistent and more dispersed, then LW females are favored. Individuals with intermediate wings, wing musculature, and maturation times are less successful in both contiguous and noncontiguous habitats. There are, in effect, two adaptive peaks. Furthermore, when living at low densities, SW females can remain at their natal site and still find food even in sparse habitats, but at higher densities LW morphs are increasingly favored regardless of habitat. The observed proportion of LW females in the population is thought to reflect the strength of selection for dispersal due to differences in the degree of habitat permanence, the abundance of the food plant, and the density of planthoppers (Figure 2.8).

Selection acts differently on males: males that track the distribution of females are favored. LW males are not able to mate in the contiguous habitat because SW males win at male–male competition and dominate the mating opportunities (Langellotto *et al.* 2000). In the sparse, noncontiguous habitat, however, SW males are unable to locate females. So, when the food plant is sparsely distributed, wings are needed to locate the dispersed females, and when planthopper densities are high, wings are favored to escape

deteriorating conditions along with the females (Langellotto and Denno 2001) (Figure 2.8). At intermediate densities wings are not useful either for locating females or habitat escape so SW forms are favored. The observed frequencies of the LW morph, then, reflect the strength of both natural selection (acting on the ability of males to escape habitat deterioration) and sexual selection (through male–male competition and nonrandom mating).

The planthopper example illustrates one important evolutionary process that may act to maintain alternative allocation phenotypes in a population. When animals are living in a patchwork of interspersed niches (or temporally varying habitats), selection favors multiple phenotypes that are specialized for exploiting resources in each niche (including the ability to make use of relevant information). No one intermediate phenotype is as effective as the extremes in exploiting the resource or at finding mates. Habitat persistence and the negative effects of density (density dependence) interact resulting in disruptive selection for morphs specialized for each habitat. The success of one morph does not depend on the other (i.e., they are not frequency dependent); each morph simply exploits the habitat to which it is adapted. The two morphs should occur at frequencies that match the resource availability in the different niches (Seger and Brockmann 1987).

The two views of disruptive selection discussed above depend on whether individuals are exploiting the same or different resources (Halama and Reznick 2001), i.e., whether they compete with one another or not. If they are vying for the same resource, then frequency-dependent

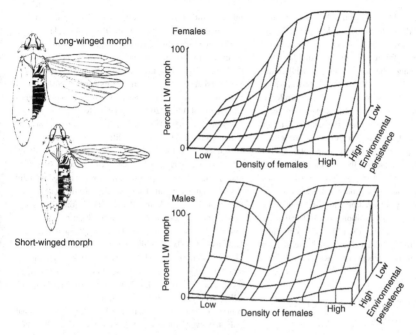

Figure 2.8 Model for the evolution of wing-morph frequency in the planthopper *Prokelisia dolus*. Long-winged (LW) and short-winged (SW) morphs differ in the length of the hind wings. The proportion of LW morphs in the population is affected by both population density and habitat persistence (habitats may deteriorate during the life of the individual). In females, LW forms are favored when persistence is low (long wings allow females to escape a deteriorating habitat), particularly at high population densities. However, in males, which are tracking the distribution of females, wings are needed at low densities to locate the dispersed females; at high densities wings are needed to get to new habitats along with the females. At intermediate densities, however, wings are less valuable either for locating females or escaping deteriorating conditions so the frequency of the LW form is reduced. (Redrawn from Denno 1994, Denno *et al.* 1985.)

selection and density dependence are often involved because the success of individuals of one morph often depends on the frequency of its own or the other morph in the population. If the two morphs are using different resources, then only density dependence (within a morph) is a likely factor. Both explanations apply to ARTs. If, for example, multiple female phenotypes exist in a population with assortative mating, then multiple niches exist for males to exploit. This could happen if females were found in different habitats (e.g., near the natal site or some distance away) and different male traits were required to locate those females (e.g., flight-capable vs. flightless) (Figure 2.8). It is also possible that variation in female behavior (including polymorphism in female preference traits) will have consequences for male phenotypes

including ARTs (Henson and Warner 1997, Alonzo and Warner 2000a, Brooks and Endler 2001, Jones 2002, Morris *et al.* 2003) (see Chapter 18).

Similarly, the two views of disruptive selection also apply to alternative life-history tactics since some involve competition over the same resource and frequency dependence, and others involve different resources. For example, some insects (Seger 1983), such as the mud-daubing wasp *Trypoxylon politum*, have a partially bivoltine life-history pattern: individuals that emerge in the spring produce some offspring that enter diapause and overwinter (delayed development) and some that pupate and emerge later in the same summer to produce a second generation (direct development) (Seger and Brockmann 1987, Brockmann and Grafen 1992, Brockmann 2004). Direct and delayed development is likely controlled by maternal cues (Tauber *et al.* 1986, McWatters and Saunders 1997). Partial bivoltinism has been explained as a bet-hedging strategy against an early winter freeze (diapausing individuals are protected from freezing whereas pupating or adult individuals are not: Taylor 1980, 1986) or other catastrophe for one or the other tactic. However, frequency dependence may be involved because if all individuals choose to overwinter, then selection will favor those that emerge later in the same summer and take advantage of the abundant available resources (Seger and Brockmann 1987). This means that the success of the two life histories (direct and delayed development) depends in part on their frequencies in the population. The same may hold for other alternative tactics. In the planthopper example, to the extent that females are competing for oviposition sites and males are competing for the same females, the LW and SW morphs may be seen as both alternative life-history tactics and ARTs in which frequency dependence, as well as density dependence, is involved.

2.3 EVOLUTION OF UNDERLYING MECHANISMS

The mechanisms of sex allocation include genetic polymorphism, a developmental switch caused by individual or environmental conditions, or individual phenotypic plasticity, or some combination of the three (Clutton-Brock and Albon 1982) (Box 2.1). Similar diversity can be found in other alternative phenotypes. Genetic polymorphisms have been described in a number of cases (Shuster and Wade 2003): the left- and right-jawed morphs of the scale-eating cichlid (one locus, two alleles: Hori 1993) (Figure 2.3); the two female damselfly morphs (one locus, two alleles sex-limited: Johnson 1964, Cordero 1990); and independent and satellite male morphs of the ruff *Philomachus pugnax* (sex-limited, single-locus, autosomal gene), which differ in the color of the display feathers (and other traits) that are used during lek courtship (Hugie and Lank 1997, Widemo 1998, Bachman and Widemo 1999, Lank *et al.* 1999) (see Box 16.1). Genetic differences have also been found in five species of fish with ARTs (Taborsky 1999), in the three male morphs of the isopod *Paracerceis sculpta* (Shuster 1989) (see Chapter 9), and in the three male morphs of the side-blotched lizard *Uta stansburiana* (Sinervo and Zamudio 2001, Sinervo *et al.* 2001) (see Chapter 12) where males differ in throat colors, size, territorial behavior, and other characters. In some cases maternal effects are known to

influence the expression of a genetic polymorphism and its associated phenotype (Tauber and Tauber 1992, Hews *et al.* 1997). In other cases environmental conditions strongly influence the expression of the genetic polymorphism (Taborsky 1999, West-Eberhard 2003, Karplus 2005).

Threshold switches from one morph to another (with few intermediates) underlie many alternative phenotypes (Shuster and Wade 2003, West-Eberhard 2003; but see Tomkins *et al.* 2005) (Figure 2.2). They are often associated with hormonal differences or differences in sensitivity to hormones (Bass 1996, Moore *et al.* 1998, Nijhout 1999, Goodson and Bass 2000, Knapp 2004) (see Chapters 5 and 7), which has been proposed to explain environmentally sensitive (and combined environmental and genetic) sex determination (Kraak and Pen 2002). In the wing polymorphism of the cricket *Gryllus rubens* (Figure 2.7), the SW morph has a pulse of juvenile hormone (JH) activity in the latter part of the last juvenile instar that the LW morph does not have (Zera and Tiebel 1989, Zera 1999, Zera and Huang 1999). If the pulse of JH remains below a specific threshold level, then cells in the hindwing pads proliferate and the LW phenotype develops. Similar threshold mechanisms control insect castes (Nijhout and Wheeler 1982), seasonal polyphenism, partial bivoltinism (Dingle and Winchell 1997, Nijhout 1999), and horned and hornless morphs in male beetles (Emlen and Nijhout 1999, 2000, 2001, Emlen 2001) (Chapters 5 and 7). Threshold changes in JH also control age polyethism in honeybees (the mechanism that organizes different worker tasks within a colony) and are associated with changes in gene expression in the brain that affect behavior (Whitfield *et al.* 2003). Threshold mechanisms show heritable variation and selection acts on variation in thresholds as it does on other traits (Hazel *et al.* 1990, 2004, Roff and Shannon 1993, Emlen 1996, Roff 1996, Shuster and Wade 2003, Tomkins and Brown 2004, Tomkins *et al.* 2004; Unrug *et al.* 2004) (see Chapter 5).

A special case of condition-dependent alternative phenotypes is frequency-dependent choice of tactics (Dominey 1984), a condition-dependent decision rule that specifies tactics according to their relative frequency in the population (Brockmann and Dawkins 1979). For example, spadefoot toads live in temporary ponds and have two larval morphs, omnivorous and carnivorous (Pfennig 1992). The omnivorous form has smaller eyes and tail and a larger gut than the carnivorous form. When the proportions of the two types are perturbed, the pond quickly returns to the equilibrium level maintained in the control (nonmanipulated) ponds. Since this occurs over a matter of days, it means that the larvae are responding to the frequency of the two morphs through phenotypic plasticity. This of course means that the animal has some mechanism for detecting the proportion (and not just the density) of the two morphs in the ponds. This condition-dependent rule will mimic frequency-dependent selection.

What might favor a conditional tactic over a genetic polymorphism? This problem has been addressed in the sex-allocation literature by asking when environmental sex determination is favored over genetic sex determination (Bull 1983, Gross 1996). Environmental sex determination is favored (a) when an individual's fitness as a male or female is strongly influenced by environmental conditions (Trivers and Willard 1973), (b) when environmental variation is large (Bull 1983), (c) when the individual has little control over the environment it will experience, and (d) when the individual grows up in an environment away from its parents and (e) in an environment that its parents do not choose (Charnov and Bull 1977, Janzen and Paukstis 1991). (f) Kraak and Pen (2002) also emphasize the importance of environmental sex determination in allowing individuals control over the sex ratio they produce (a similar argument has been used to explain the selective advantage of haplodiploidy: Bull 1983). Selection favors such control when, for example, a sex-ratio distorter or a conflict exists (Cook 2002), or when the mating structure of the population varies (Sabelis et al. 2002). These same arguments can be applied to selection favoring plastic control of alternative life-history patterns (Lessells 1991) or other alternative allocation phenotypes rather than a genetic polymorphism (recognizing that combinations of the two are likely if not the rule). For example, Hazel et al. (1990) and Unrug et al. (2004) argue that when the fitness returns from adopting alternative tactics change with status, then thresholds are likely to evolve. Because the thresholds shift with environment (such as density), large amounts of genetic variation for the exact position of the threshold are maintained.

One mechanism that has remained controversial in the alternative-strategy literature is whether animals ever choose between alternatives at random, i.e., whether animals use mixed strategies (Brockmann and Dawkins 1979, Brockmann et al. 1979, Dawkins 1980, 1982, Dominey 1984, Shuster and Wade 1991, 2003, West-Eberhard 2003). Gross (1996) asserts that no evidence exists for such a pattern and that one would not expect such a pattern since selection would always favor animals using any information they have that would correlate fitness with tactic (Neff and Sherman 2002, West and Herre 2002). This is certainly true, but what if the animal does not have reliable information? Such is often the case in sex allocation (Williams 1979). If individuals knew the instantaneous sex ratio of the population into which they were placing offspring, then selection would favor biasing offspring toward the rarer sex, but in general individuals are not privy to such information. Under this situation (and in cases like the left- and right-jawed cichlids), selection favors individuals producing offspring randomly at the equilibrium sex (or morph) ratio (normally 50 : 50 since the expectation of success through son and daughter or left- and right-jawed morph is equal) (Box 2.1, Figure 2.3). In fact, one would expect mixed strategies to exist in any alternative phenotype (just as they occur in sex allocation) whenever mutually exclusive choices exist; individuals have little or no reliable information about the relationship between tactic and fitness and when fitness is frequency dependent (Brockmann 2001, Hazel et al. 2004, Plaistow et al. 2004, Lee 2005). West-Eberhard (1979) and Moran (1992) make similar points when they argue that condition-independent systems are expected when individuals have no way to assess the appropriateness of switching from one alternative to another. Modeling shows that conditional strategies often do not entirely replace pure strategies, and many populations may be made up of combinations of conditional and pure or mixed strategies (Hazel et al. 2004, Plaistow et al. 2004). Such combinations are well known in the sex-allocation literature, including gynodioecy (Seger and Stubblefield 2002) and the phally polymorphism of pulmonate snails (Doums et al. 1998).

2.4   EVOLUTION OF TACTIC FLEXIBILITY

Alternative phenotypes are sometimes flexible (i.e., individuals switch between sexes, life-history pathways, or reproductive tactics as an adult) and sometimes inflexible with the individual retaining one or the other alternative phenotype throughout its adult lifespan. It would seem advantageous to be able to respond to available information and switch to the appropriate phenotype, yet many species do not. In sex allocation such adult flexibility (hermaphroditism) is favored under several conditions. Self-compatible hermaphrodites are more effective at colonizing unoccupied habitats: by reducing the cost of sex, they are able to colonize more quickly than animals that put effort into male production (Charnov 1982). Self-incompatible hermaphrodites are far more common; they pay the full cost of sex and must seek out mates, so when is such a pattern favored? In general, this form of hermaphroditism is favored when the fitness of hermaphrodites is greater than for separate sexes (dioecy) (Charnov 1982). For plants and immobile animals, the greater success of hermaphrodites may result from diminishing returns on investment in male function since there are only a set number of females available to a sessile male (Charnov et al. 1976). If the cost of switching sexes is low, then hermaphroditism is favored, but if switching requires many physiological, morphological, or organizational changes (as often occurs when there is strong mating competition), then dioecy will be more successful. As with dioecy, inflexible alternative phenotypes in the adult allow for specialization, i.e., the development of complex suites of traits with color, size, morphological, physiological, behavioral, and life-history differences between forms (Shuster and Wade 2003, West-Eberhard 2003). Inflexible phenotypes are favored over flexible phenotypes if the switch from one form to the other is costly in time or resources or in acquiring the information necessary to make an adaptive switch (Charnov 1982, DeWitt et al. 1998). Furthermore, when there are limits to switching, such as when the accuracy of matching phenotype to environment is constrained, then inflexible forms are favored (Moran 1992).

Similarly, if the reproductive success associated with alternative phenotypes is correlated with age, size, or environmental conditions, then selection will favor individuals capable of detecting that information and switching from one pattern to the other at the age or size or state that maximizes fitness (Shuster and Wade 2003). For example, mating tactics with developmental switches are favored by selection if different forms such as small and large competitors can take advantage of their respective sizes when performing parasitic or bourgeois behavior (Taborsky et al. 1987, Magnhagen 1992, de Fraipont et al. 1993, Alonzo and Warner 2000b). In other cases, young or small individuals may take advantage of their membership in a reproductive group that may serve different functions (Taborsky 1984, Rood 1990, Haydock et al. 1996, Balshine-Earn et al. 1998) and parasitize the reproduction of dominant individuals (Rabenold et al. 1990, Martin and Taborsky 1997, Dierkes et al. 1999, Taborsky 2001). Both possibilities are favored by an indeterminate growth pattern, i.e., by continuing growth after sexual maturation (Taborsky 1999, Wiegmann et al. 2004). Highly flexible tactics are favored if optimal mating conditions vary strongly over space and time. For example, in ten-spined sticklebacks, nesting males may take advantage of a currently attractive nest in the neighborhood and change from nest defense to sneaking and back within very short time periods (Morris 1951, 1954; for other examples see Barlow 1967, Chan and Ribbink 1990). If the costs of changing tactics between a pure growth tactic and reproductive function are relatively low during ontogeny, an opportunistic parasitic tactic may be employed by individuals that are still too small to perform the bourgeois role successfully; this pattern is particularly widespread in fishes (see Taborsky 1994 for review) (Figure 2.2). These examples point to the temporal availability and predictability of mating opportunities and the degree of specialization needed to be successful as important factors in the evolution of alternative reproductive tactics (Shuster and Wade 2003).

2.5 EVOLUTION OF DIMORPHISM

Why do alternative phenotypes show the particular form they do? Darwin (1871) pointed out that differences between the sexes in secondary sexual characters may be due to natural selection or sexual selection through male–male competition or female choice. He also recognized male choice, which might result in males mating with the more vigorous females, and traits used by males to "secure" the female as two additional factors favoring sexual dimorphism. Similarly, alternative mating and life-history tactics are adaptations whose evolution will be influenced both by natural and sexual selection. As with sexual size dimorphism (Badyaev and Hill 2000, Badyaev et al. 2000, 2001), selection on alternative phenotypes will act on both phenotypes (not just the subordinate or parasitic tactic), and the intensity and nature of selection can change over the life of the individual. Differences in size between the sexes or between morphs are due to differences in the rates, duration, and timing of growth for both phenotypes (Badyaev 2002).

Many sexually dimorphic traits show correlated selective responses due to pleiotropy and the linkage of genes affecting male and female characters, and the same should hold for morphs of one sex, such as parasitic and parental male sunfish, or morphs of both sexes, such as winged and wingless crickets or planthoppers (Figures 2.7 and 2.8). Linkage restricts the rate at which sexual dimorphism can evolve relative to that of the average phenotype of the two sexes (Lande 1980) and could have the same effect on the evolution of other alternative phenotypes. Nevertheless, under weak natural selection with relative fitness constant over time, the two sexes (or morphs) will evolve differences in optima to the point where they are partially or strongly sex or morph limited (Via and Lande 1985, Roff and Fairbairn 1993, Reeve and Fairbairn 2001, West-Eberhard 2003). This means that understanding the evolution of phenotype dimorphism requires answers to two questions: first, what are the evolutionary processes and selective pressures that create bimodal peaks of fitness (discussed in Section 2.2 above) (Smith and Girman 2000); and second, how are genetic covariances broken up in correlated traits so that the separate evolution of the alternative phenotypes is possible?

The nature of the underlying genetic architecture of traits in alternative phenotypes has been studied for the wing polymorphism of sand crickets. Genetic correlations have been found between male and female traits (Roff and Fairbairn 1993) and between traits of the two morphs: the percent of LW in a family and the duration of the calling song (Crnokrak and Roff 1995, 1998). This means that if selection were to favor LW males, for example, there would be a correlated drop in calling duration among the LW males. Since longer calling duration is associated with an increased ability to attract females, this trade-off would slow selection on the trait. This would also be the case for other traits that were correlated with LW. In this way genetic correlations could have an important effect on the evolution of dimorphism in polymorphic phenotypes, including alternative reproductive tactics.

2.6 EVOLUTION OF TACTIC FREQUENCIES

In the sex-allocation literature, the evolution of morph frequencies, i.e., sex-ratio theory, is well developed, so it is surprising that the literature on ARTs or alternative life-histories rarely attempts to predict the frequencies of alternative phenotypes in populations (Gross 1991a, b, Hori 1993, Tomkins and Brown 2004, Tomkins et al. 2004). The framework for sex-ratio theory is given in Box 2.1: when the average success through sons is equal to the success through daughters (equal reproductive value for sons and daughters), then selection favors a 1 : 1 equilibrium sex ratio; if the population should depart from that equilibrium, it will return due to the action of frequency-dependent selection (Seger and Stubblefield 2002). The frequencies of other alternative phenotypes can be explained in much the same way when the same principles (equal reproductive value and frequency-dependent selection) are operating, as in the left- and right-jawed cichlid morphs that are maintained at 1 : 1 (Hori 1993) (Figure 2.3). Some species of fig wasps have

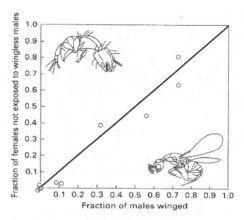

Figure 2.9 Morph ratios in fig wasps (Agaonidae). Males of some species of fig wasps are dimorphic: some males have small heads and can fly, whereas others are flightless and have enormous mandibles that are used in fighting. The wingless fighting males remain inside the fig in which they were born, fight other wingless males, and mate with females that are emerging inside the same host fig; winged males leave the fig and mate with females that are dispersing away from their natal figs. For ten species of fig wasps, there is a good relationship between the fraction of males in the population that are winged and the fraction of females leaving their natal fig before mating. Since these females will be mated by winged males, the equality of the two fractions (straight line in the figure) implies equal mating success for the two morphs. (Redrawn from Hamilton 1979.)

dimorphic males: some males are winged and disperse from their natal fig (Figure 2.9), whereas others are nondispersing and wingless with large heads and mandibles suitable for fighting other males (Bean and Cook 2001). There is a strong relationship between the fraction of males in a population that are winged and the fraction of females leaving their natal fig before mating (Cook et al. 1997) (Figure 2.9). These results support the hypothesis that there is frequency-dependent selection and equal mating success for the two morphs (Hamilton 1979, Herre 2001, Greeff 2002).

Sex ratios, however, are not always 1 : 1 (Charnov 1982). Sex-ratio theory is based on a number of assumptions, and if those assumptions are violated, then 1 : 1 sex ratios are not predicted (Frank 1987, Bull and Charnov 1988, Seger and Stubblefield 2002). Some of these exceptions (West and Herre 2002) may provide insight into the evolution of tactic ratios or the frequencies of alternative phenotypes.

2.6.1 Differential costs

The first assumption of sex-allocation theory is that resources are allocated equally to sons and daughters, and resources that are put into one sex could just as easily be put into the other sex. If one sex is cheaper to produce than the other, then equal investment will result in unequal numbers (Charnov 1982, Brockmann and Grafen 1992). So, for example, if sons are smaller and cheaper to produce, then there should be larger numbers of sons than daughters in the population so that the investment in the two sexes is equal. Sons may be cheaper to produce if, for example, they require less food than a daughter, if food limits the clutch size produced or the animal's ability to mature eggs (Rosenheim et al. 1996).

When alternative allocation phenotypes are based on maternal investment strategies, then these cost-ratio arguments from sex-ratio theory should apply. One excellent example is Dawson's burrowing bee, *Amegilla dawsoni* (Alcock 1996a, b). Female bees dig nests at aggregated sites, provision their nests with pollen, and lay an egg on the accumulated stores, which is then eaten by the developing larva. Males come in two discrete sizes, majors and minors (see Box 8.1). Large males (majors) fly little, patrol the nesting area, fight viciously with other males (and the larger males are more successful at fighting), and dig out virgin females and copulate with them as they emerge from the ground (Alcock 1996c). Small males (minors) emerge earlier and fly much of the time, patrolling the periphery of the nesting area and nearby areas and mating with previously unmated females (Alcock 1997a, b). The two types of males are the product of female allocation decisions: majors are provisioned with twice the food that minors receive (Alcock 1996d). This means that one would expect a minor-biased population-morph ratio and indeed there are about twice as many minors as majors (frequencies vary between sites from 2 : 1 to 4 : 1 minors : majors) (Alcock 1996d). Although majors appear to have much higher mating success than minors, majors also incur much higher costs and have shorter life spans due to male-male combat and increased predation (Alcock 1996a). The system is both density dependent, since minors do better at low densities (Alcock 1997a), and frequency dependent, since either morph would do better when rare (Box 8.1). Using sex-ratio theory, one must treat the success of major and minor male production as an investment strategy by the mother, taking into account the differences in cost (Dawkins 1980, Danforth and Neff 1992). As with the results from many sex-ratio studies

(Brockmann and Grafen 1992), however, this system has turned out to be very complex with interacting effects from changes in life histories (such as emergence times) and seasonal changes in sex ratios and the availability of food. Also, because large females are more likely to produce majors and females, whereas small females tend to specialize on minors, major and minor production could be a maternal conditional tactic (Tomkins et al. 2001).

2.6.2 Mode of inheritance and control of morph frequencies

The second assumption of sex-ratio theory is that sex determination is by Mendelian inheritance. We now know that a variety of factors other than the individual or its parents influence sex (Stouthamer et al. 2002). For example, sex-ratio distorters can be found on the nuclear genome as meiotic drive systems. A distorter may be a gene on the Y-chromosome that produces all male broods (Hamilton 1967) or any of a number of small, endocytoplasmic bacteria such as *Wolbachia*, which are passed to offspring through maternal inheritance and can completely alter sex-ratio patterns (Rigaud 1997, Partridge and Hurst 1998, Werren and Beukeboom 1998, Hurst and Randerson 2002). *Wolbachia*, which is found in a wide variety of insects and crustaceans, for example, can cause highly female-biased broods by killing all male brood (the bacteria is not passed on by male offspring) or by causing males to develop into females. It is interesting to note that in populations with *Wolbachia* where infected females are producing highly female-biased sex ratios, the uninfected population produces highly male-biased sex ratios. This is exactly what one would expect based on frequency-dependent selection and sex-ratio theory. As far as we know, no one has considered the possibility that maternally inherited factors might influence the frequencies of alternative reproductive or life-history tactics. We think this is possible and would be most likely to occur when the probability of transmission of the maternally inherited factor was more likely through one tactic than through the other, as might occur if one tactic were highly polygynous and the other monogamous.

2.6.3 Conflict over morph frequencies

A third assumption of sex-ratio theory is that sex-allocation decisions are maternal investment decisions. If, instead, there is some degree of control by other individuals, then a different sex ratio is predicted. For example, in haplodiploid species, selection favors males that can bias offspring sex ratios toward daughters (Brockmann and Grafen 1992). In an ant colony, if the queen controls offspring sex ratios, then a 1:1 sex ratio is expected, but if workers control sex ratios, then a 3:1 sex ratio is expected when the queen has mated once (due to the asymmetric degrees of relatedness that occur under haplodiploidy: Hamilton 1972, Trivers and Hare 1976, Alexander and Sherman 1977, Nonacs 1986, Boomsma and Grafen 1990, Boomsma 1991). Although conflict of this sort is well recognized in the sex-allocation literature, it is often overlooked when considering alternative mating and life-history tactics (see Chapter 18). The frequencies of alternative reproductive and life-history tactics should differ depending on whether they are under the control of the individual possessing them or others and on the amount of information available to each party (Mueller 1991, Sündstrom et al. 1996).

2.6.4 Specializing on a morph

A fourth assumption of sex-ratio theory is that each parent makes a decision between producing sons or daughters, but in some species one part of the population specializes in producing only one sex (referred to as split sex ratios: Grafen 1986). For example, if worker bees can assess the number of times the queen has mated, then one would expect colonies with singly mated queens to make more daughters and those with multiply mated queens to make more sons (Seger and Stubblefield 2002). Queenless colonies of bumble-bees produce males whereas queenright colonies produce a female-biased sex ratio presumably in response to frequency-dependent selection (Beekman and van Stratum 1998). The frequencies of alternative reproductive and life-history tactics may be affected when individuals specialize on producing one morph (split morph ratios), as may be the case in Dawson's burrowing bee (see Section 2.6.1).

2.6.5 Equal reproductive value for alternative morphs

The most basic assumption of sex-ratio theory is that fitness through sons is equal to that through daughters and frequency dependence is operating that stabilizes sex ratios at 1:1 (Fisher 1930). Sex-allocation theory has identified several conditions under which this assumption is violated (Hamilton 1967, Seger and Stubblefield 2002): (a) if there is competition between a philopatric parent and his/her

offspring (local resource competition; favors overproduction of the dispersive sex), (b) if offspring contribute to parental care (local resource enhancement or repayment; favors overproduction of the helping sex), or (c) if sons compete with brothers to mate with sisters (local mate competition; favors a female-biased sex ratio). For example, in some species of fig wasps, males emerge first and mate with the emerging females (Frank 1983). In species where they are mating with their sisters, the sex ratio is highly female biased (one son can mate with many sisters), but when several female foundresses have laid eggs in one fig and the sons of several females compete for mates, then the sex ratio approaches 1:1 (Frank 1985, Werren 1987, Herre 2001, Cremer and Heinze 2002, Pienaar and Greeff 2003).

Are there cases where the reproductive values of ARTs are not equal? For example, in a species where females produce two male morphs, one that only competes with his brothers for matings with his sisters (e.g., wingless, large-headed morph) and one that disperses, then sex-ratio theory would predict that females should produce fewer of the nondispersing morph (Greeff 1996, 1998). However, as with sex-allocation decisions, investment in alternative morphs depends, in part, on the information available to females (Stubblefield and Seger 1990, Flanagan and West 1998). In much of sex-allocation research (West and Herre 2002), and in studies of other alternative phenotypes, individuals are assumed to have complete information when we make predictions about the frequencies of alternative phenotypes. Clearly, this will not always be the case. Information availability needs to be a part of the equation in understanding sex as well as morph ratios.

The assumption of equal reproductive value through sons and daughters may be violated in another important way – overlapping generations (Seger 1983). For example, in the pipe-organ mud-daubing wasp *Trypoxylon politum*, some males of the first generation (overwintering, delayed development) live long enough to mate with second-generation (direct developing) females. This means that first-generation males have higher reproductive value than first-generation females or second-generation males (Brockmann and Grafen 1992), which favors females and produces a male bias in the first generation (biased by the extent of the overlap between first-generation males and second-generation females) and a female bias in the second generation. A similar effect might occur in alternative reproductive tactics if, for example, one mating tactic has overlapping generations and the other does not. Although the effect of overlapping generations has not been considered when calculating the frequencies of ARTs, sex-ratio theory suggests that one needs to consider the total reproductive value of each tactic to understand the evolution of tactic ratios.

2.6.6 Condition dependence and morph frequencies

Sex ratios, like ARTs, are often condition dependent (Trivers and Willard 1973, Charnov 1979). For example, some parasitic wasps lay sons in small hosts and daughters in large hosts. This is because fitness curves cross: the fitness of daughters is higher when they are reared from large hosts than when reared from small hosts, whereas the fitness of sons is less affected by host size (King 1992). Selection favors condition-dependent rules in individual females that take advantage of information about the relationship between sex-allocation decisions and fitness. If this sex-allocation rule for some reason results in too many sons (i.e., a departure from the equilibrium sex ratio), then frequency-dependent selection would favor a different female decision rule that produces more daughters, thus returning the population to the evolutionarily stable state (ESS) (Charnov et al. 1981) (Box 2.1).

The interaction between condition- and frequency-dependent effects in ARTs (Box 2.3) can be modeled in exactly the same way as condition-dependent sex ratios (Box 2.1). However, when condition dependence and frequency dependence are involved, then the success of alternative allocation phenotypes is not always equal (Hazel et al. 1990, Calsbeek et al. 2002, Shuster and Wade 2003), although there is a unique ESS switch point to which the population returns when perturbed (Gross and Repka 1995, 1998a, b, Repka and Gross 1995). This means that it is pointless to add up the average success of alternative tactics and expect them to have equal fitness in any natural population (Brockmann 2001). However, this model can be tested by perturbing the ratio of tactics in the population to determine whether fitness changes as predicted and whether the switch point evolves in the predicted direction (examples from sex allocation include Charnov et al. 1981, Conover and van Voorhees 1990, Sinervo et al. 2000, Horth and Travis 2002). Tomkins et al. (2004) and Radwan et al. (2002) have shown that there is heritable variation and differences between populations in the switch point, and Unrug et al. (2004) have shown heritability of the switch point of the conditional strategy found in the soil mite Sancassania berlesei (Box 2.1), a species with two male morphs. Similarly, Tomkins and Brown (2004) have demonstrated selection on a threshold switch point for a male

dimorphism in the European earwig Forficula auricularia (Figure 8.6), and Emlen (1996) and Unrug et al. (2004) selected on the switch point for dimorphic males in the horned beetle Onthophagus acuminatus (Figure 5.7).

2.6.7 The evolution of morph frequencies

This has not been an exhaustive list of the factors influencing either sex ratios or the frequencies of other alternative allocation phenotypes, but we have identified some important factors to take into consideration. Clearly, as with sex ratios, the evolution of tactic ratios is an extremely complex problem that involves a wide array of variables (e.g., cost ratios, overlapping generations, conflict). Nonetheless, understanding the evolution of relative frequencies for alternative phenotypes is certainly as important as understanding the evolution of sex ratios for sex allocation. Much of the success of sex-ratio theory has come from developing specific models that make clear, quantitative predictions that can be tested against empirical data. A similar approach should be developed for the study of alternative phenotype ratios (West and Herre 2002).

2.7 OVERVIEW

Alternative allocation phenotypes are the product of "decisions" by individuals to follow different paths for allocating their limited resources or time to a particular course of action. The underlying mechanisms controlling these decisions are subject to selection. Often, one decision path has higher fitness than the other and thus invades the population (resulting in only one tactic played by all individuals of one sex with continuous variation in individual tactics), but under some circumstances selection maintains in a population more than one route to fitness, i. e., alternative tactics. This occurs when fitness curves cross and individuals of different tactics compete for the same limited resource or when individuals of different tactics exploit different resources. Fitness curves cross when environmental conditions are such that one tactic is favored under some situations and the other tactic under other situations or when the success of a tactic depends on the individual's age or condition or when the tactics are frequency or density dependent or when some combination of these effects occurs. In general, condition-dependent mechanisms evolve when the individual making the decision has information about the correlation between tactic and fitness; if they do not, then selection should favor a mechanism controlling decision rules that is less responsive to environmental input.

A wide array of environmental, social, and individual factors, including density and frequency dependence, interact to influence the relative success of alternative tactics. But this is not all that is involved with the evolution of alternative tactics, because if phenotypes showed continuous variation, we would not think of them as "alternatives." Alternative tactics evolve when there is disruptive selection, i.e., when there is selection against intermediate phenotypes. This may occur when there is intense intraspecific competition (where the intermediates pay the price of extreme traits but do not derive the benefits) or when the environment is heterogeneous with more than one adaptive peak.

Insights into the evolution of alternative reproductive tactics and alternative life histories can be gained by comparison with one particularly well-studied alternative allocation phenotype, sex allocation. Insights from sex-allocation theory for understanding alternative tactics include the importance of considering and evaluating the effects of frequency dependence, overlapping generations and reproductive value, the availability of information to individuals about the relationship between tactics and fitness, the costs of producing individuals that use different tactics, and the mechanisms and processes controlling the expression of tactics. The study of life histories as applied to alternative reproductive tactics teaches us that we must consider trade-offs, nonlinear costs and benefits, and the effects of a heterogeneous and temporally variable environment. These have been little considered in the study of ARTs. We argue that by melding approaches, by examining mechanisms as well as evolution, by combining different modeling approaches, and by considering a new array of factors known to affect other allocation phenotypes, we can come to a better understanding of the complex outcomes and interactions in the evolution of alternative allocation phenotypes.

References

Alcock, J. 1996a. Male size and survival: the effects of male combat and bird predation in Dawson's burrowing bees, *Amegilla dawsoni*. *Ecological Entomology* 21, 309–316.

Alcock, J. 1996b. Site fidelity and homing ability of males of Dawson's burrowing bee (*Amegilla dawsoni*) (Apidae, Anthophorini). *Journal of the Kansas Entomological Society* 69, 182–190.

Alcock, J. 1996c. The relation between male body size, fighting, and mating success in Dawson's burrowing bee, *Amegilla dawsoni* (Apidae, Apinae, Anthophorini). *Journal of Zoology (London)* 239, 663–674.

Alcock, J. 1996d. Provisional rejection of three alternative hypotheses on the maintenance of a size dichotomy in males of Dawson's burrowing bee, *Amegilla dawsoni* (Apidae, Apinae, Anthophorini). *Behavioral Ecology and Sociobiology* 39, 181–188.

Alcock, J. 1997a. Competition from large males and the alternative mating tactics of small males of Dawson's burrowing bees (*Amegilla dawsoni*) (Apidae, Apinae, Anthophorini). *Journal of Insect Behavior* 10, 99–114.

Alcock, J. 1997b. Small males emerge earlier than large males in Dawson's burrowing bee (*Amegilla dawsoni*) (Hymenoptera: Anthophorini). *Journal of Zoology (London)* 242, 453–462.

Alexander, R. D. and Sherman, P. W. 1977. Local mate competition and parental investment in social insects. *Science* 196, 494–500.

Alonzo, S. H. and Warner, R. R. 2000a. Dynamic games and field experiments examining intra- and intersexual conflict: explaining counterintuitive mating behavior in a Mediterranean wrasse, *Symphodus ocellatus*. *Behavioral Ecology* 11, 56–70.

Alonzo, S. H. and Warner, R. R. 2000b. Female choice, conflict between the sexes and the evolution of male alternative reproductive behaviours. *Evolutionary Ecological Research* 2, 149–170.

Andres, J. A., Sanchez-Guillen, R. A., and Rivera, A. C. 2002. Evolution of female colour polymorphism in damselflies: testing the hypotheses. *Animal Behaviour* 63, 677–685.

Angilletta, M. J., Wilson, R. S., Navas, C. A., and James, R. S. 2003. Tradeoffs and the evolution of thermal reaction norms. *Trends in Ecology and Evolution* 18, 234–240.

Bachman, G. and Widemo, F. 1999. Relationships between body composition, body size and alternative reproductive tactics in a lekking sandpiper, the ruff (*Philomachus pugnax*). *Functional Ecology* 13, 411–416.

Badyaev, A. V. 2002. Growing apart: an ontogenetic perspective on the evolution of sexual size dimorphism. *Trends in Ecology and Evolution* 17, 369–378.

Badyaev, A. V. and Hill, G. E. 2000. The evolution of sexual dimorphism in the house finch. 1. Population divergence in morphological covariance structure. *Evolution* 54, 1784–1794.

Badyaev, A., Hill, G. E., Stoehr, A. M., Nolan, P. M., and McGraw, K. J. 2000. The evolution of sexual size dimorphism in the house finch. 2. Population divergence in relation to local selection. *Evolution* 54, 2134–2144.

Badyaev, A. V., Whittingham, L., and Hill, G. E. 2001. The evolution of sexual size dimorphism in the house finch. 3. Developmental basis. *Evolution* 55, 176–189.

Balshine-Earn, S., Neat, F. C., Reid, H., and Taborsky, M. 1998. Paying to stay or paying to breed? Field evidence for direct benefits of helping behavior in cooperatively breeding fish. *Behavioral Ecology* 9, 432–438.

Barlow, G. W. 1967. Social behavior of a South American leaf fish, *Polycentrus schomburgkii*, with an account of recurring pseudofemale behavior. *American Midland Naturalist* 78, 215–234.

Barnard, C. J. 1984. *Producers and Scroungers: Strategies of Exploitation and Parasitism.* London: Croom Helm.

Barnard, C. J. and Sibly, R. M. 1981. Producers and scroungers: a general model and its applications to captive flocks of house sparrows. *Animal Behaviour* 29, 543–550.

Bass, A. H. 1996. Shaping brain sexuality: varying reproductive tactics of plainfin midshipman fish have neural correlates. *American Scientist* 84, 352–363.

Bean, D. and Cook, J. M. 2001. Male mating tactics and lethal combat in the nonpollinating fig wasp *Sycoscapter australis*. *Animal Behaviour* 62, 535–542.

Beekman, M. and van Stratum, P. 1998. Bumblebee sex ratios: why do bumblebees produce so many males? *Proceedings of the Royal Society of London B* 265, 1535–1543.

Boomsma, J. J. 1991. Adaptive colony sex ratios in primitively eusocial bees. *Trends in Ecology and Evolution* 6, 92–95.

Boomsma, J. J. and Grafen, A. 1990. Intraspecific variation in ant sex ratios and the Trivers–Hare hypothesis. *Evolution* 44, 1026–1034.

Brockmann, H. J. 2001. The evolution of alternative strategies and tactics. *Advances in the Study of Behavior* 30, 1–51.

Brockmann, H. J. 2002. An experimental approach to altering mating tactics in male horseshoe crabs (*Limulus polyphemus*). *Behavioral Ecology* 13, 232–238.

Brockmann, H. J. 2004. Variable life-history and emergence patterns of the pipe-organ mud-daubing wasp, *Trypoxylon politum* (Hymenoptera: Sphecidae). *Journal of the Kansas Entomological Society* 77, 503–527.

Brockmann, H. J. and Dawkins, R. 1979. Joint nesting in a digger wasp as an evolutionarily stable preadaptation to social life. *Behaviour* 71, 203–245.

Brockmann, H. J. and Grafen, A. 1992. Sex ratios and life-history patterns of a solitary wasp, *Trypoxylon* (*Trypargilum*) politum (Hymenoptera: Sphecidae). *Behavioral Ecology and Sociobiology* 30, 7–27.

Brockmann, H. J. and Penn, D. 1992. Male mating tactics in the horseshoe crab, *Limulus polyphemus. Animal Behaviour* 44, 653–665.

Brockmann, H. J., Grafen, A., and Dawkins, R. 1979. Evolutionarily stable nesting strategy in a digger wasp. *Journal of Theoretical Biology* 7, 473–496.

Brockmann, H. J., Colson, T., and Potts, W. 1994. Sperm competition in horseshoe crabs (*Limulus polyphemus*). *Behavioral Ecology and Sociobiology* 35, 153–160.

Brockmann, H. J., Nguyen, C., and Potts, W. 2000. Paternity in horseshoe crabs when spawning in multiple male groups. *Animal Behaviour* 60, 837–849.

Brönmark, C. and Miner, J. G. 1992. Predator-induced phenotypical change in body morphology in crucian carp. *Science* 258, 1348–1350.

Brooks, R. and Endler, J. A. 2001. Female guppies agree to differ: phenotypic and genetic variation in mate-choice behavior and the consequences for sexual selection. *Evolution* 55, 1644–1655.

Bull, J. J. 1983. *Evolution of Sex-Determining Mechanisms.* Menlo Park, CA: Benjamin Cummings.

Bull, J. J. and Charnov, E. L. 1988. How fundamental are Fisherian sex ratios? *Oxford Surveys in Evolutionary Biology* 5, 96–135.

Cade, W. H. 1981. Alternative male strategies: genetic differences in crickets. *Science* 212, 563–564.

Calsbeek, R., Alonzo, S. H., Zamudio, K., and Sinervo, B. 2002. Sexual selection and alternative mating behaviours generate demographic stochasticity in small populations. *Proceedings of the Royal Society of London B* 269, 157–164.

Chan, T.-Y. and Ribbink, A. J. 1990. Alternative reproductive behaviour in fishes, with particular reference to *Lepomis macrochira* and *Pseudocrenilabrus philander. Environmental Biology of Fishes* 28, 249–256.

Charnov, E. L. 1979. Simultaneous hermaphroditism and sexual selection. *Proceedings of the National Academy of Sciences of the United States of America* 76, 2480–2484.

Charnov, E. L. 1982. *The Theory of Sex Allocation.* Princeton, NJ: Princeton University Press.

Charnov, E. L. 1986. An optimisation principle for sex allocation in a temporally varying environment. *Heredity* 56, 119–121.

Charnov, E. L. and Bull, J. 1977. When is sex environmentally determined? *Nature* 26, 828–830.

Charnov, E. L., Maynard Smith, J., and Bull, J. J. 1976. Why be an hermaphrodite? *Nature* 263, 125–126.

Charnov, E. L., Hartogh, R. L. L.-D., Jones, W. T., and van den Assem, J. 1981. Sex ratio evolution in a variable environment. *Nature* 289, 27–33.

Clutton-Brock, T. H. and Albon, S. D. 1982. Parental investment in male and female offspring in mammals. In King's College Sociobiology Group (eds.) *Current Problems: Sociobiology*, pp. 223–247. Cambridge, UK: Cambridge University Press.

Clutton-Brock, T. H., Guinness, F. E., and Albon, S. D. 1982. *Red Deer: Behavior and Ecology of Two Sexes*. Chicago, IL: University of Chicago Press.

Clutton-Brock, T. H., Albon, S. D., and Guinness, F. E. 1986. Great expectations: dominance, breeding success and offspring sex ratios in red deer. *Animal Behaviour* 34, 460–471.

Clutton-Brock, T. H., Albon, S. D., and Guinness, F. E. 1988. Reproductive success in male and female red deer. In T. H. Clutton-Brock (ed.) *Reproductive Success*, pp. 403–418. Chicago, IL: University of Chicago Press.

Collins, J. P. and Cheek, J. E. 1983. Effect of food and density on development of typical and cannibal salamander larvae in *Ambystoma tigrinum nebulosum*. *American Zoologist* 23, 77–84.

Conover, D. and van Voorhees, D. 1990. Evolution of a balanced sex ratio by frequency-dependent selection in a fish. *Science* 250, 1556–1558.

Cook, J. M. 2002. Sex determination in invertebrates. In I. C. W. Hardy (ed.) *Sex Ratios: Concepts and Research Methods*, pp. 178–194. Cambridge, UK: Cambridge University Press.

Cook, J. M., Compton, S. G., Herre, E. A., and West, S. A. 1997. Alternative mating tactics and extreme male dimorphism in fig wasps. *Proceedings of the Royal Society of London B* 264, 747–754.

Cordero, A. 1990. The inheritance of female polymorphism in the damselfly *Ischnura graellsii* (Rambur) (Odonata: Coenagrionidae). *Heredity* 64, 341–346.

Cordero, A. 1992. Density-dependent mating success and colour polymorphism in females of the damselfly *Ischnura graellsii* (Odonata: Coenagrionidae). *Journal of Animal Ecology* 61, 769–780.

Cordero, A. and Egido, F. J. 1998. Mating frequency, population density and female polychromatism in the damselfly *Ischnura graellsii*: an analysis of four natural populations. *Etologia* 6, 61–67.

Cordero, A., Carbone, S. S., and Utzeri, C. 1998. Mating opportunities and mating costs are reduced in androchrome

female damselflies, *Ischnura elegans* (Odonata). *Animal Behaviour* 55, 185–197.

Cremer, S. and Heinze, J. 2002. Adaptive production of fighter males: queens of the ant *Cardiocondyla* adjust the sex ratio under local mate competition. *Proceedings of the Royal Society of London B* 269, 417–422.

Crews, D. 1993. The organizational concept and vertebrates without sex chromosomes. *Brain, Behavior, and Evolution* 42, 202–214.

Crnokrak, P. and Roff, D. A. 1995. Fitness differences associated with calling behaviour in the two wing morphs of male sand crickets, *Gryllus firmus*. *Animal Behaviour* 50, 1475–1481.

Crnokrak, P. and Roff, D. A. 1998. The genetic basis of the trade-off between calling and wing morph in males of the cricket *Gryllus firmus*. *Evolution* 52, 1111–1118.

Danforth, B. N. and Desjardins, C. A. 1999. Male dimorphism in *Perdita portalis* (Hymenoptera, Andrenidae) has arisen from preexisting allometric patterns. *Insectes Sociaux* 46, 18–28.

Danforth, B. N. and Neff, J. L. 1992. Male polymorphism and polyethism in *Perdita texana* (Hymenoptera: Andrenidae). *Annals of the American Entomological Society* 85, 616–626.

Darwin, C. 1871. *The Descent of Man, and Selection in Relation to Sex*. London: John Murray.

Dawkins, R. 1980. Good strategy or evolutionarily stable strategy. In G. W. Barlow and S. Silverberg (eds.) *Sociobiology: Beyond Nature/Nurture*, pp. 331–367. Boulder, CO: Westview Press.

Dawkins, R. 1982. *The Extended Phenotype*. San Francisco, CA: W. H. Freeman.

de Fraipont, M., Fitzgerald, G. J., and Guderley, H. 1993. Age-related differences in reproductive tactics in the three-spined stickleback, *Gasterosteus aculeatus*. *Animal Behaviour* 46, 961–968.

Denno, R. F. 1994. The evolution of dispersal polymorphism in insects: the influence of habitats, host plants and mates. *Researches on Population Ecology* 36, 127–135.

Denno, R. F., Douglas, L. W., and Jacobs, D. 1985. Effects of crowding and host plant nutrition on a wing-dimorphic planthopper. *Ecology* 67, 116–123.

Denno, R. F., Roderick, G. K., Peterson, M. A., *et al.* 1996. Habitat persistence underlies the intraspecific dispersal strategies of planthoppers. *Ecological Monographs* 66, 389–408.

DeWitt, T. J., Sih, A., and Wilson, D. S. 1998. Costs and limits of phenotypic plasticity. *Trends in Ecology and Evolution* 13, 77–81.

Dierkes, P., Taborsky, M., and Kohler, U. 1999. Reproductive parasitism of broodcare helpers in a cooperatively breeding fish. *Behavioral Ecology* 10, 510–515.

Dingle, H. and Winchell, R. 1997. Juvenile hormone as a mediator of plasticity in insect life histories. *Archives of Insect Biochemistry and Physiology* 35, 359–373.

Dodson, S. 1989. Predator-induced reaction norms. *BioScience* 39, 447–452.

Dominey, W. J. 1980. Female mimicry in male bluegill sunfish: a genetic polymorphism? *Nature* 284, 546–548.

Dominey, W. J. 1984. Alternative mating tactics and evolutionarily stable strategies. *American Zoologist* 24, 385–396.

Doums, C., Viard, F., and Jarne, P. 1998. The evolution of phally polymorphism. *Biological Journal of the Linnean Society* 64, 273–296.

Eadie, J. M. and Fryxell, J. M. 1992. Density dependence, frequency dependence, and alternative nesting strategies in goldeneyes. *American Naturalist* 140, 621–641.

Emlen, D. J. 1996. Artificial selection on horn length–body size allometry in the horned beetle *Onthophagus acuminatus*. *Evolution* 50, 1219–1230.

Emlen, D. J. 2001. Costs and the diversification of exaggerated animal structures. *Science* 291, 1534–1536.

Emlen, D. J. and Nijhout, H. F. 1999. Hormonal control of male horn length dimorphism in the horned beetle *Onthophagus taurus. Journal of Insect Physiology* 45, 45–53.

Emlen, D. J. and Nijhout, H. F. 2000. The development and evolution of exaggerated morphologies in insects. *Annual Review of Entomology* 45, 661–708.

Emlen, D. J. and Nijhout, H. F. 2001. Hormonal control of male horn length dimorphism in *Onthophagus taurus* (Coleoptera: Scarabaeidae): a second critical period of sensitivity to juvenile hormone. *Journal of Insect Physiology* 47, 1045–1054.

Fisher, R. A. 1930. *The Genetical Theory of Natural Selection*. Oxford, UK: Oxford University Press.

Flanagan, K. E. and West, S. A. 1998. Local mate competition, variable fecundity and information utilization in a parasitoid. *Animal Behaviour* 56, 191–198.

Foote, C. J., Brown, G. S., and Wood, C. C. 1997. Spawning success of males using alternative mating tactics in sockeye salmon, *Oncorhynchus nerka. Canadian Journal of Fisheries and Aquatic Sciences* 54, 1785–1795.

Frank, S. 1983. A hierarchical view of sex-ratio patterns. *Florida Entomologist* 66, 42–75.

Frank, S. A. 1985. Hierarchical selection theory and sex ratios. 2. On applying the theory, and a test with fig wasps. *Evolution: International Journal of Organic Evolution* 39, 949–964.

Frank, S. A. 1987. Individual and population sex allocation patterns. *Theoretical Population Biology* 31, 47–74.

Gadgil, M. 1972. Male dimorphism as a consequence of sexual selection. *American Naturalist* 106, 574–579.

Giraldeau, L.-A. and Caraco, T. 2000. *Social Foraging Theory*. Princeton, NJ: Princeton University Press.

Giraldeau, L.-A. and Livoreil, B. 1998. Game theory and social foraging. In L. Dugatkin and H. K. Reeve (eds.) *Game Theory and Animal Behavior*, pp. 16–37. Oxford, UK: Oxford University Press.

Goodson, J. L. and Bass, A. H. 2000. Forebrain peptides modulate sexually polymorphic vocal circuitry. *Nature* 403, 769–772.

Grafen, A. 1986. Split sex ratios and the evolutionary origins of eusociality. *Journal of Theoretical Biology* 122, 95–121.

Greeff, J. M. 1996. Alternative mating strategies, partial sibmating and split sex ratios in haplodiploid species. *Journal of Evolutionary Biology* 9, 855–869.

Greeff, J. M. 1998. Local mate competition, sperm usage and alternative mating strategies. *Evolutionary Ecology* 12, 627–628.

Greeff, J. M. 2002. Mating system and sex ratios of a pollinating fig wasp with dispersing males. *Proceedings of the Royal Society of London B* 269, 2317–2323.

Greene, E. 1989. A diet-induced developmental polymorphism in a caterpillar. *Science* 243, 643–646.

Greene, E. 1999. Phenotypic variation in larval development and evolution: polymorphism, polyphenism, and developmental reaction norms. In B. Hall and M. Wake (eds.) *The Origin and Evolution of Larval Forms*, pp. 379–410. New York: Academic Press.

Gross, M. R. 1984. Sunfish, salmon, and the evolution of alternative reproductive strategies and tactics in fishes. In R. Wooton and G. Potts (eds.) *Fish Reproduction: Strategies and Tactics*, pp. 55–75. London: Academic Press.

Gross, M. R. 1985. Disruptive selection for alternative life histories in salmon. *Nature* 313, 47–48.

Gross, M. R. 1991a. Evolution of alternative reproductive strategies: frequency-dependent sexual selection in male bluegill sunfish. *Philosophical Transactions of the Royal Society of London B* 332, 59–66.

Gross, M. R. 1991b. Salmon breeding behavior and life history evolution in changing environments. *Ecology* 72, 1180–1186.

Gross, M. R. 1996. Alternative reproductive strategies and tactics: diversity within sexes. *Trends in Ecology and Evolution* 11, 92–97.

Gross, M. R. and Repka, J. 1995. Inheritance and the conditional strategy. *American Zoologist* 24, 385–396.

Gross, M. R. and Repka, J. 1998a. Game theory and inheritance in the conditional strategy. In L. Dugatkin and H. K. Reeve (eds.) *Game Theory and Animal Behavior*, pp. 168–187. Oxford, UK: Oxford University Press.

Gross, M. R. and Repka, J. 1998b. Stability with inheritance in the conditional strategy. *Journal of Theoretical Biology* 192, 445–453.

Halama, K. J. and Reznick, D. N. 2001. Adaptation, optimality, and the meaning of phenotypic variation in natural populations. In S. H. Orzack and E. Sober (eds.) *Adaptationism and Optimality*, pp. 242–272. Cambridge, UK: Cambridge University Press.

Hamilton, W. D. 1967. Extraordinary sex ratios. *Science* 156, 477–488.

Hamilton, W. D. 1972. Altruism and related phenomena, mainly in social insects. *Annual Review of Ecology and Systematics* 3, 193–232.

Hamilton, W. D. 1979. Wingless and fighting males in fig wasps and other insects. In M. S. Blum and N. A. Blum (eds.) *Sexual Selection and Reproductive Competition in Insects*, pp. 167–220. New York: Academic Press.

Haydock, J., Parker, P. G., and Rabenold, K. N. 1996. Extra pair paternity uncommon in the cooperatively breeding bicolored wren. *Behavioral Ecology and Sociobiology* 38, 1–16.

Hazel, W. N., Smock, R., and Johnson, M. D. 1990. A polygenic model for the evolution and maintenance of conditional strategies. *Proceedings of the Royal Society of London B* 242, 181–188.

Hazel, W., Smock, R., and Lively, C. M. 2004. The ecological genetics of conditional strategies. *American Naturalist* 163, 888–900.

Henson, S. A. and Warner, R. R. 1997. Male and female alternative reproductive behaviors in fishes: a new approach using intersexual dynamics. *Annual Review of Ecology and Systematics* 28, 571–592.

Herre, E. A. 2001. Selective regime and fig wasp sex ratios. In S. H. Orzack and E. Sober (eds.) *Adaptationism and Optimality*, pp. 191–217. Cambridge, UK: Cambridge University Press.

Hews, D. K., Thompson, C. W., Moore, I. T., and Moore, M. C. 1997. Population frequencies of alternative male phenotypes in tree lizards: geographic variation and common-garden rearing studies. *Behavioral Ecology and Sociobiology* 41, 371–380.

Hori, M. 1993. Frequency-dependent natural selection in the handedness of scale-eating cichlid fish. *Science* 260, 216–219.

Horth, L. and Travis, J. 2002. Frequency-dependent numerical dynamics in mosquitofish. *Proceedings of the Royal Society of London B* 269, 2239–2247.

Hugie, D. M. and Lank, D. B. 1997. The resident's dilemma: a female choice model for the evolution of alternative mating strategies in lekking male ruffs. *Behavioral Ecology* 8, 218–225.

Hurst, L. D. and Randerson, J. P. 2002. Parasitic sex puppeteers. *Scientific American* 286, 56–61.

Hutchings, J. A. and Myers, R. A. 1988. Mating success of alternative maturation phenotypes in male Atlantic salmon, *Salmo salar*. *Oecologia* 75, 169–174.

Hutchings, J. A. and Myers, R. A. 1994. The evolution of alternative mating strategies in variable environments. *Evolutionary Ecology* 8, 256–268.

Isvaran, K. and St. Mary, C. M. 2003. When should males lek? Insights from a dynamic state variable model. *Behavioral Ecology* 14, 876–886.

Janzen, F. J. and Paukstis, G. L. 1991. Environmental sex determination in reptiles: ecology, evolution and experimental design. *Quarterly Review of Biology* 66, 149–179.

Johnson, C. 1964. The inheritance of female dimorphism in the damselfly, *Ischnura damula*. *Genetics* 49, 513–519.

Jones, A. G. 2002. The evolution of alternative cryptic female choice strategies in age-structured populations. *Evolution* 56, 2530–2536.

Kaitala, A., Kaitala, V., and Lundberg, P. 1993. A theory of partial migration. *American Naturalist* 142, 59–81.

Karplus, I. 2005. Social control of growth in *Macrobrachium rosenbergii* (De Man): a review of prospects for future research. *Aquaculture Research* 36, 238–254.

King, B. H. 1992. Sex ratio manipulation by parasitoid wasps. In D. L. Wrensch and M. A. Ebber (eds.) *Evolution and Diversity of Sex Ratio in Insects and Mites*, pp. 418–441. New York: Chapman and Hall.

Knapp, R. 2004. Endocrine mediation of vertebrate male alternative reproductive tactics: the next generation of studies. *Integrative and Comparative Biology* 43, 658–668.

Kraak, S. B. M. and Pen, I. 2002. Sex-determining mechanisms in vertebrates. In I. C. W. Hardy (ed.) *Sex Ratios: Concepts and Research Methods*, pp. 158–177. Cambridge, UK: Cambridge University Press.

Kruuk, L. E. B., Clutton-Brock, T. H., Albon, S. D., Pemberton, J. M., and Guinness, F. E. 1999. Population density affects sex ratio variation in red deer. *Nature* 399, 459–462.

Lande, R. 1980. Sexual dimorphism, sexual selection and adaptation in polygenic characters. *Evolution* 34, 292–305.

Langellotto, G. A. and Denno, R. F. 2001. Benefits of dispersal in patchy environments: mate location by males of a wing-dimorphic insect. *Ecology* 82, 1870–1878.

Langellotto, G. A., Denno, R. F., and Ott, J. R. 2000. A trade-off between flight capability and reproduction in males of wing-dimorphic insects. *Ecology* 81, 865–875.

Lank, D. B., Coupe, M., and Wynne-Edwards, K. E. 1999. Testosterone-induced male traits in female ruffs (*Philomachus pugnax*): autosomal inheritance and gender differentiation. *Proceedings of the Royal Society of London B* 266, 2323–2330.

Lee, J. S. F. 2005. Alternative reproductive tactics and status-dependent selection. *Behavioral Ecology* 16, 566–570.

Lessells, C. M. 1991. The evolution of life histories. In J. R. Krebs and N. B. Davies (eds.) *Behavioral Ecology: An Evolutionary Approach*, pp. 32–68. Oxford, UK: Blackwell Scientific.

Lloyd, D. G. 1987. Parallels between sexual strategies and other allocation strategies. In S. C. Stearns (ed.) *The Evolution of Sex and Its Consequences*, pp. 263–281. Basel: Birkhauser Verlag.

Lucas, J. R., Howard, R. D., and Palmer, J. G. 1996. Callers and satellites: chorus behavior in anurans as a stochastic dynamic game. *Animal Behaviour* 51, 501–518.

Magnhagen, C. 1992. Alternative reproductive behavior in the common goby, *Pomatoschistus microps*: an ontogenic gradient. *Animal Behaviour* 44, 182–184.

Martin, E. and Taborsky, M. 1997. Alternative male mating tactics in a cichlid, *Pelvicachromis pulcher*: a comparison of reproductive effort and success. *Behavioral Ecology and Sociobiology* 41, 311–319.

Mathias, A. and Kisdi, E. 2002. Adaptive diversification of germination strategies. *Proceedings of the Royal Society of London B* 269, 151–155.

McWatters, H. G. and Saunders, D. S. 1997. Inheritance of the photoperiodic response controlling larval diapause in the blow fly, *Calliphora vicina*. *Journal of Insect Physiology* 43, 709–717.

Moczek, A. P. 2003. The behavioral ecology of threshold evolution in a polyphenic beetle. *Behavioral Ecology* 14, 841–854.

Moczek, A. P., Hunt, J., Emlen, D. J., and Simmons, L. W. 2002. Threshold evolution in exotic populations of a polyphenic beetle. *Evolutionary Ecology Research* 4, 587–601.

Mole, S. and Zera, A. J. 1992. Differential allocation of resources underlies the dispersal–reproduction trade-off in the wing-dimorphic cricket, *Gryllus rubens*. *Oecologia* 93, 121–127.

Mole, S. and Zera, A. J. 1994. Differential resource consumption obviates a potential flight–fecundity trade-off in the sand cricket (*Gryllus firmus*). *Functional Ecology* 8, 573–580.

Moore, M. C., Hews, D. K., and Knapp, R. 1998. Hormonal control and evolution of alternative male phenotypes: generalizations of models for sexual differentiation. *American Zoologist* 38, 133–151.

Moran, N. 1992. The evolutionary maintenance of alternative phenotypes. *American Naturalist* 139, 971–989.

Morris, D. 1951. Homosexuality in the ten-spined stickleback (*Pygosteus pungitius* (L.)). *Behaviour* 4, 233–261.

Morris, D. 1954. The causation of pseudofemale and pseudomale behaviour: a further comment. *Behaviour* 7, 46–56.

Morris, M. R., Nicoletto, P. F., and Hesselman, E. 2003. A polymorphism in female preference for a polymorphic male trait in the swordtail fish *Xiphophorus cortezi*. *Animal Behaviour* 65, 45–52.

Mueller, U. G. 1991. Haplodiploidy and the evolution of facultative sex ratios in a primitively eusocial bee. *Science* 254, 442–444.

Neff, J. L. and Sherman, P. W. 2002. Decision making and recognition mechanisms. *Proceedings of the Royal Society of London B* 269, 1435–1441.

Nijhout, H. F. 1999. Hormonal control in larval development and evolution: insects. In B. K. Hall and M. H. Wake (eds.) *The Origin and Evolution of Larval Forms*, pp. 217–254. New York: Academic Press.

Nijhout, H. F. and Wheeler, D. E. 1982. Juvenile hormone and the physiological basis of insect polymorphisms. *Quarterly Review of Biology* 57, 109–133.

Nonacs, P. 1986. Ant reproductive strategies and sex allocation theory. *Quarterly Review of Biology* 61, 1–21.

Partridge, L. and Hurst, L. D. 1998. Sex and conflict. *Science* 281, 2003–2008.

Penn, D. and Brockmann, H. J. 1995. Age-biased stranding and righting in male horseshoe crabs, *Limulus polyphemus*. *Animal Behaviour* 49, 1531–1539.

Pfennig, D. W. 1992. Polyphenism in spadefoot toad tadpoles as a locally adjusted evolutionarily stable strategy. *Evolution* 46, 1408–1420.

Philipp, D. P. and Gross, M. R. 1994. Genetic evidence for cuckoldry in bluegill *Lepomis macrochirus*. *Molecular Ecology* 3, 563–569.

Pienaar, J. and Greeff, J. M. 2003. Maternal control of offspring sex and male morphology in the *Otitesella* fig wasps. *Journal of Evolutionary Biology* 16, 244–253.

Piersma, T. and Drent, J. 2003. Phenotypic flexibility and the evolution of organismal design. *Trends in Ecology and Evolution* 18, 228–231.

Pigliucci, M. 2001. *Phenotypic Plasticity: Beyond Nature and Nurture*. Baltimore, MD: Johns Hopkins University Press.

Plaistow, S., Johnstone, R. A., Colegrave, N., and Spencer, M. 2004. Evolution of alternative mating tactics: conditional versus mixed strategies. *Behavioral Ecology* 15, 534–542.

Rabenold, P. P., Rabenold, K. N., Piper, W. H., Haydock, J., and Zack, S. W. 1990. Shared paternity revealed by genetic analysis in cooperatively breeding tropical wrens. *Nature* 348, 538–540.

Radwan, J., Unrug, J., and Tomkins, J. L. 2002. Status-dependence and morphological trade-offs in the expression of a sexually selected character in the mite, *Sancassania berlesei*. *Journal of Evolutionary Biology* 15, 744–752.

Reeve, J. P. and Fairbairn, D. J. 2001. Predicting the evolution of sexual size dimorphism. *Journal of Evolutionary Biology* 14, 244–254.

Repka, J. and Gross, M. R. 1995. The evolutionarily stable strategy under individual condition and tactic frequency. *Journal of Theoretical Biology* 176, 27–31.

Rigaud, T. 1997. Inherited microorganisms and sex determination of arthropod hosts. In S. L. O'Neill, A. A. Hoffmann, and J. H. Werren (eds.) *Influential Passengers: Inherited Microorganisms and Arthropod Reproduction*, pp. 81–101. Oxford, UK: Oxford University Press.

Robertson, H. 1985. Female dimorphism and mating behaviour in a damselfly, *Ischnura ramburi*: females mimicking males. *Animal Behaviour* 33, 805–809.

Roff, D. A. 1996. The evolution of threshold traits in animals. *Quarterly Review of Biology* 71, 3–35.

Roff, D. A. and Fairbairn, D. J. 1993. The evolution of alternate morphologies: fitness and wing morphology in male sand crickets. *Evolution* 47, 1572–1584.

Roff, D. A. and Shannon, P. 1993. Genetic and ontogenetic variation in behavior: its possible role in the maintenance of genetic variation in the wing dimorphism of *Gryllus firmus*. *Heredity* 71, 481–487.

Rood, J. P. 1990. Group-size, survival, reproduction, and routes to breeding in dwarf mongooses. *Animal Behaviour* 39, 566–572.

Rosenheim, J. A., Nonacs, P., and Mangel, M. 1996. Sex ratios and multifaceted parental investment. *American Naturalist* 148, 501–535.

Sabelis, M. W., Nagelkerke, C. J., and Breeuwer, J. A. J. 2002. Sex ratio control in arrhenotokous and pseudo-arrhenotokous mites. In I. C. W. Hardy (ed.) *Sex Ratios: Concepts and Research Methods*, pp. 235–253. Cambridge, UK: Cambridge University Press.

Schlichting, C. D. and Pigliucci, M. 1998. *Phenotypic Evolution*. Sunderland, MA: Sinauer Associates.

Seger, J. 1983. Partial bivoltinism may cause alternating sex-ratio biases that favour sociality. *Nature* 301, 59–62.

Seger, J. and Brockmann, H. J. 1987. What is bet-hedging? In P. H. Harvey and L. Partridge (eds.) *Oxford Surveys of Evolutionary Biology*, pp. 182–211. Oxford, UK: Oxford University Press.

Seger, J. and Stubblefield, J. W. 2002. Models of sex ratio evolution. In I. C. W. Hardy (ed.) *Sex Ratios: Concepts and Research Methods*, pp. 2–25. Cambridge, UK: Cambridge University Press.

Shuster, S. M. 1989. Male alternative reproductive strategies in a marine isopod crustacean (*Paracerceis sculpta*): the use of genetic markers to measure differences in fertilization success among a-, B-, and g-males. *Evolution* 43, 1683–1698.

Shuster, S. M. and Wade, M. J. 1991. Equal mating success among male reproductive strategies in a marine isopod. *Nature* 350, 608–610.

Shuster, S. M. and Wade, M. J. 2003. *Mating Systems and Strategies*. Princeton, NJ: Princeton University Press.

Sinervo, B. and Zamudio, K. 2001. The evolution of alternative reproductive strategies: fitness differential, heritability, and genetic correlation between the sexes. *Journal of Heredity* 92, 198–205.

Sinervo, B., Svensson, E., and Comendant, T. 2000. Density cycles and an offspring quantity and quality game driven by natural selection. *Nature* 406, 985–988.

Sinervo, B., Bleay, C., and Adamopoulou, C. 2001. Social causes of correlational selection and the resolution of a heritable throat color polymorphism in a lizard. *Evolution* 55, 2040–2052.

Sirot, L. K. and Brockmann, H. J. 2001. Costs of sexual interactions to females in Rambur's forktail damselfly, *Ischnura ramburi* (Zygoptera: Coenagrionidae). *Animal Behaviour* 61, 415–424.

Sirot, L. K., Brockmann, H. J., Marinis, C., and Muschett, G. 2003. Maintenance of a female-limited polymorphism in *Ischnura ramburi* (Zygoptera: Coenagrionidae). *Animal Behaviour* 66, 763–775.

Skúlason, S. and Smith, T. B. 1995. Resource polymorphisms in vertebrates. *Trends in Ecology and Evolution* 10, 366–370.

Smith, T. B. and Girman, D. J. 2000. Reaching new adaptive peaks: evolution of alternative bill forms in an African finch. In T. A. Mousseau, B. Sinervo, and J. A. Endler (eds.) *Adaptive Genetic Variation in the Wild*, pp. 139–156. Oxford, UK: Oxford University Press.

Smith, T. B. and Skúlason, S. 1996. Evolutionary significance of resource polymorphisms in fish, amphibians and birds. *Annual Review of Ecology and Systematics* 27, 111–134.

Stearns, S. C. 1992. *The Evolution of Life Histories*. Oxford, UK: Oxford University Press.

Stouthamer, R., Hurst, G. D. G., and Breeuwer, J. A. J. 2002. Sex ratio distorters and their detection. In I. C. W. Hardy (ed.) *Sex Ratios: Concepts and Research Methods*, pp. 195–217. Cambridge, UK: Cambridge University Press.

Stubblefield, J. W. and Seger, J. 1990. Local mate competition with variable fecundity: dependence of offspring sex ratios on information utilization and mode of male production. *Behavioral Ecology* 1, 68–80.

Sündstrom, L., Chapuisat, M., and Keller, L. 1996. Conditional manipulation of sex ratios by ant workers: a test of kin selection theory. *Science* 274, 993–996.

Sword, G. A. 1999. Density-dependent warning coloration. *Nature* 397, 217–218.

Sword, G. A. Simpson, S. J., El Hadi, O. T. M., and Wilps, H. 2000. Density-dependent aposematism in the desert locust. *Proceedings of the Royal Society of London B* 267, 63–68.

Taborsky, M. 1984. Broodcare helpers in the cichlid fish *Lamprologus brichardi*, their costs and benefits. *Animal Behaviour* 32, 1236–1252.

Taborsky, M. 1994. Sneakers, satellites, and helpers: parasitic and cooperative behavior in fish reproduction. *Advances in the Study of Behavior* 23, 1–100.

Taborsky, M. 1998. Sperm competition in fish: "bourgeois" males and parasitic spawning. *Trends in Ecology and Evolution* 13, 222–227.

Taborsky, M. 1999. Conflict or cooperation: what determines optimal solutions to competition in fish reproduction? In V. C. Almada, R. Oliveira, and E. J. Gonçalves (eds.) *Behaviour and Conservation of Littoral Fishes*, pp. 301–343. Lisbon: Instituto Superior de Psicologia Aplicada.

Taborsky, M. 2001. The evolution of bourgeois, parasitic and cooperative reproductive behaviors in fishes. *Journal of Heredity* 92, 100–110.

Taborsky, M., Hudde, B., and Wirtz, P. 1987. Reproductive behaviour and ecology of *Symphodus* (*Crenilabrus*) *ocellatus*,

a European wrasse with four types of male behaviour. *Behaviour* 102, 82–117.

Tauber, C. A. and Tauber, M. J. 1992. Phenotypic plasticity in *Chrysoperla*: genetic variation in the sensory mechanism and in correlated reproductive traits. *Evolution* 46, 1754–1773.

Tauber, M. L., Tauber, C. A., and Masaki, S. 1986. *Seasonal Adaptations of Insects*. Oxford, UK: Oxford University Press.

Taylor, F. 1980. Optimal switching to diapause in relation to the onset of winter. *Theoretical Population Biology* 18, 125–133.

Taylor, F. 1986. The fitness function associated with diapause induction in arthropods. 2. The effects of fecundity and survivorship on the optimum. *Theoretical Population Biology* 30, 93–110.

Thomaz, D., Beall, E., and Burke, T. 1997. Alternative reproductive tactics in Atlantic salmon: factors affecting mature parr success. *Proceedings of the Royal Society of London B* 264, 219–226.

Thorpe, J. E. and Morgan, R. I. G. 1978. Parental influence on growth rate smolting rate and survival in hatchery reared juvenile salmon *Salmo salar*. *Journal of Fish Biology* 13, 549–556.

Thorpe, J. E. and Morgan, R. I. G. 1980. Growth rate and smolting rate of progeny of male atlantic salmon parr *Salmo salar*. *Journal of Fish Biology* 17, 451–460.

Thorpe, J. E., Morgan, R. I. G., Talbot, C., and Miles, M. S. 1983. Inheritance of developmental rates in Atlantic salmon, *Salmo salar* L. *Aquaculture* 33, 119–128.

Tomkins, J. L. 1999. Environmental and genetic determinants of the male forceps length dimorphism in the European earwig *Forficula auricularia* L. *Behavioral Ecology and Sociobiology* 47, 1–8.

Tomkins, J. L. and Brown, G. S. 2004. Population density drives the local evolution of a threshold dimorphism. *Nature* 431, 1099–1103.

Tomkins, J. L., Simmons, L. W., and Alcock, J. 2001. Brood-provisioning strategies in Dawson's burrowing bee, *Amegilla dawsoni* (Hymenoptera: Anthophorini). *Behavioral Ecology and Sociobiology* 50, 81–89.

Tomkins, J. L., Lebas, N. R., Unrug, J., and Radwan, J. 2004. Testing the status-dependent ESS model: population variation in fighter expression in the mite *Sancassania berlesei*. *Journal of Evolutionary Biology* 17, 1377–1388.

Tomkins, J. L., Kotiaho, J. S., and LeBas, N. R. 2005. Matters of scale: positive allometry and the evolution of male dimorphisms. *American Naturalist* 165, 389–402.

Trivers, R. L. and Hare, H. 1976. Haplodiploidy and the evolution of the social insects. *Science* **191**, 249–263.

Trivers, R. L. and Willard, D. E. 1973. Natural selection of parental ability to vary the sex ratio of offspring. *Science* **179**, 90–92.

Turner, J. R. G. 1977. Butterfly mimicry: the genetical evolution of an adaptation. *Evolutionary Biology* **10**, 163–204.

Unrug, J., Tomkins, J. L., and Radwan, J. 2004. Alternative phenotypes and sexual selection: can dichotomous handicaps honestly signal quality? *Proceedings of the Royal Society of London B* **271**, 1401–1406.

van Gossum, H., Stoks, R., and De Bruyn, L. 2001. Reversible frequency-dependent switches in male mate choice. *Proceedings of the Royal Society of London B* **268**, 83–85.

Via, S. 1994. The evolution of phenotypic plasticity: what do we really know? In L. Real (ed.) *Ecological Genetics*, pp. 35–57. Princeton, NJ: Princeton University Press.

Via, S. and Lande, R. 1985. Genotype–environment interaction and the evolution of phenotypic plasticity. *Evolution* **39**, 505–522.

Walker, T. J. and Sivinski, J. M. 1986. Wing dimorphism in field crickets (Orthoptera: Gryllidae). *Annals of the Entomological Society of America* **79**, 84–90.

Waltz, E. C. 1982. Alternative mating tactics and the law of diminishing returns: the satellite threshold model. *Behavioral Ecology and Sociobiology* **10**, 75–83.

Waltz, E. C. and Wolf, L. L. 1984. By Jove! Why do alternative mating tactics assume so many different forms? *American Zoologist* **24**, 333–343.

Waltz, E. C. and Wolf, L. L. 1988. Alternative mating tactics in male white-faced dragonflies (*Leucorhinia intacta*): plasticity of tactical options and consequences for reproductive success. *Evolutionary Ecology* **2**, 205–231.

Werren, J. H. 1987. Labile sex ratios in wasps and bee: life history influences the ratio of male and female offspring. *BioScience* **37**, 498–506.

Werren, J. H. and Beukeboom, L. W. 1998. Sex determination, sex ratios, and genetic conflict. *Annual Review of Ecology and Systematics* **29**, 233–262.

West-Eberhard, M. J. 1979. Sexual selection, social competition, and evolution. *Proceedings of the American Philosophical Society* **123**, 222–234.

West-Eberhard, M. J. 2003. *Developmental Plasticity and Evolution*. Oxford, UK: Oxford University Press.

West, S. A. and Herre, E. A. 2002. Using sex ratios: why bother? In I. C. W. Hardy (ed.) *Sex Ratios: Concepts and Research Methods*, pp. 399–413. Cambridge, UK: Cambridge University Press.

Wheeler, D. E. 1991. The developmental basis of worker caste polymorphism in ants. *American Naturalist* **138**, 1218–1238.

Whitfield, C. W., Cziko, A. M., and Robinson, G. E. 2003. Gene expression profiles in the brain predict behavior in individual honey bees. *Science* **302**, 296–299.

Widemo, F. 1998. Alternative reproductive strategies in the ruff, *Philomachus pugnax*: a mixed ESS? *Animal Behaviour* **56**, 329–336.

Wiegmann, D. D., Angeloni, L. M., Baylis, J. R., and Newman, S. P. 2004. Negative maternal or paternal effects on tactic inheritance under a conditional strategy. *Evolution* **58**, 1530–1535.

Williams, G. C. 1979. The question of adaptive sex ratio in outcrossed vertebrates. *Proceedings of the Royal Society of London B* **205**, 567–580.

Zera, A. J. 1999. The endocrine genetics of wing polymorphism in *Gryllus*: critique of recent studies and state of the art. *Evolution* **53**, 973–977.

Zera, A. J. and Huang, Y. 1999. Evolutionary endocrinology of juvenile hormone esterase; functional relationship with wing polymorphism in the cricket, *Gryllus firmus*. *Evolution* **53**, 837–847.

Zera, A. J. and Rankin, M. A. 1989. Wing dimorphism in *Gryllus rubens*: genetic basis of morph determination and fertility differences between morphs. *Oecologia* **80**, 249–255.

Zera, A. J. and Tiebel, K. C. 1989. Differences in juvenile hormone esterase activity between presumptive macropterous and brachypterous *Gryllus rubens*: implications for the hormonal control of wing polymorphism. *Journal of Insect Physiology* **35**, 7–17.

Zhao, Z. and Zera, A. J. 2002. Differential lipid biosynthesis underlies a tradeoff between reproduction and flight capability in a wing-polymorphic cricket. *Proceedings of the National Academy of Sciences of the United States of America* **99**, 16829–16834.

3 · Phylogenetic analysis of alternative reproductive tactics: problems and possibilities

VÍTOR C. ALMADA AND JOANA I. ROBALO

CHAPTER SUMMARY

We present an outline of the potential that a phylogenetic approach may bring to the study of alternative reproductive tactics (ARTs) and discuss some of the difficulties and methodological problems that must be addressed if we are to apply the phylogenetic method successfully. We illustrate the principles presented by applying them to three selected examples. Specifically, based on fish studies, ARTs are, at least in some fish groups, evolutionarily unstable and rarely become incorporated as a fixed trait of a lineage at a rank as high as a family. Despite this instability, it is common for a given lineage to give rise recurrently to very similar forms of ARTs. Our results illustrate the wide spectrum of potential studies that can be enriched by a phylogenetic perspective.

3.1 BACKGROUND

3.1.1 The role of phylogenetic studies in the behavioral sciences

When Tinbergen (1963) formulated his famous four levels of explanation in ethology – causation, development, evolution, and function – the study of patterns in the evolution of behavior was explicitly made one of the central aims of the study of animal behavior. Each behavior pattern has an evolutionary history, and often it is possible to trace its origins, identifying the behavior that existed prior to the emergence of the new patterns. The pioneering work of Konrad Lorenz, who used behavior patterns to help to clarify the taxonomy of ducks, illustrates the use of behavioral phylogenies in an outstanding way (Lorenz 1941).

Obviously, evolution is a historical process. Except for organisms with very short generation times, we cannot replicate it under experimental conditions. Thus, the best we can expect to achieve is to develop hypotheses about the changes in pre-existing behavior patterns through which the new behavior patterns came into existence and the possible causes of the presumed changes. This limitation does not make evolution less interesting or less rigorous than other branches of biology. It simply means that our best achievement will only be a hypothesis, albeit a plausible one.

Phylogeny plays a crucial role in the study of behavioral evolution. Phylogenetic analysis aims to uncover the relationships among biological lineages. It tries to answer questions like:

- What are the degrees of relatedness among different groups of organisms?
- What groups of organisms shared the same common ancestor?
- What was the sequence of branching events that gave rise to the relationships among the species that descend from a given ancestor?

With the help of information about the events of the past, either fossils or molecular clocks, phylogenetic studies also deal with the attempt to date the evolutionary events that are assumed to have shaped the relationships among living forms as we know them.

When we state that phylogeny is crucial to the study of the evolution of behavior, it is useful to remember that they are not equivalent. To study the evolution of behavior is to trace the origins of a given behavior. It is also an attempt to identify the possible causes underlying the change of a pre-existing (primitive) behavior into its new (derived) state. On the other hand, to study the evolution of lineages involves formulating hypotheses about the history of the organisms themselves. Every hypothesis on the origin of a given behavior must be based on our knowledge of the history of the lineage in which the behavior changed. In other words, the understanding of behavioral evolution depends on the quality and accuracy of our knowledge of the relationships among the organisms in whose history such change in

Alternative Reproductive Tactics, ed. Rui F. Oliveira, Michael Taborsky, and H. Jane Brockmann. Published by Cambridge University Press.
© Cambridge University Press 2008.

behavior took place. Moreover, the evidence on which our inferences about behavioral evolution are based must be independent of the information that we use to work out the phylogeny of the group of organisms with which we are dealing. If we fail to keep this distinction in mind, we risk incurring a logical fallacy: suppose we use our reconstruction of the evolution of behavior to support our views on the relationships among organisms, and afterwards we use the relationships so inferred to support our reconstruction of the history of the behavior! We would have used the same data as our conclusion and as the evidence supporting it. Excellent presentations of the use of the comparative method and its application in behavioral studies may be found in Brooks and McLennan (1991), Harvey and Pagel (1991), and Martins (1996).

Behavioral ecologists have become increasingly familiar with phylogeny, for a different reason: one of the ways to try to detect associations between habitat traits and behavioral characters is to compare organisms that live in different environments. One interesting situation is when unrelated organisms that have colonized similar environments show similar behavior. The evidence will be even more compelling if organisms that share the same common ancestor become adapted to contrasting environments and also show contrasting behavior. In both cases, we have evidence suggesting that specific behavioral traits may have adaptive value in specific environments (but see Gould and Vrba [1982] and Almada and Santos [1995] for alternative, nonadaptive explanations).

This kind of comparative study requires phylogenetic information. If we want to compare descendants of a common ancestor, we have to uncover the phylogeny of the group with which we are working, and if we want to access the statistical significance of a given comparison, we must be sure that our data points are truly independent. Suppose that we compare 12 species of fish that breed in territories and 12 other species of fish that are mass spawners. Suppose also that we show that sneakers are found in a much larger proportion of the territorial species than in the mass spawners. We would be tempted to conclude that the existence of territorial males competing for favorable sites may exclude weaker or smaller males from gaining territories and that sneaking evolved as a way for weaker individuals to circumvent their limitations. Do our data support such a hypothesis? If the 12 territorial fish species are closely related phylogenetically, they may not represent 12 independent instances of the evolution of sneaking. They may share territoriality and sneaking simply because they

descended from the same ancestor. Indeed, in many cases, speciation does not imply, of necessity, changes in most traits. On the contrary, the species descended from a given ancestor tend to show similarities in their phenotypes and even in their habitats that simply reflect their common origin. If nothing has happened to cause changes, a whole lineage may keep the ecology and set of phenotypic traits of its ancestor. In such a situation, we cannot count the 12 fish species as representing 12 independent instances of evolutionary change, but only as a single one. Only a sound phylogenetic background for the organisms we are comparing will enable us to identify the number of distinct lineages available for our comparative study.

In conclusion, phylogeny is important because it provides the framework to study the origin of new behavior patterns, to control for the lack of independence, and to assist in designing comparative studies where lineages with contrasting behavior patterns must be compared.

3.1.2   The ongoing revolution in phylogenetics

The study of phylogeny is as old as the idea of evolution. In the last four decades, however, the study of phylogeny has advanced in such an unprecedented way and has reached such a new level of methodological thoroughness that we can speak of a true revolution. This revolution stemmed from three sources. The first was the understanding, especially after the work of Hennig (1965, 1966), that when trying to uncover monophyletic groups (all the life forms that descended from a common ancestor) only shared derived traits must be considered. This was an important clarification because the descendants of a given ancestor often share many characters that did not originate in the course of the history of that specific lineage but are rather much older and thus are shared with members of many other lineages. Having four legs is not a good trait to characterize mammals because four legs certainly preceded the origin of mammals. This simple appreciation prevented much confusion and forced biologists to search explicitly for traits that help to identify members of monophyletic groups instead of mingling organisms of distinct origins simply because they are similar in many primitive traits.

The second advance in the phylogenetic revolution came from incorporating molecular data into phylogenetic studies. Beginning with the use of protein electrophoresis and comparative immunology, the use of molecular data has resulted in decisive advances in developing rigorous and reliable phylogenies, particularly with the advent of efficient and relatively inexpensive techniques to amplify and sequence DNA. Polymerase chain reaction (PCR) and DNA sequencing have made it possible to use directly the information contained in the DNA sequences, instead of depending upon the expression of those sequences. Nowadays, studies that use thousands of base pairs and analyze data from different genes are increasingly common.

The third advance came from the whole new assortment of informatic and statistical tools now available to build phylogenetic trees and to evaluate the results, even with very large datasets. Without the greatly enhanced computational power that is now available, the analysis of massive DNA sequence data would not have been possible. On the other hand, a large amount of work is being invested in making phylogenetic inference increasingly rigorous by exploring the use of sophisticated statistical tools to evaluate the support for each conclusion and to access the likelihood of different hypotheses. The discussion of the different methods of phylogenetic inference is beyond the scope of this chapter but the reader may find good introductions in Harvey *et al.* (1996), Page and Holmes (1998), and Hall (2004). Fortunately many of the new, powerful tools used in phylogenetic inference and character mapping have been implemented by software packages which are freely available on the World Wide Web, together with many support materials and forums where phylogenetic issues are discussed.[1]

The preceding is intended to give the less familiar reader a first glimpse of the new possibilities now available for phylogenetic studies. The ongoing advance in our ability to generate increasingly accurate and probable hypotheses about the past is already one important achievement of modern biology.

3.1.3 Mapping behavioral character evolution in phylogenies

Once a phylogenetic hypothesis is formulated, we are in a position to map a given character or set of characters on the phylogenetic tree. For instance, this means that we can evaluate how many times and in what branches of the tree a given kind of change took place, how frequent its reversion was from a derived state to the pre-existing condition, and to what extent two or more characters tended to co-evolve

repeatedly, indicating a possible causal link in their evolution. In a similar way, we can ask what characters were present before changes in other characters took place and if those pre-existing characters may have facilitated changes in other characters. Finally, we may reconstruct the probable ecological characteristics of a given lineage, thus formulating hypotheses on past environment and possible selective pressures that promoted the evolution of a given trait in a given group of organisms.

This mapping of the evolution of characters on phylogenetic trees is becoming increasingly accurate, and statistical approaches like maximum likelihood and Bayesian inference (among other strategies) are used to evaluate different hypotheses on a quantitative basis. Again, software is freely available to perform character mapping.[2] Papers like Goodwin *et al.* (1998), Wilson *et al.* (2003), Koblmüller *et al.* (2004), Peach and Rouse (2004), and Haase (2005) provide the reader with interesting examples of the application of phylogenetic information in the study of behavioral evolution.

To end this methodological introduction, we would like to comment on two important issues. The first is statistical. When only one or a few instances of a given evolutionary change have taken place in an entire group of organisms, little can be said about the generality of the emerging results. Chance and the hazards of historical processes may not be ruled out. On the other hand, if a group is rich in recurrent instances of similar changes, we may get patterns that can be assessed statistically and bear some generality. Our second methodological remark is concerned with the proper identification of the units that are to be compared. In many cases, ARTs have been defined functionally. Suppose that we state that a sneaker male intrudes on the nest where a spawning is taking place. In terms of functional consequences, this statement may be clear and sufficient. If, however, in different animal groups, the behavior of sneakers is different – is determined by different causal factors or involves different underlying neurophysiological machinery – then phenomena that look the same functionally may not be directly comparable. The adequacy and pertinence of our comparisons will depend greatly on our understanding of the behavior patterns effectively involved in ARTs, the stimuli and internal factors that control them, and the underlying machinery that produces them and ensures their development.

---

[1] Examples of relevant links are Phylogeny Programs (http://evolution.genetics.washington.edu/phylip/software.html), Evoldir (Evolution Directory, http://evol.mcmaster.ca/brian/evoldir.html), and David Posada's Lab (http://darwin.uvigo.es/).

[2] For example, MacClade, http://macclade.org/macclade.html, and Mesquite, http://mesquiteproject.org/mesquite/mesquite.html.

3.2 SOME EXAMPLES USING FISH DATA

Fishes are an ideal group in which to study the evolution of ARTs. Their extraordinary diversity of lineages (Nelson 2006) and the numerous examples of the independent evolution of ARTs (Taborsky 1994, 1998, 1999, 2001; see also Chapter 10) provide an uncommonly rich source of comparative data.

Taborsky (1994, 1998, 1999, 2001) listed a wide spectrum of ARTs in fish that range from sneaking and female mimicry to simple intrusions of a territorial male in the territory of a neighbor during spawning. This diversity means that ARTs encompass very different phenomena. At one extreme, there are ARTs in which males differ profoundly in their ontogenetic pathway, physiology, anatomy, and size. At the other, there are ARTs that are mere variations of behavior that a male can adopt on a "moment-to-moment" basis according to the circumstances and opportunities. For instance, while sneakers are usually much smaller and frequently lack male secondary sex characters, large males with fully developed secondary sex characters may also engage in ARTs. A territorial male may leave his nest to attempt fertilizations in the territory of his neighbor, returning subsequently to his own nest.

Due to this extreme diversity and the large variation in the detail of the behavioral descriptions available in the literature, we have not attempted to map all the different forms of ARTs separately. We have adopted the simple criterion of considering that male ARTs were found when there was "male simultaneous reproductive parasitism," as adopted by Taborsky (1994). Because in most fish the eggs are fertilized after spawning by the males releasing sperm into the water, most ARTs in fish involve two or more males releasing sperm simultaneously. The occurrence of ARTs in females is not analyzed in this chapter.

3.2.1 Patterns of the evolution of ARTs in blennies, wrasses, and salmonids

In Figures 3.1, 3.2, and 3.3, inferred phylogenetic relationships of Blenniidae, Labrini, and Salmonidae are depicted. In each phylogenetic tree, the occurrence of ARTs was mapped using Mesquite v. 4.5.2 (Maddison and Maddison 2005). For the sources and methods employed in phylogenetic reconstructions see the legends of the figures. The data on ARTs are from Taborsky (1994) unless stated otherwise. The three phylogenies correspond to three independent lineages whose ancestors diverged many tens of millions of years ago, if not hundreds of millions of years ago (Nelson 1994), and thus these lineages are independent.

The first interesting feature that emerges when the blenniid and the Labrini phylogenies are inspected is the fact that ARTs occur at the tips of branches or nearly so. This means that in none of these instances have ARTs become a permanent feature of a deeper clade comprising many genera. In the blennids, ARTs evolved two or three times in accordance with the reconstruction depicted in Figure 3.1. ARTs evolved in *Parablennius parvicornis* and at least in another branch of the tree. Either ARTs were present in the common ancestor of *Salaria* and *Scartella* or evolved separately in the ancestor of *Salaria* and in one of the members of *Scartella* (*Scartella cristata*). There are no reports of ARTs in *Scartella carboverdiana* and other members of the genus not included in the tree. It is still unknown whether this absence reflects lack of ethological information or true absence of ARTs. With the data presently available, it is impossible to decide between these alternatives. A less likely hypothesis involves the presence of ARTs in the common ancestor of *Salaria* and *Scartella* and its subsequent loss in some *Scartella* species. These uncertainties are instructive because they illustrate how the inclusion of more taxa in the phylogenies and a more thorough ethological dataset will gradually reduce the number of plausible hypotheses.

In the Labrini (Figure 3.2), ARTs likely evolved twice, once in the common ancestor of *Tautogolabrus* and *Ctenolabrus* and another in the common ancestor of *Symphodus* and *Centrolabrus*. If the absence of ARTs in various *Symphodus* is confirmed, it will represent one or more events of secondary loss.

In the salmonids (Figure 3.3), we hypothesize that male parasitic tactics evolved in an ancestor that gave rise to the sister clades *Salmo* and *Oncorhynchus/Salvelinus*.

After presenting the patterns of the evolution of ARTs depicted in the figures, it is important to stress that both phylogenetic information and behavioral descriptions vary dramatically even among closely related groups. For instance, in the blenniids, genus *Scartella*, we have two species in the phylogenetic tree and behavioral information for only one, but six species are recognized within the genus. In the case of *Parablennius*, we have a single species where ARTs are known, and ten species were included in the phylogenetic tree. The genus, however, although probably not monophyletic, contains 26 species. In the Labrini there are genera that are very well studied, both behaviorally and

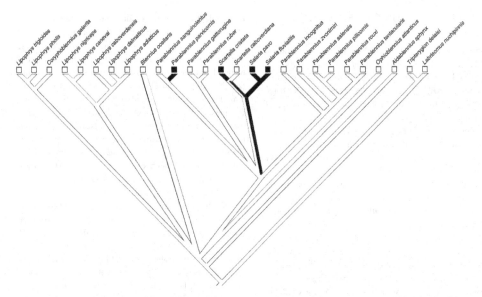

Figure 3.1 Mapping of the evolution of male ARTs in Atlanto Mediterranean blenniids. Phylogeny is based on the combined 12S–16S mitochondrial rDNA fragments adapted from Almada *et al.* (2005b). *Labrisomus nuchipinnis* and *Tripterygion delaisi* were used as outgroups. Data concerning *Parablennius sanguinolentus* from Taborsky (1994) have proven to be from *P. parvicornis*

(Almada *et al.* 2005a). The presence of ARTs in blenniids not listed in Taborsky (1994) are from *Salaria fluviatilis* (Neat *et al.* 2003a), *S. pavo* (Ruchon *et al.* 1995, Gonçalves *et al.* 1996, 2003, Oliveira *et al.* 2001a), and *Scartella cristata* (Neat *et al.* 2003b). The presence of male parasitism is marked in black.

phylogenetically. All species of *Symphodus* and *Centrolabrus* were included in the phylogenetic tree, both species of *Centrolabrus* were studied behaviorally, and ten species of *Symphodus* were carefully studied ethologically. In the salmonids, a family that is well known for the presence of male ARTs, the coverage is still fragmentary. We have data on the presence of ARTs in two of the three species of *Salmo* included in the phylogeny but there are about 28 valid species in this genus. We have data on ARTs for seven species of *Oncorhynchus*, eight species were included in Figure 3.3, but 14 species are recognized (all information on numbers of species is from Froese and Pauly 2005). We stress these limitations of the data analyzed for two reasons. The addition of more taxa in the future may change our views on the phylogenetic relationships within the groups that are still poorly sampled. The absence of information on the presence of ARTs in large numbers of species of a given taxon may have different causes: either the fish were ethologically studied in sufficient detail to make us confident that ARTs are absent or they were simply not studied or

ARTs escaped the attention of the researchers. Distinguishing among these possibilities is, for most cases, impossible. Without good quality standards for sampling species, both behaviorally and phylogenetically, all conclusions will be risky and subject to future modifications.

3.2.2 Instability and recurrent evolution of ARTs

The results outlined above support two main conclusions:

(1) Male ARTs evolved repeatedly in many independent fish lineages.

(2) ARTs evolved near the tips of the phylogenies, involving one or a few genera in each case. This observation seems to indicate that ARTs are evolutionarily short-lived, never becoming permanent features of higher taxa at the level of families or higher.

In addition, although the information is still fragmentary, ARTs were probably lost secondarily, even in some species

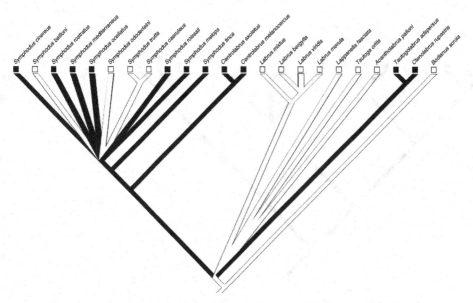

Figure 3.2 Mapping of the evolution of male ARTs in wrasses of the tribe Labrini. Phylogeny is adapted from Hanel *et al.* (2002) based on maximum likelihood analysis of mitochondrial 16S rDNA and control region sequences. *Bodianus scrofa* was used as the outgroup. In accordance with Almada *et al.* (2002), the species *Centrolabrus trutta* and *C. caeruleus* were assigned to the genus *Symphodus* and the species *S. melanocercus* to the genus *Centrolabrus*. The presence of male parasitism is marked in black.

of groups where they predominate. Short life, secondary loss, and recurrent emergence mean that male ARTs are behavior characters that, in fish, are labile on an evolutionary timescale, being relatively easy to acquire and to lose.

It is interesting to note that the incidence of ARTs may vary even within a single species. In the Atlantic salmon *Salmo salar*, the southern populations – those that are nearest to the highest temperature limits of the species – show a much higher incidence of precocious males than northern populations. This means that the patterns of distribution of sizes and ages at maturity of males and females and the incidence of male ARTs vary strongly with latitude (Moran and Garcia-Vazquez 1998, Martinez *et al.* 2000). In some populations the incidence of precocious males may be almost zero (Willson 1997, Garcia-Vazquez 2001).

In the blenniids in the genus *Salaria*, the differences between the sister species *S. pavo* and *S. fluviatilis* in the pattern of occurrence of ARTs are also remarkable. While in *S. pavo* there are populations with two clearly distinct male types: large, territorial males and small, female-like sneakers (Ruchon *et al.* 1995, Gonçalves *et al.* 1996, 2003, Oliveira

et al. 2001a), in *S. fluviatilis* the degree of dimorphism among males seems to be less marked. Smaller males often show less clearly marked secondary sex traits, but usually not so reduced as in *S. pavo* (Neat *et al.* 2003b). Apparently, some populations of *S. pavo* that are now under investigation also lack the extreme variation between male types previously described for this species (R. Oliveira, personal communication).

This inherent evolutionary instability in the occurrence of ARTs is also demonstrated when detailed genetic DNA fingerprinting of adults and offspring is performed so that the contribution of distinct types of males to the next generation can be assessed. In salmonids, there is wide variation between species and between populations of the same species in the degree of success of small males in fertilizing eggs (Blanchefield *et al.* 2003).

Sunfish, although being only distantly related to the groups discussed here, provide classical examples of species with ARTs (Gross and Charnov 1980, Gross 1982, 1984). In some sunfish species, there are territorial males, sneakers, and males that engage in female mimicry to gain access to

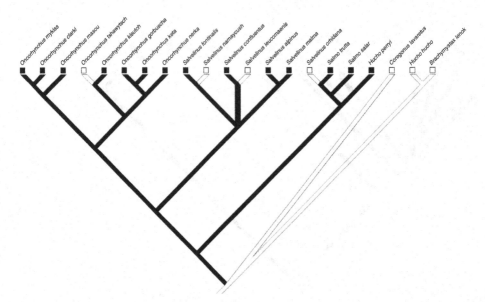

Figure 3.3 Mapping of the evolution of male ARTs in a partial phylogeny of salmonids, adapted from Crespi and Fulton (2004) and based on Bayesian analysis of 16 mitochondrial and eight nuclear genes combined. *Coregonus lavaretus*, *Brachymystax lenok*, and *Hucho perryi* were used as outgroups. The species referred to as *Salmo henshawi* and *S. gairdneri* in Taborsky (1994) are currently named *Oncorhynchus clarki* and *O. mykiss*, respectively. The ARTs in salmonids not listed in Taborsky (1994) are from *Salvelinus malma* (Hino et al. 1990), *S. confluentus* (McPhail and Baxter 1996), and *Oncorhynchus masou* (Koseki and Maekawa 2002). The presence of male parasitism is marked in black.

the eggs that females are spawning in the nests of territorial males. In other species of the genus *Lepomis*, although some nonterritorial males are present in the nest site, their contribution to the next generation is very limited (e.g., *Lepomis auritus*: DeWoody et al. 1998). The available data indicate that, in this genus, the contribution of small nonterritorial males varies widely among species (DeWoody et al. 2000, Neff 2001, Mackiewicz et al. 2002). Finally, even the quality of the habitat where the juveniles develop may influence the relative proportion and breeding success of males with different tactics in fish as different as Atlantic salmon and smallmouth bass (Wiegmann et al. 1997, Garant et al. 2003).

If the reader considers the literature reviewed in Chapter 10, the same pattern is detectable on a broader scale: examples of the occurrence of ARTs are scattered in many distinct taxa, but each taxon usually contains other species that lack ARTs. A similar inspection of other chapters of this book will confirm that the pattern we are outlining here is not restricted to fishes but very likely reflects a general rule.

How do we explain evolutionary instability? Perhaps this pattern is to be expected by the nature of ARTs. ARTs mean that a population is dimorphic or even polymorphic with regard to the diversity of phenotypes that a given sex, the males in our case, expresses to achieve reproduction. Sneaking by small, inconspicuous, and nonaggressive males, female mimicry, and intrusion into the territories of neighbors are all means to overcome or minimize the limitations that other males impose as a result of their size, degree of aggression, and so forth. As "alternative ways" to achieve egg fertilization, such tactics are likely to be extremely sensitive to social and ecological conditions in different ways. This reasoning seems to apply equally to species where each ART is fixed for the life of a male and to those where they can change with age, size, social, or ecological circumstances. This sensitivity to social and ecological conditions may also apply to situations where different ARTs have equal pay-off, as well as to those in which one or more tactics are adopted only by younger or weaker males that are making "the best of a bad job." The demographic structure of the population will typically be affected by factors like opportunities for growth, differential mortality of distinct size classes, or developmental stages and sex ratio. All these variables may, in turn, affect the operational sex ratio (density of members of each sex), which, among many other factors, may influence the balance between costs and benefits of ARTs. In species in which reproduction depends on the acquisition of nest sites, their availability may also affect drastically the opportunities a male has to establish a breeding territory and fertilize eggs (e.g., Oliveira et al. 1999). Even the variations in the reproductive tactics of the opposite sex may affect the outcomes of different ARTs. All the taxonomic chapters in this book underscore the plurality of social and ecological variables that may affect ARTs and their relative success. Thus, the fixation of multiple tactics in a high-level clade will be unlikely in most lineages.

Regardless of the adaptive value of different ARTs in various contexts, environmental factors may affect ARTs by changing the physiology of sexual maturation. The higher prevalence of precocious males in Atlantic salmon in warmer waters illustrates this type of environmental variable. Regardless of why at higher temperatures more males mature without migrating to the sea, this type of variation means that in some conditions two or more tactics are present, while in other conditions only a single tactic prevails. This effect of environment on development may itself be adaptive, or it may reflect an unavoidable peculiarity of the physiological mechanisms underlying sexual development in a given taxon. Be it adaptive or not, this variability of maturational trajectories in male development will also contribute to making the occurrence of ARTs a feature of some scattered members of their respective clades.

The recurrent evolution of ARTs in fish also raises very interesting questions. The species of *Salaria*, *Parablennius parvicornis*, and *Scartella cristata* in the blenniids are a case in point. The same type of ART evolved independently in the ancestors of the three species. In the three cases, sneakers are younger then territorials and will shift to territoriality later in life. The two tactics differ not only in behavior and fish size and age but also in anatomy and physiology. Sneakers lack the secondary sex characters of territorial males, namely the glands on the two spines of the anal fins. Their testicular glands are smaller, although their testes are proportionally larger than those of territorials (Santos 1995, Santos et al. 1996, Oliveira et al. 1999, 2001a, Neat et al. 2003a, b). As far as is known, sneakers and

territorials also differ in their hormonal profiles, the androgens of territorials showing a higher 11-ketotestosterone/testosterone ratio than in sneakers (Oliveira et al. 2001a, b). These similarities are impressive in their details and may have been caused by similar changes in the underlying physiological mechanisms.

These recurrent patterns of evolution raise several very interesting questions. While traditionally similarities were usually attributed to homologies or convergent evolution, phylogenetic studies reveal an increasing number of examples of parallel evolution. The same change occurs repeatedly in closely related taxa. In these cases, the phenotypes that served as the starting point for a given change are themselves very similar, the similarities between lineages may be very detailed. Similar lineages have undergone similar changes. What determines such parallel evolution? In this case, phylogenetic studies can provide a stimulating question, but other disciplines must help to answer it. Two closely related groups may give rise to similar changes, because being similar in ecology and behavior, the selective pressures that may promote a given type of adaptive change may be met by both lineages. On the other hand, many similar changes can be facilitated in some groups of closely related species because the mechanisms underlying sexual development are the same and thus they are prone to undergo similar transformations. Even the sensory modalities involved in courtship and mating and the details of the behavior patterns employed by a given group may themselves determine the nature of male ARTs that are more likely to evolve in its members. Chapters 10 and 11 provide excellent illustrations of this idea. A typical parasitic male fish attempts to approach a spawning female to release sperm in the vicinity of the eggs. A male parasitic frog or toad often takes advantage of the calls of other males to intercept females that are moving towards the caller. A male salamander or newt may interfere with the fertilization process of other males, either by placing his own spermatophores on top of those of the other male or by mimicking the tactile stimuli provided by a female leading the competitor to move away, a behavior that in the presence of a true female would help her to place the cloaca over a spermatophore. Apparently in fish, frogs, and salamander lineages, forms of male parasitic behavior "characteristic" of their respective groups evolved repeatedly. Many of the peculiarities of male parasitic behavior of each group were strongly influenced in their evolution, by the particular patterns of sexual behavior of the group and the stimuli involved in it.

To give a final example, consider the following: for a species to produce small sneaker males that are younger and lack the secondary sex traits of territorial males but that have fully developed testes, it is necessary that the maturation of gonads and the development of secondary male traits be physiologically uncoupled. This uncoupling is certainly easier in some lineages than in others, depending on their developmental physiology. In a similar vein, we have many hypotheses about the adaptive advantages of changing sex but in vertebrates, regular sex change seems to be an easy process only in teleosts.

The plurality of hypotheses outlined above will only be evaluated with hard data on ecology, physiology, behavior, and developmental biology and not simply by phylogeny. Invoking a vague notion like phylogenetic inertia, will, in our view, add little to our understanding of this problem. Phylogeny, although not providing the answers, will help to pinpoint areas of study that would otherwise easily escape the attention of researchers.

3.3 PROSPECTS FOR THE FUTURE

The fish examples presented above have shown how, even with many gaps in our ethological and phylogenetic knowledge, mapping the occurrence of ARTs in phylogenies has helped to raise interesting questions that go beyond the limits of phylogeny. As ethological surveys become more exhaustive and phylogenetic inferences become more robust and detailed, we may hope to address a whole array of new questions. To what extent were some forms of ARTs precursors of others? For instance, was the intrusion of floater males on the territories of others the starting point for the evolution of specialized small sneakers? Was there any tendency for female mimicry and sneaking to evolve in association? And if so, in any particular temporal sequence? Do the ARTs that are fixed for life tend to evolve from conditional ARTs? All these questions are amenable to phylogenetic analysis when the necessary information is available.

Another interesting question that phylogeny may help to answer is the following. With the phylogenies of a few fish groups, it is difficult to evaluate to what extent the incidence of independent instances of the evolution of ARTs is high or low in the clades analyzed. As more lineages are studied in this perspective, we will improve our assessments of the background probability of the evolution of a new instance of ARTs in a given lineage. With this information on "typical rates" or "ranges of rates" of emergence, we will be able to identify groups that are exceptional because they show unusually high or low levels of emergence of ARTs. Then, we will be able to consider those "exceptional" groups and attempt to relate their pattern of ARTs evolution with peculiar features of their ecology, behavior, physiology, and developmental biology.

This chapter has presented the reader with more prospects than results. We hope, however, that it will help to stimulate more active research in the phylogenetic analysis of ARTs. We believe we have shown that, at its present level of development, phylogenetic analysis provides a set of sophisticated and rigorous tools to uncover the patterns of the evolution of ARTs. When such patterns are identified, many more questions will be raised.

Acknowledgments

This chapter was much improved by the criticisms and suggestions of two referees. The contributions of Dr. Nick Goodwin were especially helpful. The authors also want to express their gratitude to Dr. Jane Brockmann for her very careful revision, which contributed much to the quality of the text. We are also indebted to Dr. André Levy for his assistance with the software Mesquite.

References

Almada, V. C. and Santos, R. S. 1995. Parental care in the rocky intertidal: a case study of adaptation and exaptation in Mediterranean and Atlantic blennies. *Reviews in Fish Biology and Fisheries* 5, 23–37.

Almada, F., Almada, V. C., Domingues, V., Brito, A., and Santos, R. S. 2005a. Molecular validation of the specific status of *Parablennius sanguinolentus* and *Parablennius parvicornis* (Pisces: Blenniidae). *Scientia Marina* 69, 519–523.

Almada, F., Almada, V., Guillemaud, T., and Wirtz, P. 2005b. Phylogenetic relationships of the north-eastern Atlantic and Mediterranean blenniids. *Biological Journal of the Linnean Society* 86, 283–295.

Almada, V., Almada, F., Henriques, M., Santos, R. S., and Brito, A. 2002. On the phylogenetic affinities of *Centrolabrus trutta* and *Centrolabrus caeruleus* (Perciformes: Labridae) to the genus *Symphodus*: molecular, meristic and behavioural evidences. *Arquipélago: Life and Marine Sciences* 19A, 85–92.

Blanchefield, P. J., Ridgway, M. S., and Wilson, C. C. 2003. Breeding success of male brook trout (*Salvelinus fontinalis*) in the wild. *Molecular Ecology* 12, 2417–2428.

Brooks, D. R. and McLennan, D. A. 1991. *Phylogeny, Ecology, and Behavior: A Research Program in Comparative Biology.* Chicago, IL: University of Chicago Press.

Crespi, B. J. and Fulton, M. J. 2004. Molecular systematics of Salmonidae: combined nuclear data yields a robust phylogeny. *Molecular Phylogenetics and Evolution* **31**, 658–679.

DeWoody, J. A., Fletcher, D. E, Wilkins, S. D., Nelson, W. S., and Avise, J. C. 1998. Molecular genetic dissection of spawning, parentage, and reproductive tactics in a population of redbreast sunfish *Lepomis auritus*. *Evolution* **52**, 1802–1810.

DeWoody, J. A., Fletcher, D. E., Mackiewicz, M., Wilkins, S. D., and Avise, J. C. 2000. The genetic mating system of spotted sunfish (*Lepomis punctatus*): mate numbers and the influence of male reproductive parasites. *Molecular Ecology* **9**, 2119–2128.

Froese, R. and Pauly, D. (eds.) 2005. *FishBase.* Available online at www.fishbase.org, version (05/2005).

Garant, D., Dodson, J. J., and Bernatchez, L. 2003. Differential reproductive success and heritability of alternative reproductive tactics in wild Atlantic salmon (*Salmo salar* L.). *Evolution* **57**, 1133–1141.

Garcia-Vazquez, E., Moran, P., Martinez, J. L., et al. 2001. Alternative mating strategies in Atlantic salmon and brown trout. *Journal of Heredity* **92**, 146–149.

Gonçalves, D., Fagundes, T., and Oliveira, R. 2003. Reproductive behaviour of sneaker males of the peacock blenny. *Journal of Fish Biology* **63**, 528–532.

Gonçalves, E. J., Almada, V. C., Oliveira, R. F., and Santos, A. J. 1996. Female mimicry as a mating tactic in males of the blenniid fish *Salaria pavo*. *Journal of the Marine Biological Association of the United Kingdom* **76**, 529–538.

Goodwin, N. B., Balshine-Earn, S., and Reynolds, J. D. 1998. Evolutionary transitions in parental care in cichlid fish. *Proceedings of the Royal Society of London B* **265**, 2265–2272.

Gould, S. J. and Vrba, E. S. 1982. Exaptation: a missing term in the science of form. *Paleobiology* **8**, 4–15.

Gross, M. R. 1982. Sneakers, satellites and parentals: polymorphic mating strategies in North American sunfishes. *Zeitschrift für Tierpsychologie* **60**, 1–26.

Gross, M. R. 1984. Sunfish, salmon, and the evolution of alternative reproductive strategies and tactics in fishes. In R. Wootton and G. Potts (eds.) *Fish Reproduction: Strategies and Tactics*, pp. 55–75. London: Academic Press.

Gross, M. R. and Charnov, E. L. 1980. Alternative male life histories in bluegill sunfish. *Proceedings of the National*

Academy of Sciences of the United States of America **77**, 6937–6940.

Haase, M. 2005. Rapid and convergent evolution of parental care in hydrobiid gastropods from New Zealand. *Journal of Evolutionary Biology* **18**, 1076–1086.

Hall, B. G. 2004. *Phylogenetic Trees Made Easy*, 2nd edn. Sunderland, MA: Sinauer Associates.

Hanel, R., Westneat, M., and Sturmbauer, C. 2002. Phylogenetic relationships, evolution of broodcare behaviour, and geographic speciation in the wrasse tribe Labrini. *Journal of Molecular Evolution* **55**, 776–789.

Harvey, P. H. and Pagel, M. D. 1991. *The Comparative Method in Evolutionary Biology.* Oxford, UK: Oxford University Press.

Harvey, P. H., Leigh Brown, A. J., Maynard Smith, J., and Nee, S. (eds.) 1996. *New Uses for New Phylogenies.* Oxford, UK: Oxford University Press.

Hennig, W. 1965. Phylogenetic systematics. *Annual Review of Entomology* **10**, 97–116.

Hennig, W. 1966. *Phylogenetic Systematics.* Urbana, IL: University of Illinois Press.

Hino, T., Maekawa, K., and Reynolds, J. B. 1990. Alternative male mating behaviours in landlocked Dolly Varden (*Salvelinus malma*) in south-central Alaska. *Journal of Ethology* **8**, 13–20.

Koblmüller, S., Salzburger, W., and Sturmbauer, C. 2004. Evolutionary relationships in the sand-dwelling cichlid lineage of Lake Tanganyika suggest multiple colonization of rocky habitats and convergent origin of biparental mouthbrooding. *Journal of Molecular Evolution* **58**, 79–96.

Koseki, Y. and Maekawa, K. 2002. Differential energy allocation of alternative male tactics in masu salmon (*Oncorhynchus masou*). *Canadian Journal of Fisheries and Aquatic Sciences* **59**, 1717–1723.

Lorenz, K. 1941. Vergleichende Bewegungsstudien an Anatinen. *Journal für Ornithologie (Ergänzungsband 3)* **89**, 194–294.

Mackiewicz, M., Fletcher, D. E., Wilkins, S. D., DeWoody, J. A., and Avise, J. C. 2002. A genetic assessment of parentage in a natural population of dollar sunfish (*Lepomis marginatus*) based on microsatellite markers. *Molecular Ecology* **11**, 1877–1883.

Maddison, W. P. and Maddison, D. R. 2005. *Mesquite: A Modular System for Evolutionary Analysis*, version 1.06. Available online at http://mesquiteproject.org.

Martinez, J.L., Moran, P., Perez, J., *et al.* 2000. Multiple paternity increases effective size of southern Atlantic salmon populations. *Molecular Ecology* **9**, 293–298.

Martins, E.P. (ed.) 1996. *Phylogenies and the Comparative Method in Animal Behavior.* Oxford, UK: Oxford University Press.

McPhail, J.D. and Baxter, J.S. 1996. *A Review of Bull Trout (Salvelinus confluentus) Life-History and Habitat Use in Relation to Compensation and Improvement Opportunities,* Fisheries Management Report No. 104. Vancouver, BC.

Moran, P. and Garcia-Vazquez, E. 1998. Multiple paternity in Atlantic salmon: a way to maintain genetic variability in relicted populations. *Journal of Heredity* **89**, 551–553.

Neat, F.C., Lengkeek, W., Westerbeek, E.P., Laarhoven, B., and Videler, J.J. 2003a. Behavioural and morphological differences between lake and river populations of *Salaria fluviatilis. Journal of Fish Biology* **63**, 374–387.

Neat, F.C., Locatello, L., and Rasotto, M.B. 2003b. Reproductive morphology in relation to alternative male reproductive tactics in *Scartella cristata. Journal of Fish Biology* **62**, 1381–1391.

Neff, B.D. 2001. Genetic paternity analysis and breeding success in bluegill sunfish (*Lepomis macrochirus*). *Journal of Heredity* **92**, 111–119.

Nelson, J.S. 2006. *Fishes of the World,* 4th edn. New York: John Wiley.

Oliveira, R.F., Almada, V.C., Forsgren, E., and Gonçalves, E.J. 1999. Temporal variation in male traits, nesting aggregations and mating success in the peacock blenny, *Salaria pavo. Journal of Fish Biology* **54**, 499–512.

Oliveira, R.F., Almada, V.C., Gonçalves, E.J., Forsgreen, E., and Canário, A.V.M. 2001a. Androgen levels in males and social interactions in breeding males of the peacock blenny. *Journal of Fish Biology* **58**, 897–908.

Oliveira, R.F., Canário, A.V.M., and Grober, M.S. 2001b. Male sexual polymorphism, alternative reproductive tactics and androgens in combtooth blennies (Pisces: Blenniidae). *Hormones and Behavior* **40**, 266–275.

Page, R.D.M. and Holmes, E.C. 1998. *Molecular Evolution: A Phylogenetic Approach.* Oxford, UK: Blackwell Scientific.

Peach, M.B. and Rouse, G.W. 2004. Phylogenetic trends in the abundance and distribution of pit organs of elasmobranches. *Acta Zoologica* **85**, 233–244.

Ruchon, F., Laugier, T., and Quignard, J.P. 1995. Alternative male reproductive strategies in the peacock blenny. *Journal of Fish Biology* **47**, 826–840.

Santos, R.S. 1995. Anatomy and histology of secondary sexual characters, gonads and liver of the rock-pool blenny (*Parablennius sanguinolentus parvicornis*) of the Azores. *Arquipélago: Life and Marine Sciences* **13A**, 21–38.

Santos, R.S., Hawkins, S.J., and Nash, R.D.M. 1996. Reproductive phenology of the Azorean rock-pool blenny (*Parablennius sanguinolentus parvicornis*), a fish with alternative mating tactics. *Journal of Fish Biology* **48**, 842–858.

Taborsky, M. 1994. Sneakers, satellites, and helpers: parasitic and cooperative behavior in fish reproduction. *Advances in the Study of Behavior* **23**, 1–100.

Taborsky, M. 1998. Sperm competition in fish: "bourgeois" males and parasitic spawning. *Trends in Ecology and Evolution* **13**, 222–227.

Taborsky, M. 1999. Conflict or cooperation: what determines optimal solutions to competition in fish reproduction? In R. Oliveira, V. Almada, and E. Gonçalves (eds.) *Behaviour and Conservation of Littoral Fishes,* pp. 301–349. Lisbon: Instituto Superior de Psicologia Aplicada.

Taborsky, M. 2001. The evolution of bourgeois, parasitic, and cooperative reproductive behaviors in fishes. *Journal of Heredity* **92**, 100–110.

Tinbergen, N. 1963. On aims and methods of ethology. *Zeitschrift für Tierpsychologie* **20**, 410–433.

Wiegmann, D.D., Baylis, J.R., and Hoff, M.H. 1997. Male fitness, body size and timing of reproduction in smallmouth bass, *Micropterus dolomieui. Ecology* **78**, 111–128.

Willson, M.F. 1997. *Variation in Salmonid Life Histories: Patterns and Perspectives,* Research Paper No. 498. United States Department of Agriculture Forest Service, Pacific Northwest Research Station.

Wilson, A.B., Ahnesjö, I., Vincent, A.C.J., and Meyer, A. 2003. The dynamics of male brooding, mating patterns and sex roles in pipefishes and seahorses (Family Syngnathidae). *Evolution* **57**, 1374–1386.

4 · Modeling alternative mating tactics as dynamic games

JEFFREY R. LUCAS AND RICHARD D. HOWARD

CHAPTER SUMMARY

Alternative reproductive tactics may result from various causal mechanisms. This is relevant for the theoretician because the mathematical approach used to address the evolution of alternative mating tactics will be affected by the causal basis of the differential expression of these behavior patterns between (and within) individuals. In this chapter, we restrict our focus to alternative male mating tactics that are strictly controlled by short-term behavioral decisions. Based on a variation of the Lucas and Howard (1995) dynamic game-theory model, we show that a detailed understanding of five properties of a system with alternative reproductive tactics is important in understanding the evolutionary trade-offs associated with the choice among alternative mating tactics. These properties include (1) physiological or morphological state and how state is affected by the tactic chosen, (2) environmental conditions, (3) frequency- and (4) density-dependent attributes of the pay-offs derived from each tactic, and (5) time constraints that either directly affect the expression of a mating tactic or affect the pay-offs derived from those tactics. These five properties should be considered simultaneously, and we demonstrate how this can be done within the framework of a dynamic game. The model is extended to consider the evolution of graded signals. Our model suggests that the prediction of Proulx *et al.* (2002) that older males should have more honest signals is sensitive to assumptions made about environmental conditions and time constraints on future success. We end with a discussion of the level of detail that should be built into models.

4.1 INTRODUCTION

Nothing in life is simple. A basic decision when modeling any biological phenomenon is to ignore complexities or incorporate them. Our initial attempts to model alternative mating tactics (ARTs) in anurans as a dynamic game

incorporated many of the complexities of caller/satellite dynamics (Lucas and Howard 1995, Lucas *et al.* 1996). We quickly realized that, despite a huge literature on mating behavior and alternative male mating tactics in anurans (see reviews in Gerhardt and Huber 2002, Shuster and Wade 2003), no single study existed that provided all the information necessary to parameterize our model. Furthermore, certain components of the model (e.g., many aspects of female behavior) were poorly known for all species. As a result, our primary objective in this chapter is to call for more complete empirical studies of alternative mating tactics, particularly for species where the choice of an alternative mating tactic is behaviorally mediated. Our approach is to provide a fairly synoptic view of certain types of male mating tactics and to illustrate why we need more information on various aspects of mating behavior. We begin by describing how alternative mating behavior patterns are classified in terms of their underlying causation. We then narrow the scope of our investigation to consider one class of alternative mating tactics, those in which behaviors are dynamically regulated by each male, and focus on caller/satellite tactics.

4.2 UNDERLYING BASES FOR ARTs: GENETIC, DEVELOPMENTAL, AND BEHAVIORAL

Shuster and Wade (2003) described three general classes of alternative mating behavior. One class represents mating *strategies* that are simple Mendelian traits that breed true (i.e., are 100% transmitted from parent to offspring), and the other two represent mating *tactics* that are controlled, at least in part, by environmental factors or environmentally induced physiological factors. These classes can be differentiated by the timescale over which they develop (Shuster and Wade 2003). As we point out later in this section, the classification is incomplete because a number of mating

Alternative Reproductive Tactics, ed. Rui F. Oliveira, Michael Taborsky, and H. Jane Brockmann. Published by Cambridge University Press.
© Cambridge University Press 2008.

tactics fall under multiple classes. Nonetheless, the classification is useful because it at least underscores potential
differences in causal mechanisms governing the expression
of these traits.

Mating *strategies* result from genetic differences between
individuals, and they are therefore fixed in an individual at
birth. The best example of this is the existence of three male
morphs of the isopod *Paracerceis sculpta* (Shuster and
Guthrie 1999). In this species, genetic differences determine the difference between territorial males, satellite
males, and males that show a third, intermediate tactic.

Two classes of mating *tactics* have been described. One
occurs when males undergo a developmental switch at some
ontogenetic stage that determines their use of a specific
mating tactic when they are adults. Such tactics most likely
result from a genotype × environment interaction, as both
genetic and environmental factors underlie trait expression
(Taborsky 1998, Garant *et al.* 2002). For example, rapidly
growing males develop into a sneaker morph, whereas more
slowly growing males become dominant morphs in both
coho salmon (Gross 1984, 1985) and Atlantic salmon
(Hutchings and Myers 1994). Similarly, horn morphology
in male dung beetles and the correlated alternative mating
tactics employed by these adults are determined by larval
feeding history (Moczek *et al.* 2002). Theoretically, the
extent to which the switch is genetically controlled will
affect the relative dynamics of the evolution of these traits.
The basis of the switch will also determine, in part, how
these systems are modeled.

The second class of alternative mating tactics includes
males that can switch rapidly between tactics. In this chapter,
we refer to tactics in this class as being under behavioral
control. For example, green tree frogs can switch from
satellites to callers in a matter of seconds (Perrill *et al.* 1978).

This classification follows Shuster and Wade (2003). A
similar classification has been discussed by Taborsky (1998).
Taborsky (1998) enumerated three dimensions of ARTs:
determination, plasticity, and *selection*. Determination refers
to whether the ART is controlled strictly by genetic differences between individuals, by a genotype × environment
interaction, or by prevailing environmental conditions.
Plasticity describes whether the ART is fixed for life, or
changes once during ontogeny, or changes multiple times
on a momentary timescale. Selection stipulates whether
alternative traits stabilize at equal fitness or whether they
reflect a disparity in quality between individuals. Under this
terminology, *strategies* are fixed, genetically determined
ARTs. *Tactics* are environmentally determined and can

either be plastic or fixed (i.e., as a result of a genotype ×
environment interaction). While the classifications by
Shuster and Wade (2003) and Taborsky (1998) are similar,
Taborsky (1998) explicitly describes more complex origins
of ARTs than those implied by Shuster and Wade's (2003)
three categories. For example, *determination* may include
both genetic and environmental inputs.

Strikingly different mathematical approaches are used to
study these three classes of alternative mating behavior.
Mendelian genetics is used to model alternative mating
strategies that are determined by one or a few loci and that
breed true. This entails an analysis of the reproductive fitness contributed by each allele that codes for a specific
mating strategy (e.g., Shuster and Wade 2003). Developmentally based tactics are better studied by using life-
history theory (e.g., Roff 1992, Stearns 1992, Charnov 1993)
or, more narrowly, the theory of reaction norms (Schlichting
and Pigliucci 1998). For alternative mating strategies (i.e.,
strictly Mendelian traits), we expect equal fitness of individuals expressing each strategy. If the alternative tactic is a
developmental phenomenon, understanding the basis of the
developmental switch is critical. For example, assume that
the tactic employed by males results from differences among
individuals in juvenile growth rate. If the tactic employed by
a male results from genetically determined differences
among individuals in juvenile growth rate, then we might
expect equal fitness across tactics. However, Gross and
Repka (1998) showed that when individuals that express
different ARTs do not breed true, unequal fitness of the
different morphs could be stable. (Simply put: if the most
successful morph generates offspring that express the less
successful morph phenotypes, then both morphs can be
maintained in the population irrespective of differences
between morphs in lifetime reproductive success.) Also, if
developmental rates are determined by stochastic components in the environment such that any individual can express
any of the possible growth rates exhibited in the population,
then there is no expectation of equal fitness (see Dawkins
1980, Gross 1996). In either case, we would think of the
mating tactic as a general rule: if growth rate is x, then
become a satellite/sneaker; if growth rate is y, then become a
territory owner. This rule may show some variation between
habitats; if so, it should be treated as a reaction norm. The
question then becomes: which rule is evolutionarily stable,
in the sense that it cannot be invaded by an alternative rule
(see Gross and Repka 1998)?

Finally, when alternative mating tactics are under
behavioral control, the problem becomes one of economic

decision-making. Game theory (Maynard Smith 1982, Parker 1984, Dugatkin and Reeve 1998) is one approach that could be used to study this class of mating tactics. A more robust but more complex approach is dynamic game theory (Houston and McNamara 1999, Clark and Mangel 2000). Dynamic games involve two components: dynamic optimization and game-theoretic pay-offs. The dynamic optimization component of the model has several functions. It acknowledges the potential for changes in some state variable to affect the pay-off associated with the choice of any given mating tactic. State variables can include physiological states (e.g., energy levels: Lucas and Howard 1995; or sperm storage levels: Harris and Lucas 2002), or some morphological states such as size (Skubic et al. 2004).

Dynamic optimization also considers the effect of time horizons on the pay-off to any given mating tactic. For example, a male near the end of his life can "afford" to expend relatively excessive amounts of energy on advertisement or territorial defense because little future reproduction is sacrificed with an excessive expenditure. In contrast, a young male may be selected to be more conservative in his expenditure if this reduced expenditure protects large expected future reproductive benefits (e.g., Lucas and Howard 1995; also see Clark 1994).

The game-theoretic component of a dynamic game acknowledges the role of both frequency- and density-dependent pay-offs on the evolution of behavior (Houston and McNamara 1987, Lucas and Howard 1995). Indeed, a critical component of the evolution of alternative mating tactics is the fact that the pay-off to any given tactic (e.g., territoriality) is affected by the frequency (and often their density) of tactics played by other members of the population (Dawkins 1980, Maynard Smith 1982, Parker 1984). The ability to combine complex state-dependent and temporal-dependent pay-offs with frequency- and density-dependent pay-offs makes for an extremely powerful theoretical approach to mating systems.

Two additional points are worth mentioning. One is that while genetic polymorphisms are explicitly considered when investigating alternative strategies using a Mendelian approach, genetic polymorphism is also implicit when studying mating tactics that are under developmental or behavioral control (Grafen 1984). That is, for any trait to be of evolutionary interest, genetic variation underlying trait differences must be involved. The phenotype influenced by genetic differences may be influenced by environmental conditions experienced during ontogeny (reflecting a genotype × environment interaction) or may be sufficiently

plastic to change instantaneously with changing social conditions (reflecting a short-term behavioral response). In nature, selection favors the best genetic option of the ones available. In modeling, one solves for the optimal solution and implicitly assumes that the genetic variation in the population was sufficient eventually to settle on this solution. Strictly speaking, a dynamic game begins with a monomorphic population with a single state- and time-dependent tactic into which competing strategies are introduced, and the tactic that remains is one that cannot be invaded by a mutant playing any alternative tactic. As Mayr (1983) and Grafen (1984) noted some time ago, our "black boxing" of genetics using optimization techniques (Grafen's "phenotypic gambit") may not be appropriate in all cases, but it has proved to be a surprisingly reasonable approach in most studies that employ it.

The second point is that not all mating systems can be easily characterized as solely under genetic, developmental, or behavioral control (see Taborsky 1998, 2001). For example, side-blotched lizards typically exhibit three different, genetically determined strategies: territorial male; nonterritorial, female-guarding male; and a female-mimic male (Sinervo and Lively 1996). However, the mate-guarding male can alternatively develop into a female mimic depending on the availability of females (Sinervo et al. 2001). In Atlantic salmon, rapidly growing young males may become sexually mature early in life. These males, known as "parr," remain in fresh water rather than migrating out to sea to continue development. Parr are a fraction of the size of anadromous males and employ a sneak alternative tactic to gain fertilization success. However, because this species is iteroparous, parr may subsequently migrate out to sea and return as territorial anadromous males (see review by Fleming and Reynolds 2004). Similarly, plumage patterns in the ruff are heritable (Lank et al. 1995, 1999). Dark-collared birds defend small mating territories on a lek; white-collared males can act as sneaker males when they dart onto a territory and mate with females, but they can also court females that arrive on a lek. Hybrid models are required in all three of these examples. For example, we could treat the ruff system as a game played within a game: white-collared birds play a dynamic game against other white-collared birds and choose their mating tactic accordingly. However, this game is nested within a genetic game played by white-collared birds against dark-collared birds. Two-level dynamic games have been described by Alonzo and Warner (2000a, b). These may provide some insight into the design of hybrid models.

Gross and Repka (1998) provide an analytical solution to the evolution of condition–dependent, developmental switching rules where tactics do not breed true. This latter analytical method is preferable to the multilevel dynamic games described by Alonzo and Warner (2000a, b) because the model is easier to interpret, but the dynamic game approach is the only complete solution available for complex systems with behaviorally regulated ARTs (see Section 4.4).

4.3   THE DYNAMIC GAME-THEORY APPROACH

4.3.1   Behaviorally regulated traits

We now narrow our discussion to modeling behaviorally regulated tactics. Some authors have suggested that behavioral regulation of alternative male mating tactics is ubiquitous (e.g., Gross 1996), although there is some debate about the prevalence of this class of mating behavior (e.g., Shuster and Wade 2003). What is clear from the literature is that behaviorally regulated alternative mating tactics are common in many mating systems and probably truly ubiquitous in some. Thus, an analysis of the theoretical aspects of behavioral regulation of mating tactics is highly relevant to our understanding of their evolution.

Our model entails optimal decision-making; that is, in any given time interval, an animal chooses to perform any one of the alternative behavior patterns in its repertoire. The particular behavior chosen has two consequences: an immediate fitness pay-off to the individual if it reproduces and a change in its future reproductive success. The change in future reproductive success is caused by changes in physiological state (such as a reduction in energy level, size, or sperm stores) and by changes in mortality risk (for example, through predation or starvation) incurred when expressing a chosen behavior. These future pay-offs should in turn affect the current decision. Thus, each decision has cascading effects into the future by affecting physiological state and mortality risk, and these cascading effects will, in turn, affect choice between alternative decisions at any given time. This pattern of temporal cascading is dynamic optimization (Houston and McNamara 1999, Clark and Mangel 2000).

Ecological conditions will dictate, in part, how far into the future the temporal cascade extends as a factor influencing a decision. For example, if predator density is high or food abundance is low, then the "time horizon" of the cascade's effect will be relatively short. However, time horizons are complex, multidimensional phenomena.

Consider a situation in which an animal faces starvation because of low food abundance. The risk of starvation could result from three different thresholds (see discussion of the "lazy L" in Stephens and Krebs 1986): (1) a constant, immediate risk of starvation if energy stores fall below some threshold; (2) a daily threshold if the animal requires energy stores to survive a period when feeding is not possible (e.g., at night for a diurnal species); and (3) a seasonal time horizon if sufficient stored reserves are required to survive for long periods such as winter. Each of these thresholds could simultaneously influence any given decision, and the relative importance of each threshold varies with time of day and season. In addition, the animal's decision is also influenced by its current energetic state and a host of other conditions. At first glance, such complexity seems too great to handle, but the beauty of dynamic optimization is that dynamic programming makes it fairly easy to model multidimensional thresholds.

In addition to ecological conditions affecting decisions, the presence of conspecifics competing for the same food or mates means that the decisions of others will influence an individual's choice of behavior. This is where the game part of dynamic game theory is important in that the pay-offs to any decision will, in part, be affected by the frequency or density of occurrence of the behavior in the population.

In sum, dynamic games involve a cascading feedback between an individual's behavioral choices and its physiological state, and pay-offs to the decisions are affected by frequency- and density-dependent trade-offs. The algorithm used to find the evolutionarily stable state (ESS) has two parts (see Houston and McNamara 1987): a backward iteration (or dynamic program) and a forward iteration (or simulation). In our example, we start with some initial guess about the number of callers and satellites of each age class on each night of the season. We then use stochastic dynamic programming (Houston and McNamara 1999, Clark and Mangel 2000) to find the best strategy that a single male should play against this population. We then use a simulation to determine the composition of a chorus composed entirely of these mutants. This two-part process is repeated until the best mutant tactic is identical with the tactic shown by the rest of the population. This tactic is the ESS.

4.3.2   Empirical issues

To illustrate the utility of the dynamic game approach to investigating alternative mating tactics, we will concentrate on caller/satellite interactions in anurans. Callers expend energy advertising for females, while satellites act as reproductive parasites by intercepting females attracted to calling males. Given space constraints, we will describe the model in general terms here. Details can be found in Lucas and Howard (1995; also see Lucas *et al.* 1996) and in the Appendix to this chapter (Section 4.5).

Caller/satellite interactions include all of the features mentioned above for a dynamic game: males can switch between each tactic, sometimes within an evening (Perrill *et al.* 1978); calling is energetically expensive (Taigen and Wells 1985, Grafe *et al.* 1992, Cherry 1993) and involves a risk of predation (Howard 1978, Ryan *et al.* 1981); and the pay-offs to each tactic are frequency-dependent (Arak 1988) and most likely density-dependent (Ryan *et al.* 1981, Dyson *et al.* 1992, Wagner and Sullivan 1992). To model caller/satellite tactics using dynamic game theory requires information on five general properties: an individual's physiological state, prevailing environmental conditions, frequency- and density-dependent pay-offs to each tactic, and time constraints. All five properties are best considered simultaneously rather than singly because they interact with each other. Below, we discuss the relevance of each property and their relationships to each other. We will make several points based on results derived from the Lucas and Howard (1995) model. We assume that males have a repertoire of four behavior patterns: calling, acting as a satellite, leaving the chorus to forage, and leaving the chorus to hide in a refuge. We model a population in which the breeding season is at most 50 days long, contains 1000 males (summed over all age classes) on the first day of the breeding season, and consists of males whose energetic stores can be arbitrarily divided into 30 intervals. For simplicity, we assume that there are two classes of males (1-year-olds and 2-year-olds). Although data on the effect of male age on mating success are rarely reported for anurans and morphological correlates of age such as body size may only distinguish first-time breeders from older males (e.g., Halliday and Verrell 1988), we assume that calling 1-year-old males attract only 70% as many females as calling 2-year-old males (e.g., Howard 1981). Initially, we assume that the reproductive rate of satellites of both ages is 50% of the reproductive rate of 2-year-old males (e.g., Miyamoto and Cane 1980, Sullivan 1982, Tejedo 1992). Overwinter survival for 1-year-old males is assumed to be dependent on the energy reserves of the male at the end of the season, with a maximum survival probability of 0.75 (e.g., Clarke 1977, Howard 1984, Caldwell 1987). The values used are roughly based on empirical estimates from several anuran species but are certainly not

meant to be representative of all anurans (Lucas and Howard 1995).

(1) PHYSIOLOGICAL STATE

Continuous chorus attendance by male anurans is usually limited to a few consecutive nights (e.g., Dyson *et al.* 1992, Murphy 1994a, Given 2002) with male condition declining with longer chorus tenure (Murphy 1994b, Judge and Brooks 2001, Given 2002). Murphy (1994b) and Marler and Ryan (1996) showed experimentally that chorus tenure is significantly influenced by energetic state (but see Green 1990, Judge and Brooks 2001). Bevier (1997) has shown that glycogen levels in trunk muscle tend to decrease more rapidly in species where males have high calling rates. Because satellite males do not call, there should be significant energetic differences between caller and satellite mating tactics. The bulk of evidence points to energetics being an important component in chorus attendance, but few studies provide quantitative information on this point. Obviously, we need to know the energetic consequences of each decision. However, even a thorough knowledge of the dynamics of a male's physiological state will not give us a complete understanding of the evolution of alternative mating tactics, in part because environmental conditions should also affect the evolution of these traits.

(2) ENVIRONMENTAL CONDITIONS

Environmental conditions can influence behavioral decisions in several ways. We will illustrate this point with an example from our original model (Lucas and Howard 1995). We consider two environmental conditions: the number of days remaining in the breeding season and the degree to which climatic conditions are favorable for breeding. For the latter, we assume that female arrival rate to a chorus is partly a function of weather (e.g., rain).

Our model generates the following predictions. When environmental conditions are often conducive for high female-arrival rate, chorus formation will be promoted. Under these conditions, 2-year-old males with high energy stores should stay in the chorus and call and 2-year-old males with low energy stores should leave the chorus and forage (Figure 4.1A). However, the threshold level at which 2-year-old males leave and forage declines as the season progresses. Thus, a male's energy stores should influence whether it enters a chorus, but the effect of energetic state on a male's mating behavior is most critical early in the season when the time horizon for future mating opportunities is relatively long.

(A)

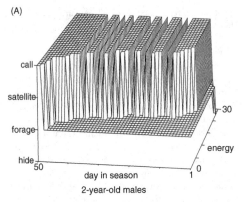

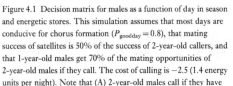

2-year-old males

(B)

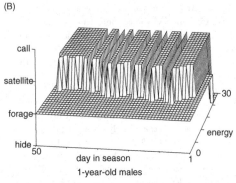

1-year-old males

Figure 4.1 Decision matrix for males as a function of day in season and energetic stores. This simulation assumes that most days are conducive for chorus formation ($P_{goodday} = 0.8$), that mating success of satellites is 50% of the success of 2-year-old callers, and that 1-year-old males get 70% of the mating opportunities of 2-year-old males if they call. The cost of calling is -2.5 (1.4 energy units per night). Note that (A) 2-year-old males call if they have high energy stores; (B) 1-year-old males choose to become satellites if they have high energy stores; the threshold energy at which 2-year-old males call decreases as the season progresses; and throughout the middle of the season all males are predicted to leave the chorus periodically in order to forage. Note that the lower right corner of this figure represents the beginning of the breeding season (day 1) and represents males with the lowest energy reserves (0).

Predictions differ for 1-year-old males: under the same conditions, 1-year-old males are expected to use satellite tactics throughout the breeding season (Figure 4.1B). The exclusive use of the satellite tactic by younger males is caused by several factors. Satellite tactics are assumed to require less energy than calling, and extended periods of environmentally favorable days for chorusing puts a premium on energetic efficiency. Furthermore, the pay-offs for calling differ: 1-year-old males attract fewer females by calling than 2-year-old males do. The net result is that 1-year-old males should weigh future reproductive success more strongly than 2-year-old males, and they should therefore choose a more conservative tactic than 2-year-olds. The more conservative strategy chosen by 1-year-olds causes the mass threshold for leaving the lek to increase as the season progresses – a trend opposite to that seen in 2-year-olds. Thus 1-year-old males in marginal condition at the end of the season should avoid the costs of entering the chorus; whereas 2-year-old males in marginal condition should accept these costs because these older males have less to lose if they die.

Regardless of age, males are predicted to move in and out of the chorus within a period of 4 to 7 days. Males are predicted to leave choruses because, under our assumptions, they will starve if they call continuously for more than 10 days, even if they begin with full fat stores. However, males are predicted

to leave choruses well before they face these energetic constraints. Males leave earlier than expected based on energetic considerations because frequency- and density-dependent pay-offs should contribute to the coherence of a chorus.

(3 AND 4) FREQUENCY- AND DENSITY-DEPENDENCE

Our model suggests that low-energy-state males are forced to leave every few days because they need to avoid starvation by foraging. The loss of these males from the chorus, in turn, potentially reduces the value of chorus attendance by males with relatively high energy states. The high-energy-state males leave because we have assumed that predation risk is both frequency (i.e., lower for satellites) and density dependent, and that female arrival rate is chorus-size dependent. The net result is what appears to be a pulsing chorus because males move in and out of the chorus without any change in environmental conditions. Note that the departure of low-energy-state males reduces the tendency for any other males to enter the chorus. This pulsing will be reinforced by an entrainment of energy states of males in the population because many males will be foraging or entering the chorus at the same time, and the modal energy state shown by males in the population will therefore cycle along with the pulsing of the chorus.

(A)

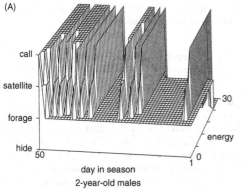

day in season
2-year-old males

(B)

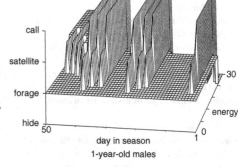

day in season
1-year-old males

Figure 4.2 Decision matrix for males as a function of day in season and energetic stores. This simulation assumes the same factors listed in Figure 4.1 and additionally assumes that the cost of calling is increased 40% over the cost assumed in Figure 4.1. (A) Note that 2-year-old males call if they have high energy stores, convert to being satellites at intermediate energy stores, and leave the chorus to forage at low energy stores. (B) The same is true of 1-year-old males, except that 1-year-old males do not call at the end of the season. Also, all males spend more days during the season foraging thus limiting the total number of days the chorus is active.

Numerous other factors could also influence frequency- or density-dependent trade-offs when pay-offs are measured in terms of increased mating success and reduced predation risk. For example, the pay-off to utilizing either the calling tactic or the satellite tactic should depend on the percentage of males in the chorus that are currently using each tactic (which should, in turn, be a function of the age/ size distribution of males in a population). The density of calling males should be positively correlated with the arrival rate of females and negatively correlated with the risk of predation each male might experience. However, merely stipulating the sign of these correlations is insufficient; the form of both functions (i.e., linear, accelerating, dampening) can be critical.

The relevance of each factor we have discussed thus far can be demonstrated with the following example. Physiological studies indicate that the cost of calling varies considerably across species (see review in Gerhardt and Huber 2002). For species with a particularly high cost of calling, we predict that chorus attendance should decrease dramatically in 2-year-old males (compare Figure 4.2A with Figure 4.1A). Early in the breeding season, 1-year-old males should enter a chorus as satellite males if they have low energy stores and they should call only if they have high energy stores. In contrast, 1-year-old males should use the satellite tactic exclusively toward the end of the season, even if they have high energy stores (Figure 4.2B). Of course, the use of

the satellite tactic by 1-year-old males is only viable if 2-year-old males call. In other words, frequency-dependent trade-offs associated with the caller/satellite decision determine critical components of chorus dynamics.

The above predictions assume that favorable environmental conditions prevail during the breeding season. In this case, energetic constraints will limit chorus attendance, and density- and frequency-dependent pay-offs will dictate how the population responds to these energetic constraints. However, unfavorable environmental conditions are predicted to eliminate both of these effects. If rain is less frequent during the breeding season, we predict that 2-year-old males should call on the few days that rain does occur (data not shown). Because a succession of rainy days should be uncommon, 2-year-old males never have the option of remaining in a chorus for periods long enough to jeopardize their energy stores. Thus, infrequent rains effectively shorten the time horizon associated with the energetic cost of chorus attendance. Males will only chorus for short intervals of time and will then have sufficient time to recoup energetic expenditures before it rains again. No pulsing of the chorus is expected as seen under more favorable and continuous breeding conditions.

When conditions favorable for breeding are rare during the season, energetic constraints should also be less relevant for 1-year-old males. These males should maximize their chances of mating on favorable days by calling rather than using the satellite tactic. The 1-year-old males should only adopt the satellite tactic early in the season, and only if they have a long breeding future.

Clearly, to understand the dynamics of chorus activity, we need to know the energetic consequences of each alternative behavior, but there are a host of frequency- and density-dependent factors that could also exert effects. In particular, these factors should affect the cohesiveness of a chorus and thereby override energetic effects. Finally, all of these effects are moderated by environmental conditions.

(5) TIME CONSTRAINTS

The last property needed to parameterize dynamic games is time constraints. As explained above, energetic constraints should have less effect on 2-year-old males than 1-year-old males as the breeding season progresses, because mating opportunities end with the current year for 2-year-old males; in contrast, 1-year-old males may survive for another year. If 1-year-old males survive the subsequent winter, prospects for mating success will be high in their next breeding season, as then they will be 2 years old. Thus, 1-year-old males have a longer time horizon than 2-year-old males. This should reduce the correlation between energetic thresholds and time of season and should cause 1-year-olds to choose an energetically conservative strategy all year long (see Figure 4.1). More generally, the reproductive consequences of most mating behavior patterns are likely to be affected by a variety of limited time horizons. A complete understanding of these time horizons is important in our characterization of behaviorally mediated mating tactics.

4.3.3   Incorporating all five factors: an example

Counterintuitive predictions can result from our model under specific parameter levels of the five factors we have outlined. In general, calling males and satellite males have an uneasy truce. Males call to attract females and the presence of callers makes the satellite tactic viable. If the cost of chorus attendance is sufficiently high for 1-year-old males, they should only enter a chorus as satellites; however, if callers suffer a significant reduction in mating success because of the presence of many satellite males and if the risk of predation is also high, then 2-year-old males that would otherwise call should not enter the chorus at all (Figure 4.3A). The consequence is no chorusing for extended periods of time. Callers stay away because the prospects of obtaining a mate are too low and chances of predation too high; satellite males stay away because there are no callers to parasitize (Figure 4.3B). Importantly, our simulation showed that every male would have a higher fitness if all individuals called in a chorus early in the season; however, this is not an ESS because it can be invaded by a male who plays satellite at least some of the time. Instead, the only ESS is for no one to call until they essentially run out of time in the season. At the end of the season,

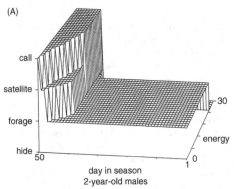

(A)

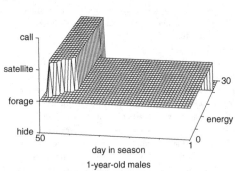

(B)

Figure 4.3 Decision matrix for males as a function of day in season and energetic stores. This simulation assumes the same factors listed in Figure 4.1, except that the mating success of satellites is 80% of the success of callers. Note that (A) 2-year-old males change

to the satellite tactic if their energy stores are low, but otherwise call; (B) 1-year-old males become satellites if they have high enough energy stores. In addition, the total duration of the chorus is only 9 days.

2-year-old males have no alternative but to enter the chorus because they will not survive to breed again. The presence of 2-year-old males with high energy loads that choose to call provides a viable option for 2-year-old males with low energy stores to act as satellites. In addition, the presence of calling 2-year-old males makes it viable for 1-year-old males to enter the chorus. Whether these males call or use the satellite tactic will depend on the frequency- and density-dependent trade-offs when in competition with calling 2-year-old males. Under the conditions we simulated here, they should always enter as satellites. Thus, the joint effect of time constraints, environmental factors, game aspects, and energetic constraints all dictate the expression of mating behavior patterns in this example.

4.3.4   Graded signals: an extension of the model

Although caller/satellite roles differ qualitatively, alternative mating tactics may also involve quantitative decisions such as the timing of when males display, and these quantitative decisions can be the basis for mating polymorphisms in a population (e.g., Boyko *et al.* 2004). Although these polymorphisms may be more subtle than qualitatively different roles, dynamic game theory can be used to investigate decision-making in these situations and thus increase our appreciation of the evolution of mating systems. Indeed, qualitative and quantitative differences are not mutually exclusive categories of behavior. For example, males exhibiting a qualitatively distinct class of signaling behavior may nonetheless show quantitative variation in the intensity of this signal. Below we extend our original caller/satellite model to illustrate how we can use dynamic game theory to study quantitative signal variation nested within qualitatively different mating tactics.

Starting with Zahavi (1975), research on signal design has focused on the factors that contribute to the evolutionary stability of signals that honestly advertise the quality of a mate (Grafen 1990, Maynard Smith 1991, Johnston and Grafen 1992). The consensus is that for signals shared among nonrelatives, the signal needs to be expensive to produce and the relative fitness consequences of producing an enhanced signal must be greater for low-quality males than for high-quality males. The models assume that females cannot detect male quality directly but can only infer quality from male signals – hence the issue of honesty and the potential for dishonest signaling (i.e., low-quality males providing high-quality signals).

In virtually all mating systems with alternative mating strategies/tactics, some males indicate their quality to females by providing information in their signals. In caller/satellite systems, calling males can vary acoustic properties of their signals such as call amplitude, rate, or duration, and females appear to prefer more exaggerated calls (reviewed by Gerhardt and Huber 2002). We know of no studies that have investigated whether the presence of satellite males affects any of these call properties of calling males.

Given the greater cost of producing a louder, longer, or more frequent call, it is assumed that these call properties can provide honest advertisements of calling-male quality. However, models used to study the evolution of honest signaling have primarily been static game-theory or genetic models (Grafen 1990, Maynard Smith 1991, Johnstone and Grafen 1993, Johnstone 2000, Gintis *et al.* 2001). While game-theoretic models have contributed significantly to our understanding of signal evolution, they leave out a potentially critical component of signal cost. In particular, the fitness consequence of investing energy on mating advertisements may change dynamically for signalers, as we have illustrated above. For example, an energetic expenditure early in a breeding season may have greater fitness consequences than an identical expenditure at the end of a breeding season. How do these dynamic components of a signaling system affect the evolution of the signal?

Proulx *et al.* (2002) showed that game-theoretic models with age structure generate different predictions than static models because young individuals have more to lose from risky signaling than do old individuals. As a result, we expect young individuals to be more conservative than old individuals when signaling; thus, we expect the signals from old individuals to be a more honest representation of quality compared to signals from young males. Although Proulx *et al.* (2002) incorporated age effects in their model, time-dependent changes in state were not explicitly treated. Dynamic games provide a mechanism to do just this and therefore can be used to evaluate the conclusions of Proulx *et al.* (2002) more completely.

To address these issues, we modified our original model to incorporate a graded call signal. We assume that the cost of an exaggerated signal falls under the category of "receiver-independent costs" (Vehrencamp 2000). That is, signaling cost is independent of the target receiver's response, as would be expected if the primary cost of producing a signal is either energetic or a risk of attracting predators. We assume that call exaggeration affects three aspects of a male's reproductive success:

(1) Relative attractiveness to a female. Assume that this is a monotonic increasing function of call investment ("intensity") but with diminishing returns. We simulate this function as

$$\text{RelAttractiveness(intensity)} = 0.4 + (5.21 \times (\text{intensity} - 1)^{0.5})/9. \quad (4.1)$$

(2) Energetic cost of call production. We modeled this relationship as a linear function of intensity following data cited in Gerhardt and Huber (2002):

$$\text{COST(intensity)} = 0.5 + (\text{intensity} - 1)/3. \quad (4.2)$$

(3) Relative risk of predation. We modeled the effect of call intensity on caller predation risk using an accelerating function of intensity:

$$\text{ProbPRED(intensity)} = 0.6112 + (\text{intensity} - 1)^2/9. \quad (4.3)$$

To simplify the analysis, we used four levels of call intensity (1–4). Increasing the number of levels to eight did not alter predictions, however. For all three functions, the coefficients were set such that there was a mean of 1.0 in the effect (attractiveness, energetic cost, or predation risk) for an equal weighting of the four intensity levels. Note that the functions described above are used as multipliers of the background attractiveness, energetic cost, and predation risk used in the original model (see Appendix, Section 4.5).

Results show firstly that the general properties of our original model are not altered if calling males use graded calls. For example, in environments conducive for chorus formation (i.e., probability that the environment is appropriate for a chorus on any given day, $P_{\text{goodday}} = 0.8$), males are expected to show pulses of mating activity punctuated by 1- to 2-day intervals where all males leave the chorus to eat. This pulsing is not shown in drier environments (e.g., $P_{\text{goodday}} = 0.4$). Also, not surprisingly, 1-year-old males will tend to act as satellites, and 2-year-old males should tend to call. The degree to which the satellite tactic is employed will depend on the relative success of satellites compared to callers. These are important results, because they suggest that our original results are robust to minor modifications of the model.

Secondly, our results partially support and extend the conclusions of Proulx et al. (2002). The relative shape of the intensity functions will strongly influence the results from the model. Due to space limitations, however, we will not explore this aspect of the model here. Instead, we will use Eqs. (4.1) to (4.3) to illustrate a few points about honest

signaling. If mate availability for satellites is only 40% of that for the mean caller and if the environment is favorable on most days ($P_{\text{goodday}} = 0.6$), then both 1-year-old and 2-year-old males call, but calling intensity increases with energy stores on a given day and call intensity changes over the course of the breeding season for both age groups. However, 1-year-old males should tend to call at lower intensities toward the end of the season (Figure 4.4A), whereas 2-year-old males should call at higher intensities at the end of the season (Figure 4.4B). Thus, as Proulx et al. (2002) suggest, one could conclude that calls from 2-year-old males appear to be more honest than those from 1-year-old males in that older males' calling intensity is a better reflection of their immediate condition compared to the younger males. However, this conclusion is correct only at the end of the breeding season. At the beginning of the season, a male's calling intensity may provide little information to the female about the quality of the calling male.

Our model suggests that the logic of Proulx et al. (2002) is sound under certain conditions: at the end of a long season (50 days) in a fairly high-quality environment ($P_{\text{goodday}} = 0.6$). If the conditions are even more conducive for chorus formation ($P_{\text{goodday}} = 0.8$), temporal trends in the intensity of calling by 1-year-old males reverse: they call more intensely at the end of the season (Figure 4.4C), as do 2-year-old males (Figure 4.4D). Paradoxically, the season is expected to be shorter for the more favorable conditions because satellite pressure drives callers from the chorus. However, the final stable solution is for all males to call! By eliminating the satellite option, we can show that the "ghost" of satellite pressure is the primary factor generating this pattern. When this happens, males call throughout the season (data not shown). Paradoxical results notwithstanding, the shortened season reduces future reproductive success for 1-year-old males and this in turn causes them to give relatively honest signals toward the end of the breeding season. Thus, the conclusion about age dependency in call honesty is sensitive to assumptions made about environmental conditions and time constraints on future reproductive success. Indeed, a change in a single parameter, such as the quality of climatic conditions, can change qualitative predictions about calling intensity. These results illustrate the value of dynamic game theory. Compared to static game-theory models, these more complex models provide a robust method of calculating future reproductive success and incorporating estimates of future reproductive success into predictions about mating tactics. Of course, this ability to measure future reproductive success comes at a

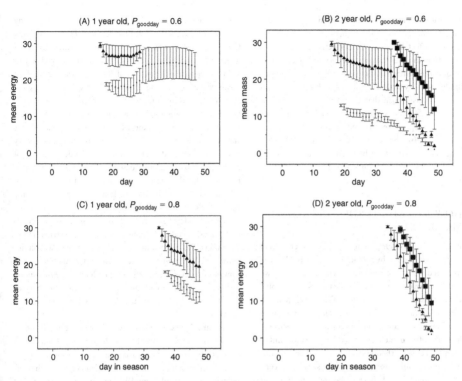

Figure 4.4 Mean predicted energy reserves as a function of time-in-season (day) and call intensity for 1- and 2-year-old males in our modeled population. Note that unlike Figures 4.1 to 4.3, this figure represents the profile of the population that results from the decision matrix, not the decision matrix itself. Panels (A) and (B) represent environments where the probability of a day favorable for chorus formation is 0.6; (C) and (D) represent environments where this probability is 0.8. The different symbols represent different call intensities: small circle = intensity 2, medium triangles = intensity 3, and large boxes = intensity 4. No male is predicted to call at the lowest intensity (1), and no males are predicted to act as satellites under these conditions. If no symbol is plotted for a given day, no chorus is predicted to form on that day (e.g., no chorus forms for the first 34 days in (C) and (D)). Note that 1-year-old males in (A) give less intense calls toward the end of the season, shifting in intensity from a mix of level 2 and 3 intensities around day 20 to only intensity 2 after day 30. The 1-year-old males in (C) tend to give more intense calls as the season progresses, with males giving intensity 3 calls over a broader range of body masses toward the end of the season. The latter trend is shown by 2-year-old males for both environments ((B) and (D)). Also, on any given day, low-energy males always give less intense calls than high-energy males if more than one intensity is produced.

cost: the empirical description of a mating system needs to incorporate all five of the factors we listed above.

Dynamic games provide a mechanism for understanding the effect of each of the five critical properties discussed in this chapter. This point is underscored by the limited results we have shown here in our model of graded signals. We assumed that females prefer 2-year-old males, but that male reproductive success is also a function of calling intensity. Should intensity honestly indicate male quality? The answer is yes, but only under limited circumstances. Our results provide evidence that only 2-year-old males should call at maximal intensity. Thus, females should be able to distinguish at least a subset of the highest-quality (older) males from lower-quality (young) males. However, our model assumes that females do not discriminate between males of different energetic states within an age class. Nonetheless, energetic state is predicted to have a profound effect on calling intensity because a male's energetic state will affect his future reproductive success. As a result, variation in calling intensities based on energetic state can blur the distinction between calling properties of high- and low-quality males. The point is that factors such as genetics or parasite load that are relevant to a female in her choice of males may be masked by variation in other factors such as energy stores that reflect stochastic events in an organism's life (see Stephens and Krebs 1986). Models such as the ones we have described here provide us with an important tool for understanding the dynamics of these decisions.

4.4 HOW DETAILED SHOULD A MODEL BE?

As in our earlier papers, our goal here is to aid researchers in prioritizing data collection by providing insights on the type of data required to analyze alternative mating tactics. As we have discussed above, alternative mating tactics are complex phenomena in that at least five general factors are involved in their evolution (an individual's physiological state, prevailing environmental conditions, frequency- and density-dependent pay-offs to each tactic, and time constraints). Such interplay between theoretical models and empiricism begs the following philosophical question: should we construct complex models of behavior that push the limits of our ability to collect data, or should we construct simple-models of behavior? We suggest that both approaches are necessary (also see Hilborn and Mangel 1997).

From our perspective, simple models perform a function different from that of more complex models. For example, the hawk/dove game (Maynard Smith 1982) revolutionized the study of behavior by introducing the concept of frequency-dependent pay-offs. This is a perfect example of a simple model that caused us to think about behavior in a new way.

A number of models of the evolution of mating behavior follow this pattern. State-independent game theoretic models provide a single-focus view of these systems. Examples include Waltz (1982) and Arak (1988), who evaluated caller/satellite decisions based on the relative attractiveness of nearby males. More complex, three-player games have also been described. Hamilton and Dill (2002) considered a game with three male strategies (resource owner, satellite, and floater), whereas Hugie and Lank (1997) modeled lekking in ruffs where they considered two

male strategies (satellite and territorial males) playing against one another and against females. Hugie and Lank (1997) showed that female choice may constrain all territorial males to have satellites, despite the fact that this reduces reproductive success of the territorial males. This result is analogous to our description of systems where the "ghost" of satellite parasites should keep males away from the chorus, despite the fact that their reproductive success would be higher if they did enter the chorus to call. Gross and Repka (1998) published a very different game-theory model that showed that condition-dependent choice of alternative mating tactics could be stable without equal fitness of the alternative mating morphs if there is partial inheritance of behavior (i.e., male morph a produces some fraction, $p < 1$, of a offspring).

These models have proved to be important in helping us understand specific components of mating systems. However they are, by definition, incomplete, and we cannot know whether their predictions are robust unless a more complex model is developed. Brodin (2000) and Pravosudov and Lucas (2001) provide an example of problems that can arise from models that are too sketchy in their depiction of behavior.

As our caller/satellite dynamic game illustrates, complex models give us some insight into the subtle and sometimes counterintuitive outcomes that can result from the interactions between a myriad of factors that regulate behavior. Indeed, only complex models can give us a way to put this myriad of factors into focus. But what level of complexity is sufficient? Dynamic programs that ignore frequency-dependent pay-offs can offer a partial solution where aspects of games played between males (or between males and females) are simply fixed. Examples of this approach are Fraizer's (1997) analysis of alternative mating tactics in digger wasps and Skubic et al.'s (2004) model of reproductive parasitism by subordinate helpers in a cooperatively breeding cichlid fish. Harris and Lucas (2002) showed that a dynamic program of sperm competition was broadly compatible with the game-theoretical models of Parker (1990), but revealed implicit assumptions about environmental factors that could invalidate predictions from the game-theoretical models when the assumptions are not met.

Dynamic games take this complexity one step further. But even with dynamic games, the level of complexity varies between models. In this chapter, we describe dynamic games played between males. A model derived from our previous work (Lucas et al. 1996) by McCauley et al. (2000) showed that broadly similar predictions could be generated with a slightly scaled-down version of our model. McCauley *et al.* (2000) argue that we do not need such a complex model. The problem is that this statement is meaningless without the one-to-one comparison between models. Even these models ignore multilevel games, which are certainly possible in the evolution of ARTs when female choice affects the pay-offs to males choosing among alternative reproductive tactics, but where the frequency of male tactics in turn affects the pay-offs to females in their choice of reproductive behavior. These multilevel games are described by Alonzo and Warner (2000b, c). Indeed, Alonzo and Warner (2000b) showed that only their most complex multilevel dynamic game explained observed relationships in a game where female Mediterranean wrasses choose spawning behavior, sneaker males choose when to join a nest, and nesting males choose if and when to desert a nest. Of course, complex models may be so complicated that it is difficult to ensure that they are correct. Grafen's (1990) classic model of honest signaling is a case in point (Siller 1998).

Nonetheless, the point we are trying to make in this chapter is that details matter in developing a sound predictive basis for the evolution of behavior. This is particularly true of behavior patterns as complex as alternative mating tactics. We have discussed the fact that the five general properties of behaviorally regulated alternative mating tactics have not been simultaneously incorporated into any study to our knowledge. We hope this short review has given students of mating behavior a good reason to expand their studies to include this broader view of their systems.

4.5 APPENDIX: THE MODEL

This Appendix is largely derived from Lucas and Howard (1995). The original model assumed that calling males had only one option when they enter a chorus, to call at a fixed intensity. We provide one important extension of the original model to allow for graded call intensities.

4.5.1 The original model

We model male anuran mating decisions as a stochastic dynamic game, using an algorithm suggested by Houston and McNamara (1987, 1988). We will first briefly outline the model; we then discuss each part of the model in detail.

Males are assumed to choose among four different behavior patterns: call, satellite, forage, or hide in a predator-safe refuge. The latter two are performed away from the chorus, and the first two are performed in the chorus. We assume that the decision is made once per day and commits the male to a given course of action for a full day. Each decision is assumed to result in a specified change in energy reserves (i.e., energy is the "state variable") and predation risk. Energetic expenditure, and therefore starvation risk, is assumed to be highest for calling males and lowest for hiding males. For males in the chorus, predation risk is assumed to decrease with chorus size; predation risk is also assumed to be generally higher for males in the chorus than for foraging or hiding males.

The choice among behavior patterns is assumed to be based on lifetime mating success. Male mating success is assumed to be a function of (1) the mating tactic chosen by a male, (2) the degree of competition between males, (3) male age, and (4) the arrival rate of females into the chorus. Female arrival rate, in turn, is a function of day in the breeding season, environmental quality (e.g., amount of precipitation), and the size and composition of the chorus. We simplify male age to allow for two age categories, 1-year-old males (males in their first year after sexual maturation) and 2-year-old males (males returning to breed in their second year of adulthood).

We seek a state- and time-dependent strategy that is evolutionarily stable or resistant to invasion by a mutant strategy. The algorithm we used to find the ESS has two parts – a backward iteration (or dynamic program) and a forward iteration (or simulation). We start with some initial guess about the number of callers and satellites of each age class on each night of the season. We then find the best strategy (i.e., the one that maximizes lifetime mating success) a single male should play against this population using stochastic dynamic programming (Houston and McNamara 1999, Clark and Mangel 2000). The dynamic program essentially identifies the best mutant strategy that could invade the population. The optimal strategy is calculated for all possible combinations of environmental state, energetic state, day in season, and male age. We then use a simulation to determine the composition of a chorus composed entirely of these mutants. This two-part process is then repeated until the best mutant strategy is identical with the strategy shown by the rest of the population. This strategy is the evolutionarily stable strategy (ESS: Parker 1984; or more specifically, the Nash equilibrium); that is, the strategy that when played by the entire population cannot be invaded by a single mutant playing some alternative strategy.

It usually takes about five to ten iterations for the algorithm to find the ESS, when one exists. However, there are conditions where no ESS is found (see Houston and

McNamara 1987, Lucas *et al.* 1996). In these cases, we present the results of the model after 50 iterations. An alternative approach for unstable models is to allow for partial "invasion" of the mutant strategy into the background population, instead of having the mutant completely replace the background population (McNamara *et al.* 1997; see Boyko *et al.* 2004 for an example). We have found that the solution generated in this manner is nearly identical to the solutions generated from our model when it is unstable because the instability is generated from only a few unstable matrix elements (i.e., combinations of time, state, and environment). We therefore retain the simpler algorithm used in our original paper.

The dynamic program solves for the best mutant strategy assuming that expected lifetime reproductive success (LRS) is maximized. LRS, in turn, is affected by survivorship and mating success.

SURVIVORSHIP

Starvation rate on any night during the breeding season is taken as a function of the level of energetic reserves and is modeled using an incomplete beta function (note: throughout this appendix, square brackets indicate that the variable is a function of the bracketed terms):

$$\mu_{st}[e] = 1 - I_e[a_e, b_e], \qquad (4.4)$$

where $I_e[a_e, b_e]$ = an incomplete beta function of relative energy state e with arguments a_e and b_e. The incomplete beta function is a cumulative distribution function of some variable ranging from 0 (at $e = 0$) to 1 (at $e = 1$). Here e is the fraction of maximal energy reserves carried by a male at the beginning of any given night. The incomplete beta function is similar in shape to a cumulative normal distribution, except it has the biological realism of finite tails (see Figure 1 in Lucas and Howard 1995).

To run the dynamic program, energetic state and time are divided into discrete intervals. Time is broken into intervals of 1 day. We divided energetic state into a series of 30 intervals, and assume that the result of each chosen behavior is a stochastic change in state. Thus if the current state is ε (which ranges from 0 to 30), then $\Delta\varepsilon_i \pm \sigma_{\varepsilon_i}$ is the per-day change in state caused by the choice of behavior i. We assume a normal frequency distribution of energy states for each age class of males in the population on the first day of the mating season, with $\mu = 25.5$ and $\sigma^2 = 2.86$ (note: this is relevant only for the forward iteration).

We assume that overwinter mortality (the probability of dying at any time from the end of one breeding season until the beginning of the next breeding season) is also a function of the energetic state of a male at the end of the season (e):

$$\mu_{ow}[e] = 1 - I_e[a_w, b_w] \times \gamma_{ow} \qquad (4.5)$$

where γ_{ow} = maximum overwinter survivorship.

Predation rate in a chorus is assumed to vary as a function of chorus size and satellite frequency:

$$\mu_{call}[\text{date, eq}] = \frac{\beta_p \times (1 - C[\text{date, eq}])}{1 + \beta_s \times \frac{S[\text{date, eq}]}{C[\text{date, eq}]}} \qquad (4.6)$$

where β_p = maximum probability of a predation event in a caller's territory and date = number of days since the breeding season started. This relationship assumes that predators locate callers, either acoustically or using movement cues, and thus primarily cue on signals emitted by the caller (e.g., Howard 1978, Ryan *et al.* 1981, Perrill and Magier 1988). C [date, eq] is the relative number of calling males on any given day of the breeding season, taken as a fraction of the maximal possible number of males (i.e., the number of males with 100% survival rate), assuming environmental quality "eq" (see below). Thus, C [date, eq] reflects both a reduction in chorus size caused by mortality and the proportion of males in the chorus that are calling. Similarly, S [date, eq] is the relative number of satellites in a chorus. The numerator in Eq. (4.6) is the probability of a predation event in a calling male's territory. This is assumed to be a linear function of chorus size. The denominator accounts for the fact that satellites can "share" the risk of predation. Here β_s is the risk to a satellite of being killed relative to the risk to a caller. We assume that $\beta_s < 1$. The denominator, therefore, is the effective number of individuals that can be preyed upon in a territory. The reciprocal of this number is the probability that the caller is killed when an attack occurs. The predation risk to a satellite on the territory is taken as a fraction of the risk to callers:

$$\mu_{sat}[\text{date, eq}] = \beta_s \times \mu_{call}[\text{date, eq}]. \qquad (4.7)$$

We assume that predation rates on foragers (μ_{forage}) and on males hiding in a refuge (μ_{hide}) are lower than those on males in a chorus, and that both μ_{forage} and μ_{hide} are constant.

MATING SUCCESS

There are two components to LRS: the mating success on a given night ("current mating success") and expected future reproductive success. Current mating success, which is nonzero for only callers and satellites, is a function of three variables: time in season, environmental quality, and chorus size. For simplicity, we assume that lifetime mating success is equivalent to LRS. This part of the model would have to be altered to include the seasonally adjusted value of a mating in species where the size and survivorship of clutches varies seasonally (e.g., Morin *et al.* 1990). Our model could also be easily extended to adjust the value of matings for phenomena such as size-assortative mating for which there is a higher fitness pay-off per mating for larger males.

The seasonal female-availability function assumes that there is some maximum number of females that could potentially arrive on a given day in the breeding season. This number increases through the first part of the year and decreases thereafter and is modeled using concatenated incomplete beta functions of day in season, "date" (see Figure 1 in Lucas and Howard 1995):

$$\Phi[\text{date}] = \begin{cases} I_{\tau 1}[a_\sigma, b_\sigma] & \text{if date} < T_{\max}/2 \\ I_{\tau 2}[a_\sigma, b_\sigma] & \text{otherwise} \end{cases} \quad (4.8)$$

where $\tau 1 = 2 \times \text{date}/T_{\max}$,

$\tau 2 = 2 \times (1 - \text{date}/T_{\max})$,

a_σ, b_σ = arguments of incomplete beta function,

$T_{\max}$ = last possible date that females could arrive.

In many species, female arrival rates are correlated with environmental variables such as rainfall or temperature, with females typically arriving on warm, rainy evenings (e.g., Robertson 1986, Telford and Dyson 1990, Ritke *et al.* 1992, Tejedo 1992). (Note that in all cases, our measure of female arrival rate is the rate per calling male.) We combine these environmental variables into a single variable representing environmental quality ("eq"), and assume that female arrival rate is a linear function of environmental quality:

$$\rho[\text{eq}] = \begin{cases} 1 - \text{eq}/4 & \text{if eq} \leq 4 \\ 1 & \text{otherwise} \end{cases} \quad (4.9)$$

and eq ranges from 0 (highest quality) to 4 (lowest quality). We assume that eq increments by $+1$ (when eq < 4 and by 0 otherwise) with some fixed probability, $1 - P_{\text{goodday}}$, on each day of the breeding season, and reverts to eq $= 0$ with probability P_{goodday}. This is analogous to rain (i.e., a "good" day) immediately increasing the availability of females and to female arrival rate decreasing with the number of days since the last rain. However, "rain" (when eq $= 0$) is meant to correspond to the suite of environmental factors that promote high female arrival rates into the chorus. Most of the chorus activity occurs on favorable days (i.e., low eq); therefore to simplify the discussion of the model results, we only present results from favorable days.

Finally, we assume that the arrival rate of females into a male's territory increases with chorus size:

$$\xi[\text{date, eq}] = \gamma\xi 1 \times (C[\text{date, eq}] \times 2 + \gamma\xi 2 \times C[\text{date, eq}]^2) \quad (4.10)$$

where $\gamma\xi 1$ = maximum female arrival rate; $\gamma\xi 2$ = constant. If $\gamma\xi 2$ is positive, this function is accelerating (concave up), and if it is negative, the function is decelerating (concave down) (see Figure 2A in Lucas and Howard 1995). If there are no calling males in the population, we assume that a "mutant" (and lone) caller represents a relative chorus size of 10^{-4}.

The net female arrival rate into an average caller's territory is

$$F[\text{date, eq}] = \xi[\text{date, eq}] \times \rho[\text{eq}] \times \Phi[\text{date}]. \quad (4.11)$$

This is the average female arrival rate across all territories. However, 1-year-old males may experience a lower mating success than old males. For simplicity, we will assume that the net effect of this age difference is that 1-year-old males "attract" fewer females into their territories than 2-year-old males and subsume any type of competition for mates into this age effect. The age difference can be accounted for as follows:

Assuming that any given 1-year-old male is able to attract some fraction ($\gamma 1$) of the number of females a 2-year-old male will attract, and assuming that $\rho 1[\text{date}]$ is the proportion of callers that are 1-year-old males, then 2-year-old males will attract females at a rate

$$v_2[\text{date, eq}] = \frac{F[\text{date, eq}]}{1 - \rho_1[\text{date}](1 - \gamma_1)} \quad (4.12)$$

and 1-year-old males will attract females at the following rate:

$$v_1[\text{date, eq}] = v_2[\text{date, eq}] \times \gamma 1. \quad (4.13)$$

In addition to age, the frequency of satellites in the chorus will affect mating rates. This is because females arriving in the territory are shared among the caller and satellites in the territory; thus, the current mating success of the caller is the rate at which females are attracted to the territory on a given night of the season, diminished by the rate at which satellites intercept females:

$$Ms_{\text{call}}[\text{age, date, eq}] = \frac{v_{\text{age}}[\text{date, eq}]}{1 + \gamma_{\text{sat}} \times \frac{S[\text{date, eq}]}{C[\text{date, eq}]}} \quad (4.14)$$

where γ_{sat} = the ability of a single satellite to obtain mates, taken as a fraction of the ability of a caller to obtain mates (we assume that $\gamma_{\text{sat}} < 1$).

The current mating success of a satellite is assumed to be age independent:

$$Ms_{sat}[age, date, eq] = \frac{F[date, eq] \times \gamma_{sat}}{1 + \gamma_{sat} \times \frac{S[date, eq]}{C[date, eq]}}. \quad (4.15)$$

The current mating success of foraging males and hiding males is zero:

$$Ms_{forage}[age, date, eq] = Ms_{hide}[age, date, eq] = 0. \quad (4.16)$$

TOTAL LIFETIME REPRODUCTIVE SUCCESS

The total reproductive success of a male equals current mating success plus expected future reproductive success:

$$
\begin{aligned}
&PO_{behav}[age, date, eq, \varepsilon] \\
&= Ms_{behav}[age, date, eq] + (1 - \mu_{st}[\varepsilon]) \\
&\times (1 - \mu_{behav}[date, eq]) \times \{P_{goodday} \times \{P_c \\
&\times \left\{ \sum_{\Delta\varepsilon = -30}^{30} P_{\Delta\varepsilon|behave} \times PO^*[age, date \right. \\
&\left. + 1, eq = 0, \varepsilon + \Delta\varepsilon] \right\} + (1 - P_c) \\
&\times \left\{ \sum_{\Delta\varepsilon = -30}^{30} P_{\Delta\varepsilon|behave} \times PO^*[age, T_{max} + 1, eq = 0, \varepsilon + \Delta\varepsilon] \right\} \} \\
&+ (1 - P_{goodday}) \times \{P_c \times \left\{ \sum_{\Delta\varepsilon = -30}^{30} P_{\Delta\varepsilon|behave} \right. \\
&\left. \times PO^*[age, date + 1, eq + 1, \varepsilon + \Delta\varepsilon] \right\} \\
&+ (1 - P_c) \times \left\{ \sum_{\Delta\varepsilon = -30}^{30} P_{\Delta\varepsilon|behave} \right. \\
&\left. \times PO^*[age, T_{max} + 1, eq + 1, \varepsilon + \Delta\varepsilon] \right\} \} \} \quad (4.17)
\end{aligned}
$$

where

$Ms_{behav}[age, date, eq] =$ current pay-off if the male exhibits behavior "behav",

$\mu_{st}[\varepsilon], \mu_{behav}[date] =$ mortality induced by starvation (a function of energetic state, ε) and mortality induced by predation conditional on behavior "behav" being exhibited,

$P_{goodday} =$ probability of a good day or a day conducive to chorus formation (eq = number of days since the last good day),

$P_c =$ probability that the mating season will continue at least another day,

$P_{\Delta\varepsilon|behave} =$ probability that energetic state is increased by $\Delta\varepsilon$, given that behavior "behav" is exhibited,

$PO^* =$ optimal pay-off for the sequence of decisions made for the rest of the male's life, starting on day = date + 1 (if the season lasts that long) or day = $T_{max} + 1$ (if the season ends), on eq = 0 (if it is a "good day") or eq + 1 (if it is not a "good day"), and at energy state $\varepsilon + \Delta\varepsilon$.

4.5.2 Graded signals: an extension of the model

We modified the model described above by allowing calling males to vary the intensity of their calls. In effect, this increased the number of mating options each male had because we treated each level of call intensity as a separate mating tactic that could be employed by a male at any time in the breeding season. We assume that the cost of an exaggerated signal falls under the category of "receiver-independent costs" (Vehrencamp 2000). That is, signaling cost is independent of the target receiver's response, as would be expected if the primary cost of producing a signal is either energetic or a risk of attracting predators. We assume that call exaggeration affects three aspects of a male's reproductive success and all of these variables are taken to be a function of call investment, here defined as "intensity" (see Section 4.3.4 for equations):

(1) Relative attractiveness to a female:

RelAttractiveness[intensity].

(2) Relative energetic cost of call production:

COST[intensity].

(3) Relative risk of predation:

ProbPRED[intensity].

The actual levels of male attractiveness are the product of the relative attractiveness and mating success (Eq. 4.14):

$$
\begin{aligned}
&MSactual_{call}[age, date, eq, intensity] \\
&= MScall[age, date, eq] \times RelAttractiveness[intensity].
\end{aligned}
\quad (4.18)
$$

Similarly the predation risk is the product of the relative risk and caller mortality (4.6):

$$
\begin{aligned}
&\mu actual_{call}[date, eq, intensity] \\
&= \mu_{call}[date, eq] \times ProbPRED[intensity]
\end{aligned}
\quad (4.19)
$$

and the energetic cost of calling is the sum of the relative cost of calling and the original value used for the cost of calling:

$$COSTactual[intensity] = \Delta\varepsilon_i \pm \sigma_\varepsilon \\ + COST[intensity]. \quad (4.20)$$

Acknowledgments

We thank Adam Boyko, Ben Fanson, Sarah Humfeld, Michael Taborsky, and Denise Zielinski for helpful comments on earlier drafts of the manuscript.

References

Alonzo, S. H. and Warner, R. R. 2000a. Allocation to mate guarding or increased sperm production in a Mediterranean wrasse. *American Naturalist* 156, 266–275.

Alonzo, S. H. and Warner, R. R. 2000b. Dynamic games and field experiments examining intra- and intersexual conflict: explaining counterintuitive mating behavior in a Mediterranean wrasse, *Symphodus ocellatus*. *Behavioral Ecology* 11, 56–70.

Alonzo, S. H. and Warner, R. R. 2000c. Female choice, conflict between the sexes and the evolution of male alternative reproductive behaviors. *Evolutionary Ecology Research* 2, 149–170.

Arak, A. 1988. Callers and satellites in the natterjack toad: evolutionarily stable decision rules. *Animal Behaviour* 36, 416–432.

Bevier, C. R. 1997. Utilization of energy substrates during calling activity in tropical frogs. *Behavioral Ecology and Sociobiology* 41, 343–352.

Boyko, A. R., Gibson, R. M., and Lucas, J. R. 2004. How predation risk affects the temporal dynamics of avian leks: greater sage grouse vs. golden eagles. *American Naturalist* 163, 154–165.

Brodin, A. 2000. Why do hoarding birds gain fat in winter the wrong way? Suggestions from a dynamic model. *Behavioral Ecology* 11, 27–39.

Caldwell, J. P. 1987. Demography and life history of two species of chorus frogs (Anura: Hylidae) in South Carolina. *Copeia*, 114–127.

Charnov, E. L. 1993. *Life History Invariants: Some Explanations of Symmetry in Evolutionary Ecology*. Oxford, UK: Oxford University Press.

Cherry, M. I. 1993. Sexual selection in the raucous toad, *Bufo rangeri*. *Animal Behaviour* 45, 359–373.

Clark, C. W. 1994. Antipredator behavior and the asset-protection principle. *Behavioral Ecology* 5, 159–170.

Clark, C. W. and Mangel, M. 2000. *Dynamic State Variable Models in Ecology: Methods and Applications*. New York: Oxford University Press.

Clarke, R. D. 1977. Postmetamorphic survivorship of Fowler's toad, *Bufo woodhousei fowleri*. *Copeia*, 594–597.

Dawkins, R. 1980. Good strategy or evolutionary stable strategy? In G. W. Barlow and J. Silverberg (eds.) *Sociobiology: Beyond Nature/Nurture?*, pp. 331–367. Boulder, CO: Westview Press.

Dugatkin, L. A. and Reeve, H. K. 1998. *Game Theory and Animal Behavior*. Oxford, UK: Oxford University Press.

Dyson, M. L., Passmore, N. I., Bishop, P. J., and Henzi, S. P. 1992. Male behavior and correlates of mating success in a natural population of African painted reed frogs (*Hyperolius marmoratus*). *Herpetologica* 48, 236–246.

Fleming, I. A. and Reynolds, J. D. 2004. Salmonid breeding systems. In A. P. Hendry and S. C. Stearns (eds.) *Evolution Illuminated: Salmon and Their Relatives*, pp. 264–294. Oxford, UK: Oxford University Press.

Fraizer, T. 1997. A dynamic model of mating behavior in digger wasps: the energetics of male–male competition mimic size-dependent thermal constraints. *Behavioral Ecology and Sociobiology* 41, 423–434.

Garant, D., Fontaine, P.-M., Good, S. P., Dodson, J. J., and Bernatchez, L. 2002. The influence of male parental identity on growth and survival of offspring in Atlantic salmon (*Salmo salar*). *Evolutionary Ecology Research* 4, 537–549.

Gerhardt, H. C. and Huber, F. 2002. *Acoustic Communication in Insects and Anurans*. Chicago, IL: University of Chicago Press.

Gintis, H., Smith, E. H., and Bowles, S. 2001. Costly signalling and cooperation. *Journal of Theoretical Biology* 213, 103–119.

Given, M. F. 2002. Interrelationships among calling effort, growth rate, and chorus tenure in *Bufo fowleri*. *Copeia*, 979–987.

Grafe, T. U., Schmuck, R., and Linsemair, K. E. 1992. Reproductive energetics of the African reed frogs, *Hyperolius viridiflavus* and *Hyperolius marmoratus*. *Physiological Zoology* 65, 153–171.

Grafen, A. 1984. Natural selection, kin selection and group selection. In J. R. Krebs and N. B. Davies (eds.) *Behavioural Ecology: An Evolutionary Approach*, pp. 62–84. Oxford, UK: Blackwell Scientific.

Grafen, A. 1990. Biological signals as handicaps. *Journal of Theoretical Biology* 144, 517–546.

Green, A. J. 1990. Determinants of chorus participation and the effects of size, weight and competition on advertisement calling in the tungara frog, *Physalemus pustulosus* (Leptodactylidae). *Animal Behaviour* 39, 620–638.

Gross, M. R. 1984. Sunfish, salmon, and the evolution of alternative reproductive strategies and tactics in fishes. In G. Potts and R. J. Wooton (eds.) *Fish Reproduction: Strategies and Tactics in Fishes*, pp. 55–75. London: Academic Press.

Gross, M. R. 1985. Alternative breeding strategies in male salmon. *Nature* 313, 47–48.

Gross, M. R. 1996. Alternative reproductive strategies and tactics: diversity within sexes. *Trends in Ecology and Evolution* 11, 92–98.

Gross, M. R. and Repka, J. 1998. Stability with inheritance in the conditional strategy. *Journal of Theoretical Biology* 192, 445–453.

Halliday, T. R. and Verrell, P. A. 1988. Body size and age in amphibians and reptiles. *Journal of Herpetology* 22, 253–265.

Hamilton, I. M. and Dill, L. M. 2002. Three-player social parasitism games: implications for resource defense and group formation. *American Naturalist* 159, 670–686.

Harris, W. E. and Lucas, J. R. 2002. A state-based model of sperm allocation in a group-breeding salamander. *Behavioral Ecology* 13, 705–712.

Hilborn, R. and Mangel, M. 1997. *The Ecological Detective: Confronting Models with Data*. Princeton, NJ: Princeton University Press.

Houston, A. I. and McNamara, J. M. 1987. Singing to attract a mate: a stochastic dynamic game. *Journal of Theoretical Biology* 129, 57–68.

Houston, A. I. and McNamara, J. M. 1988. Fighting for food: a dynamic version of the hawk–dove game. *Evolutionary Ecology* 2, 51–64.

Houston, A. I. and McNamara, J. M. 1999. *Models of Adaptive Behaviour*. Cambridge, UK: Cambridge University Press.

Howard, R. D. 1978. The evolution of mating strategies in bullfrogs, *Rana catesbeiana*. *Evolution* 32, 850–871.

Howard, R. D. 1981. Male age–size distribution and male mating success in bullfrogs. In R. D. Alexander and D. W. Tinkle (eds.) *Natural Selection and Social Behavior: Recent Research and New Theory*, pp. 61–77. New York: Chiron Press.

Howard, R. D. 1984. Alternative mating behaviors of young bullfrogs. *American Zoologist* 24, 397–406.

Hugie, D. M. and Lank, D. B. 1997. The resident's dilemma: a female choice model for the evolution of alternative mating strategies in lekking male ruffs (*Philomachus pugnax*). *Behavioral Ecology* 8, 218–225.

Hutchings, J. A. and Myers, R. A. 1994. The evolution of alternative mating strategies in variable environments. *Evolutionary Ecology* 8, 256–268.

Johnstone, R. A. 2000. Conflicts of interest in signal evolution. In Y. Espmark, T. Amundsen, and G. Rosenqvist (eds.) *Animal Signals: Signalling and Signal Design in Animal Communication*, pp. 465–485. Trondheim, Norway: Tapir Academic Press.

Johnstone, R. A. and Grafen, A. 1992. The continuous Sir Philip Sidney Game: a simple model of biological signalling. *Journal of Theoretical Biology* 156, 215–234.

Johnstone, R. A. and Grafen, A. 1993. Dishonesty and the handicap principle. *Animal Behaviour* 46, 759–764.

Judge, K. A. and Brooks, R. J. 2001. Chorus participation by male bullfrogs, *Rana catesbeiana*: a test of the energetic constraint hypothesis. *Animal Behaviour* 62, 849–861.

Lank, D. B., Smith, C. M., Hanotte, O., Burke, T., and Cooke, F. 1995. Genetic polymorphism for alternative mating behaviour in lekking male ruff. *Nature* 378, 59–62.

Lank, D. B., Coupe, M., and Wynne-Edwards, K. E. 1999. Testosterone-induced male traits in female ruffs (*Philomachus pugnax*): autosomal inheritance and gender differentiation. *Proceedings of the Royal Society of London B* 266, 2323–2330.

Lucas, J. R. and Howard, R. D. 1995. On alternative reproductive tactics in anurans: dynamic games with density and frequency dependence. *American Naturalist* 146, 365–397.

Lucas, J. R., Howard, R. D., and Palmer, J. G. 1996. Callers and satellites: chorus behaviour in anurans as a stochastic dynamic game. *Animal Behaviour* 51, 501–518.

Marler, C. A. and Ryan, M. J. 1996. Energetic constraints and steroid hormone correlates of male calling behaviour in the túngara frog. *Journal of Zoology* 240, 397–409.

Maynard Smith, J. 1982. *Evolution and the Theory of Games*. Cambridge, UK: Cambridge University Press.

Maynard Smith, J. 1991. Honest signalling: the Philip Sidney Game. *Animal Behaviour* 42, 1034–1035.

Mayr, E. 1983. How to carry out the adaptationist program. *American Naturalist* 121, 324–334.

McCauley, S. J., Bouchard, S. S., Farina, B. J., et al. 2000. Energetic dynamics and anuran breeding phenology: insights from a dynamic game. *Behavioral Ecology* 11, 429–436.

McNamara, J. M., Webb, J. N., Collins, E. J., Szekely, T., and Houston, A. I. 1997. A general technique for computing evolutionary stable strategies based on errors in decision-making. *Journal of Theoretical Biology* **189**, 211–225.

Miyamoto, M. M. and Cane, J. H. 1980. Behavioral observations of noncalling males in Costa Rican *Hyla ebraccata*. *Biotropica* **12**, 225–227.

Moczek, A. P., Hunt, J., Emlen, D. J., and Simmons, L. W. 2002. Threshold evolution in exotic populations of a polyphenic beetle. *Evolutionary Ecology Research* **4**, 587–601.

Morin, P. J., Lawler, S. P., and Johnson, E. A. 1990. Ecology and breeding phenology of larval *Hyla andersonii*: the disadvantages of breeding late. *Ecology* **71**, 1590–1598.

Murphy, C. G. 1994a. Chorus tenure of male barking treefrogs, *Hyla gratiosa*. *Animal Behaviour* **48**, 763–777.

Murphy, C. G. 1994b. Determinants of chorus tenure in barking treefrogs (*Hyla gratiosa*). *Behavioral Ecology and Sociobiology* **34**, 285–294.

Parker, G. A. 1984. Evolutionarily stable strategies. In J. R. Krebs and N. B. Davies (eds.) *Behavioural Ecology: An Evolutionary Approach*, pp. 30–61. Oxford, UK: Blackwell Scientific.

Parker, G. A. 1990. Sperm competition games: raffles and roles. *Proceedings of the Royal Society of London B* **242**, 120–126.

Perrill, S. A. and Magier, M. 1988. Male mating behavior in *Acris crepitans*. *Copeia*, 245–248.

Perrill, S. A., Gerhardt, H. C., and Daniel, R. 1978. Sexual parasitism in the green treefrog (*Hyla cinerea*). *Science* **200**, 1179–1180.

Pravosudov, V. V. and Lucas, J. R. 2001. A dynamic model of energy management in small food-caching passerine birds. *Behavioral Ecology* **12**, 207–218.

Proulx, S. R., Day, T., and Rowe, L. 2002. Older males signal more reliably. *Proceedings of the Royal Society of London B* **269**, 2291–2299.

Ritke, M. E., Babb, J. G., and Ritke, M. K. 1992. Temporal patterns of reproductive activity in the gray treefrog (*Hyla chrysoscelis*). *Journal of Herpetology* **26**, 107–111.

Robertson, J. G. M. 1986. Male territoriality, fighting and assessment of fighting ability in the Australian frog *Uperoleia rugosa*. *Animal Behaviour* **34**, 763–772.

Roff, D. A. 1992. *The Evolution of Life Histories: Theory and Analysis*. New York: Chapman and Hall.

Ryan, M. J., Tuttle, M. D., and Taft, L. K. 1981. The costs and benefits of frog chorusing behavior. *Behavioral Ecology and Sociobiology* **8**, 273–278.

Schlichting, C. D. and Pigliucci, M. 1998. *Phenotypic Evolution: A Reaction Norm Perspective*. Sunderland, MA: Sinauer Associates.

Shuster, S. M. and Guthrie, E. E. 1999. The effects of temperature and food availability on adult body length in natural and laboratory populations of *Paracerceis sculpta* (Holmes), a Gulf of California isopod. *Journal of Experimental Marine Biology and Ecology* **233**, 269–284.

Shuster, S. M. and Wade, M. J. 2003. *Mating Systems and Mating Strategies*. Princeton, NJ: Princeton University Press.

Siller, S. 1998. A note on errors in Grafen's strategic handicap models. *Journal of Theoretical Biology* **195**, 413–417.

Sinervo, B. and Lively, C. M. 1996. The rock–paper–sissors game and the evolution of alternative male strategies. *Nature* **380**, 240–243.

Sinervo, B., Bleay, C., and Adamopoulou, C. 2001. Social causes of correlational selection and the resolution of a heritable throat color polymorphism in a lizard. *Evolution* **55**, 2040–2052.

Skubic, E., Taborsky, M., McNamara, J. M., and Houston, A. I. 2004. When to parasitize? A dynamic optimization model of reproductive strategies in a cooperative breeder. *Journal of Theoretical Biology* **227**, 487–501.

Stearns, S. C. 1992. *The Evolution of Life Histories*. Oxford, UK: Oxford University Press.

Stephens, D. W. and Krebs, J. R. 1986. *The Theory of Foraging Behavior*. Princeton, NJ: Princeton University Press.

Sullivan, B. K. 1982. Male mating behaviour in the Great Plains toad (*Bufo cognatus*). *Animal Behaviour* **30**, 939–940.

Taborsky, M. 1998. Sperm competition in fish: "bourgeois" males and parasitic spawning. *Trends in Ecology and Evolution* **13**, 222–227.

Taborsky, M. 2001. The evolution of bourgeois, parasitic, and cooperative reproduction behaviors in fishes. *Journal of Heredity* **92**, 100–110.

Taigen, T. L. and Wells, K. D. 1985. Energetics of vocalization by an anuran amphibian (*Hyla versicolor*). *Journal of Comparative Physiology B* **155**, 163–170.

Tejedo, M. 1992. Large male mating advantage in natterjack toads, *Bufo calamita*: sexual selection or energetic constraints? *Animal Behaviour* **44**, 557–569.

Telford, S. R. and Dyson, M. L. 1990. The effect of rainfall on interclutch interval in painted reed frogs (*Hyperolius marmoratus*). *Copeia*, 644–648.

Vehrencamp, S. L. 2000. Handicap, index, and conventional signal elements of bird song. In Y. Espmark, T. Amundsen, and G. Rosenqvist (eds.) *Animal Signals: Signalling and Signal Design in Animal Communication*, pp. 277–300. Trondheim, Norway: Tapir Academic Press.

Wagner Jr., W. E. and Sullivan, B. K. 1992. Chorus organization in the Gulf Coast toad (*Bufo valliceps*): male and female behavior and the opportunity for sexual selection. *Copeia*, 647–658.

Waltz, E. C. 1982. Alternative mating tactics and the law of diminishing returns: the satellite threshold model. *Behavioral Ecology and Sociobiology* **10**, 75–83.

Zahavi, A. 1975. Mate selection: a selection for a handicap. *Journal of Theoretical Biology* **53**, 205–214.

Part II
Proximate mechanisms of alternative reproductive tactics

5 · The roles of genes and the environment in the expression and evolution of alternative tactics

DOUGLAS J. EMLEN

CHAPTER SUMMARY

In many animal populations, individuals may develop into any of several alternative phenotypes (e.g., guarding and sneaking male forms). Occasionally, the phenotype adopted by an individual depends entirely on the presence of a specific allele(s). More typically, it depends on the environment: individuals encountering one set of conditions produce one phenotype, individuals encountering a different set of conditions produce an alternative – often strikingly different – phenotype. Facultatively adopted alternative tactics comprise unusually tractable and intuitive forms of developmental phenotypic plasticity, and their underlying regulatory mechanisms clearly illustrate how genes and the environment can interact to control animal development. Here I review the basic components of these regulatory mechanisms to show how alternative trajectories of development are coupled with the specific environmental conditions that animals encounter. Explicit consideration of these underlying mechanisms provides a useful framework for thinking about heritable variation in tactic expression and for considering more precisely how animal alternative tactics evolve. I illustrate this integration of developmental and evolutionary perspectives using an insect example (horned and hornless male beetles), but analogous processes regulate tactic expression in other arthropods and in vertebrates.

5.1 INTRODUCTION

Expression of alternative reproductive tactics (ARTs) is often exquisitely sensitive to the environment – tactic expression is "phenotypically plastic." Ambient abiotic conditions, population density, the relative sizes or status of rival individuals, and the relative frequency of expressed alternatives all can influence the tactic adopted by an animal: individuals developing under one set of conditions express one tactic; genetically similar (e.g., sibling) individuals exposed to a different set of conditions express an alternative tactic (Figure 5.1).

Plastic mechanisms of tactic expression mean that variation among individuals in which tactic they express is often overwhelmingly influenced by environmental conditions. Indeed, genetic studies characterizing this variation typically find low to negligible heritabilities (e.g., Radwan 1993, Emlen 1994, Tomkins 1999, Kurdziel and Knowles 2002, Cremer and Heinze 2003). However, this does not mean that these phenotypes are "nongenetic"; merely, that we must shift our focus before we can meaningfully appreciate how these animal characteristics vary and how they evolve. Specifically, we must begin to consider the underlying developmental mechanisms that regulate expression of the alternative tactics. This is no different from stating that, as with any phenotypically plastic trait, the character that evolves is the *reaction norm*, rather than, or in addition to, the end phenotype per se. However, with alternative tactics, we can take this several steps farther.

Alternative reproductive tactics are members of a special class of phenotypically plastic traits called "threshold traits" (West-Eberhard 1989, 1992, 2003, Hazel *et al.* 1990, Gross 1996, Roff 1996, Gross and Repka 1998, Brockmann 2001). Expression of one or the other tactic depends on the conditions encountered by animals during development, assessed relative to an internal threshold of sensitivity. Threshold mechanisms have proven especially amenable to physiological studies of development, and we now have rigorous working models from a number of animal systems for how this class of reaction norms works (see Zera and Denno 1997, Nijhout 1999, Hartfelder and Emlen 2005 for recent reviews). Because of this, it is now possible to appreciate how genetic changes in specific components of these developmental processes might contribute to evolutionary modifications of the tactic phenotypes and the conditional patterns of their expression.

Alternative Reproductive Tactics, ed. Rui F. Oliveira, Michael Taborsky, and H. Jane Brockmann. Published by Cambridge University Press.
© Cambridge University Press 2008.

Figure 5.1 Reaction norm diagram for facultatively expressed alternative tactics. Individuals encountering conditions above a critical threshold level express one phenotype (tactic 1); individuals encountering conditions below this threshold express an alternative phenotype (tactic 2). The environment relevant to tactic expression varies across taxa but can include abiotic factors such as photoperiod or diet quality, social factors like population density or relative status, or aspects of the individual itself, such as growth or body size.

In this chapter, I build on recent advances in the study of insect development to illustrate how facultative expression of ARTs arises, and I use this information to suggest how these mechanisms are likely to evolve. The premise for this chapter is that explicit consideration of the developmental processes regulating expression of ARTs provides an exciting new picture of how these extraordinary phenotypes evolve.

5.2   ALTERNATIVE REPRODUCTIVE TACTICS AS THRESHOLD TRAITS

Several features of ARTs are essential to an appreciation of their mechanisms of expression. Firstly, the alternatives are mutually exclusive – they are not expressed at the same time in the same individual. Secondly, each individual has the genetic potential to adopt either tactic, and these alternatives are conditionally expressed. Conditional expression generates repeatable associations between the occurrence of a tactic (or, rather, of individuals expressing a particular tactic) and relevant selective situations or circumstances. Finally, tactic expression is regulated by a threshold. Threshold mechanisms uncouple gene expression of the two tactics, permitting evolutionary changes in one tactic to occur relatively independently from evolution of the alternative tactic – i.e., threshold mechanisms facilitate divergent evolution of the tactic alternatives (West-Eberhard 1989, 1992, 2003). Populations of these species are often dimorphic with respect to the tactics, with few, if any, intermediate phenotypes, and are consistent with an abrupt switch between the tactic alternatives (e.g., alternative male

morphs in amphipods: Clark 1997, Kurdziel and Knowles 2002; mites: Radwan et al. 2002; earwigs: Tomkins 1999; bees: Danforth 1991, Kukuk 1996; wasps: O'Neill and Evans 1983; ants: Cremer and Heinze 2003; beetles: Eberhard 1982, Siva-Jothy 1987, Rasmussen 1994, Iguchi 1998; and buntings: Greene et al. 2000).

At its most basic, tactic expression involves an assessment of circumstance, a translation of this information into an internal circulating signal (e.g., a hormone), and a comparison of levels of this circulating signal with an internally specified threshold level. Through this basic process, one of several alternative patterns of gene expression is initiated, eventually resulting in the expression of one or the other alternative tactic phenotype.

The "decision" to adopt a particular tactic can be reversible or irreversible (reviewed in Brockmann 2001). Reversible alternatives mean that an individual can go back and forth between tactics (e.g., "calling" versus "searching" male crickets: Zuk and Simmons 1997, or "egg-guarding" versus "egg-dumping" female lace bugs: Tallamy et al. 2002. Although they still cannot express both tactics at the same time (these are still alternatives), they can, and often do, adopt both during their lifetime. With irreversible alternatives, individuals adopt one or the other tactic and remain fixed with this option for their adult lifetime (e.g., "dispersing" versus "fighting" male ants: Cremer and Heinze 2003; and bees: Kukuk 1996).

In all cases, these "decision" mechanisms involve a threshold. However, reversible and irreversible mechanisms generally differ in the developmental timing of the decision and in the delay between the decision event (defined as the assessment of circumstance relative to threshold) and expression of the tactic phenotypes. Reversible mechanisms typically occur during the adult stage and involve little delay. This permits rapid tracking of environmental heterogeneity and allows animals to switch easily back and forth between alternatives.

Irreversible mechanisms generally occur much earlier in development, and these involve a much longer delay between the decision and expression of the final phenotype. In insects, irreversible mechanisms for alternative tactics typically occur prior to or during metamorphosis (e.g., in the final larval/nymphal stage: Wheeler and Nijhout 1981, Zera and Tiebel 1988, Nijhout 1994, Cnaani et al. 2000; or during the pupal period: Rountree and Nijhout 1995, Brakefield et al. 1998). Earlier decisions permit much more significant downstream adjustments to animal development. Combined with the capacity of many insects (and some vertebrates) to undergo radical metamorphic reorganizations of the phenotype (from larval to adult forms), these "pre-metamorphic" switches can generate impressive morphological, physiological, and behavioral differences among tactic alternatives.

5.3 ANATOMY OF A THRESHOLD MECHANISM

Although the relative timing of reversible and irreversible tactic-switching events differs, many features of their mechanisms, as well as their implications for heritability and evolution, are similar. In this section I describe the basic physiological ingredients of a typical developmental threshold mechanism. In the following section (5.4), I relate these ingredients to natural sources of genetic variation and to likely trajectories of alternative tactic evolution.

Importantly, although the details of threshold mechanisms are likely to vary from taxa to taxa, the basic components of this process can be generalized. I illustrate these components with an invertebrate irreversible example. However, most of the components of this threshold mechanism apply to reversible insect alternative tactics and to vertebrate alternative tactics as well, and the genetic and evolutionary implications of these mechanisms apply to all threshold traits.

Upstream from the threshold itself are the sensory structures of the organism. (For this chapter, "upstream" and "downstream" refer to events occurring prior to, and after the respective fate (i.e., the tactic) of an individual has been determined.) In insects this includes an arsenal of receptors to light, heat, touch, smell, and stretch, as well as the neural and neuroendocrine organs that filter and process this information (e.g., Chapman 1982). These structures dictate which elements of the environment are perceived and how this information is communicated to the endocrine organs (e.g., via the neuroendocrine system). Although I do not discuss these upstream parts of the process here, they constitute an important component of any plastic developmental response mechanism. They are, in effect, the first line of response: if an individual cannot detect a change in its surroundings, it cannot respond to that change, and any neurological filters or amplifiers of signals will affect the detection capabilities of that animal.

Once animals have detected relevant cues from their external or internal environments, this information is communicated to developing tissues by a *hormone signal* (Figure 5.2A). Levels of this circulating hormone signal

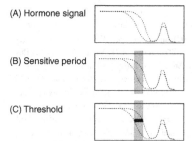

(A) Hormone signal

(B) Sensitive period

(C) Threshold

Figure 5.2 Anatomy of a threshold mechanism. (A) Most threshold mechanisms incorporate a hormone signal – an endocrine factor that provides the link between conditions occurring outside the animal and the internal tissues that enable a response. Levels of the endocrine signal are sensitive to circumstance (e.g., crowding, nutrition, growth/body size), resulting in concentrations that are higher in some individuals than in others. (B) Cells are responsive to this signal during specific sensitive periods, when levels of the hormone signal are assessed relative to (C) a genetically specified threshold of sensitivity. Thresholds reflect the critical concentration of hormone needed to elicit an all-or-none response. Animals typically have a default pattern of development, and levels of hormone above (or below) the threshold during the sensitive period "reprogram" tissues to an alternative pattern of development. Tissue reprogramming often involves interactions with secondary hormones, and/or transcription factors, and can involve complex tactic-specific patterns of downstream gene expression. For all panels, hormone concentration is shown on the vertical axis and time on the horizontal axis.

covary with aspects of the environment encountered by developing animals: some situations result in high concentrations, other situations generate low concentrations, and cells within developing structures "read" their environment through detection of levels of this circulating signal. In this way, information from the outside world is communicated to the relevant tissues as variation in the levels of this circulating hormone signal.

In many insect alternative tactics, this hormone signal is juvenile hormone (JH). For example, crowding affects the level of JH in crickets (Zera and Tiebel 1988, Zera et al. 1989), planthoppers (Bertuso et al. 2002), and locusts (Injeyan and Tobe 1981, Botens et al. 1997), and levels of dietary tannins affect JH concentrations in *Nemoria* caterpillars (Greene 1989, 1999). Other aspects of diet affect this hormone level in bugs (Rankin and Riddiford 1977), bumble-bees (Röseler et al. 1981, Strambi et al. 1984), stingless bees (Velthius 1976, Hartfelder and Engels

1998), honeybees (Asencot and Lensky 1976, Dogra *et al.* 1977, Hartfelder 1990, Rachinsky and Hartfelder 1990, Schulz *et al.* 2002), ants (Ono 1982, Wheeler 1991), earwigs (Rankin *et al.* 1997), termites (Lenz 1976), and beetles (Emlen and Nijhout 1999, 2001), and in all of these cases, levels of JH regulate the expression of condition-sensitive threshold traits. Although the identity of the signal hormone can vary from taxa to taxa (e.g., some butterflies use ecdysone [Koch and Bückmann 1987, Rountree and Nijhout 1995, Brakefield *et al.* 1998], or browning factors [Awiti and Hidaka 1982, Starnecker and Hazel 1999], and crabs appear to use methyl farnesoate [Laufer and Ahl 1995]), the function of the endocrine signal is always the same: this hormone signal provides a mechanistic link between external environmental cues and the internal cells and tissues that enable a response (Nijhout 1994, 1999).

The second component of a threshold mechanism is the *sensitive period* – a critical physiological window when target tissues are capable of responding to the circulating endocrine signal (Figure 5.2B). This is the period when hormone levels are assessed and when animals commit to one or the other developmental pathway. Sensitive periods are thought to coincide with the expression of appropriate hormone receptor proteins in the target tissues: cells in these tissues are sensitive to the endocrine signal when, and only when, they have active receptors present for that signal (reviewed in Nijhout 1994, 1999).

Sensitive periods can vary greatly in when they occur and in their length. For example, sensitive periods for elaborate, irreversible alternative tactics generally occur during the final larval or nymphal instar, or the pupal period, and often only last for a small portion of this stage (e.g., Wheeler and Nijhout 1981, Zera and Tiebel 1988, Nijhout 1994, Rountree and Nijhout 1995, Brakefield *et al.* 1998, Ayoade *et al.* 1999, Cnaani *et al.* 2000, Emlen and Nijhout 2001). Sensitive periods for reversible alternative tactics typically occur after animals are adults (e.g., Cusson *et al.* 1994, Robinson and Vargo 1997, Sullivan *et al.* 2000, Scott *et al.* 2001, Tallamy *et al.* 2002), and these sensitive periods can be very long: cells may be continually sensitive to the hormone but activated only when a brief hormone pulse occurs. For example, females of the burying beetle initiate reproductive behavior only after a brief pulse of JH, and this pulse is itself initiated by discovery of a carcass suitable for use as larval provision (Scott *et al.* 2001). Long sensitive periods permit animals to modify their behavior rapidly in response to unpredictable and infrequent, but critical, events.

Shifts in the timing of sensitive periods constitute an important and dynamic way in which complex developmental responses to specific environmental stimuli are coordinated (Nijhout 1999). Convergence on the same sensitive period can bring a suite of traits under a common regulatory control; divergence of sensitive periods can facilitate independent regulation. Thus, changes in the timing of sensitive periods can cause tissues to respond in concert with, or independent from, other animal tissues. It can also affect which tactic gets expressed: shifts in the timing of a sensitive period can cause cell sensitivities to coincide with, or to miss, pulses of signal hormone expression (Nijhout 1999), and this can lead to individual differences in the propensity to express a particular tactic.

The third component of these mechanisms is the *threshold* (Figure 5.2C). Target cells have a threshold of sensitivity to the signal hormone during the sensitive period. Tissues generally have a default developmental pathway, and levels of signal hormone that are above (or, in some cases, below) a critical response threshold initiate a switch to an alternative pattern of development (reviewed in Nijhout 1994, 1999). If the timing and duration of the sensitive period are determined by *when* appropriate hormone receptors are expressed, then the threshold of sensitivity can be interpreted as *how many* receptors are expressed. In fact, both the number of receptors and the binding affinities of those receptors can affect the sensitivity of target cells to the endocrine signal. In essence, a threshold of sensitivity refers to a critical concentration of hormone at which downstream physiological, biochemical, and transcription cascades are initiated.

Consequently, the general elements of a developmental threshold mechanism involve a circulating hormone signal, a sensitive period when target tissues express receptors for this signal, and an all-or-none response cascade that is activated when levels of the hormone signal fall above (or below) a critical threshold concentration. In all cases, the developmental, physiological, and behavioral responses associated with the switch between tactic alternatives are thought to result from altered patterns of gene expression, and very often this is mediated by *secondary signals* that act as transcription factors and coordinate downstream cascades of gene expression.

In many insects (e.g., crickets: Zera *et al.* 1989; bees: Hartfelder *et al.* 2002; beetles: Emlen and Nijhout 1999, 2001), one important secondary signal is thought to be ecdysone (in adult honeybees, octopamine may also serve as a secondary signal: Schulz *et al.* 2002). Ecdysone is a hormone known to act as a transcription factor and known to initiate major changes in patterns of gene expression (Lepesant and Richards 1989, Andres et al. 1993, Cherbas and Cherbas 1996). Furthermore, pulses of ecdysone have been shown to interact with circulating levels of JH such that levels of JH above or below a threshold determine which of several ecdysone receptors are activated (Riddiford et al. 1999) and which of several sets of downstream genes are expressed (reviewed in Bollenbacher 1988, Gilbert 1989, Berger et al. 1992, Nijhout 1994, Riddiford 1994, 1996, Gilbert et al. 1996, Truman and Riddiford 1999, 2002). Although the best characterized of these interactions all involve the metamorphic transformations from larva to adult (where animals switch abruptly from expressing larva-specific genes to pupa- and then adult-specific genes), recent advances in the study of polyphenisms – of facultative alternative life histories, age polyethisms and castes in social insects, and reproductive tactics – suggest that similar interactions drive these threshold traits as well (Nijhout 1999, Hartfelder and Emlen 2005).

I have summarized these components of a threshold mechanism in Figure 5.2, and it is probably safe to state that these basic ingredients underlie the expression of most – if not all – alternative reproductive tactics (for vertebrate parallels, see reviews by Denver 1997, Moore et al. 1998, Foran and Bass 1999, Oliveira et al. 2001). Explicit consideration of the threshold mechanism reveals several avenues by which alternative tactics could evolve: each of the components described above is likely to vary heritably in natural populations, and genetic changes in these components delineate plausible trajectories for tactic evolution.

5.4 EVOLUTION OF A THRESHOLD MECHANISM

Biologists have long known that threshold traits could evolve (e.g. Weismann 1875, Merrifield and Pouldton 1899, Uvarov 1921, Süffert 1924), and both theoretical and empirical studies clearly illustrate that developmental thresholds can have significant levels of heritable genetic variation and can respond rapidly to selection (Lively 1986, West-Eberhard 1989, 1992, 2003, Hazel et al. 1990, Moran 1992, Morooka and Tojo 1992, Roff 1994a, b, 1996, 1998, Zera and Zhang 1995, Denno et al. 1996, Gu and Zera 1996, Fairbairn and Yadlowski 1997, Krebs and Loeschcke 1997, Roff et al. 1997, Gross and Repka 1998, De Moed et al. 1999). Consideration of the underlying mechanisms regulating expression of these threshold traits takes this one

step farther and reveals three principle ways in which threshold traits evolve (West-Eberhard 1992). I discuss each of these in detail and then illustrate them with an example.

5.4.1 Evolution of sensory apparatus and cues

Effective expression of alternative tactics requires an accurate match between phenotype and environment (e.g., Levins 1968, Lively 1986, Hazel et al. 1990, Moran 1992, West-Eberhard 1989, 2003, Gross 1996, Brockmann 2001). The better the match between tactic expression and selective circumstance, the better, on average, those genotypes will perform. Genetic changes in animal sensory systems – the detection of stimuli and the filtering or amplification of responses to those stimuli – can alter individual sensitivity to environmental cues, and this can lead to evolutionary changes in at least two aspects of tactic expression.

Firstly, by changing the relationship between external conditions and circulating levels of the hormone signal, modifications to animal sensory systems can shift the sensitivity of animals to an existing cue. Secondly, genetic changes in animal sensory systems can alter the types of cues utilized, and populations and species regularly differ in the particular cues they use to trigger tactic expression. For example, green lacewings (Chrysoperla carnea) facultatively switch between direct development and reproductive diapause, and some populations respond specifically to seasonal changes in photoperiod, while others respond to both photoperiod and larval prey type (Tauber and Tauber 1992). Geometrid moth caterpillars (Nemoria spp.) switch between alternative larval morphologies, and some species rely entirely on dietary cues (levels of tannins in the plant tissues they feed on), while others rely on a combination of dietary cues and colored light reflected from their surroundings (Greene 1989, 1999). Similarly, many butterfly species form pupae that match the color of their substrates (e.g., green vs. brown), and, depending on the species, caterpillars switch between these pupal alternatives in response to photoperiod (West et al. 1972), relative humidity (Smith 1978), the color of reflected light (Wiklund 1972, Smith 1978), or the texture of their background substrate (Hazel 1977, Hazel and West 1979, 1996). Finally, facultative wing expression in crickets is cued by temperature (Ghouri and McFarlane 1958), photoperiod (Tanaka et al. 1976), population density/crowding (Zera and Tiebel 1988), or diet (McFarlane 1962) – again, depending on the species (Harrison 1979).

Consequently, numerous components of "upstream" sensory detection processes can vary genetically within populations (e.g., the types of cues detected, sensitivity of individuals to these cues, and the relationship between detected cues and circulating levels of the hormone signal), and all comprise viable avenues for the evolution of animal alternative tactics. Collectively, they determine the nature of the cues utilized and the specific conditions in which tactics are expressed.

5.4.2 Evolution of the threshold

Another avenue for evolutionary changes in tactic expression is the threshold mechanism itself. Empirical studies of threshold evolution abound (e.g., Harrison 1979, Denno *et al.* 1986, 1996, Roff 1986, 1996, Tauber and Tauber 1992, Zera and Zhang 1995, Emlen 1996, Fairbairn and Yadlowski 1997, Tomkins 1999), though only relatively recently has it been possible to begin to interpret threshold evolution in the context of developmental mechanism – as the result of genetic changes to specific components of the developmental regulatory process (Zera and Zhang 1995, Dingle and Winchell 1997, Fairbairn and Yadlowski 1997, Zera and Denno 1997, Emlen 2000, Moczek and Nijhout

2002). Numerous components of threshold mechanisms can contribute to evolutionary changes in tactic expression (e.g., Figure 5.3), and I illustrate a few of these below.

First there is the hormone signal. Levels of this circulating signal will depend on upstream sensory processes and how these affect the rates of biosynthesis and degradation of the hormone. In the case of JH, biosynthesis is regulated by the corpora allata – small neurosecretory organs located beside the insect brain (Bounhiol 1938, Wigglesworth 1940, Schooley and Baker 1985, Tobe and Stay 1985, Nijhout 1994). Corpora allata do not store JH, so levels of hormone production are directly proportional to the sizes of these organs (Tobe and Pratt 1974, Feyereisen 1985, Rachinsky and Hartfelder 1990). Corpora allata size may be sensitive to nutrition, body size, status, or other aspects of larval circumstance (e.g., Wang 1965, Lüscher 1972, Wirtz 1973, Asencot and Lensky 1976, Lenz 1976, Dogra *et al.* 1977, Ulrich and Rembold 1983, de Wilde 1985, Rembold 1985, Rachinsky and Hartfelder 1990), and genetic changes in the relative growth of this organ, by affecting the rate of JH production, can alter circulating levels of the hormone signal.

Once secreted, JH is broken down by a variety of enzymes, the most important of which is JH esterase (Hammock 1985, Zera and Tiebel 1988). Levels of JH

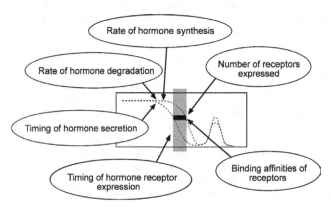

Figure 5.3 Filling in the "black box." Genetic variation for any of a multitude of components of threshold mechanisms can contribute to evolutionary changes in tactic expression. For example, variation in the rates of hormone synthesis or degradation can affect the levels of hormone signal during the sensitive period, variation in the numbers or types of receptors expressed can affect the critical hormone concentration needed to elicit a response, and variation in the timing of receptor expression can determine whether pulses of hormone coincide with, or miss,

the sensitive period. Each of these components is itself the product of numerous genes and gene products (e.g., the rate of hormone degradation will depend on rates of synthesis and degradation of enzymes, the binding affinities of those enzymes, and rates of synthesis and degradation of cofactors that affect the action of these enzymes). Consequently, threshold mechanisms regulating tactic expression are polygenic, have large quantities of additive genetic variation, and evolve readily in response to selection.

esterase are also sensitive to changes in larval diet and growth (Rachinsky and Hartfelder 1990, Browder *et al.* 2001, Tu and Tatar 2003), implicating this enzyme in the modulation of JH titers to match larval conditions or circumstance (i.e., JH esterase may couple levels of the signal hormone with perceived stimuli from the larval environment). In the cricket *Gryllus rubens*, winged and wingless individuals differ in circulating levels of JH during the sensitive period (Zera *et al.* 1989), and this difference appears to result entirely from tactic-specific differences in levels of JH esterase activity: animals reared under solitary conditions have higher levels of JH esterase (and, as a result, remove circulating JH faster) than animals reared under crowded conditions (Zera and Tiebel 1989, Zera and Tobe 1990, Zera and Holtmeier 1992; for a vertebrate parallel involving the enzyme aromatase, see Schlinger *et al.* 1999).

Importantly, populations contain genetic as well as environmental variation for JH esterase levels (e.g., Gu and Zera 1996, Roff *et al.* 1997), and genetic changes in the expression of this enzyme can lead to rapid evolutionary shifts in developmental thresholds (Zera and Zhang 1995, Fairbairn and Yadlowski 1997). Consequently, the interplay between biosynthesis and degradation ultimately determines the levels of the hormone signal at any specific time. Animals display highly stereotyped species- and situation-specific temporal profiles of hormone levels (Nijhout 1994), but genetic modifications in the organs that secrete the hormone, or in the enzymes that remove it, permit these hormone titer profiles to evolve in response to natural or artificial selection.

Tissues respond to the hormone signal during the sensitive period, defined by the physiological stage when target cells express receptors appropriate to the hormone (Nijhout 1994, 1999). Often there are several different forms of this receptor (e.g., Talbot *et al.* 1993, Jindra *et al.* 1996, Riddiford 1996, Nijhout 1999, Riddiford *et al.* 1999), and each can initiate a different downstream biochemical and transcription cascade (Riddiford *et al.* 1999, Hodin and Riddiford 2000). Genetic variation for how many hormone receptors are produced, for the amino acid sequences of the binding sites of those receptors, or for the secondary or tertiary conformational forms of the receptors can all lead to shifts in the sensitivity of target cells to the circulating endocrine signal, resulting in altered patterns of alternative tactic expression. Consequently, numerous components of these regulatory mechanisms are likely candidates for genetic changes in thresholds of sensitivity to the environment and hence for the evolution of animal ARTs.

5.4.3 Evolution of downstream processes

Binding of sufficient hormone to target cell receptors activates all-or-none downstream response cascades, often via expression of secondary hormones or transcription factors. Modifications to the biochemical properties of, or levels of expression of, any of the downstream cascades affect the phenotypes of the tactics themselves. Recent advances in molecular genetic techniques provide exciting resolution to the nature of these downstream genetic-patterning cascades and provide first glimpses of the types of genes involved in tactic-specific differences in behavior and morphology (reviewed in Evans and Wheeler 2001).

Honeybees exhibit two types of tactic alternatives: an irreversible morphological switch between queen and worker development, and then, within adult workers, a temporal switch between colony tasks (from nursing within the nest to foraging outside the nest). Both of these mechanisms involve thresholds of sensitivity to hormones and JH is the principal signal hormone in each case (e.g., Rachinsky and Hartfelder 1990, Rembold *et al.* 1992, Robinson and Vargo 1997, Rachinsky *et al.* 2000, Sullivan *et al.* 2000, Pearce *et al.* 2001). Queen and worker honeybee phenotypes differ dramatically in gene expression downstream of the developmental threshold, and the genes involved in these tactic differences range from metabolic enzymes, to transcription factors, to factors involved in cell signaling (Evans and Wheeler 1999, 2000, 2001, Hepperle and Hartfelder 2001). Nursing and foraging bees also differ in the downstream expression of at least 19 genes, again including transcription factors (Grozinger *et al.* 2003).

Similar advances have been made for downstream genetic patterning cascades in termites, which facultatively develop into either a soldier or a worker caste. Miura *et al.* (1999) recently identified a gene expressed specifically in the mandibular glands of soldiers, but not workers, and subsequent studies have found numerous transcription factor-, structural-, and enzyme-coding genes that differ in expression between soldiers and workers (Scharf *et al.* 2003).

Abouheif and Wray (2002) explored expression patterns of six genes involved in the patterning and development of insect wings and found large differences between winged and wingless castes in ants. Furthermore, by comparing these same patterning networks in several related ant species, they showed that *different points* in the patterning cascade underlay caste differences in the different species, providing remarkable insight into the genetic potential for independent evolution of alternative phenotypes. Many of these same genetic patterning networks are now thought to regulate facultative expression of eyespot patterns on the wings of butterflies (Brakefield *et al.* 1996, 1998, Keys *et al.* 1999, Weatherbee *et al.* 1999, Brunetti *et al.* 2001, Beldade and Brakefield 2002), and as I suggest below, they may be involved in the expression of beetle horns as well.

In summary, we now have identified several measurable aspects of mechanism that may contribute to additive genetic variation in the expression of alternative animal phenotypes. Although by no means complete, this list should begin to provide tangible underpinnings to the vague concept of "heritable quantitative genetic variation" in thresholds. This should also illustrate that condition sensitivity (e.g., plasticity) requires sophisticated developmental response mechanisms, which themselves comprise multiple signals, receptors, and biochemical pathways, and all of these involve multiple genes and precise timing and levels of gene expression. Plastic animal phenotypes are most definitely not "nongenetic" and modern views of how these processes work reveal a staggering potential for adaptive evolution.

5.5 AN EXAMPLE: ALTERNATIVE MALE MATING TACTICS IN HORNED BEETLES

The beetles I discuss here are dung beetles in the genus *Onthophagus* (Coleoptera: Scarabaeidae). There are over 2000 species in this genus, and these species live on every continent except Antarctica, in habitats ranging from tropical forest to desert. When these beetles find dung, they dig tunnels into the soil below (Halffter and Edmonds 1982, Cambefort 1991, Cambefort and Hanski 1991). Females dig the primary tunnels, and they spend several days inside each tunnel pulling dung fragments down to the ends and stashing them as provision for their young (Fabre 1899, Halffter and Edmonds 1982, Emlen 1997a, Hunt and Simmons 1998, Moczek and Emlen 2000, Hunt *et al.* 2002).

Male beetles guard the entrances to these tunnels and are sometimes able to monopolize access to the females inside (Figure 5.4A). In the two species best studied to date (*O. acuminatus* and *O. taurus*), large males wield a pair of horns that extend from the back of their heads, and large males with long horns are very effective at guarding tunnels: both large body size and long horn lengths significantly improve male fighting performance (Emlen 1997a, Moczek and Emlen 2000). Within guarding males, there is a positive, linear relationship between horn length and fertilization success (Hunt and Simmons 2001).

Small males are not efficient at guarding tunnels and instead employ an alternative tactic: they attempt to sneak into tunnels and mate with females on the sly (Figure 5.4A). This can entail slipping directly past the larger, guarding male, or it can involve digging a side tunnel that bypasses the guarding male entirely (Emlen 1997a, Moczek and Emlen 2000). Horns do not aid small males, and there is no relationship between either body size or horn length and

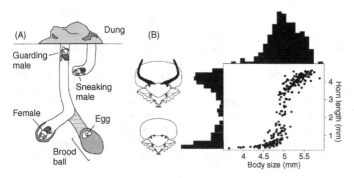

Figure 5.4 Alternative reproductive tactics in male *Onthophagus* beetles. (A) Large males fight to guard entrances to tunnels containing females; smaller males sneak into guarded tunnels and mate with females on the sly. (B) Large and small males differ in morphology as well as behavior. Males growing larger than a threshold body size produce a pair of curved horns that extend from the base of the head, whereas males not attaining this body size produce only rudimentary horns. Populations typically exhibit continuous variation for body size (top histogram) but are dimorphic for variation in male horn length (side histogram). (From Emlen 1997b and Emlen and Nijhout 1999, with permission.)

fertilization success by the sneaking male (Hunt and Simmons 2001).

The reproductive biology of these beetles and, in particular, the alternative behavioral tactics utilized by males to secure access to females generates very different selection on large and small males (Hunt and Simmons 2001). Long horns are favored in relatively large males who guard tunnel entrances; horns are not favored in smaller males who adopt the alternative tactic and sneak into tunnels (Emlen 1997a, 2000, Moczek and Emlen 2000, Hunt and Simmons 2001).

Developmentally, this is exactly what these animals do: large males produce horns, smaller males do not. Static samples of adult males from natural populations of either *O. acuminatus* or *O. taurus* have "broken" or sigmoid scaling relationships between the length of male horns and body size (Figure 5.4B). This results in natural populations being dimorphic for male horn length, with large numbers of males with full horns, large numbers of males with only rudimentary horns, and very few animals with intermediate morphologies (Emlen 1994, Hunt and Simmons 1997, 2001, Moczek and Emlen 1999).

As with most studied ARTs, the tactics of these beetles are facultatively expressed. Growth of the male horns depends on the nutritional environment encountered by animals as they develop – i.e., horn lengths are phenotypically plastic (Emlen 1994, 1996, 1997b, 2000, Hunt and Simmons 1997, 2002, Moczek 1998, 2002, Moczek and Emlen 1999, Kotiaho *et al.* 2003). Experiments perturbing the nutritional environment of larvae predictably influence the developmental fate of animals: male larvae fed large amounts of food all develop into large adults with long horns, whereas sibling males fed small food amounts mature as small adults without horns (Emlen 1994, 1997b).

Furthermore, we now know that horn development is regulated by a threshold. Males growing larger than a threshold body size grow horns according to a default pattern (extensive local proliferation of epidermal cells, resulting in two long horns at the base of the head). Males smaller than this threshold size are reprogrammed during their final larval instar and grow horns according to a different pattern (minimal proliferation of these cells, resulting in only rudimentary horns: Emlen and Nijhout 1999, 2001, Emlen 2000).

Experiments to date (measuring ecdysteroid titers, perturbing nutrition, and JH) suggest the following mechanism. Juvenile hormone acts as the signal hormone. Levels of JH appear to covary with larval diet so that by the end of the feeding period they are higher in small animals than in

(A) Large males

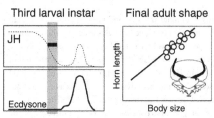

(B) Small males

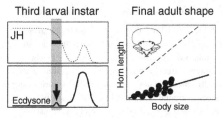

Figure 5.5 Model for the endocrine regulation of male horn expression in *Onthophagus taurus*. By the middle of the third larval instar, large and small males differ in circulating levels of juvenile hormone (JH): large males have lower concentrations than smaller males. JH levels are assessed during a brief sensitive period immediately prior to the cessation of feeding. (A) Relatively large males have JH concentrations below the critical threshold at this time. Cells in the developing horns of these individuals undergo a burst of rapid proliferation, and these larvae mature into adult males with fully developed horns (open circles). (B) Small male larvae have JH concentrations above the threshold during the sensitive period, and these animals experience a brief pulse of a second hormone, ecdysone. Ecdysone is known to initiate cascades of gene expression, and this tactic-specific pulse appears to reprogram the fate of horn cells so that they undergo only minimal proliferation. These small males mature into adults with only rudimentary horns (closed circles). (From Emlen and Allen 2004, with permission.)

larger animals (Figure 5.5). During a brief sensitive period that occurs as larvae finish feeding prior to metamorphosis, male larvae with JH levels above a threshold level (i.e., small animals) are reprogrammed to a hornless developmental trajectory by a small pulse of a secondary signal, ecdysone (Emlen and Nijhout 1999, 2001) (Figure 5.5B). This pulse of ecdysone only occurs in females and small males, and it appears to prevent significant horn growth, possibly by affecting the sensitivity of horn cells to JH during a second, later, sensitive period (not shown). The result is that large and small males end up developing along very different trajectories.

Thus, hormones translate individual patterns of overall growth into one of two specific fates – with or without horns – and this process involves a signal hormone (JH), a sensitive period (the end of the larval feeding period), a threshold of sensitivity to JH, and an interaction between levels of JH and a second hormone, ecdysone. With this as a backdrop, we can now revisit the three avenues of alternative tactic evolution and begin to consider how beetle horns and beetle horn dimorphism might evolve.

5.5.1   Evolution of sensory apparatus and cues

We still know very little about the sensory mechanisms of developing *Onthophagus* larvae, particularly those involved in detecting body size. Several features of this process are clear, however. First, animals develop in isolation within brood balls – masses of dung provisioned by the parents and buried below ground (Figure 5.4A). Consequently, detection of individual growth and overall body size must occur internally, rather than from repeated encounters with rivals during development.

Second, variation in the amount of food available to developing larvae affects their body size and can determine whether individuals mature above or below the critical threshold for horn growth (Emlen 1994, 1997b, 2000, Hunt and Simmons 1997, 2002, Moczek 1998, 2002, Moczek and Emlen 1999). Variation in food amount does not affect the threshold size itself, nor does it alter the final relationship between male horn length and body size (Emlen 1994, 1997b). Interestingly, variation in the *quality* of larval diet does affect the threshold, shifting the resulting relationship between horn length and body size (Emlen 1997b, Moczek 2002).

Thus, larvae appear to respond differently to specific aspects of their nutritional environment, and some of these dietary cues modify the sensitivity of the threshold mechanism to variation in individual status or body size. Further studies will be needed to elucidate the precise cues relevant to horn expression and to explore whether variation in these processes contributes to evolutionary changes in tactic expression.

5.5.2   Evolution of the threshold

We have a much clearer picture of how the threshold evolves. Numerous aspects of the threshold mechanism are likely to show heritable differences between individuals

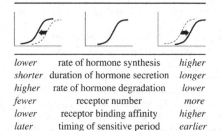

lower	rate of hormone synthesis	*higher*
shorter	duration of hormone secretion	*longer*
higher	rate of hormone degradation	*lower*
fewer	receptor number	*more*
lower	receptor binding affinity	*higher*
later	timing of sensitive period	*earlier*

Figure 5.6 Predicted consequences of mutations altering the threshold regulating beetle horn expression, illustrating multiple possible avenues for threshold evolution.

within populations, and any of these could contribute to evolutionary shifts in the threshold body size separating horned from hornless males (Figures 5.3 and 5.6). For example, levels of receptor expression by epidermal cells in the nascent horns would affect the concentration of JH needed to elicit a physiological response, as would shifts in the relationship between circulating JH and nutrition or growth (i.e., changes in the levels of JH associated with particular nutritional environments). Similarly, shifts in the timing of either the drop in JH or the sensitive period when JH levels are assessed could affect the body sizes small enough to elicit a response and hence the threshold associated with the switch between tactics.

Both artificial selection and common-garden, breeding experiments revealed extensive standing levels of heritable genetic variation for this developmental threshold (Emlen 1996, Moczek *et al.* 2002), and populations and species routinely differ in this threshold as well (Emlen 1996, 2000, Moczek *et al.* 2002, Moczek and Nijhout 2002) (Figure 5.7). This suggests that evolutionary changes in the threshold size for horn expression constitute a common and important avenue for evolution of alternative male tactics in Onthophagine beetles, permitting both the relative horn lengths and the tactic frequencies to be modified in response to changes in local conditions (such as changes in the overall population density or in the relative costs and benefits of the two tactics: Emlen 1996, 1997b, 2000, Simmons *et al.* 1999, Emlen and Nijhout 2000, Tomkins and Simmons 2000, Moczek *et al.* 2002, Moczek 2003).

One recent study combined information on the endocrine mechanism of horn expression with a comparison of genetically divergent populations in order to explore how evolutionary shifts in the threshold were brought about

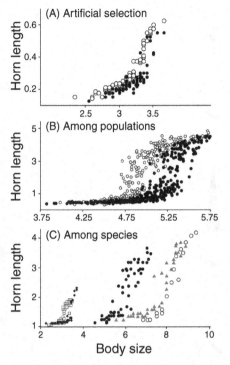

(A) Artificial selection

Horn length

0.6
0.4
0.2

2.5   3   3.5

(B) Among populations

Horn length

5
3
1

3.75   4.25   4.75   5.25   5.75

(C) Among species

Horn length

4
3
2
1

2   4   6   8   10

Body size

Figure 5.7 Evolution of a threshold. (A) Within-population variation. Artificial selection for relative horn length in *O. acuminatus* shifted the threshold body size in seven generations (final-generation animals from up and down lines shown). (B) Among-population variation. North Carolina (open circles) and Western Australia (closed circles) populations of the introduced species *O. taurus* diverged genetically for their threshold in the approximately 30 years post-introduction. (C) Among-species variation. Related *Onthophagus* species frequently differ in the threshold body size separating horned from hornless males (left to right: *O. acuminatus*, *O. marginicollis*, *O. striatulus*, *O. batesi*, and *O. incensus*). (Data from Emlen 1996, Moczek *et al.* 2002, Moczek and Nijhout 2002.)

(Moczek and Nijhout 2002). This study applied topical doses of the JH analog methoprene to animals from two populations known to differ in their horn development threshold (North Carolina and Western Australia populations of the introduced species *O. taurus*) and demonstrated that these populations had diverged in *both* the timing of a hormone sensitive period *and* in the sensitivity of cells to JH. The North Carolina populations were sensitive for a

shorter time and to smaller concentrations than Western Australia populations (Moczek and Nijhout 2002). This study illustrates how genetic variation for a variety of components of these plastic developmental response mechanisms can contribute to evolutionary changes in developmental thresholds and to the expression of alternative tactics.

5.5.3 Evolution of downstream processes

In this beetle example, the most conspicuous difference between male tactic alternatives involves the relative amounts of growth of a morphological structure – the horns (Figure 5.8). Beetle horns form as outgrowths of the epidermis that result from local bursts of cell proliferation at the base of the male head (Emlen and Nijhout 1999). Horn cells proliferate at the same time and in response to the same endocrine signals as other imaginal (imago = adult) structures, including the wings, eyes, and genitalia. Recent advances in *Drosophila* developmental genetics have identified how insect imaginal structures arise and how their sizes and shapes are regulated. Here I use this information as a foundation for exploring how beetle horns might evolve.

Cell proliferation in developing imaginal structures (e.g., in the legs, wings, and genitalia) is controlled by cascades of interacting transcription factors and signal molecules (reviewed in Cohen 1993, Serrano and O'Farrell 1997). These genetic patterning cascades result in local diffusion of signals within the epidermal fields that will form each adult structure (Lawrence and Struhl 1996, Day and Lawrence 2000). Partially overlapping gradients of these signals dictate the form of the developing structure by specifying precise regions – domains – within the growing structure. This subdivides the contiguous sheet of epidermal cells along three axes (anterior–posterior, dorsal–ventral, and proximal–distal) and results in a spatially explicit "map" of cell positions (reviewed in Cohen 1993, Serrano and O'Farrell 1997, Day and Lawrence 2000, Johnston and Gallant 2002).

Interactions between these local signals stimulate proliferation within portions of the growing structure (Peifer *et al.* 1991, Campbell *et al.* 1993, Struhl and Basler 1993, Basler and Struhl 1994, Irvine and Wieschaus 1994, Johnston and Schubiger 1996, Johnston and Gallant 2002). Importantly, changes in the concentrations of specific signals, or in the sensitivities of cells to those signals, can change the relative proportions of domains within the limb-epidermal field and, in so doing, can alter the shape

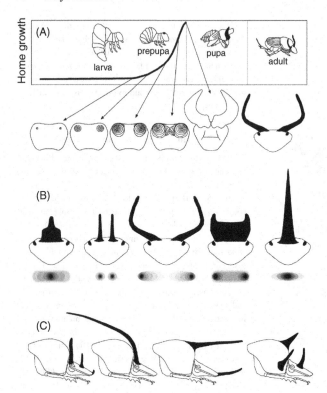

Figure 5.8 Evolution of downstream genetic patterning processes. Large and small male *O. taurus* differ in the relative amount of proliferation that occurs in the cells that will form the horns. (A) Horn growth is concentrated at the end of the larval period (during the "prepupal" stage) and coincides with growth of the other imaginal (adult) structures (e.g., eyes, wings, genitalia). Horns form as local evaginations of the larval epidermis, but these tubes of epidermis remain trapped inside the larval head capsule as they grow, so they fold on themselves into dense concentric rings that expand to their full length when the animal pupates. The tips of these developing horns are predicted to act as local sources for gene products that diffuse into the surrounding epithelium and specify a distal, or outgrowth, fate for these regions (see text).

Genetic modifications to the concentrations of these signals (e.g., epidermal growth factor-receptor [EGFR], *Distalless* [*Dll*] or *aristaless* [*al*]) would affect the size of the outgrowth and thus the length of the horn. (B) Modifications to the shape of the signal sources and the resulting diffusion gradients could lead to evolutionary changes in the shapes of the horns (proposed diffusion gradients shown below each horn type; left to right: *O. nuchicornis*, *O. asperulus*, *O. taurus*, *O. capella*, and *O. tersidorsis*). (C) Changes in the physical location of these signal sources could underlie evolutionary changes in the location of horns (e.g., back, middle, or front of the head, and central or lateral pronotum; left to right: *O. demarzi*, *O. raffayi*, *O. pentacanthus*, and *O. brooksi*). (Part (A) from Emlen and Allen 2004, with permission.)

and form of the resulting structure (e.g., Neumann and Cohen 1996, Niwa *et al.* 2000, Adachi-Yamada and O'Connor 2002, Martín-Castellanos and Edgar 2002). Thus, mutations in any of the patterning genes can subtly alter the final proportions of developing structures, and these patterning cascades constitute a likely mechanism for

evolutionary changes in animal shape (e.g., Weatherbee *et al.* 1999, Jockusch *et al.* 2000, Niwa *et al.* 2000, Keisman *et al.* 2001, Abouheif and Wray 2002, Stern 2003, Emlen and Allen 2004).

In *Drosophila*, the genes *Wingless* (*Wg*) and *Decapentaplegic* (*Dpp*) are the primary organizers of these developing imaginal structures (Serrano and O'Farrell 1997, Johnston and Gallant 2002). Patterns of *Wg* and *Dpp* expression determine where the axis of outgrowth will form, in part by activating signaling of the epidermal growth factor-receptor (EGFR) in the cells that will form the distal-most tip of the final structure (Campbell 2002, Galindo *et al.* 2002). In all invertebrate taxa studied to date, the distal tips of developing appendages act as local sources for EGFR, which forms steep concentration gradients along the distal-to-proximal axis of these structures (with the highest concentration occurring at the distal tip: Campbell 2002, Galindo *et al.* 2002). Gradients of Wg, Dpp, and EGFR activity coordinate the expression of specific downstream genes in concentric rings along the outgrowth axis (Campbell 2002, Galindo *et al.* 2002), and two of these genes – *Distalless* (*Dll*) and *aristaless* (*al*) – play especially prominent roles in specifying the distal extremes of developing structures (Campbell *et al.* 1993, Basler and Struhl 1994, Diaz-Benjumea *et al.* 1994, Lecuit and Cohen 1997, Campbell 2002, Galindo *et al.* 2002). Changes in levels of expression of any of these genes can affect the relative sizes of a structure, and homologs of many of these genes do the same thing for distal regions of developing vertebrate limbs (Panganiban *et al.* 1997, Martin 1998, Campbell 2002). Consequently, Wg, Dpp, and EGFR signaling, as well as the expression of *Distalless*, and *aristaless*, are all likely candidates for evolutionary changes in beetle horn morphology, with levels of expression predicted to be positively correlated with horn size. In Figure 5.8 I illustrate how simple changes in expression patterns of these genes could lead to evolutionary changes in the shape of beetle horns (Figure 5.8B) and also to changes in the physical location of horns on the beetle body (e.g., base of the head, front of the head, or thorax: Figure 5.8C).

Two recent studies now suggest that these limb-patterning genes are expressed in growing beetle horns, and that altered activities of this pathway may contribute to dimorphism in patterns of horn growth. Moczek and Nagy (2005) showed that Dll and al proteins are present in the horns of *Onthophagus taurus* and *O. nigriventris*, and that the relative sizes of their respective domains of expression differ between horned males, hornless males, and females. Similarly, Laura Corley, Quenna Szafran, Ian Dworkin and I showed that *Wg* and *Dpp* also are expressed in developing beetle horns (*O. nigriventris*), and that their relative levels of expression (measured as mRNA transcript abundances) are significantly greater in horned males than in either hornless males or females (D. J. Emlen *et al.*, unpublished data).

Consequently, the limb-patterning pathway may be involved with the downstream mechanisms responsible for generating male dimorphism (and sexual dimorphism) in horn expression, and recently this pathway has been implicated in the evolution of horn shape and physical location as well (Emlen *et al.* 2006).

A second promising pathway is the insulin-signaling pathway, which regulates overall rates of cell proliferation by controlling the process of protein synthesis (Edgar 1999, Kawamura *et al.* 1999, Bryant 2001). Levels of both insulin and insulin-dependent growth factors are sensitive to larval nutrition (Kawamura *et al.* 1999, Bryant 2001, Britton *et al.* 2002, Ikeya *et al.* 2002, Nijhout and Grunert 2003), and this physiological pathway interacts with cells in the developing structures to control tissue growth.

Cells in developing imaginal structures express receptors for insulin, and their sensitivity to this circulating signal couples their growth with the nutritional environment of larvae (Chen *et al.* 1996, Brogiolo *et al.* 2001, Bryant 2001). Genetic changes in levels of insulin-receptor expression can alter the growth of specific structures (Leevers *et al.* 1996, Huang *et al.* 1999, Weinkove *et al.* 1999), and by altering the relationship between larval nutrition and trait growth, these mutations change the *relative sizes* of these traits (i.e., their final size relative to the overall body size or status of the animal: Mirth *et al.* 2005; Shingleton *et al.* 2005). The insulin pathway is now considered to be the most promising mechanism for the development and evolution of allometry in insects, since it couples the sizes of developing structures with nutritional environments and growth (Emlen and Allen 2004; Shingleton *et al.* 2007).

Because changes in patterns of insulin signaling alter trait allometry (e.g., Mirth *et al.* 2005; Shingleton *et al.* 2005), this pathway is also a candidate "downstream" mechanism for dimorphism in the expression of beetle horns (horned and hornless individuals differ in horn allometry: Emlen and Allen 2004) (Figure 5.5). In the first beetle species to be examined (*O. nigriventris*), the horn discs of horned and hornless individuals differ significantly in their relative levels of insulin receptor (InR) expression (Emlen *et al.* 2006), and we now suspect that male dimorphism (and sexual dimorphism) in this species involves a truncation of insulin pathway activity in the horn disks of small males and females.

In conclusion, a suite of genes and gene products interacts to control the form and final size of adult morphological structures. Bringing the expression of any of these genes under the regulation of a physiological threshold can couple major changes in limb shape and/or size with larval exposure to environmental conditions – i.e., it can bring expression of morphological structures into the suite of phenotypic characteristics associated with conditionally expressed alternative tactics. Examination of the pathways themselves reveals a multitude of ways that the morphology of beetle horns could evolve, a pattern consistent with the extraordinary diversity in weapon form exhibited by extant species of this genus.

5.6 HORMONES, THRESHOLDS, AND GENETIC CORRELATIONS AMONG TRAITS

Identifying the types of mutations that would affect an endocrine response mechanism is informative for an additional reason: it can reveal developmental sources of linkage, or genetic correlation, among traits. For example, many hormones coordinate a large number of developmental and physiological events (e.g., growth, sexual maturity, egg production, aging), and genetic changes in circulating levels of these hormones or in cellular sensitivities to the hormone can affect the expression of multiple traits ("hormonal pleiotropy": Ketterson and Nolan 1992, Finch and Rose 1995, Sinervo and Svensson 1998, Zera and Harshman 2001, Flatt and Kawecki 2004). In *Drosophila melanogaster*, mutations in the insulin-like receptor gene (*InR*) alter circulating levels of JH, and genetic variation at this locus contributes to a negative genetic correlation between female fecundity and adult lifespan (Tatar and Yin 2001, Tatar *et al.* 2001a, b, 2003). Similarly, *Met* is a gene that affects JH sensitivity in target cells (Shemshedini and Wilson 1990, Pursley *et al.* 2000), and a recent study by Flatt and Kawecki (2004) found large, pleiotropic consequences of genetic variation at this locus, contributing to genetic correlations between development time, the onset of reproduction, and both early and late fecundity. In both of these examples, shared utilization of a circulating hormone signal (JH) appears to have contributed to genetic correlations among life-history traits, and knowledge of the mechanism made it possible to predict both the direction and relative magnitude of the resulting trait associations.

Beetle horns are often correlated with other morphological traits, such as eyes, wings, antennae, and genitalia (Nijhout and Emlen 1998, Emlen 2000, 2001, Moczek and Nijhout 2004), and this also appears to result from shared utilization of circulating signals. During the period of horn growth (the prepupal period: Figure 5.8A), the epidermal cells that will form the horns are sensitive to levels of JH (Emlen and Nijhout 1999, Moczek and Nijhout 2002) and probably also to levels of nutrients, insulin, and growth factors (Emlen and Allen 2004). Many other morphological structures grow at this same time. Proliferation in these other structures is similarly affected by circulating hormones, nutrients, insulin, and growth factors (reviewed in Stern and Emlen 1999, Emlen and Allen 2004), and herein lies the potential for trait interactions: if proliferating organs compete with other organs for access to any of these circulating signals or nutrients, then mutations affecting the relative growth of one of the structures could have negative pleiotropic consequences for growth of other structures.

Although the relevant signal(s) has not yet been identified, it is already clear that horn growth in some *Onthophagus* species is negatively correlated with growth of other structures, and perturbations to the growth of one of these traits (e.g., the horns) alters the relative sizes of the other traits (Nijhout and Emlen 1998, Emlen 2000, 2001, Moczek and Nijhout 2004). In at least one of these cases (horns vs. eyes), the resulting genetic correlation appears to have constrained the independent evolution of these traits, leading to evolutionary losses of horns and also to major changes in the type of horn produced (Emlen *et al.* 2005). Consequently, shared utilization of circulating endocrine or other signals can generate nonrandom associations among different traits, and this can affect the subsequent evolution of the involved traits.

Not all mutations are expected to have such extensive pleiotropic effects, however (reviewed in Stern 2000). Genetic modifications to spatially localized patterning networks, such as those expressed only in a specific cell type or body region (e.g., within a limb field), often will only affect the trait in question (e.g., Raff 1996, Raff and Sly 2000). Separation of processes in time can also reduce correlations among traits: if tissues respond to a hormone signal during different sensitive periods, then the potential exists for them to evolve relatively independently (Nijhout 1994, 1999).

Finally, the most striking implication of developmental thresholds for the evolution of animal phenotypes is that they *minimize genetic correlations between tactics* (e.g., Cheverud 1984, 1996, Bonner 1988, West-Eberhard 1992, 2003, Raff 1996, Wagner 1996, Nijhout 1999, Raff and Sly 2000). By partially uncoupling gene expression of the alternatives (e.g., Evans and Wheeler 1999, 2000, 2001, Miura *et al.* 1999, Miura 2001, Grozinger *et al.* 2003, Scharf *et al.* 2003), threshold mechanisms permit tactics to evolve along strikingly independent trajectories (West-Eberhard

1989, 1992, 2003). The general pattern that emerges is that mutations in genes involved with a hormone signal or with cellular sensitivities to this signal (i.e., with the threshold mechanism itself) often have large pleiotropic effects on other aspects of morphology, life history, and behavior, whereas mutations in the downstream pathways regulated by a threshold often do not – these effects are much more likely to be confined to the trait in question, and in particular, to just one of the two tactics.

The threshold mechanism generating male dimorphism in beetle horn expression illustrates this well. This mechanism uncouples the development of large and small males, and as a result, the alternative tactics have evolved along very different trajectories: selection for altered horn morphology has led to profound genetic changes in the shape and size of beetle horns (e.g., Emlen 2001, Emlen *et al.* 2005) (Figure 5.8), and this radiation of form has occurred with little or no corresponding changes to the phenotype of the alternative tactic.

5.7 SUMMARY

The expression of beetle horns is condition-sensitive, with overwhelming influences of larval nutrition and negligible heritabilities for traditional variables like horn length. In this respect, alternative reproductive tactics in male beetles are typical of alternative tactics in general – plastic, conditionally expressed, phenotypic alternatives expressed within one sex of a single species. Extreme condition-sensitivity of tactic expression does not imply an absence of genetic variation. The purpose of this chapter has been to describe the basic physiological processes that confer condition-sensitivity to tactic expression – the mechanism – and to use this knowledge of mechanism to illustrate how genes contribute to both the expression and evolution of animal alternative tactics.

Alternative reproductive tactics evolve in three major ways. First, the sensory structures that detect and respond to external circumstances can evolve, leading to shifts in the relative sensitivities of individuals to specific stimuli and to changes in the types of stimuli utilized. Second, the physiological response mechanism that translates perceived stimuli into altered patterns of gene expression, tissue growth, and behavior, can evolve. This is the threshold mechanism, and genetic changes in these response thresholds can change the conditions under which tactics are expressed. Finally, genetic changes in downstream regulatory pathways can alter the tactics themselves, leading to

within-tactic changes in morphology or behavior that are expressed at least partially independently from the tactic alternatives.

Why incorporate a developmental perspective? As we improve our understanding of the components of these developmental mechanisms – hormones, sensitive periods, receptors, and their interactions – we learn *how* evolutionary changes in phenotype are brought about. We begin to identify the types of mutations that would affect these mechanisms and the consequences of these genetic changes for the resulting phenotypes. Even a superficial understanding of development can bring improved resolution to the processes, both past and present, that have shaped the evolution of animal alternative tactics.

Acknowledgments
I thank H. J. Brockmann, K. L. Bright, and an anonymous reviewer for helpful comments on this manuscript, and the National Science Foundation (IBN-0092873) for funding the research.

References
Abouheif, E. and Wray, G. A. 2002. Evolution of the genetic network underlying wing polyphenism in ants. *Science* 297, 249–252.

Adachi-Yamada, T. and O'Connor, M. B. 2002. Morphogenetic apoptosis: a mechanism for correcting discontinuities in morphogen gradients. *Developmental Biology* 251, 74–90.

Andres, A. J., Fletcher, J. C., Karim, F. D., and Thummel, C. S. 1993. Molecular analysis of the initiation of metamorphosis: a comparative study of *Drosophila* ecdysteroid-regulated transcription. *Developmental Biology* 160, 388–404.

Asencot, M. and Lensky, Y. 1976. The effect of sugars and juvenile hormone on the differentiation of the female honeybee larvae (*Apis mellifera* L.) to queens. *Life Sciences* 18, 693–700.

Awiti, L. R. and Hidaka, T. 1982. Neuroendocrine mechanisms involved in pupal color dimorphism in swallowtail *Papilio xuthus*. *Insect Science Applications* 3, 181–192.

Ayoade, O., Morooka, S., and Tojo, S. 1999. Enhancement of short wing formation and ovarian growth in the genetically defined macropterous strain of the brown planthopper, *Nilaparvata lugens*. *Journal of Insect Physiology* 45, 93–100.

Basler, K. and Struhl, G. 1994. Compartment boundaries and the control of *Drosophila* limb pattern by *hedgehog* protein. *Nature* 368, 208–214.

Beldade, P. and Brakefield, P. M. 2002. The genetics and evo-devo of butterfly wing patterns. *Nature Reviews Genetics* 3, 442–452.

Berger, E. M., Goudie, K., Klieger, L., and DeCato, R. 1992. The juvenile hormone analogue, methoprene, inhibits ecdysterone induction of small heat shock protein gene expression. *Developmental Biology* 151, 410–418.

Bertuso, A. G., Morooka, S., and Tojo, S. 2002. Sensitive periods for wing development and precocious metamorphosis after precocene treatment of the brown planthopper, *Nilaparvata lugens*. *Journal of Insect Physiology* 48, 221–229.

Bollenbacher, W. E. 1988. The interendocrine regulation of larval–pupal development in the tobacco hornworm, *Manduca sexta*: a model. *Journal of Insect Physiology* 34, 941–947.

Bonner, J. T. 1988. *The Evolution of Complexity by Means of Natural Selection.* Princeton, NJ: Princeton University Press.

Botens, F. F. W., Rembold, H., and Dorn, A. 1997. Phase-related juvenile hormone determinations in field catches and laboratory strains of different *Locusta migratoria* subspecies. In S. Kawashima and S. Kikuyama (eds.) *Advances in Comparative Endocrinology,* pp. 197–203. Bologna, Italy: Monduzzi.

Bounhiol, J. J. 1938. Recherches expérimentales sur le déterminisme de la métamorphose chez Lépidoptères. *Biologie Bulletin (Suppl.)* 24, 1–199.

Brakefield, P. M., Gates, J., Keys, D., *et al.* 1996. Development, plasticity and evolution of butterfly eyespot patterns. *Nature* 384, 236–242.

Brakefield, P. M., Kesbeke, F., and Koch, P. B. 1998. The regulation of phenotypic plasticity of eyespots in the butterfly *Bicyclus anynana*. *American Naturalist* 152, 853–860.

Britton, J. S., Lockwood, W. K., Li, L., Cohen, S. M., and Edgar, B. A. 2002. *Drosophila*'s insulin/PI3-kinase pathway coordinates cellular metabolism with nutritional conditions. *Developmental Cell* 2, 239–249.

Brockmann, H. J. 2001. The evolution of alternative strategies and tactics. *Advances in the Study of Behavior* 30, 1–51.

Brogiolo, W., Stocker, H., Ikeya, T., *et al.* 2001. An evolutionarily conserved function of the *Drosophila* insulin receptor and insulin-like peptides in growth control. *Current Biology* 11, 213–221.

Browder, M. H., D'Amico, L. J., and Nijhout, H. F. 2001. The role of low levels of juvenile hormone esterase in the metamorphosis of *Manduca sexta*. *Journal of Insect Science* 1, 1–4.

Brunetti, C., Selegue, J. E., Monterio, A., *et al.* 2001. The generation and diversification of butterfly eyespot patterns. *Current Biology* 11, 1578–1585.

Bryant, P. J. 2001. Growth factors controlling imaginal disc growth in *Drosophila*. In G. Bock, G. Cardew, and J. A. Goode (eds.) *The Cell Cycle and Development,* pp. 182–199. New York: John Wiley.

Cambefort, Y. 1991. Biogeography and evolution. In I. Hanski and Y. Cambefort (eds.) *Dung Beetle Ecology,* pp. 51–68. Princeton, NJ: Princeton University Press.

Cambefort, Y. and Hanski, I. 1991. Dung beetle population biology. In I. Hanski and Y. Cambefort (eds.) *Dung Beetle Ecology,* pp. 36–50. Princeton: Princeton University Press.

Campbell, G. 2002. Distalization of the *Drosophila* leg by graded EGF-receptor activity. *Nature* 418, 781–785.

Campbell, G., Weaver, T., and Tomlinson, A. 1993. Axis specification in the developing *Drosophila* appendage: the role of wingless, *decapentaplegic*, and the homeobox *aristaless*. *Cell* 74, 1113–1123.

Chapman, R. F. 1982. *The Insects: Structure and Function.* London: Hodder and Stoughton.

Chen, C., Jack, J., and Garofalo, R. S. 1996. The *Drosophila* insulin receptor is required for normal growth. *Endocrinology* 137, 846–856.

Cherbas, P. and Cherbas, L. 1996. Molecular aspects of ecdysteroid hormone action. In L. I. Gilbert, J. R. Tata, and B. G. Atkinson (eds.) *Metamorphosis: Postembryonic Reprogramming of Gene Expression in Amphibian and Insect Cells,* pp. 175–222. San Diego, CA: Academic Press.

Cheverud, J. M. 1984. Quantitative genetics and developmental constraints on evolution by selection. *Journal of Theoretical Biology* 110, 155–172.

Cheverud, J. M. 1996. Developmental integration and the evolution of pleiotropy. *American Zoologist* 36, 44–50.

Clark, R. A. 1997. Dimorphic males display alternative reproductive strategies in the marine amphipod *Jassa marmorata* Holmes (Corophioidea: Ischyroceridae). *Ethology* 103, 531–553.

Cnaani, J., Robinson, G. E., and Hefetz, A. 2000. The critical period for caste determination in *Bombus terrestris* and its juvenile hormone correlates. *Journal of Comparative Physiology A* 186, 1089–1094.

Cohen, S. M. 1993. Imaginal disc development. In M. Bate and A. Martinez-Arias (eds.) *The Development of Drosophila melanogaster,* pp. 747–841. New York: Cold Spring Harbor Laboratory Press.

Cremer, S. and Heinze, J. 2003. Stress grows wings: environmental induction of winged dispersal males in *Cardiocondyla* ants. *Current Biology* 13, 219–223.

Cusson, M., Tobe, S., and McNeil, J. 1994. Juvenile hormones: their role in the regulation of the pheromonal communication system of the armyworm moth, *Pseudaletia unipuncta*. *Archives of Insect Biochemistry and Physiology* 25, 329–345.

Danforth, B. N. 1991. The morphology and behavior of dimorphic males in *Perdita portalis* (Hymenoptera: Andrenidae). *Behavioral Ecology and Sociobiology* 29, 235–247.

Day, S. J. and Lawrence, P. A. 2000. Measuring dimensions: the regulation of size and shape. *Development* 127, 2977–2987.

De Moed, G. H., Kruitwagen, C. L. J. J., De Jong, G., and Scharloo, W. 1999. Critical weight for the induction of pupariation in *Drosophila melanogaster*: genetic and environmental variation. *Journal of Evolutionary Biology* 12, 852–858.

Denno, R. F., Douglass, L. W., and Jacobs, D. 1986. Effects of crowding and host plant nutrition on a wing dimorphic planthopper. *Ecology* 67, 116–123.

Denno, R. F., Roderick, G. K., Peterson, M. A., *et al.* 1996. Habitat persistence underlies intraspecific variation and dispersal strategies of planthoppers. *Ecological Monographs* 66, 389–408.

Denver, R. J. 1997. Environmental stress as a developmental cue: corticotropin-releasing hormone is a proximate mediator of adaptive phenotypic plasticity in amphibian metamorphosis. *Hormones and Behavior* 31, 169–179.

Diaz-Benjumea, F. J., Cohen, B., and Cohen, S. M. 1994. Cell interaction between compartments establishes the proximal–distal axis of *Drosophila* legs. *Nature* 372, 175–179.

Dingle, H. and Winchell, R. 1997. Juvenile hormone as a mediator of plasticity in insect life histories. *Archives of Insect Biochemistry and Physiology* 35, 359–373.

Dogra, G. S., Ulrich, G. M., and Rembold, H. 1977. A comparative study of the endocrine system of the honeybee larvae under normal and experimental conditions. *Zeitschrift für Naturforschung*, 32C, 637–642.

Eberhard, W. G. 1982. Beetle horn dimorphism: making the best of a bad lot. *American Naturalist* 119, 420–426.

Edgar, B. A. 1999. From small flies come big discoveries about size control. *Nature Cell Biology* 1, E191–E193.

Emlen, D. J. 1994. Environmental control of horn length dimorphism in the beetle *Onthophagus acuminatus* (Coleoptera: Scarabaeidae). *Proceedings of the Royal Society of London B* 256, 131–136.

Emlen, D. J. 1996. Artificial selection on horn length–body size allometry in the horned beetle *Onthophagus acuminatus* (Coleoptera: Scarabaeidae). *Evolution* 50, 1219–1230.

Emlen, D. J. 1997a. Alternative reproductive tactics and male dimorphism in the horned beetle *Onthophagus acuminatus* (Coleoptera: Scarabaeidae). *Behavioral Ecology and Sociobiology* 41, 335–341.

Emlen, D. J. 1997b. Diet alters male horn allometry in the beetle *Onthophagus acuminatus* (Coleoptera: Scarabaeidae). *Proceedings of the Royal Society of London B* 264, 567–574.

Emlen, D. J. 2000. Integrating development with evolution: a case study with beetle horns. *BioScience* 50, 403–418.

Emlen D. J. 2001. Costs and the diversification of exaggerated animal structures. *Science* 291, 1534–1536.

Emlen, D. J. and Allen, C. E. 2004. Genotype to phenotype: physiological control of trait size and scaling in insects. *Integrative and Comparative Biology* 43, 617–634.

Emlen, D. J. and Nijhout, H. F. 1999. Hormonal control of male horn length dimorphism in the horned beetle *Onthophagus taurus*. *Journal of Insect Physiology* 45, 45–53.

Emlen, D. J. and Nijhout, H. F. 2000. The development and evolution of exaggerated morphologies in insects. *Annual Review of Entomology* 45, 661–708.

Emlen, D. J. and Nijhout, H. F. 2001. Hormonal control of male horn length dimorphism in the dung beetle *Onthophagus taurus* (Coleoptera: Scarabaeidae): a second critical period of sensitivity to juvenile hormone. *Journal of Insect Physiology* 47, 1045–1054.

Emlen, D. J., Marangelo, J., Ball, B., and Cunningham, C. W. 2005. Diversity in the weapons of sexual selection: horn evolution in the beetle genus *Onthophagus*. *Evolution* 59, 1060–1084.

Emlen, D. J., Szafran, Q., Corley, L. S., and Dworkin, I. 2006. Candidate genes for the development and evolutionary diversification of beetle horns. *Heredity* (in press).

Evans, J. D. and Wheeler, D. E. 1999. Differential gene expression between developing queens and workers in the honey bee, *Apis mellifera*. *Proceedings of the National Academy of Sciences of the United States of America* 96, 5575–5580.

Evans, J. D. and Wheeler, D. E. 2000. Expression profiles during honeybee caste determination. *Genome Biology* 2.1, pp. research0001.1–research0001.6. Available online at: http://genomebiology.com/2000/2/1/research/0001.1.

Evans, J. D. and Wheeler, D. E. 2001. Gene expression and the evolution of insect polyphenisms. *BioEssays* 23, 62–68.

Fabre, J. H. 1899. *Souvenirs Entomologiques* Excerpts translated by A. T. de Mattos in *More Beetles* (1922) London: Hodder and Stoughton.

Fairbairn, D. J. and Yadlowski, D. E. 1997. Coevolution of traits determining migratory tendency: correlated response of a critical enzyme, juvenile hormone esterase, to selection on wing morphology. *Journal of Evolutionary Biology* 10, 495–513.

Feyereisen, R. 1985. Radiochemical assay for juvenile hormone III biosynthesis *in vitro*. *Methods in Enzymology* 111, 530–539.

Finch, C. E. and Rose, M. R. 1995. Hormones and the physiological architecture of life history evolution. *Quarterly Review of Biology* 10, 1–52.

Flatt, T. and Kawecki, T. J. 2004. Pleiotropic effects of *methoprene-tolerant* (*Met*), a gene involved in juvenile hormone metabolism, on life history traits in *Drosophila melanogaster*. *Genetica* 122, 141–160.

Foran, C. M. and Bass, A. H. 1999. Preoptic GnRH and AVT: axes for sexual plasticity in teleost fish. *General and Comparative Endocrinology* 116, 141–152.

Galindo, M. I., Bishop, S. A., Greig, S., and Couso, J. P. 2002. Leg patterning driven by proximal–distal interactions and EGFR signaling. *Science* 297, 256–259.

Ghouri, A. S. K. and McFarlane, J. E. 1958. Occurrence of a macropterous form of *Gryllodes sigillatus* (Walker) (Orthoptera:Gryllidae) in laboratory culture. *Canadian Journal of Zoology* 36, 837–838.

Gilbert, L. I. 1989. The endocrine control of molting: the tobacco hornworm, *Manduca sexta*, as a model system. In J. Koolman (ed.) *Ecdysone: From Chemistry to Mode of Action*, pp. 448–471. Stuttgart, Germany: Thieme-Verlag.

Gilbert, L. I., Rybczynski, R., and Tobe, S. 1996. Endocrine cascade in insect metamorphosis. In L. I. Gilbert, J. R. Tata, and B. G. Atkinson (eds.) *Metamorphosis: Postembryonic Reprogramming of Gene Expression in Amphibian and Insect Cells*, pp. 60–108. San Diego, CA: Academic Press.

Greene, E. 1989. A diet induced developmental polymorphism in a caterpillar. *Science* 243, 643–646.

Greene, E. 1999. Phenotypic variation in larval development and evolution: polymorphism, polyphenism, and developmental reaction norms. In B. K. Hall and M. H. Wake (eds.) *The Origin and Evolution of Larval Forms*, pp. 379–410. San Diego, CA: Academic Press.

Greene, E., Lyon, B. E., Muehter, V. R., *et al.* 2000. Disruptive sexual selection for plumage colouration in a passerine bird. *Nature* 407, 1000–1003.

Gross, M. R. 1996. Alternative reproductive strategies and tactics: diversity within sexes. *Trends in Ecology and Evolution* 11, 92–98.

Gross, M. R. and Repka, J. 1998. Stability with inheritance in the conditional strategy. *Journal of Theoretical Biology* 192, 445–453.

Grozinger, C. M., Sharabash, N. M., Whitfield, C. W., and Robinson, G. E. 2003. Pheromone-mediated gene expression in the honey bee brain. *Proceedings of the National Academy of Sciences of the United States of America* 100, 14519–14525.

Gu, X. and Zera, A. J. 1996. Quantitative genetics of juvenile hormone esterase, juvenile hormone binding and general esterase activity in the cricket, *Gryllus assimilis*. *Heredity* 76, 136–142.

Halffter, G. and Edmonds, W. G. 1982. *The Nesting Behavior of Dung Beetles (Scarabaeidae): An Ecological and Evolutive Approach*, Publication No. 10. Mexico City: Instituto de Ecologica.

Hammock, B. D. 1985. Regulation of juvenile hormone titer: degradation. In G. A. Kerkut and L. I. Gilbert (eds.) *Comprehensive Insect Physiology, Biochemistry, and Pharmacology*, vol. 7, pp. 431–472. New York: Pergamon Press.

Harrison, R. G. 1979. Flight polymorphism in the field cricket *Gryllus pennsylvanicus*. *Oecologia* 40, 125–132.

Hartfelder, K. 1990. Regulatory steps in caste development of eusocial bees. In W. Engels (ed.) *Social Insects: An Evolutionary Approach to Castes and Reproduction*, pp. 245–264. Heidelberg, Germany: Springer-Verlag.

Hartfelder, K. and Emlen, D. J. 2005. Endocrine control of insect polyphenism. In L. I. Gilbert, K. Iatrou, and S. S. Gill (eds.) *Comprehensive Molecular Insect Science*, vol. 3, *Endocrinology*, pp. 651–703. Boston, MA: Elsevier.

Hartfelder, K. and Engels, W. 1998. Social insect polymorphism: hormonal regulation of plasticity in development and reproduction in the honeybee. *Current Topics in Developmental Biology* 40, 45–77.

Hartfelder, K., Bitondi, M. M. G., Santana, W. C., and Simões, Z. L. P. 2002. Ecdysteroid titers and reproduction in queens and workers of the honey bee and of a stingless bee: loss of ecdysteroid function at increasing levels of sociality? *Insect Biochemistry and Molecular Biology* 32, 211–216.

Hazel, W. N. 1977. The genetic basis of pupal colour dimorphism and its maintenance by natural selection in *Papilio polyxenes* (Papilionidae: Lepidoptera). *Heredity* 38, 227–236.

Hazel, W. N. and West, D. A. 1979. Environmental control of pupal colour in swallowtail butterflies (Lepidoptera: Papilionidae: *Battus philenor* (L.) and *Papilio polyxenes* Fabr.) *Ecological Entomology* 4, 393–408.

Hazel, W. N. and West, D. A. 1996. Pupation site preference and environmentally-cued pupal colour dimorphism in the swallowtail butterflies *Papilio polyxenes* Fabr. (Lepidoptera: Papilionidae). *Biological Journal of the Linnean Society* 57, 81–87.

Hazel, W. N., Smock, R., and Johnson, M. D. 1990. A polygenic model for the evolution and maintenance of conditional strategies. *Proceedings of the Royal Society of London B* 242, 181–187.

Hepperle, C. and Hartfelder, K. 2001. Differentially expressed regulatory genes in honey bee caste development. *Naturwissenschaften* 88, 113–116.

Hodin, J. and Riddiford, L. M. 2000. Different mechanisms underlie phenotypic plasticity and interspecific variation for a reproductive character in Drosophilids (Insecta: Diptera). *Evolution* 54, 1638–1653.

Huang, H., Potter, C. J., Tao, W., *et al.* 1999. PTEN affects cell size, cell proliferation and apoptosis during *Drosophila* eye development. *Development* 126, 5365–5372.

Hunt, J. and Simmons, L. W. 1997. Patterns of fluctuating asymmetry in beetle horns: an experimental examination of the honest signaling hypothesis. *Behavioral Ecology and Sociobiology* 41, 109–114.

Hunt, J. and Simmons, L. W. 1998. Patterns of parental provisioning covary with male morphology in a horned beetle (*Onthophagus taurus*) (Coleoptera: Scarabaeidae). *Behavioral Ecology and Sociobiology* 42, 447–451.

Hunt, J. and Simmons, L. W. 2001. Status-dependent selection in the dimorphic beetle *Onthophagus taurus*. *Proceedings of the Royal Society of London B* 268, 2409–2414.

Hunt, J. and Simmons, L. W. 2002. The genetics of maternal care: direct and indirect genetic effects on phenotype in the dung beetle *Onthophagus taurus*. *Proceedings of the National Academy of Sciences of the United States of America* 99, 6828–6832.

Hunt, J., Simmons, L. W., and Kotiaho, J. S. 2002. A cost of maternal care in the dung beetle *Onthophagus taurus*? *Journal of Evolutionary Biology* 15, 57–64.

Iguchi, Y. 1998. Horn dimorphism of *Allomyrina dichotoma septentrionalis* (Coleoptera: Scarabaeidae) affected by larval nutrition. *Annals of the Entomological Society of America* 91, 845–847.

Ikeya, T., Galic, M., Belawat, P., Nairz, K., and Hafen, E. 2002. Nutrient-dependent expression of insulin-like peptides from neuroendocrine cells in the CNS contributes to growth regulation in *Drosophila*. *Current Biology* 12, 1293–1300.

Injeyan, H. S. and Tobe, S. S. 1981. Phase polymorphism in *Schistocerca gregaria*: assessment of juvenile hormone synthesis in relation to vitellogenesis. *Journal of Insect Physiology* 27, 203–210.

Irvine, K. D. and Wieschaus, E. 1994. *fringe*, a boundary-specific signaling molecule, mediates interactions between dorsal and ventral cells during *Drosophila* wing development. *Cell* 79, 595–606.

Jindra, M., Malone, F., Hiruma, K., and Riddiford, L. W. 1996. Developmental profiles and ecdysteroid regulation of the mRNAs for two ecdysone receptor isoforms in the epidermis and wings of the tobacco hornworm, *Manduca sexta*. *Developmental Biology* 180, 258–272.

Jockusch, E. L., Nulsen, C., Newfeld, S. J., and Nagy, L. M. 2000. Leg development in flies versus grasshoppers: differences in *dpp* expression do not lead to differences in the expression of downstream components of the leg patterning pathway. *Development* 127, 1617–1626.

Johnston, L. A. and Gallant, P. 2002. Control of growth and organ size in *Drosophila*. *BioEssays* 24, 54–64.

Johnston, L. A. and Schubiger, G. 1996. Ectopic expression of *wingless* in imaginal discs interferes with *decapentaplegic* expression and alters cell determination. *Development* 122, 3519–3529.

Kawamura, K., Shibata, T., Saget, O., Peel, D., and Bryant, P. J. 1999. A new family of growth factors produced by the fat body and active on *Drosophila* imaginal disc cells. *Development* 126, 211–219.

Keisman, E. L., Christiansen, A. E., and Baker, B. S. 2001. The sex determination gene *doublesex* regulates the A/P organizer to direct sex-specific patterns of growth in the *Drosophila* genital imaginal disc. *Developmental Cell* 1, 215–225.

Ketterson, E. D. and Nolan, V. 1992. Hormones and life histories: an integrative approach. *American Naturalist* 140, S33–S62.

Keys, D. N., Lewis, D. L., Selegue, J. E., *et al.* 1999. Recruitment of a *hedgehog* regulatory circuit in butterfly eyespot evolution. *Science* 283, 532–534.

Koch, P. B. and Bückmann, D. 1987. Hormonal control of seasonal morphs by the timing of ecdysteroid release in *Araschnia levana* L. (Nymphalidae: Lepidoptera). *Journal of Insect Physiology* 33, 823–829.

Kotiaho, J. S., Simmons, L. W., Hunt, J., and Tomkins, J. L. 2003. Males influence maternal effects that promote sexual selection: a quantitative genetic experiment with dung beetles *Onthophagus taurus*. *American Naturalist* **161**, 852–859.

Krebs, R. A. and Loeschcke, V. 1997. Estimating heritability in a threshold trait: heat-shock tolerance in *Drosophila buzzatii*. *Heredity* **79**, 252–259.

Kukuk, P. F. 1996. Male dimorphism in *Lasioglossum (Chialictus) hemichalceum*: the role of larval nutrition. *Journal of the Kansas Entomological Society* **69**, 147–157.

Kurdziel, J. P. and Knowles, L. L. 2002. The mechanisms of morph determination in the amphipod *Jassa*: implications for the evolution of alternative male phenotypes. *Proceedings of the Royal Society of London B* **269**, 1749–1754.

Laufer, H. and Ahl, J. S. B. 1995. Mating behavior and methyl farnesoate levels in male morphotypes of the spider crab, *Libinia emarginata* (Leach). *Journal of Experimental Marine Biology and Ecology* **193**, 15–20.

Lawrence, P. A. and Struhl, G. 1996. Morphogens, compartments, and pattern: lessons from *Drosophila*? *Cell* **85**, 951–961.

Lecuit, T. and Cohen, S. M. 1997. Proximal–distal axis formation in the *Drosophila* leg. *Nature* **388**, 139–145.

Leevers, S. J., Weinkove, D., MacDougall, L. K., Hafen, E., and Waterfield, M. D. 1996. The *Drosophila* phosphoinositide 3-kinase Dp110 promotes cell growth. *EMBO Journal* **15**, 6584–6594.

Lenz, M. 1976. The dependence of hormone effects in termite caste determination on external factors. In M. Lüscher (ed.) *Phase and Caste Determination in Insects: Endocrine Aspects*, pp. 73–89. Oxford, UK: Pergamon Press.

Lepesant, J.-A. and Richards, G. 1989. Ecdysteroid-regulated genes. In J. Koolman (ed.) *Ecdysone: From Chemistry to Mode of Action*, pp. 355–367. Stuttgart, Germany: Thieme-Verlag.

Levins, R. 1968. *Evolution in Changing Environments*. Princeton, NJ: Princeton University Press.

Lively, C. M. 1986. Canalization versus developmental conversion in a spatially variable environment. *American Naturalist* **128**, 561–572.

Lüscher, M. 1972. Environmental control of juvenile hormone (JH) secretion and caste differentiation in termites. *Comparative Endocrinology* (**Supplement**) **3**, 509–514.

Martin, G. R. 1998. The roles of FGFs in the early development of vertebrate limbs. *Genes and Development* **12**, 1571–1586.

Martín-Castellanos, C. and Edgar, B. A. 2002. A characterization of the effects of *Dpp* signaling on cell

growth and proliferation in the *Drosophila* wing. *Development* **129**, 1003–1013.

McFarlane, J. E. 1962. Effect of diet and temperature on wing development of *Gryllodes sigillatus* (Walk.) (Orthoptera: Gryllidae). *Annales de la Société Entomologique de Québec* **7**, 28–33.

Merrifield, F. and Pouldton, E. B. 1899. The color relation between the pupae *of Papilio machaon, Pieris napai* and many other species, and the surroundings of the larvae preparing to pupate, etc. *Transactions of the Entomological Society of London* **1899**, 369–433.

Mirth, C., Truman, J. W., and Riddiford, L. M. 2005. The role of the prothoracic gland in determining critical weight for metamorphosis in *Drosophila melanogaster*. *Current Biology* **15**, 1796–1807.

Miura, T. 2001. Morphogenesis and gene expression in the soldier-caste differentiation of termites. *Insectes Sociaux* **48**, 216–223.

Miura, T., Kamikouchi, A., Sawata, M., *et al.* 1999. Soldier caste-specific gene expression in the mandibular glands of *Hodotermopsis japonica* (Isoptera: Termopsidae). *Proceedings of the National Academy of Sciences of the United States of America* **96**, 13874–13879.

Moczek, A. P. 1998. Horn polyphenism in the beetle *Onthophagus taurus*: larval diet quality and plasticity in parental investment determine adult body size and male horn morphology. *Behavioral Ecology* **9**, 636–642.

Moczek, A. P. 2002. Allometric plasticity in a polyphenic beetle. *Ecological Entomology* **27**, 58–67.

Moczek, A. P. 2003. The behavioral ecology of threshold evolution in a polyphenic beetle. *Behavioral Ecology* **14**, 841–854.

Moczek, A. P. and Emlen, D. J. 1999. Proximate determination of male horn dimorphism in the beetle *Onthophagus taurus* (Coleoptera: Scarabaeidae). *Journal of Evolutionary Biology* **12**, 27–37.

Moczek, A. P. and Emlen, D. J. 2000. Male horn dimorphism in the scarab beetle, *Onthophagus taurus*: do alternative reproductive tactics favour alternative phenotypes? *Animal Behaviour* **59**, 459–466.

Moczek, A. P., Nagy, L. M. 2005. Diverse developmental mechanisms contribute to different levels of diversity in horned beetles. *Evolution and Development* **7**, 175–185.

Moczek, A. P. and Nijhout, H. F. 2002. Developmental mechanisms of threshold evolution in a polyphenic beetle. *Evolution and Development* **4**, 252–264.

Moczek, A. P. and Nijhout, H. F. 2004. Trade-offs during the development of primary and secondary sexual traits in a horned beetle. *American Naturalist* 163, 184–191.

Moczek, A. P., Hunt, J., Emlen, D. J., and Simmons, L. W. 2002. Threshold evolution in exotic populations of a polyphenic beetle. *Evolutionary Ecology Research* 4, 587–601.

Moore, M. C., Hews, D. K., and Knapp, R. 1998. Hormonal control and evolution of alternative male phenotypes: generalizations of models for sexual differentiation. *American Zoologist* 38, 133–152.

Moran, N. A. 1992. The evolutionary maintenance of alternative phenotypes. *American Naturalist* 139, 971–989.

Morooka, S. and Tojo, S. 1992. Maintenance and selection of strains exhibiting specific wing form and body colour under high density conditions in the brown planthopper *Nilaparvata lugens* (Homoptera: Delphacidae). *Applied Entomology and Zoology* 27, 445–454.

Neumann, C. J. and Cohen, S. M. 1996. Distinct mitogenic and cell fate specification functions of *wingless* in different regions of the wing. *Development* 122, 1781–1789.

Nijhout, H. F. 1994. *Insect Hormones*. Princeton, NJ: Princeton University Press.

Nijhout, H. F. 1999. Control mechanisms of polyphenic development in insects. *BioScience* 49, 181–192.

Nijhout, H. F. and Emlen, D. J. 1998. Competition among body parts in the development and evolution of insect morphology. *Proceedings of the National Academy of Sciences of the United States of America* 95, 3685–3689.

Nijhout, H. F. and Grunert, L. W. 2003. Bombyxin is a growth factor for wing imaginal disks in Lepidoptera. *Proceedings of the National Academy of Sciences of the United States of America* 99, 15446–15450.

Niwa, N., Inoue, Y., Nozawa, A., *et al.* 2000. Correlation of diversity of leg morphology in *Gryllus bimaculatus* (cricket) with divergence in *dpp* expression pattern during leg development. *Development* 127, 4373–4381.

Oliveira, R. F., Canario, A. V. M., and Grober, M. S. 2001. Male sexual polymorphism, alternative reproductive tactics, and androgens in combtooth blennies (Pisces: Blennidae). *Hormones and Behavior* 40, 266–275.

O'Neill, K. M. and Evans, H. E. 1983. Alternative male mating tactics in *Bembicinus quinquespinosus* (Hymenoptera: Sphecidae): correlations with size and color variation. *Behavioral Ecology and Sociobiology* 14, 39–46.

Ono, S. 1982. Effect of juvenile hormone on the caste determination in the ant *Pheidole fervida* Smith

(Hymenoptera: Formicidae). *Applied Entomology and Zoology* 17, 1–7.

Panganiban, G., Irvine, S. M., Lowe, C., *et al.* 1997. The origin and evolution of animal appendages. *Proceedings of the National Academy of Sciences of the United States of America* 94, 5162–5166.

Pearce, A. N., Huang, Z. Y., and Breed, M. D. 2001. Juvenile hormone and aggression in honey bees. *Journal of Insect Physiology* 47, 1243–1247.

Peifer, M., Rauskolb, C., Williams, M., Riggleman, B., and Wieschaus, E. 1991. The segment polarity gene *armadillo* interacts with the *wingless* signalling pathway in both embryonic and adult pattern formation. *Development* 111, 1029–1043.

Pursley, S., Ashok, M., and Wilson, T. G. 2000. Intracellular localization and tissue specificity of the *Methoprene-tolerant* (*Met*) gene product in *Drosophila melanogaster*. *Insect Biochemistry and Molecular Biology* 30, 839–845.

Rachinsky, A. and Hartfelder, K. 1990. Corpora allata activity, a prime regulating element for caste-specific juvenile hormone titre in honey bee larvae (*Apis mellifera carnica*). *Journal of Insect Physiology* 36, 189–194.

Rachinsky, A., Strambi, C., Strambi, A., and Hartfelder, K. 2000. Caste and metamorphosis: hemolymph titers of juvenile hormone and ecdysteroids in last instar honeybee larvae. *General and Comparative Endocrinology* 79, 31–38.

Radwan, J. 1993. The adaptive significance of male polymorphism in the acarid mite *Caloglyphus berlesei*. *Behavioral Ecology and Sociobiology* 33, 201–208.

Radwan, J., Unrug, J., and Tomkins, J. L. 2002. Status-dependence and morphological trade-offs in the expression of a sexually selected character in the mite, *Sancassania berlesei*. *Journal of Evolutionary Biology* 15, 744–752.

Raff, R. A. 1996. *The Shape of Life: Genes, Development, and the Evolution of Animal Form*. Chicago, IL: University of Chicago Press.

Raff, R. A. and Sly, B. J. 2000. Modularity and dissociation in the evolution of gene expression territories in development. *Evolution and Development* 2, 102–113.

Rankin, M. A. and Riddiford, L. M. 1977. Hormonal control of migratory flight in *Oncopeltus fasciatus*: the effects of the corpus cardiacum, corpus allatum and starvation on migration and reproduction. *General and Comparative Endocrinology* 33, 309–321.

Rankin, S. M., Chambers, J., and Edwards, J. P. 1997. Juvenile hormone in earwigs: roles in oogenesis, mating, and maternal behaviors. *Archives of Insect Biochemistry and Physiology* 35, 427–442.

Rasmussen, J. L. 1994. The influence of horn and body size on the reproductive behavior of the horned rainbow scarab beetle *Phanaeus difformis* (Coleoptera: Scarabaeidae). *Journal of Insect Behavior* 7, 67–82.

Rembold, H. 1985. Sequence of caste differentiation steps in *Apis mellifera*. In J. A. L. Watson, B. M. Okot-Kotber, and C. H. Noirot (eds.) *Caste Differentiation in Social Insects*, pp. 347–359. New York: Pergamon Press.

Rembold, H., Czoppelt, C., Grüne, M., *et al.* 1992. Juvenile hormone titers during honey bee embryogenesis and metamorphosis. In B. Mauchamp, F. Couillaud, and J. C. Baehr (eds.) *Insect Juvenile Hormone Research*, pp. 37–43. Paris: INRA.

Riddiford, L. M. 1994. Cellular and molecular actions of juvenile hormone. 1. General considerations and premetamorphic actions. *Advances in Insect Physiology* 24, 213–274.

Riddiford, L. M. 1996. Molecular aspects of juvenile hormone action in insect metamorphosis. In L. I. Gilbert, J. R. Tata, and B. G. Atkinson (eds.) *Metamorphosis: Postembryonic Reprogramming of Gene Expression in Amphibian and Insect Cells*, pp. 223–253. San Diego, CA: Academic Press.

Riddiford, L. M., Hiruma, K., Lan, Q., and Zhou, B. 1999. Regulation and role of nuclear receptors during larval molting and metamorphosis of Lepidoptera. *American Zoologist* 39, 736–746.

Robinson, G. E. and Vargo, E. L. 1997. Juvenile hormone in adult eusocial Hymenoptera: gonadotropin and behavioral pacemaker. *Archives of Insect Biochemistry and Physiology* 35, 559–583.

Roff, D. A. 1986. The genetic basis of wing dimorphism in the sand cricket, *Gryllus firmus* and its relevance to the evolution of wing dimorphism in insects. *Heredity* 57, 221–231.

Roff, D. A. 1994a. Evolution of dimorphic traits: effect of directional selection on heritability. *Heredity* 72, 36–41.

Roff, D. A. 1994b. The evolution of dimorphic traits: predicting the genetic correlation between environments. *Genetics* 136, 395–401.

Roff, D. A. 1996. The evolution of threshold traits in animals. *Quarterly Review of Biology* 71, 3–35.

Roff, D. A. 1998. Evolution of threshold traits: the balance between directional selection, drift and mutation. *Heredity* 80, 25–32.

Roff, D. A., Stirling, G., and Fairbairn, D. J. 1997. The evolution of threshold traits: a quantitative genetic analysis of the physiological and life-history correlates of wing dimorphism in the sand cricket. *Evolution* 51, 1910–1919.

Röseler, P.-F., Röseler, I., and van Honk, C. G. J. 1981. Evidence for inhibition of corpora allata activity in workers of *Bombus terrestris* by a pheromone from the queen's mandibular glands. *Experientia* 37, 348–351.

Rountree, D. B. and Nijhout, H. F. 1995. Hormonal control of a seasonal polyphenism in *Precis coenia* (Lepidoptera: Nymphalidae). *Journal of Insect Physiology* 41, 987–992.

Scharf, M. E., Wu-Scharf, D., Pittendrigh, B. R., and Bennett, G. W. 2003. Caste- and development-associated gene expression in a lower termite. *Genome Biology* 4, R62.

Schlinger, B. A., Greco, C., and Bass, A. H. 1999. Aromatase activity in the hindbrain vocal control region of teleost fish: divergence among males with alternative reproductive tactics. *Proceedings of the Royal Society of London B* 266, 131–136.

Schooley, D. A. and Baker, F. C. 1985. Juvenile hormone biosynthesis. In G. A. Kerkut and L. I. Gilbert (eds.) *Comprehensive Insect Physiology, Biochemistry and Pharmacology*, vol. 7, pp. 363–389. New York: Pergamon Press.

Schulz, D. J., Sullivan, J. P., and Robinson, G. E. 2002. Juvenile hormone and octopamine in the regulation of division of labor in honey bee colonies. *Hormones and Behavior* 42, 222–231.

Scott, M. P., Trumbo, S. T., Neese, P. A., Bailey, W. D., and Roe, R. M. 2001. Changes in biosynthesis and degradation of juvenile hormone during breeding by burying beetles: a reproductive or social role? *Journal of Insect Physiology* 47, 295–302.

Serrano, N. and O'Farrell, P. H. 1997. Limb morphogenesis: connections between patterning and growth. *Current Biology* 7, R186–R195.

Shemshedini, L. and Wilson, T. G. 1990. Resistance to juvenile hormone and insect growth regulator in *Drosophila* is associated with altered cytosolic juvenile hormone-binding protein. *Proceedings of the National Academy of Sciences of the United States of America* 87, 2072–2076.

Shingleton, A. W., Das, J., Vinicius, L., and Stern, D. L. 2005. The temporal requirements for insulin signaling during development in *Drosophila*. *PLoS Biology* 3, 1607–1617.

Shingleton, A., Frankino, A., Flatt, T., Nijhout, H. F., and Emlen, D. J. 2007. Size and shape: the developmental regulation of static allometry in insects. *BioEssays* 29, 536–548.

Simmons, L. W., Tomkins, J. L., and Hunt, J. 1999. Sperm competition games played by dimorphic male beetles. *Proceedings of the Royal Society of London B* 266, 145–150.

Sinervo, B. and Svensson, E. 1998. Mechanistic and selective causes of life history trade-offs and plasticity. *Oikos* 83, 432–442.

Siva-Jothy, M. T. 1987. Mate securing tactics and the cost of fighting in the Japanese horned beetle *Allomyrina dichotoma* L. (Scarabaeidae). *Journal of Ethology* 5, 165–172.

Smith, A. G. 1978. Environmental factors influencing pupal colour determination in Lepidoptera. 1. Experiments with *Papilio polytes*, *Papilio demoleus* and *Papilio polyxenes*. *Proceedings of the Royal Society of London B* 200, 295–329.

Starnecker, G. and Hazel, W. N. 1999. Convergent evolution of neuroendocrine control of phenotypic plasticity in pupal colour in butterflies. *Proceedings of the Royal Society of London B* 266, 2409–2412.

Stern, D. L. S. 2000. Evolutionary developmental biology and the problem of variation. *Evolution* 54, 1079–1091.

Stern, D. L. S. 2003. The Hox gene *Ultrabithorax* modulates the shape and size of the third leg of *Drosophila* by influencing diverse mechanisms. *Developmental Biology* 256, 355–366.

Stern, D. L. S. and Emlen, D. J. 1999. The developmental basis for allometry in insects. *Development* 126, 1091–1101.

Strambi, A., Strambi, C., Röseler, P.-F., and Röseler, I. 1984. Simultaneous determination of juvenile hormone and ecdysteroid titers in the hemolymph of bumblebee prepupae (*Bombus hypnorum* and *B. terrestris*). *General and Comparative Endocrinology* 55, 83–88.

Struhl, G. and Basler, K. 1993. Organizing activity of *Wingless* protein in *Drosophila*. *Cell* 72, 527–540.

Süffert, F. 1924. Bestimmungsfaktoren des Zeichnungsmusters beim Saisondimorphismus von *Araschnia levana-prorsa*. *Biologisches Zentralblatt* 44, 173–188.

Sullivan, J. P., Jassim, O., Fahrbach, S. E., and Robinson, G. E. 2000. Juvenile hormone paces behavioral development in the adult worker honey bee. *Hormones and Behavior* 37, 1–14.

Talbot, W. S., Swyryd, E. A., and Hogness, D. 1993. *Drosophila* tissues with different metamorphic responses to ecdysone express different ecdysone receptor isoforms. *Cell* 73, 1323–1337.

Tallamy, D. W., Monaco, E. L., and Pesek, J. D. 2002. Hormonal control of egg dumping and guarding in the lace bug, *Gargaphia solani* (Hemiptera: Tingidae). *Journal of Insect Behavior* 15, 467–475.

Tanaka, S., Matsuka, M., and Sakai, T. 1976. Effect of change in photoperiod on wing form in *Pteronemobius taprobanensis* (Orthoptera: Gryllidae). *Applied Entomology and Zoology* 11, 27–32.

Tatar, M. and Yin, C.-M. 2001. Slow aging during insect reproductive diapause: why butterflies, grasshoppers and flies are like worms. *Experimental Gerontology* 36, 723–738.

Tatar, M., Chien, S. A., and Priest, N. K. 2001a. Negligible senescence during reproductive diapause in *Drosophila melanogaster*. *American Naturalist* 158, 248–258.

Tatar, M., Kopelman, A., Epstein, D., *et al.* 2001b. A mutant *Drosophila* insulin receptor homolog that extends life-span and impairs neuroendocrine function. *Science* 292, 107–110.

Tatar, M., Bartke, A., and Antebi, A. 2003. The endocrine regulation of aging by insulin-like signals. *Science* 299, 1346–1351.

Tauber, C. A. and Tauber, M. J. 1992. Phenotypic plasticity in *Chrysoperla*: genetic variation in the sensory mechanisms and in correlated reproductive traits. *Evolution* 46, 1754–1773.

Tobe, S. S. and Pratt, G. E. 1974. The influence of substrate concentrations on the rate of insect juvenile hormone biosynthesis by corpora allata of the desert locust *in vitro*. *Biochemical Journal* 144, 107–113.

Tobe, S. S. and Stay, B. 1985. Structure and regulation of the corpus allatum. *Advances in Insect Physiology* 18, 305–432.

Tomkins, J. L. 1999. Environmental and genetic determinants of the male forceps length dimorphism in the European earwig *Forficula auricularia* L. *Behavioral Ecology and Sociobiology* 47, 1–8.

Tomkins, J. L. and Simmons, L. W. 2000. Sperm competition games played by dimorphic male beetles: fertilization gains with equal mating access. *Proceedings of the Royal Society of London B* 267, 1547–1553.

Truman, J. W. and Riddiford, L. M. 1999. The origins of insect metamorphosis. *Nature* 401, 447–452.

Truman, J. W. and Riddiford, L. M. 2002. Endocrine insights into the evolution of metamorphosis in insects. *Annual Review of Entomology* 47, 467–500.

Tu, M.-P. and Tatar, M. 2003. Juvenile diet restriction and the aging and reproduction of adult *Drosophila melanogaster*. *Aging Cell* 2, 327–333.

Ulrich, G. M. and Rembold, H. 1983. Caste-specific maturation of the endocrine system in the female honey bee larva. *Cell and Tissue Research* 230, 49–55.

Uvarov, B. P. 1921. A revision of the genus *Locusta* L. (= *Pachytylus* Fieb.), with a new theory as to periodicity and migrations of locusts. *Bulletin of Entomological Research* 12, 135–163.

Velthius, H. H. W. 1976. Environmental, genetic and endocrine influences in stingless bee caste determination.

In M. Lüscher (ed.) *Phase and Caste Determination in Insects: Endocrine Aspects*, pp. 35–53. New York: Pergamon Press.

Wagner, G. P. 1996. Homologues, natural kinds, and the evolution of modularity. *American Zoologist* 36, 36–43.

Wang, D.-I. 1965. Growth rates of young queen and worker honeybee larvae. *Journal of Apicultural Research* 4, 3–5.

Weatherbee, S. D., Nijhout, H. F., Grunert, L. W., *et al.* 1999. *Ultrabithorax* function in butterfly wings and the evolution of insect wing patterns. *Current Biology* 9, 109–115.

Weinkove, D., Neufeld, T., Twardzik, T., Waterfield, M., and Leevers, S. J. 1999. Regulation of imaginal disc cell size, cell number and organ size by *Drosophila* class IA phosphinositide 3-kinase and its adaptor. *Current Biology* 9, 1019–1029.

Weismann, A. 1875. *Studien zur Deszendenztheorie* vol., 1, *Über den Saisondimorphismus der Schmetterlinge.* Leipzig, Germany: Engelmann.

West, D. A., Snelling, W. N., and Herbeck, T. A. 1972. Pupal color dimorphism and its environmental control in *Papilio polyxenes asterias* Stoll (Lep. Papilionidae). *Journal of the New York Entomological Society* 80, 205–211.

West-Eberhard, M. J. 1989. Phenotypic plasticity and the origins of diversity. *Annual Review of Ecology and Systematics* 20, 249–278.

West-Eberhard, M. J. 1992. Behavior and evolution. In P. R. Grant and H. S. Horn (eds.) *Molds, Molecules and Metazoa: Growing Points in Evolutionary Biology*, pp. 57–75. Princeton, NJ: Princeton University Press.

West-Eberhard, M. J. 2003. *Developmental Plasticity and Evolution.* Oxford, UK: Oxford University Press.

Wheeler, D. E. 1991. The developmental basis of worker caste polymorphism in ants. *American Naturalist* 138, 1218–1238.

Wheeler, D. E. and Nijhout, H. F. 1981. Soldier determination in ants: new role for juvenile hormone. *Science* 213, 361–363.

Wigglesworth, V. B. 1940. Local and general factors in the development of "pattern" in *Rhodnius prolixus* (Hemiptera). *Journal of Experimental Biology* 17, 180–200.

Wiklund, C. 1972. Pupal colour polymorphism in *Papilio machaon* L. in response to wavelength of light. *Naturwissenschaften* 59, 219.

Wilde, J. de 1985. Extrinsic control of caste differentiation in the honey bee (*Apis mellifera* L.) and in other Apidae. In

J. A. L. Watson, B. M. Okot-Kotber, and C. H. Noirot (eds.) *Caste Differentiation in Social Insects*, pp. 361–369. New York: Pergamon Press.

Wirtz, P. 1973. Differentiation in the honeybee larva. *Mededelingen van de Landbouwhogeschool Wageningen* 73–75, 1–66.

Zera, A. J. and Denno, R. F. 1997. Physiology and ecology of dispersal polymorphism in insects. *Annual Review of Entomology* 42, 207–230.

Zera, A. J. and Harshman, L. G. 2001. The physiology of life history trade-offs in animals. *Annual Review of Ecology and Systematics* 32, 95–126.

Zera, A. J. and Holtmeier, C. L. 1992. In vivo and in vitro degradation of juvenile hormone-III in presumptive long-winged and short-winged *Gryllus rubens*. *Journal of Insect Physiology* 38, 61–74.

Zera, A. J. and Tiebel, K. C. 1988. Brachypterizing effect of group rearing, juvenile hormone III and methoprene in the wing-dimorphic cricket, *Gryllus rubens*. *Journal of Insect Physiology* 34, 489–498.

Zera, A. J. and Tiebel, K. C. 1989. Differences in juvenile hormone esterase activity between presumptive macropterous and brachypterous *Gryllus rubens*: implications for the hormonal control of wing polymorphism. *Journal of Insect Physiology* 35, 7–17.

Zera, A. J. and Tobe, S. S. 1990. Juvenile hormone-III biosynthesis in presumptive long-winged and short-winged *Gryllus rubens*: implications for the endocrine regulation of wing dimorphism. *Journal of Insect Physiology* 36, 271–280.

Zera, A. J. and Zhang, C. 1995. Evolutionary endocrinology of juvenile hormone esterase in *Gryllus assimilis*: direct and correlated responses to selection. *Genetics* 141, 1125–1134.

Zera, A. J., Strambi, C., Tiebel, K. C., Strambi, A., and Rankin, M. A. 1989. Juvenile hormone and ecdysteroid titers during critical periods of wing determination in *Gryllus rubens*. *Journal of Insect Physiology* 35, 501–511.

Zuk, M. and Simmons, L. W. 1997. Reproductive strategies of the crickets (Orthoptera: Gryllidae). In J. C. Choe and B. J. Crespi (eds.) *Mating Systems in Insects and Arachnids*, pp. 89–109. Cambridge, UK: Cambridge University Press.

6 · Neuroendocrine mechanisms of alternative reproductive tactics: the chemical language of reproductive and social plasticity

ANDREW H. BASS AND PAUL M. FORLANO

CHAPTER SUMMARY

The wide range of variation in reproductive tactics displayed among teleost fishes has provided a rich source of natural experiments for investigating the neural mechanisms of alternative reproductive tactics (ARTs). These studies have mainly focused on identifying the location and extent of neuropeptide-containing cells in the forebrain's preoptic area (POA), in part, because of the well-established influence of these neurons on reproductive mechanisms. We first review the ARTs of teleost species that have served as model systems for investigating the neural mechanisms of reproductive plasticity and then the general organization of the POA of vertebrates. Comparative surveys then show how life-history trajectories and reproductive tactics vary with inter- and intrasexual dimorphisms in the size and number of POA neurons that synthesize either arginine vasotocin (AVT) or gonadotropin-releasing hormone (GnRH). The emerging evidence for the potential role of neurosteroids in mechanisms of reproductive plasticity inclusive of ARTs is then considered before concluding with a listing of a suite of neuroendocrinological traits that may provide proximate mechanisms essential to the widespread evolution of ARTs among teleost fish.

6.1 INTRODUCTION: DIVERGENT LIFE-HISTORY TRAJECTORIES

A major theme that continues to emerge from many studies of the neural mechanisms of ARTs is the uncoupling of gonadal and neurobiological traits that provides for the adaptable patterning of suites of mechanisms between alternative behavioral phenotypes (Bass 1992). We briefly discuss the life-history patterns that can give rise to alternative reproductive/ behavioral morphs of the major study species discussed in this review to provide some background for a comparative survey of neural mechanisms.

Teleost fishes exhibit a remarkable range of reproductive phenotypes (e.g., see Taborsky 1994). Alternative male reproductive morphs among teleosts may originate from any one of several developmental trajectories (see Foran and Bass 1999 for a more complete discussion) (Figure 6.1). In some species, like midshipman fish and sunfish (reviews: Gross 1991, Bass 1996), alternative male morphs become fixed and males will follow one of two nonoverlapping developmental pathways (shown in Figure 6.1A as type I or type II males: nomenclature after Bass and Marchaterre 1989). Thus, type I and II males differ in a large suite of traits. Type I males delay the onset of maturity to invest in larger body size and, in the case of midshipman fish, a vocal motor system that functions in the production of advertisement calls used in courtship and agonistic calls used in territorial defense (Box 6.1). Sunfish have comparable male morphs, although there is no information on possible morph divergence in vocal traits (sunfish are also sonic: Gerald 1971). Conditional mating tactics (Figure 6.1B), like those described for some gobies (Mazzoldi et al. 2000), pupfish (e.g., Leiser and Itzkowitz 2004) and type I male midshipman fish (Lee and Bass 2004), have males that show reversible, social-context-dependent changes in reproductive status between territorial (T) and sneaking, nonterritorial (NT) morphs. For sex/role-changing fish such as the bluehead wrasse (review: Godwin et al. 2003), either initial-phase (IP) males or females transform permanently into territorial, terminal-phase (TP) males (Figure 6.1C). Thus, one individual experiences sequential life-history stages that, by contrast, are separated between individuals in species like midshipman and sunfish

Alternative Reproductive Tactics, ed. Rui F. Oliveira, Michael Taborsky, and H. Jane Brockmann. Published by Cambridge University Press.
© Cambridge University Press 2008.

Life-history patterns of ARTs in teleost fishes

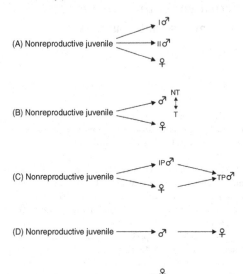

Figure 6.1 Life-history patterns for teleost fish showing
alternative reproductive tactics and strategies (see Gross 1996,
Brockmann 2001 for discussion of terminology). (A) For
gonochoristic species (juveniles are either male or female), there
are two distinct male phenotypes shown here as type I and type II
males that represent terminally differentiated life-history
trajectories. (B) Conditional strategies can be represented by
individuals that exhibit reversible changes between a territorial
(T) and nonterritorial (NT) status. (C) For sex/role-changing
species (sequential hermaphrodites) that show female-to-male
transformations (protogyny), initial-phase males (IP) and
females can transform into terminal-phase males (TP).
(D) For sequential hermaphrodites with male-to-female sex
change (protandry), a monogamous male can become the
dominant female in a social group. (E) Simultaneous
hermaphrodites exhibit serial sex change, and repeatedly switch
from male to female phenotypes. (Adapted from Foran and Bass
1999.)

(Figure 6.1A). Individuals in yet other sex-changing species
like anemonefish (Godwin *et al.* 2003) may show permanent
male-to-female sex reversal (Figure 6.1D). Lastly, serially
sex-changing fish like gobies (review: Cole 1990) switch back
and forth between the sexes (Figure 6.1E).

Box 6.1 Vocal behavior and motor system of
midshipman fish

Midshipman fish have a pair of muscles (sm) attached to
the lateral walls of their swim bladder (sb), as shown here
in a line drawing of a midshipman fish (Figure 6.2A).
The synchronous contraction of the sonic muscles leads
to the production of sounds. Type I male midshipman
fish produce long-duration (more than 1 hour), multi-
harmonic calls known as "hums" (Figure 6.2B, a segment of a continuous hum recorded from a nest at
16.1 °C). Midshipman fish, and the closely related
toadfishes, have a vocal control network as depicted here
in a sagittal view of the brain and anterior spinal cord.
The vocal motor network (Figure 6.2C) includes vocal-
acoustic integration centers (VAC) at forebrain (f),
midbrain (m), and hindbrain (h) levels (Bass *et al.* 1994,
Goodson and Bass 2002). Auditory input is provided to
each VAC by way of auditory nuclei positioned at
hindbrain, midbrain, and forebrain levels (see Bass *et al.*
2000, Goodson and Bass 2002). A hindbrain–spinal vocal
pacemaker circuit (shaded region) includes a column of
pacemaker neurons positioned ventrolateral to the sonic
motor nucleus that innervates the sonic muscles via
ventral, sonic occipital nerve roots (Bass and Baker 1990,
Bass *et al.* 1994, 1996). A ventral medullary nucleus
provides for extensive coupling of the pacemaker–sonic
circuit across the midline (Bass *et al.* 1994, 1996).
The contraction rate of the sonic muscles is directly
determined by the rhythmic output of the pacemaker–
motor neuron circuit. This output is easily recorded in a
neurophysiological preparation and is known as a fictive
vocalization because its temporal properties directly
establish the temporal features of natural calls such as
the fundamental frequency and duration (Bass and Baker
1990). Hence, this preparation provides a simple model
for investigating the effects of hormones and other
neurochemicals on the neural substrates of vocal
behavior in a vertebrate.

6.2 NEURAL MECHANISMS OF ARTs: THE CHEMICAL LANGUAGE OF THE PREOPTIC AREA

Before launching into a survey of the diversity of the preoptic area (POA) phenotypes among teleosts, we will first
consider the general organization of the POA to provide a
more general context for understanding why this region of

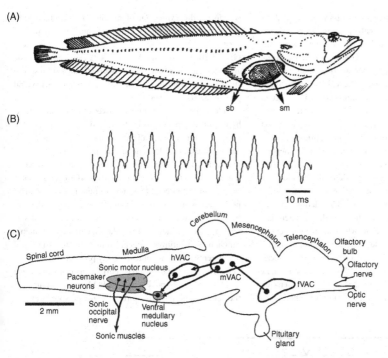

Figure 6.2 Overview of local behavior and motor system of midshipman fish. (A) Portrait showing position of swim bladder (sb) and sonic muscle (sm) at level of the pectoral fin. (B) Oscillogram record of segment of the "hum" advertisement call of a type I male. (C) Sagittal view of brain and spinal cord showing nuclei that form a central vocal–auditory network. See Box 6.1 for details.

the vertebrate brain plays an essential role in coordinating the divergent neural mechanisms that underlie the performance of any reproductive-related behavior. The term POA–anterior hypothalamus has often been used interchangeably with the term POA alone. For the purposes of this review, we consider the POA and anterior hypothalamus as a single functional unit, the POA, for two reasons. First, the POA and anterior hypothalamus share a common developmental origin (Puelles 2001). Second, while many of the neuropeptide-containing neurons in teleosts are located in brain nuclei identified as part of the POA (e.g., see Bass and Grober 2001), the homologous cell groups of tetrapods (e.g., the paraventricular and supraoptic nuclei) are typically identified as part of the anterior hypothalamus (e.g., Moore and Lowry 1998, Puelles 2001).

One context in which to frame the functional organization of the POA of teleosts and vertebrates in general is its central location within a neurochemically rich "core" of the brain as recognized by Nieuwenhuys et al. (1989). While Nieuwenhuys and colleagues discuss this concept within the context of a mammalian limbic system, we can apply it to nonmammals as well, especially given the conserved organization of the POA across vertebrate classes (see Butler and Hodos 1996, Meek and Nieuwenhuys 1998). Core regions, like the POA, lie adjacent to the brain's ventricular spaces and contain neuronal populations that synthesize a wide range of neuropeptides, concentrate androgens and estrogens, and are generally implicated in the control of homeostatic and social behavior patterns (Nieuwenhuys et al. 1989). A laterally positioned "paracore" region at brainstem levels is especially rich in monoamines (serotonin and catecholamines) and interconnected with the core region. Together, the core and paracore regions form a neuroendocrine "axis" in the brain.

Herbert (1993) articulates a similar organizational pattern for neuropeptide-containing cell groups and further points out an added degree of complexity afforded by interactions between different peptide systems and between peptides and steroids. Peptide interactions may involve either multiple peptides acting on a single target or one peptide system acting upon another in a somewhat hierarchical fashion. Moreover, individual brain nuclei may have multiple peptides that influence a wide range of peripheral and central structures and, in turn, the related behavior patterns. Finally, steroid hormones may affect all of these targets via one or more peptide systems. Herbert (1993) proposes that the different neuropeptide systems "function as chemical coding systems organizing patterns of adaptive responses to defined demands. ... The structure and diversity of peptides raises the possibility that there may be some predictable relation between individual composition and function ... that is, there is a chemical 'code' or 'language' in which defined functions are encoded into interpretable sequences in amino acids." One of the long-term goals of continuing neuroendocrinological studies of species with ARTs should be to show how different neuropeptides (and steroids) are operating either independently or in concert with one another to coordinate the expression of a suite of characters (both neural and nonneural) leading to the performance of ARTs (also see Goodson and Bass 2001, Perry and Grober 2002, Rose and Moore 2002). Such a pluralistic approach is essential to a neuroethological research strategy that aims to explain the existence of behavioral phenotypes (Bass 1998).

The POA exerts an influence over other organ systems by way of its connections to the somatic motor system, the visceral motor system, and the pituitary gland (Figure 6.3; also see Markakis 2002). The somatic motor system includes motor neurons in the brain and spinal cord that directly innervate skeletal muscle. By contrast, the central motor neurons of the visceral (autonomic) motor system contact peripheral motor neurons in autonomic ganglia that, in turn, innervate either glands or the smooth muscle of visceral organs. The adrenal medulla is a modified autonomic ganglion that utilizes catecholamines (epinephrine and norepinephrine) as its neurosecretory products. The POA's linkage to the pituitary gland is central to its neuroendocrine function. Multiple populations of POA neurons innervate the anterior and posterior pituitary (adenohypophysis and

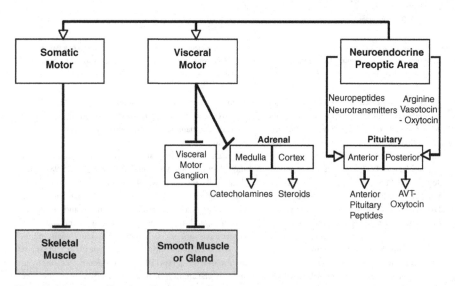

Figure 6.3 Schematic overview of somatic motor, visceral motor, and neuroendocrine systems. Steroids released from the adrenal cortex include glucocorticoids and mineralocorticoids. Catecholamines released from adrenal medulla include epinephrine and norepinephrine. See Bentley (1998) for more details of anterior pituitary peptides, the arginine vasotocin–oxytocin family of neuropeptides, and other neuropeptides and transmitters produced by neurons in the preoptic area.

neurohypophysis respectively). POA neurons synthesize peptides that influence the activity of anterior pituitary secretory cells. These secretory cells synthesize and release peptidergic hormones into the circulation that target organs throughout the body including the adrenal gland that releases corticosteroids (Bentley 1998). Most vertebrates have a blood portal system that transports the neurosecretory products of the POA to the pituitary; teleosts lack this portal system and instead have axons that directly terminate in the pituitary (Peter and Fryer 1983). POA neurons also synthesize the family of arginine vasotocin (AVT)-like peptides that are directly released into the posterior pituitary that, like the anterior pituitary, interfaces with the circulation.

6.3 DIVERGENT GONADOTROPIN-RELEASING HORMONE AND ARGININE VASOTOCIN PHENOTYPES

Studies of the neural mechanisms of ARTs have largely focused on the forebrain's POA, in part, because of its neuroendocrine functions (Section 6.2) and more general influence on a wide range of reproductively related behavior patterns (Nelson 1998, Pfaff *et al.* 2002). Several reviews of teleosts with ARTs show how the size and number of neuropeptide-containing neurons within the POA vary with developmental trajectories and reproductive tactics (Foran and Bass 1999, Bass and Grober 2001, Grober and Bass 2002; see Goodson and Bass 2001, Rhen and Crews 2002, Rose and Moore 2002 for more general reviews of vertebrates). Tables 6.1 and 6.2 summarize this information for gonadotropin-releasing hormone (GnRH)- and arginine vasotocin (AVT)-containing neurons. (Earlier versions of these tables appeared in Foran and Bass [1999] and Bass and Grober [2001] but have been updated here for studies published up to 2003.)

The POA of teleosts includes several subdivisions. We, as many others, follow the nomenclature of Braford and Northcutt (1983) and recognize a POA with an anterior parvocellular nucleus, a posterior parvocellular nucleus, and a magnocellular nucleus that is further divided into small (parvocellular), medium (magnocellular), and large (gigantocellular) cell regions. A retinal-recipient, suprachiasmatic nucleus is identified at the level of posterior parvocellular nucleus. This pattern of POA organization is highly conserved across teleosts (see references in Bass and Grober

2001). The POA transitions into the anterior hypothalamus, which shows extensive interspecific variation in its organization across teleosts; see Braford and Northcutt (1983) for a comparative discussion.

Of particular relevance here are neurons that synthesize neurochemicals that are members of the nine-amino-acid family of arginine vasopressin (AVP) like neuropeptides and the ten-amino-acid family of gonadotropin-releasing hormones (GnRH). As in mammals, there are a large number of other peptides synthesized in the POA (review: Meek and Nieuwenhuys 1998). Arginine vasotocin (AVT) and isotocin are the teleost homologs of, respectively, mammalian AVP and oxytocin; they are mainly found in the magnocellular nucleus. AVT is considered the ancestral peptide; hence, our reference to the AVT-like family. Among teleosts, neurons containing GnRH (homolog of mammalian luteinizing-hormone-releasing hormone) are mainly located within the anterior parvocellular nucleus. AVT, isotocin, and GnRH neurons have a similar distribution across diverse teleost groups, although the pattern of axonal trajectories and terminal fields may vary (see Goodson and Bass 2000a, Goodson *et al.* 2003 for AVT and Lethimonier *et al.* 2004 for GnRH).

6.3.1 Gonadotropin-releasing hormone

Gonadotropin-releasing hormone (GnRH)-containing neurons release their contents into the anterior pituitary where they regulate the release of gonadotropins (luteinizing and follicle-stimulating hormones, gonadotropic hormones I and II in teleosts) that, in turn, influence gonadal size and steroidogenesis during either sexual maturation or adulthood. Given the POA's direct input to the anterior pituitary in teleosts, changes in GnRH–ir (see below) neuron activity may be more rapidly reflected in blood gonadotropin levels than in other vertebrates with a hypophyseal portal system.

Teleosts have two major populations of GnRH neurons in the forebrain. One population is within the ganglion of the terminal nerve (TN) that is positioned either within the olfactory bulb and nerve or at the junction of the olfactory bulb and telencephalon. A second GnRH population is within the POA. Studies in the dwarf gourami show that only the POA cells project to the pituitary (Oka and Ichikawa 1990), whereas TN neurons have widespread projections throughout the forebrain and do not provide input to the pituitary (Oka and Matsushima 1993). Individual

Table 6.1. *Sexual dimorphisms of POA-GnRH neurons among teleost fish with ARTs*

Species	Male morph(s)	Male life history	Male GSI[a]	Cell size	Cell number	mRNA	Reference
Plainfin midshipman, *Porichthys notatus*	Type I (territorial/courting) and type II (nonterritorial/noncourting) male morphs	Permanent, early diverging developmental trajectories	TII > TI	TI > TII and female	TI = TII = female	Not available	Grober et al. 1994
Swordtails, *Xiphophorus maculatus*	"Small" (S) and "large" (L) male morphs	Permanent, early diverging developmental trajectories	S > L	S = L	S > L	Not available	Halpern-Sebold et al. 1986
Bluehead wrasse, *Thalassoma bifasciatum*	Terminal phase (TP) and initial phase (IP) males	Single, permanent, sex/role change for IP male and females into TP male	IP > TP	IP = TP = female	TP > IP and female	Not available	Grober and Bass 1991
Anemonefish *Amphiprion melanopus*	One reproductive male and several nonreproductives (NR)	Permanent, one-time, adult male-to-female sex change	R > NR	Female > R and NR male[b]	R > NR and female	Not available	Elofsson et al. 1997
Ballan wrasse, *Labrus berggylta*	Male defends harem of females	Permanent, one-time, adult female-to-male sex change		Males postspawning > males prespawning and females	Male > female[c]	Not available	Elofsson et al. 1999

[a] Gonosomatic index (gonad weight/body weight).

[b] Explained by differences in body size.

[c] Explained by differences in body size among males only; thus, no difference in cell number between males and females of same body size.

Table 6.2. *Sexual dimorphisms of POA–AVT neurons among teleost fish with ARTs*

Species	Male morph(s)	Male life history	Male GSI[a]	Cell size	Cell number	mRNA density	Reference
Plainfin midshipman, *Porichthys notatus*	Type I (territorial/courting/nest-guarding) and type II (non-territorial/noncourting) male morphs	Permanent, early diverging trajectories	TII > TI	TI and female > TII[b]	TII = TI = female	Not available	Foran and Bass 1998
Saddleback wrasse, *Thalassoma duperrey*	Terminal phase (TP) and initial phase (IP) males	Single, permanent, change; IP and females to TP	IP > TP	TP > IP and female	TP > IP and female	TP > IP, female[c]	Grober 1998
Bluehead wrasse, *Thalassoma bifasciatum*	Terminal phase (TP) and initial phase (IP) males	Single, permanent, change; IP and females to TP	IP > TP	Not available	TP > female[d]	TP > IP> female[e]	Godwin *et al.* 2000
Marine goby, *Trimma okinawae*	Territorial males	Reversible sex change	—	Female > male	Not available	Not available	Grober and Sunobe 1996
Bluebanded goby, *Lythrypnus dalli*	Nesting males	Permanent, one-time, female-to-male change	—	Male > female	Not available	Not available	Reavis and Grober 1999
Peacock blenny, *Salaria pavo*	Females and sneak males (SM) court nest-holding males (NM)	SM transforms into NM	SM > NM	SM = NM; Female > SM, NM	SM = NM; Female < SM, NM	SM and female > NM[f]	Grober *et al.* 2002
Rock-pool blenny, *Parablennius sanguinolentus parvicornis*	Territorial, nesting males (NM) and territorial, sneak males (SM)	SM transforms into NM	SM > NM	SM = NM = female	SM = NM = female	Not available	Miranda *et al.* 2003

[a] Gonosomatic index (gonad weight/body weight).
[b] Explained by differences in body size.
[c] Number and size of mRNA cells.
[d] Explained by differences in body size.
[e] mRNA density (expression levels/cell measured as number of grains/cell averaged across all cells).
[f] mRNA density (expression levels/cell measured as number of grains/cell averaged across all cells).

GnRH–TN neurons also show rhythmic firing properties, which led Oka and Matsushima (1993) to propose that GnRH–TN neurons might have widespread functions as a neuromodulator. As discussed below, neuroanatomical studies have used either immunocytochemical methods to detect the presence of the peptide or in situ hybridization histochemistry for identifying neuropeptide mRNA transcripts. When discussing immunocytochemically detected, neuropeptide-containing (i.e., immunoreactive-like, ir) neurons, it is important to keep in mind that increases in either cell size or number may reflect either increased synthesis or decreased release of the peptides, while decreases in the magnitude of those parameters may reflect either decreased synthesis or increased release.

To our knowledge, the first studies of POA organization in species with ARTs were on platyfish by Schreibman and colleagues who used this species not to study ARTs per se but rather as a model to establish the temporal relationship between the onset of sexual maturation and changes in the morphology of pituitary gonadotropes and GnRH neurons (review: Schreibman and Magliulo-Cepriano 2002). Platyfish have "large" and "small" males that are analogous, respectively, to the type I and II males shown in Figure 6.1A. This is also the one group of teleosts with ARTs for which there is strong evidence that the morphs are genetically determined. Immunocytochemical studies showed a correlation between the onset of sexual maturation and changing GnRH–POA phenotype (Halpern-Sebold et al. 1986). Thus, the small, earlier-maturing males had more GnRH neurons than the large males. Consistent with these results, studies across a wide range of species have since shown that, in general, GnRH dimorphisms are associated with differences in relative gonad size and reproductive tactic (Table 6.1). Thus, the male morph with larger gonad mass/body mass ratio (GSI) generally has either larger or more GnRH–POA neurons. This same morph is also typically the courting, territorial, and/or aggressive morph.

It is not possible in the space available to review many of the studies on GnRH phenotypes summarized in Table 6.1 (but see Foran and Bass 1999 and Bass and Grober 2001 in the context of ARTs, and Okuzawa and Kobayashi 1999 for studies in salmon in the context of spawning migrations). Also of interest to the general study of plasticity in POA phenotypes have been studies of GnRH neurons in the cichlid fish *Astatotilapia (Haplochromis) burtoni*, where males can reversibly transform from a reproductive to a nonreproductive condition (Box 6.2).

Box 6.2   GnRH neuronal plasticity in cichlids

Several investigations have explored the relationship between GnRH–ir and mRNA expression in *Astatotilapia burtoni* and an individual's social status as either a nonterritorial/nonreproductive (NT) male or a territorial/courting (T) male. Davis and Fernald (1990) first showed an increase in the size (but not number) of GnRH–ir neurons in the POA that was paralleled by increasing gonad size as males transitioned from NT to T status (also see Hofmann and Fernald 2000 for similar changes in somatostatin-containing POA neurons). Subsequent studies identified three different forms of GnRH in *A. burtoni*: GnRH1, GnRH2, and GnRH3 in, respectively, the POA, the midbrain, and the TN (review: Fernald and White 1999); eight forms have been identified among teleosts (see Lethimonier et al. 2004). Only GnRH1–ir and GnRH1 mRNA expression varies with NT/T status. White et al. (2002) investigated the relationship between social status and relative gonad size, levels of GnRH1 mRNA, and size of GnRH–ir neurons in the POA. Levels of GnRH mRNA expression, GnRH–ir neuron size, and gonad size were positively correlated with status; all parameters were greater in magnitude among T males. When NT males were placed in a social situation that allowed them to adopt a T status, there was an increase in GnRH1 mRNA levels and GnRH–ir neuron size, whereas males that were induced to transform from T to NT status showed the opposite trends. Behavioral changes (measured as levels of aggression) were observed after 1 day among NT males that were on a "social ascent" to being T males; their behavior resembled that of T males after 2 weeks had elapsed, although their GnRH traits resembled those of T males after just 1 week. For T-to-NT males that were on a "social decline," T males behaved like NT males after just 1 day, although their GnRH traits did not resemble those of NT males until after 3 weeks. White et al. (2002) suggest that unstable social conditions might explain the temporal disparities between the rate of change of GnRH traits and behavior.

6.3.2   Arginine vasotocin

The arginine vasotocin (AVT)-like family of neuropeptides includes 12 different peptides among vertebrates and two among invertebrates (Bentley 1998). Recall that AVT and isotocin are the teleost homologs, respectively, of mammalian AVP and oxytocin. The evolution of the AVT-like peptides with only four variants that differ by one or two amino acids from AVT is more conserved than that of the oxytocin-like peptides with eight variants that differ by one to three amino acids from oxytocin. Among mammals, AVP and oxytocin modulate a wide variety of social (e.g., parental care, courtship, aggression) and nonsocial (e.g., hibernation) behavior patterns (see Goodson and Bass 2001). The behavioral functions of AVP are often associated with males and those of oxytocin with females (e.g., see Insel and Young 2000); comparable dichotomies are becoming apparent among nonmammals (reviews: Goodson and Bass 2001, Rose and Moore 2002). Across species, AVT/AVP's facilitatory influence on courtship behavior is fairly consistent. However, AVT's influence on aggression is more dependent on the social system in question, namely either a territorial or nonterritorial species; in general, AVT is inhibitory in the former and facilitatory in the latter (see Goodson and Bass 2001 for extended discussion). There are few behavioral or neuroendocrinological studies of isotocin (but see below).

A complete understanding of the functional significance of divergent patterns of neuropeptide expression will depend, in part, on explanations at a neurophysiological level of analysis. By way of example, we review studies of male morph-specific effects of AVT and isotocin on fictive calling in midshipman fish (see Rose and Moore 2002 for comparable studies of the neural substrates of mating behavior in salamanders). Midshipman fish have two male morphs, types I and II (Figure 6.1A), which follow divergent growth trajectories (Bass et al. 1996) and reproductive tactics (Brantley and Bass 1994, Bass 1996) (Figure 6.4). Territorial type I males build nests under rocky shelters in the intertidal zone along the northwestern coast of the United States and Canada and then court females with a long-duration (more than 1 hour) advertisement call known as a "hum." Type I males also produce a long-duration, repetitive series of brief (millisecond) "grunts" during nest defense (Brantley and Bass 1994, Bass et al. 1999). Type II males neither build nests nor acoustically court females but rather attempt to steal fertilizations from type I males by either sneaking into their nest or by satellite spawning from a nest's periphery. Recent studies also show, however, that small, type I males may also show behavioral plasticity and sneak-spawn (Lee and Bass 2004). Thus, type I male midshipman fish show a combination of the ART patterns illustrated in Figures 6.1A and 6.1B, which highlights once

again the wide range of phenotypic plasticity among reproductive morphs across teleosts. Type II males, as females, infrequently produce low-amplitude grunts that have so far been documented only in a nonspawning context (Brantley and Bass 1994).

Neuroanatomical and neurophysiological studies in midshipman fish and the closely related toadfishes have delineated a vocal control network that leads to sound production (Bass and McKibben 2003) (Box 6.1). There are intrasexual dimorphisms in many vocal traits that parallel the divergence in vocal behavior patterns between type I and II males (Bass 1996, Bass and McKibben 2003). This includes differences in the size of AVT–ir neurons in the POA (Foran and Bass 1998) (Table 6.2). The descending vocal motor system interfaces with central AVT and oxytocin-like pathways at multiple levels of the central nervous system (Goodson and Bass 2000a, 2002, Goodson et al. 2003). Goodson and Bass (2000b) showed male, morph-specific patterns of vocal motor activity with microinjections of either AVT or isotocin into vocally active sites of the anterior hypothalamus (part of the fVAC depicted in Figure 6.2). Of particular advantage to studies in midshipman fish (and the closely related toadfishes) is the ability to record "fictive" vocalizations from ventral occipital nerve roots that represent the rhythmic activity of a vocal pacemaker circuit in the caudal hindbrain and rostral spinal cord (Bass and Baker 1990) (Box 6.1). Fictive calls predict the most salient temporal features of natural calls, namely fundamental frequency and duration. Hence, this preparation provided the opportunity to assess how neuropeptides modulate the output of a central pattern generator that is directly translated into a naturally occurring social behavior, i.e., vocalizations. AVT and isotocin influenced both fictive call initiation and duration; there was no influence on fundamental frequency (although there are inter- and intrasexual dimorphisms in this parameter: Bass and Baker 1990). AVT inhibits, and the appropriate antagonists facilitate, fictive calling in type I males, whereas isotocin has no effects. By contrast, only isotocin and its appropriate antagonists have significant and parallel effects on vocal activity in both type II males and females. The midshipman studies show that (1) there are both inter- and intrasexual divergences in the efficacy of AVT-like peptides in modulating the neural substrates of a behavior (also see Bastian et al. 2001 for another demonstration of male–female differences in a weakly electric fish), (2) forebrain neuropeptides can modulate vocal motor patterning (as in other vertebrate groups: see Goodson and Bass 2001), (3)

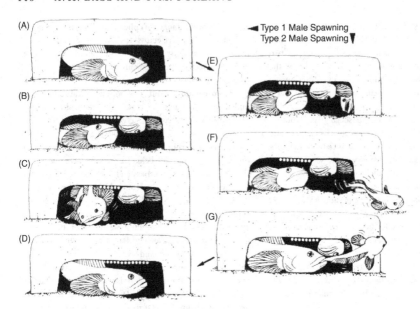

Figure 6.4 Alternative reproductive tactics in the plainfin midshipman fish, *Porichthys notatus*. Plainfin midshipman fish readily reproduce during the breeding season when moved from their nests in the intertidal zone to aquaria with flow-through seawater (Brantley and Bass 1994, Lee and Bass 2004). Type I males will take up residence under an artificial rocky shelter – for example, a portion of a cement block as shown in this schematic overview (A) that summarizes the studies of Brantley and Bass (1994; see Bass 1996 for photographs of nests in the intertidal zone). Type I males acoustically court females with a hum advertisement call (see Box 6.1) after nightfall. After a female enters the nest and remains to spawn, the male will cease to hum. Females deposit their eggs on the surface of the nest's interior (B).

Eggs have an adhesive disk that attaches them to the surface. The male rolls and quivers as he releases sperm near each egg as they are deposited one at a time on the nest's surface by the female (C). After a female releases all of her eggs, she will leave the nest and the type I male remains to guard the eggs (D). The type I male will then court other females on subsequent nights. When present, type II males will either enter a nest and sneak spawn (far right, E) or remain along the periphery of the nest and attempt to satellite spawn by fanning their sperm into the nest's interior (far right, F). Territorial type I male attacks satellite spawning type II males (G). Under some conditions, small nonterritorial type I males will sneak-spawn (Lee and Bass 2004). (Adapted from Brantley and Bass 1994.)

AVT's action as an inhibitory substance in the territorial male morph is consistent with studies in birds showing a similar neuropeptide-behavioral phenotype (see Goodson and Bass 2001), and (4) males with a female-like behavioral trait (in this case, a vocalization) converge with females in the neurochemical mechanism that leads to modulation of that behavior's central pattern generator. Together, the results emphasized once again that the uncoupling of gonadal and behavioral sex from neural mechanisms leads to an evolutionarily adaptable patterning of these traits (Bass 1992, 1996).

Recent additions to the comparative literature on patterns of AVT–ir and AVT mRNA expression among species with ARTs include studies of the lagoon-dwelling peacock blenny, *Salaria pavo*, and the Azorean rock-pool blenny, *Parablennius sanguinolentus parvicornis* (Table 6.2; also see Chapter 7). *Salaria pavo* females show behavioral role reversal in that they are the reproductive morph that courts; smaller and younger nonnesting males sneak-spawn by mimicking female courtship behavior to gain access to the nest of larger males. Sneaker males transform into nesting males (analogous to the transformation of initial-phase males into terminal-phase males in wrasses (see Oliveira et al. 2001) (Figure 6.1C). AVT–ir cell number is smaller in females compared to either male morph (which are equal: Grober et al. 2002). By contrast, AVT–ir cell size is larger in females than either male morph. Variation in either cell size or number cannot be explained by the divergence in body size among the reproductive morphs. AVT mRNA density (grain counts per neuron) is greater in the POA of either females or sneaker males compared to nest-holding males. Thus, while the pattern of AVT–ir traits is sex specific, the pattern of AVT mRNA expression is consistent with similar courtship tactics by females and sneaker males. The same AVT–ir pattern is not observed in *P. s. parvicornis* that also has nesting and nonnesting/sneaker male morphs (Miranda *et al.* 2003). Although sneaker males in both species transform into nesting males, there are important species differences. Unlike *S. pavo*, territorial/nest-holding *P. s. parvicornis* males court females and *P. s. parvicornis* sneaker/satellite males help to defend territories (although sneaker males also transform into nesting males in this species: see Oliveira *et al.* 2001). There are no significant differences in either AVT–ir cell size or number in the POA among all three reproductive morphs. However, significant differences are found for the ratio of either cell size or number to body mass (as in midshipman fish: Foran and Bass 1998). Thus, the smaller, nonnesting males (like type II midshipman) have a larger ratio of AVT–ir cell number/body mass than either nesting males (like type I midshipman) or females, whereas nonnesting males and females have a larger ratio of AVT–ir cell size/body mass than nesting males (AVT mRNA density was not reported for this blenniid). As with midshipman (Foran and Bass 1998), which they generally resemble in the pattern of male morph tactics, the results in the blenny suggest that AVT–ir cell number develops prior to the onset of sexual maturation and the differences in the cell size or number/body mass ratios may indicate a much higher concentration of AVT per gram body mass.

Black *et al.* (2004) showed changes in the number of putative isotocin-containing neurons in the POA during the process of sex reversal in the bluebanded goby, *Lythrypnus dalli* (we say putative because these authors used an antibody that recognizes the closely related oxytocin peptide: see Goodson *et al.* [2003] for comparable methodology). This species exhibits one-time, permanent adult female-to-male sex change; males have fewer isotocin–ir neurons than females (there were no significant differences in cell size). A previous study for this species showed that males and females have a similar number of AVT–ir neurons in the POA, although the neurons are larger in males (see Table 6.2).

Several studies of the bluehead wrasse, *Thalassoma bifasciatum*, have investigated the relationship between patterns of neuropeptide expression and social status. The

bluehead wrasse has been a focus of study since Grober and Bass (1991) first reported inter- and intrasexual differences in its GnRH–POA phenotype (Table 6.1). Since that time, several reports have also investigated AVT–POA phenotypes. Very briefly, the bluehead wrasse has IP and TP males (Figure 6.1C). TP males are highly territorial and aggressively compete for sole access to females. Some TP males are nonterritorial floaters (Semsar *et al.* 2001). IP males either group spawn or sneak-spawn with a territorial TP male and female. Either adult females or IP males can be induced to transform into TP males by removing territorial TP males from a reef. If all IP males and TP males are removed, the largest females transform into TP males and adopt TP male-like behavior. AVT promotes courtship behavior in either TP or nonterritorial TP males but only increases aggression in the nonterritorial TP males (Semsar *et al.* 2001). This is consistent with the general pattern of AVT's involvement in promoting courtship behavior, whereas its effects on aggression vary with territorial status (Goodson and Bass 2001). The increased aggression among AVT-treated, nonterritorial TP males is consistent with the overproduction of aggressive behavior that might be critical to their becoming territorial.

The first study of POA–AVT mRNA levels in wrasses showed that TP males, IP males, and sex-reversed females had significantly higher levels than females and that levels were four times greater in sex-changing females than other females after just 2–3 days following removal of TP males from a reef (Godwin *et al.* 2000) (see Table 6.2 for similar results in another wrasse, *T. duperrey*). Recently, Semsar and Godwin (2002) tested the effects of social, gonadal, and hormonal status on the AVT–POA phenotype of *T. bifasciatum* (also see Godwin *et al.* 2000). They first wanted to know if the size of AVT–ir neurons and AVT mRNA content would change in sex-changing females that were socially dominant compared to subordinate females, regardless of their gonadal status (i.e., either intact or ovariectomized). Transformation to a TP male phenotype was correlated with significant increases in both AVT mRNA signal and the size of AVT–ir somata (only in the PMg, the gigantocellular portion of the magnocellular nucleus of the preoptic area); only the changes in neuron size were gonadally dependent. Consistent with this, castration of TP males had no effect on their AVT mRNA phenotype although AVT–ir somata in the PMg were larger, again suggesting a gonadal effect on AVT peptide expression. Together, these studies show how social environment may influence AVT phenotype in sex-changing fish. At the same time, however, these studies show a mismatch between AVT mRNA and AVT–ir patterns that is somewhat perplexing but presumably related to steroid secretion by the gonad (also see earlier described study of the peacock blenny).

Perry and Grober (2002) suggest for bluehead wrasse that glucocorticoids regulate changes in the brain and gonad linked to the upregulation of AVT. At least in trout, there are glucocorticoid receptors throughout the neuroendocrine regions of the brain, including both the parvocellular and magnocellular nuclei of the POA (Teitsma et al. 1997, 1998). These glucocorticoid receptors are colocalized with GnRH neurons in the caudal telencephalon/anterior POA (Teitsma et al. 1999). Evidence in mammals shows that glucocorticoids modulate AVP mRNA and its receptor in the hypothalamus and forebrain (see Goodson and Bass 2001). Thus, glucocorticoids may be promising candidates that would translate social and other environmental cues to changes in neuropeptide expression involved in proximate mechanisms of behavior in alternative male phenotypes.

While there is not enough space here to discuss the many other elegant studies of AVT expression in teleost fish, the reader is urged to consider the work of Urano and colleagues on neuronal AVT and isotocin mRNA expression and immunoreactivity in chum salmon (*Oncorhynchus keta*) across different life-history stages (review: Urano et al. 1994). Although these studies mainly define the relationship between AVT and isotocin expression and the osmotic challenges linked to the migration from freshwater to saltwater environments, several studies reveal expression patterns linked to reproductive status (e.g., Ota et al. 1996, 1999, Hiraoka et al. 1997). Two other recent neurophysiological studies provide new insights into the neurosecretory function of the teleost POA. Saito and Urano (2001) showed separately synchronized patterns of electrical activity between the AVT and isotocin neurons in an in vitro preparation of the POA of rainbow trout, while Saito et al. (2003) have shown that GnRH can affect the oscillatory activity of AVT neurons. This work also begins to address the interaction between neuropeptide systems that we discussed earlier.

A number of studies in anuran amphibians have identified intersexual dimorphisms in brain AVT phenotypes (review: Boyd 1994). Of particular relevance here is the report of Marler et al. (1999) on the relationship between forebrain AVT–ir and ARTs in the cricket frog (*Acris crepitans*). Cricket frogs have calling males that court females and noncalling, satellite males that try to intercept females moving toward calling males. Intraperitoneal AVT injections increased calling among males engaged in agonistic encounters (as in other anurans: see Marler et al. 1999). AVT's facilitation of aggressive calling is consistent with such a role in nonterritorial species (see earlier comments). Calling males also had smaller AVT–ir neurons in the ventral forebrain's nucleus accumbens and less dense AVT–ir (i.e., labeled neuronal processes) in the region adjacent to nucleus accumbens. The role of nucleus accumbens in either a vocalization or reproductive context is apparently not known.

6.4 NEUROSTEROIDS AND AROMATASE

To our knowledge, there are no studies that address the organizational mechanisms responsible for fixed, alternative male phenotypes in fishes. Although studies from salmon, bluegill sunfish, and platyfish suggest a genetic role (Gross 1996), this still does not address the underlying mechanisms. Although teleosts with fixed alternative phenotypes have diandric males that can be distinguished by multiple traits including GnRH and AVT brain phenotypes (see Tables 6.1 and 6.2), evidence of how dimorphic neural circuitry can lead to dimorphic behavior remains undefined for most species. One exception has been the vocal motor circuit of midshipman fish. In midshipman, the sonic motor nucleus (SMN) that innervates sonic swimbladder muscles is inter- and intrasexually dimorphic. Thus, individual motor neurons comprising the nucleus and total SMN volume itself is larger in type I males compared to type II males and females (Bass and Baker 1990, Bass et al. 1996). Sonic motor neuron size is also an androgen-sensitive trait (Bass 1995; also see Brantley et al. 1993a). In all vertebrates, sex steroids organize neural substrates important in sex-specific reproductive behavior (review: De Vries and Simerly 2002). In this regard, midshipman fish provide an ideal model to examine the influence of neurosteroids as proximate mechanisms that influence the development and maintenance of dimorphic male brain structures that directly control divergent reproductive tactics. Neurosteroids "include both neuroactive compounds produced de novo and steroids metabolized to neuroactive compounds in the brain but derived from circulating precursors" (Compagnone and Mellon 2000). Here, we focus on the conversion of testosterone to estradiol by aromatase.

Activity levels of brain aromatase appear to be conserved throughout vertebrates; highest levels are consistently localized in forebrain areas known to control sexual behavior and reproduction (review: Balthazart and Ball 1998). Aromatase affects the development of sexually dimorphic brain nuclei (reviews: Beyer 1999, Burke *et al.* 1999). To date, studies in teleosts have localized aromatase using specific antibodies and mRNA probes in midshipman (Forlano *et al.* 2001), trout (Menuet *et al.* 2003), zebrafish (Goto-Kazeto *et al.* 2004, Menuet *et al.* 2005), and silversides (protein only: Strobl-Mazzulla *et al.* 2005). As expected, these studies identified aromatase in the POA and throughout the hypothalamus, but unexpectedly, as first shown in midshipman fish, aromatase–ir was localized to radial glial cells along ventricular zones throughout the brain. In midshipman, the SMN is enshrouded with aromatase–ir cells and fibers, contains high levels of aromatase mRNA, and probably accounts for most of the aromatase activity found in the hindbrain and rostral spinal cord (Schlinger *et al.* 1999, Forlano *et al.* 2001; also see Pasmanik and Callard 1985).

Both type I and type II males have aromatase expression in the vocal regions of the brain, although activity levels are significantly higher in type II males (Schlinger *et al.* 1999), and mRNA expression is significantly higher in the SMN (but not POA) in type II males (see also Forlano and Bass 2005a) (Figure 6.5). Thus, aromatase likely has divergent functions in the vocal hindbrain of adult male midshipman. Estradiol has rapid, modulatory effects on the vocal output of type I males (Remage-Healey and Bass 2004), and, therefore, local estradiol production may function to modulate vocal signaling in type I males. Among type II males, aromatase may also largely bind or convert testosterone to estradiol to prevent circulating androgens from reaching androgen-sensitive circuitry (Schlinger *et al.* 1999). Forlano *et al.* (2005) demonstrated estrogen receptor alpha mRNA in the sonic motor nucleus of type I males. The absence of membrane-bound or nuclear estrogen receptor in the SMN of type IIs would support the differential function of neurosteroids between male morphs.

While type I male midshipman alone have detectable levels of 11-ketotestosterone, type II males and females have similarly higher testosterone levels than type I males (Brantley *et al.* 1993b, Knapp *et al.* 1999, Sisneros *et al.* 2004). Our results suggest that, like some other vertebrates (e.g., see Balthazart and Ball 1998, Gelinas *et al.* 1998), testosterone can both upregulate aromatase expression (Forlano and Bass 2005b) and masculinize the sonic motor system (Bass 1995). We hypothesized that relative levels of aromatase expression in and around the SMN may function to prevent its transformation by circulating testosterone to a

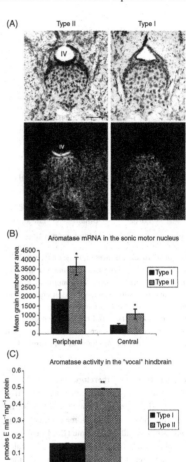

Figure 6.5 Intrasexual differences in brain aromatase expression in the two male midshipman fish phenotypes. (A) Type I and type II males of similar lengths show differences in aromatase mRNA expression at the level of the dimorphic sonic motor nucleus (SMN). Brightfield (top) and darkfield (bottom) visualizations of in situ hybridization show strongest signal at the dorsal periphery of the nucleus which contacts the fourth ventricle (IV). Scale bar = 200 µm for all micrographs. (B) Quantification of mRNA silver grains shows significantly higher levels of expression in both peripheral ($P = 0.029$) and central regions ($P = 0.020$) of the nucleus in type II males ($n = 5$) compared to type I males ($n = 7$) (see Forlano and Bass 2005a for methods). (C) Compared to type I males ($n = 5$), type II males ($n = 5$) have significantly higher levels of aromatase activity in hindbrain–spinal regions that contain the dimorphic vocal circuitry ($P < 0.0001$). (After Schlinger *et al.* 1999.)

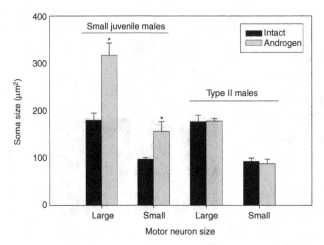

Figure 6.6 Effect of androgen treatment on sonic motor neuron size. Both large and small motor neurons within the sonic motor nucleus show a significant increase in size after implantation with androgens (testosterone proprionate) for 8–9 weeks in small juvenile males ($n = 5$ and 3 respectively for intact and androgen-treated animals, $P = 0.004$ and 0.36 for large and small motor neurons, respectively); however, the same treatment has no effect on type II males ($n = 6$ and 5 respectively for intact and androgen-treated animals, $P = 0.965$ and 0.698) (A. Bass, B. Horvath, and M. Marchaterre, unpublished observations). Changes in juvenile males parallel an increase in sonic muscle fiber number and diameter (Brantley et al. 1993a; also see for method of hormone treatment); see Bass et al. (1996) for age classification and quantification of motor neuron size. Other studies show that 11-ketotestosterone also does not induce a transformation of the type II male vocal motor phenotype (Lee and Bass 2005).

type I male phenotype and therefore may be a key mechanism in both generating and maintaining alternative male phenotypes in this species (Schlinger et al.1999). In support of this, the SMN of type II male midshipman treated with testosterone will not become type I male-like, although the same treatment given to small juvenile males that have not yet adopted a type I male growth trajectory (see Bass et al. 1996) will lead to a type I male-like phenotype (Figure 6.6). Also, type II males castrated and implanted with testosterone will show an upregulation of aromatase mRNA in and around the SMN as well as in other brain areas (Forlano and Bass 2001) (Figure 6.7). This positive feedback of testosterone on brain aromatase may function as a buffering system to regulate the amount of circulating steroid reaching specific brain nuclei. The localization of aromatase in radial glial cells lining the ventricle throughout the brain (Forlano et al. 2001) allows for direct exchange of neurosteroids between the brain, cerebrospinal fluid, and circulatory system and may account for a source of circulating estrogen in both type I (Sisneros et al. 2004) and type II (J. Sisneros, P. Forlano,

R. Knapp, and A. Bass, unpublished data) males, thus altering the overall hormonal milieu of the animal.

One hypothesis for a mechanism that may influence the ontogeny of alternative male phenotypes in midshipman fish stems from studies that demonstrate differential expression of steroidogenic enzymes around the time of sexual differentiation. Aromatase activity and gene expression appear to be specific to female gonadal tissue, while the enzymes needed to make 11-oxygenated androgens are found only in male gonadal tissue, as demonstrated in studies using genetic female and male rainbow trout (Baroiller et al. 1999). Thus, differences between a type II male and a female at early stages in development may simply be due to the absence of gonadal production of estradiol in type II males. However, while a type II male testis may produce testosterone, it may have little or no 11β-hydroxylase (11β-H) or 11β-hydroxysteroid dehydrogenase (11β-HSD) that would be needed to make 11-ketotestosterone, the more potent teleost androgen (see Brantley et al. 1993b, Knapp 2004). Thus, sex differentiation in midshipman may be the result of gonadal aromatase expression. The divergence and

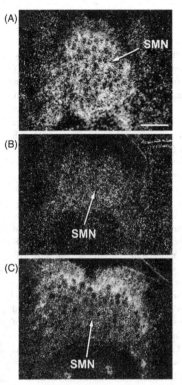

Figure 6.7 Effect of androgen treatment on aromatase expression in the sonic motor nucleus (SMN) in type II males (P. Forlano and A. Bass, unpublished observations). (A) Intact, type II male shows abundant mRNA expression in the SMN (in situ hybridization methods after Forlano and Bass 2005a, b). (B) Castration results in a large reduction in aromatase mRNA expression. (C) Castration with testosterone implant induces a dramatic upregulation of aromatase mRNA in the SMN, especially around the periphery (castration and hormone treatment methods after Brantley et al. 1993a). Notice that the hybridization signal clearly surrounds motor neuron somata. For visualization of aromatase-ir glial cells in this pattern, see Forlano et al. (2001). Scale bar = 150 μm.

differentiation of male phenotypes may then be the result of differential expression of aromatase or the androgenic enzymes 11β-H or 11β-HSD in the brain. During ontogeny, the type I morph may be the "default" developmental pathway if aromatase levels are low or absent in the hindbrain–spinal vocal motor regions. One method to test this hypothesis is to inhibit aromatase activity during a critical developmental window before developmental trajectories are adopted. If aromatase is inhibited during an androgen-sensitive window, all type I males should result. Aromatase may, in fact, ultimately function to organize gonadal and neural substrates to determine a fixed developmental pathway and maintain a certain male phenotype in the midshipman fish as well as in other vertebrates that show sexual polymorphisms in brain and behavior.

Differences in aromatase levels at a critical period may modify the hormonal milieu (e.g., the ratio of testosterone to estradiol) which, in turn, may determine male phenotype. At the same time, levels of brain aromatase gene expression may be either inherited or induced by environmental (including social) factors (see Schlinger et al. 2001). Brain aromatase levels in tilapia (Oreochromis niloticus) were approximately twofold higher in genetic females compared to males during sexual differentiation, and temperature-induced masculinization of females induced a threefold decrease in aromatase activity in the brain along with a decrease in the gonad. Genetic males reared at the same temperature that masculinized females also showed a decrease in brain aromatase activity (D'cotta et al. 2001; also see Tsai et al. 2003). Now that it is established that aromatase gene expression is thermosensitive in at least some fishes, perhaps its lability may also be affected by other environmental factors such as social interactions.

Sequential hermaphrodites by definition change sex during adulthood and therefore do not appear to have a true organizational period during early development as seen in gonochoristic fishes and other vertebrates. Therefore, the classical concepts of hormonal organization and activation do not necessarily apply to this group (see Crews 1993). Several studies suggest that either an increase in 11-keto-testosterone or a decrease in estradiol or a combination of both may induce sex change in protogynous fishes – species with female-to-male transformations (Cardwell and Liley 1991, Grober et al. 1991, Kroon and Liley 2000, Bhandari et al. 2004). In support of this, several studies have shown a significant decrease in gonadal aromatase mRNA during protogynous sex change in Thalassoma duperrey and Epinephelus coioides (Morrey et al. 1998, Zhang et al. 2004). Thus, a downregulation of the aromatase gene seems necessary to enable male differentiation. Conversely, elevated aromatase activity levels in gonads, elevated plasma estradiol levels, and decreased plasma 11-ketotestosterone levels were associated with protandry (male-to-female sex change) in the black porgy, Acanthopagrus schlegeli (Chang and Lin 1998). In another protandrous fish, Amphiprion *melanopus*, the estradiol/11-ketotestosterone ratio also showed a clear increase during sex change (Godwin and Thomas 1993). The importance of brain aromatase during sex change was first elucidated by experiments with *A. schlegeli*, which exist as functional males during the first 2 years and then change to female in the third year. Lee *et al.* (2001) supplemented the diet of 2-year-old males for 9 months with aromatase inhibitors. Compared to controls, treatment with the inhibitor significantly downregulated aromatase activity in all brain areas (fore-, mid-, and hindbrain) and pituitary but not in the gonad, and all treated fish remained as functional males. Treated males also showed increased levels of plasma luteinizing hormone and 11-ketotestosterone and an induction of spermiation (also see Lee *et al.* 2002). Thus, inhibition of brain aromatase blocked the natural sex change in this species. Although other studies have induced sex change in protogynous and bidirectional sex-changing fishes using aromatase inhibitors (Kroon and Liley 2000, Bhandari *et al.* 2004, Kroon *et al.* 2005), changes in the brain were not investigated. Since adult sex change in several fishes appears to be under social control, and changes in behavior may occur within minutes to hours in the absence of gonads (Godwin *et al.* 1996), endogenous steroids in the brain may initiate the cascade of events that lead to changes in gonad structure and circulating steroids. Recent evidence from studies in the protogynous bluebanded goby, *Lythrypnus dalli*, supports this hypothesis. Females had brain aromatase activity that was about seven times higher than males. Within hours of sex change to male, female brain aromatase activity decreased by over 40%, while aggressive behavior increased significantly (Black *et al.* 2005).

Additional evidence for the role of aromatase in sexual plasticity comes from studies of temperature-sensitive sex determination in fish (see above, Kitano *et al.* 1999; review: Devlin and Nagahama 2002), reptiles (Crews and Bergeron 1994, Jeyasuria and Place 1998, Crews *et al.* 2001), and amphibians (Kuntz *et al.* 2003). Jeyasuria and Place (1998) demonstrated in the diamondback terrapin that aromatase is transcribed in the brain well before the temperature-sensitive period of embryonic development at both male and female temperatures. However, in females, there is a switch to lower aromatase in the brain while concurrently increasing aromatase transcripts in the putative ovary. In males, brain aromatase levels rise exponentially. Thus, in temperature-dependent sex determination in reptiles, two different forms of aromatase may establish a feedback system linked to the environment in order to ensure proper timing and expression of aromatase in different tissues for sex differentiation. Studies in the leopard gecko demonstrate that the endocrinology, brain morphology, and behavior of adults are dependent on embryonic incubation temperature (reviews: Crews 1998, Rhen and Crews 2002). Compared to males incubated in male-biased temperature, males from female-biased temperatures are more sexually active and less aggressive toward females, have higher estrogen levels and lower testosterone levels, and have greater metabolic capacity in brain areas associated with sexual behavior (i.e., POA). In contrast, males from male-biased temperatures have a higher metabolic capacity in areas of the brain associated with agonistic behavior (i.e., septum, anterior hypothalamus). Evidence from studies in other species of reptiles suggests that temperature determines gonadal sex by influencing sex steroid metabolizing enzymes (i.e., aromatase) during embryonic development (reviews: Crews 1996, Crews *et al.* 2001). Thus, it is probable that temperature directly or indirectly (via a thermosensitive factor) affects brain aromatase levels that, in turn, organize the brain toward a particular phenotype.

Although the effects of steroid hormones on neuropeptide systems have been investigated (see Goodson and Bass 2001), few studies have investigated the interaction of neuropeptides and neurosteroids. Thus, many studies have shown that AVT/AVP systems are sensitive to testosterone. However, in gonadectomized rats, estradiol, but not dihydrotestosterone (DHT, a non-aromatizeable androgen like 11-ketotestosterone), is effective at upregulating AVP mRNA in the medial amygdala and the bed nucleus of the stria terminalis (DeVries *et al.* 1994, Wang and DeVries 1995). Furthermore, studies in quail show that aromatization of testosterone before hatching organizes the sexually dimorphic AVT sensitivity to testosterone in adults (Panzica *et al.* 1998). In the bullfrog, the AVT receptor is sensitive to estradiol and DHT in the amygdala, septum, and habenula, but only androgen sensitive in more posterior dimorphic areas (Boyd 1997). In the midshipman fish model, there are several regions of overlap between aromatase and AVT–ir, as well as estrogen receptor alpha (ERα), especially within the AVT-sensitive vocal motor pathway (e.g., within the anterior hypothalamus and the periaqueductal gray: see Goodson and Bass 2000a, Forlano *et al.* 2001, 2005). Brain aromatase may function in these areas to regionally regulate steroid concentrations reaching AVT neurons/receptors, which in turn may contribute to inter- and intrasexual dimorphism in AVT content and vocal motor sensitivity.

Lastly, catecholaminergic inputs could also have a significant effect on brain aromatase regulation because both dopamine and norepinephrine can alter adenyl cyclase activity and therefore cyclic AMP. Cyclic AMP is known to upregulate aromatase activity in gonadal and other nonneuronal tissue but to inhibit aromatase in the brain, and evidence exists for a cyclic AMP-responsive element on the aromatase gene in both neuronal and nonneuronal tissue (Lephart 1996, Balthazart and Ball 1998). In midshipman, high aromatase and ERα expression overlap with tyrosine hydroxylase immunoreactive (TH–ir) somata in several brain regions, including preoptic and hypothalamic regions that are integration sites for auditory and vocal processing, and dense TH–ir fibers terminate in the aromatase-rich sonic motor nucleus. Thus, aromatase in TH–ir areas suggests another mechanism through which neuroestrogens could modulate variation in vocal–auditory physiology and behavior (see Forlano et al. 2005 for more discussion).

6.5 CONCLUDING COMMENTS: NEUROENDOCRINOLOGICAL TRAITS SUPPORTING ALTERNATIVE REPRODUCTIVE TACTICS IN TELEOSTS

We propose that at least three neuroendocrinological traits may support the widespread evolution of ARTs, and more generally reproductive and social plasticity, among teleost fishes.

Trait 1: Direct input of neuropeptide-containing (e.g., GnRH and AVT) neurons to the pituitary gland. A direct preoptic–pituitary pathway that bypasses a hypophyseal blood portal system may allow for a more rapid change in blood gonadotropin levels.

Trait 2: Abundant brain aromatase. Given the demonstrated role for aromatase in primary sexual differentiation, an aromatase-dependent mechanism may lead to intrasexual dimorphisms as well. That the brain is the site of abundant aromatase synthesis and activity and thus potentially the major source of brain estrogen, emphasizes both the primacy of the brain (see Francis 1992) and possibly of neurosteroids in general in directing events leading to social and reproductive plasticity.

Trait 3: 11-ketotestosterone. A review of androgens in teleosts with male dimorphisms showed that (a) 11-ketotestosterone was the principal circulating steroid

in the courting/territorial male morph, and that (b) 11-ketotestosterone was a more potent androgen than testosterone in the induction of male secondary sex characteristics (Brantley et al. 1993a, b). Studies completed since that review have essentially supported this conclusion (e.g., Lee et al. 2001). As discussed here, the ratio of 11-ketotestosterone levels to estradiol levels (as regulated by aromatase) may provide a key mechanism leading to the adoption of alternative male phenotypes in gonochoristic species and to either sex- or role-reversal in hermaphroditic species.

Acknowledgments
Research support during preparation of this manuscript was provided by NSF (IBN 9987341, 0516748) and NIH (DC00092) grants to AHB and a NIMH predoctoral training grant (5T32MH15793) to PMF. Thanks to T. Natoli for help with preparation of the manuscript, to the editors for their invitation to prepare this review, and to R. Oliveira and two anonymous referees for many helpful comments. All references are up to date as of 2005 when this chapter was originally completed.

References
Balthazart, J. and Ball, G. 1998. New insights into the regulation and function of brain estrogen synthase (aromatase). *Trends in Neuroscience* 21, 243–248.

Baroiller, J.-F., Guiguen, Y., and Fostier, A. 1999. Endocrine and environmental aspects of sex differentiation in fish. *Cell and Molecular Life Sciences* 55, 910–931.

Bass, A. H. 1992. Dimorphic male brains and alternate reproductive tactics in a vocalizing fish. *Trends in Neuroscience* 15, 139–145.

Bass, A. H. 1995. Alternative life history strategies and dimorphic males in an acoustic communication system. In F. W. Goetz and P. Thomas (eds.) *Proceedings of the 5th International Symposium on the Reproductive Physiology of Fish*, pp. 258–260. Port Aransas, TX: Marine Science Institute, University of Texas at Austin.

Bass, A. H. 1996. Shaping brain sexuality. *American Scientist* 84, 352–363.

Bass, A. H. 1998. Behavioral and evolutionary neurobiology: a pluralistic approach. *American Zoologist* 38, 97–107.

Bass, A. H. and Baker, R. 1990. Sexual dimorphisms in the vocal control system of a teleost fish: morphology of physiologically identified cells. *Journal of Neurobiology* 21, 1155–1168.

Bass, A. H. and Grober, M. S. 2001. Social and neural modulation of sexual plasticity in teleost fish. *Brain, Behavior and Evolution* 57, 293–300.

Bass, A. H. and Marchaterre, M. A. 1989. Sound-generating (sonic) motor system in a teleost fish (*Porichthys notatus*): sexual polymorphism in the ultrastructure of myofibrils. *Journal of Comparative Neurology* 286, 141–153.

Bass, A. H. and McKibben, J. R. 2003. Neural mechanisms and behaviors for acoustic communication in teleost fish. *Progress in Neurobiology* 69, 1–26.

Bass, A. H., Marchaterre, M. A., and Baker, R. 1994. Vocal–acoustic pathways in a teleost fish. *Journal of Neuroscience* 14, 4025–4039.

Bass, A. H., Horvath, B. J., and Brothers, E. B. 1996. Nonsequential developmental trajectories lead to dimorphic vocal circuitry for males with alternative reproductive tactics. *Journal of Neurobiology* 30, 493–504.

Bass, A. H., Bodnar, D. A., and Marchaterre, M. A. 1999. Complementary explanations for existing phenotypes in an acoustic communication system. In M. Hauser and M. Konishi (eds.) *Neural Mechanisms of Communication*, pp. 493–514. Cambridge, MA: MIT Press.

Bass, A. H., Bodnar, D. A., and Marchaterre, M. A. 2000. Midbrain acoustic circuitry in a vocalizing fish. *Journal of Comparative Neurology* 419, 505–531.

Bastian, J., Schniederjan, S., and Nguyenkim, J. 2001. Arginine vasotocin modulates a sexually dimorphic communication behavior in the weakly electric fish *Apteronotus leptorhynchus. Journal of Experimental Biology* 204, 1909–1923.

Bentley, P. J. 1998. *Comparative Vertebrate Endocrinology*, 3rd edn. Cambridge, UK: Cambridge University Press.

Beyer, C. 1999. Estrogen and the developing mammalian brain. *Anatomy and Embryology* 199, 379–390.

Bhandari, R. K., Higa, M., Nakamura, S., and Nakamura, M. 2004. Aromatase inhibitor induces complete sex change in the protandrous honeycomb grouper (*Epinephelus merra*). *Molecular Reproduction and Development* 67, 303–307.

Black, M. P., Reavis, R. H., and Grober, M. S. 2004. Socially induced sex change regulates forebrain isotocin in *Lythrypnus dalli. Neuroreport* 15, 185–189.

Black, M. P., Balthazart, J., Baillien, M., and Grober, M. S. 2005. Socially induced and rapid increases in aggression are inversely related to brain aromatase activity in a sex-changing fish, *Lythrypnus dalli. Proceedings of the Royal Society of London B* 272, 2337–2344.

Boyd, S. K. 1994. Arginine vasotocin facilitation of advertisement calling and call phonotaxis in bullfrogs. *Hormones and Behavior* 28, 232–240.

Boyd, S. K. 1997. Brain vasotocin pathways and the control of sexual behaviors in the bullfrog. *Brain Research Bulletin* 44, 345–350.

Braford Jr., M. R., and Northcutt, R. G. 1983. Organization of the diencephalon and pretectum of the ray-finned fishes. In R. E. Davis and R. G. Northcutt (eds.) *Fish Neurobiology*, vol. 2, pp. 117–164. Ann Arbor, MI: University of Michigan Press.

Brantley, R. K. and Bass, A. H. 1994. Alternative male spawning tactics and acoustic signaling in the plainfin midshipman fish, *Porichthys notatus. Ethology* 96, 213–232.

Brantley, R. K., Marchaterre, M. A., and Bass, A. H. 1993a. Androgen effects on vocal muscle structure in a teleost fish with inter- and intra-sexual dimorphism. *Journal of Morphology* 216, 305–318.

Brantley, R. K., Wingfield, J. C., and Bass, A. H. 1993b. Sex steroid levels in *Porichthys notatus*, a fish with alternative male reproductive tactics, and a review of the hormonal bases for male dimorphism among teleost fishes. *Hormones and Behavior* 27, 332–347.

Brockmann, H. J. 2001. The evolution of alternative strategies and tactics. *Advances in the Study of Behavior* 30, 1–51.

Burke, K. A., Kuwajima, M., and Sengelaub, D. R. 1999. Aromatase inhibition reduces dendritic growth in a sexually dimorphic rat spinal nucleus. *Journal of Neurobiology* 38, 301–312.

Butler, A. B. and Hodos, W. 1996. *Comparative Vertebrate Neuroanatomy*. New York: Wiley-Liss.

Cardwell, J. R. and Liley, N. R. 1991. Hormonal control of sex and color change in the stoplight parrotfish, *Sparisoma viride. General and Comparative Endocrinology* 81, 7–20.

Chang, C.-F. and Lin, B.-Y. 1998. Estradiol-17β stimulates aromatase activity and reversible sex change in protandrous black porgy, *Acanthopagrus schlegeli. Journal of Experimental Zoology* 280, 165–173.

Cole, K. S. 1990. Patterns of gonad structure in hermaphroditic gobies (Teleostei: Gobiidae). *Environmental Biology of Fishes* 28, 125–142.

Compagnone, N. A. and Mellon, S. H. 2000. Neurosteroids: biosynthesis and function of the novel neuromodulators. *Frontiers in Neuroendocrinology* 21, 1–56.

Crews, D. 1993. The organizational concept and vertebrates without sex chromosomes. *Brain, Behavior and Evolution* 42, 202–214.

Crews, D. 1996. Temperature-dependent sex determination: the interplay of steroid hormones and temperature. *Zoological Science* 13, 1–13.

Crews, D. 1998. On the organization of individual differences in sexual behavior. *American Zoologist* 38, 118–132.

Crews, D. and Bergeron, J. M. 1994. Role of reductase and aromatase in sex determination in the red-eared slider (*Trachemys scripta*) turtle with temperature-dependent sex determination. *Journal of Endocrinology* 143, 279–289.

Crews, D., Fleming, A., Willingham, E., Baldwin, R., and Skipper, J. K. 2001. Role of steroidogenic factor 1 and aromatase in temperature-dependent sex determination in the red-eared slider turtle. *Journal of Experimental Zoology* 290, 597–606.

Davis, M. R. and Fernald, R. D. 1990. Social control of neuronal soma size. *Journal of Neurobiology* 21, 1180–1189.

D'cotta, H., Fostier, A., Guiguen, Y., Govoroun, M., and Baroiller, J.-F. 2001. Aromatase plays a key role during normal and temperature-induced sex differentiation of tilapia *Oreochromis niloticus*. *Molecular Reproduction and Development* 59, 265–276.

Devlin, R. H. and Nagahama, Y. 2002. Sex determination and sex differentiation in fish: an overview of genetic, physiological, and environmental influences. *Aquaculture* 208, 191–364.

DeVries, G. J. and Simerly, R. 2002. Anatomy, development, and functions of sexually dimorphic neural circuits in the mammalian brain. In D. Pfaff, A. P. Arnold, A. M. Etgen, S. E. Fahrbach, and R. T. Rubin (eds.) *Hormones, Brain, and Behavior*, vol. 4, pp. 137–191. San Diego, CA: Academic Press.

DeVries, G. J., Wang, Z., Bullock, N. A., and Numan, S. 1994. Sex differences in the effects of testosterone and its metabolites on vasopression mRNA levels in the bed nucleus of the stria terminalis of rats. *Journal of Neuroscience* 14, 1789–1794.

Elofsson, U., Winburg, S., and Francis, R. C. 1997. Number of preoptic GnRH-immunoreactive cells correlates with sexual phase in a protandrously hermaphroditic fish, the dusky anemonefish (*Amphiprion melanopus*). *Journal of Comparative Physiology* 181, 484–492.

Elofsson, U., Winburg, S., and Nilsson, G. E. 1999. Relationships between sex and the size and number of forebrain gonadotropin-releasing hormone-immunoreactive neurons in the ballan wrasse (*Labrus berggylta*), a protogynous hermaphrodite. *Journal of Comparative Neurology* 410, 158–170.

Fernald, R. D. and White, R. B. 1999. Gonadotropin-releasing hormone genes: phylogeny, structure, and functions. *Frontiers in Neuroendocrinology* 20, 224–240.

Foran, C. M. and Bass, A. H. 1998. Preoptic AVT immunoreactive neurons of a teleost fish with alternative reproductive tactics. *General and Comparative Endocrinology* 111, 271–282.

Foran, C. M. and Bass, A. H. 1999. Preoptic GnRH and AVT: axes for sexual plasticity in teleost fish. *General and Comparative Endocrinology* 116, 141–152.

Forlano, P. M. and Bass, A. H. 2001. Sex steroid modulation of brain aromatase mRNA expression in a vocal fish. *Society for Neuroscience Abstracts* 27, 1081.

Forlano, P. M. and Bass, A. H. 2005a. Seasonal plasticity of brain aromatase mRNA expression in glia: divergence across sex and vocal phenotypes. *Journal of Neurobiology* 65, 37–49.

Forlano, P. M. and Bass, A. H. 2005b. Steroid regulation of brain aromatase expression in glia: female preoptic and vocal motor nuclei. *Journal of Neurobiology* 65, 50–58.

Forlano, P. M., Deitcher, D. L., Myers, D. A., and Bass, A. H. 2001. Anatomical distribution and cellular basis for high levels of aromatase activity in the brain of teleost fish: aromatase enzyme and mRNA expression identify glia as source. *Journal of Neuroscience* 21, 8943–8955.

Forlano, P. M., Deitcher, D. L., and Bass, A. H. 2005. Distribution of estrogen receptor alpha mRNA in the brain and inner ear of a vocal fish with comparisons to sites of aromatase expression. *Journal of Comparative Neurology* 483, 91–113.

Francis, R. C. 1992. Sexual lability in teleosts: developmental factors. *Quarterly Review of Biology* 67, 1–18.

Gelinas, D., Pitoc, G. A., and Callard, G. V. 1998. Isolation of a goldfish brain cytochrome P450 aromatase cDNA: mRNA expression during the seasonal cycle and after steroid treatment. *Molecular and Cellular Endocrinology* 138, 81–93.

Gerald, J. W. 1971. Sound production during courtship in six species of sunfish (Centrarchidae). *Evolution* 25, 75–87.

Godwin, J. R. and Thomas, P. 1993. Sex change and steroid profiles in the protandrous anemonefish *Amphiprion melanopus* (Pomacentridae, teleostei). *General and Comparative Endocrinology* 91, 144–157.

Godwin, J. R., Crews, D., and Warner, R. R. 1996. Behavioral sex change in the absence of gonads in a coral reef fish. *Proceedings of the Royal Society of London B* 263, 1683–1688.

Godwin, J., Sawby, R., Warner, R. R., Crews, D., and Grober, M. S. 2000. Hypothalamic arginine vasotocin mRNA abundance variation across sexes and with sex change in a coral reef fish. *Brain, Behavior and Evolution* **55**, 74–84.

Godwin, J., Luckenbach, J. A., and Borski, R. J. 2003. Ecology meets endocrinology: environmental sex determination in fishes. *Evolution and Development* **5**, 40–49.

Goodson, J. L. and Bass, A. H. 2000a. Vasotocin innervation and modulation of vocal–acoustic circuitry in the teleost *Porichthys notatus*. *Journal of Comparative Neurology* **422**, 363–379.

Goodson, J. L. and Bass, A. H. 2000b. Forebrain peptide modulation of sexually polymorphic vocal motor circuitry. *Nature* **403**, 769–772.

Goodson, J. L. and Bass, A. H. 2001. Social behavior functions and related anatomical characteristics of vasotocin/vasopressin systems in vertebrates. *Brain Research Reviews* **35**, 246–265.

Goodson, J. L. and Bass, A. H. 2002. Forebrain and midbrain vocal–acoustic complexes: intraconnectivity and descending vocal motor pathways. *Journal of Comparative Neurology* **448**, 298–321.

Goodson, J. L., Evans, A. K., and Bass, A. H. 2003. Putative isotocin distributions in sonic fish: relation to vasotocin and vocal–acoustic circuitry. *Journal of Comparative Neurology* **462**, 1–14.

Goto-Kazeto, R., Kight, K. E., Zohar, Y., Place, A. R., and Trant, J. M. 2004. Localization and expression of aromatase mRNA in adult zebrafish. *General and Comparative Endocrinology* **139**, 72–84.

Grober, M. S. 1998. Socially controlled sex change: integrating ultimate and proximate levels of analysis. *Acta Ethologica* **1**, 3–17.

Grober, M. S. and Bass, A. H. 1991. Neuronal correlates of sex/role change in labrid fishes: LHRH-like immunoreactivity. *Brain, Behavior and Evolution* **38**, 302–312.

Grober, M. S. and Bass, A. H. 2002. Life history, neuroendocrinology, and behavior in fish. In D. Pfaff, A. P. Arnold, A. M. Etgen, S. E. Fahrbach, and R. T. Rubin (eds.) *Hormones, Brain and Behavior*, vol. 2, pp. 331–347. San Diego, CA: Academic Press.

Grober, M. S. and Sunobe, T. 1996. Serial adult sex change involves rapid and reversible changes in forebrain neurochemistry. *Neuroreport* **7**, 2945–2949.

Grober, M. S., Jackson, I. M. D., and Bass, A. H. 1991. Gonadal steroids affect LHRH preoptic cell number in a sex-role changing fish. *Journal of Neurobiology* **22**, 734–741.

Grober, M. S., Fox, S. H., Laughlin, C., and Bass, A. H. 1994. GnRH cell size and number in a teleost fish with two male reproductive morphs: sexual maturation, final sexual status and body size allometry. *Brain, Behavior and Evolution* **43**, 61–78.

Grober, M. S., George, A. A., Watkins, K. K., Carneiro, L. A., and Oliveira, R. 2002. Forebrain AVT and courtship in fish with male alternative reproductive tactics. *Brain Research Bulletin* **57**, 423–425.

Gross, M. R. 1991. Evolution of alternative reproductive strategies: frequency-dependent selection in male bluegill sunfish. *Philosophical Transactions of the Royal Society of London B* **332**, 59–66.

Gross, M. 1996. Alternative reproductive tactics and strategies: diversity within sexes. *Trends in Ecology and Evolution* **11**, 92–97.

Halpern-Sebold, L., Schreibman, M. P., and Margolis-Nunno, H. 1986. Differences between early- and late-maturing genotypes of the platyfish (*Xiphophorus maculatus*) in the morphometry of their immunoreactive leuteinizing hormone releasing hormone-containing cells: a developmental study. *Journal of Experimental Zoology* **240**, 245–257.

Herbert, J. 1993. Peptides in the limbic system: neurochemical codes for co-ordinated adaptive responses to behavioural and physiological demand. *Progress in Neurobiology* **41**, 723–791.

Hiraoka, S., Ando, H., Ban, M., Ueda, H., and Urano, A. 1997. Changes in expression of neurohypophysial hormone genes during spawning migration in Chum salmon, *Oncorhynchus keta*. *Journal of Molecular Endocrinology* **18**, 49–55.

Hofmann, H. A. and Fernald R. D. 2000. Social status controls somatostatin neuron size and growth. *Journal of Neuroscience* **20**, 4740–4744.

Insel, T. and Young, L. J. 2000 The neurobiology of attachment. *Nature Reviews Neuroscience* **2**, 129–136.

Jeyasuria, P. and Place, A. R. 1998. Embryonic brain–gonadal axis in temperature-dependent sex determination of reptiles: a role for P450 aromatase (CYP19). *Journal of Experimental Zoology* **281**, 428–449.

Kitano, T., Takamune, K., Koyayashi, T., Nagahama, Y., and Abe, S.-I. 1999. Suppression of P450 aromatase gene expression in sex-reversed males produced by rearing genetically female larvae at a high water temperature during a period of sex differentiation in the Japanese flounder (*Paralichthys olivaceus*). *Journal of Molecular Endocrinology* **23**, 167–176.

Knapp, R. 2004. Endocrine mediation of vertebrate male alternative reproductive tactics: the next generation of studies. *Integrative and Comparative Biology* 43, 658–668.

Knapp, R., Wingfield, J. C., and Bass, A. H. 1999. Steroid hormones and paternal care in the plainfin midshipman fish (*Porichthys notatus*). *Hormones and Behavior* 35, 81–89.

Kroon, F. J. and Liley, N. R. 2000. The role of steroid hormones in protogynous sex change in the blackeye goby, *Coryphopterus nicholsii* (Teleostei: Gobiidae). *General and Comparative Endocrinology* 118, 273–283.

Kroon, F. J., Munday, P. L., Wescott, D. A., Hobbs, J. P., and Liley, N. R. 2005. Aromatase pathway mediates sex change in each direction. *Proceedings of the Royal Society of London B* 272, 1399–1405.

Kuntz, S., Chesnel, A., Duterque-Coquillaud, M., *et al.* 2003. Differential expression of P450 aromatase during gonadal sex differentiation and sex reversal of the newt *Pleurodeles walti. Journal of Steroid Biochemistry and Molecular Biology* 84, 89–100.

Lee, J. S. F. and Bass, A. H. 2004. Does the "exaggerated" morphology preclude plasticity to cuckoldry? A test in the midshipman fish, *Porichthys notatus. Naturwissenschaften* 91, 338–341.

Lee, J. S. F. and Bass, A. H. 2005. Differential effects of 11-ketotestosterone on dimorphic traits in a teleost with alternative male reproductive morphs. *Hormones and Behavior* 47, 523–531.

Lee, Y.-H., Du, J.-L., Yueh, W.-S., *et al.* 2001. Sex change in the protandrous black porgy, *Acanthopagrus schegeli*: a review in gonadal development, estradiol, estrogen receptor, aromatase activity and gonadotropin. *Journal of Experimental Zoology* 290, 715–726.

Lee, Y.-H., Yueh, W.-S., Du, J.-L., Sun, L.-T., and Chang, C.-F. 2002. Aromatase inhibitors block natural sex change and induce male function in the protandrous black porgy, *Acanthopagrus schegeli* Bleeker: possible mechanism of natural sex change. *Biology of Reproduction* 66, 1749–1754.

Leiser, J. K. and Itzkowitz, M. 2004. Changing tactics: dominance, territoriality, and the responses of "primary" males to competition from conditional breeders in the variegated pupfish (*Cyprinodon variegatus*). *Behavioral Processes* 66, 119–130.

Lephart, E. D. 1996. A review of brain aromatase cytochrome P450. *Brain Research Reviews* 22, 1–26.

Lethimonier, C., Madigou, T., Munoz-Cueto, J. A., Lareyre, J. J., and Kah, O. 2004. Evolutionary aspects of GnRHs,

GnRH neuronal systems and GnRH receptors in teleost fish. *General and Comparative Endocrinology* 135, 1–16.

Markakis, E. A. 2002. Development of the neuroendocrine hypothalamus. *Frontiers in Neuroendocrinology* 23, 257–291.

Marler, C. A., Boyd, S. K., and Wilczynski, W. 1999. Forebrain arginine vasotocin correlates of alternative mating strategies in frogs. *Hormones and Behavior* 36, 53–61.

Mazzoldi, C., Scaggiante, M., Ambrosin, E., and Rasotto, M. B. 2000. Mating system and alternative male mating tactics in the grass goby *Zosterisessor ophiocephalus* (Teleostei: Gobiidae). *Marine Biology* 137, 1041–1048.

Meek, J. and Nieuwenhuys, R. 1998. Holosteans and teleosts. In R. Nieuwenhuys, H. J. ten Donkelaar, and C. Nicholson (eds.) *The Central Nervous System of Vertebrates*, pp. 759–937. New York: Springer-Verlag.

Menuet, A., Anglade, I., Le Guevel, R., *et al.* 2003. Distribution of aromatase mRNA and protein in the brain and pituitary of female rainbow trout: comparison with estrogen receptor. *Journal of Comparative Neurology* 462, 180–193.

Menuet, A., Pellegrini, E., Brion, F., *et al.* 2005. Expression and estrogen-dependent regulation of the zebrafish brain aromatase gene. *Journal of Comparative Neurology* 485, 304–320.

Miranda, J. A., Oliveira, R. F., Carneiro, L. A., Santos, R., and Grober, M. S. 2003. Neurochemical correlates of male polymorphism and alternative reproductive tactics in the Azorean rock-pool blenny, *Parablennius parvicornis. General and Comparative Endocrinology* 132, 183–189.

Moore, F. L. and Lowry, C. A. 1998. Comparative neuroanatomy of vasotocin and vasopressin in amphibians and other vertebrates. *Comparative Biochemistry and Physiology* 119, 251–260.

Morrey, G. E., Kobayashi, T., Nakamura, M., Grau, E. G., and Nagahama, Y. 1998. Loss of gonadal P450 aromatase mRNA corresponds with the de-differentiation of the ovary in the protogynous wrasse, *Thalassoma duperrey. Experimental Zoology* 281, 507–508.

Nelson, R. 1998. *Behavioral Neuroendocrinology*. Sunderland, MA: Sinauer Associates.

Nieuwenhuys, R., Veening, J. G., and Van Domburg, P. 1989. Core and paracores: some new chemoarchitectural entities in the mammalian neuraxis. *Acta Morphologica Neerlando-Scandinavica* 26, 131–163.

Oka, Y. and Ichikawa, M. 1990. Gonadotropin-releasing hormone (GnRH) immunoreactive system in the brain of the dwarf gourami (*Colisa lalia*) as revealed by light microscopic immunocytochemistry using a monoclonal antibody to common amino acid sequence of GnRH. *Journal of Comparative Neurology* 300, 511–522.

Oka, Y. and Matsushima, T. 1993. Gonadotropin-releasing hormone (GnRH)-immunoreactive terminal nerve cells have intrinsic rhythmicity and project widely in the brain. *Journal of Neuroscience* 13, 2161–2176.

Okuzawa, K. and Kobayashi, M. 1999. Gonadotropin-releasing hormone neuronal systems in the teleostean brain and functional significance. In P. Rao and P. Kluwer (eds.) *Neural Regulation in the Vertebrate Endocrine System*, pp. 85–100. New York: Plenum Press.

Oliveira, R. F., Canario, A. V. M., and Grober, M. S. 2001. Male sexual polymorphism, alternative reproductive tactics, and androgens in combtooth blennies (Pisces: Blenniidae). *Hormones and Behavior* 40, 266–275.

Ota, Y., Ando, H., Ban, M., Ueda, H., and Urano, A. 1996. Sexually different expression of neurohypophysial hormone genes in the preoptic nucleus of pre-spawning Chum salmon. *Zoological Science* 13, 593–601.

Ota, Y., Ando, H., Ueda, H., and Urano, A. 1999. Seasonal changes in expression of neurohypophysial hormone genes in the preoptic nucleus of immature female Masu salmon. *General and Comparative Endocrinology* 116, 31–39.

Panzica, G. C., Castagna, C., Viglietti-Panzica, C., *et al.* 1998. Organizational effects of estrogens on brain vasotocin and sexual behavior in quail. *Journal of Neurobiology* 37, 684–699.

Pasmanik, M. and Callard, G. V. 1985. Aromatase and 5α-reductase in the teleost brain, spinal cord and pituitary gland. *General and Comparative Endocrinology* 60, 244–251.

Perry, A. N. and Grober, M. S. 2002. A model for social control of sex change: interactions of behavior, neuropeptides, glucocorticoids, and sex steroids. *Hormones and Behavior* 43, 31–38.

Peter, R. E. and Fryer, J. N. 1983. Endocrine functions of the hypothalamus of actinopterygians. In R. E. Davis and R. G. Northcutt (eds.) *Fish Neurobiology*, vol. 2, pp. 165–201. Ann Arbor, MI: University of Michigan Press.

Pfaff, D. W., Arnold, A. P., Etgen, A. M., Fahrbach, S. E., and Rubin, R. T. 2002. *Hormones, Brain and Behavior*. San Diego, CA: Academic Press.

Puelles, L. 2001. Brain segmentation and forebrain development in amniotes. *Brain Research Bulletin* 55, 695–710.

Reavis, R. H. and Grober, M. S. 1999. An integrative approach to sex change: social, behavioural and neurochemical changes in *Lythrypnus dalli* (Pisces). *Acta Ethological* 2, 51–60.

Remage-Healey, L. and Bass, A. H. 2004. Rapid, hierarchical modulation of vocal patterning by steroid hormones. *Journal of Neuroscience* 24, 5892–5900.

Rhen, T. and Crews, D. 2002. Variation in reproductive behavior within a sex: neural systems and endocrine activation. *Journal of Neuroendocrinology* 14, 517–531.

Rose, J. D. and Moore, F. L. 2002. Behavioral neuroendocrinology of vasotocin and vasopressin and the sensorimotor processing hypothesis. *Frontiers in Neuroendocrinology* 23, 317–341.

Saito, D. and Urano, A. 2001. Synchronized periodic Ca^{2+} pulses define neurosecretory activities in magnocellular vasotocin and isotocin neurons. *Journal of Neuroscience* 21, RC178.

Saito, D., Hasegawa, Y., and Urano, A. 2003. Gonadotropin-releasing hormones modulate electrical activity of vasotocin and isotocin neurons in the brain of rainbow trout. *Neuroscience Letters* 351, 107–110.

Schlinger, B. A., Greco, C., and Bass, A. H. 1999. Aromatase activity in hindbrain vocal control region of a teleost fish: divergence among males with alternative reproductive tactics. *Proceedings of the Royal Society of London B* 266, 131–136.

Schlinger, B. A., Soma, K., and London, S. E. 2001. Neurosteroids and brain sexual differentiation. *Trends in Neuroscience* 24, 429–431.

Schreibman, M. P. and Magliulo-Cepriano, L. 2002. Differentiation/maturation of centers in the brain regulating reproductive function in fishes. In D. Pfaff, A. P. Arnold, A. M. Etgen, S. E. Fahrbach, and R. T. Rubin (eds.) *Hormones, Brain, and Behavior*, vol. 4, pp. 303–323. San Diego, CA: Academic Press.

Semsar, K. and Godwin, J. 2002. Social influences on the arginine vasotocin system status are independent of gonads in a sex-changing fish. *Journal of Neuroscience* 23, 4386–4393.

Semsar, K., Kandel, F. L., and Godwin, J. 2001. Manipulations of the AVT system shift social status and related courtship and aggressive behavior in the bluehead wrasse. *Hormones and Behavior* 40, 21–31.

Sisneros, J. A., Forlano, P. M., Knapp, R., and Bass, A. H. 2004. Seasonal variation of steroid hormone levels in an intertidal-nesting fish, the vocal plainfin midshipman. *General and Comparative Endocrinology* 136, 101–116.

Strobl-Mazzulla, P. H., Moncaut, N. P., Lopez, G. C., *et al.* 2005. Brain aromatase from pejerrey fish (*Odontesthes bonariensis*): cDNA cloning, tissue expression, and immunohistochemical localization. *General and Comparative Endocrinology* **143**, 21–32.

Taborsky, M. 1994. Sneakers, satellites, and helpers: parasitic and cooperative behavior in fish reproduction. *Advances in the Study of Behavior* **23**, 1–100.

Teitsma, C. A., Bailhache, B., Anglade, I., *et al.* 1997. Distribution and expression of glucocorticoid receptor mRNA in the forebrain of the rainbow trout. *Neuroendocrinology* **66**, 294–304.

Teitsma, C. A., Anglade, I., Toutirais, G., *et al.* 1998. Immunohistochemical localization of glucocorticoid receptors in the forebrain of the rainbow trout (*Oncorhynchus mykiss*). *Journal of Comparative Neurology* **401**, 395–410.

Teitsma, C. A., Anglade, I., Lethimonier, C., *et al.* 1999. Glucocorticoid receptor immunoreactivity in neurons and pituitary cells implicated in reproductive functions in rainbow trout: a double immunohistochemical study. *Biology of Reproduction* **60**, 642–650.

Tsai, C. L., Chang, S. L., Wang, L. H., and Chao, T. Y. 2003. Temperature influences the ontogenetic expression of aromatase and oestrogen receptor mRNA in the developing tilapia (*Oreochromis mossambicus*) brain. *Journal of Neuroendocrinology* **15**, 97–102.

Urano, A., Kurokawa, K., and Hiraoka, S. 1994. E xpression of the vasotocin and isotocin gene family in fish. In N. Sherwood and C. L. Hew (eds.) *Fish Physiology*, vol. 13, *Molecular Aspects of Hormonal Regulation in Fish*, pp. 101–132. San Diego, CA: Academic Press.

Wang, Z. and DeVries, G. J. 1995. Androgen and estrogen effects on vasopressin messenger RNA expression in the medial amygdaloid nucleus in male and female rats. *Journal of Neuroendocrinology* **7**, 827–831.

White, S. A., Nguyen, T., and Fernald, R. D. 2002. Social regulation of gonadotropin-releasing hormone. *Journal of Experimental Biology* **205**, 2567–2581.

Zhang, Y., Zhang, W., Zhang, L., *et al.* 2004. Two distinct cytochrome P450 aromatases in the orange-spotted grouper (*Epinephelus coioides*): cDNA cloning and differential mRNA expression. *Journal of Steroid Biochemistry and Molecular Biology* **92**, 39–50.

7 · Hormones and alternative reproductive tactics in vertebrates

RUI F. OLIVEIRA, ADELINO V. M. CANÁRIO, AND ALBERT F. H. ROS

CHAPTER SUMMARY

The wide diversity of alternative tactics of reproduction found among vertebrates offers a unique opportunity to study the endocrine mechanisms underlying the phenotypic variation of reproductive traits. Here, we first assess the existing conceptual frameworks on the mechanisms underlying the expression of alternative reproductive tactics (ARTs) by reviewing the available data on hormone levels in alternative phenotypes and on the effects of hormone manipulations in different vertebrate taxa. We then highlight recent studies that have opened new avenues of research on the neuroendocrine basis of ARTs, such as the use of functional genomics to study differential gene expression between morphs. Finally, we stress the need to integrate the study of ARTs with the mechanisms underlying the expression of alternative phenotypes and with functional studies of ARTs. Only such an integrative approach will allow a comprehensive understanding of the evolution and development of ARTs.

7.1 INTRODUCTION

7.1.1 Setting the scene

According to the classic paradigm of the endocrine control of vertebrate reproduction, the hypothalamus–pituitary–gonadal (HPG) axis controls gonadal maturation, the expression of secondary sexual characters, and reproductive behavior (Figure 7.1A). However, in some species there are males in which gonadal maturation and sperm production are dissociated from the expression of behavioral and morphological male traits (i.e., secondary sexual characters). They are males with male alternative reproductive tactics (ARTs), and they offer unique opportunities to study the proximate mechanisms of reproduction (Figure 7.1B). ARTs are also valuable models for the study of the causal

mechanisms underlying individual variation in reproduction since within-sex variation in reproductive traits can be studied without the confounding effects of gender (Moore 1991, Godwin and Crews 2002).

Historically, typological classifications of ARTs have been based on the evolutionary processes underlying their expression (e.g., genetic polymorphisms vs. conditional tactics, Gross 1996; or Mendelian strategies vs. developmental strategies vs. behavioral strategies, Shuster and Wade 2003). In this chapter we will use a classification based on observed patterns of ARTs that does not require knowledge of their underlying processes (e.g., genetic vs. conditional strategies). The classification scheme is modified from that proposed by other authors (Caro and Bateson 1986, Moore 1991, Taborsky 1994, Moore et al. 1998, Brockmann 2001). We will consider alternative reproductive phenotypes as fixed if the individuals adopt one of the tactics for their entire lifetime or as plastic if individuals change their reproductive tactic. Within plastic ART phenotypes, we will distinguish between irreversible sequential patterns, when individuals switch from one tactic to another at a particular moment in their lifetime, and reversible patterns, when individuals can change back and forth between patterns (Moore 1991, Moore et al. 1998, Brockmann 2001) (see Figure 1.1).

A number of reviews on the proximate mechanisms of ARTs have been published lately, but each has a different focus from the present chapter. Moore and co-authors (1998) develop a conceptual framework for the role of hormones on tactic differentiation, Rhen and Crews (2002) provide an overview of mechanisms involved in ARTs in different vertebrate taxa, Knapp (2003) proposes a new generation of studies more focused on target tissues than on circulating levels of hormones, and Oliveira (2005) and Oliveira and co-authors (2005) focus on mechanisms operating in fish ARTs. So what can be added by another chapter on the causal mechanisms of ARTs?

Alternative Reproductive Tactics, ed. Rui F. Oliveira, Michael Taborsky, and H. Jane Brockmann. Published by Cambridge University Press.
© Cambridge University Press 2008.

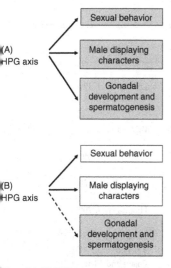

Figure 7.1 (A) Different reproductive traits share a common underlying causal agent (e.g., testosterone); (B) in species with ARTs a dissociation between the different traits may occur resulting in a phenotypic mosaic that can express both male and female traits (e.g., sneaker males that mimic female behavior and morphology in order to achieve fertilizations).

This chapter has two main objectives. The first is to present an exhaustive revision of the available data on hormone levels in alternative phenotypes and on the effects of hormone manipulations in different vertebrate taxa. This will provide the basis for the assessment of existing conceptual frameworks on the mechanisms underlying the expression of ARTs. The second objective is to highlight recent studies that have opened new avenues of research on the physiological basis of ARTs and its implications for understanding the evolution of ARTs (e.g., the study of differential hormonal-mediated costs of alternative phenotypes and the field of functional genomics to study differential gene expression between morphs).

7.1.2   Who's in the ARTs ark?

We will address only male ARTs since they are the most common and best-studied cases at a proximate level. In contrast to other recent reviews of ARTs, we also include species with cooperative breeding in which parentage is

shared between breeders and helpers (e.g., acorn woodpecker, *Melanerpes formicivorus*: Haydock *et al.* 2001), in which there are behavioral observations of breeding attempts with the female of the pair by helpers (e.g., bell miner, *Manorina melanophrys*: Poiani and Fletcher 1994; but see Conrad *et al.* 1998), and in which helpers are non-breeders in their home group but attempt extra-pair copulations (EPC) with other group females (e.g., superb fairy-wrens, *Malurus cyaneus*: Mulder *et al.* 1994). In these cases we consider helping to be an alternative tactic to achieve breeding. According to these criteria we have included in our analyses the cooperative breeding species listed in Table 7.1. It should be noted that the use of these criteria assumes that observed mating episodes result in reproductive output, which may not always be the case. In contrast, we have discarded other cooperative breeding species for which detailed hormonal data are available when paternity analyses have revealed that the species are genetically monogamous (e.g., Florida scrub-jay, *Aphelocoma coerulescens*: Schoech *et al.* 1991, 1996, Quinn *et al.* 1999; red-cockaded woodpecker, *Picoides borealis*: Haig *et al.* 1994, Khan *et al.* 2001). In the white-browed sparrow weaver, *Plocepasser mahali* (Wingfield *et al.* 1991), for which there are hormone data for both breeders and helpers, the information on the helpers' behavior suggests that they do not try to sneak copulations (J. C. Wingfield, personal communication), and therefore this species was not included. Finally, there are species for which the available information regarding the reproduction of helpers is dubious or indirect. In the pied kingfisher (*Ceryle rudis*), two types of helpers occur: primary helpers that are off-spring of the breeding pair and secondary helpers that are unrelated to breeders (Reyer 1980, 1984). Primary helpers have small, immature gonads and have lower testosterone levels than both male breeders and secondary helpers, and thus are not able to fertilize eggs (Reyer *et al.* 1986). In contrast, secondary helpers, which have mature gonads, sometimes fight with the breeder male to get access to the female of the pair (Reyer *et al.* 1986). Therefore, even without parentage data, we decided to consider secondary helping of the pied kingfisher as an ART and have included it in the analysis.

Two cooperatively breeding rodents in which helpers do not achieve reproductive success were also included, as they might be seen as special cases of ARTs: the naked mole-rat (*Heterocephalus glaber*) and the Mongolian gerbil (*Meriones unguiculatus*). In both cases subordinate individuals acting as helpers are incapable of direct reproduction and are

Table 7.1. *Cooperative breeding species in which helpers also breed*

Species	Evidence for breeding in helpers (reproductive success of helpers)	References
Fish		
Princess of Burundi, *Neolamprologus brichardi*	Genetic (10.8% of offspring)	Dierkes *et al.* 1999
Birds		
Seychelles warbler, *Acrocephalus sechellensis*	Genetic (15% of offspring)	Richardson *et al.* 2001
Mexican scrub-jay, *Aphelocoma coerulescens*	Genetic (low)	Bowen *et al.* 1995
Acorn woodpecker, *Melanerpes formicivorus*	Genetic (approx. 25% of offspring)	Haydock *et al.* 2001
Australian magpie, *Gymnorhina tibicen*	Genetic (high; up to 82% of extra-group paternity)	Hughes *et al.* 2003
Azure-winged magpie, *Cyanopica cyanus*	Behavioral (high)	De la Cruz *et al.* 2003; Valencia *et al.* 2003
Bell miner, *Manorina melanophrys*	Behavioral/genetic (genetic data indicates very low success)	Poiani and Fletcher 1994; Conrad *et al.* 1998
Superb fairy-wren, *Malurus cyaneus*	Genetic (within-group = 2.2%; extra-group = 76%)	Mulder *et al.* 1994
Pied kingfisher, *Ceryle rudis*	Behavioral (low)	Reyer *et al.* 1986
Harris's hawk, *Parabuteo unicinctus*	Behavioral (low)	Dawson and Mannan 1991
Mammals		
Ring-tailed lemur, *Lemur catta*	Behavioral (high)	Sauther 1991; Sussman 1991
Common marmoset, *Callithrix jacchus*	Behavioral/genetic (genetic data indicates very low success within the group)	Digby 1999; Nievergelt *et al.* 2000
Alpine marmot, *Marmota marmota*	Genetic (only subordinate helpers)	U. Bruns and W. Arnold, unpublished data in Dierkes *et al.* 1999
Dwarf mongoose, *Helogale parvula*	Genetic (24% of offspring)	Keane *et al.* 1994
Meerkat, *Suricata suricatta*	Genetic (low)	Griffin *et al.* 2003
Gray wolf, *Canis lupus*	Behavioral (low)	Creel 2005
African wild-dog, *Lycaon pictus*	Behavioral/genetic (low)	Girman *et al.* 1997; Creel and Creel 2002

obligate helpers, and thus their fitness is entirely indirect (Clark and Galef 2000, Faulkes and Bennett 2001) (see Box 7.1). In these two cases, it can be argued that helping is a conditional strategy, without which these individuals would have zero fitness.

In summary, this chapter will cover not only the usual ARTs but also the cooperative breeders that fit the conditions described above.

7.2 PROFILES OF ALTERNATIVE REPRODUCTIVE PHENOTYPES

In general, two alternative modes or tactics of reproduction can be found in species with male ARTs: a conventional or bourgeois tactic or an alternative or parasitic tactic. Whereas bourgeois males invest resources to attract mates (e.g., differentiation of morphological ornaments; expression

Box 7.1   Obligatory helping as an alternative reproductive tactic

In cooperatively breeding animals, it is usual that reproduction is monopolized by some group members resulting in a high within-group reproductive skew. Kin selection theory may explain indirect benefits for nonbreeding individuals that act as helpers in these groups, while direct benefits such as queuing to take over the breeding position when it is vacant have been advocated (see Solomon and French 1997). There are two extreme cases of obligatory helping that have been described among cooperatively breeding mammals: the naked mole-rat (*Heterocephalus glaber*) and the Mongolian gerbil (*Meriones unguiculatus*). In these two cases it can be argued that since their inclusive fitness equals their indirect fitness (i.e., the only chance that nonbreeding individuals have during their whole lifespan to get copies of their genes into the next generation is by helping kin to reproduce), individuals that specialize in alloparenting and/or helping behavior patterns can be seen as adopting an alternative tactic.

The naked mole-rat fits the eusociality definition derived from insects, since division of labor is present in the colony among the nonbreeding helpers, which is based on body size (Lacey and Sherman 1991). A single female, the "queen," is sexually active breeding with up to three breeding males (Bennett and Faulkes 2000). The queen controls the reproductive physiology of both sexes, maintaining the reproductive suppression of their subordinate colony mates (Faulkes and Abbott 1997). There is

also evidence for the existence of castes, with a disperser morph among males and a morphologically distinct "queen" (O'Riain *et al.* 1996, 2000b). In addition, this mating system with high rates of inbreeding leads to a genetic structure similar to insect haplodiploidy, with intra-colony relatedness coefficients as high as 0.8, which is greater than the 0.75 achieved by the haplodiploid system (Reeve *et al.* 1990). This system seems to have evolved due to high costs of dispersal, and most subordinate individuals spend their whole lives as nonbreeding colony defenders.

In Mongolian gerbils male fetuses vary in their intra-uterine positions, and this variation is reflected in adult testosterone levels. Males gestated between two males (2M males) have higher testosterone levels when adults than their brothers that were gestated between two females (2F males) (Clark *et al.* 1992b). This intrauterine position has a major impact in the development of male sex characters and sexual behavior: 2F males have reduced bulbocavernosus muscle mass (involved in penile erection) and alterations in their copulatory and scent-marking behavior, achieving a lower reproductive success than their 2M siblings (Clark *et al.* 1990, 1992a). Conversely, 2F males express more paternal behavior than the 2M males (Clark *et al.* 1998). Among 2F males some individuals that have extremely low levels of circulating testosterone (similar to those of females) show no interest in receptive females, failing to impregnate them when they are paired. Therefore, nonbreeding 2F males are incapable of direct reproduction and are obligate helpers (Clark and Galef 2000).

---

of visual, chemical, or acoustic courtship signals; defense of breeding territories) (see Chapter 1 and Taborsky 1997), parasitic males, in contrast, exploit the investment made by the bourgeois males to get access to mates (e.g., female mimicry, sneaking, satellite) (see Chapter 1 and Taborsky 1997). Therefore, the traits selected in the two male types are usually divergent. In bourgeois males, traits related with mate attraction and monopolization will be favored by selection, while in parasitic males, traits that increase the probability of stealing fertilizations from bourgeois males will prevail. This disruptive selection acting on a constellation of phenotypic traits may result in the creation of phenotypic mosaics in which both male and female traits are expressed in the same individual, as is the case with parasitic males that mimic female morphology and behavior to get access to fertilization events (e.g., female mimicry in sneaker males of the peacock

blenny, *Salaria pavo*: Gonçalves *et al.* 1996, Gonçalves *et al.* 2005). In this example, the expression of male reproductive behavior and male secondary sex characters become dissociated from the differentiation of a functional male gonad. Classically, male sexual differentiation involves the action of androgens (e.g., testosterone), which, in a cascade of events, promote the masculinization of different body parts (see Box 7.2 on sexual differentiation in vertebrates). However, ARTs offer the possibility to gain insight into the proximate mechanisms underlying sexual differentiation, since in the parasitic tactic, gonadal maturation and spermatogenesis can be dissociated from the expression of behavioral and morphological male traits (Figure 7.1). The decoupling of different male traits in parasitic males may be achieved by different means (e.g., by variation in the local micro-environments in target tissues, as a result of differential

Box 7.2   Sex determination in vertebrates

What determines sex in an individual starts with a blue-print laid out in the genetic material organized in chromosomes, referred to as *genetic sex*. In most verte-brates, sex chromosomes contain the most important genes required for the developing gonad to differentiate according to the genetic plan into an ovary or a testis, referred to as *gonadal sex*. As the gonads develop they start to secrete hormones that will act on the urogenital system, central nervous system, and external features to promote the secondary sexual characteristics originating what we recognize from behavior and appearance as the *phenotypic sex*.

During early development two urogenital ridges along the entire length of the dorsal body wall originate from the vertebrate mesoderm; the mid portion of these ridges differentiates into a single genital ridge from which a bipotential gonad originates. The urinary and reproductive systems are therefore closely associated, and in more primitive vertebrates, they share common ducts.

In eutherian (placental) mammals, maleness is deter-mined by the Y-chromosome being present in normal indi-viduals. This chromosome contains one-third of the number of genes present in the X-chromosome, some inactive, and includes *SRY* (Sex determining Region on Y). SRY protein acts on the bipotential gonad to initiate a cascade of gene expression leading to the development of the testis (Morrish and Sinclair 2002). One of the essential factors expressed specifically in the testis differentiation pathway is *SOX9*, an autosomal gene also involved in cartilage and bone forma-tion. Both *SOX9* and *SRY* are thought to have derived from *SOX3*, located in the X-chromosome. As soon as a testis is formed, Sertoli cells start secreting antimüllerian hormone (AMH), which inhibits the differentiation of Müllerian ducts into female reproductive tract structures (fallopian tubes, uterus, and part of the vagina), and Leydig cells secrete testosterone, which promotes the differentiation of the Wolffian ducts into seminiferous tubules, vas deferens, and seminal vesicle. However, for the differentiation of the external genitalia (prostate, scrotum, and penis), testosterone needs to be converted to 5α-dihydrotestosterone through the action of 5α-reductase.

In the female differentiation pathway, *SRY* is absent and DAX1, the product of a gene located in the X-chromosome, is thought to inhibit *SOX9* expression and therefore inhibit the male differentiation pathway (Swain *et al.* 1998). The expression of DAX1 itself is upregulated by WNT4, a factor that is also essential for Müllerian duct formation and steroidogenesis (Mizusaki *et al.* 2003). In the mammalian female, differentiation of the ovary and external genitalia proceeds without the intervention of sex steroid hormones, which led to the notion that female differentiation is "passive." However, it is, like the male pathway, an active process in which failure in one step can lead to partial or total phenotypic sex reversal.

Phenotypic sex reversal can happen as a result of gene duplication, deletion, inversion, or mutations, which ori-ginate a higher or lower formation of gene product. This is the concept of sex related to gene dosage (number of copies of a gene), which is thought to be the ancestral form of sex determination. For example, any of these conditions ori-ginate a female phenotype in XY individuals: absence of *SRY*, two copies of *DAX1*, one copy of *SOX9*, or one copy of *SF1* (steroidogenic factor 1, a factor required for ster-oidogenesis). Three copies of *SOX9* in XX individuals will also originate a male phenotype. In marsupial mammals, gonadal sex is also determined by the presence of a Y-chromosome, but the development of female pouch versus male scrotum depends on X-chromosome dosage (Vaiman and Pailhoux 2000).

Sex determination mechanisms evolve rapidly, and this has resulted in the independent development of sex chromosomes throughout the vertebrates. The monotremes (egg-laying mammals) appear to have a hybrid between the mammalian XY chromosome system and the avian WZ/ZZ system (Grutzner *et al.* 2004). In birds WZ/ZZ sex chromosomes are universal (female heterogamety). Male and female heterogamety is present in reptiles, amphibians, and fish. Environmental sex determination (ESD) is com-mon in reptiles, but it is also present in amphibians and fish. Parthenogenesis has been reported in reptiles and fish, and polygenic systems are present in several fish species (Kraak and Pen 2002).

Only mammals, except monotremes, have the master sex determining gene *SRY*. In other species only in medaka fish (*Oryzias latipes*) has a master sex-determining gene been found – *DMY*, related to *DMRT1* (also important in the male sex-differentiation pathway) (Matsuda *et al.* 2002, Nanda *et al.* 2002). However, it is absent in some popula-tions of the same species and other fishes (Volff *et al.* 2003).

Other than *SRY*, it appears that most of the above fac-tors indicated as important in mammalian sex differentiation are also present and are expressed at the appropriate time during development in nonmammalian vertebrates, which may indicate common mechanisms (Smith and Sinclair 2004). However, unlike in mammals, in birds and in other vertebrates, steroids are required for the development of the female pathway – androgens promote testicular development and estrogens ovarian development. Thus, the non-mammalian female gonad expresses aromatase, which converts testosterone to estradiol-17β inducing its feminization (Sarre et al. 2004).

The most common form of ESD is through the action of incubation temperature (TSD). The temperature at which embryos are incubated influences the activity of steroidogenic enzymes, in particular aromatase. The inhibition of aromatase leads to the accumulation of testosterone and masculinization, while optimum temperatures for aromatase activity favor the ratio of estrogen to androgen and feminization (Pieau and Dorizzi 2004). Socially induced ESD will ultimately influence steroidogenic enzymes to promote sex change in fishes (Devlin and Nagahama 2002).

Figure 7.2 shows a schematic representation of the sex determination pathway in mammals.

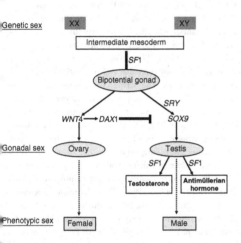

Figure 7.2 Schematic representation of the sex determination pathway in mammals.

expression of receptors or of differential levels of activity of steroidogenic enzymes that modulate the availability of the active hormone) (see Section 7.6).

7.3 PROXIMATE CAUSES OF PHENOTYPIC PLASTICITY: NEURAL AND ENDOCRINE MECHANISMS

7.3.1 Neural structural reorganization versus biochemical switching

Structural reorganization and biochemical switching have been recognized as the major mechanisms underlying behavioral plasticity (Zupanc and Lamprecht 2000). Structural reorganization of neural networks underlying behavior may include processes such as neurogenesis, synaptogenesis, apoptosis, and changes in the dendritic structure of neurons that lead to the differentiation of new neural circuits. These processes are not necessarily restricted to early developmental phases, since adult neurogenesis, for example, has been demonstrated to occur in a variety of vertebrates including humans (Alvarez-Buylla and Lois 1995, Zupanc 2001, Ming and Song 2005). Neural structural reorganization leads to changes in the properties of the networks and therefore in their behavioral output. Functional changes in neural networks activity may also be achieved by alterations of glia cells. For example, changes in astrocyte volume may alter the area of neuronal membrane that is juxtaposed in adjacent neurons. Therefore, glial withdrawal (which can be induced by water deprivation) could increase the area of contact between neurons, potentially leading to an increased excitability of these cells (Zupanc and Lamprecht 2000). In summary, structural reorganization can occur at different life-history stages and involves the modification of the structure of neurons and/or glial cells. As a result, behavioral changes that depend on this mechanism are expected to be slow, long-lasting, and drastic.

In contrast, biochemical switching involves the modulation of synaptic transmission within circuits that are not being rearranged. The main neuromodulators that have been identified include catecholamines, serotonin, and neuropeptides (Zupanc and Lamprecht 2000). Since neuropeptides and catecholamines can be released in a non-synaptic fashion, they may act on larger areas of the central nervous system by diffusion, which would allow them to influence more than one behavioral system at a time. Biochemical switching is thus a mechanism that allows for reversible behavioral output and underlies faster, gradual, or transient changes (Zupanc and Lamprecht 2000).

These potential neural mechanisms underlying pheno-typic plasticity have a parallel in hormonal mechanisms: structural (re)organization of neural circuits can be influenced by organizational effects of hormones during well-defined, sensitive periods in the life of an individual, while biochemical switches can be driven by activational effects of hormones on central pathways underlying behavior (for a review on organizational vs. activational effects of hormones in vertebrates see Arnold and Breedlove 1985).

Therefore, it is predicted that reversible tactics that require rapid and transient changes in neural activity are mediated by biochemical switches influenced by hormones in an activational fashion, whereas fixed and sequential tactics, which involve, in the first case, an organization of the phenotype early in the development or, in the second case, a post-maturational reorganization of the phenotype, are mediated by structural reorganization of neural networks. Concomitantly, the role of hormones in the expression of the different types of tactics should differ: organizational (or reorganizational) effects should be associated with fixed and sequential tactics, activational effects with reversible tactics.

7.3.2 Organizational versus activational effects of hormones

The action of hormones, in particular sex steroids, on behavior has been classically divided into activational and organizational effects. Activational effects are transient and occur throughout the lifespan of the individual, while organizational effects are long-lasting and occur early in ontogeny, typically during a critical period of development (Arnold and Breedlove 1985). This dichotomy of sex hormone action was initially proposed by Phoenix and co-authors (1959) and assumes that activational effects act through the activation of neural circuits that are already present, whereas organizational effects require the organization of new neural circuits at critical periods during development.

The use of the dichotomy between activational and organizational effects of hormones has also been proposed by Moore (1991) as a conceptual framework for the hormonal basis of ART, and it is known as the *relative plasticity hypothesis*. The rationale behind this hypothesis is that the effects of hormones in the differentiation of alternative reproductive tactics are equivalent to their effects in primary sex differentiation (Moore 1991). Thus, by making a distinction between fixed alternative phenotypes (in which individuals adopt one of the tactics for their entire life) and

flexible alternative phenotypes (in which individuals may switch tactics during their lifetime), Moore (1991) proposed an organizational-like role for hormones in the former case and an activational-like role in the latter case. Two predictions can then be extracted from this hypothesis (Moore 1991). (1) In species with plastic ARTs, hormone levels should differ between adult alternative morphs; in species with fixed ARTs, adult hormone profiles should be similar among alternative morphs, except when morphs experience different social environments (Moore 1991). (2) In species with plastic ARTs, hormone manipulations should be effective in adults but not during early development (activational effect); in fixed ARTs hormone manipulations should be effective during early development but not in adults (organizational effect). More recently, a second generation of the relative plasticity hypothesis has been proposed (Moore et al. 1998). This revised version emphasizes the distinction between reversible and irreversible phenotypes among plastic tactics and between conditional and unconditional fixed tactics. Accordingly, the plastic, reversible tactics would be the true equivalents of activational effects of hormones, and thus, the original predictions of the relative plasticity hypothesis would only apply to this type of alternative tactic. The plastic, irreversible (i.e., sequential) ARTs would represent a post-maturational reorganization effect, in which the phenotypic outcome would be produced immediately (Moore et al. 1998). Thus, hormone differences needed to differentiate the two alternative phenotypes need not be permanent and may only be present during the transitional phase. Among the fixed ARTs, the distinction between conditional and unconditional fixed tactics has no consequences for the predictions concerning the endocrine mechanisms of ARTs, with organizational actions being predicted in both cases (Moore et al. 1998). Thus, the predictions of Zupanc and Lamprecht (2000) for the neural mechanisms underlying phenotypic plasticity and those of the relative plasticity hypothesis are in good agreement (Table 7.2).

7.3.3 Endocrine candidates: sex hormones, glucocorticoids, and neuropeptides

Sex steroids, glucocorticoids, and neuropeptides emerge as candidates to play a major role in the differentiation and maintenance of alternative reproductive morphs. As mentioned above, sex steroids have an essential role in sexual differentiation and in the control of male reproduction in vertebrates (e.g., Dixon 1998, Wilson et al. 2002, Nelson

Table 7.2. *Neural and hormonal mechanisms of alternative reproductive tactics in vertebrates*

ART type	Neural mechanism (Zupanc and Lamprecht 2000)	Hormonal mechanism following the relative plasticity hypothesis v.1 (Moore 1991)	Hormonal mechanism following the relative plasticity hypothesis v.2 (Moore et al. 1998)
Fixed	Structural organization	Organizational effect	Organizational effect (post-maturational)
Sequential	Structural reorganization	Activational effect	Organizational effect
Reversible	Biochemical switching	Activational effect	Activational effect

2005). In particular androgens participate in the differentiation of primary and secondary sex characters, in the expression of reproductive behavior, in the feedback regulation of the hypothalamus and pituitary, and in spermatogenesis (Nelson 2005, Oliveira 2005). These pivotal roles in reproduction make them the preferential target for studies of endocrine correlates of male ARTs. However, as discussed below, the development of male ARTs is likely to be influenced by the neuroendocrine system in addition to gonadal steroids.

Glucocorticoids play an important role as mediators of interindividual variation in social behavior. One classic example of such an effect is provided by a series of studies on the relationship between social status and cortisol levels among free-living male olive baboons (*Papio anubis*) in an African national park (Sapolsky 1983, Sapolsky and Ray 1989, Virgin and Sapolsky 1997). In stable social hierarchies, dominant males have lower basal cortisol concentrations than do subordinates, but these differences disappear at times of social instability when all males show elevated basal cortisol levels and suppressed cortisol responsiveness to stress (Sapolsky 1983). Moreover, within high- and low-ranking males, individuals adopting different behavioral profiles also share different endocrine profiles. Among dominant males, only those with a high degree of social skill (e.g., those that are able to distinguish between threatening and neutral interactions with rivals and therefore more likely to initiate fights in the first but not in the latter case) had lower basal cortisol titers. Dominant males lacking these skills had cortisol levels as high as subordinates (Sapolsky and Ray 1989). Also among low-ranking males, a subset of individuals with high rates of consortships had higher cortisol levels than subordinates who had high rates of surreptitious copulations. This might reflect the stress experienced by the former subset of subordinates, which adopt a precocious strategy of open reproductive competition with the dominant males (Virgin and Sapolsky 1987). Overall, these studies suggest that glucocorticoid profiles are associated with distinctive behavioral styles. Moreover, glucocorticoids can interact with the HPG axis and thus modulate the expression of reproductive traits (Sapolsky et al. 2000).

Finally, studies of two forebrain neuropeptide systems may help us to understand the differentiation of ARTs: gonadotropin-releasing hormone (GnRH) and arginine vasopressin (AVP; or arginine vasotocin [AVT] in non-mammalian vertebrates). GnRH plays a central role in the control of vertebrate reproduction by orchestrating the functioning of the HPG axis (Parhar 2002) and AVP/AVT influences the expression of social behavior patterns, including courtship behavior, in a wide range of vertebrates (Goodson and Bass 2001). Since both neuropeptide systems have been reviewed in the light of ARTs (Foran and Bass 1999, Bass and Grober 2001) and will be addressed in a separate chapter in this volume (see Chapter 6), we will limit this review to the evidence for the involvement of sex steroids and glucocorticoids in ARTs in the next two sections.

7.4 SEX HORMONES AND ARTs: THE RELATIVE PLASTICITY HYPOTHESIS AND BEYOND

7.4.1 Testing the relative plasticity hypothesis: the first prediction

In order to look for associations between patterns of circulating sex hormone levels (i.e., gonadotropins, androgens, estrogens, and progestogens) and the expression of alternative reproductive morphs in the different classes of vertebrates, we have surveyed the published literature (see Table 7.3).

Table 7.3. *Comparison of hormone levels between alternative reproductive tactics in male vertebrates*

CLASS/Family/ Species	Alternative phenotypes	Intrasexual dimorphism	Pituitary tropic hormones	Androgens	Estrogens	Progestogens	Glucocorti- coids	References
TELEOSTEI								
Batrachoididae								
Lusitanian toadfish, *Halobatrachus didactylus*	Fixed?	+	?	KT: B>P T: B=P	E2: B=P	B=P	?	Modesto and Canário 2003a
Plainfin midshipman, *Porichthys notatus*	Fixed: type I calling vs. type II noncalling males	+	?	KT: B>P T: B<P	E2: B=P	?	?	Brantley *et al.* 1993b
Blenniidae								
Peacock blenny, *Salaria pavo*	Sequential: nest-holders vs. female-mimic sneakers	+	?	KT: B>P[a] T: B>P[a]	?	?	?	Oliveira *et al.* 2001b
Rock-pool blenny, *Parablennius parvicornis*	Sequential: nest-holders vs. satellites	+	?	KT: B>P T: B=P	?	?	?	Oliveira *et al.* 2001c
Centrarchidae								
Bluegill sunfish, *Lepomis macrochirus*	Fixed: parentals vs. sneakers and satellites	+	?	KT: B>P T: B=P	?	?	?	Kindler *et al.* 1989
Longear sunfish, *Lepomis megalotis*	Fixed?	+	?	KT: B>P T: B=P	?	?	F: B<P	Knapp 2003
Cichlidae								
Princess of Burundi, *Neolamprologus pulcher*	Sequential: breeders vs. helpers	–	?	KT: B=P[b] T: B=P[b]	?	?	?	Oliveira *et al.* 2003
Mozambique tilapia,	Reversible: territorial courting vs.	+	?	uKT: B>Pu T: B>P	?	uP: B>P	?	Oliveira *et al.* 1996, Oliveira

Species	Pattern			KT / T	E2 / B		Reference
Oreochromis mossambicus	nonterritorial female mimics						and Almada 1998a
St. Peter's fish, *Sarotherodon galilaeus*	Reversible: monogamous vs. polygynous males	−	?	KT: B=P; T: B=P	B=P	?	Ros et al. 2003
Labridae							
Corkwing wrasse, *Symphodus melops*	Fixed: territorial vs. female mimics	+	?	KT: B>P; T: B<P	E2: B<P	?	Uglem et al. 2000, 2002
Rainbow wrasse, *Coris julis*	Sequential: initial-phase vs. terminal-phase males	+	?	KT: B>P[c]	?	?	Reinboth and Becker 1984
Saddleback wrasse, *Thalassoma duperrey*	Sequential: initial-phase vs. terminal-phase males	+	?	KT: B>P; T: B=P	E2: B=P; B=P	?	Hourigan et al. 1991
Poeciliidae							
Sailfin molly, *Poecilia velifera*	Reversible: large, courting vs. small, noncourting males	−	?	KT: B=P[b]; T: B=P[b]	?	?	R. F. Oliveira, D. M. Gonçalves, and I. Schlupp, unpubl. data
Salmonidae							
Atlantic salmon, *Salmo salar*	Fixed: mature parr vs. anadromous males	+	?	KT: B>P[d]; T: B<P[d]	B=P[d]	?	Mayer et al. 1990
Scaridae							
Stoplight parrotfish, *Sparisoma viride*	Sequential: initial-phase vs. terminal-phase males	+	?	KT: B>P; T: B>P	E2: B<P	?	Cardwell and Liley 1991
Serranidae							
Belted sunfish, *Serranus subligarius*	Reversible: streakers vs. pair spawners in a simultaneous hermaphrodite	−	?	KT: B>P	B>P	?	Cheek et al. 2000

Table 7.3. (cont.)

CLASS/Family/Species	Alternative phenotypes	Intrasexual dimorphism	Pituitary tropic hormones	Androgens	Estrogens	Progestogens	Glucocorti-coids	References
AMPHIBIA								
Bufonidae								
Great plains toad, *Bufo cognatus*	Reversible: callers vs. satellites	−	?	DHT: B = P T: B = P	?	?	B: B > P	Leary *et al.* 2004
Woodhouse's toad, *Bufo woodhousii*	Reversible: callers vs. satellites	−	?	DHT: B = P T: B = P	?	?	B: B > P	Leary *et al.* 2004
Ranidae								
Bullfrog, *Rana catesbeiana*	Reversible: callers vs. satellites	−	?	DHT: B < P T: B < P	?	?	B: B > P	Mendonça *et al.* 1985
REPTILIA								
Iguanidae								
Marine iguana, *Amblyrhynchus cristatus*	Sequential: territorial vs. satellite vs. sneaker	+	?	T: B > P1 = P2	?	?	?	Wikelski *et al.* 2005
Gekkonidae								
Leopard gecko, *Eublepharis macularis*	Fixed: males from male-biased incubation temperature vs. males from female-biased incubation temperature (more sexually active and less aggressive)	+(males from male-biased incubation temperature with larger snout–vent length)	?	T: B = P DH T: B = P	E2: B < P	?	B: B = P	Flores *et al.* 1994, Tousignant and Crews 1995, Crews *et al.* 1998
Phrynosomatidae								
Tree lizard, *Urosaurus ornatus*	Fixed: territorial vs. nomadic rovers/sedentary satellites	+(dewlap color: orange with blue spot, orange)	?	T: B = P T: P₁ < P₂	?	?	B: B = P, P₁ > P₂	Thompson and Moore 1992, Moore *et al.* 1998, Knapp *et al.* 2003

Side-blotched lizard, *Uta stansburiana*	Fixed: ultraterritorial vs. territorial vs. nonterritorial female mimics	+(throat color: orange, blue and yellow morphs)	?	?	T: B > P$_1$ = P$_2$	?	Sinervo et al. 2000
Colubridae							
Red-sided garter snake, *Thamnophis sirtalis parietalis*	Sequential: he-males vs. she-males	+(skin lipids)	?	?	T: B < P	?	Mason and Crews 1985
AVES							
Brown-headed cowbird, *Molothrus ater*	Reversible: paired vs. unpaired (EPC)	–	?	?	T: B > P	?	Dufty and Wingfield 1986
House finch, *Carpodacus mexicanus*	Fixed: dominant and no parental care vs. subordinate and parental care	+(plumage: dull vs. redder)	?	?	T: B < P	?	Duckworth et al. 2004
White-throated sparrow, *Zonotrichia albicollis*	Fixed: dominant singing vs. subordinate and less singing	+(plumage: white stripe vs. tan stripe)	?	?	T: B > P	?	Maney et al. 2005
Pied flycatcher, *Ficedula hypoleuca*	Reversible: polyterritorial vs. home-territorial	–	LH: B > P	?	T: B > P	?	Silverin and Wingfield 1982
Seychelles warbler, *Acrocephalus sechellensis*	Sequential: primary males vs. subordinate males (cooperative breeding, EGC)	–	?	?	T: B > P	?	Crommenacker et al. 2004
Florida scrub-jay, *Aphelocoma coerulescens*	Sequential: nest-owners vs. non-nesting (EPC)	–	?	?	T: B = P	?	Vleck and Brown 1999
Acorn woodpecker, *Melanerpes formicivorus*	Sequential: breeder vs. helper (cobreeder)	–	?	?	T: B = P	?	Koenig and Dickinson, this volume
Australian magpie, *Gymnorhina tibicen*	Sequential/reversible: breeding adults vs.	+(plumage between adult	?	?	T: B1 = P1 B2 > P2	?	Schmidt et al. 1991

Table 7.3. (*cont.*)

CLASS/Family/Species	Alternative phenotypes	Intrasexual dimorphism	Pituitary tropic hormones	Androgens	Estrogens	Progestogens	Glucocorticoids	References
	nonbreeding adults vs. breeding subadults vs. nonbreeding subadults	and subadult morphs)	LH: B1=P1 B2>P2	DHT: B1=P1 B2>P2				
Azure-winged magpie, *Cyanopica cyanus*	Reversible: breeder vs. helper (EPC)	–	?	T: B=P	?	?	?	De la Cruz *et al.* 2003
Bell miner, *Manorina melanophrys*	Sequential: breeder vs. helper (EPC)	–	?	T: B>P	?	?	?	Poiani and Fletcher 1994
Superb fairy-wren, *Malurus cyaneus*	Sequential: dominant group breeder vs. paired vs. helper (EGC)	–	?	T: B>P1=P2	?	?	?	Peters *et al.* 2001
Pied kingfisher, *Ceryle rudis*	Sequential: breeders vs. primary helpers vs. secondary helpers	–	LH: B=P1=P2	T: B=P2>P1	?	?	?	Reyer *et al.* 1986
Harris's hawk, *Parabuteo unicinctus*	Sequential: adult breeders vs. adult-plumaged helpers vs. juvenal-plumaged helpers	+(plumage)	LH: B=P1>P2	T: B=P1>P2	E2: B=P1=P2	?	B: B=P1=P2	Mays *et al.* 1991
MAMMALIA								
Hominidae								
Human, *Homo sapiens*	Reversible: polygynously married Swahili men vs. monogamously married Swahili men	–	?	sT: AT1 > AT2[e] ?	?	?	?	Gray 2003

144

Species	Mating system	Secondary trait	Hormones	T		F	References
Chimpanzee, *Pan troglodytes schweinfurthii*	Reversible: dominant vs. subordinate (surreptitious matings)	–	?	$_u$T: B > P $_f$; T: B > P	?	$_u$F: B > P	Muehlenbein et al. 2004, Muller and Wrangham 2004a, 2004b
Orang-utan, *Pongo* spp.	Sequential: territorial flanged males (courting with long call vocalization) vs. nomadic unflanged males (forced copulations)	+(facial flanges, long, thick hair and a vocal throat sac in territorial males)	$_u$LH: B > P; $_u$FSH: B = P; $_u$GH: B > P; $_u$TSH: B = P	$_u$T: B > P; $_u$DHT: B > P	?	$_u$F: B = P	Maggioncalda et al. 1999, 2000, 2002
Cercopithecidae							
Mandrill, *Mandrillus sphinx*	Reversible: social fatted males vs. solitary nonfatted males (sneak matings in group incursions)	+(brilliant red and blue skin in the face and fat rumps in alfa social males)	?	T: B > P	?	?	Setchell and Dixson 2001
Olive baboon, *Papio anubis*	Sequential: dominant (maintain consortships) vs. subordinate (stolen copulations)	–	?	T: B > P $_f$	?	F: B > P $_f$	Sapolsky 1983, Virgin and Sapolsky 1997, Ray and Sapolsky 1992, Rose et al. 1971
Rhesus macaque, *Macaca mulatta*	Plastic: dominant vs. subordinate (surreptitious matings)	–	?	T: B > P	?	?	Barrett et al. 2002
Japanese macaque, *Macaca fuscata*	Plastic: dominant (multiday consortships with estrous female) vs. subordinate (forced copulations)	–	?	$_f$T: B = P	?	$_d$F: B > P	
Cebidae							
Mantled howling monkey, *Alouatta palliata*	Sequential: dominant group mating vs. subordinate (opportunistic mating)	–	?	$_f$T: B > P	?	?	Jones 1995, Zucker et al. 1996

145

Table 7.3. (cont.)

CLASS/Family/Species	Alternative phenotypes	Intrasexual dimorphism	Pituitary tropic hormones	Androgens	Estrogens	Progestogens	Glucocorticoids	References
Tufted capuchin monkey, *Cebus paella nigritus*	Reversible: dominant (courtship followed by multiple mounts) vs. subordinate (rapid copulation)	—	?	$_f$T: B = P	?	?	$_f$F: B = P	Lynch *et al.* 2002, Lynch Alfaro 2005
Callitrichidae								
Common marmoset, *Callithrix jacchus*	Sequential: dominant breeder vs. subordinate helper (copulation with extragroup females during intergroup encounters)	—	?	T: B = P	?	?	?	Abbott and Hearn 1978, Baker *et al.* 1999, Digby 1999
Indridae								
Sifaka, *Propithecus verreauxi*	Plastic: dominant (scent marking and mate guarding) vs. subordinate	—	?	$_f$T: B > P	?	?	?	Kraus *et al.* 1999
Lemuridae								
Ring-tailed lemur, *Lemur catta*	Plastic: dominant vs. subordinate (group transfer?)	—	?	$_f$T: B > P	?	?	?	Cavigelli and Pereira 2000, Gould 2005
Cricetidae								
Mongolian gerbil, *Meriones unguiculatus*	Fixed (intrauterine position): 2M males vs. 2F males (reduced copulatory behavior and scent marking; alloparenting)	+(reduced bulbocavernosus muscle mass)	?	T: B > P	?	?	?	Clark and Galef 2000

Taxon	Alternative phenotype	Morphology		Androgens		Glucocorticoids	Reference
Sciuridae Alpine marmot, *Marmota marmota*	Sequential: dominant territorial vs. subordinate kin vs. subordinate nonkin	—	?	T: B = P1 > P2	?	GC: B = P2 > P1	Arnold and Dittami 1997
Bathyersidae Naked mole-rat, *Heterocephalus glaber*	Fixed: colony defenders vs. dispersers	+(higher girth-to-body length and more fat in the neck region)	LH: B < P	?	?	?	O'Riain et al. 1996
	Plastic: within-colony breeders vs. nonbreeders	—	?	uT: B > P	?	uF: B = P	Clarke and Faulkes 1998
Elephantidae African elephant, *Loxodonta africana*	Reversible: musth vs. sexually active without musth	—	?	fEA: B > P	?	fGC: B = P	Ganswindt et al. 2005
Rhinocerotidae White rhino, *Ceratotherium simum*	Sequential: territorial vs. nonterritorial (satellite)	—	?	fT: B > P	?	?	Rachlow et al. 1998
Equidae Plains zebra, *Equus burchelli*	Sequential: harem breeding vs. bachelor	—	?	uT: B > P	?	?	Chaudhuri and Ginsberg 1990
Grevy's zebra, *Equus grevyi*	Sequential: territorial vs. bachelor	—	?	uT: B > P	?	?	Chaudhuri and Ginsberg 1990
Shetland pony, *Equus caballus*	Sequential: harem stallion vs. bachelor	—	?	T: B > P	?	?	McDonnell and Murray 1995
Misaki feral horse, *Equus caballus*	Sequential: harem stallion vs. bachelor	—	?	T: B > P	?	?	Khalil et al. 1998
Przewalski horse, *Equus ferus przewalski*	Sequential: harem stallion vs. bachelor	—	?	fT: B > P fEA: B > P	fE2: B > P	?	Schwarzenberger et al. 2004

Table 7.3. (cont.)

CLASS/Family/Species	Alternative phenotypes	Intrasexual dimorphism	Pituitary tropic hormones	Androgens	Estrogens	Progestogens	Glucocorticoids	References
Bovidae								
Plains bison, *Bison bison*	Sequential: tending vs. nonguarding	−	?	$_f$T: B > P	?	?	?	Mooring et al. 2004
Bighorn sheep, *Ovis canadensis*	Sequential: tending vs. coursing/blocking	+	?	$_f$T: B > P	?	?	?	Pelletier et al. 2003
Cervidae								
Impala, *Aepyceros melampus*	Sequential: territorial vs. bachelor	−	?	T: B > P	?	?	?	Illius et al. 1983
Viverridae								
Dwarf mongoose, *Helogale parvula*	Sequential: dominant breeder vs. subordinate helper	−	?	$_u$T: B = P	?	?	?	Creel and Waser 1994, Creel et al. 1992
Meerkat, *Suricata suricatta*	Sequential: dominant breeder vs. subordinate helper	−	LH: B = P	$_f$T: B = P$_u$ T: B = P	?	?	?	O'Riain et al. 2000a, Moss et al. 2001
Felidae								
African lion, *Panthera leo*	Sequential: dominant vs. subordinate	+(darker mane vs. lighter mane)	?	T: B > P	?	?	?	West and Packer 2002
Canidae								
Gray wolves, *Canis lupus*	Sequential: dominant vs. subordinate helper	−	?	?	?	?	$_f$F: B = P	Sands and Creel 2004
African wild dog, *Lycaon pictus*	Sequential: dominant vs. subordinate	−	?	$_f$T: B > P	?	?	$_f$F: B > P	Creel et al. 1997

148

Phocidae								
Harbor seal, *Phoca vitulina*	Sequential: larger hauled-out alone vs. medium-sized with low fidelity to haul-out site	—	?	T: B=P	?	?	?	Coltman *et al.* 1999
Weddell seal, *Leptinychotes weddelli*	Sequential: territorial vs. nonterritorial	—	?	T: B > P	?	?	F: B > P	Bartsch *et al.* 1992

References: For each species we give the reference that has reported the relative hormone levels of the alternative tactics. In cases in which the endocrine data is given in a paper that does not mention the ARTs, we also cite a paper that documents the occurrence of alternative tactics in that species (e.g., one paper reports that dominant males of a given species have higher testosterone levels than subordinates and an independent paper suggests that subordinate males of that same species use an alternative mating tactic).

Hormone abbreviations: LH, luteinizing hormone; FSH, follicle stimulating hormone; GH, growth hormone; TSH, thyroid stimulating hormone; T, testosterone; KT, ketotestosterone; DHT, dihydrotestosterone; EA, epiandrosterone; E2, estradiol; P, progestogen (can vary across taxa, e.g., progesterone in mammals and various kinds among teleost fish); F, cortisol; B, corticosterone; GC, antibody used had higher cross reactivity with more than one glucocorticoid.

Hormone prefixes: u, urinary levels; f, fecal levels; s, salivary levels.

Alternative tactic abbreviations: B, bourgeois tactic; P, parasitic tactic (P1 and P2 are used when there is more than one alternative tactic); EPC, extra-pair copulation; EGC, extra-group copulation.

[a] Testicular androgen levels.

[b] Steroid levels in fish holding water.

[c] In vitro gonadal production from [14C] T incubation.

[d] Values for late summer, when GSI values peak.

[e] In some species alternative tactics occur that do not match the functional classification of bourgeois vs. parasitic males, e.g., polygynously vs. monogamously married Swahili men; in these cases we have used AT1 and AT2 as abbreviations for alternative tactic 1 and alternative tactic 2.

[f] Only present in stable hierarchies.

149

A clear association exists between androgen levels and the expression of one of the alternative reproductive tactics (Table 7.3). For the majority of the species, the conventional morph has higher levels of androgens than the alternative morphs, but in many other cases, there are no significant differences in androgens between the two alternative morphs, and in some cases the parasitic males may even have higher androgen levels than the bourgeois males (Table 7.3). How can such variability be explained?

Could this variability be explained by the first prediction of Moore's reproductive plasticity hypothesis – that hormone profiles should differ in plastic adult morphs but not in fixed ones?

Unfortunately, the relative plasticity hypothesis is flawed. Androgen levels not only influence behavior (and thus can be expected to play an activational role in species with plastic ARTs), but they can also be influenced by the social environment in which the animal lives (Wingfield et al. 1990, Oliveira et al. 2002, Oliveira 2004). This means that any conclusions derived from finding different levels of androgens in alternative reproductive morphs (either fixed or plastic) are suspect. Moore (1991) argued that in fixed ARTs, adult hormone profiles should be similar among alternative male phenotypes, except when alternative morphs experience different social environments (see also Thompson and Moore 1992). Therefore, positive associations, negative associations, and even the lack of an association between androgen levels and the ART type are to be expected. As a result, the study of androgen levels in species with plastic ARTs is far more informative. In fact, among plastic species androgen levels should differ between the alternative morphs, and any negative result (lack of difference) cannot be explained by differential influences of the social environment on the androgen levels of the alternative phenotypes. Thus, the most robust estimate of this prediction is to compute the percentage of plastic species in which there are no differences in circulating levels between the bourgeois and the parasitic morph. In order to make this exercise easier and to avoid potential phylogenetic bias (i.e., bias introduced by some patterns being more characteristic of some vertebrate classes than others), the raw data from Table 7.3 were reorganized into contingency tables for each vertebrate class (the data for amphibians and reptiles were pooled into a single table owing to the low number of species for which endocrine data on ARTs are available) (Tables 7.4 through Table 7.7). In these tables, the shaded background cells represent cases that support the first prediction of the relative plasticity hypothesis and the white background cells

Table 7.4. *Test of the first prediction of the relative plasticity hypothesis in fish*

Androgen levels	ART type	
	Fixed	Plastic
Bourgeois > Parasitic	Plainfin midshipman Lusitanian toadfish Bluegill sunfish Corkwing wrasse Atlantic salmon	Peacock blenny Rock-pool blenny Stoplight parrotfish Rainbow wrasse Saddleback wrasse Mozambique tilapia Belted sunfish
Bourgeois = Parasitic		Princess of Burundi St. Peter's fish Sailfin molly
Bourgeois < Parasitic		

Table 7.5. *Test of the first prediction of the relative plasticity hypothesis in reptiles and amphibians*

Androgen levels	ART type	
	Fixed	Plastic
Bourgeois > Parasitic	Side-blotched lizard	Marine iguana
Bourgeois = Parasitic	Tree lizard	Great plains toad Woodhouse's toad
Bourgeois < Parasitic		Bullfrog Red-sided garter snake

represent those that reject it. The tables illustrate that by using this conservative estimate from the relative plasticity hypothesis, we cannot explain 30% of the occurrences of plastic ARTs in fish, 40% of those in amphibians and reptiles, 54.5% of the plastic ART cases in birds, and 19.4%

Table 7.6. *Test of the first prediction of the relative plasticity hypothesis in birds*

Androgen levels	ART type	
	Fixed	Plastic
Bourgeois > Parasitic		Brown-headed cowbird
		Pied flycatcher
		Seychelles warbler
		Bell miner
		Superb fairy-wren
Bourgeois = Parasitic		Mexican scrub-jay
		Acorn woodpecker
		Australian magpie
		Azured magpie
		Pied kingfisher
		Harris's hawk
Bourgeois < Parasitic	House finch	

Table 7.7. *Test of the first prediction of the relative plasticity hypothesis in mammals*

Androgen levels	ART type	
	Fixed	Plastic
Bourgeois > Parasitic	Mongolian gerbils	Human
		Chimpanzee
		Orang-utan
		Mandrill
		Olive baboon
		Rhesus monkey
		Mantled howling monkey
		Sifaka
		Ring-tailed lemur
		Alpine marmot
		Naked mole-rat
		African elephant
		White rhino
		Plain zebra
		Grevy's zebra
		Shetland pony
		Misaky feral horse
		Przewalski horse
		Plains bison
		Bighorn sheep
		Impala
		African lion
		African wild dog
		Harbor seal
		Weddell seal
Bourgeois = Parasitic		Japanese monkey
		Common marmoset
		Tufted capuchin monkey
		Dwarf mongoose
		Meerkat
Bourgeois < Parasitic		

of mammalian plastic ARTs. This means that the model can potentially explain over 80% of the ART cases in mammals, where sex is genetically determined, males are the hetero-gametic sex, and the expression of their secondary sexual characteristics is androgen dependent. Among other vertebrate classes, where the mechanisms of primary sex determination vary from those present in eutherian mammals, the model loses its predictive power. In birds, females are the heterogametic sex and the expression of male ornaments, a typical bourgeois trait, is, in most cases, not androgen dependent (e.g., male breeding plumage: Owens and Short 1995; but see Kimball and Ligon 1999). In amphibians, reptiles, and fish, primary sex determination mechanisms are more labile and open to influences from the environment, such as temperature or the social context (environmental sex determination, ESD), even though sex chromosomes may be present (Crews 1998). For example, genetic sex determination (GSD) mechanisms in fish, which are present in approximately half the species that have been studied using cytogenetical data, are very diverse. They range from polygenic systems to systems with dominant sex-determining factors, to sex chromosomes with either heterogametic males (XY) or females (ZW) (Devlin and Nagahama 2002). Interestingly, the number of species that display male heterogamety is twice the number of those

with female heterogamety (Devlin and Nagahama 2002), a fact that could, to a degree, explain why fish appear as the second best fit of the model. In summary, an association between the mechanisms of sex determination operating in each animal class and the role of sex hormones on the expression of ARTs seems to be present, which in turn suggests that differences between alternative reproductive morphs within a sex are based on the same mechanisms that generate sex differences within a species (Godwin and Crews 2002). Crews (1998) already pointed out the relationship between the sex-determination mechanism and the type of ART displayed, suggesting that species with fixed tactics should have GSD, whereas species with plastic tactics should have either GSD or ESD (but see Oliveira 2005 for a review of this issue among teleost fish yielding different results). The parallels between the processes of sex differentiation (i.e., males vs. females) and the differentiation of discrete alternative reproductive phenotypes within the same sex further support a role for sex steroids in the differentiation of intrasexual alternative phenotypes.

How can we explain species with fixed ARTs in which androgen levels differ between the alternative phenotypes? As mentioned above differences in sex hormone levels between alternative reproductive male types might not reflect different hormone profiles due to an activational effect on the expression of the bourgeois tactic, but rather might reflect the responsiveness of these hormones to the expression of the tactic itself (Thompson and Moore 1992). That is, they are a consequence and not a cause of the expression of alternative mating tactics. This can be the case if the alternative phenotypes experience different social environments, which is very likely since by definition bourgeois males defend resources to get access to mates and thus are expected to face higher levels of social challenges than parasitic males. For example, in the peacock blenny, nest-holder males show an increase in androgen levels during the breeding season that is positively correlated with an increase in sneaking attempts to which they are exposed (Oliveira et al. 2001a). In only three cases does the parasitic tactic have a higher testosterone level than the bourgeois tactic: the house finch (*Carpodacus mexicanus*), the bullfrog (*Rana catesbeiana*), and the red-sided garter snake (*Thamnophis sirtalis parietalis*). In the house finch, the dull and less ornamented males are dominant over redder males, but the redder males pair earlier and provide more parental care than the dull males (Duckworth et al. 2004). In addition, the higher testosterone levels found in free-living, dull males are probably the result of dull males having a higher motivation to access food resources and are not a direct cause for the differentiation of alternative phenotypes (Duckworth et al. 2004). In the bullfrog, the lower levels of androgens present in calling (bourgeois) males have been interpreted as a stress-related cost due to frequent combat to defend territories (Mendonça et al. 1985). In the red-sided

garter snake, higher androgen levels in recently emerged she-males (which is a phase through which apparently all males go after emerging from winter dormancy: Shine et al. 2000) can be a consequence of the twofold higher mating activity that they experience compared to conventional males (Mason 1992).

Data on progestogens are available for six species with ARTs, all of them teleosts (Table 7.3). Interestingly, progestogens are never higher in the parasitic morph than in the bourgeois morph (they are higher in the bourgeois males than in the parasitic males in two species, and no differences are present in the other four species). However, the progestogen(s) measured varied from species to species. For example 17,20β,21-trihydroxy-4-pregen-3-one (17,20β21P), 17,20α-dihydroxy-4-pregen-3-one (17,20αP), and 17,20βP were measured in the Lusitanian toadfish (Modesto and Canário 2003a); 17,20αP and 17,20βP were assayed in the Mozambique tilapia (Oliveira et al. 1996); 17,20β21P and 17,20βP were determined in the belted sunfish (Cheek et al. 2000); whereas only 17,20βP has been monitored in the saddleback wrasse (Hourigan et al. 1991), in the St. Peter's fish (Ros et al. 2003), and in the Atlantic salmon (Mayer et al. 1990). The available data suggest that 17,20β21P in the toadfish, 17,20βP in the saddleback wrasse, and 17,20βP in the Atlantic salmon may play a role in male reproduction (e.g., spermiation). In the Mozambique tilapia, territorial males have higher levels of both 17,20αP and 17,20βP than nonterritorial, female-mimicking males, but only a 17,20αP increase in the plasma concentration in the presence of females when courtship behavior is expressed by the males (Oliveira et al. 1996), suggesting that 17,20αP may play a major role in spawning behavior and/or spermiation in this species. In the belted sandfish, 17,20β21P rather than 17,20βP seems to be associated with male reproductive behavior (Cheek et al. 2000). In summary, progestogens appear to be associated with the expression of bourgeois reproductive traits, but for most species it is difficult to disentangle potential effects of progestogens on male courtship behavior from effects on spermiation. It is also interesting to note that in the tree lizard, a species with fixed ARTs determined early in ontogeny (see Section 7.4.2), progesterone peaks twice during the critical period, and on both occasions the levels are bimodal at the population level, suggesting a potential involvement of progesterone on morph differentiation (Moore et al. 1998). This is further supported by the fact that approximately 90% of the individuals that received a single injection of progesterone on the day of hatching differentiated into the bourgeois morph (Moore et al. 1998). Future studies should examine the role of progestogens on the expression of ARTs.

Estrogens have also been measured in alternative morphs of five teleosts and in one mammal. Among fish estradiol titers are never higher in the bourgeois morph (they are lower in two cases and equal in the other three; see Table 7.3). In contrast, fecal estrogen levels are significantly higher in stallions than in bachelor males of Przewalski horses (Table 7.3). However, it should be stressed that, in all cases, estrogen levels are almost always very low, suggesting that high circulating estrogen levels are incompatible with the expression of the bourgeois tactic, at least among teleost fish.

Finally, data are available on luteinizing hormone (LH) for seven species (four birds and three mammals). One of the cases for which an LH level is available is an interesting type of ART in which a dispersive morph has been described in naked mole-rats (see Box 7.1). Since it is not clear that the colony defenders are playing a bourgeois tactic and the dispersers a parasitic tactic, no clear prediction can be made for this case; however, it has been found that dispersers exhibit higher LH circulating concentrations than colony defenders (O'Riain et al. 1996). In the remaining six cases in which the adopted functional dichotomy bourgeois-vs.-parasitic tactic seems to be valid, LH levels are never lower in the bourgeois morph (it is higher in two cases and similar in the other four) than in the parasitic morph. In all of these cases, LH perfectly mirrors the differences in androgen levels between morphs (Table 7.3). Therefore, a direct involvement of LH in the differentiation of alternative tactics is not plausible, and the most parsimonious hypothesis for its action upon morph differentiation is through sex steroids.

7.4.2 Testing the relative plasticity hypothesis: the second prediction

As mentioned above, according to the second prediction of the relative plasticity hypothesis, in species with fixed ARTs, hormone manipulations should only be effective early in development (i.e., should have organizational effects), whereas in species with plastic ARTs, the exogenous administration of hormones should be effective in adults (Moore 1991, Moore et al. 1998). Unlike the first prediction, the second prediction does not suffer from epistemological flaws and provides, therefore, a stronger test for the assessment of the relative plasticity hypothesis. Unfortunately, hormone

levels of alternative phenotypes have been manipulated in only 12 species (see Table 7.8 for a survey of the available literature on hormone manipulations in species with ARTs).

In only one case, the tree lizard, have the effects of early administration of androgens to males of a species with fixed ARTs been evaluated. Males treated with testosterone implants the day they hatched developed into the orange-blue morph in a significantly higher proportion than sham-operated males. Conversely, males castrated at the same age preferentially developed into the orange phenotype (Hews et al. 1994). These data support an organizational effect of androgens in the expression of tree lizard ARTs and suggest a well-defined critical period for this effect in the ontogeny of the species. Tree lizard males begin to express their color morphs between days 60 and 90 post-hatching (Moore et al. 1998). Testosterone implants on day 1 and on day 30 were effective in directing morph differentiation, while those performed on day 60 had no effect, indicating the presence of a critical period that ends between day 30 and day 60 post-hatching (Hews and Moore 1996). Another case demonstrating that early exposure to hormones manipulates the expression of ARTs is the Mongolian gerbil. In this species an intrauterine position effect has been described in which males gestated between two females (2F males) have lower testosterone levels when adults than their brothers gestated between two male fetuses (2M males) (see Box 7.1). Some of the 2F males that display exceptionally low levels of circulating testosterone (i.e., similar to those of females) do not express male sexual behavior when exposed to females in oestrus but, in contrast, overexpress allopaternal behavior. Therefore, the early exposure to androgens determines the tactic adopted by male Mongolian gerbils, with some 2F males becoming asexual and obligate helpers (Clark and Galef 2000). These two examples strongly support a straightforward organizational effect of androgens on the development of fixed alternative phenotypes.

The evidence compiled for hormone manipulations in adulthood yields much less clear results (Table 7.9). Of the 11 species that have been studied, only five support Prediction 2. Of the five supportive cases, in two of them (one reptile and one cooperatively breeding bird), the administration of testosterone to the parasitic morph of species with plastic ARTs induced a tactic switch (see Tables 7.8 and 7.9). In a third case, the inhibition of testosterone production reduced the sexual activity of juvenile males that tried to steal copulations in Soay sheep, Ovis aries (Stevenson and Bancroft 1995). In the other two cases, there was no effect of the administration of testosterone on the

Table 7.8. *Effects of hormone manipulations on adult alternative (i.e., parasitic) reproductive morphs in vertebrates*

CLASS/Species	Alternative phenotypes	Hormone manipulation (manipulated sex type)	Effects on behavior	Effects on morphology	References
TELEOSTEI					
Plainfin midshipman	Fixed	KT (type II males)	–	+ (sonic muscle)	Lee and Bass 2004
Peacock blenny	Sequential	KT implant (sneakers)	–/+ (no effect on bourgeois behavior but inhibits sneaking behavior)	+ (anal gland and genital papillae)	Oliveira *et al.* 2001d
Rock-pool blenny	Sequential	KT and MT implants (satellite males)	–/+ (MT treatment increases the time satellite males spend in independent nests)	+ (anal gland and genital papillae)	Oliveira *et al.* 2001e
Sailfin molly	Reversible	MT injections (small males)	–	Not determined	R. F. Oliveira, I. Schlupp, D. M. Gonçalves, and A. V. M. Canário, unpubl. data
REPTILES					
Marine iguana	Sequential	Testosterone IP injection (satellite males)	+ (establishment of temporary territories)	–	Wikelski *et al.* 2005
		Testosterone IP injection (sneaker males)	+ (leave female groups and start behaving like satellite males)	–	Wikelski *et al.* 2005
		Androgen receptor blocker + aromatase inhibitor IP injection (territorial males)	+ (decrease head-bob patrolling, territory size and number of females in the territory)	–	Wikelski *et al.* 2005
Tree lizard	Fixed	Castration (neonatal males)	Not determined	+ (increased the frequency of orange males as adults)	Hews *et al.* 1994
		Testosterone implant (neonatal males)	Not determined	+ (increased the frequency of orange-blue males as adults)	Hews *et al.* 1994
		Testosterone implant (males at days 1, 30, and 60 after hatching)	Not determined	+ (d1 and d30 treatments increased and d60 had no effect on the frequency of orange-blue males as adults critical period)	Hews and Moore 1996

154

Side-blotched lizard	Fixed	Progesterone injection (neonatal males)	Not determined	+ (increased the frequency of orange-blue males as adults)	Moore et al. 1998
		Testosterone implants (blue and yellow morphs)	+ (increase endurance [measured in a treadmill], activity, home-range size, and control over female territories)	–	DeNardo and Sinervo 1994b, Sinervo et al. 2000
AVES					
Ruff, *Philomachus pugnax*	Fixed	Testosterone implants (satellite males)	– (satellite males did not express territorial behavior but increased satellite behavior)	Not determined	D. B. Lank, unpubl. data in Rhen and Crews 2002
		Castration (independent males)	?	+ (lack growth of male displaying feathers)	Van Cordt and Junge 1936 in Lank et al. 1999
		Testosterone implants (females)	+ (male courtship)	+ (increase in body mass and development of male display feathers)	Lank et al. 1999
House finch	Fixed?	Testosterone implants (males)	+ (increased dominance)	–	Duckworth et al. 2004
Azure-winged magpie	Plastic	Testosterone implants (males at the beginning of the breeding season)	– (likelihood of becoming a helper or a breeder)	–	De la Cruz et al. 2003
Superb fairy-wren	Plastic	Testosterone implants (helpers)	+ (increases helper courtship behavior towards own female)	–	Peters et al. 2002
MAMMALS					
Soay sheep, *Ovis aries*	Plastic	Inhibition of testosterone production by medroxy-progesterone acetate injections (juveniles that attempt copulations by harassing consorting pairs)	– P (reduces harassing behavior)	–	Stevenson and Bancroft 1995

For scientific names of each species and information on their taxonomy consult Table 7.3. +, transformations towards the bourgeois phenotype; –, no transformations towards the bourgeois phenotype; –/+, partial transformation towards the bourgeois phenotype; – P, inhibition of parasitic tactic.

155

Table 7.9. *Testing the second prediction of the relative plasticity hypothesis (shaded cells represent cases that support the prediction)*

Manipulation of androgen levels in parasitic males		ART type	
		Fixed	Plastic
Early in development	Effective	Tree lizards	
	No effects		
In adults	Effective	Side-blotched lizards	Marine iguanas
		House finch	Superb fairy-wren
			Soay sheep
	No effects	Plainfin midshipman	Peacock blenny
		Ruff	Rock-pool blenny
			Sailfin molly
			Azure-winged magpie

parasitic morph of the "fixed" type species (one fish and one lek-breeding bird; see Tables 7.8 and 7.9). Of the six cases that do not support the second prediction, two correspond to positive effects of testosterone administration in "fixed" species (one lizard and one bird), and the other four to absence of effects of testosterone administration in "plastic" species (three fish and one cooperatively breeding bird) (see Tables 7.8 and 7.9). Therefore, overall, the validity of the second prediction of the relative plasticity hypothesis is only present in 50% of the species studied so far. Unfortunately, in the vertebrate taxa for which the hypothesis is probably most adequately applied, the mammals, there is only one species for which data are available (and it supports the hypothesis).

Interestingly, of all the hormone manipulations performed on vertebrates with the objective of unraveling the physiological mechanisms of ARTs, only in one case (the marine iguana) has the reversibility of the transformation from parasitic to bourgeois male in "plastic" species been tested. In the experiment, territorial males were implanted with an androgen receptor blocker (flutamide) together with an aromatase inhibitor (1,4,6-androstatrien-3,17-dione; ATD) in order to block the direct (i.e., testosterone acting on an androgen receptor) and indirect (i.e., testosterone being aromatized into estradiol, which would activate the behavior) effects of testosterone on the expression of bourgeois behavior (Wikelski *et al.* 2005). Treated males decreased the expression of their territorial behavior, had their territories reduced in size, and suffered a decrease in the number of females present on their territories, but they did not develop the full expression of parasitic behavior. These results

suggest that the blockage of androgens in bourgeois males can reduce the expression of bourgeois behavior but cannot induce a tactic change to a parasitic morph in a "plastic" species with sequential tactics. This conforms to the expectation that plasticity in alternative morphs should only be permissible in directions that correspond to normal sexual differentiation (i.e., parasitic males can transform into bourgeois males but not the reverse).

In summary, although the relative plasticity hypothesis provides a tentative conceptual framework for the study of the hormonal basis of ARTs and has been elegantly developed (Moore *et al.* 1998), it does not seem to apply across vertebrate taxa. One of the major reasons for this mismatch may reside in the fact that this hypothesis, derived from the organizational paradigm of mammalian sex differentiation, is not common to other vertebrate classes and, in particular, is not found in those with labile sex-determining mechanisms.

7.4.3 Beyond the relative plasticity hypothesis: the "making of" alternative phenotypes

It is also important to be able to distinguish whether alternative phenotypes diverge only in terms of behavioral traits, or if they also differ in the expression of morphological traits. Since behavior is often more labile than morphology and anatomy, the mechanisms underlying the expression of behavioral variation are expected to be more flexible than those underlying morphological and anatomical variations. It follows that alternative reproductive tactics that only involve differences in behavior should differ in the activation of different neural substrates but not necessarily display different hormonal profiles. In contrast, alternative reproductive phenotypes that also show a divergence in morphological traits (i.e., intrasexual polymorphisms), in which the differentiation of sexual characters between the alternative morphs needs a whole-organism control system, are expected to have different hormone profiles to account for these differences. It could be argued that differences in hormone levels should only be present at the period of the differentiation of the tactic, if their effects were to be organizational. However, there are several pieces of evidence suggesting that androgen-dependent traits, typical of bourgeois males, need continuous exposure to androgens to be maintained. For example, in adults androgens inhibit the shrinkage of motorneurons in the spinal nucleus of the bulbocavernosus that controls penile erection in rodents (Breedlove and Arnold 1981, Forger et al. 1992, Watson et al. 2001). Also, castration induces the regression and exogenous administration of androgens restores the development of sonic muscles in vocalizing male fish (Brantley et al. 1993a; but see Modesto and Canário 2003b). The hypothesis that androgens may play differential roles in the differences between male morphs across different phenotypic traits (i.e., behavioral, morphological, and gonadal) will be discussed below.

HORMONES AND DIFFERENCES BETWEEN
ALTERNATIVE PHENOTYPES IN SECONDARY SEX
CHARACTERS

Since androgens play a major role in the induction of secondary sex characters in male vertebrates (Nelson 2005), differences in androgen levels among morphs may be of little importance in species with alternative tactics lacking major tactic-specific morphological specializations (such as the expression of male secondary sex characters in bourgeois males). Among the species displaying ARTs and intrasexual dimorphism, 100% of the fish, 66.6% of the reptiles, 75% of the birds, and 100% of the mammals (i.e., 90.9% of all studied species) displayed significant differences in circulating androgen levels, with the bourgeois morphs having consistently higher levels than those of the parasitic males (Table 7.3).

Recently, the association between the degree of phenotypic specialization of the alternative tactics and the magnitude of the difference in androgen levels between alternative male types was investigated among teleost fish (Oliveira 2005). In all species for which androgen levels are known and for which the ART involves a morphological intrasexual dimorphism (apart from differences in body

size), the levels of 11-ketotestosterone (KT, the most potent androgen in fish) are higher in the bourgeois than in the parasitic male, irrespective of the type of ART displayed (Oliveira 2005). This suggests a parallel to the androgen correlates of sex-changing fish, in which androgens may play a major role in morphological differentiation during sex change but are not essential for behavioral sex change (Godwin et al. 1996, Grober 1998, Reavis and Grober 1999). These results, together with the data presented here, suggest a major role for androgens in the differentiation of morphological traits typical of the bourgeois tactic.

HORMONES AND DIFFERENCES BETWEEN
ALTERNATIVE PHENOTYPES IN REPRODUCTIVE
BEHAVIOR

In species with reversible ARTs without morphological modifications, changes in the activity of neural pathways underlying the behavioral changes are to be expected rather than differences in androgen levels (Zupanc and Lamprecht 2000; see Section 7.3.1). This could explain, for example, the lack of differences in KT levels between polygynous and monogamous males in the St. Peter's fish (Ros et al. 2003) and between callers and satellites in toads (Leary et al. 2004). Hence, reversible ARTs lacking intrasexual dimorphisms may have been emancipated from a sex-differentiation mechanism ruled by sex hormones. In this respect, it is interesting to note that in the peacock blenny, where sneaker males mimic female courtship behavior, castrated sneakers (that mimic females) continue to exhibit female courtship (D. M. Gonçalves, J. Alpedrinha, and R. F. Oliveira, unpublished data), indicating that gonadal steroids are not crucial for the behavioral expression of the parasitic tactic in this species.

HORMONES AND DIFFERENCES BETWEEN
ALTERNATIVE PHENOTYPES IN GONADAL
ALLOCATION

For a large number of species with ARTs, in particular among fish, the parasitic morph has relatively larger gonads, a phenomenon which has been explained by the sperm competition hypothesis (Taborsky 1998). This is intriguing from a physiological perspective since androgens are also involved in spermatogenesis. There are several possible explanations for this paradox.

(1) In the particular case of teleost fish, KT and testosterone (T) have different roles in the control of spermatogenesis: KT stimulates germ cell proliferation

and maturation, and T is involved in the negative feedback mechanisms needed to control KT-dependent spermatogenesis. Thus, a balance between T and KT is critical for the control of spermatogenesis (Schulz and Miura 2002). A plot of the KT to T ratio as a function of the relative size of the gonad (GSI) shows that in species in which the magnitude of the ratio between bourgeois and parasitic is larger, there is a smaller difference in GSI (Oliveira 2005). This means that a higher GSI among parasitic males is associated with a lower KT : T ratio, which allows them to have larger testis without a linked expression of bourgeois male secondary sex characters and behavior (Oliveira 2005).

(2) In the case of other vertebrates, a potential alternative explanation is differential density of gonadal receptors among morphs, so that the gonads of parasitic males may become particularly reactive to the same levels of gonadotrophic hormones when compared with those of bourgeois males.

(3) In vertebrates direct innervation of the gonads has been demonstrated, and this might allow for an alternative route for controlling gonadal function in alternative phenotypes. In all vertebrates, both afferent and efferent neural connections between the gonad and the hypothalamus have been described, with the efferent fibers terminating on steroidogenic cells of the gonad (for references see Crews 1993). Moreover, de-innervation of the gonad causes gonadal atrophy whereas the electrical stimulation of these fibers induces variations in gonadal steroid secretion and sperm release (Demski 1987, Damber 1990). Thus, a private channel between the brain and the gonads is present that might allow for a control of gonadal activity in parasitic males independent of the systemic action of the HPG axis.

In summary, the relative importance of different physiological mechanisms for the differentiation of tactic-specific traits might vary among behavioral, morphological, and gonadal traits. If this occurs in species with ARTs, it would challenge the classic paradigm of androgens controlling, in a whole-organism fashion, the expression of the entire set of reproductive characters that distinguish each tactic.

7.5   STRESS, GLUCOCORTICOID LEVELS, AND ARTs

One of the axioms of the current ART theory is that alternative morphs have a lower competitive ability and therefore

a subordinate status if in direct competition with bourgeois morphs. Dominance relationships are also known to have a differential effect on glucocorticoid (GC) levels, and for a long time it was assumed that circulating concentrations of a subordinate's GCs should be higher than those of dominant individuals and that these differences should mediate the effects of social rank on reproductive physiology (Creel 2005). This belief has led to the concept of social status as almost synonymous with stress for subordinates in a social group. This concept was built on a logical inference using three independent pieces of evidence: (a) in staged fights both winners and losers experience an increase in circulating levels of GCs, but there is a higher magnitude in the loser's response; (b) GCs suppress the HPG axis; and (c) social stress leads to the suppression of reproduction in subordinates (for references see Creel 2005). However, it has become increasingly clear that in most free-living species, either there is no difference in GC levels according to social status, or there is a trend for dominant males to have higher circulating levels of GCs than subordinates (Creel 2001, 2005, Abbott et al. 2003; however, these reviews included only bird and mammalian studies). In fact, the winner–loser effects on GC levels do not predict differences between dominant and subordinate individuals in free-living groups that conform to different social systems, and there is no parsimonious argument that allows one to predict whether dominants or subordinates are more stressed in the wild. While dominants are expected to face the stressful situation of having to fight harder and at higher rates to keep their status, subordinates, in turn, are exposed to the stress of repeated defeats (although in the wild they can often spatially avoid being exposed to dominant individuals or even take the option of dispersal) (Creel 2005). Based on a meta-analysis of rank differences in cortisol levels among primates, Abbott and co-authors (2003) proposed that two conditions should explain the relationship between social status and GC levels. According to this analysis, subordinates should have higher GC titers than dominants (1) when subjected to higher rates of stressors, either physical (e.g., food availability, exposure to predators and to pathogens, likelihood of facing aggressive challenges) or psychological (e.g., control access to resources, exposure to aggression, establish stable and predictable social relations) or (2) when they experience decreased opportunities of social support.

According to the rationale proposed by Creel (2005), in species with ARTs, the bourgeois morph, characterized by its investment in the monopolization of access to mates, should face more social challenges and therefore would be expected to have higher circulating levels of GC than the parasitic morph. However, an analysis of Table 7.3 does not support this prediction. In fact, the three possible relationships between GC levels and ART type are present: of the 16 species studied so far, levels are higher in bourgeois males in 37.5% of the cases, are higher in the parasitic morph in 12.5% of the cases, and there are no differences in the remaining 50%. Moreover, the differences in GC levels are independent of the type of ART expressed (fixed vs. plastic; ACTUS – simulation statistics for contingency tables with low expected values – $P < 0.05$), indicating that the first prediction of the relative plasticity hypothesis also does not conform to the available data on GCs. However, this result should be taken with caution since in most studies only basal levels were reported. Glucocorticoids act through a dual receptor system where two receptor types are present in target tissues: type I receptors (or mineralocorticoid) and type II receptors (or glucocorticoid) (de Kloet et al. 1993). Since type I receptors have a higher affinity for glucocorticoids than type II, at baseline levels most GCs are bound to type I receptors. This receptor subtype mediates permissive actions of GCs (i.e., actions that are already present before the stressor and that prime the stress defenses of the organism). When GCs increase in response to a stressor and type I receptors become saturated, then there is a binding shift towards type II receptors, which mediate suppressive actions of GCs mainly outside the HPA axis, such as reproductive suppression (Sapolsky et al. 2000). The disruption of the HPG axis by glucocorticoids can be achieved by several different mechanisms, namely by decreasing both the hypothalamic release of GnRH, and the LH secretion from the pituitary, as well as by reducing the gonadal responsiveness to LH and the local density of LH receptors (Sapolsky et al. 2000). For a clearer picture of a potential role of GCs on ARTs, we need to look for differences in GC responses to challenges between alternative tactics and to confirm that the dual GC receptor system described in mammals is also present in the other vertebrate classes. Below we illustrate some known examples of the involvement of GCs on the expression of ARTs in different vertebrate taxa.

In the tree lizard, *Urosaurus ornatus*, two fixed reproductive phenotypes exist: territorial males display an orange dewlap with a blue spot (orange-blue males), and non-territorial males have an orange dewlap (Moore 1991, Moore et al. 1998). Within the orange morph, the males may switch between a sedentary satellite tactic and a nomadic tactic, depending on the environmental conditions they

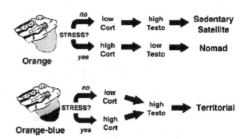

Figure 7.3 Proposed endocrine mechanism for tactic switching in the tree lizard. Males have a two-step reaction to stress. Both morphs increase their corticosterone levels in response to a stressor. However, orange-blue male testosterone levels are corticosterone resistant, while testosterone levels of orange males are sensitive to suppression by corticosterone. Therefore, orange-blue males express territorial behavior independently of exposure to stress, whereas orange males switch their tactic from satellite (with low corticosterone) to nomad (with high corticosterone) depending on the environmental conditions. (Reprinted with permission from Knapp et al. 2003.)

face, thus representing plastic ARTs (Moore et al. 1998). The corticosterone response to stress seems to be the key factor triggering this switch within the orange morph (see Figure 7.3). In harsh conditions, corticosterone levels increase causing a decrease in testosterone concentrations, which leads to a lack of site attachment (cf. DeNardo and Sinervo 1994a, b) and a concomitant switch from the satellite to the nomadic tactic (Figure 7.3). Apparently the orange-blue males are resistant to testosterone suppression by corticosterone, and thus, independently of the environmental conditions, continue to express the territorial tactic (Knapp et al. 2003).

In amphibians the *energetics–hormone vocalization* model has been proposed (Emerson 2001, Emerson and Hess 2001), which aims to explain transitions in vocal production (i.e., calling vs. noncalling) in anurans. It proposes that elevated levels of corticosterone due to the energetic demands of calling behavior inhibit androgen production which inhibits calling. Data are available for three anuran species with noncalling satellite males (Table 7.3). In two of these species, the Woodhouse and the Great Plains toads, although corticosterone levels are higher in the calling morph, there are no differences between morphs in androgen levels. These findings are contrary to a suppression of the HPG axis by increased levels of corticosterone in calling males and support the occurrence of direct effects of

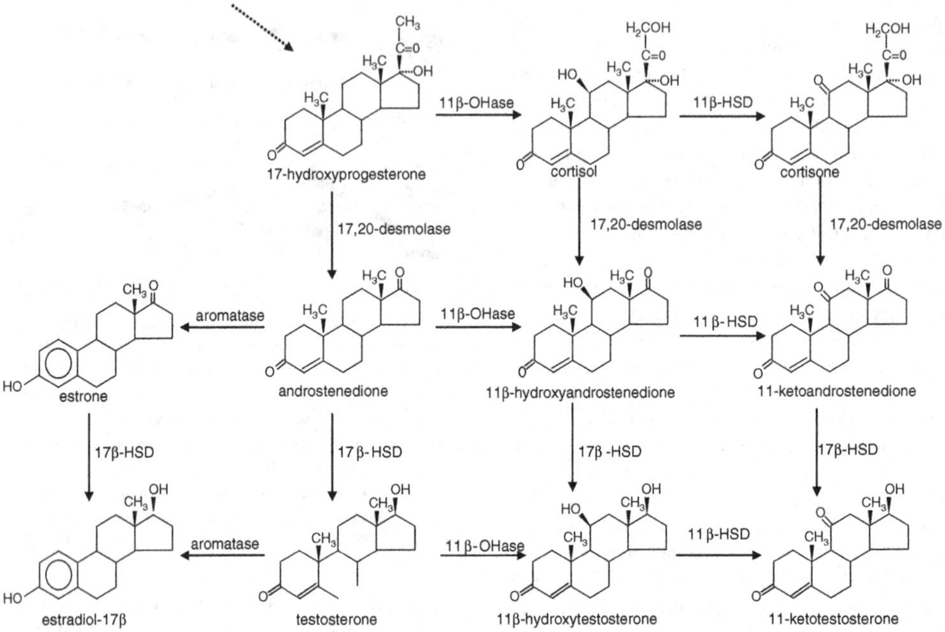

Figure 7.4 Steroidogenic pathways illustrating the similarities in the enzymes involved in androgen and glucocorticoid metabolism.

corticosterone on vocal brain nuclei that control calling behavior (Leary et al. 2004).

Finally, a model for the differentiation of alternative phenotypes in teleost fish based on glucocorticoid–androgen interactions has been proposed by Knapp and co-workers (Knapp et al. 2002, Knapp 2003). Since the same enzymes that participate in the synthesis of KT are also involved in the synthesis (11β-OHase = 11β-hydroxylase) and inactivation (11β-HSD = 11β-hydroxysteroid dehydrogenase) of GCs (see Figure 7.4), it is proposed that reciprocal competitive inhibition can regulate the activity of these enzymes (Knapp 2003). Consequently, in species with plastic ARTs, reciprocal inhibition creates the possibility that these enzymes may mediate the transduction of social into endocrine signals that will modulate the adoption of a certain ART (Knapp 2003). This model assumes that parasitic males have higher cortisol levels than bourgeois males, as a result of aggressive interactions among the two morphs. Competitive inhibition of 11β-OHase and/or 11β-HSD would yield lower levels of KT in parasitic males and result in an accumulation of T. The increased T could then be available to the enzyme aromatase for estrogen production. Therefore, higher levels of aromatase activity are predicted in parasitic males, as has been observed in midshipman fish type II males (Schlinger et al. 1999). A potential pitfall of this model is the assumption of competition for cortisol and KT production. However, this is only expected if occurring in the same tissue (i.e., gonad or adrenals). Data on cortisol levels in teleost species with ART are only available for the longear sunfish, where parasitic males have both higher levels of cortisol and lower levels of KT than bourgeois males, suggesting that parasitic males may have a lower activity of 11β-HSD both in the interrenal glands and in the testes relative to bourgeois males (Knapp 2003). A similar model has been independently proposed by Perry and Grober (2003) to explain the social modulation of sex change in sequential hermaphroditic teleosts. This model is supported by the fact that in the bidirectional, socially induced, sex-changing goby Gobiodon histrio, a glucocorticoid responsive element has been identified in the promoter region of the aromatase gene CYP19A1 (gonadal isoform) that could allow GC to act as an upregulatory transcription factor, ultimately promoting estrogen synthesis responsible for male-to-female sex change (Gardner et al. 2005). Thus cortisol could play a pivotal role when subordinate males change back to females as a response to the stress of competition with dominant males (Munday and Jones 1998).

In vertebrates other than the teleosts, where KT is not present, a role for these steroidogenic enzymes is still possible. In mammals 11β-HSD plays a major role at the intracellular level in regulating the availability of GC to glucocorticoid receptors. This enzyme has two isoforms with different activities. Whereas 11β-HSD2 catalyzes the irreversible inactivation of GCs, leading to the formation of 11-keto-steroids (i.e., cortisone from cortisol and 11-dehydrocorticosterone from corticosterone), 11β-HSD1 can promote both the inactivation or the activation (by reduction of the 11-ketosteroids) of GCs (de Kloet *et al.* 1998, Sapolsky *et al.* 2000). In Leydig cells, 11β-HSD activity modulates the availability of intracellular GC to the type II receptors that in turn inhibit testosterone production (Gao *et al.* 1996a, b). Therefore, differential expression of the two isoforms in different tissues between alternative morphotypes can be a mechanism that explains intrasexual variation in the expression of reproductive traits. The lizard, anuran, and teleost examples illustrate the fact that GCs seem to act in the expression of alternative tactics, but their exact role may depend on the social systems and on particular GC mechanisms present in different taxa (e.g., the duality of GC receptors present in mammals).

7.6 BEYOND HORMONE PROFILES: FOCUSING ON TARGET TISSUES

The decoupling of different male traits in alternative reproductive phenotypes may be achieved by mechanisms other than differences in hormone levels, namely by varying the local microenvironment in the different target tissues. This could result from differential expression of receptors or differential levels of activity of catabolic enzymes that modulate the availability of the active hormone to specific targets (e.g., 11β-OHase and 11β-HSD, which metabolize testosterone into KT, are key steps in the expression of male secondary sex characters, in spermatogenesis, and in the modulation of the expression of reproductive behavior in male teleosts: Borg 1994). This focus on target tissues, when studying the mechanisms of intrasexual variation in reproduction, has rarely been used. One rare example of such an approach is a study on the relative levels of brain steroid receptors between alternative reproductive phenotypes in the protogynous wrasse *Halichoeres trimaculatus*. In this species it was found that by using competitive reverse transcriptase–polymerase chain reaction, the levels of androgen receptor (AR) transcripts were significantly higher in the brain of terminal-phase males (bourgeois tactic) than in initial-phase males (parasitic tactic) (Kim *et al.* 2002). No other significant differences in gene expression were observed, either for AR in the gonads or for estrogen receptor (ER) in the brain and in the gonads. Thus, by regulation of the expression of AR in specific tissues (by varying AR density in different tissues such as brain vs. gonad) of bourgeois males (in this case terminal-phase males), the sensitivity to circulating androgen levels in specific targets (the brain) can be increased, and the effects of androgens compartmentalized (Ketterson and Nolan 1999). This mechanism hypothetically makes it possible to activate the expression of an androgen-dependent reproductive behavior in bourgeois males without having the associated costs of increasing spermatogenesis or expressing a sex character, since the androgen action can be independently modulated at each compartment (brain vs. gonad vs. morphological secondary sex character).

Another level at which the availability of steroid hormones to target tissues can be differentially modulated between alternative phenotypes is through steroid-binding globulins (SBGs). SBGs can regulate the availability of circulating steroids to target tissues, since only the free (unbound) fraction is biologically active. To our knowledge, there is only one published study in vertebrates that documents differences in binding capacity of an SBG among alternative morphs (Jennings *et al.* 2000). In the tree lizard two SBGs have been identified: one with a high affinity to androgens and estradiol (i.e., a typical sex-hormone-binding globulin), and another with a high affinity to androgens, progesterone, and corticosterone, thus named androgen–glucocorticoid–steroid-binding globulin (AGBG: Jennings *et al.* 2000). Whereas the capacity of the former SBG does not differ between the two morphs, the AGBG capacity is much larger in the orange-blue males, resulting in higher levels of free (i.e., unbound) corticosterone in the orange morph (Jennings *et al.* 2000). Consequently, testosterone levels in the orange morph are more sensitive to negative feedback by corticosterone, especially during periods of stress (e.g., staged male–male encounters: Knapp and Moore 1996, 1997). Thus, at least for tree lizards, SBGs can act as mediators of the environmental effects on the differentiation and expression of alternative morphs. Further studies focusing on target tissues are thus a major avenue for future research in this area.

7.7   ARTs IN THE GENOMICS ERA: A HOLISTIC APPROACH TO THE PROXIMATE MECHANISMS OF ARTs

Functional genomics tools now provide a new approach to understanding the proximate mechanisms of ARTs. Using microarray technology, the activity of large sets of genes (thousands) can be monitored simultaneously in key tissues (e.g., brain, gonads). It is therefore possible to identify genes and regulatory networks that are consistently upregulated or downregulated in each morph. These differentially expressed genes are then taken as likely candidates involved in the expression of the alternative morphotypes (Hofmann 2003). Only two studies have been published that used microarray techniques to study alternative phenotypes. In the honeybee (*Apis mellifera*), workers socially regulate the division of labor, with younger individuals acting as hive workers and older individuals as foragers. The transition between these two alternative (sequential) phenotypes is associated with differential gene expression in 39% of the approximately 5500 genes tested (Whitfield *et al.* 2003), indicating a link between different profiles of brain gene expression and the occurrence of behavioral plasticity. In a second study, the only one of a vertebrate species, gene expression profiles were compared between sneaker males and immature juveniles (of the same age) of the Atlantic salmon, *Salmo salar* (Aubin-Horth *et al.* 2005). Males that will reproduce as sneakers do not migrate to the sea and attain sexual maturity earlier (1–3 years old) than migratory males that return later to the breeding grounds as large, anadromous individuals (3–7 years old) (Fleming 1998). Thus, the immature males represent the anadromous phenotype before migration, and they are the same age as the sneakers (in order to avoid age-related differences in gene expression). A differential expression of 15% of the 2917 genes tested has been detected between the sneaker and the juvenile immature males (Aubin-Horth *et al.* 2005). Most of the upregulated genes in sneakers are associated with reproduction and associated processes (e.g., gonadotropins, growth hormone, prolactin, and POMC genes), and the upregulated genes in immature males are mainly associated with somatic growth (e.g., genes involved in transcription regulation and protein synthesis, folding, and maturation). These differences reflect, at the cellular level, the life history trade-off between reproduction and growth that is found in these two alternative phenotypes (Aubin-Horth *et al.* 2005). Interestingly, genes involved in neural

plasticity (e.g., genes coding for synaptic function and for cell-adhesion glycoproteins that have been implicated in memory formation) and neural signaling (i.e., genes coding for nitric oxide synthesis, a neurotransmitter involved in the regulation of neuropeptide action) were upregulated in sneakers suggesting that the expression of this tactic might be particularly demanding at the level of cognition (Aubin-Horth *et al.* 2005). This approach not only allows us to confirm predictions of differential gene expression between alternative phenotypes, in processes that are a priori expected to differ between alternative morphs (e.g., reproduction vs. growth), but it also enables the detection of differences in gene expression between morphs in unsuspected biological processes (e.g., neural plasticity).

7.8   DIFFERENTIAL COSTS IN ENDOCRINE-MEDIATED ARTs

The study of the physiological mechanisms underlying the expression of ARTs may also shed light on the evolutionary mechanisms involved, since from a functional point of view, the potential benefits of high androgen levels for the fitness of the individuals adopting the bourgeois tactic have to outweigh the costs associated with keeping those levels high for long periods. Androgens facilitate the physiology and behavior related to high intra- and intersexual competition typical of the bourgeois tactic. The required extra energetic resources needed for the expression of exaggerated secondary sexual characters and agonistic behavior patterns might have consequences for the allocation of energy to other functions. Especially when animals are constrained in their opportunities to increase energy uptake or when gains in reproduction are high, it may pay to evolve a mechanism that facilitates the expression of sexual traits, while downregulating other energetically expensive functions. This trade-off might explain why, in many species, androgens seem to suppress immunity (Folstad and Karter 1992, Wedekind and Folstad 1994). There is evidence indicating that humoral and cellular immunocompetence are costly (e.g., Martin *et al.* 2003) and trade off with reproduction (Sheldon and Verhulst 1996, Deerenberg *et al.* 1997, Norris and Evans 2000, Cichoń *et al.* 2001).

Few studies have addressed the differential costs in immunocompetence for alternative morphs due to different hormonal profiles of alternative tactics. In the corkwing wrasse (*Symphodus melops*), despite the fact that sneaker males differ from nest-holders in androgen levels (Uglem *et al.* 2002), no relationship has been found between male reproductive tactics and leukocyte count (Uglem *et al.* 2001). In ruffs, there are no differences among morphs in humoral immunity but territorial males have higher cell-mediated immunity than satellites (Lozano and Lank 2004).

We have recently started to address this issue using the rock-pool blenny (*Parablennius parvicornis*) and the peacock blenny (*Salaria pavo*). In both species, bourgeois males exhibit both parental and territorial behavior, which does not allow them to forage far from their nest sites. In contrast, parasitic males do not have such constraints on energy uptake during the breeding season, and, as a result, nest-holder males of both species suffer a dramatic decrease in body condition not experienced by parasitic males (Gonçalves and Almada 1997). We therefore tested whether the expression of alternative male tactics has consequences at the level of immunocompetence in these two blennies. In salmonids, androgen treatment decreases antibody production by lymphocytes and may even kill them by apoptosis (Slater *et al.* 1995, Slater and Schreck 1997). Interestingly, a specific androgen receptor has been detected in these leukocytes (rainbow trout, *Oncorhynchus mykiss*, and chinook salmon, *Oncorhynchus tshawytsha*: Slater *et al.* 1995, Slater and Schreck 1998). We therefore focused our studies on the relative number of lymphocytes (i.e., leukocytes responsible for the production of specific antibodies) and on antibody production in response to a challenge with a nonpathogenic antigen. In accordance with expectation, lymphocyte count (in both species) and antibody responsiveness (in the rock-pool blenny) were found to be higher in parasitic males than in bourgeois males (Ros *et al.* 2006; A. F. H. Ros and R. F. Oliveira, unpublished data) (Figure 7.5). This suggests that alternative morphs differ in their capacity to mount "specific" immune responses. Moreover, since lymphocyte numbers are negatively correlated to body size (Figure 7.5), and since competitive ability of the males increases with body size (Oliveira *et al.* 2000), it is plausible that in larger animals, relatively more energy is traded off with immunity than in smaller animals.

7.9 CONCLUSIONS AND PROSPECTS FOR FUTURE RESEARCH

We have summarized the effects on ARTs of different hormones (mainly androgens and glucocorticoids) at both the organizational and the activational levels. However, these effects vary from species to species in a fashion that is not consistent with the type of ART expressed, as predicted by the relative plasticity hypothesis. In particular, in the case of sex steroids, it is conceivable that the expression of a given tactic requires that androgens reach a threshold level for the expression of the bourgeois traits. But, above that threshold, further variations in androgen levels are not associated with the expression of the tactic and may merely reflect the social environment faced by individuals following different tactics. We have also shown that androgens are more relevant for the differentiation of morphological traits than of behavioral traits, which implies that differences in androgen levels between alternative tactics are more likely when the ART involves an intrasexual dimorphism. This difference between ARTs with and without associated variation in the expression of morphological traits is thus a point that should not be neglected in future studies. Another point that needs to be stressed here is that in order to understand the mechanisms of ARTs more research effort is needed focusing on the processes of hormone action at the target tissues, since they may vary between alternative tactics. Most of the work conducted so far is based on correlations of circulating levels of hormones in individuals following alternative tactics and on hormone manipulations in different adult morphs.

At the conceptual level, the views on the role that hormones play in the control of behavior have been changing with time. Two major changes have occurred in

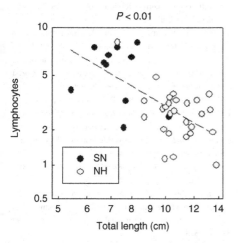

Figure 7.5 Preliminary results on the relationship between lymphocyte counts, total length, and alternative reproductive tactics in the peacock blenny. SN, sneaker males; NH, nest-holder males. (A. F. H. Ros, unpublished data.)

recent years. Hormones have been seen classically as causal agents of behavior of the type one-hormone-one-behavior relationship. This view has been supported mainly by studies of castration and hormone-replacement therapy that showed that a behavior was abolished by castration and restored by exogenous administration of androgens (Nelson 2005). Currently a probabilistic approach to the effects of hormones on behavior has been adopted and hormones are seen as facilitators of behavior rather than as determinant factors (Simon 2002). Accordingly, hormones may increase or decrease the probability of the expression of a given behavior by acting as neuromodulators on the neural pathways underlying that behavioral pattern. Second, there has been a recognition that the social environment feeds back to influence hormone levels (Wingfield *et al.* 1990, Oliveira 2004), which is seen as an adaptive mechanism through which individuals may adjust their motivation according to the social context they are facing. This indicates a two-way type of interaction between hormones and behavior. Accordingly, hormones (e.g., androgens) are viewed as playing a key role as endocrine mediators of the effects of social context on the expression of social behavior. These new views of the role hormones play in the control of behavior should be incorporated in future studies on the endocrine basis of ARTs.

Finally, the strengths of the comparative approach in understanding the proximate mechanisms of intrasexual variation in reproductive behavior should be stressed. It is a valuable tool for various reasons. First, it promotes the development of a conceptual framework to explain these phenomena that is not species centered. One major problem in this area is that a lot of research effort has been invested in only a reduced number of species, so that the information available for these few species has great detail but tends to be extrapolated as valid to the vertebrates as a whole. Therefore, the collection of data on different species exhibiting alternative tactics contributes to the awareness that similar functional phenomena may have different underlying mechanisms and promotes the search for commonalities among species. In turn, these prompt the generation of hypotheses that organize the observed variation and thus contribute to the development of a framework that explains the evolution of proximate mechanisms underlying alternative tactics.

Acknowledgments

We would like to thank all the people who have collaborated in the studies that have been conducted in both laboratories on ARTs and a number of colleagues who have provided thoughtful discussions on this topic. They certainly have helped to shape our views on this subject. They are in alphabetical order: Vitor Almada, João Alpedrinha, Eduardo Barata, Luis Carneiro, Inês Domingues, Teresa Fagundes, David Gonçalves, Emanuel Gonçalves, Matthew Grober, João Saraiva, and Mariana Simões. We thank also Jane Brockmann, Michael Taborsky, and Rosemary Knapp for their comments on earlier versions of this manuscript that contributed to its improvement. The research from the RFO laboratory described in this review was supported by a series of grants from the Fundação para a Ciência e a Tecnologia (FCT). The writing of this chapter was directly funded both by the Pluriannual Program of FCT (UIandD 331/2001) and by the FCT research grant POCTI/BSE/38395/2001. Finally, we would like to express our gratitude to our families for tolerating our love of science and blennies.

References

Abbott, D. H. and Hearn, J. P. 1978. Physical, hormonal and behavioral aspects of sexual development in the marmoset monkey, *Callithrix jacchus. Journal of Reproduction and Fertility* 53, 155–166.

Abbott, D. H., Keverne, E. B., Bercovitch, F. B., *et al.* 2003. Are subordinates always stressed? A comparative analysis of rank differences in cortisol levels among primates. *Hormones and Behavior* 43, 67–82.

Alvarez-Buylla, A. and Lois, C. 1995. Neuronal stem cells in the brain of adult vertebrates. *Stem Cells* 13, 263–272.

Arnold, A. B. and Breedlove, S. M. 1985. Organizational and activational effects of sex steroids on brain and behavior: a reanalysis. *Hormones and Behavior* 19, 469–498.

Arnold, W. and Dittami, J. 1997. Reproductive suppression in male alpine marmots. *Animal Behaviour* 53, 53–66.

Aubin-Horth, N., Landry, C. R., Letcher, B. H., and Hofmann, H. 2005. Alternative life histories shape brain gene expression profiles in males of the same population. *Proceedings of the Royal Society of London B* 272, 1655–1662.

Baker, J. V., Abbott, D. H., and Saltzman, W. 1999. Social determinants of reproductive failure in male common marmosets housed with their natal family. *Animal Behaviour* 58, 501–513.

Barrett, G. M., Shimizu, K., Bardi, M., Asaba, S., and Mori, A. 2002. Endocrine correlates of rank, reproduction, and female-directed aggression in male Japanese macaques (*Macaca fuscata*). *Hormones and Behavior* 42, 85–96.

Bartsch, S. S., Johnston, S. D., and Siniff, D. B. 1992. Territorial behaviour and breeding frequency of male Weddell seals (*Leptinychotes weddelli*) in relation to age, size, and concentration of serum testosterone and cortisol. *Canadian Journal of Zoology* 70, 680–692.

Bass, A. H. and Grober, M. S. 2001. Social and neural modulation of sexual plasticity in teleost fish. *Brain, Behavior and Evolution* 57, 293–300.

Bennett, N. C. and Faulkes, C. G. 2000. *African Mole-Rats: Ecology and Eusociality*. Cambridge, UK: Cambridge University Press.

Borg, B. 1994. Androgens in teleost fishes. *Comparative Biochemistry and Physiology C* 109, 219–245.

Bowen, B. S., Koford, R. R., and Brown, J. L. 1995. Genetic evidence for undetected alleles and unexpected parentage in the gray-breasted jay. *Condor* 97, 503–511.

Brantley, R. K., Marcharterre, M. A., and Bass, A. H. 1993a. Androgen effects on vocal muscle structure in a teleost fish with inter- and intra-sexual dimorphisms. *Journal of Morphology* 216, 305–318.

Brantley, R. K., Wingfield, J. C., and Bass, A. H. 1993b. Sex steroid levels in *Porichthys notatus*, a fish with alternative reproductive tactics, and a review of the hormonal bases for male dimorphism among teleost fishes. *Hormones and Behavior* 27, 332–347.

Breedlove, S. M. and Arnold, A. P. 1981. Sexually dimorphic motor nucleus in the rat lumbar spinal cord: response to adult hormone manipulation, absence in androgen-insensitive rats. *Brain Research* 225, 297–307.

Brockmann, H. J. 2001. The evolution of alternative strategies and tactics. *Advances in the Study of Behavior* 30, 1–51.

Cardwell, J. R. and Liley, N. R. 1991. Hormonal control of sex and color change in the stoplight parrotfish, *Sparisoma viride*. *General and Comparative Endocrinology* 81, 7–20.

Caro, T. M. and Bateson, P. 1986. Organization and ontogeny of alternative tactics. *Animal Behaviour* 34, 1483–1499.

Cavigelli, S. A. and Pereira, M. E. 2000. Mating season aggression and fecal testosterone levels in male ring-tailed lemurs (*Lemur catta*). *Hormones and Behavior* 37, 246–255.

Chaudhuri, M. and Ginsberg, J. R. 1990. Urinary androgen concentrations and social status in two species of free ranging zebra (*Equus burchelli* and *E. grevyi*). *Journal of Reproduction and Fertility* 88, 127–133.

Cheek, A. O., Thomas, P., and Sullivan, C. V. 2000. Sex steroids relative to alternative mating behaviors in the simultaneous hermaphrodite *Serranus subligarius* (Perciformes: Serranidae). *Hormones and Behavior* 37, 198–211.

Cichoń, M., Dubiec, A., and Chadzińska, M. 2001. The effect of elevated reproductive effort on humoral immune function in collared flycatcher females. *Acta Oecologica* 22, 71–76.

Clark, M. M. and Galef Jr., B. C. 2000. Why some male Mongolian gerbils may help at the nest: testosterone, asexuality and alloparenting. *Animal Behaviour* 59, 801–806.

Clark, M. M., Malenfant, S. A., Winter, D. A., and Galef Jr., B. G. 1990. Fetal uterine position affects copulation and scent marking by adult gerbils. *Physiology and Behavior* 47, 301–305.

Clark, M. M., Tucker, L., and Galef Jr., B. G. 1992a. Stud males and dud males: intrauterine position effects on the success of male gerbils. *Animal Behaviour* 43, 215–221.

Clark, M. M., vom Saal, F. S., and Galef Jr., B. G. 1992b. Fetal intrauterine position correlates with endogenous testosterone levels in adult male Mongolian gerbils. *Physiology and Behavior* 51, 957–960.

Clark, M. M., Vonk, J. M., and Galef Jr., B. G. 1998. Intrauterine position, parenting and nest-site attachment in male Mongolian gerbils. *Developmental Psychobiology* 32, 177–181.

Clarke, F. M. and Faulkes, C. G. 1998. Hormonal and behavioural correlates of male dominance and reproductive status in captive colonies of the naked mole-rat, *Heterocephalus glaber*. *Proceedings of the Royal Society of London B* 47, 83–91.

Coltman, D. W., Bowen, W. D., and Wright, J. M. 1999. A multivariate analysis of phenotype and paternity in male harbor seals, *Phoca vitulina*, at Sable Island, Nova Scotia. *Behavioral Ecology* 10, 169–177.

Conrad, K. F., Clarke, M. F., Robertson, R. J., and Boag, P. T. 1998. Paternity and the relatedness of helpers in the cooperatively breeding bell miner. *Condor* 100, 343–349.

Creel, S. 2001. Social dominance and stress hormones. *Trends in Ecology and Evolution* 16, 491–497.

Creel, S. 2005. Dominance, aggression, and glucocorticoid levels in social carnivores. *Journal of Mammalogy* 86, 255–264.

Creel, S. and Creel, N. M. 2002. *The African Wild Dog: Behavior, Ecology and Evolution*. Princeton, NJ: Princeton University Press.

Creel, S. and Waser, P. M. 1994. Inclusive fitness and reproductive strategies in dwarf mongooses. *Behavioral Ecology* 5, 339–348.

Creel, S., Creel, N., Wildt, D., and Monfort, S. L. 1992. Behavioural and endocrine mechanisms of reproductive suppression in Serengeti dwarf mongooses. *Animal Behaviour* 43, 231–245.

Creel, S., Creel, N., Mills, M., and Monfort, S. 1997. Rank and reproduction in cooperatively breeding African wild dogs: behavioral and endocrine correlates. *Behavioral Ecology* 8, 298–306.

Crews, D. 1993. The organizational concept and vertebrates without sex chromosomes. *Brain, Behavior and Evolution* 42, 202–214.

Crews, D. 1998. On the organization of individual differences in sexual behavior. *American Zoologist* 38, 118–132.

Crews, D., Sakata, J., and Rhen, T. 1998. Developmental effects on intersexual and intrasexual variation in growth and reproduction in a lizard with temperature-dependent sex determination. *Journal of Comparative Physiology C* 119, 229–241.

Crommenacker, J. van de, Richardson, D. S., Groothuis, T. G. G., et al. 2004. Testosterone, cuckoldry risk and extra-pair opportunities in the Seychelles warbler. *Proceedings of the Royal Society of London B* 271, 1023–1031.

Damber, J. E. 1990. The effect of guanethidine treatment of testicular blood flow and testosterone production in rats. *Experientia* 46, 486–487.

Dawson, J. W. and Mannan, W. 1991. Dominance hierarchies and helper contributions in Harris' hawks. *Auk* 108, 649–660.

Deerenberg, C., Apanius, V., Daan, S., and Bos, N. 1997. Reproductive effort decreases antibody responsiveness. *Proceedings of the Royal Society of London B* 264, 1021–1029.

De Kloet, E. R., Oitzl, M. S., and Joels, M. 1993. Functional implications of brain corticosteroid receptor diversity. *Cellular and Molecular Neurobiology* 13, 433–455.

De Kloet, E. R., Vreugdenhil, E., Oitzl, M. S., and Joëls, M. 1998. Brain corticosteroid receptor balance in health and disease. *Endocrine Reviews* 19, 269–301.

de la Cruz, C., Solís, E., Valencia, J., Chastel, O., and Sorci, G. 2003. Testosterone and helping behavior in the azure-winged magpie (*Cyanopica cyanus*): natural covariation and an experimental test. *Behavioral Ecology and Sociobiology* 55, 103–111.

Demski, L. 1987. Diversity in reproductive patterns and behavior in fishes. In D. Crews (ed.) *Psychobiology of Reproductive Behavior: An Evolutionary Perspective*, pp. 1–27. Englewood Cliffs, NJ: Prentice-Hall.

DeNardo, D. F. and Sinervo, B. 1994a. Effects of corticosterone on activity and home-range size of free-ranging male lizards. *Hormones and Behavior* 28, 53–65.

DeNardo, D. F. and Sinervo, B. 1994b. Effects of steroid hormone interactions on activity and home-range size of male lizards. *Hormones and Behavior* 28, 273–287.

Devlin, R. H. and Nagahama, Y. 2002. Sex determination and sex differentiation in fish: an overview of genetic, physiological, and environmental influences. *Aquaculture* 208, 191–364.

Dierkes, P., Taborsky, M., and Kohler, U. 1999. Reproductive parasitism of broodcare helpers in a cooperatively breeding fish. *Behavioral Ecology* 10, 510–515.

Digby, L. J. 1999. Sexual behavior and extragroup copulations in a wild population of common marmosets (*Callithrix jacchus*). *Folia Primatologica* 70, 136–145.

Dixon, A. F. 1998. *Primate Sexuality: Comparative Studies of the Prosimians, Monkeys, Apes, and Humans*. Oxford, UK: Oxford University Press.

Duckworth, R. A., Mendonça, M. T., and Hill, G. E. 2004. Condition-dependent sexual traits and social dominance in the house finch. *Behavioral Ecology* 15, 779–784.

Dufty Jr., A. M. and Wingfield, J. C. 1986. The influence of social cues on the reproductive endocrinology of male brown-headed cowbirds: field and laboratory studies. *Hormones and Behavior* 20, 222–234.

Emerson, S. B. 2001. Male advertisement calls: behavioral variation and physiological processes. In M. J. Ryan (ed.) *Anuran Communication*, pp. 36–44. Washington, DC: Smithsonian Institution Press.

Emerson, S. B. and Hess, D. I. 2001. Glucocorticoids, androgens, testis mass, and the energetics of vocalizations in breeding male frogs. *Hormones and Behavior* 39, 59–69.

Faulkes, C. G. and Abbott, D. H. 1997. Proximate mechanisms regulating a reproductive dictatorship: a single dominant female controls male and female reproduction in colonies of naked mole-rats. In N. G. Solomon and J. A. French (eds.) *Cooperative Breeding in Mammals*, pp. 302–334. Cambridge, UK: Cambridge University Press.

Faulkes, C. G. and Bennett, N. C. 2001. Family values: group dynamics and social control of reproduction in African mole-rats. *Trends in Ecology and Evolution* 16, 184–190.

Fleming, I. A. 1998. Pattern and variability in the breeding system of Atlantic salmon, with comparisons to other salmonids. *Canadian Journal of Fisheries and Aquatic Sciences* 55, 59–76.

Flores, D., Tousignant, A., and Crews, D. 1994. Incubation temperature affects the behavior of adult leopard geckos (*Eublepharis macularius*). *Physiology and Behavior* 55, 1067–1072.

Folstad, I. and Karter, A. J. 1992. Parasites, bright males, and the immunocompetence handicap. *American Naturalist* **139**, 603–622.

Foran, C. M. and Bass, A. H. 1999. Preoptic GnRH and AVT: axes for sexual plasticity in teleost fish. *General and Comparative Endocrinology* **116**, 141–152.

Forger, N. G., Fishman, R. B., and Breedlove, S. M. 1992. Differential effects of testosterone metabolites upon the size of sexually dimorphic motoneurons in adulthood. *Hormones and Behavior* **26**, 204–213.

Ganswindt, A., Rasmussen, H. B., Heistermann, M., and Hodges, J. K. 2005. The sexually active states of free-ranging male African elephants (*Loxodonta africana*): defining musth and non-musth using endocrinology, physical signals, and behavior. *Hormones and Behavior* **47**, 83–91.

Gao, H. B., Ge, R. S., Lakshmi, V., Marandici, A., and Hardy, M. P. 1996a. Hormonal regulation of oxidative and reductive activities of 11β-hydroxysteroid dehydrogenase in rat Leydig cells. *Endocrinology* **138**, 156–161.

Gao, H. B., Shan, L. X., Monder, C., and Hardy, M. P. 1996b. Supression of endogenous corticosterone levels *in vivo* increases the steroidogenic capacity of purified rat Leydig cells *in vitro*. *Endocrinology* **137**, 1741–1718.

Gardner, L., Anderson, T., Place, A. R., Dixon, B., and Elizur, A. 2005. Sex change strategy and the aromatase genes. *Journal of Steroid Biochemistry and Molecular Biology* **94**, 395–404.

Girman, D. J., Mills, M. G. L., Geffen, E., and Wayne, R. K. 1997. A genetic analysis of social structure and dispersal in African wild dogs (*Lycaon pictus*). *Behavioral Ecology and Sociobiology* **40**, 187–198.

Godwin, J. and Crews, D. 2002. Hormones, brain and behavior in reptiles. In D. W. Pfaff, A. P. Arnold, A. M. Etgen, S. E. Farbach, and R. T. Rubin (eds.) *Hormones, Brain and Behavior*, vol. 2, pp. 649–798. New York: Academic Press.

Godwin, J., Crews, D., and Warner, R. R. 1996. Behavioural sex change in the absence of gonads in a coral reef fish. *Proceedings of the Royal Society of London B* **263**, 1683–1688.

Gonçalves, D. M., Matos, R., Fagundes, T., and Oliveira, R. F. 2005. Do bourgeois males of the peacock blenny, *Salaria pavo*, discriminate females from female-mimicking sneaker males? *Ethology* **111**, 559–572.

Gonçalves, E. J. and Almada, V. C. 1997. Sex differences in resource utilization by the peacock blenny. *Journal of Fish Biology* **51**, 624–633.

Gonçalves, E. J., Almada, V. C., Oliveira, R. F., and Santos, A. J. 1996. Female mimicry as a mating tactic in males of the blenniid fish *Salaria pavo*. *Journal of the Marine Biological Association of the UK* **76**, 529–538.

Goodson, J. and Bass, A. H. 2001. Social behaviour functions and related anatomical characteristics of vasotocin/vasopressin systems in vertebrates. *Brain Research Reviews* **35**, 246–265.

Gould, L. 2005. Variation in fecal testosterone levels, intermale aggression, dominance rank and age during mating and post-mating periods in wild adult male ring-tailed lemurs (*Lemur catta*). *American Journal of Physical Anthropology* Suppl. **40**, 108.

Gray, P. B. 2003. Marriage, parenting and testosterone variation among Kenyan Swahili men. *American Journal of Physical Anthropology* **122**, 279–286.

Griffin, A. S., Pemberton, J. M., Brotherton, P. N. M., *et al.* 2003. A genetic analysis of breeding success in the cooperative meerkat (*Suricata suricatta*). *Behavioral Ecology* **14**, 472–480.

Grober, M. S. 1998. Socially controlled sex change: integrating ultimate and proximate levels of analysis. *Acta Ethologica* **1**, 3–17.

Gross, M. R. 1996. Alternative reproductive strategies and tactics: diversity within sexes. *Trends in Ecology and Evolution* **11**, 92–98.

Grutzner, F., Rens, W., Tsend-Ayush, E., *et al.* 2004. In the platypus a meiotic chain of ten sex chromosomes shares genes with the bird Z and mammal X chromosomes. *Nature* **432**, 913–917.

Haig, S. M., Walters, J. R., and Plissner, J. H. 1994. Genetic evidence for monogamy in the cooperatively breeding red-cockaded woodpecker. *Behavioral Ecology and Sociobiology* **34**, 295–303.

Haydock, J., Koenig, W. D., and Stanback, M. T. 2001. Shared parentage and incest avoidance in the cooperatively breeding acorn woodpecker. *Molecular Ecology* **10**, 1515–1525.

Hews, D. K. and Moore, M. C. 1996. A critical period for the organization of alternative male phenotypes of tree lizards by exogenous testosterone? *Physiology and Behavior* **60**, 425–429.

Hews, D. K., Knapp, R., and Moore, M. C. 1994. Early exposure to androgens affects adult expression of alternative male types in tree lizards. *Hormones and Behavior* **28**, 96–115.

Hofmann, H. A. 2003. Functional genomics of neural and behavioral plasticity. *Journal of Neurobiology* **54**, 272–282.

Hourigan, T. F., Nakamura, N., Nagahama, Y., Yamauchi, K., and Grau, E. G. 1991. Histology, ultrastructure, and in vitro steroidogenesis of the testes of two male phenotypes of the protogynous fish, *Thalassoma duperrey* (Labridae). *General and Comparative Endocrinology* 83, 193–217.

Hughes, J. M., Mather, P. B., Toon, A., et al. 2003. High levels of extra-group paternity in a population of Australian magpies *Gymnorhina tibicen*: evidence from microsatellite analysis. *Molecular Ecology* 12, 3441–3450.

Illius, A. W., Haynes, N. B., Lamming, G. E., et al. 1983. Evaluation of LH-RH stimulation of testosterone as an index of reproductive status in rams and its application in wild antelope. *Journal of Reproduction and Fertility* 68, 105–112.

Jennings, D. H., Moore, M. C., Knapp, R., Matthews, L., and Orchinik, M. 2000. Plasma steroid-binding globulin mediation of differences in stress reactivity in alternative male phenotypes in tree lizards, *Urosaurus ornatus*. *General and Comparative Endocrinology* 120, 289–299.

Jones, C. B. 1995. Alternative reproductive behaviors in the mantled howler monkey (*Alouatta palliata* Gray): testing Carpenter's hypothesis. *Boletín de Primatología Latina* 5, 1–5.

Keane, B., Waser, P. M., Creel, S. R., et al. 1994. Subordinate reproduction in dwarf mongooses. *Animal Behaviour* 47, 65–75.

Ketterson, E. D. and Nolan Jr., V. 1999. Adaptation, exaptation, and constraint: a hormonal perspective. *American Naturalist* 154, S4–S25.

Khalil, A. M., Murakami, N., and Kaseda, Y. 1998. Relationship between plasma testosterone concentrations and age, breeding season and harem size in Misaki feral horses. *Journal of Veterinary Medical Science* 60, 643–645.

Khan, M. Z., McNabb, F. M. A., Walters, J. R., and Sharp, P. J. 2001. Patterns of testosterone and prolactin concentrations and reproductive behavior of helpers and breeders in the cooperatively breeding red-cockaded woodpecker (*Picoides borealis*). *Hormones and Behavior* 40, 1–13.

Kim, S. J., Ogasawara, K., Park, J. G., Takemura, A., and Nakamura, M. 2002. Sequence and expression of androgen receptor and estrogen receptor gene in the sex types of protogynous wrasse, *Heliochoeres trimaculatus*. *General and Comparative Endocrinology* 127, 165–173.

Kimball, R. T. and Ligon, J. D. 1999. Evolution of avian plumage dichromatism from a proximate perspective. *American Naturalist* 154, 182–193.

Kindler, P. M., Philipp, D. P., Gross, M. R., and Bahr, J. M. 1989. Serum 11-ketotestosterone and testosterone concentrations associated with reproduction in male bluegill

(*Lepomis macrochirus*: Centrarchidae). *General and Comparative Endocrinology* 75, 446–453.

Knapp, R. 2003. Endocrine mediation of vertebrate male alternative reproductive tactics: the next generation of studies. *Integrative and Comparative Biology* 43, 658–668.

Knapp, R. and Moore, M. C. 1996. Male morphs in tree lizards, *Urosaurus ornatus*, have different delayed hormonal responses to aggressive encounters. *Animal Behaviour* 52, 1045–1055.

Knapp, R. and Moore, M. C. 1997. Male morphs in tree lizards have different testosterone responses to elevated levels of corticosterone. *General and Comparative Endocrinology* 107, 273–279.

Knapp, R., Carlisle, S. L., and Jessop, T. S. 2002. A model for androgen–glucocorticoid interactions in male alternative reproductive tactics: potential roles for steroidogenic enzymes. *Hormones and Behavior* 41, 475.

Knapp, R., Hews, D. K., Thompson, C. W., Ray, L. E., and Moore, M. C. 2003. Environmental and endocrine correlates of tactic switching by nonterritorial male tree lizards (*Urosaurus ornatus*). *Hormones and Behavior* 43, 83–92.

Kraak, S. B. M. and Pen, I. R. 2002. Sex ratios: concepts and research methods. In Hardy, I. C. W. (ed.) *Sex Determining Mechanisms in Vertebrates*, pp. 158–177. Cambridge, UK: Cambridge University Press.

Kraus, C., Heistermann, M., and Kappeler, P. M. 1999. Physiological supression of sexual function of subordinate males: a subtle form of intrasexual competition among male sifakas (*Propithecus verreauxi*)? *Physiology and Behavior* 66, 855–861.

Lacey, E. A. and Sherman, P. W. 1991. Social organization of naked mole-rat colonies: evidence for division of labour. In P. W. Sherman, J. U. M. Jarvis, and R. D. Alexander (eds.) *The Biology of the Naked Mole-Rat*, pp. 275–336. Princeton, NJ: Princeton University Press.

Lank, D. B., Coupe, M., and Wynne-Edwards, K. E. 1999. Testosterone-induced male traits in female ruffs (*Philomachus pugnax*): autosomal inheritance and gender differentiation. *Proceedings of the Royal Society of London B* 266, 2323–2330.

Leary, C. J., Jessop, T. S., Garcia, A. M., and Knapp, R. 2004. Steroid hormone profiles and relative body condition of calling and satellite toads: implications for proximate regulation of behavior in anurans. *Behavioral Ecology* 15, 313–320.

Lee, J. S. F. and Bass, A. H. 2004. Effects of 11-ketotestosterone on brain, sonic muscle, and behavior in type-II midshipman fish. *Hormones and Behavior* 46, 115–116.

Lozano, G. A. and Lank, D. B. 2004. Immunocompetence and testosterona-induced condition traits in male ruffs (*Philomachus pugnax*). *Animal Biology* 54, 315–329.

Lynch, J. W., Ziegler, T. E., and Strier, K. B. 2002. Individual and seasonal variation in fecal testosterone and cortisol in wild male tufted capuchin monkeys, *Cebus apella nigritus*. *Hormones and Behavior* 41, 275–287.

Lynch Alfaro, J. W. 2005. Male mating strategies and reproductive constraints in a group of wild tufted capuchin monkeys, *Cebus apella nigritus*. *American Journal of Primatology* 67, 313–328.

Maggioncalda, A. N., Sapolsky, R. M., and Czekala, N. M. 1999. Reproductive hormone profiles in captive male orangutans: implications for understanding developmental arrest. *American Journal of Physical Anthropology* 109, 19–32.

Maggioncalda, A. N., Czekala, N. M., and Sapolsky, R. M. 2000. Growth hormone and thyroid stimulating hormone concentrations in captive male orangutans: implications for understanding developmental arrest. *American Journal of Primatology* 50, 67–76.

Maggioncalda, A. N., Czekala, N. M., and Sapolsky, R. M. 2002. Male orangutan subadulthood: a new twist on the relationship between chronic stress and developmental arrest. *American Journal of Physical Anthropology* 118, 25–32.

Maney, D. L., Erwin, K. L., and Goode, C. T. (2005). Neuroendocrine correlates of behavioral polymorphism in white-throated sparrows. *Hormones and Behavior* 48, 196–206.

Martin II , L. B., Scheuerlein, A., and Wikelski, M. 2003. Immune activity elevates energy expenditure of house sparrows: a link between direct and indirect costs? *Proceedings of the Royal Society of London B* 270, 153–158.

Mason, R. T. 1992. Reptilian pheromones. In C. Gans and D. Crews (eds.) *Hormones, Brain and Behavior*, vol. 18, *Biology of the Reptilia*, pp. 114–228. Chicago, IL: University of Chicago Press.

Mason, R. T. and Crews, D. 1985. Female mimicry in garter snakes. *Nature* 316, 59–60.

Matsuda, M., Nagahama, Y., Shinomiya, A., *et al.* 2002. DMY is a Y-specific DM-domain gene required for male development in the medaka fish. *Nature* 417, 559–563.

Mayer, I., Lundqvist, H., Berglund, I., *et al.* 1990. Seasonal endocrine changes in Baltic salmon, *Salmo salar*, immature parr and mature male parr. 1. Plasma levels of five androgens, 17α-hydroxy-20β-dihydroprogesterone, and 17β-estradiol. *Canadian Journal of Zoology* 68, 1360–1365.

Mays, N. A., Vleck, C. M., and Dawson, J. 1991. Plasma luteinizing hormone, steroid hormones, behavioral role, and nest stage in cooperatively breeding harris' hawks (*Parabuteo unicinctus*). *Auk* 108, 619–637.

McDonnell, S. M. and Murray, S. C. 1995. Bachelor and harem stallion behavior and endocrinology. *Biology of Reproduction Monographs* 1, 577–590.

Mendonça, M. T., Licht, P., Ryan, M. J., and Barnes, R. 1985. Changes in hormone levels in relation to breeding behavior in male bullfrogs (*Rana catesbeiana*) at the individual and population levels. *General and Comparative Endocrinology* 58, 270–279.

Ming, G.-L. and Song, H. 2005. Adult neurogenesis in the mammalian central nervous system. *Annual Reviews in Neuroscience* 28, 223–250.

Mizusaki, H., Kawabe, K., Mukai, T., *et al.* 2003. *Dax-1* (dosage-sensitive sex reversal-adrenal hypoplasia congenita critical region on the X chromosome, gene 1) gene transcription is regulated by *wnt4* in the female developing gonad. *Molecular Endocrinology* 17, 507–519.

Modesto, T. and Canário, A. V. M. 2003a. Morphometric changes and sex steroid levels during the annual reproductive cycle of the Lusitanian toadfish, *Halobatrachus didactylus*. *General and Comparative Endocrinology* 131, 220–231.

Modesto, T. and Canário, A. V. M. 2003b. Hormonal control of swimbladder sonic muscle dimorphism in the Lusitanian toadfish *Halobatrachus didactylus*. *Journal of Experimental Biology* 206, 3467–3477.

Moore, M. C. 1991. Application of organization-activation theory to alternative male reproductive strategies: a review. *Hormones and Behavior* 25, 154–179.

Moore, M. C., Hews, D. K., and Knapp, R. 1998. Hormonal control and evolution of alternative male phenotypes: generalizations of models for sexual differentiation. *American Zoologist* 38, 133–151.

Mooring, M. S., Patton, M. L., Lance, V. A., *et al.* 2004. Fecal androgens of bison bulls during the rut. *Hormones and Behavior* 46, 392–398.

Morrish, B. C. and Sinclair, A. H. 2002. Vertebrate sex determination: many means to an end. *Reproduction* 124, 447–457.

Moss, A. M., Clutton-Brock, T. H., and Monfort, S. L. 2001. Longitudinal gonadal steroid excretion in free-living male and female meerkats (*Suricata suricatta*). *General and Comparative Endocrinology* 122, 158–171.

Muehlenbein, M. P., Watts, D. P., and Whitten, P. 2004. Dominance rank and fecal testosterone levels in adult male chimpanzees (*Pan troglodytes schweinfurthii*) at Ngogo,

Kibale National park, Uganda. *American Journal of Primatology* **64**, 71–82.

Mulder, R. A., Duna, P. O., Cockburn, A., Lazenby-Cohen, K. A., and Howell, M. J. 1994. Helpers liberate female fairy-wrens from constraints on extra-pair mate choice. *Proceedings of the Royal Society of London B* **255**, 223–229.

Muller, M. N. and Wrangham, R. W. 2004a. Dominance, aggression and testosterone in wild chimpanzees: a test of the "challenge hypothesis." *Animal Behaviour* **67**, 113–123.

Muller, M. N. and Wrangham, R. W. 2004b. Dominance, cortisol and stressing wild chimpanzees (*Pan troglodytes schweinfurthii*). *Behavioral Ecology and Sociobiology* **55**, 332–340.

Munday, P. and Jones, G. 1998. Bi-directional sex change in a coral-dwelling goby. *Behavioral Ecology and Sociobiology* **43**, 371–377.

Nanda, I., Kondo, M., Hornung, U., *et al.* 2002. A duplicated copy of *DMRT1* in the sex-determining region of the Y chromosome of the medaka, *Oryzias latipes. Proceedings of the National Academy of Sciences of the United States of America* **99**, 11778–11783.

Nelson, R. J. 2005. *An Introduction to Behavioral Endocrinology*, 3rd edn. Sunderland, MA: Sinauer Associates.

Nievergelt, C. M., Digby, L. J., Ramiakrishnan, U., and Woodruff, D. S. 2000. Genetic analysis of group composition and breeding system in a wild common marmoset (*Callithrix jacchus*) population. *International Journal of Primatology* **21**, 1–20.

Norris, K. and Evans, M. R. 2000. Ecological immunity: life history trade-offs and immune defense in birds. *Behavioral Ecology* **11**, 19–20.

Oliveira, R. F. 2004. Social modulation of androgens in vertebrates: mechanisms and function. *Advances in the Study of Behavior* **34**, 165–239.

Oliveira, R. F. 2005. Neuroendocrine mechanisms of alternative reproductive tactics in fish. In K. A. Sloman, R. W. Wilson, and S. Balshine (eds.) *Fish Physiology*, vol. 24, *Behavior and Physiology of Fish*, pp. 297–357. New York: Elsevier.

Oliveira, R. F. and Almada, V. C. 1998. Mating tactics and male–male courtship in the lek-breeding cichlid *Oreochromis mossambicus. Journal of Fish Biology* **52**, 1115–1129.

Oliveira, R. F., Almada, V. C., and Canario, A. V. M. 1996. Social modulation of sex steroid concentrations in the urine of male cichlid fish *Oreochromis mossambicus. Hormones and Behavior* **30**, 2–12.

Oliveira, R. F., Miranda, J. S., Carvalho, N., *et al.* 2000. Male mating success in the Azorean rock-pool blenny: the effects of body size, male behaviour and nest characteristics. *Journal of Fish Biology* **57**, 1416–1428.

Oliveira, R. F., Almada, V. C., Gonçalves, E. J., Forsgren, E., and Canario, A. V. M. 2001a. Androgen levels and social interactions in breeding males of the peacock blenny. *Journal of Fish Biology* **58**, 897–908.

Oliveira, R. F., Canario, A. V. M., and Grober, M. S. 2001b. Male sexual polymorphism, alternative reproductive tactics and androgens in combtooth blennies (Pisces: Blenniidae). *Hormones and Behavior* **40**, 266–275.

Oliveira, R. F., Canário, A. V. M., Grober, M. S., and Santos, R. S. 2001c. Endocrine correlates of alternative reproductive tactics and male polymorphism in the Azorean rock-pool blenny, *Parablennius sanguinolentus parvicornis. General and Comparative Endocrinology* **121**, 278–288.

Oliveira, R. F., Carneiro, L. A., Gonçalves, D. M., Canario, A. V. M., and Grober, M. S. 2001d. 11-ketotestosterone inhibits the alternative mating tactic in sneaker males of the peacock blenny, *Salaria pavo. Brain, Behavior and Evolution* **58**, 28–37.

Oliveira, R. F., Carneiro, L. A., Canário, A. V. M., and Grober, M. S. 2001e. Effects of androgens on social behaviour and morphology of alternative reproductive males of the Azorean rock-pool blenny. *Hormones and Behavior* **39**, 157–166.

Oliveira, R. F., Hirschenhauser, K., Carneiro, L. A., and Canario, A. V. M. 2002. Social modulation of androgens in male teleost fish. *Comparative Biochemistry and Physiology B* **132**, 203–215.

Oliveira, R. F., Hirschenhauser, K., Canario, A. V. M., and Taborsky, M. 2003. Androgen levels of reproductive competitors in a cooperatively breeding cichlid. *Journal of Fish Biology* **63**, 1615–1620.

Oliveira, R. F., Ros, A. F. H., and Gonçalves, D. M. 2005. Intra-sexual variation in male reproduction in teleost fish: a comparative approach. *Hormones and Behavior* **48**, 430–439.

O'Riain, M. J., Jarvis, J. U. M., and Faulkes, C. G. 1996. A dispersive morph in the naked mole-rat. *Nature* **380**, 619–621.

O'Riain, M. J., Bennett, N. C., Brotherton, P. N. M., McIlrath, G., and Clutton-Brock, T. 2000a. Reproductive suppression and inbreeding avoidance in wild populations of cooperatively breeding meerkats (*Suricata suricatta*). *Behavioral Ecology and Sociobiology* **48**, 471–477.

O'Riain, M. J., Jarvis, J. U. M., Alexander, R., Buffenstein, R., and Peeters, C. 2000b. Morphological castes in a vertebrate.

Proceedings of the National Academy of Sciences of the United States of America **97**, 13194–13197.

Owens, I. P. F. and Short, R. 1995. Hormonal basis of sexual dimorphism in birds: implications of new theories of sexual selection. *Trends in Ecology and Evolution* **10**, 44–47.

Parhar, I. 2002. Cell migration and evolutionary significance of GnRH subtypes. *Progress in Brain Research* **141**, 3–17.

Pelletier, F., Bauman, J., and Festa-Bianchet, M. 2003. Fecal testosterone in bighorn sheep (*Ovis canadensis*): behavioral and endocrine correlates. *Canadian Journal of Zoology* **81**, 1678–1684.

Perry, A. N. and Grober, M. S. 2003. A model for social control of sex change: interactions of behavior, neuropetides, glucocorticoids, and sex steroids. *Hormones and Behavior* **43**, 31–38.

Peters, A., Astheimer, L. B., and Cockburn, A. 2001. The annual testosterone profile in cooperatively breeding superb fairy-wrens, *Malurus cyaneus*, reflects their extreme infidelity. *Behavioral Ecology and Sociobiology* **50**, 519–527.

Peters, A., Cockburn, A., and Cunningham, R. 2002. Testosterone treatment suppresses paternal care in superb fairy-wrens, *Malurus cyaneus*, despite their concurrent investment in courtship. *Behavioral Ecology and Sociobiology* **51**, 538–547.

Phoenix, C., Goy, R., Gerall, A., and Young, W. 1959. Organizing action of prenatally-administered testosterone proprionate on the tissues mediating mating behavior in the female guinea pig. *Endocrinology* **65**, 369–382.

Pieau, C. and Dorizzi, M. 2004. Oestrogens and temperature-dependent sex determination in reptiles: all is in the gonads. *Journal of Endocrinology* **181**, 367–377.

Poiani, A. and Fletcher, T. 1994. Plasma levels of androgens and gonadal development of breeders and helpers in the bell miner (*Manorina melanophrys*). *Behavioral Ecology and Sociobiology* **34**, 31–41.

Quinn, J. S., Woolfenden, G. E., Fitzpatrick, J. W., and White, B. N. 1999. Multi-locus DNA fingerprinting supports genetic monogamy in Florida scrub-jays. *Behavioral Ecology and Sociobiology* **45**, 1–10.

Rachlow, J. L., Berkeley, E. V., and Berger J. 1998. Correlates of male mating strategies in white rhinos (*Ceratotherium simum*). *Journal of Mammalogy* **79**, 1317–1324.

Ray, J. C. and Sapolsky, R. M. 1992. Styles of male social behavior and their endocrine correlates among high-ranking wild baboons. *American Journal of Primatology* **28**, 231–250.

Reavis, R. H. and Grober, M. S. 1999. An integrative approach to sex change: social, behavioural and neurochemical changes in *Lythrypnus dalli* (Pisces). *Acta Ethologica* **2**, 51–60.

Reeve, H. K., Westneat, D. F., Noon, W. A., Sherman, P. W., and Aquadro, C. F. 1990. DNA "fingerprinting" reveals high levels of inbreeding in colonies of the eusocial naked mole-rat. *Proceedings of the National Academy of Sciences of the United States of America* **87**, 2496–2500.

Reinboth, R. and Becker, B. 1984. In vitro studies on steroid metabolism by gonadal tissues from ambisexual teleosts. 1. Conversion of 14-C testosterone by males and females of the protogynous wrasse *Coris julis* L. *General and Comparative Endocrinology* **55**, 245–250.

Reyer, H.-U. 1980. Flexible helper structure as an ecological adaptation in the pied kingfisher (*Ceryle rudis*). *Behavioral Ecology and Sociobiology* **6**, 219–227.

Reyer, H.-U. 1984. Investment and relatedness: a cost/benefit analysis of breeding and helping in the pied kingfisher (*Ceryle rudis*). *Animal Behaviour* **32**, 1163–1178.

Reyer, H.-U., Dittami, J., and Hall, M. R. 1986. Avian helpers at the nest: are they psychologically castrated? *Ethology* **71**, 216–228.

Rhen, T. and Crews, D. 2002. Variation in reproductive behaviour within a sex: neural systems and endocrine activation. *Journal of Neuroendocrinology* **14**, 517–531.

Richardson, D. S., Jury, F. L., Blaakmer, K., Komdeur, J., and Burke, T. 2001. Parentage assignment and extra-group paternity in a cooperative breeder: the Seychelles warbler (*Acrocephalus sechellensis*). *Molecular Ecology* **10**, 2263–2273.

Ros, A. F. H., Canario, A. V. M., Couto, E., Zeilstra, I., and Oliveira, R. F. 2003. Endocrine correlates of intra-specific variation in the mating system of the St. Peter's fish (*Sarotherodon galilaeus*). *Hormones and Behavior* **44**, 365–373.

Ros, A. F. H., Bouton, N., Santos, R. S., and Oliveira, R. F. 2006. Alternative male reproductive tactics and the immuncompetence handicap in the Azorean rock-pool blenny, *Parablennius parvicornis*. *Proceedings of the Royal Society of London B* **273**, 901–909.

Rose, R. M., Holaday, J. W., and Bernstein, I. S. 1971. Plasma testosterone, dominance rank and aggressive behaviour in male rhesus monkeys. *Nature* **231**, 366–368.

Sands, J. and Creel, S. 2004. Social dominance, aggression and faecal glucocorticoid levels in a wild population of wolves, *Canis lupus*. *Animal Behaviour* **67**, 387–396.

Sapolsky, R. M. 1983. Endocrine aspects of social instability in the olive baboon (*Papio anubis*). *American Journal of Primatology* **5**, 365–379.

Sapolsky, R. M. and Ray, J. 1989. Styles of dominance and their physiological correlates among wild baboons. *American Journal of Primatology* 18, 1–13.

Sapolsky, R. M., Romero, L. M., and Munck, A. U. 2000. How do glucocorticoids influence stress responses? Integrating permissive, suppressive, stimulatory, and preparative actions. *Endocrine Reviews* 21, 55–89.

Sarre, S. D., Georges, A., and Quinn, A. 2004. The ends of a continuum: genetic and temperature-dependent sex determination in reptiles. *BioEssays* 26, 639–645.

Sauther, M. L. 1991. Reproductive behavior of free-ranging *Lemur catta* at Beza Mahafaly Special Reserve, Madagascar. *American Journal of Physical Anthropology* 84, 463–477.

Schlinger, B. A., Greco, C., and Bass, A. H. 1999. Aromatase activity in the hindbrain and vocal control region of a teleost fish: divergence among males with alternative reproductive tactics. *Proceedings of the Royal Society of London B* 266, 131–136.

Schmidt, L. G., Bradshaw, S. D., and Follett, B. K. 1991. Plasma levels of luteinizing hormone and androgens in relation to age and breeding status among cooperatively breeding Australian magpies (*Gymnorhina tibicen* Latham). *General and Comparative Endocrinology* 83, 48–55.

Schoech, S. J., Mumme, R. L., and Moore, M. 1991. Reproductive endocrinology and mechanisms of breeding inhibition in cooperatively breeding Florida scrub jays (*Aphelocoma c. coerulescens*). *Condor* 93, 354–364.

Schoech, S. J., Mumme, R. L., and Wingfield, J. C. 1996. Delayed breeding in the cooperatively breeding Florida scrub-jay (*Aphelocoma coerulescens*): inhibition or the absence of stimulation? *Behavioral Ecology and Sociobiology* 39, 77–90.

Schulz, R. W. and Miura, T. 2002. Spermatogenesis and its endocrine regulation. *Fish Physiology and Biochemistry* 26, 43–56.

Schwarzenberger, F., Sterregaard, F., Elias, F., Baumgartner, R., and Walzer, C. 2004. Who is the boss? Endocrinological evaluation of re-introduced takhis in Takhin Tal: implications and consequences. In P. Kaczensky (ed.) *Abstracts of the 2nd International Workshop on the Re-Introduction of the Przewaski's Horse*, pp. 22–23. Takhin Tal (Mongolia): International Takhi Group.

Setchell, J. M. and Dixon, A. F. 2001. Arrested development of secondary sexual adornments in subordinate adult male mandrills (*Mandrillus sphinx*). *American Journal of Physical Anthropology* 115, 245–252.

Sheldon, B. C. and Verhulst, S. 1996. Ecological immunity: costly parasite defences and trade-offs in evolutionary ecology. *Trends in Ecology and Evolution* 11, 317–321.

Shine, R., Harlow, P., Lemaster, M. P., Moore, I. T., and Mason, R. T. 2000. The travestite serpent: why do male garter snakes court (some) other males? *Animal Behaviour* 59, 349–359.

Shuster, S. M. and Wade, M. J. 2003. *Mating Systems and Strategies*. Princeton, NJ: Princeton University Press.

Silverin, B. and Wingfield, J. C. 1982. Patterns of breeding behaviour and plasma levels of hormones in a free-living population of pied flycatchers, *Ficedula hypoleuca. Journal of Zoology* 198, 117–129.

Simon, N. G. 2002. Hormonal processes in the development and expression of aggressive behavior. In D. W. Pfaff, A. P. Arnold, A. M. Etgen, S. E. Farbach, and R. T. Rubin (eds.) *Hormones, Brain and Behavior*, vol. 1, pp. 339–392. New York: Academic Press.

Sinervo, B., Miles, D. B., Frankino, W. A., Klukowski, M., and DeNardo, D. F. 2000. Testosterone, endurance, and Darwinian fitness: natural and sexual selection on the physiological bases of alternative male behaviors in side-blotched lizards. *Hormones and Behavior* 38, 222–233.

Slater, C. H. and Schreck, C. B. 1997. Physiological levels of testosterone kill salmonid leukocytes *in vitro*. *General and Comparative Endocrinology* 106, 113–119.

Slater, C. H. and Schreck, C. B. 1998. Season and physiological parameters modulate salmonid leukocyte androgen receptor affinity and abundance. *Fish and Shellfish Immunology* 8, 379–391.

Slater, C. H., Fitzpatrick, M. S. and Schreck, C. B. 1995. Characterization of an androgen receptor in salmonid lymphocytes: possible link to androgen induced immunosuppression. *General and Comparative Endocrinology* 100, 218–225.

Solomon, N. G. and French, J. A. (eds.) 1997. *Cooperative Breeding in Mammals*. Cambridge, UK: Cambridge University Press.

Smith, C. A. and Sinclair, A. H. 2004. Sex determination: insights from the chicken. *BioEssays* 26, 120–132.

Stevenson, I. R. and Bancroft, D. R. 1995. Fluctuating trade-offs favour precocial maturity in male Soay sheep. *Proceedings of the Royal Society of London B* 262, 267–275.

Sussman, R. W. 1991. Demography and social organization of free-ranging *Lemur catta* in the Beza Mahafaly Reserve, Madagascar. *American Journal of Physical Anthropology* 84, 43–58.

Swain, A., Narvaez, S., Burgoyne, P., Camerino, G., and Lovellbadge, R. 1998. *Dax1* antagonizes *Sry* action in mammalian sex determination. *Nature* **391**, 761–767.

Taborsky, M. 1994. Sneakers, satellites, and helpers: parasitic and cooperative behavior in fish reproduction. *Advances in the Study of Behavior* **23**, 1–100.

Taborsky, M. 1997. Bourgeois and parasitic tactics: do we need collective, functional terms for alternative reproductive behaviours? *Behavioral Ecology and Sociobiology* **41**, 361–362.

Taborsky, M. 1998. Sperm competition in fish: "bourgeois" males and parasitic spawning. *Trends in Ecology and Evolution* **13**, 222–227.

Thompson, C. W. and Moore, M. C. 1992. Behavioral and hormonal correlates of alternative reproductive strategies in a polygynous lizard: tests of the relative plasticity and challenge hypotheses. *Hormones and Behavior* **26**, 568–585.

Tousignant, A. and Crews, D. 1995. Incubation temperature and gonadal sex affect growth and physiology in the leopard gecko (*Eublepharis macularius*), a lizard with temperature-dependent sex determination. *Journal of Morphology* **224**, 159–170.

Uglem, I., Rosenqvist, G., and Schioler Wasslavik, H. 2000. Phenotypic variation between dimorphic males in corkwing wrasse (*Symphodus melops* L.). *Journal of Fish Biology* **57**, 1–14.

Uglem, I., Galloway, T. F., Rosenqvist, G., and Folstad, I. 2001. Male dimorphism, sperm traits and immunology in the corkwing wrasse (*Symphodus melops* L.). *Behavioral Ecology and Sociobiology* **50**, 511–518.

Uglem, I., Mayer, I., and Rosenqvist, G. 2002. Variation in plasma steroids and reproductive traits in dimorphic males of corkwing wrasse (*Symphodus melops* L.). *Hormones and Behavior* **41**, 396–404.

Vaiman, D. and Pailhoux, E. 2000. Mammalian sex reversal and intersexuality: deciphering the sex-determination cascade. *Trends in Genetics* **16**, 488–494.

Valencia, J., de la Cruz, C., and González B. 2003. Flexible helping behaviour in the azure-winged magpie. *Ethology* **109**, 545–558.

Virgin, C. E. and Sapolsky, R. M. 1997. Styles of male social behavior and their endocrine correlates among low-ranking baboons. *American Journal of Primatology* **42**, 25–39.

Vleck, C. M. and Brown, J. L. 1999. Testosterone and social and reproductive behaviour in Aphelcoma jays. *Animal Behaviour* **58**, 943–951.

Volff, J.-N., Kondo, M., and Schartl, M. 2003. Medaka dmY/dmrt1Y is not the universal primary sex-determining gene in fish. *Trends in Genetics* **19**, 196–199.

Watson, N. V., Freeman, L. M., and Breedlove, S. M. 2001. Neuronal size in the spinal nucleus of the bulbocavernosus: direct modulation by androgen in rats with mosaic androgen insensitivity. *Journal of Neuroscience* **21**, 1062–1066.

Wedekind, C. and Folstad, I. 1994. Adaptive and non-adaptive immunosuppression by sex hormones. *American Naturalist* **143**, 936–938.

West, P. M. and Packer, C. 2002. Sexual selection, temperature and the lion's mane. *Science* **297**, 1339–1343.

Whitfield, C. W., Cziko, A. M., and Robinson, G. E. 2003. Gene expression profiles in the brain predict behavior in individual honey bees. *Science* **302**, 296–299.

Wikelski, M., Steiger, S. S., Gall, B., and Nelson, K. N. 2005. Sex, drugs, and mating role: testosterone-induced phenotype-switching in Galapagos marine iguanas. *Behavioral Ecology* **16**, 260–268.

Wilson, J. D., Leihy, M. W., Shaw, G., and Renfree, M. B. 2002. Androgen physiology: unsolved problems at the millennium. *Molecular and Cellular Endocrinology* **198**, 1–5.

Wingfield, J. C., Hegner, R. E., Dufty, A. M., and Ball, G. F. 1990. The "challenge hypothesis": theoretical implications for patterns of testosterone secretion, mating systems, and breeding strategies. *American Naturalist* **136**, 829–846.

Wingfield, J. C., Hegner, R. E., and Lewis, D. M. 1991. Circulating levels of luteinizing hormone and steroid hormones in relation to social status in the cooperatively breeding white-browed sparrow weaver, *Plocepasser mahali*. *Journal of Zoology* **225**, 43–58.

Zucker, E. L., O'Neil, J. A. S., and Harrison, R. M. 1996. Fecal testosterone values for free-ranging male mantled howling monkeys (*Alouatta palliata*) in Costa Rica. IPS/ASP 1996 Congress Abstracts, p. 112.

Zupanc, G. K. H. 2001. A comparative approach towards the understanding of adult neurogenesis. *Brain, Behavior and Evolution* **58**, 246–249.

Zupanc, G. K. H. and Lamprecht, J. 2000. Towards a cellular understanding of motivation: structural reorganization and biochemical switching as key mechanisms of behavioral plasticity. *Ethology* **106**, 467–477.

Part III
Taxonomic reviews of alternative reproductive tactics

8 · Alternative reproductive tactics in insects

H. JANE BROCKMANN

CHAPTER SUMMARY

Discrete, alternative reproductive tactics (ARTs) occur in most orders of insects, across a wide array of mating systems, and during all steps in the reproductive process (locating a mate, gaining access to him/her, copulating, and post-copulatory behavior). ARTs for mate searching are particularly common and often involve a division between high-investment or high-risk but sedentary, nondispersing tactics and low-investment or low-risk but active, dispersing, searching tactics with longer-range movements. ARTs in insects are often associated with intense sexual selection and include individuals that avoid costly or high-risk intrasexual interactions, parasitizing the costly investment of others, or circumventing intersexual interactions and mate conflict. Two or more tactics can arise in a population when opportunities for success are discrete and require different and mutually exclusive behavior or morphology. Suites of distinctive, correlated traits arise through condition-dependent, developmental switches and when there is selection against individuals with intermediate traits (disruptive selection). ARTs are maintained in populations by frequency dependence and equality of fitness among tactics or as condition- or environment-dependent alternatives that maximize individual fitness.

8.1 THE PROBLEM

Discrete, alternative reproductive tactics (ARTs) within one sex and one population occur in most groups of insects (Table 8.1). For example, nonflying, large-headed, fighting males live in the nests of some female halictid (Kukuk and Schwarz 1988) and andrenid bees (Danforth 1991b) and mate with females inside the nest just prior to oviposition; while smaller, flying males with normal-looking heads and mandibles mate with females outside the nest as they forage on flowers (Danforth 1991b) (Figure 8.1). In this case ARTs are correlated with distinct morphological differences

between male morphs, but in other cases no obvious differences exist. In the meloid beetle *Tegrodera aloga*, males usually engage females in a long, face-to-face courtship and mounting follows a specific signal from the female (Pinto 1975). Some males, however, "short-circuit" the process by mounting abruptly from behind and attempting to copulate forcefully without prior display. Such discrete, alternative tactics are an evolutionary puzzle because if one form was just a little less successful (on average) than the other, then it should be eliminated from the population through natural selection. Nonetheless, in many species ARTs are maintained over long periods at relatively stable frequencies. Often they are maintained in spite of the fact that one form appears to be less successful than the other(s). What are the evolutionary processes that maintain such variation? What are the selective pressures that favor two discrete tactics rather than one continuously variable tactic? What are the underlying genetic, developmental, and physiological mechanisms that result in two (or more) different forms of one sex? Why do the tactics take the particular forms they do? What maintains the population at particular frequencies of tactics? In this chapter I will begin to address these issues by documenting some of the remarkable diversity of ARTs that can be found among insects.

Many schemes exist for organizing the diversity of ARTs (e.g., Alcock 1979b, Cade 1979b, Waltz and Wolf 1984, Taborsky 1997, Brockmann 2001, Shuster and Wade 2003). Some are based on mechanisms, such as separating genetic polymorphism from condition-dependent tactics (Austad 1984, Gross 1996, Tomkins 1999) or separating ontogenetic switch from behavioral plasticity (Fincke 1985), but since most ARTs involve both genetic and environmental components, this is not a particularly useful or practical means of categorizing patterns (Brockmann 2001). Other methods of categorizing ARTs separate out those with frequency-dependent effects from those without (Davies 1982, Austad 1984). However, it is now clear that

Alternative Reproductive Tactics, ed. Rui F. Oliveira, Michael Taborsky, and H. Jane Brockmann. Published by Cambridge University Press.
© Cambridge University Press 2008.

Table 8.1. *Alternative reproductive tactics in male insects*[a]

Order	Family	Species	Tactic[b]		Morphological correlates	Tactic type[c]	Reference
			Tactic 1	Tactic 2			
Odonata – Anisoptera (dragonflies)	Libellulidae	*Leucorrhinia intacta*	T	NT, S	No	C	Waltz and Wolf 1984, 1988, Wolf and Waltz 1993
		Plathemis lydia	T	NT, S	No	CD, CA	Campanella and Wolf 1974
		Nannophya pygmaea	T	NT, Sk	No	CD	Tsubaki and Ono 1986
		Sympetrum parvulum	T, G♀	NT, G♀/NG	No	CD	Uéda 1979
Odonata – Zygoptera (damselflies)	Calopterygidae	*Mnais nawai, M. pruinosa,*	T, G♀	NT, NG, S	Yes: 1 = larger, orange wings; 2 = smaller, pale wings	P	Nomakuchi and Higashi 1996, Higashi and Nomakuchi 1997, Tsubaki et al. 1997
		M. costalis	T	NT, S, Sk	Yes: 1 = larger, orange wings; 2 = smaller, pale wings	P, CA, CF	Hooper et al. 1999, Tsubaki 2003
		Calopteryx maculata	T, G♀, C♀	NT, S, NG, NC, Sk	No	CA, CD	Waage 1973, 1979, Forsyth and Montgomerie 1987
		Calopteryx haemorrhoidalis	T, C♀	NT, S, NC	No	CD	Cordero 1999
	Coenagrionidae	*Enallagma hageni*	W	NT, S, G♀	No	CSu	Fincke 1982, 1984, 1985, 1986
	Megapodagrionidae	*Paraphlebia quinta*	T	Sa	Yes: 1 = black-winged; 2 = hyaline-winged	P	Gonzalez-Soriano and Cordoba-Aguilar 2003
Dermaptera (earwigs)	Forficulidae	*Forficula auricularia*, several related species and genera	F	NF	Yes: 1 = larger, macrolabic; 2 = smaller, brachylabic	CS	Lamb 1976, Briceño and Eberhard 1987, Eberhard and Gutiérrez 1991, Moore and Wilson 1993, Radesäter and Halldórsdóttir 1993, Tomkins and Simmons 1996, Tomkins 1999, Forslund 2003

178

Orthoptera – Caelifera	Acrididae (grasshoppers)	*Melanoplus sanguinipes*	A, C♀	NC	No	CF	Belovsky *et al.* 1996
		Ligurotettix coquilletti	T	NT, Sa	No	CD	Greenfield and Shelly 1985, 1989, Shelly and Greenfield 1985, 1989,
	Pneumoridae (bladder grasshopper)	*L. planum*	T	NT	No	C	Shelly *et al.* 1987
		Bullacris membracioides	A (♂ approach)	NA, S	Yes: 1 = larger, wings; 2 = small, wingless,	P	Alexander and van Staaden 1989, van Staaden and Römer 1997
Orthoptera – Ensifera	Tettigoniidae (katydids)	*Orchelimum nigripes*	A	NA, Sa	No	CD, CA	Feaver 1983
		Elephantodeta nobilis	A	NA, Sa	No	CD	Bailey and Field 2000
	Anostostomatidae (wetas)	*Hemideina crassidens, H. maori*	T	S	Yes: 1 = larger; 2 = smaller (mature earlier)	CH	Koning and Jamieson 2001, Gwynne and Jamieson 1998, Leisnham and Jamieson 2004
	Gryllotalpidae (mole crickets)	*Scapteriscus acletus, S. vicinus*	A	Sa	No	C	Forrest 1983
	Gryllidae (crickets)	*Gryllus integer, G. pennsylvanicus, G. campestris*	A	NA, S	No	P	Hedrick 1988, French and Cade 1989, Cade and Cade 1992
		Gryllus texensis	A	NA, S	No	P, CD	Cade 1979b, 1981, Cade and Wyatt 1984, Rowell and Cade 1993, Bertram 2002
		Gryllus rubens, G. firmus	A	A,S	Yes: 1 = larger, short-winged; 2 = smaller, long-winged	P	Walker and Sivinski 1986, Crnokrak and Roff 1995, 1998
		Anurogryllus arboreus	A	A,S	No	C	Walker 1980, 1983
Thysanoptera (thrips)	Phlaeothripidae	*Hoplothrips karnyi*	F, G♀	S, Sk	Yes: 1 = larger, wingless males with larger fore-femora and thorax, nondispersing; 2 = winged, dispersing	P, CF	Crespi 1988a, c
		Elaphrothrips tuberculatus	F, G♀	NF, NG, Sk	Yes: 1 = larger	CD	Crespi 1986, 1988b
	Delphacidae (planthoppers)	*Prokelisia dolus*	F	S	Yes: 1 = larger, short-winged; 2 = smaller, long-winged	P	Langellotto *et al.* 2000

Table 8.1. (*cont.*)

Order	Family	Species	Tactic[b]		Morphological correlates	Tactic type[c]	Reference
			Tactic 1	Tactic 2			
Hemiptera – Auchenorrhyncha							
Hemiptera – Heteroptera	Gerridae (water striders)	*Aquarius elongatus*	T, C♀, G♀	NT, S, NC	Yes: 1 = larger, longer midlegs for fighting	P	Hayashi 1985, Andersen 1993, 1996, Krupa and Sih 1993
		Aquarius remigis	G♀	NG, S	Yes: 1 = larger	CS	Rubenstein 1984, Kaitala and Dingle 1993, Lauer 1996, Lauer *et al.* 1996
		Gerris odonto-gaster, G. incognitus	C♀	NC	No	P	Arnqvist 1989, 1992a, b, 1997
		Limnoporus dissortis, L. notabilis, L. rufoscutellatus	T, A	NT, S	Yes: 1 = larger	CS	Vepsäläinen and Nummelin 1985, Spence and Wilcox 1986
	Lygaeidae (chinch bugs)	*Cavelerius saccharivorus*	F	S	Yes: 1 = larger, short-winged; 2 = smaller, long-winged	P	Fujisaki 1992
	Rhopalidae (soapberry bug)	*Jadera haematoloma*	G♀	NG, S	No	P	Carroll 1993, Carroll and Corneli 1995
Coleoptera	Scarabaeidae (dung beetles)	*Onthophagus binodis, O. taurus, O. acuminatus, Ageopsis nigricollis, Podischnus agenor, Heliocopris domi-nus, Phanaeus*	G♀, IP	NG, S, NI, Sk	Yes: 1 = larger, horns; 2 = smaller, hornless, emerge earlier	P, CF	Eberhard 1982, 1987, Cook 1987, 1988, 1990, Eberhard and Gutiérrez 1991, Emlen 1994, 1996, 1997a, b, Joseph 1994, Rasmussen 1994, Moczek and Emlen 2000, Tomkins and Simmons 2000, Rowland 2003

180

	difformis, *Xylotrupes* spp.					
	Allomyrina dichotoma	T, F	NT, NF, S	Yes: 1 = larger, horns; 2 = smaller, hornless	P	Siva-Jothy 1987
Lucanidae (Atlas beetles)	*Chalcosoma caucasus, C. atlas*	G♀, IP	NG, S, NI, Sk	Yes: 1 = larger, horns; 2 = smaller, hornless, emerge earlier	P, CF	Kawano 1995
Silphidae (burying beetles)	*Nicrophorus orbicollis, N. vespilloides*	T, A, G♀	A	Yes: 1 = larger, find carrion; 2 = smaller males, do not search for carrion	CS	Eggert 1992, Beeler *et al.* 1999
Cerambycidae (longhorn beetles)	*Trachyderes mandibularis*	T	NT, S	Yes: 1 = larger, larger mandibles	CS	Goldsmith 1985a, b, 1987, Goldsmith and Alcock 1993
	Tetraopes tetraophthalmus	F	NF, S	Yes: 1 = larger, sedentary; 2 = smaller, dispersing	CD	Lawrence 1986, 1987
Brentidae	*Brenthus anchorago*	G♀	NG, Sk	Yes: 1 = larger	P	Johnson 1982
Staphylinidae (rove beetles)	*Leistotrophus versicolor*	F	NF	Yes: 1 = larger; 2 = smaller, ♀ mimic	P	Forsyth and Alcock 1990
	Oxyporus stygicus	T	NT, S	Yes: 1 = larger, larger mandibles	P	Hanley 2000
Tenebrionidae (fungus beetles)	*Bolitotherus cornutus*	T	NT, S	Yes: 1 = larger, larger horns	P	Brown and Bartalon 1986, Conner 1989
Bruchidae (seed beetles)	*Callosobruchus maculatus*	W	S	Yes: 1 = flightless, lacks pubescence, darker, matures quickly, sedentary; 2 = flight, pubescence, delayed maturity, disperse	CD, CF	Utida 1972
Meloidae (blister beetles)	*Tegrodera aloga*	C♀	NC	No	–	Pinto 1975
Mecoptera Panorpidae (scorpionflies)	*Panorpa* sp., *P. vulgaris*	C♀	NC	Yes: 1 = produce salivary secretion or hunt insect as nuptial gift	CF	Thornhill 1979b, 1980, 1981, Bockwinkel and Sauer 1994, Sauer *et al.* 1998

181

Table 8.1. (cont.)

Order	Family	Species	Tactic[b]		Morphological correlates	Tactic type[c]	Reference
			Tactic 1	Tactic 2			
	Bittacidae (hanging flies)	*Hylobittacus apicalis*	C♀	NC	Yes: 1 = produce salivary secretion or hunt insect as nuptial gift	CF	Thornhill 1979a, b, 1984
Diptera – Nematocera	Chironomidae (midges)	*Tokunagayusurika akamusi*	L	S	Yes: 1 = larger (swarming)	P, CT	Kon *et al.* 1986, Takamura 1999
		Chaoborus flavicans	L	S	Yes: 1 = larger (swarming)	P	McLachlan and Neems 1989
	Ceratopogonidae (blood-sucking midges)	*Culicoides nebeculosus, C. riethi*	L	S			Downes 1955
Diptera – Brachycera	Scatophagidae (dung flies)	*Scatophaga stercoraria*	T, G♀	NT, S	Yes: 1 = larger	CD	Parker 1970, 1974, Borgia 1980, 1982
	Dryomyzidae	*Dryomyza anilis*	T	NT, S, G♀	Yes: 1 = larger	P, C?	Otronen 1984a, b
	Coelopidae (seaweed flies)	*Coelopa frigida, C. nebularum*	C♀	NC	Yes: 1 = larger	P	Dunn *et al.* 1999
	Phoridae	*Puliciphora borinquenensis*	IT	NI	Yes: 1 = older	CA	Miller 1984
	Drosophilidae (fruitflies)	*Drosophila melanogaster*	T	NT	No	CD	Hoffmann and Cacoyianni 1990
	Neriidae (cactus flies)	*Odontoloxozus longicornis*	T, G♀, C♀	NT, S	No	CE, CF	Mangan 1979
	Empididae (dancing flies)	*Empis tessellata,*	H, IG	NH, IG	Yes: 1 = higher quality	CF	Preston-Mafham 1999
		E. opaca	S	W	No	CT	Maier and Waldbauer 1979

Order	Family	Species					Reference
	Syrphidae (hoverflies)	*Mallota posticata, Somula decora, Spilomyia hamifera*	L, C♀	T, NC	No	CT	Smith and Prokopy 1980
	Tephritidae (true fruitflies)	*Rhagoletis pomonella*	L, A, C♀	T, NA, NC	No	CT	Prokopy and Hendrichs 1979, Hendrichs *et al.* 1994, Warburg and Yuval 1997
		Ceratitis capitata	T, W	H	No	P	Catts 1979
	Gasterophilidae (bot flies)	*Gasterophilus intestinalis*	?	?	Yes	P	Gomez *et al.* 2003
Siphonaptera (fleas)	Ctenophthalmidae	*Ctenophthalmus apertus personatus*	H	S	No	CT	Scott 1974a
Lepidoptera (butterflies and moths)	Nymphalidae	*Poladryas minuta*	G♀	S	No	CD	Alcock 1994
		Chlosyne californica	T	S	No	C	Hernández and Benson 1998
		Heliconius sara	T	NT, S	No	P, CT	Kemp 2001
		Hypolimnas bolina	H	NT, S	No	CT	van Dyck *et al.* 1997, van Dyck and Matthysen 1998, van Dyck and Wiklund 2002
	Satyridae	*Pararge aegeria*	T	S	Yes: 1 = lighter color, perch more, fight, bask; 2 = darker, longer flights	CE	Wickman 1988
		Lasiommata megera		NT, S	No	C	Courtney and Parker 1985
	Lycaenidae	*Taracus theophrastus*	C♀, A	NC, NA, Sk	No. 2 may use ♀mimicry	C	Field and Keller 1993
Hymenoptera – Apocrita	Braconidae	*Cotesia rubecula*	F	S	No		Ramirez and Marsh 1996
	Ichneumonidae	*Psenobolus ficarius*	F	S	Yes: 1 = wingless; 2 = winged	P	Salt 1952
		Gelis corruptor	F	S	Yes: 1 = wingless with larger mandibles, nondispersing 2 = winged, dispersing	P	Jousselin *et al.* 2004
	Pteromalidae	*Philotrypesis* spp.		S	–	–	

183

Table 8.1. (cont.)

Order	Family	Species	Tactic[b]		Morphological correlates	Tactic type[c]	Reference
			Tactic 1	Tactic 2			
	Chalcididae	*Otitesella pseudoserrata, O. longicauda, O. rotunda*	F	S, W	Yes: 1 = large, wingless; 2 = small, winged	P, CH	Greef and Ferguson 1999, Pienaar and Greeff 2003a, b, Moore *et al.* 2004
					Yes: 1 = "religiosa" form, flightless, larger with massive jaws, extremely aggressive; 2 = "digitata" form, smaller, dispersing		
		Melittobia chalybii, M. australica	F	NF, S	Yes: 1 = larger, both wingless	P	Schmieder 1933, Freeman and Ittyeipe 1982
		Pseudidarnes minerva	NF	S	Yes: 1 = smaller, wingless, nondispersing; 2 = dispersing	P	Cook *et al.* 1997
		Trichogramma sembludis	F	S	Yes: 1 = wingless with larger mandibles, nondispersing; 2 = winged, dispersing	P	Salt 1937
		Sycoscapter australis	F	S	Yes: 1 = wingless, large mandibles, nondispersing; 2 = dispersing	P	Bean and Cook 2001
	Bethylidae	*Cephalonomia gallicola*, other *Cephalonomia* species	F	S	Yes: 1 = smaller, wingless with larger mandibles, nondispersing; 2 = dispersing	P	Kearns 1934, Evans 1963
	Sphecidae (digger wasps)	*Bembecinus quinquespinosus*	W, F, IT	S, NI	Yes: 1 = larger, more yellow	P	O'Neill and Evans 1983a, O'Neill *et al.* 1989, O'Neill 2001
		Sphecius grandis	T	NT, Sa, S	Yes: 1 = larger	P	Hastings 1989
		Bembix rostrata	W, F, IT	S, NT	Yes: 1 = larger	P	Schöne and Tengö 1981

184

Family	Species			Dimorphism		References
	Philanthus bicinctus, P. basilaris, P. zebratus, P. triangulum	T, A	NT, Sa, NA	Yes: 1 = smaller	P	Alcock *et al.* 1978, Evans and O'Neill 1978, O'Neill and Evans 1981, 1983b, O'Neill 1983
	Philanthus bicinctus	T, A	S	Yes: 1 = larger	P	Alcock 1979b, Gwynne 1980
Colletidae	*Tachytes tricinctus*	T	NT, S	Yes: 1 = larger	P	Elliott and Elliott 1992
Halictidae (sweat bees)	*Hylaeus alcyoneus*	T	S	Yes: 1 = larger	P	Alcock and Houston 1987
	Lassioglossum hemichalceum, L. erythrurum	F	NF, S	Yes: 1 = flightless with larger head and jaws, nondispersing; 2 = winged, dispersing	P	Houston 1970, Kukuk and Schwarz 1987, 1988, Kukuk 1996
Andrenidae (short-tongued bees)	*Perdita texana, P. portalis*	F	S	Yes: 1 = flightless, larger head and jaws and atrophied wing muscles, reduced eyes, nondispersing; 2 = winged, dispersing	P	Danforth 1991a, b, Neff and Danforth 1991, Danforth and Neff 1992, Danforth and Desjardins 1999
Megachilidae (leafcutter bees)	*Anthidium maculosum, A. manicatum*	T	NT, Sa	Yes: 1 = larger	P	Alcock *et al.* 1977a, Alcock 1979b, Severinghaus *et al.* 1981, Starks and Reeve 1999
	Osmia rufa, Hoplitis anthocopoides	T	NT, Sa	Yes: 1 = larger		Eickwort 1977
Anthophoridae (digging bees)	*Amegilla dawsoni, Centris pallida*	W, G♀	S	Yes: 1 = larger	P	Alcock *et al.* 1977b, Alcock 1996a, b, c, Alcock 1997a, b, Tomkins *et al.* 2001, Simmons *et al.* 2000
Apidae (social bees)	*Nannotrigona postica*	?	?	Yes: 1 = larger	P	Bego and Camargo 1984
Pompilidae (spider wasps)	*Cryptocheilus* spp.	W, G♀	S	Yes: 1 = flattened, elongate morph; 2 = female-like morph	P	Alcock 1979a, Day 1984
Vespidae (wasps)	*Polistes dominulus*	T	NT, S	Yes: 1 = larger	C	Beani and Turillazzi 1988
	Euodynerus foraminatus	T	NT	Yes: 1 = larger	P	Cowan 1979, 1981
		W, T	S, NT	Yes: 1 = larger		Groddeck *et al.* 2004

185

Table 8.1. (*cont.*)

Order	Family	Species	Tactic[b]		Morphological correlates	Tactic type[c]	Reference
			Tactic 1	Tactic 2			
		Ceramius fonscolombei					
		Synagris cornuta	F, G♀	S	Yes: 1 = larger, tusks on head; 2 = no tusks	P	Longair 2004
	Formicidae (ants)	*Cardiocondyla obscurior, C. nuda, C. wroughtoni, C. minutior*	F, G♀	S	Yes: 1 = wingless, ergatoid, more elongate mandibles, nondispersing, spermatogenesis continues as adults; 2 = winged, dispersing, cease spermatogenesis	P	Kinomura and Yamauchi 1987, Yamauchi and Kawase 1992, Heinze *et al.* 1993, 1998, Cremer and Heinze 2002, Cremer *et al.* 2002, Heinze *et al.* 2004
		Hypoponera opacior, H. punctatissima, H. bondroiti	F, G♀	S	Yes: 1 = larger, wingless, nondispersing; 2 = smaller, winged, dispersing, ♀mimic?	P	Hamilton 1979, Yamauchi *et al.* 1996, Foitzik *et al.* 2002
		Formica exsecta, F. sanguinea	G♀	S	Yes: 1 = larger, nondispersing; 2 = smaller, dispersing	P	Agosti and Hauschteck-Jungen 1987, Fortelius *et al.* 1987

[a] This compilation is meant to illustrate the diversity of species that show ARTs and the many kinds of ARTs that have evolved in insects (the table is incomplete). The information presented in this table is based on descriptions from the literature. Inclusion in the table depends on the author describing male behavior as alternative tactics, strategies or morphs. When an order is missing from the table (Ephemeroptera, Plecoptera, Grylloblattodea, Embioptera, Phasmatodea, Zoraptera, Psocoptera, Phthiraptera, Megaloptera, Raphidioptera, Neuroptera, Strepsiptera, Trichoptera), it means that I was not able to find an example of male ARTs in this insect group.

[b] Tactic. I have assigned Tactic 1 as the tactic with the higher investment. Tactics 1 and 2: A, advertises for females (females approach); C♀, courts females; F, fights other males over females within confined space; G♀, guards females; H, intercepts females by hilltopping or waiting at a particular location and intercepting dispersed females; IG, male invests in nuptial gift; IP, male invests in paternal care of offspring; IT, male invests in female transport; L, mating occurs in lek or swarm; NA, does not advertise; NC, does not court; NF, does not fight; NG, does not guard; NI, no investment; NT, not territorial; S, searches for females over long distances or transient; Sa, satellite of territorial or advertising male; Sk, sneaks copulations on territories or where males are guarding females; T, territorial or guards site where females are found; W, waits for females at emergence or oviposition sites or other resources; ?, not known.

[c] Tactic type: C, conditional; CA, conditional on age; CD, conditional on density; CE, conditional on environmental conditions; CF, conditional on foraging success; CH, conditional on harem size or operational sex ratio; CS, conditional on size; CSu, conditional on success; CT, conditional on time of day or season; P, permanent; –, not known.

186

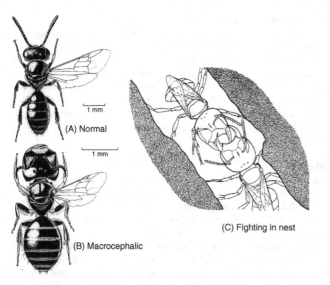

(A) Normal

(B) Macrocephalic

(C) Fighting in nest

Figure 8.1 Males of the ground-nesting halictid bee *Lasioglossum erythrurum* occur in two forms: (A) a male that is similar to that of other species of bees and (B) a large, flightless, macrocephalic male with hypertrophied mandibles and reduced wings and thoracic musculature. Normal males leave the natal nest to search for females whereas (C) the macrocephalic males fight other males and mate with females within the nest. ((A) and (B) from Houston (1970); (C) from Kukuk and Schwarz (1988). (Reproduced with permission.)

frequency dependence may operate together with condition dependence in many systems (Gross 1996, Walker and Cade 2003). Still another system for categorizing ARTs is based on whether tactics are permanent over the adult lifespan of the individual or whether individuals change at a particular point such as at a certain size or age (switch) or whether individuals can change flexibly back and forth throughout adult life (Alcock 1979b, Caro and Bateson 1986, Waltz and Wolf 1988, Taborsky 1997, Brockmann 2001). While this scheme is a useful way to categorize initial observations and to address some questions that we might ask about ARTs, it does not reveal function and the likely selective pressures favoring the evolution of ARTs. In this chapter, I will propose functional categories for insect ARTs. Such categories provide a means of understanding the common ecological and evolutionary factors favoring ARTs, although they can be faulty if insufficient information is available.

8.2 PATTERNS OF ARTs IN INSECTS

For most male insects, reproductive behavior involves four steps: locating a female, gaining access to her, copulating with her, and engaging in post-copulatory behavior that increases the male's chance that his sperm will fertilize the female's eggs. Female reproductive behavior can be divided into similar steps that include finding a mate, choosing or gaining access to that mate, acquiring and choosing sperm, and fertilizing and laying eggs. (Post-laying paternal or maternal care and social nesting are outside the scope of this chapter.) Continuous variation in reproductive characters (behavior, morphology, etc.) is common but here we are seeking to understand special cases in which the variation in reproductive characters constitutes consistent, discretely different ways of achieving the same functional end, i.e., alternative ways of completing one of the steps involved in reproduction.

One tactic that can be found during any reproductive step is sexual mimicry. Subordinate males show a variety of techniques for getting around or deflecting the aggressive behavior of dominant males and one of the most common is to look, act, or smell like a female, i.e., sexual mimicry (e.g., cockroaches: Wendelken and Barth 1985; beetles: Peschke 1987, Forsyth and Alcock 1990). In the scorpionfly *Hylobittacus apicalis*, males capture arthropod prey that are used as nuptial gifts, a costly and risky behavior (Thornhill 1976). The male advertises his prey with pheromones and when a female arrives, he gives the prey over to the female who feeds on it during copulation. Females will not mate if males have no prey and they copulate longer when they have larger prey. A male without a nuptial gift may approach an advertising male, and, acting like a female, grab the rival's prey and then use the pirated item to attract a female of his own (Thornhill 1979a). Deceptive practices like sexual mimicry are particularly likely to be displayed by individuals using low-investment or low-risk tactics that get around the high-investment decisions of others.

Female sexual mimicry occurs in a number of species of butterflies, dragonflies, and damselflies (Nielsen and Watt 2000, Sherratt 2001, Svensson et al. 2005). In the damselfly *Ischnura ramburi*, females of one morph (andromorphs) are physically similar to males, are often treated as though they were males, and act like males when approached. Females of the other morph (gynomorph) look different from males, are always recognized and treated like females by males, and do not act like males when approached (Robertson 1985, Sirot et al. 2003). The advantage to male mimicry is that females can avoid mating or mating attempts by males (Robertson 1985), which are known to be costly (Sirot and Brockmann 2001). Gynomorphs avoid mating attempts by being less active and more cryptic; andromorphs by looking and acting like males. However, since males learn to recognize andromorphic females, the system is frequency dependent, like other mimicry systems (Fincke 2004). The disadvantage is apparently that at very high frequencies when their crypsis is broken, andromorph fitness is more affected by mating attempts than is gynomorph fitness (Sirot and Brockmann 2001).

8.2.1 ARTs associated with locating mates

Insects show three general types of mate-locating behavior: some (usually males) remain in one place and fight for the mates with whom they were born, others advertise and attract mates, and still others (usually males) travel to where mates can be found. Alternative tactics have been described for each type or combination of types in male insects. Finding mates is not usually a problem for females and female–female competition for mates appears to be rare, but selection nonetheless acts on females to locate mates efficiently and to reduce the costs associated with finding mates, which may on occasion favor the evolution of female ARTs.

REMAIN IN THE NATAL GROUP

When males are born into a situation where many conspecifics compete for females, selection sometimes favors males that disperse and sometimes favors males that remain in the nest (e.g., various species of fig wasps: Herre et al. 1997; various species of *Melittobia*, a parasitic chalcid wasp: Consoli and Vinson 2002). On occasion, however, discrete alternative tactics arise within one species with some males staying and some dispersing (see Figure 2.9).

(A) FIGHTER VS. DISPERSER

Fighter males tend to evolve in species where many females are sequestered with a few males or, as Hamilton (1979) called it, a "seraglio" (see also Figures 2.9, 8.1). Clearly, if an alternative, dispersing tactic is to be favored, however, some receptive females must be available in dispersed locations. The fighter/disperser mating pattern is found in a number of species of parasitic insects such as the tiny bethylid wasp *Cephalonomia gallicola*, which parasitizes cigarette beetles (Kearns 1934). This species has dimorphic males, some wingless with large mandibles and some winged that look like females. Males emerge first and chew their way into nearby cocoons, fertilizing the females before they emerge. Multiple fighter males are attracted to female cocoons and vicious, damaging fights occur, even between close relatives. Winged males do not fight but disperse and mate with females that emerge in nests without wingless males. Dimorphic fighter males also occur in the thrips *Hoplothrips karnyi* (Crespi 1988a), which live in colonies that vary in size from a few individuals to hundreds located on shelf fungi where they feed on mycelia. Under dense rearing conditions, some males develop into a large fighting morph whereas others develop into a small, nonfighting morph; under low density conditions all are small, nonfighters. Females oviposit onto communal egg masses in crevices under bark that the large, wingless males defend using their enlarged forelegs and abdomens (Figure 8.2). The smaller males (winged or wingless) may mate with females away from the egg mass or attempt to sneak a mating within the guarded area. Although large males have much higher mating success inside the nest than smaller, nondefending males, the large males cannot disperse when conditions deteriorate. In both of these examples, the frequency of fighter males within a nest or colony depends on the availability of mating opportunities within the nest as compared with outside: when there are few receptive, dispersing females then fighters are more common (Figure 2.9).

(A)

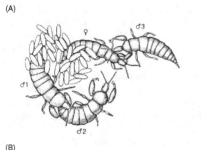

(B)

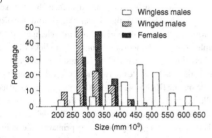

Figure 8.2 (A) Males of the colonial, polygynous thrips *Hoplothrips karnyi* (Thysanoptera) around a communal egg mass and female oviposition area. Males 1 and 2 are large, wingless males with enlarged fore-femora that guard oviposition sites, mate with females that come to lay on the communal egg mass, and fight with other males. In the figure, Male 1 is stabbing Male 2. Male 3 is a small, nonguarding male (most of which are winged) that usually mates with females away from the egg mass. Sometimes (shown in the figure) nonguarding males make incursions onto the egg mass where they attempt copulations with females and avoid the guarding males by running away. (B) Frequency distribution of the sizes of the two types of males and females. (From Crespi 1988c, reprinted with kind permission of Springer Science and Business Media.)

(B) GUARDER VS. DISPERSER

Fighting is not the only outcome when males are sequestered with females. Nonfighting dimorphic males are found in the ant *Hypoponera opacior* (Foitzik *et al.* 2002). In this species wingless males remain within their natal colony, sit on top of the cocoons of unemerged queens, insert their genitalia, and remain in copula with cocooned queens for up to 40 h (ant copulations usually take only a few minutes), presumably a form of mate guarding. Winged males disperse to other nests and mate with queens that have not mated prior to initiating a new nest. Wingless males are found in multiple-queen nests (polygynous, polydomous),

whereas winged males occur in single-queen colonies (monogynous), a pattern that is repeated in other species of ants (Foitzik *et al.* 2002).

ADVERTISING AND ATTRACTING MATES

When males and females are dispersed, some species evolve female advertisement for mates whereas other species evolve male advertisement (Thornhill and Alcock 1983). In species such as fireflies where females actively advertise their presence to males and males compete to approach females swiftly (Lloyd 1979), I did not identify any intraspecific, alternative mate-locating tactics (although predatory females are known to use such tactics in attracting heterospecific prey: Lloyd 1980). However, in species where males advertise for females, ARTs are common.

(A) ADVERTISE VS. SEARCH, SNEAK, OR SATELLITE

In *Gryllus* crickets, males sing from burrows they have dug for use as refuges or sites for mating or female oviposition. Their calling attracts males as well as females, so males must defend their burrows from other males (Figure 2.7). Calling signals may be costly to produce (Prestwich 1994) and may increase the male's risk of attracting parasites and predators (Cade 1975, Burk 1982, Sakaluk and Belwood 1984, Hedrick and Dill 1993, Gwynne 2001). Not surprisingly, then, in some species with long-distance advertising, some males call whereas other males search for females without calling (Cade 1979b, 1980, 1981, Walker 1980, Cade and Cade 1992, Rowell and Cade 1993), a less successful but also less risky behavior. In addition, variation in female responsiveness to male calls (phonotaxis) has been found in crickets such as *G. integer* (Wagner *et al.* 1995) and *G. texensis* (Cade 1979a, Bertram 2002). In *G. texensis* the population is made up of females from different generations that differ in their phonotactic behavior due to different risk factors from seasonal predation or parasitism. Predation during phonotaxis may also alter female responses to males (Hedrick and Dill 1993) and of course the quality of the previous mate may change a female's willingness to remate (Brown 1997). The result is that some females are much more responsive to calling males than others, which means that searching for mates may be more successful than advertising under some conditions.

An extreme example of alternative mate-locating tactics can be found in the alternative reproductive tactics of the South African bladder grasshopper *Bullacris membracioides*

(A) Male

(B) Alternative
male

(C) Female

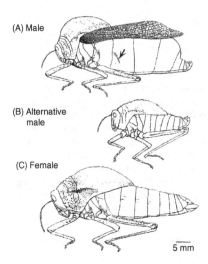

5 mm

Figure 8.3 Two male forms of the South African bladder grasshopper *Bullacris membracioides*. (A) One is a large, flight-capable morph that advertises for females over very long distances. This male morph has a stridulatory organ (indicated by the arrow) and inflated abdomen that acts as a sound resonating chamber. (B) When the male receives a response, he flies to the female and initiates courtship. Males of the much smaller, flightless morph remain on the shrub on which they emerged where they search for females, sometimes approaching and courting a female that has responded to a large male. These apterous, alternative males do not have a stridulatory organ for long-distance calling nor an inflated abdomen. (C) The wingless females mate with males in and around the shrub on which they emerged. (From van Staaden and Romer 1997.) (Reproduced with permission of the Company of Biologists.)

(Figure 8.3). The large, winged male transmits low-frequency acoustic signals that can be heard from a distance of more than 2 km (van Staaden and Römer 1997). This is accomplished with an abdominal–femoral stridulatory organ and a greatly inflated abdomen that magnifies the sound. A small, wingless alternative morph does not produce the loud, long-distance calls and does not have the femoral stridulatory organ or inflated abdomen. Winged males are extremely aggressive toward other winged males but largely ignore the smaller, wingless males. Females are wingless and never leave the plant on which they emerge; they respond to a winged male's signal with a low-intensity call that can be heard for only a few meters. Her response initiates a duet with the winged male, who flies to the female and mates (Alexander and van Staaden 1989). Wingless males also

remain on the plant on which they emerged; they fight other wingless males and respond to females that duet with winged males as well as searching out females on their own. Females do not appear to prefer one type of male over the other. There is still much to be learned about the ARTs of this species, but one prediction is that the winged males would have an advantage under low-density conditions when they can seek mates over a wider area than wingless males (see also Figure 2.8 for a similar pattern in plant-hoppers).

Parasitizing the advertising efforts of other males is taken one step further by the Australian bushcricket *Elephantodeta nobilis* (Tettigoniidae). Males and females perform duets: the male initiates the interchange with a complex call and the female responds with clicks that are precisely timed to one particular male and that male replies with further clicks, while the two move toward one another prior to mating (Bailey and Field 2000). Satellite males place a volley of clicks between parts of the duetting male's call; the effect is to change the behavior of the female's response so that she is no longer replying to the primary male alone. In two-speaker choice tests that mimic the primary and inserted satellite calls, most females approach the primary male's call, but some approach the satellite (Bailey and Field 2000) thus suggesting a benefit for satellites at least under some conditions.

(B) LEK OR SWARM VS. SEARCH OR SNEAK

In some species males form groups that attract females. In swarming chironomid midges, for example, larger males swarm while smaller individuals locate mates where they gather before entering the male swarm (McLachlan and Neems 1989). In the tephritid fly *Ceratitis capitata*, most males join groups on the bottom surface of leaves and jointly produce pheromone signals and wing fan, behavior that attracts receptive females (Prokopy and Hendrichs 1979). Some males, however, can be found nearby not producing pheromones or wing fanning, but nonetheless mating occasionally with females as they approach the signaling males. If signaling is costly, as seems likely due to increased wasp predation (Warburg and Yuval 1997), then individuals with this alternative tactic may have higher lifetime success.

TRAVELING TO WHERE MATES OCCUR

Males seek out receptive females wherever they can be found. This often means that males search for areas with valuable resources that females require, such as food or oviposition sites. In mobile, low-density populations, males may seek females on migration routes or in locations where they pass by, as in hilltopping species (Thornhill and Alcock 1983). When males invest in locating and defending receptive females or areas where females are likely to be found, selection sometimes favors individuals that use alternative mate-locating tactics (Alcock *et al.* 1978, Alcock 1979b).

In a number of species of insects, males guard resources that are required by females, such as feeding or nesting sites, and exclude conspecific males from that area (Alcock *et al.* 1978). Often only one tactic occurs but in a few species nonterritorial males seek females outside of established territories. They are referred to as patrolling, wandering, or transient males (Alcock *et al.* 1977b, Waltz and Wolf 1984, Beani and Turillazzi 1988), whereas if they move into established male territories when females are present, they are referred to as sneaker males (Waltz 1982). If males remain in long-term association with one territorial male, they are referred to as residents or satellites (Howard 1978, Wolf and Waltz 1988).

(A) TERRITORIAL VS. TRANSIENT, SATELLITE, OR SNEAK

When some individuals guard resources that attract mates, selection may favor parasitic, nonterritorial tactics. For example, in many species of odonates, males are territorial around female oviposition sites, perching on stems within a pond's perimeter and flying out to attack intruders. Pairing and oviposition occur on pond territories (Wolf and Waltz 1988). Transient, nonterritorial males perch away from the pond and make flights over the pond and adjoining areas where nonovipositing females may be found (Campanella and Wolf 1974). When pairing occurs with a nonterritorial male, the couple flies to vegetation around the pond (away from territories) and when copulation is complete they fly to the pond in tandem where the female oviposits (Wolf and Waltz 1988). This means that nonterritorial males do not pay the costs of maintaining a territory, but they may suffer reduced success due to sperm competition if a female is taken over by the male on whose territory she is ovipositing (Uéda 1979, Wolf and Waltz 1993).

Ephemeral resources or extreme environmental conditions also provide opportunities for ARTs. Male cactus flies maintain territories around oviposition sites, fighting off conspecific males with characteristic displays and occasional fights in which males bump and push one another (Mangan 1979). Females arrive at the site, probe the necrotic tissue, mate with the resident male, and oviposit. Males without territories mate with females away from the guarded oviposition sites. Territorial males are more successful than searching males, but because of the extreme environmental conditions of the desert, territorial males cannot remain at one site for long so females that have mated with nonterritorial males may lay their eggs at unguarded oviposition sites. Similar behavior is seen in other species that use ephemeral resources (Parker 1970, Borgia 1980).

In many territorial species, males that hold territories are morphologically different from nonterritorial individuals and may possess specialized structures for defense or attracting females. For example, in the water strider *Gerris elongatus*, territorial males are larger with longer midlegs, which are used for fighting (Hayashi 1985), when compared with nonterritorial males, which use nonaggressive behavior to gain access to females. Presumably the structures that make males better at fighting make them less efficient at searching. In some cases morphological adaptations other than size are associated with alternative tactics. Territorial males of the digging sphecid wasp *Bembecinus quinquespinosus* are lighter in color than nonterritorial individuals. This allows the territorial animals to be more active in the hot, exposed environment of the nesting area they are defending (O'Neill and Evans 1983a, O'Neill *et al.* 1989). The nonterritorial males are less exposed and their darker color allows them to maintain flight temperatures over longer periods during the day and season than lighter colored individuals.

Of course, male territorial vs. transient ARTs can evolve only if females are willing to mate both inside and outside territories with both territorial and nonterritorial males. It is not known in these species whether females mate with any male that comes along or whether some (perhaps larger) females prefer territorial males and some prefer nonterritorial males (or perhaps that some do not discriminate among males), which would amount to a female ART.

Alternative territorial tactics among females are known from some species of water striders. Some female *Aquarius remigis* are territorial, occupying the center of permanent pools in creeks where prey capture rates are higher, whereas other females are not territorial, living along creek margins with lower prey-capture rates (Rubenstein 1984). Only the largest females are territorial and since wingless females tend to be larger than winged individuals, wingless females are more likely to be territorial (Kaitala and Dingle 1993). Furthermore, males guard territorial females that are regularly harassed by males when left unattended, behavior that is known to be costly to females (Krupa *et al.* 1990, Krupa and Sih 1993, Lauer *et al.* 1996). This means that reproductive females may adopt one of two tactics: be territorial in the center of a pool, get more food but suffer more harassment from males, or occupy the edges of pools with less food and less harassment from males.

(B) MALE DOMINANCE VS. TRANSIENT, SATELLITE, OR SNEAK

In some species males are not territorial but they fight intensely at female emergence sites or over resources required by females (Box 8.1). For example, females of the desert cerambycid beetle *Dendrobias mandibularis* depend on feeding at sap ooze sites on desert broom plants (Goldsmith 1985a, 1987). The large major males, a distinct morph with massive mandibles, are more successful at dominating other males and mating with the females that visit the ooze sites than are minor males with their much smaller mandibles. However, unlike majors, minor males mate with females on saguaro fruits and in foliage at widely dispersed locations (Goldsmith and Alcock 1993). Guarding is costly; not only is it energy-consuming and risky from damaging intraspecific fights, but it may also expose males to greater risks from

---

Box 8.1 Multiple interacting factors favor the evolution of ARTs in solitary bees

The males of two species of ground-nesting anthophorid bees, *Centris pallida* (Alcock *et al.* 1977b) and Dawson's burrowing bee *Amegilla dawsoni* (Alcock 1997a), show alternative mating tactics. Large males (majors) fight other males and patrol near the ground where virgin females are emerging (Alcock *et al.* 1977b) whereas smaller individuals (minors) hover near emergence areas or near flowering trees waiting for already emerged females (Alcock 1979b, 1999) (Figure 8.4). Mating in these two areas has different costs for females as well as males. Newly emerged females can be damaged by the intense male–male fights that occur and they seem to leave the emergence area as quickly as possible, thus providing mating opportunities for smaller males away from emergence sites. Despite trade-offs of this sort, it nonetheless appears that major males are more successful (Alcock 1995, 1996b, c). Yet, the two patterns are maintained stably in populations over generations (Alcock 1984, 1989, 1995). Several explanations are possible and many of them point to the possibility that average success has not been estimated correctly and that density-, condition-, and frequency-dependent effects may interact in complex ways in the evolution of ARTs in these species (Alcock *et al.* 1977b, Alcock 1997a) (Figure 8.4B). For example, minors may be favored when densities are low and females are dispersed whereas majors may do better when nests are aggregated (Alcock 1979c); bird predation is heavier on major than on minor males suggesting that under conditions of high predation, minors would be favored whereas majors would be favored under low predation (Alcock 1995, 1996a). In *A. dawsoni*, mark–recapture studies show that majors have shorter lifespans than minors (Alcock 1996a). Also, minor males (*A. dawsoni*) emerge earlier in the season and are active over a greater portion of the day than majors,

providing additional opportunities for mating (Alcock 1997b). Another possibility is that minors may be able to compete with majors through sperm competition if females mate multiply, but this appears not to be the case (Simmons *et al.* 2000). Still another possibility is that since male size is the product of female investment decisions (Alcock 1979b, c, 1999), an understanding of the factors maintaining male body size must include maternal allocation tactics and not just the relative success of the two tactics in the male offspring (Alcock 1989). Alcock (1979c) speculates that females may be adjusting the numbers of males to the prevailing nesting conditions, making more small males when densities are lower. This is a much harder problem to address (Alcock 1996b) but similar adaptations have been found in other species. Females clearly adjust the amount of provisions they supply (Alcock 1999) but even when females are manipulated by adding weights or by clipping their wings, they nonetheless produce small and large offspring in the same proportions as control females (Tomkins *et al.* 2001), so individual female effort is not affecting the numbers of small and large males they produce. Furthermore, smaller females produce more small males than larger females. There is also a seasonal effect on resource allocation decisions with females producing more large offspring early in the season and smaller offspring later (Alcock *et al.* 2005). These results suggest that condition dependence and density dependence are involved in the maintenance of the maternal investment decision in male burrowing bee tactics. Frequency dependence is also likely to be involved (particularly at high densities) since the success of the minor males is likely to be affected by their proportion in the population, i.e., the frequency of different mating tactics will affect their success. Nothing is known about the developmental mechanisms that control the expression of correlated traits in the two morphs (e.g., the role of juvenile hormone and threshold mechanisms).

(A)

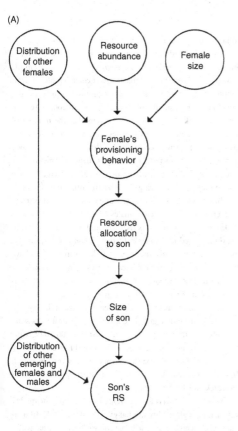

(B)

Figure 8.4 Alternative tactics in the digging bee *Centris pallida*. Small, minor males hover over the nesting area or patrol nearby areas for females whereas large, major males mate with females as they emerge, fighting off other large males. (A) A variety of factors affect the success of a male including the amount of investment provided by the mother and the distribution and abundance of other females. (B) This information suggests the hypothesis that maternal investment will depend in part on female density; females should make minor sons when densities are high since major males are more successful when densities are high and major sons when densities are high. RS, reproductive success. (Based on information in Alcock *et al.* 1977b, Alcock 1979c, 1984, 1989.)

predation. This means that nonguarding individuals may survive longer with more time to reproduce even if their daily success rate is lower.

(C) NONTERRITORIAL WAITING VS. PATROLLING: ALTERNATIVE SEARCHING TACTICS

In some species males wait for females at likely spots without being territorial, and when this occurs, selection may favor other males that search more widely for females. For example, in the nonterritorial, explosively breeding damselfly *Enallagma hageni*, some males perch and wait for females while they are ovipositing under water, whereas others search for females that have not yet mated (Fincke 1982, 1984, 1985). Perching males pull ovipositing females out of the water and mate with them, and since there is strong last-male sperm precedence, this is a worthwhile tactic for males if the female has not yet laid all her eggs. The male's behavior may also be worthwhile for a female since it may prevent her from drowning (Fincke 1986). In butterflies, males of most species either perch or wait in a likely spot for females to come by or they slowly fly over areas likely to contain females, but in a few species, both patterns occur in one population (Scott 1974a). Waiters spend most of their time perched with only brief flights whereas patrollers spend all of their time searching the habitat. Thus, a common pattern of alternative mate-locating tactics is for some individuals to focus on higher-density areas and not move around very much whereas others trapline between many different, lower-density, dispersed sites.

8.2.2   ARTs associated with choosing and gaining access to mates

Once a mate is located, individuals may have to compete for access to that mate (intrasexual selection) or engage in behavior that results in mate choice (intersexual selection). In most species copulation is preceded by some form of courtship: the male may stroke or tap the female prior to intromission (e.g., beetles), engage in courtship song (e.g., crickets), perform an aerial flight (e.g., butterflies), or produce a courtship pheromone (e.g., moths) (Thornhill and Alcock 1983). Females respond by changing the position of their abdomen to allow or prevent intromission, by releasing pheromones, or by signaling in other ways. Females may evaluate males on the basis of their displays either by rejecting them outright or by various post-copulatory mechanisms (see Section 8.2.3). In some species, however, males subvert the courtship and mate choice process by using alternative tactics, such as attempting to copulate without courtship.

In a few species, females compete for access to males. For example, in certain katydids, males turn over substantial resources during copulation and females fight for access to the highest-quality mates (Gwynne 1983, 1984). When female–female competition is high, females show adaptations such as larger size or increased sensitivity to male song (Gwynne and Bailey 1999), but female ARTs for gaining access to males have not been described. Models of ARTs point out that selection should sometimes favor females that employ alternative tactics of female choice (e.g., tend to follow less risky tactics when young) and that such tactics will affect the evolution of male characters (Jennions and Petrie 1997, Alonzo and Warner 2000, Jones 2002, Luttbeg 2004).

(A) HIGH VS. LOW INVESTMENT

When males make costly pre-copulatory investments, selection may favor individuals that parasitize these efforts. In the horned dung beetle *Onthophagus binodus*, for example (Chapter 5), large, horned males invest considerable effort in helping their mate to gather larval provisions prior to oviposition whereas small, hornless males do not cooperate with females (Cook 1987, 1990). This means that the number of brood masses produced by females is greater when a female is paired with a horned as compared with a hornless male (Cook 1988). Rather than remaining with females, hornless males search for already guarded and provisioning females with whom they mate, fathering about half the offspring (Tomkins and Simmons 2000). Hornless, minor males have larger testes and ejaculates than major males (Tomkins and Simmons 2002) but similar lifespans (Kotiaho and Simmons 2003). In another horned dung beetle *Phanaeus difformis*, both major and minor males assist the female but larger individuals have higher success in pairing because they win in male–male competition (Rasmussen 1994). The smaller, hornless, minor males use alternative tactics to gain access to females by entering a major male's burrow when he is absent, by digging a burrow around a guarding major male, or by locating females away from oviposition sites (Hanley 2000). The large, horned guarding males are more effective at fighting whereas the smaller, hornless males are more maneuverable within burrows and can slip more easily past the guarding males (Moczek and Emlen 2000).

Males may subvert female choice by providing low-quality rather than high-quality nuptial gifts (Luttbeg 2004). In dancing flies (*Empis*), females will mate only when presented with a nuptial gift. Some males present the female with a high-quality arthropod prey item on which she feeds during copulation, but other males present the female with a useless object such as fluff from a willow (Preston-Mafham 1999). Such low-quality gifts result in shorter copulations than when fresh prey are used, but copulations with low-quality gifts do not differ from those in which the female is presented with a dried out (reused) arthropod. If nuptial gifts are costly to acquire, as seems likely (Thornhill 1979b), then low-quality gifts may result in higher success than if the male tried to mate with no gift at all.

(B) COURTSHIP VS. NO COURTSHIP (FORCED COPULATION)

In species in which males usually engage in costly courtship prior to copulation, selection may favor individuals

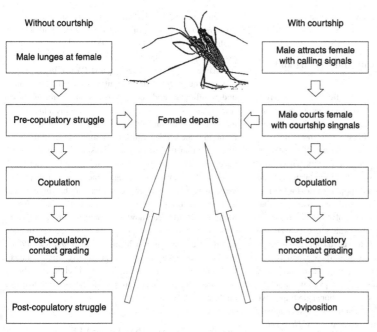

Without courtship

Male lunges at female

⬇

Pre-copulatory struggle ⇨ Female departs ⇦ Male courts female with courtship singnals

⬇

Copulation

⬇

Post-copulatory contact grading

⬇

Post-copulatory struggle

With courtship

Male attracts female with calling signals

⬇

⬇

Copulation

⬇

Post-copulatory noncontact grading

⬇

Oviposition

Figure 8.5 Courtship vs. no courtship ARTs in water striders. The picture shows a pair of water striders involved in a pre-mating struggle. The male has successfully grasped the female and attempts copulation while the female rears back on her midlegs which often causes the couple to flip over backwards. The flow diagram describes the two types of mating sequence, with and without courtship. (Reprinted from Arnqvist [1997] with permission from Cambridge University Press.)

approaching a mate without courtship (Figure 8.5). For example, in the grasshopper *Melanoplus sanguinipes* (Belovsky *et al.* 1996), some males court females by facing them and vibrating their femora; if she displays in response, the male leaps on her. Some males, however, stalk females and pounce on them from behind without displaying (Otte 1981). The female responds to a stalker's pounce by leaping away, flipping over, or prying at the male with her hind legs. These copulation attempts, which are characteristic of males that are in poor condition, are only half as successful as those that are preceded by courtship. Males that forage better are more likely to display prior to copulation, and they transfer larger spermatophores, which results in higher female fecundity. In mate-choice trials, females prefer males that forage better but when the female is not able to see the male or when he is not moving, she does not make such distinctions (Belovsky *et al.* 1996).

A similar pattern is found in *Panorpa latipennis* and *P. vulgaris*, species of scorpionfly in which males normally provide a costly nuptial gift to the female that the male advertises with pheromones (Thornhill 1979b, 1980). Nuptial gifts consist of hardened salivary secretions deposited on the substrate or of captured dead arthropods (Thornhill 1981). The size of the gift is correlated with the time the female takes to consume it, the time the male spends copulating, and the proportion of eggs he fertilizes (Thornhill 1979b). The ability to produce salivary masses is correlated with male quality, i.e., his ability to find food and win fights (Thornhill 1981, Bockwinkel and Sauer 1994, Sauer *et al.* 1998, Engqvist and Sauer 2003). A male with no gift does not advertise but lashes out at a passing female with his abdomen, grasps her in his genital forceps, secures her wings with a notal clamp, and may copulate with her while she appears to struggle (Thornhill and Sauer 1991).

These examples are clear cases where inferior males subvert the female's ability to choose quality males by adopting a nondisplaying, alternative mating tactic. Females may be forced to mate or they may choose to mate with these males based on their own condition and the costs of assessment or resisting (Luttbeg 2004).

8.2.3   ARTs associated with transferring and acquiring sperm

Males that use different tactics often show different adaptations for sperm competition (Tomkins and Simmons 2000, 2002). For example, the tactic faced with higher male–male competition should evolve more investment in larger testes and more sperm per ejaculate (Gage 1995), or more costly guarding behavior, whereas the tactic faced with reduced competition (e.g., the dispersed tactic) should evolve a longer lifespan with more opportunities for mating.

A large literature has developed demonstrating that males of many insect species have evolved structures (as in *Panorpa*), behavior, or seminal products that subvert the female choice process, forcing a female to copulate longer or oviposit sooner than she would without these products. For example, in the sagebrush cricket *Cyphoderris strepitans*, males possess a pair of spines (gin trap) that pinch the female's abdomen, and males with functional gin traps have greater success than those whose gin trap has been removed (Sakaluk *et al.* 1995). In the ground weta *Hemiandrus pallitarsis*, the female has a secondary copulatory organ that secures the male's genitalia during the transfer of a large and nutritious spermatophylax (Gwynne 2002). I found no evidence of ARTs in either male or female mate-securing tactics. Although notal organs, gin traps, and genital clamps vary in size, the variation is continuous, not dimorphic (Andersen 1996, Arnqvist 1997). Closely related species, however, differ in the presence or absence of these structures, suggesting that there might have been a time in the past when such traits were dimorphic within the species (West-Eberhard 2003).

8.2.4   ARTs associated with fertilizing and laying eggs

Females have evolved an elaborate set of post-copulatory mechanisms to affect the use of a particular male's sperm. These include manipulating the use of competing ejaculates during or after copulation so that the sperm of the higher-quality male fertilizes her eggs by removing sperm plugs and discarding or failing to transport sperm to storage organs, by failing to ovulate (Eberhard 1996); or by remating more quickly after copulating with a poorly performing male (Simmons and Gwynne 1991, Brown 1997).

In some species, females that mate with superior males invest more in their offspring than those mating with lower-quality males (Wedell 1996). Female dung flies (*Scatophaga stercoraria*) store the sperm from different males in different spermathecae (sperm storage organs) (Ward 1993). They then use the sperm from different males when ovipositing under different conditions (Ward 1998). None of these studies, however, describes a female dimorphism in cryptic female choice, although it seems possible that one will be found in some species with further study (Jones 2002).

In addition to post-copulatory displays or investment, males may guard females after mating. Such behavior may increase the proportion of eggs the male fertilizes by manipulating cryptic female choice or by preventing another male from mating with his mate (Eberhard 1996). Male insects are also known to ejaculate substances that affect female remating or oviposition, sometimes to the detriment of the female (Chapman *et al.* 1995, Rice 1996, Wolfner 2002). Although no alternative mate-conflict tactics of this sort are known, it seems likely that if these substances are expensive to produce, such alternatives might eventually be found.

(A) MATE GUARDING VS. MATE SEARCHING

After mating, a male may guard the female until she oviposits or he may fail to guard her and spend his time searching for additional females (Harari *et al.* 2003). In the territorial dragonfly *Calopteryx maculata*, most males are territorial and defend mates as they are ovipositing on the territory (Waage 1979). Guarding is highly advantageous for males because it prevents takeovers of mates with 80–100% sperm removal; it is also advantageous for females because it allows females time to oviposit without being disturbed by males. Nonterritorial or transient males do not defend females while they oviposit. Their success depends on the ability of their mate to oviposit on a guarded territory and exploit the guarding behavior of a territorial male (Waage 1979). Territorial males will sometimes defend a female that he has not mated with if she is ovipositing near his mate or when he has recently mated. Density affects the willingness of males to use a guarding tactic (Uéda 1979, Waltz and Wolf 1988). For example, at low densities transient, nonterritorial males of the dragonfly *Sympetrum parvulum* guard females during oviposition, much as territorial males do; but at high densities, a nonterritorial male remains in tandem with a female, holding her with his claspers as she oviposits (Uéda 1979). This much more energy-intensive form of guarding has the effect of reducing the number of takeovers.

(B) THEFT OF INVESTMENT

A few cases exist where high post-copulatory investment provides an opportunity for noninvesting males. For example, in a very small phorid fly, males mate with pheromone-advertising females (Miller 1984). After mating the male picks up the wingless female and transports her to a new oviposition site (males scout for and learn the location of these sites before mating). Alternatively, males may wait at the oviposition sites and mate with females that have already been transported, thus saving the considerable effort and time associated with transport (Miller 1984). The transporting male may stay and guard the female or he may return to the emergence site to mate with another female.

Female–female thefts of investment are common in species with maternal care (Eickwort 1975, Field 1992). Facultative, intraspecific brood parasitism (cleptoparasitism) or egg dumping occurs in a wide array of species that provision offspring, an activity that is often dangerous as well as energy- and time-consuming (Kurczewski and Spofford 1998). For example, in solitary, sphecid wasps, particularly those nesting in dense aggregations such as the pipe-organ mud-daubing wasp, *Trypoxylon politum* (Brockmann 1980), females open brood cells that have been completed by other females, remove the egg laid by another female, lay an egg of their own, and reseal the nest (Field 1992). In a species of lace bug (Hemiptera: Tingidae, *Gargaphia solani*) and a treehopper (Membracidae, *Publilia concave*) with maternal care, some females lay eggs in the nests of other females and leave without providing care whereas others engage in a lengthy period of egg guarding (Tallamy and Horton 1990, Zink 2003). In these cases brood parasitism is the more successful tactic as long as hosts are common (i.e., it is frequency dependent). Female burying beetles *Necrophorus vespilloides* fight over carcasses that are essential for rearing young and they remain with and feed their brood until pupation. Conspecific females occasionally sneak a few eggs onto a carcass which are then reared by the host female (Müller *et al.* 1990). This is a case of females making the best of a bad situation when they fail to compete successfully for a valuable resource. Sphecid digger wasps also brood parasitize conspecifics by removing prey from the nests of other females as an alternative to hunting for prey on their own (Alexander 1986, Field 1989, 1992,

Villalobos and Shelly 1996, Kurczewski and Spofford 1998). Females also steal prey that have been left momentarily outside the nest (Brockmann 1985) or attack females as they arrive at the nesting area with prey (Villalobos and Shelly 1996). Females also take over the nests of other females, which saves the time and energy required for digging (Brockmann and Dawkins 1979, Field 1992).

Female social Hymenoptera show a wide array of alternative reproductive tactics. In some species of vespid social wasps, for example, some females join (adopt) the nests of other females to join nests that are more mature. The joiner is less likely to build or provision on her new nest than the original foundresses and instead she seems to wait on the nest for an opportunity to take over the worker brood that will then care for her eggs (Nonacs and Reeve 1995, Starks 1998, 2001). Alternative, colony-founding tactics also occur in a few species of ants, and differences in female tactics may affect male mating opportunities. Most species of ants initiate colonies in one of two ways: independently by lone females that rely completely on body reserves during the initial period of colony founding, or by young queens joining already established colonies where they are assisted by workers from the adopting colony. A few species use both tactics (*Cardiocondyla batesii, Leptothorax rugatulus, Solenopsis geminata*) and in most of these, the queens are dimorphic (Rüppell and Heinze 1999, Heinze *et al.* 2002). In *L. rugatulus*, for example, large females with more fat and greater flying ability (macrogynes) initiate nests independently (Rüppell et al. 1998) whereas small females (microgynes) are adopted by their natal colony (Rüppell *et al.* 2001). The two sizes are thought to be specializations for their different nest-initiating tactics. The result is that macrogynes occur in colonies with only a single queen and microgynes in colonies with multiple queens. This may lead to different patterns of mating, with selection for some males to remain inside polygynous nests (fighter males) and some to disperse.

8.3   DISCUSSION

8.3.1   Diversity of ARTs in insects

ARTs are found in most orders of insects (Table 8.1) but they are much more common in some groups than in others. Hymenoptera and Coleoptera seem to be the champions of ARTs, although this may be because there are more studies of mating behavior in these groups. Particularly remarkable is the paucity of ARTs among Hemiptera (except water striders) and Lepidoptera (Scott 1974b), which are both well-studied groups. One partial explanation might be that parental care is more common in Hymenoptera and Coleoptera than in other groups (Trumbo 1996), thus providing more opportunities for male parasitic behavior and female brood parasitism. Although ARTs are found in a wide range of species, they do not seem to occur in all species even within a small, well-studied group such as fig wasps. This suggests that ARTs are labile traits that evolve only under special conditions. A number of factors seem to favor the simultaneous expression of alternative tactics within one population.

Male ARTs may occur at any step in the reproductive process but seem particularly common as mate-searching tactics (Table 8.1). When females mate in more than one location and when those locations are mutually exclusive (i.e., searching in one prevents an individual from searching in another), then at least some males may have higher success by searching alternative patches. This is particularly true if conventional sites are already occupied or have a high operational sex ratio (OSR). In fact, one expects males to distribute themselves among different patches where females are found according to an ideal free distribution (Parker 1978, Milinski and Parker 1991). If mating opportunities exist in a patch, one expects at least a few males even at low-density sites. If females at low-density sites are of lower quality, however, then one expects males to be present but in lower numbers than predicted based strictly on a simple, ideal free distribution (Sutherland and Parker 1985, Parker and Sutherland 1986).

Some ARTs are associated with morphological differences and some are not (Table 8.1). In most cases authors have looked for obvious differences such as the presence or absence of wings or substantial size differences between individuals that behave differently. However, physical differences may be subtle. For example, some long-winged crickets histolyze their flight muscles and are unable to fly (Zera and Denno 1997). Subtle differences in metabolism or lipid biosynthesis (Zera and Harshman 2001, Crnokrak and Roff 2002, Zhao and Zera 2002) may underlie some tactics. Therefore, considerable care should be taken when describing the suites of characters associated with behavioral differences between tactics.

Identifying ARTs requires judgment by the observer and my categories (Table 8.1) reflect authors' descriptions. For example, in the milkweed beetle *Tetraopes tetraophthalmus*, the data clearly show that size variation is continuous (Lawrence 1987), but the males are behaviorally

dichotomous with some individuals remaining in high-density patches where they have to fight for mates and others dispersing to less dense patches; it is the smaller, less competitive (in the dense patch) males that are more likely to disperse. Hence, discrete, alternative behavioral options are correlated with size in this species. A broad definition of ARTs means that Table 8.1 is just a sampling of insect ARTs. Additional species may be found in more specialized accounts (Alcock *et al.* 1978, Hamilton 1979, Thornhill and Alcock 1983, Field 1992, Eberhard 1996, Arnqvist 1997, Herre *et al.* 1997, Danforth and Desjardins 1999, Gwynne 2001, O'Neill 2001, Shuster and Wade 2003).

8.3.2   Sedentary, high-investment vs. dispersing, low-investment ARTs

Many male and female ARTs seem to involve a division between high-investment or high-risk but sedentary, non-dispersing tactics and low-investment or low-risk but active, dispersing, searching tactics with longer-range movements (Table 8.2). These broad patterns are often associated with morphological specializations (Table 8.1). Dispersing males are often smaller, with longer wings and larger wing muscles than nondispersing males that remain at high density sites. Nondispersing males are usually larger, with smaller wings and reduced musculature and enhanced structures adapted for fighting other males such as hypertrophied mandibles. The same pattern holds for females; females that disperse are smaller, have longer wings, greater wing musculature, often delayed maturation, larger amounts of fat, and reduced fecundity (Roff 1986, Mole and Zera 1992, Zera and Denno 1997). One might also expect wing length to body mass and aspect ratio differences for sedentary vs. dispersing tactics as well as different adaptations for sperm competition.

If we were not focusing on ARTs, it would be easy to interpret many of the patterns we have been discussing in quite a different light, i.e., as alternative life-history patterns for dispersing and nondispersing (Table 8.2). For example, the speckled wood butterfly *Pararge aegeria* has a color polymorphism in which pale males are territorial and darker males search for females over wider areas. The dark-winged males are adapted for searching since they engage in longer flights interspersed by more infrequent periods of basking and dark animals heat up more quickly than pale ones when basking (van Dyck and Matthysen 1998). The pale, territorial males perch and bask in sunspots more often between much shorter and more frequent fights, so they

Table 8.2. *Functional categories for ARTs*[a]

Type of sexual selection	Nature of high-investment resource	High-investment tactic	Tactics for avoiding high-investment decisions		
			Tactic for opting out of high investment	Tactic for parasitizing investment of others	ARTs that involve a pattern of nondispersing vs. dispersing tactics
Male ARTs					
Intrasexual selection Male–male competition	Aggregation of unmated females	Fighter	Disperser	Satellite or sneaker (nonfighting)	Fighter or nonfighter (sedentary) vs. disperser
	Resource required by females	Territorial	Transient searcher (patrolling)	Satellite or sneaker	Sedentary territorial vs. ranging patroller
		Male dominance	Transient searcher (patrolling)	Satellite or sneaker	Male dominance (sedentary) vs. transient
	Receptive female	Mate guarding (pre- or post-copulation)	Mate searching	Takeover guarding	Sedentary guarder vs. long-distance searcher
	Females mate with males as they are encountered	Waiting hilltopping		Patrolling or traplining	Sedentary waiter vs. long-distance patroller
Intersexual selection Female choice and mate conflict	Females attracted to an advertising male	Advertise	Search for females not attracted by advertiser	Satellite or sneak in on advertiser	Sedentary advertiser vs. longer-distance searcher
	Females mate with males in groups	Lek or swarm	Search outside group	Satellite or sneak on lek	Sedentary lek or swarm vs. longer-distance searcher

Table 8.2. (*cont.*)

Type of sexual selection	Nature of high-investment resource	High-investment tactic	Tactics for avoiding high-investment decisions		ARTs that involve a pattern of nondispersing vs. dispersing tactics
			Tactic for opting out of high investment	Tactic for parasitizing investment of others	
	Females mate based on nuptial gift	Provide high-quality gift	Low-cost replacement gift; forced copulation	Thieve nuptial gifts	
	Females mate based on courtship displays	Use courtship	No courtship (forced copulation)	Take over after courtship by another	
	Females mate based on paternal investment	Provide investment or care	Search for matings not requiring investment	Sneak in after care is provided by another	Sedentary paternal care vs. searching for opportunities
Female ARTs					
Intrasexual selection	Egg-laying opportunities at high-quality sites	Solitary species: territorial Social species: solitary foundress	Solitary: disperse Social: multiple foundress associations	Solitary: satellite or sneak eggs into nests Social: sneak eggs into nests or queuing for egg-laying position	
Female–female competition	Access to high-quality males[b] Maternal care	Fight for or guard high-quality males Provide care	Search for males willing to mate Brood parasitism	Sneak in on successful female Brood parasitism	Sedentary fighter vs. ranging searcher
Intersexual selection Male choice and mate conflict	Males mate based on finding a female; matings or mating	Conspicuous	Hiding		

attempts are costly for females[c]				
Males mate based on female quality[b]	Females display quality	Satellite of attractive female	Sneaking in on advertising female	
Males mate based on female advertisement[b]	Advertise	Search for males	Court males as they approach advertising females	Sedentary advertising vs. searching

[a] Under intrasexual selection, the table gives the nature of the resource over which males (or females) are competing; under intersexual selection, the table gives the basis on which a mating decision is made (either male or female choice). In the third to sixth columns, tactics known to occur in insects are given; the names are taken from Table 8.1 for male ARTs and from the text for female ARTs. The third column gives common high-investment tactics and the fourth and fifth columns give common associated tactics for avoiding that high investment either by opting out (taking an entirely different route) or by parasitizing the high-investment tactic. Many ARTs involve nondispersing and dispersing tactics; these are listed in the final column.

[b] Selection of this type is not known to result in ARTs in insects but is included here to illustrate the type of ART that might be found with further study.

[c] Females have alternative tactics that allow them to control mating attempts through male mimicry.

more easily maintain the temperature necessary for flight. But speckled wood butterflies have a complex life cycle (van Dyck and Wiklund 2002), with overlapping spring and summer generations that differ in the degree of melanization of the wings (spring animals are darker, which extends their active periods), in body size, and in wing loading associated with seasonal differences in distances traveled (van Dyck et al. 1997, van Dyck and Matthysen 1998). Clearly, there is a complex interplay between male ARTs and life-history patterns that cannot be easily distinguished (see also Chapter 2).

8.3.3 Sexual selection and the evolution of ARTs

Male and female ARTs are associated with many different kinds of mating systems including polygyny, polyandry, explosive breeding, monogamy, leks, and swarms. Male ARTs seem to be particularly common when males are sequestered with many females and there is intense male–male competition. Such a "seraglio" occurs in fig wasps (Hamilton 1979, Heinze and Hölldobler 1993, Greeff 2002) and Cardiocondyla ants (Cremer and Heinze 2002, 2003, Anderson et al. 2003, Heinze et al. 2004) and may sometimes result in highly dimorphic fighter vs. disperser males (Figure 8.1). Although less extreme, many other species show a similar pattern with large or fighting males remaining at the nesting, emergence, or oviposition sites and smaller, less well-armed males dispersing and searching for females, as in the beetle Dendrobias mandibularis, or the anthophorine solitary digging bees Centris pallida and Amegilla dawsoni.

ARTs are often associated with intense sexual selection. Patterns arise in response to both male–male competition and female choice (Table 8.2). ARTs arising from male–male competition are of two types: (a) opting out of costly or high-risk male–male interactions to seek mates under less competitive circumstances (e.g., fighting vs. dispersing males) or (b) noncombative parasitizing of costly male investment (e.g., dung beetles). Patterns arising in response to female choice are much less common but include (c) circumventing female choice (e.g., useless nuptial gifts) and mate conflict (e.g., noncourtship and forced copulation) or (d) parasitizing the mate-acquiring investment of other males (e.g., calling vs. noncalling). Although much less common, female ARTs are well known in insects and occur in response to (a) the parasitism of the costly maternal investment of other females (e.g., brood parasitism), (b) female–female competition for egg-laying opportunities (e.g., parasitism in social insects), or (c) conflict with males (e.g., avoiding male harassment) (Table 8.2). Although

technically possible, no known ARTs arise from female–female competition for mates.

It might seem reasonable to categorize ARTs on the basis of the different types of sexual selection, but too often it is impossible to separate adaptations for male–male competition from those associated with female choice or mate conflict. For example, some males of Europe's common earwig Forficula auricularia have a greatly enlarged terminal forceps (macrolabic morph) whereas other, smaller males have forceps only slightly larger than those of females (brachylabic morph) (Figure 8.6). When placed in competition with one another, macrolabic males are slightly more likely to mate and remain in copula longer than brachylabic males (Radesäter and Halldórsdóttir 1993). This may be because a macrolabic male is more likely to win a fight with a brachylabic male (by hitting his opponent with his forceps) and gain access to a female or because a macrolabic male is more likely to take over a female by pushing a mating male off the female or by raising a mating pair off the ground with his forceps thus dislodging the male (Radesäter and Halldórsdóttir 1993, Forslund 2003). However, males also use their forceps during courtship (Tomkins and Simmons 1998). The male taps and strokes the female's abdomen with his forceps and displays the forceps at the female's head whereupon she nibbles on them prior to soliciting copulation. Females solicit more quickly from males with larger forceps and manipulations of forceps length demonstrate that females prefer males with longer forceps (Tomkins and Simmons 1998). Clearly in this example, and in many others, intra- and intersexual selection are tightly intertwined.

8.3.4 Phylogeny of insect ARTs

Throughout this chapter, I have not made clear which alternative tactic evolves first. In the few cases that have been investigated, it has been found that either may be ancestral. For example, in Papilio butterflies with a genetic polymorphism for male mimicry, the andromorph is ancestral in some species (e.g., P. aegeus) and in other species (e.g., P. phorcas) it is derived (based on whether the andromorph is recessive or dominant: Clarke et al. 1985). In the water striders (Gerridae) (Andersen 1993) and fig wasps (Pteromalidae) (Jousselin et al. 2004), wing dimorphism can evolve from either short-winged or long-winged forms. In the andrenid bees, the male dimorphism found in Perdita portalis is derived (Danforth and Desjardins 1999). In each case a phylogenetic analysis is needed to determine whether ARTs are ancestral or derived, i.e., the direction of evolution cannot be assumed (West-Eberhard 2003).

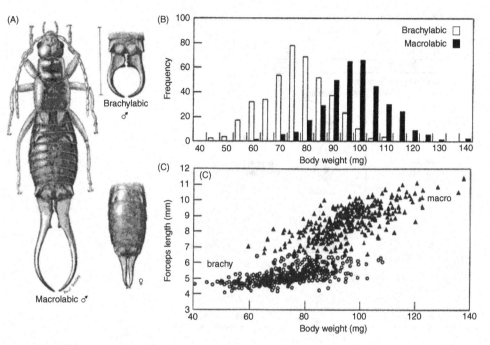

Figure 8.6 Males of the European earwig (*Forficula auricularia*, Dermaptera) occur in two forms: (A) some (macrolabic) have a large forceps at the posterior end of the body whereas others (brachlabic) have a much smaller forceps. (B) Frequency distributions of body size and forceps length for brachylabic and macrolabic male earwigs. (C) The relationship between body size and forceps length reveals a threshold mechanism underlying development into the two morphs. (Reprinted from Forslund [2003] with permission from Elsevier.)

8.3.5 Threshold mechanisms and the heritability of insect ARTs

Suites of distinctive, correlated traits may arise as a result of a developmental switch (Roff 1996) as described for horned beetles (Emlen and Nijhout 2000; see also Chapter 5). A developmental switch occurs when individuals below a particular size (or condition) at the time of pupation develop one set of characters and those above the threshold another set (Eberhard and Gutiérrez 1991, Roff 1996, Tomkins 1999) (Figure 8.7; see also Boxes 2.1 and 2.3). These switch mechanisms are often mediated by complex hormonal processes including the common insect gonadotropin, juvenile hormone (Zera 1999, Emlen and Nijhout 2001, Emlen and Allen 2004). The switch point or threshold evolves (Emlen 1996, Emlen and Nijhout 1999): if individuals are switching from one tactic to another at a size (or condition) that results in reduced fitness, then selection will favor individuals with a different switch point (Emlen 1996, Tomkins and Brown 2004, Tomkins *et al.* 2004). This mechanism has been used to explain winged and wingless morphs and extreme dimorphism in beetles, fig wasps, earwigs, and *Perdita* bees, as well as in castes (Wheeler 1991), polyethism in social Hymenoptera (Robinson 1992), and other forms of phenotypic plasticity in insects (Nijhout 1999, 2003). Threshold mechanisms have not been used to explain less extreme forms of male dimorphism, however, such as differences between territorial vs. transient or advertising vs. sneaking males. It seems possible that similar threshold mechanisms may be controlling these tactics as well.

Alternative tactics are often heritable and the threshold switch between tactics is known to evolve in most organisms that have been studied (Roff 1996, Moczek 2003, Shuster and Wade 2003), such as horned vs. hornless beetles (Figure 5.7) (Emlen 1996, Moczek *et al.* 2002, Moczek and Nijhout 2003),

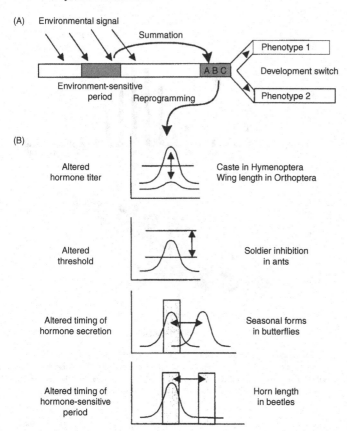

Figure 8.7 Endocrine mechanisms underlying the threshold or switch mechanisms in insects. (A) The developmental period for a typical insect. At some point during the larval stage, there is a sensitive period during which specific environmental stimuli such as temperature, photoperiod, or pheromones are received. These result in reprogramming of an endocrine mechanism just before or during metamorphosis that leads to alternative developmental pathways that result in different adult phenotypes. (B) Four kinds of hormonally controlled, developmental switching mechanisms have been identified in insects. In all cases hormones act during a tissue-specific sensitive period. Subsequent alternative developmental pathways depend on whether the hormone is above or below a threshold value during this period. (Modified from Nijhout [1999]. Copyright, American Institute of Biological Sciences, reprinted with permission.)

singing vs. nonsinging crickets (Cade 1981), wing dimorphic crickets (Walker 1987, Crnokrak and Roff 1998), short vs. long forceps in earwigs (Tomkins 1999, Tomkins et al. 2004), and male (Tsubaki 2003) and female color polymorphism in damselflies (Johnson 1964, 1966, Cordero 1990, Andres and Cordero 1999) and butterflies (Clarke et al. 1985). In relatively few insects, however, are the underlying genetics or associated natural and sexual selection processes clear.

Suites of distinctive, correlated traits may also arise in association with a genetic polymorphism. In seaweed flies (Coelopidae), size variation is maintained by a chromosomal inversion system and males of two size classes adopt different reproductive tactics (Day and Gilburn 1997). Large males (aa) gain by mounting less often but are more likely to be accepted when they try, a result of both female choice and the ability of larger males to subdue the reluctant females. Large males also live longer (Butlin and Day 1985). Small males ($\beta\beta$) ($\alpha\beta$ are intermediate in size) are more active and maneuverable and can mount more females although they are accepted less often (Weall and Gilburn 2000). However, small size has additional advantages in earlier emergence and faster development (Dunn *et al.* 1999), which can be beneficial in this species' unpredictable, shoreline habitat of rotting seaweed that is periodically swept out to sea before larvae find a safe pupation site. The evidence is that the α and β forms of the inversion have evolved independently with different correlated traits (Gilburn and Day 1994, 1996). The α inversion appears to gain its advantage primarily through sexual selection whereas the β inversion gains its advantage through viability selection. Although the evolution of correlated traits is well understood in this and some other species, little is known about the developmental pathways (e.g., threshold mechanism) that result in the differential expression of correlated traits in the different morphs of a genetic polymorphism (Brakefield *et al.* 2003).

8.3.6 Disruptive selection and the evolution of ARTs

Discrete tactics evolve by disruptive selection, i.e., selection against individuals of intermediate phenotype. For example,

in the horned beetles, hornless minor males use an entirely different tactic for acquiring females than the horned major males, and the minors with horns and majors without horns are not as successful as the two extremes (Moczek and Emlen 2000). Similarly, in the butterfly *Heliconius sara*, some males locate females as they are emerging from their pupal cases by responding to a female pheromone while other males set up territories in areas that emerged females frequent (Hernández and Benson 1998). Large males are more successful at pupal mating since several males may respond to one female; but small males are more successful at territorial mating because they are more successful in escalated aerial disputes. When selection regimes are bimodal in this way, a mechanism that channels individuals into one developmental pathway or another with few intermediates is favored. In insects this is often a threshold switch during the latter part of development (Nijhout 2003) (Figure 8.7; Chapter 5).

The evolution of dimorphic male traits has been particularly well studied in the andrenid bee *Perdita portalis* (Danforth and Desjardins 1999) (Figure 8.8). The dimorphism in *P. portalis* is extreme, with one morph being entirely flightless with reduced flight muscles, wings, and eyes but with a greatly enlarged head and mandibles (LH) (Danforth 1991b). The large-headed males remain inside their natal nest, fight other LH males to the death, and

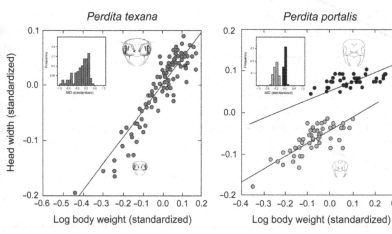

Figure 8.8 The evolution of alternative phenotypes in the andrenid bee *Perdita*. *Perdita portalis* shows two discrete and nonoverlapping phenotypes, small-headed males (shaded) and large-headed males. Other species of *Perdita*, such as *Perdita texana*, show an equally wide range of head sizes but the distribution is continuous. In *P. texana* the scaling is linear, whereas in *P. portalis* it is sigmoidal with few intermediates between the two forms, suggesting an underlying threshold mechanism for the development of the two morphs. (Redrawn from figures in Danforth and Desjardins [1999].)

mate with females inside the nest (Danforth 1991a). The small-headed, dispersing males (SH) leave their natal nest, perch on flowers, which they defend from other males, and mate with foraging females. Females of this species nest in groups of 2–29, often reusing their natal nests. There is no reproductive dominance in this communally nesting species; rather, each female constructs and provisions separate brood cells within their common nest. Most nests contain LH males but newly established nests do not have LH males, so all females mate outside the nest with SH males. In older nests 29% of emerging males are of the LH morph (Danforth 1999). A positive correlation exists between the number of females in a nest and the proportion of LH males. Female Hymenoptera control the size and sex of their offspring (in

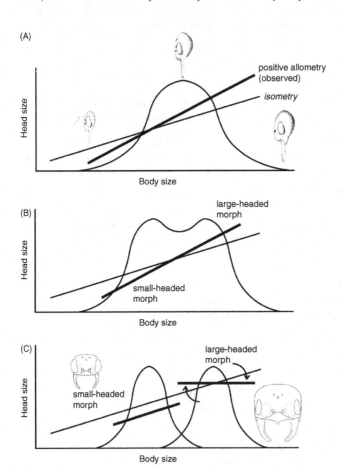

Figure 8.9 Model for the evolution of dimorphism based on morphometric data from *Perdita*. (A) The primitive condition is for males to show positive head allometry with a unimodal distribution of body sizes, as in *P. texana*. (B) Through disruptive selection, males of intermediate body size may be partially or completely eliminated. (C) Once developmentally decoupled through the addition of a new threshold size for pupation, the two male morphs are free to evolve independently in the allometric relationship between head size and body size and in the development of other correlated characters. (Modified from Danforth and Desjardins [1999] with permission from Birkhäuser-Verlag, Basel, Switzerland.)

species that are eusocial, workers may also control size but this is a communal species), so this correlation must be the product of maternal investment decisions from the preceding generation (Danforth and Desjardins 1999). Danforth (1999) and Danforth and Desjardins (1999) (Figure 8.9) speculate that increased female density, either on flowers or within nests, leads to elevated levels of male combat and hence favors increased male size with disproportionately larger heads for fighting. The result is a population with high variation in male size. Male dimorphism would then arise when males of intermediate size were less successful than those at the extremes. This might occur, for example, when there are two distinct mating opportunities available to males, such as mating in nests and mating on flowers. If these different mating "niches" favor different traits, then selection favors a threshold switch mechanism between alternatives (Wheeler 1991, Nijhout 2003) (Figure 8.7).

8.3.7 Condition dependence and the evolution of insect ARTs

Most ARTs are dependent on condition or status (Gross 1996) and a wide range of environmental- and individual-based conditions influence the expression of alternative tactics (Table 8.1). For example, in the damselfly *Calopteryx maculata*, older males switch to nonterritorial tactics (Waage 1973, 1979, Forsyth and Montgomerie 1987) presumably because their ability to hold territories has declined or the cost of holding territories has increased. In thrips, earwigs, horned beetles, and the meal moth *Plodia*, experiments have demonstrated that larval food affects the switch from small to large body size or low- to high-investment reproductive tactics (Crespi 1988a, Gage 1995, Emlen 1997b, Tomkins 1999, Hunt and Simmons 2001). Density is one of the most common factors affecting tactics. In the territorial dragonfly *Nannophya pygmaea*, males are more likely to switch to a nonterritorial satellite tactic at high densities (Tsubaki and Ono 1986). In an experimental manipulation, Cade and Cade (1992) show that male crickets (*Gryllus texensis*) call more and search less in low-density populations. This is because at low female densities, mating success is inversely correlated with time spent searching and positively correlated with time spent calling. At high densities, however, males do better when they search for females since females are encountered more readily and calling has reduced success due to increased parasitism by noncalling males. The most important factor, however, is the interplay between density, frequency dependence, and parasitism by flies that alters the benefits to

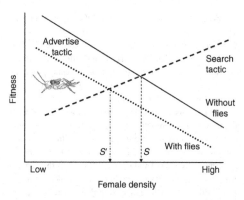

Figure 8.10 Hypothesized crossing fitness curves for the calling vs. searching tactics of crickets. In a population without parasitic flies when female density is low, males have higher success by calling than by searching, but when density rises to point *S* (crossover point), then males have higher fitness by searching. Above *S*, then, males should search and below this line they should advertise. In a population with parasitic flies, however, the fitness of the advertise tactic drops because the flies are attracted to the calling song of male crickets (advertise tactic fitness curve decreases). This decreased fitness for the advertise tactic means that the crossover point, *S'*, is at a lower density. This means that a higher proportion of the population will use the search tactic in a population with flies than in a population without flies.

the two tactics to such an extent that selection can favor the maintenance of both patterns in a population with flies and only the calling pattern in a population without flies (Walker and Cade 2003). When condition-dependent tactics exist, we expect well-adapted animals to switch from one tactic to the other at the condition that maximizes fitness (Parker 1982) (Figure 8.10). So for example, if one tactic is advantageous at low density and another at a higher density, we expect individuals to switch at the density that maximizes fitness. Individuals of many species are remarkably good at switching tactics in such a way that fitness is maximized (making the *best* of their situation). For example, in white-faced dragon-flies *Leucorrhinia intacta*, males adopt territorial vs. transient tactics based on daily conditions and the result is that males of each tactic mate in proportion to their representation in the population (Wolf and Waltz 1993).

8.3.8 Equality of fitness among ARTs

In some species such as crickets, seaweed flies, and fig wasps, alternative tactics are equally successful (Holtmeier and Zera 1993, Day and Gilburn 1997). For example, dimorphic male fig wasps have equal fitness through time (Cook *et al.* 1997) either because the males are able to perceive the number and sex of individuals maturing alongside them and adjust their developmental program accordingly (Pienaar and Greeff 2003a) or because the dimorphism is a result of a maternal decision using similar information (Pienaar and Greeff 2003b, Moore *et al.* 2004). Similarly, in the damselfly *Mnais pruinosa*, two male color morphs exist; orange-winged males are territorial around oviposition sites and guard females whereas clear-winged males are not territorial and intercept females as they arrive at territories or they take over ovipositing females (Tsubaki *et al.* 1997). Although orange-winged males have higher daily rates of success, clear-winged males live longer and suffer less from parasites and the result is no significant difference between the two tactics in lifetime reproductive success (Tsubaki and Hooper 2004).

Alternative tactics are often reported as having unequal success. It is important to point out that it is very difficult to measure accurately the lifetime reproductive success associated with each tactic for a number of reasons. First, the measure includes the average for all males that follow each tactic over their lifetimes, including the many individuals that never mate or die before they reproduce (Shuster and Wade 2003). Second, most ARTs involve opportunities for sperm competition and cryptic female choice and the outcome of these processes may be affected by local conditions (Simmons *et al.* 2004). Third, it is difficult to measure trade-offs between lifespan and fertility or mating success and lifespan (Gadgil 1972, Gadgil and Taylor 1975, Banks and Thompson 1985, Alcock 1996b, c), and it is particularly difficult to take into account differences in success in different years or seasons when different densities and different ecological conditions affect success of the two tactics unequally. Finally, some ARTs, such as the two morphs of digging bees, are maternal investment strategies (Alcock 1989) (Box 8.1). This means that females should produce large and small sons at the frequencies that maximize the mothers' long-term reproductive success, not their sons' success. Given all these problems with getting a proper measure of reproductive success, it seems likely that some studies have prematurely claimed unequal success between tactics.

Nevertheless, many studies argue that alternative tactics are not equally successful (Campanella and Wolf 1974, Eberhard 1982, Fincke 1982, 1985, Forsyth and Montgomerie 1987, Plaistow and Siva-Jothy 1996, Starks and Reeve 1999). Most ARTs are condition or status

dependent (Gross 1996): individuals follow alternative tactics in such a way that individual fitness is maximized (Box 8.1), i.e., they are making "the *best* of a bad situation" (Dawkins 1980, Wolf and Waltz 1993, Brockmann 2002). When averaged over all individuals, such conditional tactics are not necessarily equally successful (Gross 1996). Furthermore, when frequency dependence is superimposed on conditional tactics, calculations of average success are more complex.

8.3.9   Frequency-dependent selection and ARTs

Many ARTs are frequency dependent (Tomkins 1999), i.e., the success of a disperser, satellite, or sneaker depends on the proportion of that tactic in the population (regardless of density). So, for example, if the success of satellite male crickets depends on calling males to attract females, then as the proportion of satellite males increases, their success declines because there are just more satellites around the same number of calling males. Similarly, if female choice of a male morph depends on the proportion of that morph (e.g., rare-male effect), then its success will be frequency dependent (Knoppien 1985, Partridge 1988). Under negative frequency-dependent selection, the threshold switch that controls the development of tactics (Box 8.1) should evolve: if individuals switch from one tactic to another at a point that results in so many individuals of one tactic that they have reduced fitness, then selection will favor a different threshold that reduces the frequency of the excess tactic (Boxes 2.1 and 2.2).

An important and interesting feature of negative frequency-dependent selection is that it acts to maintain alternative tactics in a population. Shuster and Wade (2003) argue that frequency-dependent selection acts on condition-dependent traits in exactly the same way that it operates on other traits and will maintain tactics at a frequency such that they are equally successful. Gross and Repka (Gross 1996, Gross and Repka 1995, 1998a, b, Repka and Gross 1995) argue that when frequency dependence acts on condition-dependent traits, there is no longer an expectation of equal success for the alternative tactics. In either case, frequency-dependent selection will have the effect of maintaining alternative tactics in the population at stable frequencies. Unlike many other forms of selection, frequency-dependent effects cannot be understood just by measuring the behavior and reproductive success of individuals, but require knowledge of population frequencies and should be evaluated experimentally by manipulating those frequencies (Waltz and Wolf 1988, Brockmann 2001).

8.3.10 Economics and the evolution of ARTs in insects

Two or more tactics can arise in a population when selection favors specialized morphs or tactics that exploit discrete reproductive opportunities or niches. What creates these niches? Each alternative reproductive tactic must have some expected success if it is to be maintained in a population and intermediates must be less successful, e.g., fighter vs. disperser, guarder vs. disperser, territorial vs. nonterritorial (Figure 2.6). In many species there is good evidence that males adjust their tactics to differences in the availability of receptive females (pay-offs) and to differences in energetic costs and exposure to risks such as predators and parasites (costs) at different sites.

Females are also part of the equation and the economics of ARTs, including male ARTs, must include the interplay between male and female tactics, costs, and benefits (see Chapter 18). This point is nicely illustrated by water striders. Some males do not court females but lunge and struggle whereas others attract females with calling signals communicated through surface waves (Arnqvist 1997) (Figure 8.5). When a male pounces on a female, she responds with a vigorous pre-mating struggle that includes backwards somersaults and kicking as the male attempts to secure a pair of abdominal processes that hold the female securely (Arnqvist 1992b). The longer the processes, the more effective the male is at securing the female (Arnqvist 1992c). Males that mate without courtship, although less successful, nonetheless save the time and energy associated with advertising and guarding (Hayashi 1985, Spence and Wilcox 1986, Arnqvist 1989, 1992a, Krupa and Sih 1993); unlike courting males, the noncourting males do not show post-copulatory guarding, so they return to mate searching (Figure 8.5). In some species females possess abdominal spines, which they use to thwart unwanted mounting by males (Arnqvist and Rowe 1995, Andersen 1996). Struggles are costly to females (Arnqvist 1989), so females adjust their resistance depending on gains (e.g., sperm-depleted females are less likely to resist: Lauer 1996), mating costs (Jabloński and Vepsäläinen 1995), and other costs such as predation (Sih and Krupa 1995). This means that when females are living at high operational sex ratios with little predation, they are less likely to exert pre-mating choice (Arnqvist 1992a). Alternative tactics are affected by the interplay between male and female behavior, costs, and benefits.

Alternative reproductive opportunities or niches seem most likely to evolve under intense sexual selection (Table 8.2).

When intrasexual competition becomes very costly, selection may favor males that opt out of that escalating investment to utilize an alternative route to success, if one exists (Gadgil 1972). A common pattern is for some individuals to focus on highly competitive, higher-density areas and not move around very much and others to patrol or trapline between many different, lower-density, dispersed sites (Table 8.1). For some males the costs of male–male competition may be much higher than for others. For example, for larvae that are growing slowly or are born late in the season, the cost of emerging late at a larger size may be very high, and for these individuals the only expected success comes when they utilize an alternative smaller size or earlier emergence tactic, if one exists. This favors the evolution of condition-dependent switches from one tactic to another. So, for example, a slowly growing male may emerge at a smaller size but with other traits that make him particularly good at finding dispersed females as opposed to fighting over aggregated females. In situations like this, individuals with intermediate traits would be selected against. A similar argument can be applied to costly intersexual selection: when female choice becomes very costly (i.e., when it is extremely risky, time-consuming, or requires a very large investment for males to attract a mate), selection may favor switching to a different and less costly tactic (if fitness is available through an alternative route). This switch is more likely for some individuals and will be based on individual or environmental conditions (condition-dependent tactics). Selection will then favor correlated traits that make individuals particularly effective at each alternative tactic with selection against those with intermediate traits. A likely mechanism to achieve this divergence is a threshold switch.

ACKNOWLEDGMENTS

I wish to thank Michael Taborsky, Laura Sirot, and two anonymous reviewers for their valuable comments. Skye White drew the figures.

References

Agosti, D. and Hauschteck-Jungen, E. 1987. Polymorphism of males in *Formica exsecta* Nyl (Hym.: Formicidae). *Insectes Sociaux* **34**, 280–290.

Alcock, J. 1979a. The behavioural consequences of size variation among males of the territorial wasp *Hemipepsis ustulata* (Hymenoptera: Pompilidae). *Behaviour* **71**, 322–335.

Alcock, J. 1979b. The evolution of intraspecific diversity in male reproductive strategies in some bees and wasps. In

M. S. Blum and N. A. Blum (eds.) *Sexual Selection and Reproductive Competition in Insects*, pp. 381–402. New York: Academic Press.

Alcock, J. 1979c. The relation between female body size and provisioning behavior in the bee *Centris pallida* Fox (Hymenoptera: Anthophoridae). *Journal of the Kansas Entomological Society* 52, 623–632.

Alcock, J. 1984. Long-term maintenance of size variation in populations of *Centris pallida* (Hymenoptera: Anthophoridae). *Evolution* 38, 220–223.

Alcock, J. 1989. Size variation in the anthophorid bee *Centris pallida*: new evidence on its long-term maintenance. *Journal of the Kansas Entomological Society* 62, 484–489.

Alcock, J. 1994. Alternative mate-locating tactics in *Chlosyne californica* (Lepidoptera, Nymphalidae). *Ethology* 97, 103–118.

Alcock, J. 1995. Persistent size variation in the anthophorine bee *Centris pallida* (Apidae) despite a large male mating advantage. *Ecological Entomology* 20, 1–4.

Alcock, J. 1996a. Male size and survival: the effects of male combat and bird predation in Dawson's burrowing bees, *Amegilla dawsoni*. *Ecological Entomology* 21, 309–316.

Alcock, J. 1996b. Provisional rejection of three alternative hypotheses on the maintenance of a size dichotomy in males of Dawson's burrowing bee, *Amegilla dawsoni* (Apidae, Apinae, Anthophorini). *Behavioral Ecology and Sociobiology* 39, 181–188.

Alcock, J. 1996c. The relation between male body size, fighting, and mating success in Dawson's burrowing bee, *Amegilla dawsoni* (Apidae, Apinae, Anthophorini). *Journal of Zoology (London)* 239, 663–674.

Alcock, J. 1997a. Competition from large males and the alternative mating tactics of small males of Dawson's burrowing bees (*Amegilla dawsoni*) (Apidae, Apinae, Anthophorini). *Journal of Insect Behavior* 10, 99–114.

Alcock, J. 1997b. Small males emerge earlier than large males in Dawson's burrowing bee (*Amegilla dawsoni*) (Hymenoptera: Anthophorini). *Journal of Zoology (London)* 242, 453–462.

Alcock, J. 1999. The nesting behavior of Dawson's burrowing bee, *Amegilla dawsoni* (Hymenoptera: Anthophorini), and the production of offspring of different sizes. *Journal of Insect Behavior* 12, 363–384.

Alcock, J. and Houston, T. F. 1987. Resource defence and alternative mating tactics in the Banksia bee *Hylaeus alcynoneus* (Erichson). *Ethology* 76, 177–188.

Alcock, J. and Houston, T. F. 1996. Mating systems and male size in Australian hylaeine bees (Hymenoptera: Colletidae). *Ethology* 102, 591–610.

Alcock, J., Eickwort, G. C., and Eickwort, K. R. 1977a. The reproductive behavior of *Anthidium maculosum* (Hymenoptera: Megachilidae) and the evolutionary significance of multiple copulations by females. *Behavioral Ecology and Sociobiology* 2, 385–396.

Alcock, J., Jones, C. E., and Buchmann, S. L. 1977b. Male mating strategies in the bee *Centris pallida* Fox (Anthophoridae: Hymenoptera). *American Naturalist* 111, 145–155.

Alcock, J., Barrows, E. M., Gordh, G., et al. 1978. The ecology and evolution of male reproductive behaviour in bees and wasps. *Zoological Journal of the Linnean Society* 64, 293–326.

Alcock, J., Simmons, L. W., and Beveridge, M. 2005. Seasonal change in offspring sex and size in Dawson's burrowing bees (*Amegilla dawsoni*) (Hymenoptera: Anthophorini). *Ecological Entomology* 30, 247–254.

Alexander, A. J. and van Staaden, M. J. 1989. Alternative sexual tactics in male bladder grasshoppers (Orthoptera, Pneumoridae). In M. N. Bruton (ed.) *Alternative Life-History Styles of Animals*, pp. 261–277. Dordrecht, The Netherlands: Kluwer.

Alexander, B. 1986. Alternative methods of nest provisioning in the digger wasp *Clypeadon laticinctus* (Hymenoptera: Sphecidae). *Journal of the Kansas Entomological Society* 59, 59–63.

Alonzo, S. H. and Warner, R. R. 2000. Female choice, conflict between the sexes and the evolution of male alternative reproductive behaviours. *Evolutionary Ecology Research* 2, 149–170.

Andersen, N. M. 1993. The evolution of wing polymorphism in water striders (Gerridae): a phylogenetic approach. *Oikos* 67, 433–443.

Andersen, N. M. 1996. Ecological phylogenetics of mating systems and sexual dimorphism in water striders (Heteroptera: Gerridae). *Vie et Milieu* 46, 103–114.

Anderson, C., Cremer, S., and Heinze, J. 2003. Live and let die: why fighter males of the ant *Cardiocondyla* kill each other but tolerate their winged rivals. *Behavioral Ecology* 14, 54–62.

Andres, J. A. and Cordero, A. 1999. The inheritance of female colour morphs in the damselfly *Ceriagrion tenellum* (Odonata: Coenagrionidae). *Heredity* 82, 328–335.

Arnqvist, G. 1989. Multiple mating in a water strider: mutual benefits or intersexual conflict? *Animal Behaviour* 38, 749–756.

Arnqvist, G. 1992a. The effects of operational sex ratio on the relative mating success of extreme male phenotypes in the water strider *Gerris odontogaster* (Zett.) (Heteroptera: Gerridae). *Animal Behaviour* 43, 681–683.

Arnqvist, G. 1992b. Pre-copulatory fighting in a water strider: inter-sexual conflict or mate assessment? *Animal Behaviour* 43, 559–567.

Arnqvist, G. 1992c. Spatial variation in selective regimes: sexual selection in the water strider, *Gerris odontogaster*. *Evolution* 46, 914–929.

Arnqvist, G. 1997. The evolution of water strider mating systems: causes and consequences of sexual conflicts. In J. Choe and B. J. Crespi (eds.) *Mating Systems in Insects and Arachnids*, pp. 146–163. Cambridge, UK: Cambridge University Press.

Arnqvist, G. and Rowe, L. 1995. Sexual conflict and arms races between the sexes: a morphological adaptation for control of mating in a female insect. *Proceedings of the Royal Society of London B* 261, 123–127.

Austad, S. N. 1984. A classification of alternative reproductive behaviors and methods for field-testing ESS models. *American Zoologist* 24, 309–319.

Bailey, W. J. and Field, G. 2000. Acoustic satellite behaviour in the Australian bushcricket *Elephantodeta nobilis* (Phaneropterinae, Tettigoniidae, Orthoptera). *Animal Behaviour* 59, 361–369.

Banks, M. J. and Thompson, D. J. 1985. Lifetime mating success of females in the damselfly *Coenagrion puella*. *Animal Behaviour* 33, 1175–1183.

Bean, D. and Cook, J. M. 2001. Male mating tactics and lethal combat in the nonpollinating fig wasp *Sycoscapter australis*. *Animal Behaviour* 62, 535–542.

Beani, L. and Turillazzi, S. 1988. Alternative mating tactics in males of *Polistes dominulus* (Hymenoptera: Vespidae). *Behavioral Ecology and Sociobiology* 22, 257–264.

Beeler, A. E., Rauter, C. M., and Moore, A. J. 1999. Pheromonally mediated mate attraction by males of the burying beetle *Nicrophorus orbicollis*: alternative calling tactics conditional on both intrinsic and extrinsic factors. *Behavioral Ecology* 10, 578–584.

Bego, L. R. and Camargo, C. A. 1984. On the occurrence of giant males in *Nannotrigona (Scaptotrigona) postica* Latreille (Hymenoptera, Apidae, Meliponinae). *Boletim de Zoologia, Universidade de São Paulo* 8, 11–16.

Belovsky, G. E., Slade, J. B., and Chase, J. M. 1996. Mating strategies based on foraging ability: an experiment with grasshoppers. *Behavioral Ecology* 7, 438–444.

Bertram, S. M. 2002. Temporally fluctuating selection of sex-limited signaling traits in the Texas field cricket, *Gryllus texensis*. *Evolution* 56, 1831–1839.

Bockwinkel, G. and Sauer, K. P. 1994. Resource dependence of male mating tactics in the scorpionfly, *Panorpa vulgaris* (Mecoptera, Panorpidae). *Animal Behaviour* 94, 203–209.

Borgia, G. 1980. Sexual competition in *Scatophaga stercoraria*: size- and density-related changes in male ability to capture females. *Behaviour* 75, 185–206.

Borgia, G. 1982. Experimental changes in resource structure and male density: size-related differences in mating success among male *Scatophaga stercoraria*. *Evolution* 36, 307–315.

Brakefield, P. M., French, V., and Zwaan, B. J. 2003. Development and the genetics of evolutionary change within insect species. *Annual Review of Ecology, Evolution and Systematics* 34, 633–660.

Briceño, R. D. and Eberhard, W. G. 1987. Genetic and environmental effects on wing polymorphisms in two tropical earwigs (Dermaptera: Labiidae). *Oecologia* 74, 253–255.

Brockmann, H. J. 1980. Diversity in the nesting behavior of mud-dauber (*Trypoxylon politum* Say; Sphecidae). *Florida Entomologist* 63, 53–64.

Brockmann, H. J. 1985. Provisioning behavior of the great golden digger wasp, *Sphex ichneumoneus* (L.) (Sphecidae). *Journal of the Kansas Entomological Society* 58, 631–655.

Brockmann, H. J. 2001. The evolution of alternative strategies and tactics. *Advances in the Study of Behavior* 30, 1–51.

Brockmann, H. J. 2002. An experimental approach to altering mating tactics in male horseshoe crabs (*Limulus polyphemus*). *Behavioral Ecology* 13, 232–238.

Brockmann, H. J. and Dawkins, R. 1979. Joint nesting in a digger wasp as an evolutionarily stable preadaptation to social life. *Behaviour* 71, 203–245.

Brown, L. and Bartalon, J. 1986. Behavioral correlates of male morphology in a horned beetle. *American Naturalist* 127, 565–570.

Brown, W. D. 1997. Female remating and the intensity of female choice in black-horned tree crickets, *Oecanthus nigricornis*. *Behavioral Ecology* 8, 66–74.

Burk, T. 1982. Evolutionary significance of predation on sexually signaling males. *Florida Entomologist* 65, 90–104.

Butlin, R. K. and Day, T. H. 1985. Adult size, longevity and fecundity in the seaweed fly, *Coelopa frigida*. *Heredity* 54, 107–110.

Cade, W. H. 1975. Acoustically orienting parasitoids: fly phonotaxis to cricket song. *Science* 190, 1212.

Cade, W. H. 1979a. Effect of male deprivation on female phonotaxis in field crickets (Orthoptera: Gryllidae: *Gryllus*). *Canadian Entomologist* 111, 741–744.

Cade, W. 1979b. The evolution of alternative male reproductive strategies in field crickets. In M. Blum and

N. A. Blum (eds.) *Sexual Selection and Reproductive Competition in Insects*, pp. 343–379. New York: Academic Press.

Cade, W. H. 1980. Alternative male reproductive strategies. *Florida Entomologist* **63**, 30–44.

Cade, W. H. 1981. Alternative male strategies: genetic differences in crickets. *Science* **212**, 563–564.

Cade, W. H. and Cade, E. S. 1992. Male mating success, calling and searching behaviour at high and low densities in the field cricket *Gryllus integer*. *Animal Behaviour* **43**, 49–56.

Cade, W. H. and Wyatt, D. R. 1984. Factors affecting calling behaviour in field crickets, *Teleogryllus* and *Gryllus* (age, weight, density, and parasites). *Behaviour* **88**, 61–75.

Campanella, P. J. and Wolf, L. L. 1974. Temporal leks as a mating system in a temperate zone dragonfly (Odonata: Libellulidae). 1. *Plathemis lydia*. *Behaviour* **51**, 49–87.

Caro, T. M. and Bateson, P. 1986. Organization and ontogeny of alternative tactics. *Animal Behaviour* **34**, 1483–1499.

Carroll, S. P. 1993. Divergence in male mating tactics between two populations of the soapberry bug. 1. Guarding versus nonguarding. *Behavioral Ecology* **4**, 156–164.

Carroll, S. P. and Corneli, P. S. 1995. Divergence in male mating tactics between two populations of the soapberry bug. 2. Genetic change and the evolution of a plastic reaction norm in a variable social environment. *Behavioral Ecology* **6**, 46–56.

Catts, E. P. 1979. Hilltop aggregation and mating behavior by *Gasterophilus intestinalis* (Diptera: Gasterophilidae). *Journal of Medical Entomology* **16**, 461–464.

Chapman, T. W., Liddle, L. F., Kalb, J. M., Wolfner, M. F., and Partridge, L. 1995. Costs of mating in *Drosophila melanogaster* is mediated by male accessory gland products. *Nature* **373**, 241–244.

Clarke, C., Clarke, F. M. M., Collins, S. C., Gill, A. C. L., and Turner, J. R. G. 1985. Male-like females, mimicry and transvestism in butterflies (Lepidoptera: Papilionidae). *Systematic Entomology* **10**, 257–283.

Conner, J. 1989. Density dependent sexual selection in the fungus beetle *Bolitotherus cornutus*. *Evolution* **43**, 1378–1386.

Consoli, F. L. and Vinson, S. B. 2002. Clutch size, development and wing morph differentiation of *Melittobia digitata*. *Entomologia Experimentalis et Applicata* **102**, 135–143.

Cook, D. 1987. Sexual selection in dung beetles. 1. A multivariate study of the morphological variation in two species of *Onthophagus* (Scarabaeidae: Onthophagini). *Australian Journal of Zoology* **35**, 123–132.

Cook, D. 1988. Sexual selection in dung beetles. 2. Female fecundity as an estimate of male reproductive success in

relation to horn size, and alternative behavioural strategies in *Onthophagus binodis* Thunberg (Scarabaeidae: Onthophagini). *Australian Journal of Zoology* **36**, 521–532.

Cook, D. F. 1990. Differences in courtship, mating and postcopulatory behaviour between male morphs of the dung beetle *Onthophagus binodis* Thunberg (Coleoptera: Scarabaeidae). *Animal Behaviour* **40**, 428–436.

Cook, J. M., Compton, S. G., Herre, E. A., and West, S. A. 1997. Alternative mating tactics and extreme male dimorphism in fig wasps. *Proceedings of the Royal Society of London B* **264**, 747–754.

Cordero, A. 1990. The inheritance of female polymorphism in the damselfly *Ischnura graellsii* (Rambur) (Odonata: Coenagrionidae). *Heredity* **64**, 341–346.

Cordero, A. 1999. Forced copulations and female contact guarding at a high male density in a calopterygid damselfly. *Journal of Insect Behavior* **12**, 27–37.

Courtney, S. P. and Parker, G. A. 1985. Mating behaviour of the tiger butterfly (*Tarucus theophrastus*): competitive mate-searching when not all females are captured. *Behavioral Ecology and Sociobiology* **17**, 213–221.

Cowan, D. P. 1979. Sibling matings in a hunting wasp: adaptive inbreeding? *Science* **205**, 1403–1405.

Cowan, D. P. 1981. Parental investment in two solitary wasps *Ancistrocerus adiabatus* and *Euodynerus foraminatus* (Eumenidae: Hymenoptera). *Behavioral Ecology and Sociobiology* **9**, 95–102.

Cremer, S. and Heinze, J. 2002. Adaptive production of fighter males: queens of the ant *Cardiocondyla* adjust the sex ratio under local mate competition. *Proceedings of the Royal Society of London B* **269**, 417–422.

Cremer, S. and Heinze, J. 2003. Stress grows wings: environmental induction of winged dispersal males in *Cardiocondyla* ants. *Current Biology* **13**, 219–223.

Cremer, S., Lautenschlager, B., and Heinze, J. 2002. A transitional stage between ergatoid and winged male morph in the ant *Cardiocondyla obscurior*. *Insectes Sociaux* **49**, 221–228.

Crespi, B. J. 1986. Size assessment and alternative fighting tactics in *Elaphrothrips tuberculatus* (Insecta: Thysanoptera). *Animal Behaviour* **34**, 1324–1335.

Crespi, B. J. 1988a. Adaptation, compromise, and constraint: the development, morphometrics, and behavioral basis of a fighter-flier polymorphism in male *Hoplothrips karnyi* (Insecta: Thysanoptera). *Behavioral Ecology and Sociobiology* **23**, 93–104.

Crespi, B. J. 1988b. Alternative male mating tactics in a thrips: effects of sex ratio variation and body size. *American Midland Naturalist* **119**, 83–92.

Crespi, B. J. 1988c. Risks and benefits of lethal male fighting in the colonial, polygynous thrips *Hoplothrips karnyi* (Insecta: Thysanoptera). *Behavioral Ecology and Sociobiology* 22, 293–301.

Crnokrak, P. and Roff, D. A. 1995. Fitness differences associated with calling behaviour in the two wing morphs of male sand crickets, *Gryllus firmus*. *Animal Behaviour* 50, 1475–1481.

Crnokrak, P. and Roff, D. A. 1998. The genetic basis of the trade-off between calling and wing morph in males of the cricket *Gryllus firmus*. *Evolution* 52, 1111–1118.

Crnokrak, P. and Roff, D. A. 2002. Trade-offs to flight capability in *Gryllus firmus*: the influence of whole-organism respiration rate on fitness. *Journal of Evolutionary Biology* 15, 388–398.

Danforth, B. N. 1991a. Female foraging and intranest behavior of a communal bee, *Perdita portalis* (Hymenoptera: Andrenidae). *Annals of the Entomological Society of America* 84, 537–548.

Danforth, B. N. 1991b. The morphology and behavior of dimorphic males in *Perdita portalis* (Hymenoptera: Andrenidae). *Behavioral Ecology and Sociobiology* 29, 235–247.

Danforth, B. N. 1999. Emergence dynamics and bet hedging in a desert bee, *Perdita portalis*. *Proceedings of the Royal Society of London B* 266, 1985–1994.

Danforth, B. N. and Desjardins, C. A. 1999. Male dimorphism in *Perdita portalis* (Hymenoptera, Andrenidae) has arisen from preexisting allometric patterns. *Insectes Sociaux* 46, 18–28.

Danforth, B. N. and Neff, J. L. 1992. Male polymorphism and polyethism in *Perdita texana* (Hymenoptera: Andrenidae). *Annals of the Entomological Society of America* 85, 616–626.

Davies, N. B. 1982. Behaviour and competition for scarce resources. In King's College Sociobiology Group (eds.) *Current Problems in Sociobiology*, pp. 363–380. Cambridge, UK: Cambridge University Press.

Dawkins, R. 1980. Good strategy or evolutionarily stable strategy. In G. W. Barlow and S. Silverberg (eds.) *Sociobiology: Beyond Nature/Nurture*, pp. 331–367. Boulder, CO: Westview Press.

Day, M. C. 1984. Male polymorphism in some Old World species of *Cryptocheilus* Panzer (Hymenoptera: Pompilidae). *Zoological Journal of the Linnean Society* 80, 83–101.

Day, T. H. and Gilburn, A. S. 1997. Sexual selection in seaweed flies. *Advances in the Study of Behavior* 26, 1–49.

Downes, J. A. 1955. Observations on the swarming flight and mating of *Culicoides* (Diptera: Ceratopogonidae).

Transactions of the Royal Entomological Society of London 106, 213–236.

Dunn, D. W., Crean, C. S., Wilson, C. L., and Gilburn, A. S. 1999. Male choice, willingness to mate and body size in seaweed flies. *Animal Behaviour* 57, 847–853.

Eberhard, W. G. 1982. Beetle horn dimorphism: making the best of a bad lot. *American Naturalist* 119, 420–426.

Eberhard, W. G. 1987. Use of horns in fights by the dimorphic males of *Ageopsis nigricollis* (Coleoptera, Scarabeidae, Dynastinae). *Journal of the Kansas Entomological Society* 60, 504–509.

Eberhard, W. G. 1996. *Female Control: Sexual Selection by Cryptic Female Choice*. Princeton, NJ: Princeton University Press.

Eberhard, W. G. and Gutiérrez, E. E. 1991. Male dimorphisms in beetles and earwigs and the question of developmental constraints. *Evolution* 45, 18–28.

Eggert, A. K. 1992. Alternative male mate-finding tactics in burying beetles. *Behavioral Ecology* 3, 243–254.

Eickwort, G. 1975. Gregarious nesting of the mason bee *Hoplites anthocopoides* and the evolution of parasitism and sociality among megachilid bees. *Evolution* 29, 142–150.

Eickwort, G. 1977. Male territorial behaviour in the mason bee *Hoplites anthocopoides* (Hymenoptera: Megachilidae). *Animal Behaviour* 25, 542–554.

Elliott, N. B. and Elliott, W. M. 1992. Alternative male mating tactics in *Tachytes tricinctus* (Hymenoptera: Sphecidae, Larrinae). *Journal of the Kansas Entomological Society* 65, 261–266.

Emlen, D. J. 1994. Environmental control of horn length dimorphism in the beetle *Onthophagus acuminatus* (Coleoptera: Scarabaeidae). *Proceedings of the Royal Society of London B* 256, 131–136.

Emlen, D. J. 1996. Artificial selection on horn length–body size allometry in the horned beetle *Onthophagus acuminatus*. *Evolution* 50, 1219–1230.

Emlen, D. J. 1997a. Alternative reproductive tactics and male-dimorphism in the horned beetle *Onthophagus acuminatus* (Coleoptera: Scarabaeidae). *Behavioral Ecology and Sociobiology* 41, 335–341.

Emlen, D. J. 1997b. Diet alters male horn allometry in the beetle *Onthophagus acuminatus* (Coleoptera: Scarabaeidae). *Proceedings of the Royal Society of London B* 264, 567–574.

Emlen, D. J. and Allen, C. E. 2004. Genotype to phenotype: physiological control of trait size and scaling in insects. *Integrative and Comparative Biology* 43, 617–634.

Emlen, D. J. and Nijhout, H. F. 1999. Hormonal control of male horn length dimorphism in the horned beetle *Onthophagus taurus*. *Journal of Insect Physiology* 45, 45–53.

Emlen, D. J. and Nijhout, H. F. 2000. The development and evolution of exaggerated morphologies in insects. *Annual Review of Entomology* 45, 661–708.

Emlen, D. J. and Nijhout, H. F. 2001. Hormonal control of male horn length dimorphism in *Onthophagus taurus* (Coleoptera: Scarabaeidae): a second critical period of sensitivity to juvenile hormone. *Journal of Insect Physiology* 47, 1045–1054.

Engqvist, L. and Sauer, K. P. 2003. Influence of nutrition on courtship and mating in the scorpionfly *Panorpa cognata* (Mecoptera, Insecta). *Ethology* 109, 911–928.

Evans, H. E. 1963. A new species of *Cephalonomia* exhibiting an unusually complex polymorphism (Hymenoptera: Bethylidae). *Psyche* 70, 151–163.

Evans, H. E. and O'Neill, K. M. 1978. Alternative mating strategies in a digger wasp *Philanthus zebratus* Cresson. *Proceedings of the National Academy of Sciences of the United States of America* 75, 1901–1903.

Feaver, M. N. 1983. Pair formation in the katydid *Orchelimum nigripes* (Orthoptera: Tettigoniidae). In D. T. Gwynne and G. K. Morris (eds.) *Orthopteran Mating Systems: Sexual Competition in a Diverse Group of Insects*, pp. 205–239. Boulder, CO: Westview Press.

Field, J. 1989. Alternative nesting tactics in a solitary wasp. *Behaviour* 110, 219–243.

Field, J. 1992. Intraspecific parasitism as an alternative reproductive tactic in nest-building wasps and bees. *Biological Review* 67, 79–126.

Field, S. A. and Keller, M. A. 1993. Alternative mating tactics and female mimicry as post-copulatory mate-guarding behaviour in the parasitic wasp *Cotesia rubecula*. *Animal Behaviour* 46, 1183–1189.

Fincke, O. M. 1982. Lifetime mating success in a natural population of the damselfly, *Enallagma hageni* (Walsh) (Odonata: Coenagrionidae). *Behavioral Ecology and Sociobiology* 10, 293–302.

Fincke, O. M. 1984. Sperm competition in the damselfly *Enallagma hageni* Walsh (Odonata: Coenagrionidae): benefits of multiple mating to males and females. *Behavioral Ecology and Sociobiology* 14, 235–240.

Fincke, O. M. 1985. Alternative mate-finding tactics in a non-territorial damselfly (Odonata: Coenagrionidae). *Animal Behaviour* 33, 1124–1137.

Fincke, O. M. 1986. Underwater oviposition in a damselfly (Odonata: Coenagrionidae) favors male vigilance, and

multiple mating by females. *Behavioral Ecology and Sociobiology* 18, 405–412.

Fincke, O. M. 2004. Polymorphic signals of harassed female odonates and the males that learn them support a novel frequency-dependent model. *Animal Behaviour* 67, 833–845.

Foitzik, S., Heinze, J., Oberstadt, B., and Herbers, J. 2002. Mate guarding and alternative reproductive tactics in the ant *Hypoponera opacior*. *Animal Behaviour* 63, 597–604.

Forrest, T. G. 1983. Calling songs and mate choice in mole crickets. In D. T. Gwynne and G. K. Morris (eds.) *Orthopteran Mating Systems*, pp. 185–204. Boulder, CO: Westview Press.

Forslund, P. 2003. An experimental investigation into status-dependent male dimorphism in the European earwig, *Forficula auricularia*. *Animal Behaviour* 65, 309–316.

Forsyth, A. and Alcock, J. 1990. Female mimicry and resource defense polygyny by males of a tropical rove beetle, *Leistotrophus versicolor* (Coleoptera: Staphylinidae). *Behavioral Ecology and Sociobiology* 26, 325–330.

Forsyth, A. and Montgomerie, R. D. 1987. Alternative reproductive tactics in a territorial damselfly *Calopteryx maculata*: sneaking by older males. *Behavioral Ecology and Sociobiology* 21, 73–81.

Fortelius, W., Pamilo, P., Rosengren, R., and Sundström, L. 1987. Male size dimorphism and alternative reproductive tactics in *Formica exsecta* ants (Hymenoptera, Formicidae). *Annales Zoologici Fennici* 24, 45–54.

Freeeman, B. E. and Ittyeipe, K. 1982. Morph determination in *Melittobia*, an eulophid wasp. *Ecological Entomology* 7, 355–363.

French, B. W. and Cade, W. H. 1989. Sexual selection at varying population densities in male field crickets *Gryllus veletis* and *G. pennsylvanicus*. *Journal of Insect Behavior* 2, 105–121.

Fujisaki, K. 1992. A male fitness advantage to wing reduction in the oriental chinch bug, *Cavelerius saccharivorus* Okajima (Heteroptera: Lygaeidae). *Researches on Population Ecology* 34, 173–183.

Gadgil, M. 1972. Male dimorphism as a consequence of sexual selection. *American Naturalist* 106, 574–579.

Gadgil, M. and Taylor, C. E. 1975. Plausible models of sexual selection and polymorphism. *American Naturalist* 109, 470–472

Gage, M. J. G. 1995. Continuous variation in reproductive strategy as an adaptive response to population density in the moth *Plodia interpunctella*. *Proceedings of the Royal Society of London B* 261, 25–30.

Gilburn, A. S. and Day, T. H. 1994. Sexual dimorphism, sexual selection and the *αβ* chromosomal inversion polymorphism in the seaweek fly, *Coelopa frigida*. *Proceedings of the Royal Society of London B* **257**, 303–309.

Gilburn, A. S. and Day, T. H. 1996. The evolution of female choice when the preference and the preferred trait are linked to the same inversion system. *Heredity* **76**, 19–27.

Goldsmith, S. K. 1985a. Male dimorphism in *Dendrobias mandibularis* (Coleoptera: Cerambycidae). *Journal of the Kansas Entomological Society* **58**, 534–538.

Goldsmith, S. K. 1985b. The mating system and alternative reproductive behaviors of *Dendrobias mandibularis* (Coleoptera: Cerambycidae). *Behavioral Ecology and Sociobiology* **20**, 111–115.

Goldsmith, S. K. 1987. The mating system and alternative reproductive behavior of *Dendrobias mandibularis* (Coleoptera: Cerambycidae). *Behavioral Ecology and Sociobiology* **20**, 111–115.

Goldsmith, S. K. and Alcock, J. 1993. The mating chances of small males of the cerambycid beetle *Trachyderes mandibularis* differ in different environments (Coleoptera: Cerambycidae). *Journal of Insect Behavior* **6**, 351–360.

Gomez, M. S., Fernandez-Salvador, R., and Garcia, R. 2003. First report of Siphonaptera infesting *Microtus* (*Microtus*) *cabrerae* (Rodentia, Muridae, Arvicolinae) in Cuenca, Spain and notes about the morphologic variability of *Ctenophthalmus* (*Ctenophthalmus*) *apertus personatus* (Insecta, Siphonaptera, Ctenophthalmidae). *Parasite Journal de la Société Française de Parasitologie* **10**, 127–131.

Gonzalez-Soriano, E. and Cordoba-Aguilar, A. 2003. Sexual behaviour in *Paraphlebia quinta* Calvert: male dimorphism and a possible example of female control (Zygoptera: Megapodagrionidae). *Odonatologica* **32**, 345–353.

Greeff, J. M. 2002. Mating system and sex ratios of a pollinating fig wasp with dispersing males. *Proceedings of the Royal Society of London B* **269**, 2317–2323.

Greeff, J. M. and Ferguson, W. H. 1999. Mating ecology of the nonpollinating fig wasps of *Ficus ingens*. *Animal Behaviour* **57**, 215–222.

Greenfield, M. D. and Shelly, T. E. 1985. Alternative mating strategies in a desert grasshopper: evidence of density dependence. *Animal Behaviour* **33**, 1192–1210.

Greenfield, M. D. and Shelly, T. E. 1989. Territory-based mating systems in desert grasshoppers: effect of host plant distribution and variation. In T. W. Chapman and A. Joern

(eds.) *A Biology of Grasshoppers*, pp. 315–335. New York: Wiley Interscience.

Groddeck, J., Mauss, V., and Reinhold, K. 2004. The resource-based mating system of the Mediterranean pollen wasp *Ceramius fonscolombei* Latreille 1820 (Hymenoptera, Vespidae, Masarinae). *Journal of Insect Behavior* **17**, 397–418.

Gross, M. R. 1996. Alternative reproductive strategies and tactics: diversity within sexes. *Trends in Ecology and Evolution* **11**, 92–97.

Gross, M. R. and Repka, J. 1995. Inheritance and the conditional strategy. *American Zoologist* **24**, 385–396.

Gross, M. R. and Repka, J. 1998a. Game theory and inheritance in the conditional strategy. In L. Dugatkin and H. K. Reeve (eds.) *Game Theory and Animal Behavior*, pp. 168–187. Oxford, UK: Oxford University Press.

Gross, M. R. and Repka, J. 1998b. Stability with inheritance in the conditional strategy. *Journal of Theoretical Biology* **192**, 445–453.

Gwynne, D. T. 1980. Female defence polygyny in the bumblebee wolf, *Philanthus bicinctus* (Hymenoptera: Sphecidae). *Behavioral Ecology and Sociobiology* **7**, 213–225.

Gwynne, D. T. 1983. Male nutritional investment and the evolution of sexual differences in Tettigoniidae and other Orthoptera. In D. T. Gwynne and G. K. Morris (eds.) *Orthopteran Mating Systems: Sexual Competition in a Diverse Group of Insects*, pp. 337–366. Boulder, CO: Westview Press.

Gwynne, D. T. 1984. Sexual selection and sexual differences in mormon crickets (Orthoptera, Tettigoniidae, *Anabrus simplex*). *Evolution* **38**, 1011–1022.

Gwynne, D. T. 2001. *Katydids and Bush-Crickets: Reproductive Behavior and Evolution of the Tettigoniidae*. Ithaca, NY: Cornell University Press.

Gwynne, D. T. 2002. A secondary copulatory structure in a female insect: a clasp for a nuptial meal? *Naturwissenschaften* **89**, 125–129.

Gwynne, D. T. and Bailey, W. J. 1999. Female–female competition in katydids: sexual selection for increased sensitivity to a male signal. *Evolution* **53**, 546–551.

Gwynne, D. T. and Jamieson, I. G. 1998. Sexual selection and sexual dimorphism in a harem-polygynous insect, the alpine weta (*Hemideina maori*, Orthoptera, Stenopelmatidae). *Ethology Ecology and Evolution* **10**, 393–402.

Hamilton, W. D. 1979. Wingless and fighting males in fig wasps and other insects. In M. S. Blum and N. A. Blum (eds.) *Sexual Selection and Reproductive Competition in Insects*, pp. 167–220. New York: Academic Press.

Hanley, R. S. 2000. Mandibular allometry and male dimorphism in a group of obligately mycophagous beetles (Insecta: Coleoptera: Staphylinidae: Oxyporinae). *Biological Journal of the Linnean Society* 72, 451–459.

Harari, A. R., Landolt, P. J., O'Brien, C. W., and Brockmann, H. J. 2003. Prolonged mate guarding and sperm competition in the weevil *Diaprepes abbreviatus* (L.). *Behavioral Ecology* 14, 89–96.

Hastings, J. M. 1989. The influence of size, age, and residency status on territory defense in male western cicada killer wasps (*Sphecius grandis*, Hymenoptera: Sphecidae). *Journal of the Kansas Entomological Society* 62, 363–373.

Hayashi, K. 1985. Alternative mating strategies in the water strider *Gerris elongatus* (Heteroptera, Gerridae). *Behavioral Ecology and Sociobiology* 16, 301–306.

Hedrick, A. V. 1988. Female choice and the heritability of attractive male traits: an empirical study. *American Naturalist* 132, 267–276.

Hedrick, A. V. and Dill, L. M. 1993. Mate choice by female crickets is influenced by predation risk. *Animal Behaviour* 46, 193–196.

Heinze, J. and Hölldobler, B. 1993. Fighting for a harem of queens: physiology of reproduction in *Cardiocondyla* male ants. *Proceedings of the National Academy of Sciences of the United States of America* 90, 8412–8414.

Heinze, J., Kuhnholz, S., Schilder, K., and Hölldobler, B. 1993. Behavior of ergatoid males in the ant, *Cardiocondyla nuda*. *Insectes Sociaux* 40, 273–282.

Heinze, J., Hölldobler, B., and Yamauchi, K. 1998. Male competition in *Cardiocondyla* ants. *Behavioral Ecology and Sociobiology* 42, 239–246.

Heinze, J., Schrempf, A., Seifert, B., and Tinaut, A. 2002. Queen morphology and dispersal tactics in the ant, *Cardiocondyla batesii*. *Insectes Sociaux* 49, 129–132.

Heinze, J., Bottcher, A., and Cremer, S. 2004. Production of winged and wingless males in the ant, *Cardiocondyla minutior*. *Insectes Sociaux* 51, 275–278.

Hendrichs, J., Katsoyannos, B. I., Wornoayporn, V., and Hendrichs, M. A. 1994. Odour-mediated foraging by yellowjacket wasps (Hymenoptera: Vespidae): predation on leks of pheromone-calling Mediterranean fruit fly males (Diptera: Tephritidae). *Oecologia* 99, 88–94.

Hernández, M. I. M. and Benson, W. W. 1998. Small-male advantage in the territorial tropical butterfly *Heliconius sara* (Nymphalidae): a paradoxical strategy? *Animal Behaviour* 56, 533–540.

Herre, E. A., West, S. A., Cook, J. M., Compton, S. G., and Kjellberg, F. 1997. Fig-associated wasps: pollinators and parasites, sex-ratio adjustment and male polymorphism, population structure and its consequences. In J. Choe and B. J. Crespi (eds.) *The Evolution of Mating Systems in Insects and Arachnids*, pp. 226–239. Cambridge, UK: Cambridge University Press.

Higashi, K. and Nomakuchi, S. 1997. Alternative mating tactics and aggressive male interactions in *Mnais nawai* Yamamoto (Zygoptera: Calopterygidae). *Odontologica* 26, 159–169.

Hoffmann, A. A. and Cacoyianni, Z. 1990. Territoriality in *Drosophila melanogaster* as a conditional strategy. *Animal Behaviour* 40, 526–537.

Holtmeier, C. and Zera, A. J. 1993. Differential mating success of male wing morphs of the cricket, *Gryllus rubens*. *American Midland Naturalist* 129, 223–233.

Hooper, R. E., Tsubaki, Y., and Siva-Jothy, T. 1999. Expression of a costly, plastic secondary sexual trait is correlated with age and condition in a damselfly with two male morphs. *Physiological Entomology* 24, 364–369.

Houston, T. F. 1970. Discovery of an apparent soldier caste in a nest of a halictine bee (Hymenoptera: Halictidae), with notes on the nest. *Australian Journal of Zoology* 18, 345–351.

Howard, R. 1978. The evolution of mating strategies in bullfrogs, *Rana catesbeiana*. *Evolution* 32, 850–871.

Hunt, J. and Simmons, L. W. 2001. Status-dependent selection in the dimorphic beetle *Onthophagus taurus*. *Proceedings of the Royal Society of London B* 268, 2409–2414.

Jabłoński, P. and Vepsäläinen, K. 1995. Conflict between the sexes in the water strider, *Gerris lacustris*: a test of two hypotheses for male guarding behavior. *Behavioral Ecology* 6, 388–396.

Jennions, M. D. and Petrie, M. 1997. Variation in mate choice and mating preferences: a review of causes and consequences. *Biological Review* 72, 283–327.

Johnson, C. 1964. The inheritance of female dimorphism in the damselfly, *Ischnura damula*. *Genetics* 49, 513–519.

Johnson, C. 1966. Genetics of female dimorphism in *Ischnura demorsa*. *Heredity* 21, 453–459.

Johnson, L. K. 1982. Sexual selection in a brentid weevil. *Evolution* 36, 251–262.

Jones, A. G. 2002. The evolution of alternative cryptic female choice strategies in age-structured populations. *Evolution* 56, 2530–2536.

Joseph, K. J. 1994. Sexual dimorphism and intra-sex variations in the elephant dung beetle, *Heliocopris dominus* (Coprinae: Scarabaeidae). *Entomon* 19, 165–168.

Jousselin, E., van Noort, S., and Greeff, J.M. 2004. Labile male morphology and intraspecific male polymorphism in the *Philotrypesis* fig wasps. *Molecular Phylogenetics and Evolution* 33, 706–718.

Kaitala, A. and Dingle, H. 1993. Wing dimorphism, territoriality and mating frequency of the waterstrider, *Aquarius remegis*. *Annales Zoologici Fennici* 30, 163–168.

Kawano, K. 1995. Horn and wing allometry and male dimorphism in giant rhinoceros beetles (Coleoptera: Scarabaeidae) of tropical Asia and America. *Annals of the Entomological Society of America* 88, 92–99.

Kearns, C.W. 1934. Method of wing inheritance in *Cephalonomia gallicola* Ashmead (Bethylidae: Hymenoptera). *Annals of the Entomological Society of America* 27, 533–539.

Kemp, D.J. 2001. Investigating the consistency of mate-locating behavior in the territorial butterfly *Hypolimnas bolina* (Lepidoptera: Nymphalidae). *Journal of Insect Behavior* 14, 129–147.

Kinomura, K. and Yamauchi, K. 1987. Fighting and mating behaviors of dimorphic males in the ant *Cardiocondyla wroughtoni*. *Journal of Ethology* 5, 75–81.

Knoppien, P. 1985. Rare male mating advantage: a review. *Biological Reviews* 60, 81–117.

Kon, M., Otsuka, K., and Hidaka, T. 1986. Mating system of *Tokunagayusurika akamusi* (Diptera: Chironomidae). 1. Copulation in the air by swarming and on the ground by searching. *Journal of Ethology* 4, 49–58.

Koning, J.W. and Jamieson, I.G. 2001. Variation in size of male weaponry in a harem-defence polygynous insect, the mountain stone weta *Hemideina maori* (Orthoptera: Anostostomatidae). *New Zealand Journal of Zoology* 28, 109–117.

Kotiaho, J.S. and Simmons, L.W. 2003. Longevity cost of reproduction for males but no longevity cost of mating or courtship for females in the male-dimorphic dung beetle *Onthophagus binodis*. *Journal of Insect Physiology* 49, 817–822.

Krupa, J.J. and Sih, A. 1993. Experimental studies on water strider mating dynamics: spatial variation in density and sex ratio. *Behavioral Ecology and Sociobiology* 33, 107–120.

Krupa, J.J., Leopold, W.R., and Sih, A. 1990. Avoidance of male giant water striders by females. *Behaviour* 115, 247–253.

Kukuk, P.F. 1996. Male dimorphism in *Lasioglosum* (Chilalictus) *hemichalceum*: the role of larval nutrition. *Journal of the Kansas Entomological Society* 69, 147–157.

Kukuk, P.F. and Schwarz, M.P. 1987. Intranest behavior of the communal sweat bee *Lasioglossum* (*Chilallictus*) *erythrurum* (Hymenoptera: Halictidae). *Journal of the Kansas Entomological Society* 60, 58–64.

Kukuk, P.F. and Schwarz, M.P. 1988. Macrocephalic male bees as functional reproductives and probable guards. *Pan-Pacific Entomologist* 64, 131–137.

Kurczewski, F.E. and Spofford, M.G. 1998. Alternative nesting strategies in *Ammophila urnaria* (Hymenoptera: Sphecidae). *Journal of Natural History* 32, 99–106.

Lamb, R.J. 1976. Polymorphisms among males of the European earwig *Forficula auricularia* (Dermaptera: Forficulidae). *Canadian Entomologist* 108, 69–75.

Langellotto, G.A., Denno, R.F., and Ott, J.R. 2000. A trade-off between flight capability and reproduction in males of wing-dimorphic insect. *Ecology* 81, 865–875.

Lauer, M.J. 1996. Effect of sperm depletion and starvation on the mating behavior of the water strider, *Aquarius remigis*. *Behavioral Ecology and Sociobiology* 38, 89–96.

Lauer, M.J., Sih, A., and Krupa, J.J. 1996. Male density, female density and intersexual conflict in a stream-dwelling insect. *Animal Behaviour* 52, 929–939.

Lawrence, W.S. 1986. Male choice and competition in *Tetraopes tetraophthalmus*: effects of local sex ratio variation. *Behavioral Ecology and Sociobiology* 18, 289–296.

Lawrence, W.S. 1987. Dispersal: an alternative mating tactic conditional on sex ratio and body size. *Behavioral Ecology and Sociobiology* 21, 367–373.

Leisnham, P.T. and Jamieson, I.G. 2004. Relationship between male head size and mating opportunity in the harem-defense, polygynous tree weta *Hemideina maori* (Orthoptera: Anostostomatidae). *New Zealand Journal of Ecology* 28, 49–54.

Lloyd, J.E. 1979. Sexual selection in luminescent beetles. In M.S. Blum and N.A. Blum (eds.) *Sexual Selection and Reproductive Competition in Insects*, pp. 293–342. New York: Academic Press.

Lloyd, J.E. 1980. Male *Photuris* fireflies mimic sexual signals of their females' prey. *Science* 210, 669–671.

Longair, R.W. 2004. Tusked males, male dimorphism and nesting behavior in a subsocial afrotropical wasp, *Synagris cornuta*, and weapons and dimorphism in the genus (Hymenoptera: Vespidae: Eumeninae). *Journal of the Kansas Entomological Society* 77, 528–557.

Luttbeg, B. 2004. Female mate assessment and choice behavior affect the frequency of male mating tactics. *Behavioral Ecology* 15, 239–247.

Maier, C. T. and Waldbauer, G. P. 1979. Dual mate-seeking strategies in male syrphid flies (Diptera: Syrphidae). *Annals of the Entomological Society of America* 72, 54–61.

Mangan, R. L. 1979. Reproductive behavior of the cactus fly, *Odontoloxozus longicornis*, male territoriality and female guarding as adaptive strategies. *Behavioral Ecology and Sociobiology* 4, 265–278.

McLachlan, A. J. and Neems, R. 1989. An alternative mating system in small male insects. *Ecological Entomology* 14, 85–91.

Milinski, M. and Parker, G. A. 1991. Competition for resources. In J. R. Krebs and N. B. Davies (eds.) *Behavioral Ecology: An Evolutionary Approach*, pp. 122–147. Oxford, UK: Blackwell Scientific.

Miller, P. L. 1984. Alternative reproductive routines in a small fly, *Puliciphora borinquenensis* (Diptera: Phoridae). *Ecological Entomology* 9, 293–302.

Moczek, A. P. 2003. The behavioral ecology of threshold evolution in a polyphenic beetle. *Behavioral Ecology* 14, 841–854.

Moczek, A. P. and Emlen, D. J. 2000. Male horn dimorphism in the scarab beetle *Onthophagus taurus*: do alternative tactics favor alternative phenotypes? *Animal Behaviour* 59, 459–466.

Moczek, A. P. and Nijhout, H. F. 2003. Rapid evolution of a polyphenic threshold. *Evolution and Development* 5, 259–268.

Moczek, A. P., Hunt, J., Emlen, D. J., and Simmons, L. W. 2002. Threshold evolution in exotic populations of a polyphenic beetle. *Evolutionary Ecology Research* 4, 587–601.

Mole, S. and Zera, A. J. 1992. Differential allocation of resources underlies the dispersal-reproduction trade-off in the wing-dimorphic cricket, *Gryllus rubens*. *Oecologia* 93, 121–127.

Moore, A. J. and Wilson, P. 1993. The evolution of sexually dimorphic earwig forceps: social interactions among adults of the toothed earwig *Vostox apicedentatus*. *Behavioral Ecology* 4, 40–56.

Moore, J. C., Pienaar, J., and Greeff, J. M. 2004. Male morphological variation and the determinants of body size in two otitesselline fig wasps. *Behavioral Ecology* 15, 735–741.

Müller, J. K., Eggert, A. K., and Dressel, J. 1990. Intraspecific brood parasitism in the burying beetle, *Necrophorus vespilloides* (Coleoptera: Silphidae). *Animal Behaviour* 40, 491–499.

Neff, B. D. and Danforth, B. N. 1991. The nesting and foraging behavior of *Perdita texana* (Cresson)

(Hymenoptera: Andrenidae). *Journal of the Kansas Entomological Society* 64, 394–405.

Nielsen, M. G. and Watt, W. B. 2000. Interference competition and sexual selection promote polymorphism in *Colias* (Lepidoptera, Pieridae). *Functional Ecology* 14, 718–730.

Nijhout, H. F. 1999. Control mechanisms of polyphenic development in insects. *BioScience* 49, 181–192.

Nijhout, H. F. 2003. Development and evolution of adaptive polyphenisms. *Evolution and Development* 5, 9–18.

Nomakuchi, S. and Higashi, K. 1996. Competitive habitat utilization in the damselfly *Mnais nawai* (Zygoptera: Calopterygidae) coexisting with a related species, *Mnais pruinosa*. *Researches on Population Ecology* 38, 41–50.

Nonacs, P. and Reeve, H. K. 1995. The ecology of cooperation in wasps: causes and consequences of alternative reproductive decisions. *Ecology* 76, 953–967.

O'Neill, K. M. 1983. The significance of body size in territorial interactions of male beewolves (Hymenoptera: Sphecidae, *Philanthus*). *Animal Behaviour* 31, 404–411.

O'Neill, K. M. 2001. *Solitary Wasps: Behavior and Natural History*. Ithaca, NY: Cornell University Press.

O'Neill, K. M. and Evans, H. E. 1981. Predation on conspecific males by females of the beewolf *Philanthus basilaris* Cresson (Hymenoptera: Sphecidae). *Journal of the Kansas Entomological Society* 54, 553–556.

O'Neill, K. M. and Evans, H. E. 1983a. Alternative male mating tactics in *Bembecinus quinquespinosus* (Hymenoptera: Sphecidae): correlations with size and color variation. *Behavioral Ecology and Sociobiology* 14, 39–46.

O'Neill, K. M. and Evans, H. E. 1983b. Body size and alternative mating tactics in the beewolf *Philanthus zebratus* (Hymenoptera: Sphecidae). *Biological Journal of the Linnean Society* 20, 39–46.

O'Neill, K. M., Evans, H. E., and O'Neill, R. P. 1989. Phenotypic correlates of mating success in the sand wasp *Bembecinus quinquespinosus* (Hymenoptera: Sphecidae). *Canadian Journal of Zoology* 67, 2557–2568.

Otronen, M. 1984a. The effect of differences in body size on the male territorial system of the fly *Dryomyza anilis*. *Animal Behaviour* 32, 882–890.

Otronen, M. 1984b. Male contests for territories and females in the fly *Dryomyza anilis*. *Animal Behaviour* 32, 891–898.

Otte, D. 1981. *The North American Grasshoppers*, vol. 1, Acrididae. Cambridge, MA: Harvard University Press.

Parker, G. A. 1970. The reproductive behavior and the nature of sexual selection in *Scatophaga stercoraria*. 2. The fertilization rate and the spatial and temporal relationships of each around the site of mating and oviposition. *Journal of Animal Ecology* **39**, 205–228.

Parker, G. A. 1974. Courtship persistence and female-guarding as male time investment strategies. *Behaviour* **48**, 157–184.

Parker, G. A. 1978. Evolution of competitive mate searching. *Annual Review of Entomology* **23**, 173–196.

Parker, G. A. 1982. Phenotype-limited evolutionarily stable strategies. In King's College Sociobiology Group (eds.) *Current Problems in Sociobiology*, pp. 173–201. Cambridge, UK: Cambridge University Press.

Parker, G. A. and Sutherland, W. J. 1986. Ideal free distribution when individuals differ in competitive ability: phenotype limited ideal-free models. *Animal Behaviour* **34**, 1222–1242.

Partridge, L. 1988. The rare-male effect; what is its evolutionary significance? *Philosophical Transactions of the Royal Society of London B* **319**, 525–539.

Peschke, K. 1987. Male aggression, female mimicry and female choice in the rove beetle, *Aleochara curtula*. *Ethology* **75**, 265–284.

Pienaar, J. and Greeff, J. M. 2003a. Different male morphs of *Otitesella pseudoserrata* fig wasps have equal fitness but are not determined by different alleles. *Ecology Letters* **6**, 286–289.

Pienaar, J. and Greeff, J. M. 2003b. Maternal control of offspring sex and male morphology in the *Otitesella* fig wasps. *Journal of Evolutionary Biology* **16**, 244–253.

Pinto, J. D. 1975. Intra- and interspecific courtship behavior in blister beetles of the genus *Tegrodera* (Meloidae). *Annals of the Entomological Society of America* **68**, 275–285.

Plaistow, S. and Siva-Jothy, T. 1996. Energetic constraints and male mate-securing tactics in the damselfly *Calopteryx splendens xanthostoma* (Charpentier). *Proceedings of the Royal Society of London B* **263**, 1233–1238.

Preston-Mafham, K. G. 1999. Courtship and mating in *Empis* (*Xanthempis*) *trigramma* Meig., *E. tessellata* F. and *E.* (*Polyblepharis*) *opaca* F. (Diptera: Empididae) and the possible implications of "cheating" behaviour. *Journal of Zoology (London)* **247**, 239–246.

Prestwich, K. N. 1994. The energetics of acoustic signaling in anurans and insects. *American Zoologist* **34**, 625–643.

Prokopy, R. J. and Hendrichs, J. 1979. Mating behavior of *Ceratitis capitata* on a field-caged host tree. *Annals of the Entomological Society of America* **72**, 642–648.

Radesäter, T. and Halldórsdóttir, H. 1993. Two male types of the common earwig: male–male competition and mating success. *Ethology* **95**, 89–96.

Ramirez, W. and Marsh, P. M. 1996. Review of the genus *Psenobolus* (Hymenoptera: Braconidae) from Costa Rica, an inquiline fig wasp with brachypterous males, with descriptions of two new species. *Journal of Hymenoptera Research* **5**, 64–72.

Rasmussen, J. L. 1994. The influence of horn and body size on the reproductive behavior of the horned rainbow scarab beetle *Phanaeus difformis* (Coleoptera: Scarabaeidae). *Journal of Insect Behavior* **7**, 67–82.

Repka, J. and Gross, M. R. 1995. The evolutionarily stable strategy under individual condition and tactic frequency. *Journal of Theoretical Biology* **176**, 27–31.

Rice, W. R. 1996. Sexually antagonistic male adaptation triggered by experimental arrest of female evolution. *Nature* **381**, 232–234.

Robertson, H. 1985. Female dimorphism and mating behaviour in a damselfly, *Ischnura ramburi*: females mimicking males. *Animal Behaviour* **33**, 805–809.

Robinson, G. E. 1992. Regulation of division of labor in insect societies. *Annual Review of Entomology* **37**, 637–665.

Roff, D. A. 1986. The evolution of wing dimorphism in insects. *Evolution* **40**, 1009–1020.

Roff, D. A. 1996. The evolution of threshold traits in animals. *Quarterly Review of Biology* **71**, 3–35.

Rowe, L., Arnqvist, G., Sih, A., and Krupa, J. J. 1994. Sexual conflict and the evolutionary ecology of mating patterns: water striders as a model system. *Trends in Ecology and Evolution* **9**, 289–293.

Rowell, G. A. and Cade, W. H. 1993. Simulation of alternative male reproductive behavior: calling and satellite behavior in field crickets. *Ecological Modelling* **65**, 265–280.

Rowland, J. M. 2003. Male horn dimorphism, phylogeny and systematics of rhinoceros beetles of the genus *Xylotrupes* (Scarabaeidae: Coleoptera). *Australian Journal of Zoology* **51**, 213–258.

Rubenstein, D. I. 1984. Resource acquisition and alternative mating strategies in water striders. *American Zoologist* **24**, 345–353.

Rüppell, O. and Heinze, J. 1999. Alternative reproductive tactics in females: the case of size polymorphism in winged ant queens. *Insectes Sociaux* **46**, 6–17.

Rüppell, O., Heinze, J., and Hölldobler, B. 1998. Size-dimorphism in the queens of the North American ant *Leptothorax rugatulus* (Emery). *Insectes Sociaux* **45**, 67–77.

Rüppell, O., Heinze, J., and Hölldobler, B. 2001. Alternative reproductive tactics in the queen-size dimorphic ant *Leptothorax rugatulus* (Emery) and their consequences for genetic population structure. *Behavioral Ecology and Sociobiology* 50, 189–197.

Sakaluk, S. K. and Belwood, J. J. 1984. Gecko phonotaxis to cricket calling song: a case of satellite predation. *Animal Behaviour* 32, 659–662.

Sakaluk, S. K., Bangert, P. J., Eggert, A. K., Gack, C., and Swanson, L. V. 1995. The gin trap as a device facilitating coercive mating in sagebrush crickets. *Proceedings of the Royal Society of London B* 261, 65–72.

Salt, G. 1937. The egg parasite of *Sialis lutaria*: a study of the influence of the host upon a dimorphic parasite. *Parasitology* 29, 539–553.

Salt, G. 1952. Trimorphism in the ichneumonid parasite *Gelis corruptor*. *Quarterly Journal of Microscopy Science* 93, 453–474.

Sauer, K. P., Lubjuhn, T., Sindern, J., *et al*. 1998. Mating system and sexual selection in the scorpionfly, *Panorpa vulgaris* (Mecoptera: Panorpidae). *Naturwissenschaften* 85, 219–228.

Schmieder, R. G. 1933. The polymorphic forms of *Melittobia chalybii* Ashmead and the determining factors involved in their production (Hymenoptera: Chalcidoidea, Eulophidae). *Biological Bulletin* 65, 338–354.

Schöne, H. and Tengö, J. 1981. Competition of males, courtship behaviour and chemical communication in the digger wasp *Bembix rostrata* (Hymenoptera, Sphecidae). *Behaviour* 77, 44–66.

Scott, J. A. 1974a. Adult behavior and population biology of *Poladryas minuta*, and the relationship of the Texas and Colorado populations. *Pan-Pacific Entomologist* 50, 9–22.

Scott, J. A. 1974b. Mate-locating behavior of butterflies. *American Midland Naturalist* 91, 103–117.

Severinghaus, I. L., Kurtak, B. H., and Eickwort, G. C. 1981. The reproductive behavior of *Anthidium manicatum* (Hymenoptera: Megachilidae) and the significance of size for territorial males. *Behavioral Ecology and Sociobiology* 9, 51–58.

Shelly, T. E. and Greenfield, M. D. 1985. Alternative mating strategies in a desert grasshopper: a transitional analysis. *Animal Behaviour* 33, 1211–1222.

Shelly, T. E. and Greenfield, M. D. 1989. Satellites and transients: ecological constraints on alternative mating tactics in male grasshoppers. *Behaviour* 109, 200–221.

Shelly, T. E., Greenfield, M. D., and Downum, K. R. 1987. Variation in host plant quality: influences on the mating system of a desert grasshopper. *Animal Behaviour* 35, 1200–1209.

Sherratt, T. N. 2001. The evolution of female-limited polymorphisms in damselflies: a signal detection model. *Ecology Letters* 4, 22–29.

Shuster, S. M. and Wade, M. J. 2003. *Mating Systems and Strategies*. Princeton, NJ: Princeton University Press.

Sih, A. and Krupa, J. J. 1995. Interacting effects of predation risk and male and female density on male/female conflicts and mating dynamics of stream water striders. *Behavioral Ecology* 6, 316–325.

Simmons, L. and Gwynne, D. T. 1991. The refractory period of female katydids (Orthoptera: Tettigoniidae): sexual conflict over the remating interval? *Behavioral Ecology* 2, 276–282.

Simmons, L. W. 2001. *Sperm Competition and Its Evolutionary Consequences in Insects*. Princeton, NJ: Princeton University Press.

Simmons, L. W., Tomkins, J. L., and Alcock, J. 2000. Can minor males of Dawson's burrowing bee, *Amegilla dawsoni* (Hymenoptera: Anthophorini) compensate for reduced access to virgin females through sperm competition? *Behavioral Ecology* 11, 319–325.

Simmons, L. W., Beveridge, M., and Krauss, S. 2004. Genetic analysis of parentage within experimental populations of a male dimorphic beetle, *Onthophagus taurus*, using amplified fragment length polymorphism. *Behavioral Ecology and Sociobiology* 57, 164–173.

Sirot, L. K. and Brockmann, H. J. 2001. Costs of sexual interactions to females in Rambur's forktail damselfly, *Ischnura ramburi* (Zygoptera: Coenagrionidae). *Animal Behaviour* 61, 415–424.

Sirot, L. K., Brockmann, H. J., Marinis, C., and Muschett, G. 2003. Maintenance of a female-limited polymorphism in *Ischnura ramburi* (Zygoptera: Coenagrionidae). *Animal Behaviour* 66, 763–775.

Siva-Jothy, T. 1987. Mate securing tactics and the cost of fighting in the Japanese horned beetle, *Allomyrina dichotoma* L. (Scarabaeidae). *Journal of Ethology* 5, 165–172.

Smith, D. C. and Prokopy, R. J. 1980. Mating behavior of *Rhagoletis pomonella* (Diptera, Tephritidae). 6. Site of early season encounters. *Canadian Entomologist* 112, 585–590.

Spence, J. R. and Wilcox, R. S. 1986. The mating system of two hybridizing species of water striders (Gerridae). 2. Alternative tactics of males and females. *Behavioral Ecology and Sociobiology* 19, 87–95.

Starks, P. T. 1998. A novel "sit and wait" reproductive strategy in social wasps. *Proceedings of the Royal Society of London B* 265, 1407–1410.

Starks, P. T. 2001. Alternative reproductive tactics in the paper wasp *Polistes dominulus* with specific focus on the sit-and-wait tactic. *Annales Zoologici Fennici* 38, 189–199.

Starks, P. T. and Reeve, H. K. 1999. Condition-based alternative reproductive tactics in the wool-carder bee, *Anthidium manicatum*. *Ethology Ecology and Evolution* 11, 71–75.

Sutherland, W. J. and Parker, G. A. 1985. Distribution of unequal competitors. In R. M. Sibly and R. H. Smith (eds.) *Behavioural Ecology: Ecological Consequences of Adaptive Behaviour*, pp. 255–274. Oxford, UK: Blackwell Scientifics.

Svensson, E. I., Abbott, J., and Hardling, R. 2005. Female polymorphism, frequency dependence, and rapid evolutionary dynamics in natural populations. *American Naturalist* 165, 567–576.

Taborsky, M. 1997. Bourgeois and parasitic tactics: do we need collective, functional terms for alternative reproductive behaviours? *Behavioral Ecology and Sociobiology* 41, 361–362.

Takamura, K. 1999. Wing length and asymmetry of male *Tokunagayusurika akamusi* chironomid midges using alternative mating tactics. *Behavioral Ecology* 10, 498–503.

Tallamy, D. W. and Horton, L. A. 1990. Costs and benefits of the egg-dumping alternative in *Gargaphia* lace bugs (Hemiptera: Tingidae). *Animal Behaviour* 39, 352–359.

Thornhill, R. 1976. Sexual selection and nuptial feeding behavior in *Bittacus apicalis* (Insecta: Mecoptera). *American Naturalist* 110, 529–548.

Thornhill, R. 1979a. Adaptive female-mimicking behavior in a scorpionfly. *Science* 205, 412–414.

Thornhill, R. 1979b. Male and female sexual selection and the evolution of mating strategies in insects. In M. S. Blum and N. A. Blum (eds.) *Sexual Selection and Reproductive Competition in Insects*, pp. 81–121. New York: Academic Press.

Thornhill, R. 1980. Rape in *Panorpa* scorpionflies and a general rape hypothesis. *Animal Behaviour* 28, 52–59.

Thornhill, R. 1981. *Panorpa* (Mecoptera: Panorpidae) scorpionflies: systems for understanding resource-defense polygyny and alternative male reproductive efforts. *Annual Review of Ecology and Systematics* 12, 355–386.

Thornhill, R. 1984. Alternative female choice tactics in the scorpionfly *Hylobittacus apicalis* (Mecoptera) and their implications. *American Zoologist* 24, 367–383.

Thornhill, R. and Alcock, J. 1983. *The Evolution of Insect Mating Systems*. Cambridge, MA: Harvard University Press.

Thornhill, R. and Sauer, K. P. 1991. The notal organ of the scorpionfly (*Panorpa vulgaris*): an adaptation to coerce mating duration. *Behavioral Ecology* 2, 156–164.

Tomkins, J. L. 1999. Environmental and genetic determinants of the male forceps length dimorphism in the European earwig *Forficula auricularia* L. *Behavioral Ecology and Sociobiology* 47, 1–8.

Tomkins, J. L. and Brown, G. S. 2004. Population density drives the local evolution of a threshold dimorphism. *Nature* 431, 1099–1103.

Tomkins, J. L. and Simmons, L. W. 1996. Dimorphisms and fluctuating asymmetry in the forceps of male earwigs. *Journal of Evolutionary Biology* 9, 753–770.

Tomkins, J. L. and Simmons, L. W. 1998. Female choice and manipulations of forceps size and symmetry in the earwig *Forficula auricularia* L. *Animal Behaviour* 56, 347–356.

Tomkins, J. L. and Simmons, L. W. 2000. Sperm competition games played by dimorphic male beetles: fertilization gains with equal mating access. *Proceedings of the Royal Society of London B* 266, 1547–1553.

Tomkins, J. L. and Simmons, L. W. 2002. Measuring relative investment: a case study of testes investment in species with alternative male reproductive tactics. *Animal Behaviour* 63, 1009–1016.

Tomkins, J. L., Simmons, L. W., and Alcock, J. 2001. Brood-provisioning strategies in Dawson's burrowing bee, *Amegilla dawsoni* (Hymenoptera: Anthophorini). *Behavioral Ecology and Sociobiology* 50, 81–89.

Tomkins, J. L., Lebas, N. R., Unrug, J., and Radwan, J. 2004. Testing the status-dependent ESS model: population variation in fighter expression in the mite *Sancassania berlesei*. *Journal of Evolutionary Biology* 17, 1377–1388.

Trumbo, S. T. 1996. Parental care in invertebrates. *Advances in the Study of Behavior* 25, 3–52.

Tsubaki, Y. 2003. The genetic polymorphism linked to mate-securing strategies in the male damselfly *Mnais costalis* Selys (Odonata: Calopterygidae). *Population Ecology* 45, 263–266.

Tsubaki, Y. and Hooper, R. E. 2004. Effects of eugregarine parasites on adult longevity in the polymorphic damselfly *Mnais costalis* Selys. *Ecological Entomology* 29, 361–366.

Tsubaki, Y. and Ono, T. 1986. Competition for territorial sites and alternative mating tactics in the dragonfly, *Nannophya pygmaea* Rambur (Odonata: Libellulidae). *Behaviour* 97, 234–252.

Tsubaki, Y., Hooper, R. E., and Siva-Jothy, T. 1997. Differences in adult and reproductive lifespan in the two male forms of *Mnais pruinosa costalis* (Selys) (Odonata: Calopterygidae). *Researches on Population Ecology* 39, 149–155.

Uéda, T. D. 1979. Plasticity of the reproductive behavior in a dragonfly, *Sympetrum parvulum* Bartenoff, with reference to the social relationsihps of males and the density of territories. *Researches on Population Ecology* **21**, 135–152.

Utida, S. 1972. Density dependent polymorphism in the adult of *Callosobruchus maculatus* (Coleoptera, Bruchidae). *Journal of Stored Products Research* **8**, 111–126.

Van Dyck, H. and Matthysen, E. 1998. Thermoregulatory differences between phenotypes in the speckled wood butterfly: hot perchers and cold patrollers. *Oecologia* **114**, 326–334.

Van Dyck, H. and Wiklund, C. 2002. Seasonal butterfly design: morphological plasticity among three developmental pathways relative to sex, flight and thermoregulation. *Journal of Evolutionary Biology* **15**, 216–225.

Van Dyck, H., Matthysen, E., and Dhondt, A. A. 1997. The effect of wing colour on male behavioural strategies in the speckled wood butterfly. *Animal Behaviour* **53**, 39–51.

Van Staaden, M. J. and Römer, H. 1997. Sexual signaling in bladder grasshoppers: tactical design for maximizing calling range. *Journal of Experimental Biology* **200**, 2597–2608.

Vepsäläinen, K. and Nummelin, M. 1985. Male territoriality in the water strider *Limnoporus rufoscutellatus*. *Annales Zoologici Fennici* **22**, 441–448.

Villalobos, E. M. and Shelly, T. E. 1996. Intraspecific nest parasitism in the sand wasp *Stictia heros* (Fabr.) (Hymenoptera: Sphecidae). *Journal of Insect Behavior* **9**, 105–119.

Waage, J. K. 1973. Reproductive behavior and its relation to territoriality in *Calopteryx maculata* (Beauvois) (Odonata: Calopterygidae). *Behaviour* **47**, 240–256.

Waage, J. K. 1979. Adaptive significance of postcopulatory guarding of mates and non-mates by male *Calopteryx maculata* (Odonata). *Behavioral Ecology and Sociobiology* **6**, 147–154.

Wagner, W. E., Murray, A. M., and Cade, W. H. 1995. Phenotypic variation in the mating preferences of female field crickets, *Gryllus integer*. *Animal Behaviour* **49**, 1269–1281.

Walker, S. E. and Cade, W. H. 2003. A simulation model of the effects of frequency dependence, density dependence and parasitoid flies on the fitness of male field crickets. *Ecological Modelling* **169**, 119–130.

Walker, T. J. 1980. Reproductive behavior and mating success of male short-tailed crickets: differences within and between demes. *Evolutionary Biology* **13**, 219–260.

Walker, T. J. 1983. Mating modes and female choice in short-tailed crickets (*Anugryllus arboreus*). In D. T. Gwynne and G. K. Morris (eds.) *Orthopteran Mating Systems: Sexual*

Competition in a Diverse Group of Insects, pp. 240–267. Boulder, CO: Westview Press.

Walker, T. J. 1987. Wing dimorphism in *Gryllus rubens* (Orthoptera: Gryllidae). *Annals of the Entomological Society of America* **69**, 547–560.

Walker, T. J. and Sivinski, J. M. 1986. Wing dimorphism in field crickets (Orthoptera: Gryllidae). *Annals of the Entomological Society of America* **79**, 84–90.

Waltz, E. C. 1982. Alternative mating tactics and the law of diminishing returns: the satellite threshold model. *Behavioral Ecology and Sociobiology* **10**, 75–83.

Waltz, E. C. and Wolf, L. L. 1984. By Jove! Why do alternative mating tactics assume so many different forms? *American Zoologist* **24**, 333–343.

Waltz, E. C. and Wolf, L. L. 1988. Alternative mating tactics in male white-faced dragonflies (*Leucorhinia intacta*): plasticity of tactical options and consequences for reproductive success. *Evolutionary Ecology* **2**, 205–231.

Warburg, M. S. and Yuval, B. 1997. Effects of energetic reserves on behavioral patterns of Mediterranean fruit flies (Diptera: Tephritidae). *Oecologia* **112**, 314–319.

Ward, P. I. 1993. Females influence sperm storage and use in the yellow dung fly *Scatophaga stercoraria* (L.). *Behavioral Ecology and Sociobiology* **32**, 313–319.

Ward, P. I. 1998. A possible explanation for cryptic female choice in the yellow dung fly, *Scatophaga stercoraria* (L.). *Ethology* **104**, 97–110.

Weall, C. V. and Gilburn, A. S. 2000. Factors influencing the choice of female mate rejection strategies in the seaweed fly *Coelopa nebularum* (Diptera: Coelopidae). *Journal of Insect Behavior* **13**, 539–552.

Wedell, N. 1996. Mate quality affects reproductive effort in a paternally investing species. *American Naturalist* **148**, 1075–1088.

Wendelken, P. W. and Barth, R. H. 1985. On the significance of pseudofemale behavior in the neotropical cockroach genera *Blaberus*, *Archimandrita*, and *Byrsotria*. *Psyche* **92**, 493–503.

West-Eberhard, M. J. 2003. *Developmental Plasticity and Evolution*. Oxford, UK: Oxford University Press.

Wheeler, D. E. 1991. The developmental basis of worker caste polymorphism in ants. *American Naturalist* **138**, 1218–1238.

Wickman, P. O. 1988. Dynamics of mate-searching behaviour in a hilltopping butterfly, *Lasiommata megera* (L.): the effects of weather and male density. *Zoological Journal of the Linnean Society* **93**, 357–377.

Wolf, L. L. and Waltz, E. C. 1988. Oviposition site selection and spatial predictability of female white-faced dragonflies *Leucorrhinia intacta* (Hagen). *Journal of Ethology* **78**, 306–320.

Wolf, L. L. and Waltz, E. C. 1993. Alternative mating tactics in male white-faced dragonflies: experimental evidence for a behavioural assessment ESS. *Animal Behaviour* **46**, 325–334.

Wolfner, M. F. 2002. The gifts that keep on giving: physiological functions and evolutionary dynamics of male seminal proteins in *Drosophila*. *Heredity* **88**, 85–93.

Yamauchi, K. and Kawase, N. 1992. Pheromonal manipulation of workers by a fighting male to kill his rival males in the ant *Cardiocondyla wroughtoni*. *Naturwissenschaften* **79**, 274–276.

Yamauchi, K., Kimura, Y., Corbara, B., Kinomura, K., and Tsuji, K. 1996. Dimorphic ergatoid males and their reproductive behavior in the ponerine ant *Hypoponera bondroiti*. *Insectes Sociaux* **43**, 119–130.

Zera, A. J. 1999. The endocrine genetics of wing polymorphism in *Gryllus*: critique of recent studies and state of the art. *Evolution* **53**, 973–977.

Zera, A. J. and Denno, R. F. 1997. Dispersal polymorphism in insects: integrating physiology, genetics and ecology. *Annual Review of Entomology* **42**, 207–231.

Zera, A. J. and Harshman, L. G. 2001. The physiology of life history trade-offs in animals. *Annual Review of Ecology and Systematics* **32**, 95–126.

Zhao, Z. and Zera, A. J. 2002. Differential lipid biosynthesis underlies a tradeoff between reproduction and flight capability in a wing-polymorphic cricket. *Proceedings of the National Academy of Sciences of the United States of America* **99**, 16829–16834.

Zink, A. G. 2003. Intraspecific brood parasitism as a conditional reproductive tactic in the treehopper *Publilia concava*. *Behavioral Ecology and Sociobiology* **54**, 406–415.

9 · The expression of crustacean mating strategies

STEPHEN M. SHUSTER

CHAPTER SUMMARY

Three fundamental patterns of phenotypic expression exist for alternative mating strategies. These patterns include Mendelian strategies, developmental strategies, and behavioral strategies. Each pattern of expression is revealed by hormonal and neurological factors that regulate the timing and degree to which phenotypic differences appear; however, the nature of each regulatory mechanism depends fundamentally on its underlying mode of inheritance. The genetic architectures underlying such inheritance in turn depend on the circumstances in which mating opportunities arise, including the intensity of selection favoring distinct reproductive morphologies, and the predictability of mating opportunities within individual lifespans. This chapter concerns the nature of this variation and its possible causes, with illustrations from the Crustacea.

9.1 INTRODUCTION

Although crustaceans were among the first recorded examples of alternative mating strategies (*Orchestia darwinii*: Darwin 1874, p. 275; *Tanais* spp.: Darwin 1874, p. 262), there is currently no synthetic treatment of how such polymorphisms are expressed within this group. The apparent scarcity of reports of male polymorphism among crustaceans is unexpected given the frequency with which sexual selection has been demonstrated within this taxon (Holdich 1968, 1971, Manning 1975, Stein 1976, Thompson and Manning 1981, Knowlton 1980, Shuster 1981, Christy 1983, Hatziolos and Caldwell 1983, reviews in Salmon 1984, Koga *et al.* 1993). As explained below, when sexual selection occurs, alternative mating strategies are likely to evolve. This chapter provides an evolutionary framework for understanding the expression of alternative mating strategies, with illustrations from the Crustacea (Table 9.1). My goals are to show that in this fascinating collection of species, all known forms of alternative mating strategies are represented and opportunities for further research abound.

Several frameworks for understanding alternative mating strategies now exist (Gadgil 1972, Maynard Smith 1982, Austad 1984, Dominey 1984, Gross 1985, 1996, Lucas and Howard 1995, Gross and Repka 1998). Because these approaches have focused primarily on behavioral or developmental differences among individuals (that is, on "condition-dependent phenotypes" often called "tactics" (Box 9.1), and because such polymorphisms seldom conform to the simplifying assumptions required by game theory regarding inheritance and fitness, there has been little consensus about the theoretical and empirical approaches best suited for investigating alternative mating strategies and tactics, in the laboratory as well as in nature.

In response to this confusion, Shuster and Wade (2003; see also Hazel *et al.* 1990, Roff 1992, 1996, Sinervo 2000, 2001, Shuster 2002) explained how alternative mating strategies can be understood using conventional evolutionary genetic principles including game theory, provided that the average as well as the variance in fitness among the observed morphs is considered within quantitative analyses. This requirement is necessary because alternative mating strategies evolve in response to sexual selection, an evolutionary context in which fitness variance is often extreme. When fundamental principles are applied, the contexts in which alternative mating strategies evolve as well as the forms these adaptations assume become clear.

This chapter has three parts. First, I will explain the source of sexual selection and how it produces alternative mating strategies in the first place. Second, I will describe Levins' (1968) scheme for understanding polyphenism (the tendency for individuals to express variable phenotypes in response to environmental cues) to show why alternative mating strategies can be understood in this light (see Shuster and Wade 2003). Third, using crustacean examples, I will demonstrate how this approach predicts the

Alternative Reproductive Tactics, ed. Rui F. Oliveira, Michael Taborsky, and H. Jane Brockmann. Published by Cambridge University Press.
© Cambridge University Press 2008.

Taxon	Species	Conventional morph	Alternative morph(s)	Mode of expression	References
Branchiopoda					
Anostraca	*Eubranchipus serratus*	Guard individual females in sequence	Usurper	Behavioral	Belk 1991
Notostraca	*Triops newberryi*	Hermaphrodite	Male	Mendelian	Sassaman 1991
Conchostraca	*Eulimnadia texana*	Hermaphrodite	Male	Mendelian	Sassaman 1989, Sassaman and Weeks 1993, Weeks and Zucker 1999
Maxillipoda					
Copepoda	*Euteropina acutifrons*	Large males	Small males	Mendelian/developmental	Haq 1965, 1972, 1973, D'Apolito and Stancyk 1979, Moreira *et al.* 1983, Moreira and McNamara 1984, Stancyk and Moreira 1988
Malacostraca					
Stomatopoda	*Pseudosquilla ciliata*	Guard individual females in sequence	Usurper	Behavioral	Hatziolos and Caldwell 1983
	Gonodactylus bredini	Guard individual females in sequence	Usurper	Behavioral	Shuster and Caldwell 1989
Decapoda					
Caridea	*Alpheus armatus*	Guard individual or groups of females	Usurper/sneaker?	Behavioral	Knowlton 1980
	Argis dentata	Protandrous hermaphrodites	Primary females	Developmental?	Fréchette *et al.* 1970
	Athanas spp.	Protandrous hermaphrodites	Primary males	Developmental?	Nakashima 1987, Gherardi and Calloni 1993
	Crangon crangon	Protandrous hermaphrodites	Primary females	Developmental?	Boddeke *et al.* 1991
	Exhippolysmata sp.	Protandrous simultaneous hermaphrodites		Developmental?	Bauer 2002
	Lysmata spp.	Protandrous simultaneous hermaphrodites		Developmental	Bauer 2000, Baeza and Bauer 2004
	Macrobrachium dayanum	Guard female groups	Female mimic, sneaker	Developmental	Kuris *et al.* 1987

Table 9.1 (*cont.*)

Taxon	Species	Conventional morph	Alternative morph(s)	Mode of expression	References
	Macrobrachium idea	Guard female groups	Female mimic, sneaker	Developmental	Kuris et al. 1987
	Macrobrachium malcolmsonii	Guard female groups	Female mimic, sneaker	Developmental	Kuris et al. 1987
	Macrobrachium rosenbergii	Guard female groups	Female mimic, sneaker	Developmental	Nagamine et al. 1980, Ra'anan and Sagi 1989, Kuris et al. 1987, Barki et al. 1992, Karplus et al. 2000, Kurup et al. 2000
	Macrobrachium rosenbergii	Large females	Small females	Unknown	Harikrishnan et al. 1999
	Macrobrachium scabriculum	Guard female groups	Female mimic, sneaker	Developmental	Kuris et al. 1987
	Pandalus spp.	Protandrous hermaphrodites	Primary females	Developmental?	Charnov 1979, 1982, Bergström 1997
	Processa edulis	Protandrous hermaphrodites	Primary females	Developmental?	Noël 1976
	Rhynchocinetes typus	Guard individual females in sequence (robustus, intermedius)	Sneaker (intermedius), sperm competitor (typus)	Behavioral	Correa et al. 2000, 2003, Correa and Thiel 2003
	Thor manningi	Protandrous hermaphrodites	Primary males	Developmental?	Chace 1972, Bauer 1986
Brachyura	*Callinectes sapidus*	Guard individual females in sequence	Usurper	Behavioral/developmental	Jivoff and Hines 1998
	Chionoecetes opilio	Smooth spermatophores	Wrinkled spermatophores	Unknown	Moriyasu and Benhalima 1998
	Inachus phalangium	Mate with many females in sequence	Sperm competitors	Behavioral	Diesel 1989
	Libinia emarginata	Guard individual females in sequence (large males)	Usurper (small males)	Developmental	Sagi et al. 1994, Ahl and Laufer 1996
	Pachygrapsus transversus	Guard female groups (large males)	Sneaker (small males)	Behavioral/developmental	Abele et al. 1986
	Scopimera globosa	Guard resources required by females (males occupy burrows)	Usurpers/wanderers (small males)	Behavioral/developmental	Wada 1986, Koga 1998
	Uca spp.	Guard resources required by females (males occupy	Usurpers/wanderers	Behavioral	Salmon and Hyatt 1983, Christy and

Taxon	Species	Dominant tactic	Alternative tactic	Mechanism	References
Palinura	*Jasus edwardsii*	Mate with many females in sequence	Sperm competitor	Behavioral	Salmon 1991, Jennions and Backwell 1998, MacDiarmid and Butler 1999
Dendrobrachiata	*Sicyonia dorsalis*	Mate with many females in sequence	Sperm competitor?	Behavioral	Bauer 1992
Astacidea	*Homarus americanus*	Guard individual females in sequence	Usurper	Behavioral	Cowan and Atema 1990, Cowan 1991
Amphipoda	*Jassa falcata*	Guard individual females in sequence (thumbed males)	Sneaker (thumbless males)	Developmental	Borowsky 1985
	Jassa marmorata	Guard individual females in sequence (thumbed males)	Sneaker (thumbless males)	Developmental	Clark 1997, Kurdziel and Knowles 2002
	Microdeutopus gryllotalpa	Guard individual females in sequence (thumbed males)	Sneaker (thumbless males)	Developmental	Borowsky 1980, 1984, 1989
	Orchestia darwinii	Guard individual females in sequence (thumbed males)	Sneaker (thumbless males)	Developmental?	Darwin 1874
Isopoda	*Elaphognathia cornigera*	Guard female groups (large males)	Sneaker (small males)	Unknown	Tanaka and Aoki 1999, Tanaka 2003
	Idotea baltica	Guard individual females in sequence	Usurper	Behavioral	Jormalainen et al. 1994
	Jaera albifrons	Guard individual females in sequence (males with setose walking legs)	Unknown, males with nonsetose walking legs	Mendelian	Bocquet and Veuille 1973
	Paracerceis sculpta	Guard female groups (α-males)	Female mimic (β-males), sneaker (γ-males)	Mendelian	Shuster 1989, 1992, Shuster and Wade 1991, 2003, Shuster and Sassaman 1997, Shuster and Levy 1999, Shuster et al. 2001
	Thermosphaeroma spp.	Guard individual females in sequence	Usurper	Behavioral	Shuster 1981, Jormalainen and Shuster 1999
Tanadiacea	*Tanais* spp.	Guard individual females in sequence (large males with gnathopods)	Sneaker (small male with small gnathopods)	Developmental?	Darwin 1874

227

Box 9.1   Strategies and tactics

The term "strategy" as defined in evolutionary game theory describes a preprogrammed set of behavioral or life history characteristics (Maynard Smith 1982). Alternative mating strategies can thus be viewed as functional sets of behavior patterns or morphologies that are used by their bearers to acquire mates (Shuster 2002). An evolutionarily stable strategy (ESS: Maynard Smith 1982) is a strategy that persists in a population for one of two reasons: either the average fitness of individuals expressing the ESS *equals* that of all other strategies existing in the population or the average fitness of individuals expressing the ESS *exceeds* that of other strategies that have appeared in the population to date. If a strategy's average fitness is consistently *less* than that of other strategies, it will be removed from the population by selection (Darwin 1874, Maynard Smith 1982, Shuster and Wade 2003). *By definition*, individuals with fitness less than the population average are *selected against*. Thus, a *strategy* is an adaptation whose expression has been *shaped by selection*.

This definition implies that two further assumptions are met. First, genetic variation must underlie such traits. Heritability is required for any trait to change in frequency or be removed from a population as described above. If genetic variation is lacking; that is, if all individuals in the population are presumed to be genetically identical for a given trait (e.g., Eberhard 1979, 1982, Lucas and Howard 1995, Gross 1996), no evolutionary response to selection is possible. Second, stabilizing selection is presumed to refine trait expression. This is the process by which less-fit trait variants are eliminated by selection, more-fit trait variants reproduce, and over time, a trait's function becomes recognizable. Traits with uniformly inferior fitness are usually eliminated from populations *before* their phenotypes can become modified. And as mentioned above, no response to selection is possible unless genetic variation underlies the trait. Thus, stabilizing selection can operate only on *heritable* traits whose average fitness, relative to other similar traits, allows them to persist within the population over time. Stated differently, the average fitnesses of coexisting traits *must be equivalent*. If either of these assumptions is not met, discussions of trait evolution become meaningless.

Recent descriptions of discontinuous variation in mating phenotype have distinguished between genetically distinct "strategies" and phenotypes that represent condition-dependent "tactics" (Gross 1996, Gross and Repka 1998, Correa et al. 2003, Neff 2003, Howard et al. 2004). The term "tactic" is used to describe behavioral or morphological characteristics whose expression is contingent on environmental conditions or on the "status" of the individuals in which they appear. Status-dependent selection (SDS), the term now used to describe how selection may operate on such traits (Gross 1996, Gross and Repka 1998, Denoel et al. 2001, Hunt and Simmons 2001, Taru et al. 2002, Tomkins and Brown 2004), is presumed to allow individuals to assess their potential mating opportunities in terms of their physical condition, social status, or probability of success and then to make behavioral or developmental "decisions" that lead to greater mating success than if the choice had not been made.

According to the SDS hypothesis, dimorphic populations arise because *all* individuals choose one or another status-dependent phenotype. "Status" is presumed to translate into fitness according to a linearly increasing function, with the rate of increase greater for higher status individuals than for lower status individuals. The fitnesses of each phenotype are considered equal only at the intersection of their fitness functions, a location defined as the "switch point" (s^*) (Figure 9.1). Condition-dependent choices appear to cause much of the population to "make the best of a bad job"; that is, to experience inferior mating success compared to individuals of higher status (Eberhard 1979, 1982, Dawkins 1980). Furthermore, according to the SDS hypothesis, *all* individuals in the population are assumed to be genetically monomorphic with respect to their ability to make conditional choices (but see Gross and Repka 1998 and below). This part of the hypothesis salvages the lower fitness of males with apparently lower mating success (Y, Figure 9.1A) because, as stated above, a genetically uniform population *cannot* respond to selection. Thus, in spite of their inability to secure mates, the SDS hypothesis conveniently bends the principles of population genetics to allow inferior phenotypes to persist within populations over time.

Gross and Repka (1998) acknowledged that conditional strategies representing genetic monomorphisms are unlikely to exist due to overwhelming evidence that heritable factors influence trait expression. However, their revised model concluded that the assertions of Gross (1996) were still appropriate and that the SDS hypothesis is the best explanation for the appearance of behavioral polymorphism in nature (see also Hunt and Simmons 2001, Forslund 2003, Tomkins and Brown 2004). But two problems remain with the revised SDS approach. The first difficulty is that it presumes *from the outset* that the average fitnesses of the two tactics considered (fighters and sneakers) are *unequal* (Repka and Gross 1995, p. 28; Gross 1996, p. 93; Gross and Repka 1998, p. 170). As stated above, this premise is evolutionarily untenable.

The assumption of unequal fitnesses among morphs prevents this and related theoretical methods (Lucas and Howard 1995, Repka and Gross 1995, Gross 1996) from considering situations in which the fitnesses of the different morphs *are* equal. It also places severe limits on the potential influence inheritance can have, both on trait expression as well as on how selection may influence trait frequency within the population. Furthermore, genetic monomorphism is still presumed to exist at the switch point (Gross 1996, Gross and Repka 1998), again removing any possibility that selection can influence its position. This issue is *not an assumption of models that consider condition-dependent phenotypes as quantitative genetic polymorphisms* (e.g., Hazel et al. 1990, Roff 1996, Flaxman 2000, Shuster and Wade 2003).

The fitnesses of the two tactics *are* considered equal at the switch point. However, this is merely a consequence of how tactic fitnesses are defined – as linear relationships between phenotype and fitness that happen to have different slopes (Figure 9.1B). The notion of the switch point as it is used in this theoretical approach is inappropriate because it assumes equal fitnesses to exist *only* at the population frequencies described at the switch point. This is contrary to the principles of game theory and population genetics, which state that for polymorphism to persist within a population, the relative fitnesses of the alternative morphs must be equal at *all* population frequencies, not just those occurring at the switch point (Shuster and Wade 2003). But again, assumption of equal fitnesses at the switch point is of little evolutionary consequence anyway because, as mentioned above, genetic variation is presumed to be absent for the polymorphism at this location (Gross and Repka 1998).

The second difficulty with the Gross and Repka (1998) approach is that relationships among the parameters used to estimate the frequency and fitness of the alternative tactics, as well as the proportion of progeny of each type that are transmitted to the next generation, are constrained by the authors *in advance* of the simulations they conduct. Thus, a higher existing frequency of one tactic imposes lower possible values for recruitment and heritability of the other tactic. The apparent goal of these interwoven constraints is to make the influences of each tactic on the other frequency dependent, and, indeed, measurable narrow-sense heritability of quantitative traits does depend on the frequency of the trait within the population (Falconer 1989). However, there is no population genetic precedent for the *inheritability* of traits to rely to such a large degree on their own population frequency, their own fitness, their own rate of recruitment into the population, or on the frequency, fitness, rate of recruitment, or mode of inheritance for another alternative trait.

Contrary to the predictions of the SDS model (Gross 1996, Gross and Repka 1998), considerable evidence already exists indicating that polymorphisms in mating phenotype with flexible expression represent mixtures of evolutionarily stable strategies (e.g., a normal distribution of genetically based reaction norms: Hazel et al. 1990, Roff 1992, 1996, Schlicting and Pigliucci 1998, Flaxman 2000, Shuster and Wade 2003). These results indicate that genetic architectures allowing phenotypic flexibility can persist in populations by frequency-dependent selection, a mechanism functionally identical to the way polymorphisms controlled by Mendelian factors persist in nature. In models of frequency-dependent selection, the inheritability of traits *does not* depend on their frequency in the population as in Gross and Repka (1998). If this condition were imposed, the salient feature of frequency-dependent selection (i.e., the tendency for alternative genotypes to have high relative fitness at low population frequency and low relative fitness at high population frequency) would cease to exist.

Thus, while the term "tactic" is indeed useful for describing phenotypes that are flexible in their expression, as opposed to those controlled by more rigid (e.g., Mendelian) rules, there is no need to distinguish a "strategy," as a phenotype that is *inheritable*, from a "tactic" as a phenotype for which genetic variation is *constrained* or *nonexistent*. Both traits clearly represent adaptations, that are underlain by genetic variation, and that are maintained in populations by selection. When viewed in this light, the term "strategy" is appropriate for *all* evolved polymorphisms in reproductive behavior, regardless of how their expression is controlled.

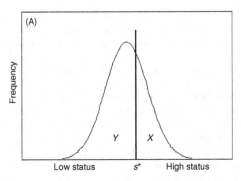

Figure 9.1 The status-dependent selection model of Gross (1996, redrawn). (A) The distribution of male phenotypes in a hypothetical population; Y indicates the fraction of the population with low status and thus which assume the sneaker phenotype, X indicates the fraction of the male population with high status and thus which assume the fighter phenotype; the position of the switchpoint, s^*, is determined by the location of the intersection of the fitness functions for sneakers and fighters in (B). (B) The fitness functions for X and Y phenotypes; slope of the X phenotype is steeper than that of the Y phenotype, at s^* the fitnesses of X and Y phenotypes are presumed to be equal.

three fundamental ways in which alternative mating strategies are expressed, as well as the types of data that may be used for further hypothesis testing. My hope is that this framework will stimulate research on crustacean mating systems, particularly field studies that quantify the source and intensity of sexual selection, as well as laboratory studies that explore the genetic architectures underlying polymorphic mating phenotypes.

9.2   SEXUAL SELECTION AND THE MATING NICHE

Darwin (1874) considered sexual selection to have evolutionary effects functionally similar to those that exist in populations with a surplus of males. He noted that "if each male secures two or more females, many males cannot pair" (Darwin 1874, p. 212). This observation is the primary reason why male and female phenotypes tend to diverge under the influence of sexual selection, and why in many sexual species, including a large number of crustaceans, males and females are sexually dimorphic in appearance.

When some males mate and others do not, a sex difference in fitness variance often appears. This occurs because of the necessary relationship between the mean and variance in male and female fitness in all sexual species (Wade 1979, Shuster and Wade 2003). Although many factors may contribute to "fitness," this concept is least confounded and most easily understood when considered in terms of offspring numbers (Wolf and Wade 2001). Because every offspring has a mother and a father (Fisher 1958), the average number of offspring per male must equal the average number of offspring per female when the sex ratio equals 1 (Wade and Shuster 2002, Shuster and Wade 2003). Also under this condition, the variance in offspring numbers for each sex, that is, the variance in fitness for each sex, will be equivalent if all males and females mate once.

However if some males mate and others do not, as is usually the case, then the average fitness of males who do not mate becomes less than the fitness of the average female, for obvious reasons. Simultaneously, the average fitness of males who do mate equals the average female fitness multiplied by the number of mates that male secures (Wade 1979, Shuster and Wade 2003). As matings by females become clustered with fewer and fewer males, the class of males with no mates and no fitness becomes increasingly larger. In contrast, the class of males who do mate becomes increasingly smaller, but these males secure an increasingly larger fraction of the total offspring produced. Whereas the total variance in female fitness remains unchanged by this process, the divergence of the male population into reproductive "haves" and "have-nots" causes the total variance in male fitness to become very large.

The magnitude of this sex difference in fitness variance provides an estimate of the strength of sexual selection $(V_{W\text{males}} - V_{W\text{females}}$: Shuster and Wade 2003). As the magnitude of this fitness difference becomes larger, sexual selection becomes increasingly intense and male character- istics that promote polygamy are transmitted dis- proportionately to the next generation. This is why, over evolutionary time, males within such populations become modified in their appearance to a greater degree than females. When fitness variance is greater in females than it is in males, as it is in sex-role-reversed species, it is also why females become modified in appearance to a greater degree than males (S. M. Shuster and M. J. Wade 2003, unpub- lished data). Among related species in which sexual selec- tion occurs, this is why the sex in which selection is more intense shows greater phenotypic diversity than the sex in which selection is weaker. Also, within individual species, this is why the sex in which selection is strongest shows a greater tendency to express alternative mating strategies (Gadgil 1972, Shuster and Wade 2003).

Why does this last relationship exist? Why should alternative mating strategies appear within the sex in which sexual selection is strongest? The answer is that alternative mating strategies readily evolve when male mating success becomes uneven among males (or uneven among females in role-reversed species). The existence of uneven mating success among males not only causes sexual selection, as observed by Darwin (1874), it also creates a "mating niche" for males engaging in unconventional mating behavior (Shuster and Wade 1991, 2003).

For example, in many species, small males invade the breeding territories of larger males by avoiding direct competition altogether. Once inside breeding territories, these "sneaker" or "satellite" males surreptitiously mate with receptive females, as occurs in isopods (*Paracerceis sculpta*: Shuster 1992), amphipods (*Microdeutopus gryllotalpa*: Borowsky 1980; *Jassa marmorata*: Clark 1997, Kurdziel and Knowles 2002) and in many decapods (e.g., freshwater prawns, *Macrobrachium* spp.: Ra'anan and Sagi 1989; sand bubbler crabs, *Scopimera globosa*, Koga *et al.* 1993; spider crabs, *Libinia emarginata*: Sagi *et al.* 1994, Ahl and Laufer 1996; and rock shrimp, *Rhynchocinetes typus*: Correa *et al.* 2003). In each of these species, stolen matings appear to provide unconventional or satellite males with only a tiny fraction of the fertilization success gained by those males that defend harems. These satellite males appear to "make the best of a bad job" (Eberhard 1979, Gross 1996). Yet in each of these examples, because unconventional males take fertil- izations away from males whose fertilization success is already disproportionately large, satellite males are more successful at siring offspring than territorial males who secure no mates at all.

Game theory and population genetic analyses agree on the conditions necessary for the invasion and persistence of evolutionarily stable strategies (Maynard Smith 1982, Crow 1986). These conditions are most easily met for males employing alternative strategies (Wade and Shuster 2004) (Box 9.2). The important relationship is this: the larger the average harem size is among conventional males, the larger the fraction of conventional males must be who cannot secure mates. Because the average fitness of conventional males includes the fitness of males who mate as well as the fitness of males who do not mate, the *larger* the average harem size is among conventional males, the *smaller* the fraction of the total fertilizations unconventional males need to acquire within harems for their average fitness to equal the average fitness of *all* conventional males combined (Shuster and Wade 2003, Wade and Shuster 2004) (Box 9.2). Although the average fitness of unconventional males seems inferior to that of conventional males, in fact, the average fitness of uncon- ventional males often *equals or exceeds* the average fitness of all conventional males (see also Shuster and Wade 2003).

9.3 THE EXPRESSION OF ALTERNATIVE MATING STRATEGIES

Levins (1968) proposed that polymorphism can persist in natural populations when selection acts in changing envir- onments. When environments change little, selection is usually weak and phenotypic tolerance is allowed; however, when environments change frequently, selection is stronger, phenotypic tolerance is impermissible, and genetic poly- morphism is expected to arise. Shuster and Wade (2003) argued that such conditions are especially likely when sexual selection occurs. Sexual selection is often extremely strong and circumstances favoring mating success are often highly variable. Thus, they proposed that sexual selection acting in variable environments will most often favor distinct pheno- types and genetic polymorphism. It is important to note that the term "genetic polymorphism" not only refers to single- locus polymorphisms with alleles that segregate according to Mendelian rules, but also describes the normal distribution of genetic factors that influence the expression of condition-dependent patterns in development or behavior (i.e., "tactics") (Shuster and Wade 2003; Box 9.1). The game- theory-inspired concepts of genetically "fixed" pure pheno- types versus genetically monomorphic "conditional" pheno- types, while useful as heuristic devices (Maynard Smith 1982, Gross 1996, Alcock 2005), make little evolutionary sense

---

Box 9.2   Jack-of-All-Harems

Shuster and Wade (2003) showed how to visualize the quantitative relationship between the intensity of sexual selection and the ease with which alternative mating strategies may evolve. If we assume that H is the average mating success of harem-holding males and that satellite males succeed in mating by invading the harems of such males, then the fitness of satellites, W_β, can be expressed as

$$W_\beta = Hs, \qquad (B9.2.1)$$

where s equals the fertilization success of satellite males within the harems of territorial males. Although the fitness of territorial males who successfully secure mates equals H, the average number of mates per male is less than H. This happens because when territorial males acquire a harem containing k females, $k - 1$, other territorial males will be unable to mate at all (Shuster and Wade 2003). To calculate the average success of territorial males *as a class*, it is necessary to consider the distribution of mates among *all* of the males in that class. Thus, the average success of *all* territorial males must be

$$W_a = R, \qquad (B9.2.2)$$

where R, the sex ratio $(= N_\female/N_\male)$, is equal to the distribution of all females over all territorial males. As Shuster and Wade (2003) showed, the condition necessary for satellite males to invade a population of territorial males is

$$W_\beta > W_a. \qquad (B9.2.3)$$

That is, the average fitness of satellite males, W_β, must exceed the fitness of territorial males, W_a. By substitution with Eqs. (B9.2.1) and (B9.2.2), this relationship can also be expressed as

$$Hs > R. \qquad (B9.2.4)$$

If the sex ratio, R, equals 1 (i.e., $Hs > 1$), then by rearrangement, the condition necessary for the invasion of

a polygynous male population by an alternative mating strategy becomes

$$s > 1/H. \qquad (B9.2.5)$$

That is, to invade a population of territorial males, satellite males must obtain a fraction of the total fertilizations in harems, s, that exceeds the reciprocal of the average harem size of *successful* territorial males. To understand this relationship, we need only imagine that the average harem-holding male mates with three females, or $H = 3$. In such circumstances, Eq. (B9.2.5) shows that satellite males need only secure mates one-third as successfully as territorial males to invade this mating system ($s = 0.333$). Thus, on average, satellite males would need only fertilize 1/3 of the clutch of each female, or sire the progeny of 1 of the 3 females in each harem, to invade the population. And, as harem size increases (as females become increasingly clustered around fewer territorial males), the invasion of alternative mating strategies becomes easier still – satellites can be even less successful within harems and still invade because the fraction, $1/H$, becomes smaller with increasing values of H (Figure 9.2).

Shuster and Wade (2003) showed that in a polygamous population, the fraction of nonmating males is $p_0 = 1 - (1/H)$. By rearrangement of this equation, we can see that $1/H = 1 - p_0$. Now, by substitution with Eq. (B9.2.5), it is clear that

$$s > 1 - p_0. \qquad (B9.2.6)$$

This relationship shows the same result as Eq. (B9.2.5) but in a slightly different way. Here, as the fraction of territorial males excluded from mating, p_0, increases, the mating success necessary for satellites to invade this mating system, s, becomes increasingly small. At equilibrium (i.e., $s = 1 - p_0$), this relationship explicitly identifies the fraction of the territorial male population that is excluded from mating, p_0, when territorial and satellite males coexist.

---

when considering how selection might shape phenotypic expression for one obvious reason: traits lacking underlying genetic variation cannot respond to selection and therefore cannot evolve (Shuster and Wade 2003).

Whether the genetic architecture underlying a phenotypic polymorphism will be Mendelian or polygenic

depends on the "environmental grain"; that is, on the relative predictability of environmental change (Levins 1968) (Box 9.3). Shuster and Wade (2003) argued that mobile organisms like animals experience environmental grain primarily on a temporal rather than on a spatial scale. Furthermore, with respect to the evolution of

Box 9.3   Fitness sets and sexual selection

Levins' (1968) proposed that the fitness of each phenotype within a population changes as environmental conditions change and that the distribution of fitness for a specific phenotype can be characterized by the average and the variance in fitness. In these terms, environmentally "tolerant" phenotypes show a broader distribution of fitness in the face of environmental change than environmentally more "sensitive" phenotypes, i.e., the variance in the fitness of a tolerant phenotype, $V_{W(\text{tolerant})}$, is larger than that of a more sensitive phenotype, $V_{W(\text{sensitive})}$ (Figure 9.3) or

$$V_{W(\text{tolerant})} > V_{W(\text{sensitive})}. \qquad (B9.3.1)$$

Because tolerance to variable environments is likely to impose fitness costs, the average fitness of the tolerant phenotype, W_{tolerant}, is less than for the sensitive type, $W_{\text{sensitive}}$ (Figure 9.3) or

$$W_{(\text{tolerant})} < W_{(\text{sensitive})}. \qquad (B9.3.2)$$

Thus, a phenotype that maintains some fitness in marginal environments will be unable to achieve the highest fitness in the more common environment, whereas a phenotype that achieves low fitness in marginal environments will achieve higher fitness in the environment for which it is specialized (Figure 9.3).

A graphical means for identifying the optimal phenotype for a particular environment is obtained by holding environmental conditions fixed and examining performance as a function of phenotype (Levins 1968, Shuster and Wade 2003). This procedure generates a curve describing the distribution of fitness for a given phenotype, i, across a limited range of environments, j. The peak of each curve identifies the optimal phenotype in each subset of environments, and because phenotypes deviating from this optimum have lower performance, fitness decreases symmetrically away from the phenotypic optimum toward zero (Figure 9.3). When the performance curves generated by the two most common environments are considered together (Figure 9.4), the phenotypic *tolerance* of a population can be quantified. Specifically, tolerance (T) is equal to $2d$, where d is the distance in phenotypic performance units from the peak of the distribution to its point of inflection.

The environmental range, E, is the difference in the average phenotypic performances in each environment

$(s_2 - s_1)$ (Figure 9.4). Approximately overlapping performance curves produced by each environment indicate a *tolerant* phenotype; that is, a phenotype whose ability to tolerate environmental change exceeds the range of conditions that usually appear within the environment. In such cases, $T > E$, and the optimal phenotype is approximately similar in each environment (Figure 9.4A). On the other hand, nonoverlapping curves (those in which $T < E$) indicate *intolerant* phenotypes. These phenotypes are favored when the range of environmental conditions is so great that a single phenotype is unable to tolerate all environmental circumstances. Thus, different phenotypes are optimal in each of the most common environments (Figure 9.4B).

When the values of the performance curve in environment 1 are plotted *against* those in environment 2, the familiar shapes of Levins' *fitness sets* appear (Figure 9.5). The similar performance curves of tolerant phenotypes generate *convex* fitness sets (Figure 9.5A), whereas nonoverlapping performance curves of intolerant phenotypes generate *concave* fitness sets (Figure 9.5B). Tolerant phenotypes can persist despite rapid changes in the environment, provided that the magnitude environmental variation, E, is small. These phenotypes experience environmental variation as an average of environment types (Levins 1968, Lloyd 1984). However, increasing the range of environmental fluctuation makes environmental tolerance more difficult. Thus, when the environmental range, E, becomes large, tolerant phenotypes, which achieve modest success across all environments, tend to go extinct and are replaced by specialists, which, while phenotypically inflexible compared to more tolerant phenotypes, can achieve higher average fitness due to their enhanced success in a particular environment.

In short, increasing the range of environmental variation intensifies selection in favor of phenotypes that are specialized for particular conditions. Given the postulated trade-off between fitness mean and variance, as selection intensity increases, specialization is favored and performance distributions must become narrower, more distinct, and therefore likely to generate concave fitness sets (Levins 1968). Thus, when environments fluctuate widely, more specialized phenotypes with higher average fitness are expected to invade populations consisting of tolerant, generalist phenotypes (Figure 9.3).

A wide range of environmental fluctuations alone can favor phenotypic specialization. However, Shuster and Wade (2003) argued that *stronger* selection is in favor of a

234   S. M. SHUSTER

Box 9.3   (Cont.)

particular phenotype, the more *narrow* the distribution of performance in a particular environment will be. Thus, a concave fitness set will arise *whenever* selection becomes intense, even if the range of environmental fluctuation remains small. This occurs because under intense selection, the performance distributions within each environmental extreme will *contract* and the variance of the fitness distribution will be *reduced* (Figure 9.6). Thus, as selection becomes more intense, fitness sets will become increasingly concave and increasingly specialized phenotypes are expected to appear (see discussions in Bradshaw 1965, Lloyd 1984, Via and Lande 1985, Lively 1986, Moran 1992, Winn 1996, Schlicting and Pigliucci 1998).

But this is only part of the story. The optimum strategy for a given environment is not determined by the shape of the fitness set alone (Levins 1968, Shuster and Wade 2003). Rather, it is the *pattern* of environmental change impinging on each fitness set that determines (1) whether polymorphism will evolve and (2) the mechanism by which phenotypes will be expressed. When environmental changes occurs slowly, with periodicity *greater than* the average lifespan, individuals tend to experience their environments as alternative conditions with proportionately large, nonlinear effects on their fitness. Environmental changes occurring more rapidly, with periodicity *less than* the average lifespan, cause linear increases or decreases in the fitness of individuals because individuals experience the environment as a succession of different developmental conditions with their fitness averaged over them. The spatial and temporal scale of environmental change is the basis of Levins' (1968) concept of environmental *grain*.

Few or no changes within an individual's lifetime constitute *coarse* environmental grain, whereas rapid changes within an individual's lifetime cause the environment to be experienced as an average, and thereby constitute *fine* environmental grain. Phenotypes showing little variation are expected to evolve when environmental fluctuation is small in magnitude (Bradshaw 1965, Levins 1968). When the environment fluctuates, fitness sets become concave and polymorphic phenotypes of several kinds are expected to evolve, depending on how organisms perceive their environment.

If the arrival of change is unpredictable, environmental grain is coarse and Mendelian polymorphisms are expected to evolve. Under these conditions, the frequencies of

genetically distinct phenotypes will depend on the probability with which each environment occurs and on the relative fitness that each phenotype obtains therein. Distinct genotypes persist when their fitnesses averaged across the environmental grain are equal (Bradshaw 1965, Levins 1968, Maynard Smith 1982, Lively 1986).

In fluctuating environments, if environmental grain is perceived as fine, then selection will favor *polyphenism* (Lloyd 1984). This variation differs from simple environmental tolerance because the fitness set is concave. That is, selection is so intense that even the most tolerant individuals cannot persist; only specialists can. Thus, selection favors individuals who are *developmentally* capable of generating more than one phenotype, over individuals developing only a single phenotype with broader tolerance (Bradshaw 1965, Levins 1968, Lively 1986, Moran 1992, Roff 1992). Polyphenism is a mechanism for tolerance of environmental variation and its existence is evidence of "adaptive plasticity" (Shuster and Wade 2003). A coarse-grained environment can be experienced as a fine-grained one by individuals who use environmental cues to predict when change will occur and adjust their developmental trajectories appropriately (Bradshaw 1965, Levins 1968, Lively 1986, Moran 1992, Roff 1992, Winn 1996).

The ability to respond to a change in one's environment represents a genotype-by-environment interaction (G×E) (Schlicting and Pigliucci 1998). The particular way in which this interaction is expressed – the way in which an individual responds to environmental change – is known as its *reaction norm*. Within populations, reaction norms tend to be normally distributed due to genetic differences among individuals (Hazel *et al.* 1990, Roff 1996). The level of adaptive plasticity is the average efficiency with which different individuals in the population respond to environmental change. Thus, the equilibrium distribution of genotypes in a population depends on the distribution of reaction norms, the distribution of environments, and the distributions of fitness for the different possible phenotypes within the population. As for any genetic polymorphism, stable phenotypic distributions (or in this case, stable distributions of reaction norms that allow plastic responses to changing environments) are expected to persist when the fitnesses of their underlying genotypes are *equal* (Hazel *et al.* 1990, Roff 1996, Flaxman 2000, Shuster and Wade 2003).

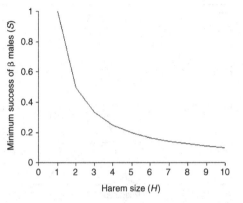

Figure 9.2 The relationship between harem size, H, and the minimum success necessary for satellite (β) males to invade a mating system consisting of territorial males. (Redrawn from Shuster and Wade 2003.)

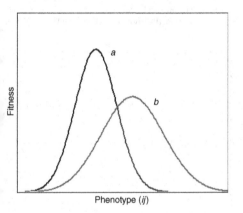

Figure 9.3 The distribution of fitness for phenotypes (reaction norms) exposed to environmental variation (after Levins 1968); each curve describes the relationship between phenotype and fitness for a given reaction norm, i, when it is expressed in environment, j; a specialized reaction norm (a) has a narrow distribution of possible phenotypes that achieve low fitness in marginal environments but achieve high fitness in a particular environment; a tolerant reaction norm (b) has a broader distribution of possible phenotypes, that obtain some fitness in marginal environments but are unable to achieve high fitness in any environment.

alternative mating strategies, they proposed that environmental grain is perceived by males in terms of (1) the existence of cues that predict mating opportunities as well

as (2) the timing of cue perception, relative to the lifespan of individual males.

With these two factors in mind, it is easy to see that the grain of the environment will be coarse if cues predicting male mating success do not exist. Such conditions may appear most often when male lifetimes are short. However, regardless of whether males are presented with few or many mating opportunities within their lifespans, when environments are unpredictable, specialists are favored and male mating behavior patterns are expected to represent Mendelian alternatives (Shuster and Wade 2003). The grain of the environment will be perceived by males as fine if environmental cues do predict the type of mating opportunities that will become available. Such conditions may exist most often when male lifetimes are long. But regardless of whether males are presented with few or many mating opportunities, strong sexual selection combined with fine-grained environments will favor the evolution of polygenic inheritance underlying the expression alternative mating strategies. In general, the expression of such traits is well explained by current models for threshold inheritance (Shuster and Wade 2003) (Box 9.4).

When environmental cues perceived early in life predict mating opportunities later in life (when the interval between the perception of the cue and mating opportunities is long relative to total male lifespan), developmental processes will prevail. Thus, abundant food may enhance growth rate, increasing a male's body size as well as his likelihood of success in combat. Food shortages, on the other hand, may decrease the probability of such success in combat and instead lead to the expression of a noncombative, default phenotype. Males who respond to environmental cues with appropriate developmental trajectories are likely to outcompete males whose genotypes resist modification when environments change, as do Mendelian alternatives.

When environmental cues predicting mating success occur immediately before mating opportunities arise (when the interval between the perception of the cue and mating opportunities is short relative to total male lifespan), behavioral processes will prevail. Thus, a particular density of mating competitors may induce some individuals to become aggressive, whereas individuals insensitive to such cues will not engage actively in the commotion of direct mating competition. Or, a particular density of females may cause some males to associate themselves with individual females to await their impending receptivity, whereas individuals insensitive to such cues may continue searching for females more immediately receptive. The

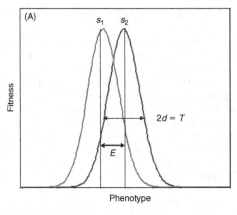

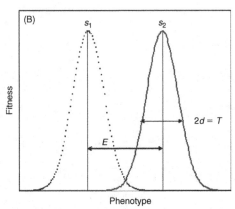

Figure 9.4 The performance (fitness) curves generated by the two most common environments considered together (grey line and black line). Tolerance (T) is equal to $2d$, where d is the distance, in phenotypic performance units, from the peak of the distribution to its point of inflection. Environmental range, E, is the difference in the average phenotypic performances in each environment ($S_2 - S_1$).

Heavily overlapping curves (A) produced by each environment indicate a tolerant phenotype, one in which $T > E$ or in which the optimal phenotype is similar in each environment. Minimally overlapping curves (B), in which $T < E$, indicate intolerant phenotypes, or more simply, that different phenotypes are optimal in each environment. (Redrawn from Schuster and Wade 2003.)

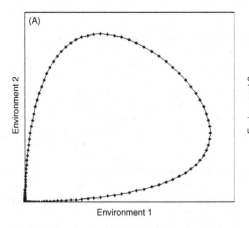

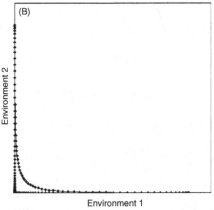

Figure 9.5 When the fitness values of each performance curve are plotted against each other fitness sets are generated; similar performance curves generate (A) convex fitness sets, whereas non-overlapping performance curves generate (B) concave fitness sets. (Redrawn from Shuster and Wade 2003.)

relative frequencies of sensitive and insensitive individuals in any population will depend on the relative success of these phenotypes over time (Box 9.3). Males who respond rapidly and appropriately to environmental cues that predict mating success in changing environments are likely to outcompete males whose genotypes resist environmental change, as well as males who cannot respond as rapidly to changes in conditions favoring mating success.

And yet, while it is widely acknowledged that genetic architectures sensitive to environmental cues can allow

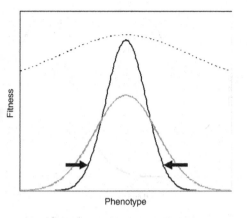

Figure 9.6 Strong stabilizing selection (dotted line) causes phenotypic distributions to contract.

males to express appropriate phenotypes in response to changing environments (Roff 1996, Schlicting and Pigliucci 1998), the evolutionary restrictions on phenotypic plasticity are seldom mentioned when variable phenotypes are observed (Shuster and Wade 2003). Phenotypic plasticity in development or in behavior is likely to evolve only if the following conditions exist:

(1) Genetic variation allowing a plastic response to changing environmental conditions must be present in the population – individuals must be genetically variable, not genetically identical.

(2) The cost of making the wrong developmental or behavioral "choice" must be high; that is, expressing an inappropriate phenotype in an environment in which it is not favored, leads to little, or more often, no reproduction at all.

(3) Circumstances favoring plasticity must occur frequently. Conditions in which a plastic response is required must be common, they must occur in a consistent way, and they must not be contingent on special circumstances (such as a uniquely debilitating

---

Box 9.4   Threshold characters

Discrete phenotypic classes within a population that fail to segregate according to Mendelian rules are often explained by threshold models of quantitative inheritance. As with most complex characters, continuous genetic variation appears to underlie threshold traits. However, a threshold of "lability" within this distribution also exists that makes trait expression discontinuous. Individuals with genotypes below the threshold express a default phenotype, whereas individuals with genotypes above the threshold express a modified phenotype (Figure 9.7).

The expression of threshold traits is not absolute. Depending on trait heritability, threshold position, and the environment, each genotype has its own probability of trait expression (Dempster and Lerner 1950, Gianola and Norton 1981). For this reason, environmental influences on threshold characters can be viewed in two ways. When the environment is constant, or reasonably so, as might exist over a period of maturation, trait expression appears as described above; genotypes above the threshold usually express the trait, trait expression becomes increasingly unlikely for genotypes below the threshold, and the population appears dimorphic. Alternative mating strat-

egies involving distinct developmental trajectories are well described by this hypothesis.

When the environment changes over shorter time-scales, few or no genotypes may express the trait at one environmental extreme, whereas at the other extreme, all or nearly all genotypes will become modified (Figure 9.8). The wider the environmental range, the greater is the proportion of the population that is likely to change. Although the probability of trait expression remains constant for each genotype, depending on the intensity of the environmental "cue" at any time, few, some, or all individuals in the population may express the trait. Alternative mating strategies involving behavioral polyphenism are well described by this hypothesis.

Threshold models may also explain age-dependent mating strategies, although contrary to current models of this phenomenon, a threshold view predicts that few males will perform both "young" and "old" mating strategies within their lifetimes (e.g., Correa et al. 2003). Instead, quantitative genetic variation is expected to predispose males to mate as satellites when young or as territorial males when old, with frequency-dependent selection maintaining the position of the threshold within the distribution of male maturation rates.

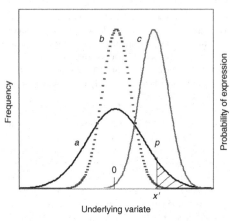

Underlying variate

Figure 9.7   The distribution of individuals expressing and not expressing a threshold trait; the solid curve a represents the distribution of the underlying heritable character, x, for the entire population; the average genotype, $x = 0$; x' represents the threshold value of x; the hatched area, p, represents the proportion of individuals exhibiting the character; curve b represents the distribution of individuals with the mean genotype in the population ($x = 0$) that are expected to express the character or not; the distribution of individuals with different genotypes expressing and not expressing a threshold trait; individuals with the average genotype, $x = 0$, curve b, have only a small probability of expressing the extreme trait, whereas individuals with genotype x', curve c, have a higher probability of trait expression. (Redrawn from Dempster and Lerner 1949.)

injury: West-Eberhard 2003; or the appearance of a uniquely compatible mate: Gowaty 1997, Tregenza and Wedell 2000).

(4) Conditions in which a plastic response is required must be experienced by a large fraction of the population.

All of these conditions must apply for phenotypic plasticity to evolve, because if they do not exist, either a response to selection will be impossible or selection on genetic factors allowing polyphenism will be weak. Clearly, phenotypic plasticity cannot evolve in the absence of genetic variation mediating a flexible developmental or behavioral response. However, phenotypic plasticity is also unlikely to evolve when circumstances favoring it are rare and highly contingent on the behavior of other individuals, or when they are experienced by only a few individuals in the population. Under these conditions, selection will be of low intensity, intermittent in its effects, and likely to influence

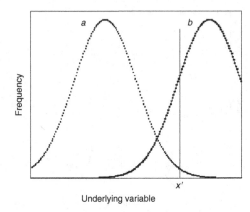

Underlying variable

Figure 9.8   The expression of a threshold character in a variable environment; solid vertical line at x' represents the threshold of trait expression; all individuals to the right of the line express the character, all individuals to the left of the line do not; curve a represents the distribution of character expression when the environmental cue stimulating the expression of the character is weak; curve b represents the distribution of expression when the cue is strong.

only a small number of individuals in the population. In combination, these factors will weaken, if not obliterate entirely, the effects of directional selection favoring adaptive phenotypic plasticity (Shuster and Wade 2003).

9.4   MENDELIAN STRATEGIES

Among crustaceans, examples of Mendelian strategies include marine isopods (Shuster and Wade 1991, Shuster and Sassaman 1997, K. Tanaka, personal communication), freshwater isopods (Bocquet and Veuille 1973), androdioecious branchiopods (Sassaman 1991, Weeks and Zucker 1999), and sequentially hermaphroditic decapods in which primary males or primary females persist (Bauer 2000, 2002).

In *Paracerceis sculpta*, a marine isopod inhabiting the northern Gulf of California, three discrete male morphotypes coexist (Figure 9.9). Phenotypic differences among males are controlled primarily by an autosomal locus of major effect (*Ams* = alternative mating strategy), whose inheritance is Mendelian and whose alleles exhibit directional dominance ($Ams^\beta > Ams^\gamma > Ams^\alpha$). The different *Ams* alleles interact with alleles at other loci, switching on distinct developmental cascades that lead to discontinuous adult phenotypes. These interactions appear to influence male as well as female phenotypes. Alleles at *Ams*, and at an additional autosomal

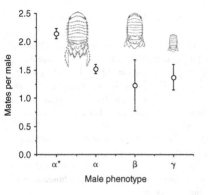

Figure 9.9 The mating success of *Paracerceis sculpta* α-, β-, and γ-males in *Leucetta lonangelensis* spongocoels between 1983 and 1985. The average harem size of mating α-males is represented by α*; α represents the average harem size of all α-males. (Redrawn from Shuster and Wade 2003.)

locus (*Tfr* = transformer), epistatically interact to radically distort family sex ratios (Shuster and Sassaman 1997, Shuster et al. 2001). This is accomplished when particular *Ams–Tfr* allelic combinations override the primary sex determination mechanism (WW = males; ZW = females: Shuster and Levy 1999) and cause individuals to mature as adults of the opposite sex (Shuster et al. 2001).

The dynamic nature of this mating system is consistent with the hypothesis that sexual selection is intense and mating opportunities for males, as well as for females, are highly variable and unpredictable from one generation to the next. At breeding sites, mating success among the male morphs varies with the number of females, as well as with the number and type of other males (Shuster 1989, 1992, Shuster and Wade 1991). However, over time, the average fitnesses of α-, β-, and γ-males are equal (average ± 95% CI; α-males: 1.52 ± 0.16, N = 452; β-males: 1.25 ± 0.86, N = 20; γ-males: 1.37 ± 0.45, N = 83: Shuster and Wade 1991, 2003) In this species, α-males defend territories in sponges; thus mating as well as nonmating α-males are identifiable. Indeed, when only the average mating success of *mating* α-males (α*) is considered, this value is significantly larger than the average mating success for β- and γ-males (2.22 ± 0.17) (Figure 9.9). This example shows why satellite males in many animal species may appear to "make the best of a bad job" when unsuccessful males cannot be identified. In reality, satellite males usually experience average fitness equal to that of all territorial males, winners as well as losers.

A genetic polymorphism in male leg morphology exists in *Jaera albifrons*, a freshwater isopod in which males guard females before mating (Bocquet and Veuille 1973). In other isopods with similar pre-copulatory behavior (*Thermosphaeroma*: Shuster 1981, Jormalainen et al. 1999), the legs of males in several species are more setose than those of females and may be useful in retaining control of mates during usurpation attempts by other males. Unfortunately, in *J. albifrons*, environmental or social factors maintaining the polymorphism have not been examined nor has the possibility that males may employ different reproductive strategies in the context of mate guarding.

However, males in species with pre-copulatory guarding are widespread within the Crustacea (Ridley 1983, Conlan 1991, Jormalainen 1998), and individual males are known to switch between mate-guarding and mate-usurping behavior. It is likely that the inheritance of mate-guarding behavior is polygenic rather than Mendelian, consistent with behavioral polyphenism (see below). And yet, the existence of Mendelian phenotypes in *J. albifrons* suggests that such genetic architectures are more widespread than is currently recognized. In such species, more detailed experiments designed to identify genetic influences on morphology as well as behavior among males are clearly needed.

A Mendelian polymorphism may also exist in the gnathiid isopod *Elaphognathia cornigera*, a species inhabiting mud banks and coral rubble in coastal regions of the western Pacific (Tanaka and Aoki 1999, Tanaka 2003). In this species, males are larger than females and possess enormous mandibles that are used to defend breeding aggregations and to encourage females to enter breeding sites. In addition to large males, small, sexually mature males coexist with territorial males in these populations (K. Tanaka, personal communication). Although the inheritance of this polymorphism is currently unknown, like *P. sculpta*, gnathiid isopods are semelparous (Upton 1987, Tanaka 2003); thus, individual lifetimes are relatively short. In addition, the pelagic praniza larvae of gnathiids, which are parasitic on fish (Roberts and Janovy 2005), seem unlikely to have opportunities to detect cues predicting their mating success until they arrive at breeding sites as adults. Such conditions could favor Mendelian inheritance of the adult male phenotype (although a developmental polymorphism is also possible; see below). Experiments are currently under way to test this hypothesis (K. Tanaka, personal communication).

Although usually not considered in discussions of alternative mating strategies, mating systems in which males persist with hermaphrodites (androdioecy) or in which females persist with hermaphrodites (gynodioecy) often represent Mendelian polymorphisms controlling the expression of alternative mating strategies (Charlesworth 1984, Charlesworth and Charlesworth 1987). In the clam shrimp *Eulimnadia texana*, males coexist with two phenotypically similar, but genetically different, types of hermaphrodites that may self-fertilize or outcross. Sex in this species is controlled by a single genetic locus (Sassaman and Weeks 1993), in which a dominant allele codes for the hermaphroditic condition (S) and a recessive allele codes for males (s). Phenotypic males are homozygous recessives (ss) whereas hermaphrodites may be homozygous (SS = monogenic) or heterozygous (Ss = amphigenic). Monogenic hermaphrodites are homozygous dominants (SS) and produce 100% monogenic progeny when selfed (SS) or 100% amphigenic progeny when outcrossed (Ss). Amphigenic hermaphrodites produce mixtures of monogenic, amphigenic, and male progeny depending on whether they self or outcross (25% monogenics, 50% amphigenics, and 25% males when selfed; 50% amphigenics and 50% males when outcrossed). The composition and the relative fitness of each genotype within each population determine the observed genotype frequencies and, despite a high frequency of selfing for *E. texana* (inbreeding coefficients ranging from 0.20 to 0.97: Sassaman 1989, Weeks and Zucker 1999), androdioecy persists in nature.

Sex determination also appears to be controlled by a simple mechanism in the notostracan *Triops newberryi* (Sassaman 1991), in which the female genotype at a single autosomal locus influences whether females produce families that are all-female or which consist of mixtures of males and females. *Triops* and *Eulimnadia* species, like many other branchiopods, inhabit temporary pools in arid regions (Sassaman 1991, Weeks and Zucker 1999). The dormant zygotes of these species can persist for years in a desiccated state and are dispersed by wind and floods. Thus, as predicted above, the lifetimes of these species are short, and conditions favoring males or hermaphrodites in newly colonized pools are likely to be unpredictable, with intense selection favoring selfing or outcrossing from one habitat to the next.

A bewildering array of mating systems appears to exist within the caridean decapods (Bauer and VanHoy 1996, Bauer 2000), and while many species appear to include some form of developmental sex change (see below), the persistence of individuals who do not change sex suggests the existence of a Mendelian polymorphism (Roff 1996,

Lively *et al.* 2000). Caridean examples include protandrous mating systems with primary males (e.g., *Thor manningi*: Chace 1972, Bauer 1986; *Athanas* spp.: Nakashima 1987, Gherardi and Calloni 1993), mating systems with protandric simultaneous hermaphroditism (PSH) (e.g., *Lysmata* spp. and possibly *Exhippolysmata*: Bauer 2000, 2002), and protandrous mating systems with primary females (e.g., *Processa edulis*: Noël 1976; *Crangon crangon*: Boddeke *et al.* 1991; *Argis dentata*: Fréchette *et al.* 1970; *Pandalus*: Charnov 1979, 1982, Bergström 1997). The persistence of distinct, nonchanging adult phenotypes in each of these mating systems strongly suggests the existence of underlying genetic variation that is maintained within the population by equal fitnesses among the recognizable morphs.

9.5 DEVELOPMENTAL STRATEGIES

Examples of developmental strategies within the Crustacea appear to include certain copepods (Haq 1972, Stancyk and Moreira 1988), a large number of decapods (Carpenter 1978, Kuris *et al.* 1987, Ra'anan and Sagi 1989, Bauer 2000, Baeza and Bauer 2004), as well as numerous amphipods and tanaids (Darwin 1874, Borowsky 1980, 1984, Conlan 1991, Clark 1997, Kurdziel and Knowles 2002).

In the harpacticoid copepod *Euteropina acutifrons*, two distinct male morphs coexist with females (Haq 1965, 1972, 1973, D'Apolito and Stancyk 1979, Moreira *et al.* 1983, Moreira and McNamara 1984, Stancyk and Moreira 1988). While earlier descriptions suggested that the male morphs differed in their developmental rates (Haq 1972, 1973), later experiments that directly examined the possible effects of paternity and temperature on morph expression (Stancyk and Moreira 1988) suggested that a combination of Mendelian and developmental polymorphisms may exist in which primary males coexist with protandrous hermaphrodites, a situation similar to that observed in many caridean shrimp (Bauer 2000).

Ra'anan and Sagi (1989) described three male morphs representing successive growth stages in freshwater prawns (*Macrobrachium rosenbergii*) (see Nagamine *et al.* 1980). Also in this species, Kuris *et al.* (1987) demonstrated that developmental trajectories among the male morphs are determined by feeding schedule and social interactions among individuals. These authors suggested that, in fact, four morphs are identifiable (see also Barki *et al.* 1992, Kurup *et al.* 2000) and that dominance hierarchies among the morphs exist, wherein removal of larger individuals induces smaller individuals to grow and in some cases assume the morphology of the missing larger class. However, not all individuals responded equally to this stimulus, as expected if males vary in their responsiveness to social and nutritional cues (Kuris et al. 1987, Karplus et al. 2000). Such variation is characteristic of traits with threshold expression (Box 9.4).

Other *Macrobrachium* species also appear to exhibit male polymorphism (*M. dayanum, M. idae, M. malcolmsonii, M. scabriculum*; reviewed in Kuris et al. 1987), and it is likely that similar growth and social stimuli influence the fitness and frequencies of the different male morphs, and thus the means by which developmental switches are favored. In *M. rosenbergii*, females are dimorphic as well (Harikrishnan et al. 1999). In all of these populations, extreme sexual dimorphism involving large size and elongated chelae in males suggests that competition for mates is intense (Wade and Shuster 2004). Moreover, relatively long-lived individuals appear to obtain information and respond appropriately with respect to their future mating opportunities and to nutritional and social cues during development (Kuris et al. 1987). Populations of the same species appear to vary in the proportions of individuals exhibiting different male morphologies (Karplus et al. 2000, Kurup et al. 2000). All of these observations are consistent with quantitative genetic inheritance of threshold traits (Roff 1996) (Box 9.4). Given that freshwater prawns provide an important food source in many countries, there is likely to be continued commercial interest in studies identifying the degree to which developmental programs can be manipulated, either by altered environments or by selection on norms of reaction (Emlen 1996).

In the rock shrimp, *Rhynchocinetes typus*, males exhibit three phenotypes of increasing size that evidently represent increasing states of maturation (*typus, intermedius, robustus*; Correa et al. 2000, 2003). The *typus* morphotype is similar in morphology to the female. The *robustus* morphotype possesses powerful chelae and elongated third maxillipeds. During development, males undergo several intermediate molts between these two morphs. Such males are classified as *intermedius*. All three male types are sexually mature and all three exhibit similar behavior when allowed to mate with females in isolation. However, in competitive situations, males established size-based dominance hierarchies (*robustus > intermedius > typus*), and males differed in their rates of interaction with females and spermatophore deposition. Subordinate males engaged in these activities more rapidly than more dominant males.

It is unclear whether these morphs represent the same level of specialization that appears to exist in *Macrobrachium* spp. or in amphiopods with two developmentally distinct

male morphs (Kurdziel and Knowles 2002). Because *typus* males appear to eventually grow into *intermedius* and *robustus* stages, Correa et al. (2000, 2003, Correa and Thiel 2003) consider the variation observed in *R. typus* to be entirely behavioral and consistent with models for condition-dependent switching of phenotypes (see below). However, populational variation in the tendency for *typus* males to follow this developmental trajectory is unknown, and unfortunately, like other marine decapods, the presence of planktonic larvae and prolonged juvenile development are likely to prevent the detailed breeding experiments necessary to identify genetic variation underlying different male phenotypes. The tendency for *typus* males to display highly specialized mating and sperm-transfer behavior associated with their small size (Correa et al. 2003) suggests that selection favoring this morphotype is strong. Also, the tendency for the population frequencies of the three male morphs to remain constant within populations and to be variable among populations suggests an underlying developmental mechanism involving threshold inheritance (Correa et al. 2003).

Darwin (1874) identified a male polymorphism in the Brazilian amphipod *Orchestia darwinii*. In this species, adult males possess gnathopods (chelae) that are either enlarged or reduced in size. Darwin added "the two male forms probably originated by some having varied in one manner and some in another; both forms having derived certain special, but nearly equal advantages, from their differently shaped organs." He also mentioned a dimorphism in *Tanais* "in which the male is represented by two distinct forms, which never graduate into each other. In the one form the male is furnished with more numerous smelling-threads, and in the other form with more powerful and more elongated chelae or pincers, which serve to hold the female" (Darwin 1874, p. 262).

Whether these dimorphisms are developmental in their expression is unclear. However, the possibility that they are is made credible by the detailed work of Borowsky (1984, 1985, 1989), Clark (1997), and Kurdziel and Knowles (2002), who have documented two sexually mature male morphs representing different growth stages in the marine amphipods *Microdeutopus gryllotalpa, Jassa falcata*, and *J. marmorata*. In each of these species, larger males ("majors" or "thumbed" males) possess enlarged gnathopods that are lacking in smaller males ("minors" or "thumbless" males). Majors vigorously defend tubes inhabited by receptive females against other majors, whereas minors tend to travel among tubes and avoid conflict. Similar mating systems are evidently widespread within the Amphipoda, and sexual dimorphism involving enlarged male gnathopods is common in taxa in which males defend females in burrows or other cavities (Conlan 1991).

Kurdziel and Knowles (2002) demonstrated that in *J. marmorata*, the polymorphism is indeed developmental, and their results are consistent with threshold models of quantitative trait inheritance (Roff 1996, Shuster and Wade 2003) (Box 9.4). Well-fed males tend to grow to large size and develop enlarged gnathopods, whereas poorly fed males do not. However, Kurdziel and Knowles' (2002) initial interpretation of their results ("Heritability analyses indicated the reproductive phenotypes do not reflect genetic differences between dimorphic males," p. 1749) is suspect given their application of a standard full-sib breeding design to investigate broad sense heritability for male phenotype (Falconer 1989). The probabilistic nature of threshold trait expression (Box 9.4) makes this approach less likely to identify a genetic component underlying male differences than the method specifically designed to detect the heritability of threshold traits (Falconer 1989, p. 300). A reanalysis of the data of Kurdziel and Knowles (2002), or a breeding design conducted using methodology appropriate for such traits, could determine whether male polymorphism in these and other amphipods involves threshold inheritance.

Tendencies for individuals within populations to undergo sex change are likely to represent a developmental strategy that evolves when individuals regularly encounter distortions in population sex ratio (Shuster and Wade 2003). The dynamics of sex change are well known in *Pandalus* shrimp, for which much of sex allocation theory was developed (Charnov 1979, 1982). Also, in the caridean shrimp *Lysmata wurdemanni* (Bauer 2000, Baeza and Bauer 2004), individuals mature as male-phase (MP) individuals and later change to female-phase (FP) individuals, which possess female external morphology but retain both male and female reproductive capacity (another example of protandric simultaneous hermaphroditism).

To examine social mediation of sex change, Baeza and Bauer (2004) reared MP individuals in both large and small social groups with different sexual and size composition. As expected, if the availability of mating opportunities for members of each sex influenced the intensity of sexual selection (Shuster and Wade 2003), these authors found that the speed of sex change was inversely related to the abundance of FP individuals in the "large group" experiment but the trait was less obvious in smaller groups. Baeza and Bauer

(2004) suggested that a more rapid change to the female phase may occur when male mating opportunities are low because the simultaneous-hermaphrodite FPs can immediately reproduce as a female while maintaining male mating capacity.

While sex allocation theory is consistent with the observations above, the theoretical framework for sex ratio equalization almost without exception involves family selection (review: Wade et al. 2003). That is, the primary source of selection on sex ratio derives from the fitness of females who bias their family sex ratios toward the minority sex, relative to that of females who either bias their family sex ratios toward the majority sex or do not bias their family sex ratios at all. This evolutionary process is extremely slow and is unlikely by itself to explain the observed dynamics in sex ratio in natural populations. Shuster and Wade (2003; see also Wade et al. 2003) proposed that in many species undergoing sex change, genetic polymorphisms may exist that mediate individual abilities to either change sex or remain unchanged, as evidently occurs in the isopod *Paracerceis sculpta* (Shuster and Sassaman 1997, Shuster et al. 2001)

A simple method for investigating this possibility involves estimating the frequency of the population that exists as a single sex. If this fraction represents an alternative mating strategy, then as explained above (see Eq. (B9.2.6), Box 9.2), their fitness relative to that of hermaprodites, s, may be used to approximate the fraction of the hermaphroditic population that is *unsuccessful* in reproducing as that sex, p_0. Thus, primary males would represent the alternative phenotype in androdioecy and in protandrous and protandric simultaneous hermaphroditic mating systems with primary males. Similarly, females would represent the alternative phenotype in gynodioecy and in protandrous mating systems with primary females.

9.6 BEHAVIORAL STRATEGIES

Within the Crustacea, examples of polymorphism in mating behavior appear, as predicted, in long-lived taxa such as stomatopods (*Pseudosquilla ciliata*: Hatziolos and Caldwell 1983; *Gonodactylus bredini*: Shuster and Caldwell 1989) and decapods (*Alpheus armatus*: Knowlton 1980; *Pachygrapsus transversus*: Abele et al. 1986; *Homarus americanus*: Cowan 1991, Cowan and Atema 1990; *Uca* spp.: Salmon and Hyatt 1983, Christy and Salmon 1991, Jennions and Backwell 1998; *Sicyonia dorsalis*: Bauer 1992; *Callinectes sapidus*: Jivoff and Hines 1998; *Scopimera globosa*: Wada 1986, Koga 1998; *Chionoecetes opilio*: Moriyasu and Benhalima 1998;

Jasus edwardsii: MacDiarmid and Butler 1999; *Rhynchocinetes typus*: Correa and Thiel 2003, Correa *et al.* 2000, 2003). Howevever, they also appear in shorter-lived taxa including amphipods (*Microdeutopus gryllotalpa*: Borowsky 1984; *Gammarus duebeni*: Dick and Elmwood 1995; *Jassa marmorata*: Clark 1997) and isopods (*Thermosphaeroma* spp.: Shuster 1981, Jormalainen and Shuster 1999; *Paracerceis sculpta*: Shuster 1992, Shuster and Arnold 2007; *Idotea baltica*: Jormalainen *et al.* 1994). In each of these cases, males, and often females as well, are highly mobile, have multiple mating opportunities within their lifetimes, and individuals can rapidly change their behavior in ways that allow them to exploit mating opportunities as they arise.

The underlying genetic architectures responsible for such variability appear to be similar to those described above for developmental strategies (Hazel *et al.* 1990, Shuster and Wade 2003). That is, genetic variation underlying quantitative traits is expected to influence the likelihood that individuals will express a particular mating behavior. In a given situation, individuals with phenotypes below the liability threshold express one set of mating behavior, whereas individuals with phenotypes above this threshold express another behavioral set (Box 9.4). In variable situations, weak stimuli will induce few individuals to perform mate-acquiring behavior. Strong stimuli, however, will cause most individuals to attempt to mate (Box 9.4; see also Shuster and Wade 2003). This "behavioral threshold" hypothesis predicts differential responsiveness to the same environmental cues among individuals within populations due to genetic differences among males. This hypothesis also predicts differential responses to different cue intensities among individuals within populations, again, due to genetic differences among males.

Behavioral strategies are expected to arise when sexual selection favors specialized mating phenotypes, as in all of the cases previously considered. In these polymorphisms the relative mating success of each phenotype is predictable within male lifetimes and the timescale for change is short; so short, in fact, that environments may change dramatically within minutes or seconds. Behavioral plasticity is expected to exclude major genes and developmental plasticity as modes of phenotypic expression when reliable cues predicting mating success are available and mating opportunities change quickly.

Pre-copulatory mate guarding is widespread among crustaceans (Ridley 1983, Jormalainen 1998). The explanation for this tendency in many species is that molting

initiates female receptivity and chemical cues present in female urine or present on females themselves prior to this molt allow males to locate, guard, and inseminate females as soon as they become receptive. Mate guarding reduces the ability of females to mate more than once; thus, a male who guards a female successfully fertilizes all of her ova. If a male unsuccessfully guards his mate, or if he leaves her in search of other females before her receptivity is complete, the male's fertilization success with that female will be eroded due to matings by other males. Sperm competition as an alternative mating strategy in crustaceans and in other taxa is discussed in more detail in Shuster and Wade (2003; see also Diesel 1989, Koga *et al.* 1993, Orensanz *et al.* 1995, Jormalainen 1998).

Males in a wide range of crustacean species that engage in mate guarding exhibit flexibility in guarding duration in response to local sex ratios, as well as in their responses to female body size, reproductive condition, parasitemia, and resistance to male guarding attempts (reviews: Shuster 1981, Ridley 1983, Jormalainen 1998, Plaistow *et al.* 2001). Variability in guarding duration in response to sex ratio shows a consistent pattern in several peracarids (Jormalainen 1998). In at least five species, males tend to shorten their average guarding durations when exposed to operational sex ratios that are female biased ($R_0 = N_{males}/N_{females} < 1$) and to lengthen their average guarding durations when sex ratios are male biased ($R_0 > 1$). Such behavioral flexibility is consistent with the hypothesis that mate guarding evolves as an adaptation to prevent multiple mating. Flexibility in mate-guarding behavior is evidently under strong sexual selection because males who guard ineffectively lose fertilizations to other males. Thus, the expression of this behavioral trait is consistent with the predictions of threshold inheritance of behavioral phenotypes (Shuster and Wade 2003).

Genetically variable characters likely to influence behavioral lability include individual sensitivities to crowding and to circulating hormone levels (Sagi *et al.* 1994, Briceno and Eberhard 1998, Borash *et al.* 2000, Peckol *et al.* 2001, Nephew and Romero 2003). Other characters likely to influence mating behavior may include heritable sensitivities to pheromone concentrations (Ferveur 1997, Giorgi and Rouquier 2002), to the density of mating competitors (Haig and Bergstrom 1995), or to the perception of mating behavior by other individuals (Shuster 1981). In the presence of a strong environmental cue, all but a few individuals are expected to express a modified behavioral phenotype. Weaker cue intensity, on the other hand, may induce few or no individuals toward behavioral change (e.g., Lively *et al.* 2000). The behavioral threshold hypothesis, like the developmental threshold hypothesis, predicts differential responsiveness to the same environmental cues among individuals within populations. Thus, the same female distributions that induce some males to assume satellite behavior are expected to cause other males to persist as territorial males, as is widely observed.

Among populations, the tendency for males to exhibit one behavior or another is also likely to vary, leading to the likelihood that different proportions of each population will express the modified behavioral phenotype at any given time. Such interpopulational variation in behavioral expression is known to anyone studying behavior in multiple populations. The explanation for this variation is the same as for developmental polymorphisms. The underlying genetic basis for behavioral expression is similar to that of most threshold characters. Moreover, observed proportions of different morphotypes within a population depend multiplicatively on the frequency and relative fitness of each type. Thus, as with Mendelian and developmental polymorphisms, behavioral polymorphism is maintained within a population because the average fitnesses of each phenotype are equal.

9.7 CONCLUSIONS

The expression, inheritance, dynamics, and persistence of alternative mating strategies in natural populations of crustaceans, like other strategies discussed in an evolutionary context, can be investigated following well-established principles from population genetics and, given certain assumptions, evolutionary game theory. Alternative mating strategies clearly evolve under intense sexual selection. Contrary to current hypotheses regarding the importance of male genetic quality or sexual conflict (review: Shuster and Wade 2003), this condition minimizes the potential influence of viability selection on morph fitness, making investigation of life-history differences among males less important for understanding the persistence of polymorphism than investigation of the intensity of sexual selection within and among morphs (Gross 1996).

In species in which sexual selection is strong, not only are alternative mating strategies expected to evolve, but the evolutionary effects of sexual selection on alternative mating phenotypes are likely to be more easily documented than for phenotypes evolving in response to natural selection. Because population sizes may be large and generation times

long, students of natural selection may fail to observe the effects of selection within their own lifetimes. In contrast, students of alternative mating strategies, because sexual selection acts so intensely on these traits, are usually able to observe the evolutionary effects of sexual selection in real time in many species.

How is this possible? Characters evolving under sexual selection, particularly polymorphic male phenotypes, are often easily recognizable. The signature of evolutionary change, as revealed by changes in morph frequency resulting from differential mating success, can be readily observed in some cases over a few days or weeks. Crustaceans are second only to the insects in their abundance and diversity among arthropods. They include economically and ecologically important species. In many major taxa, the majority of the scientific literature addresses newly described species (Brusca and Brusca 2004). If sexual selection is indeed one of the most powerful evolutionary forces known, then studies of adaptations evolving in this context are likely to reward investigators with abundant data and important new insights. The future is bright for continued studies in the evolution, persistence, and expression of mating strategies, particularly among the Crustacea.

Acknowledgments
I am grateful for comments on early drafts of this manuscript by Jane Brockmann, Rui Oliveira, Emily Omana, Con Slobodchikoff, and Brian Tritle. This research was supported by NSF Grant DBI-0243914.

References

Abele, L. G., Campanella, P. J., and Salmon, M. 1986. Natural history and social organization of the semiterrestrial grapsid crab, *Pachygrapsus transversus* (Gibbes). *Journal of Experimental Marine Biology and Ecology* **104**, 153–170.

Ahl, J. S. B. and Laufer, H. 1996. The pubertal molt in Crustacea revisited. *Invertebrate Reproduction and Development* **30**, 177–180.

Alcock, J. 2005. *Animal Behavior*, 8th edn. Sunderland, MA: Sinauer Associates.

Austad, S. N. 1984. A classification of alternative reproductive behaviors, and methods for field testing ESS models. *American Zoologist* **24**, 309–320.

Baeza, J. A. and Bauer, R. T. 2004. Experimental test of socially mediated sex change in a protandric simultaneous hermaphrodite, the marine shrimp *Lysmata wurdemanni* (Caridea: Hippolytidae). *Behavioral Ecology and Sociobiology* **55**, 544–550.

Barki, A., Karplus, I., and Goren, M. 1992. Effects of size and morphotype on dominance hierarchies and resource competition in the freshwater prawn *Macrobrachium rosenbergii*. *Animal Behaviour* **44**, 547–555.

Bauer, R. T. 1986. Sex change and life history pattern in the shrimp *Thor manningi* (Decapoda: Caridea): a novel case of partial protandric hermaphroditism. *Biological Bulletin* **170**, 11–31.

Bauer, R. T. 1992. Repetitive copulation and variable success of insemination in the marine shrimp *Sicyonia dorsalis* (Decapoda: Penaeoidea). *Journal of Crustacean Biology* **12**, 153–160.

Bauer, R. T. 2000. Simultaneous hermaphroditism in caridean shrimps: a unique and puzzling sexual system in the Decapoda. *Journal of Crustacean Biology* **20**, 116–128.

Bauer, R. T. 2002. The reproductive ecology of a protandric simultaneous hermaphrodite, the shrimp *Lysmata wurdemanni* (Decapoda: Caridea: Hippolytidae). *Journal of Crustacean Biology* **22**, 742–749.

Bauer, R. T. and VanHoy, R. 1996. Simultaneous hermaphroditism in the marine shrimp *Lysmata wurdemanni* (Caridea: Hippolytidae): an undescribed sexual system in the decapod Crustacea. *Marine Biology* **132**, 223–235.

Belk, D. 1991. Anostracan mating behavior: a case of scramble-competition polygyny. In R. T. Bauer and J. W. Martin (eds.) *Crustacean Sexual Biology*, pp. 111–125. New York: Columbia University Press.

Bergström, B. I. 1997. Do protandric pandalid shrimp have environmental sex determination? *Marine Biology* **128**, 397–407.

Bocquet C. and Veuille, M. 1973. Le polymorphisme des variants sexuels des mâles chez *Jaera (albifrons) ischiosetosa* Forsman (Isopoda: Asellota). *Archives de Zoologie Expérimentale et Générale* **114**, 111–128.

Boddeke, R., Bosschieter, J. R., and Goudswaard, P. C. 1991. Sex change, mating, and sperm transfer in *Crangon crangon* (L.). In R. T. Bauer and J. W. Martin (eds.) *Crustacean Sexual Biology*, pp. 164–182. New York: Columbia University Press.

Borash, D. J., Teoto-Nio, H., Rose, M. R., and Mueller, L. E. 2000. Density-dependent natural selection in *Drosophila*: correlations between feeding rate, development time and viability. *Journal of Evolutionary Biology* **13**, 181–187.

Borowsky, B. 1980. The pattern of tube-sharing in *Microdeutopus gryllotalpa* (Crustacea: Amphipoda). *Animal Behaviour* **28**, 790–797.

Borowsky, B. 1984. Effects of receptive females' secretions on some male reproductive behaviors in the amphipod

crustacean *Microdeutopus gryllotalpa*. *Marine Biology* **84**, 183–187.

Borowsky, B. 1985. Differences in reproductive behavior between two male morphs of the amphipod crustacean, *Jassa falcata* Montagu. *Physiological Zoology* **58**, 497–502.

Borowsky, B. 1989. The effects of residential tubes on reproductive behaviors in *Microdeutopus gryllotalpa* (Costa) (Crustacea: Amphipoda). *Journal of Experimental Marine Biology and Ecology* **8**, 117–125.

Bradshaw, A. D. 1965. Evolutionary significance of phenotypic plasticity in plants. *Advances in Genetics* **13**, 115–155.

Briceno, R. D. and Eberhard, W. G. 1998. Medfly courtship duration: a sexually selected reaction norm changed by crowding. *Ethology Ecology and Evolution* **10**, 369–382.

Brusca, R. C. and Brusca, G. J. 2004. *Invertebrates*, 2nd edn. Sunderland, MA: Sinauer Associates.

Carpenter, A. 1978. Protandry in the freshwater shrimp, *Paratya curvirostris* (Heller, 1862) (Decapoda: Atyidae), with a review of the phenomenon and its significance in the Decapoda. *Journal of the Royal Society of New Zealand* **8**, 343–358.

Chace Jr., F. A. 1972. The shrimps of the Smithsonian Bredin Caribbean expeditions, with a summary of the West Indian shallow-water shrimps (Crustacea: Natantia: Decapoda). *Smithsonian Contributions to Zoology* **998**, 1–179.

Charlesworth, D. 1984. Androdioecy and the evolution of dioecy. *Biological Journal of the Linnean Society* **23**, 333–348.

Charlesorth, D. and Charlesworth, B. 1987. Inbreeding depression and its evolutionary consequences. *Annual Review of Ecology and Systematics* **18**, 237–268.

Charnov, E. L. 1979. Natural selection and sex change in pandalid shrimp: test of a life history theory. *American Naturalist* **113**, 715–734.

Charnov, E. L. 1982. *The Theory of Sex Allocation*. Princeton, NJ: Princeton University Press.

Christy, J. H. 1983. Female choice in the resource defense mating system of the sand fiddler crab, *Uca pugilator*. *Behavioral Ecology and Sociobiology* **12**, 160–180.

Christy, J. H. and Salmon, M. 1991. Comparative studies of reproductive behavior in mantis shrimps and fiddler crabs. *American Zoologist* **31**, 329–337.

Clark, R. A. 1997. Dimorphic males display alternative reproductive strategies in the marine amphipod *Jassa marmorata* Holmes (Corophioidea: Ischyroceridae). *Ethology* 103, 531–553.

Conlan, K. E. 1991. Precopulatory mating behavior and sexual dimorphism in the amphipod Crustacea. *Hydrobiologia* 223, 255–282.

Correa, C. J. A. and Thiel, M. 2003. Population structure and operational sex ratio in the rock shrimp, *Rynchocinetes typus* (Decapoda: Caridea). *Journal of Crustacean Biology* 23, 849–861.

Correa, C. J., Baeza, A., Dupre, E., Hinojosa, I. A., and Thiel, M. 2000. Mating behaviour and fertilization success of three ontogenetic stages of male rock shrimp *Rynchocinetes typus* (Decapoda: Caridea). *Journal of Crustacean Biology* 20, 628–640.

Correa, C. J., Baeza, A., Hinojosa, I. A., and Thiel, M. 2003. Male dominance hierarchy and mating tactics in the rock shrimp, *Rhynchocinetes typus* (Decapoda: Caridea). *Journal of Crustacean Biology* 23, 33–45.

Cowan, D. F. 1991. Courtship and chemical signals in the American lobster. *Journal of Shellfish Research* 10, 284.

Cowan, D. F. and Atema, J. 1990. Moult staggering and serial monogamy in American lobsters, *Homarus americanus*. *Animal Behaviour* 39, 1199–1206.

Crow, J. F. 1986. *Basic Concepts in Population, Quantitative and Evolutionary Genetics*. San Francisco, CA: W. H. Freeman.

D'Apolito, L. M. and Stancyk, S. E. 1979. Population dynamics of *Euterpina acutifrons* (Copepoda: Harpacticoida) from North Inlet, South Carolina, with reference to dimorphic males. *Marine Biology* 54, 251–260.

Darwin, C. R. 1874. *The Descent of Man and Selection in Relation to Sex*. London: John Murray.

Dawkins, R. 1980. Good strategy or evolutionary stable strategy? In G. W. Barlow and J. Silverberg (eds.) *Sociobiology: Beyond Nature/Nurture?* pp. 331–367. Boulder, CO: Westview Press.

Dempster, E. R. and Lerner, I. M. 1950. Heritability of threshold characters. *Genetics* 35, 212–236.

Denoel, M., Poncin, P., and Ruwet, J. -C. 2001. Alternative mating tactics in the alpine newt *Triturus alpestris alpestris*. *Journal of Herpetology* 35, 62–67.

Dick, J. T. A. and Elmwood, R. W. 1995. Effects of natural variation in sex ratio and habitat structure on mate-guarding decisions in amphipods (Crustacea). *Behaviour* 133, 985–996.

Diesel, R. 1989. Structure and function of the reproductive system of the symbiotic spider crab *Inachus phalangium*

(Decapoda: Majidae): observations on sperm transfer, sperm storage, and spawning. *Journal of Crustacean Biology* 9, 266–277.

Dominey, W. J. 1984. Alternative mating tactics and evolutionary stable strategies. *American Zoologist* 24, 385–396.

Eberhard, W. G. 1979. The function of horns in *Podichnus agenor* (Dynastinae) and other beetles. In M. S. Blum and N. A. Blum (eds.) *Sexual Selection and Reproductive Competition in Insects*, pp. 231–258. New York: Academic Press.

Eberhard, W. G. 1982. Beetle horn dimorphism: making the best of a bad lot. *American Naturalist* 119, 420–426.

Emlen, D. J. 1996. Artificial selection on horn length–body size allometry in the horned beetle, *Onthophagus acuminatus* (Coleoptera: Scarabaeidae). *Evolution* 50, 1219–1230.

Falconer, D. S. 1989. *Introduction to Quantitative Genetics*, 2nd edn. New York: Longman Scientific and Technical.

Ferveur, J. -F. 1997. The pheromonal role of cuticular hydrocarbons in *Drosophila melanogaster*. *BioEssays* 19, 353–358.

Fisher, R. A. 1958. *The Genetical Theory of Natural Selection*, 2nd edn. New York: Dover.

Flaxman, S. M. 2000. The evolutionary stability of mixed strategies. *Trends in Ecology and Evolution* 15, 482–495.

Forslund, P. 2003. An experimental investigation into status-dependent male dimorphism in the European earwig, *Forficula auricularia*. *Animal Behaviour* 65, 309–316.

Fréchette, J. G., Corrivault G. W., and Couture, R. 1970. Hermaphroditisme protérandrique chez une crevette de la famille des crangonidés, *Argis dentata* Rathbun. *Le Naturaliste Canadien* 97, 805–822.

Gadgil, M. 1972. Male dimorphism as a consequence of sexual selection. *American Naturalist* 106, 574–580.

Gherardi, F. and Calloni, C. 1993. Protandrous hermaphroditism in the tropical shrimp *Athanus indicus* (Decapoda: Caridea) a symbiont of sea urchins. *Journal of Crustacean Biology* 13, 675–689.

Gianola, D. and Norton, H. W. 1981. Scaling threshold characters. *Genetics* 99, 357–364.

Giorgi, D. and Rouquier, S. 2002. Identification of V1R-like putative pheromone receptor sequences in non-human primates: characterization of V1R pseudogenes in marmoset, a primate species that possesses an intact vomeronasal organ. *Chemical Senses* 27, 529–537.

Gowaty, P. A. 1997. Sexual dialectics, sexual selection and variation in reproductive behavior. In P. A. Gowaty (ed.)

Feminism and Evolutionary Biology: Boundaries, Intersections and Frontiers, pp. 351–384. New York: Chapman and Hall.

Gross, M. R. 1985. Disruptive selection for alternative life histories in salmon. Nature 313, 47–48.

Gross, M. R. 1996. Alternative reproductive strategies and tactics: diversity within sexes. Trends in Ecology and Evolution 11, 92–97.

Gross, M. R. and Repka, J. 1998. Game theory and inheritance of the conditional strategy. In L. A. Dugatkin and H. K. Reeve (eds.) Game Theory and Animal Behavior, pp. 168–187. Oxford, UK: Oxford University Press.

Haig, D. and Bergstrom, C. T. 1995. Multiple mating, sperm competition and meiotic drive. Journal of Evolutionary Biology 8, 265–282.

Haq, S. M. 1965. Development of the copepod Euterpina acutifrons with special reference to dimorphism in the male. Proceedings of the Zoological Society of London 144, 175–201.

Haq, S. M. 1972. Breeding of Euterpina acutifrons, a harpacticoid copepod, with special reference to dimorphic males. Marine Biology 15, 221–235.

Haq, S. M. 1973. Factors affecting production of dimorphic males of Euterpina acutifrons. Marine Biology 19, 23–26.

Harikrishnan, M. B., Kurup, M., and Sankaran, T. M. 1999. Differentiation of morphotypes in female population of Macrobrachium rosenbergii (de Man) on the basis of distance function analysis. Proceedings of the 4th Indian Fisheries Forum, 24–28 November, 1996, Kochi, Kerala, pp. 65–67.

Hatziolos, M. E. and Caldwell, R. L. 1983. Role reversal in courtship in the stomatopod Pseudosquilla ciliata (Crustacea). Animal Behaviour 31, 1077–1087.

Hazel, W. N., Smock, R., and Johnson, M. D. 1990. A polygenic model of the evolution and maintenance of conditional strategies. Proceedings of the Royal Society of London B 242, 181–187.

Holdich, D. M. 1968. Reproduction, growth and bionomics of Dynamene bidentata (Crustacea: Isopoda). Journal of Zoology (London) 156, 136–153.

Holdich, D. M. 1971. Changes in physiology, structure and histochemistry during the life history of the sexually dimorphic isopod, Dynamene bidentata (Crustacea: Peracarida). Marine Biology 8, 35–47.

Howard, R. D., DeWoody, J. A., and Muir, W. M. 2004. Transgenic male mating advantage provides opportunity for trojan gene effect in a fish. Proceedings of the National Academy of Sciences of the United States of America 101, 2934–2938.

Hunt, J. and Simmons, L. W. 2001. Status-dependent selection in the dimorphic beetle Onthophagus taurus. Proceedings of the Royal Society of London B 268, 2409–2414.

Jennions, M. D. and Backwell, P. R. Y. 1998. Variation in courtship rate in the fiddler crab Uca annulipes: is it related to male attractiveness? Behavioral Ecology 9, 605–611.

Jivoff, P. and Hines, A. H. 1998. Female behaviour, sexual competition and mate guarding in the blue crab, Callinectes sapidus. Animal Behaviour 55, 589–603.

Jormalainen, V. 1998. Precopulatory mate guarding in crustaceans: male competitive strategy and intersexual conflict. Quarterly Review of Biology 73, 275–304.

Jormalainen, V. and Shuster, S. M. 1999. Female reproductive cycles and sexual conflict over precopulatory mate-guarding in Thermosphaeroma isopods. Ethology 105, 233–246.

Jormalainen, V., Merilaita, S., and Tuomi, J. 1994. Male choice and male–male competition in Idotea baltica (Crustacea, Isopoda). Ethology 96, 46–57.

Jormalainen, V., Shuster, S. M., and Wildey, H. 1999. Reproductive anatomy, sexual conflict and paternity in Thermosphaeroma thermophilum. Marine and Freshwater Behavior and Physiology 32, 39–56.

Karplus, I., Malecha, S. R., and Sagi, A. 2000. The biology and management of size variation. In M. New and W. C. Valenti (eds.) Freshwater Prawn Culture, pp. 259–289. Bet-Dagan, Israel: Agricultural Research Organization.

Knowlton, N. 1980. Sexual selection and dimorphism in two demes of a symbiotic, pair-bonding snapping shrimp. Evolution 34, 161–173.

Koga, T. 1998. Reproductive success and two modes of mating in the sand-bubbler crab Scopimera globosa. Journal of Experimental Marine Biology and Ecology 229, 197–207.

Koga, T., Henmi, Y., and Murai, M. 1993. Sperm competition and the assurance of underground copulation in the sand-bubbler crab, Scopiemera globosa (Brachyura: Ocypodidae). Journal of Crustacean Biology 13, 134–137.

Kurdziel, J. P. and Knowles, L. L. 2002. The mechanisms of morph determination in the amphipod Jassa: implications for the evolution of alternative male phenotypes. Proceedings of the Royal Society of London B 269, 1749–1754.

Kuris, A. M., Ra'anan, Z., Sagi, A., and Cohen, D. 1987. Morphotypic differentiation of male Malaysian giant prawns, *Macrobrachium rosenbergii*. *Journal of Crustacean Biology* 7, 219–237.

Kurup, B. M., Harikrishnan, M., and Sureshkumar, S. 2000. Length–weight relationship of male morphotypes of *Macrobrachium rosenbergii* (de Man) as a valid index for differentiating their developmental pathway and growth phases. *Indian Journal of Fisheries* 47, 283–290.

Levins, R. 1968. *Evolution in Changing Environments*. Princeton, NJ: Princeton University Press.

Lively, C. M. 1986. Canalization versus developmental conversion in a spatially variable environment. *American Naturalist* 128, 561–572.

Lively, C. M., Hazel, W. N., Schellenberger, M. J., and Michelson, K. S. 2000. Predator-induced defense: variation for inducibility in an intertidal barnacle. *Ecology* 81, 1240–1247.

Lloyd, D. G. 1984. Variation strategies of plants in heterogeneous environments. *Biological Journal of the Linnean Society* 21, 357–385.

Lucas, J. and Howard, R. D. 1995. On alternative reproductive tactics in anurans: dynamic games with density and frequency dependence. *American Naturalist* 146, 365–397.

MacDiarmid, A. B. and Butler, M. J. IV. 1999. Sperm economy and limitation in spiny lobsters. *Behavioral Ecology and Sociobiology* 46, 14–24.

Manning, J. T. 1975. Male discrimination and investment in *Asellus aquaticus* (L.) and *A. meridianus* Racovitsza (Crustacea: Isopoda). *Behaviour* 55, 1–14.

Maynard Smith, J. 1982. *Evolution and the Theory of Games*. New York: Cambridge University Press.

Moran, N. A. 1992. Evolutionary maintenance of alternative phenotypes. *American Naturalist* 139, 971–989.

Moreira, G. S. and McNamara, J. C. 1984. Annual variation in abundance of female and dimorphic male *Euterpina acutifrons* (Dana) (Copepoda: Harpacticoida) from the Hauraki Gulf, New Zealand. *Crustaceana* 47, 298–302.

Moreira, G. S., Yamashita, C., and McNamara, J. C. 1983. Seasonal variation in the abundance of the developmental stages of *Euterpina acutifrons* (Copepoda: Harpacticoida) from the São Sebastião Channel, southern Brazil. *Marine Biology* 74, 111–114.

Moriyasu, M. and Benhalima, K. 1998. Snow crabs, *Chionoecetes opilio* (O. Fabricius, 1788) (Crustacea: Majidae) have two types of spermatophore: hypotheses on the mechanism of fertilization and population reproductive dynamics in the southern Gulf of St. Lawrence, Canada. *Journal of Natural History* 32, 1651–1665.

Nagamine, C., Knight, A. W., Maggenti, A., and Paxman, G. 1980. Effects of androgenic gland ablation on male primary and secondary sexual characteristics in the Malaysian prawn, *Macrobrachium rosenbergii* (de Man) (Decapoda, Palaemonidae), with first evidence of induced feminization in a nonhermaphroditic decapod. *General and Comparative Endocrinology* 41, 423–441.

Nakashima, Y. 1987. Reproductive strategies in a partially protandrous shrimp, *Athanas kominatoensis* (Decapoda: Alpheidae): sex changes as the best of a bad situation for subordinates. *Journal of Ethology* 5, 145–159.

Neff, B. D. 2003. Paternity and condition affect cannibalistic behavior in nest-tending bluegill sunfish. *Behavioral Ecology and Sociobiology* 54, 377–384.

Nephew, B. C. and Romero, L. M. 2003. Behavioral, physiological, and endocrine responses of starlings to acute increases in density. *Hormones and Behavior* 44, 222–232.

Noël, P. 1976. L'évolution des caractères sexuels chez *Processa edulis* (Risso)(Decapoda: Natantia). *Vie et Milieu* 26, 65–104.

Orensanz, J. M., Parma, A. M., Armstrong, D. A., Armstrong, J., and Wardrup, P. 1995. The breeding ecology of *Cancer gracilis* (Crustacea: Decapoda: Cancridae) and the mating systems of cancrid crabs. *Journal of Zoology (London)* 235, 411–437.

Peckol, E. L., Troemel, E. R., and Bargmann, C. I. 2001. Sensory experience and sensory activity regulate chemosensory receptor gene expression in *Caenorhabditis elegans*. *Proceedings of the National Academy of Sciences of the United States of America* 98, 11032–11038.

Plaistow, S. J., Troussard, J.-P., and Cézilly, F. 2001. The effect of the acanthodephalan parasite *Pomphorhynchuys laevis* on the lipid and glycogen content of its intermediate host *Gammmarus pulex*. *International Journal of Parasitology* 31, 346–351.

Ra'anan, Z. and Sagi, A. 1989. Alternative mating strategies in male morphotypes of the freshwater prawn *Machrobrachium rosenbergii* (DeMan). *Biological Bulletin* 169, 592–601.

Repka, J. and Gross, M. R. 1995. The evolutionarily stable strategy under individual condition and tactic frequency. *Journal of Theoretical Biology* 176, 27–31.

Ridley, M. 1983. *The Explanation of Organic Diversity: The Comparative Method and Adaptations for Mating*. Oxford, UK: Oxford University Press.

Roberts, L. S. and Janovy Jr., J. 2005. *Foundations of Parasitology*, 7th edn. Boston, MA: McGraw-Hill.

Roff, D. A. 1992. *The Evolution of Life Histories*. New York: Chapman and Hall.

Roff, D. A. 1996. The evolution of threshold traits in animals. *Quarterly Review of Biology* 71, 3–35.

Sagi, A., Ahl, J. S. B., Danaee, H., and Laufer, H. 1994. Methyl farnesoate levels in male spider crabs exhibiting active reproductive behavior. *Hormones and Behavior* 28, 261–272.

Salmon, M. 1984. The courtship, aggression and mating system of a "primitive" fiddler crab (*Uca vocans*: Ocypodidae). *Transactions of the Zoological Society of London* 37, 1–50.

Salmon, M. and Hyatt, G. W. 1983. Spatial and temporal aspects of reproduction in North Carolina fiddler crabs (*Uca pugilator* Bosc). *Journal of Experimental Marine Biology and Ecology* 70, 21–43.

Sassaman, C. 1989. Inbreeding and sex ratio variation in female-biased populations of a clam shrimp, *Eulimnadia texana*. *Bulletin of Marine Science* 45, 425–432.

Sassaman, C. 1991. Sex ratio variation in female-biased populations of notostracans. *Hydrobiologia* 212, 169–179.

Sassaman, C. and Weeks, S. C. 1993. The genetic mechanism of sex determination in the conchostracan shrimp *Eulimnadia texana*. *American Naturalist* 141, 314–328.

Schlicting, C. D. and Pigliucci, M. 1998. *Phenotypic Evolution: A Reaction Norm Perspective*. Sunderland, MA: Sinauer Associates.

Shuster, S. M. 1981. Sexual selection in the Socorro isopod, *Thermosphaeroma thermophilum* (Cole and Bane) (Crustacea: Peracarida). *Animal Behaviour* 29, 698–707.

Shuster, S. M. 1989. Male alternative reproductive behaviors in a marine isopod crustacean (*Paracerceis sculpta*): the use of genetic markers to measure differences in fertilization success among α-, β-, and γ-males. *Evolution* 34, 1683–1698.

Shuster, S. M. 1992. The reproductive behaviour of α, β-, and γ-males in *Paracerceis sculpta*, a marine isopod crustacean. *Behaviour* 121, 231–258.

Shuster, S. M. 2002. Mating strategies, alternative. In M. Pagel *et al.* (eds.) *Encyclopedia of Evolution*, pp. 688–693. Oxford, UK: Oxford University Press.

Shuster, S. M. and Caldwell, R. L. 1989. Male defense of the breeding cavity and factors affecting the persistence of breeding pairs in the stomatopod, *Gonodactylus bredini* (Crustacea: Hoplocarida). *Ethology* 82, 192–207.

Shuster, S. M. and Levy, L. 1999. Sex-linked inheritance of a cuticular pigmentation marker in a marine isopod, *Paracerceis sculpta*. *Journal of Heredity* 90, 304–307.

Shuster, S. M. and Sassaman, C. 1997. Genetic interaction between male mating strategy and sex ratio in a marine isopod. *Nature* 388, 373–376.

Shuster, S. M. and Wade, M. J. 1991. Equal mating success among male reproductive strategies in a marine isopod. *Nature* 350, 606–610.

Shuster, S. M. and Wade, M. J. 2003. *Mating Systems and Strategies*. Princeton, NJ: Princeton University Press.

Shuster, S. M., Ballard, J. O. W., Zinser, G., Sassaman, C., and Keim, P. 2001. The influence of genetic and extrachromosomal factors on population sex ratio in *Paracerceis sculpta*. In R. C. Brusca and B. Kensley (eds.) *Isopod Systematics and Evolution*, vol. 13, *Crustacean Issues*, pp. 313–326. Amsterdam: Balkema.

Shuster, S. M. and Arnold, E. M. 2007. The effect of females on male–male competition in the isopod, *Paracerceis Sculpta*: a reaction norm approach to behavioral plasticity. *Journal of Crustacean Biology* 27, 417–424.

Sinervo, B. 2000. Selection in local neighborhoods, the social environment, and ecology of alternative strategies. In L. Dugatkin (ed.) *Model Systems in Behavioral Ecology*, pp. 191–226. Princeton, NJ: Princeton University Press.

Sinervo, B. 2001. Runaway social games, genetic cycles driven by alternative male and female strategies, and the origin of morphs. *Genetica* 112, 417–434.

Stancyk, S. E. and Moreira, G. S. 1988. Inheritance of male dimorphism in Brazilian populations of *Euterpina acutifrons* (Dana) (Copepoda: Harpacticoida). *Journal of Experimental Marine Biology and Ecology* 120, 125–144.

Stein, R. A. 1976. Sexual dimorphism in crayfish chelae: functional significance linked to reproductive activities. *Canadian Journal of Zoology* 54, 220–227.

Tanaka, K. 2003. Population dynamics of the sponge-dwelling gnathiid isopod *Elaphognathia cornigera*. *Journal of the Marine Biological Association of the United Kingdom* 83, 95–102.

Tanaka, K. and Aoki, M. 1999. Spatial distribution patterns of the sponge-dwelling gnathiid isopod *Elaphognathia cornigera* (Nunomura) on an intertidal rocky shore of the Isu Peninsula, southern Japan. *Crustacean Research* 28, 160–167.

Taru, M., Kanda, T., and Sunobe, T. 2002. Alternative mating tactics of the gobiid fish *Bathygobius fuscus*. *Journal of Ethology* 20, 9–12.

Thompson, D. J. and Manning, J. T. 1981. Mate selection by *Asellus* (Crustacea: Isopoda). *Behaviour* **78**, 178–187.

Tomkins, J. L. and Brown, G. S. 2004. Population density drives the local evolution of a threshold dimorphism. *Nature* **431**, 1099–1103.

Tregenza, T. and Wedell, N. 2000. Invited review: genetic compatibility, mate choice and patterns of parentage. *Molecular Ecology* **9**, 1013–1027.

Upton, N. P. D. 1987. Asynchronous male and female life cycles in the sexually dimorphic, harem-forming isopod *Paragnathia formica* (Crustacea: Isopoda). *Journal of Zoology (London)* **212**, 677–690.

Via, S. and Lande, R. 1985. Genotype–environment interactions and the evolution of phenotypic plasticity. *Evolution* **39**, 505–522.

Wada, K. 1986. Burrow usurpation and duration of surface activity in *Scopimera globosa* (Crustacea: Brachyura: Ocypodidae). *Publications of the Seto Marine Biological Laboratory* **31**, 327–332.

Wade, M. J. 1979. Sexual selection and variance in reproductive success. *American Naturalist* **114**, 742–764.

Wade, M. J. and Shuster, S. M. 2002. The evolution of parental care in the context of sexual selection: a critical reassessment of parental investment theory. *American Naturalist* **160**, 285–292.

Wade, M. J. and Shuster, S. M. 2004. Sexual selection: harem size and the variance in male reproductive success. *American Naturalist* **164**, E83–E89.

Wade, M. J., Shuster, S. M., and Demuth, J. P. 2003. Sexual selection favors female-biased sex ratios: the balance between the opposing forces of sex-ratio selection and sexual selection. *American Naturalist* **162**, 403–414.

Weeks, S. C. and Zucker, N. 1999. Rates of inbreeding in the androdioecious clam shrimp *Eulimnadia texana*. *Canadian Journal of Zoology* **77**, 1402–1408.

West-Eberhard, M. J. 2003. *Developmental Plasticity and Evolution*. Oxford, UK: Oxford University Press.

Winn, A. A. 1996. The contributions of programmed developmental change and phenotypic plasticity to within-individual variation in leaf traits in *Dicerandra linearifolia*. *Journal of Evolutionary Biology* **9**, 737–752.

Wolf, J. B. and Wade, M. J. 2001. On the assignment of fitness to parents and offspring: whose fitness is it and when does it matter? *Journal of Evolutionary Ecology* **14**, 347–358.

10 · Alternative reproductive tactics in fish

MICHAEL TABORSKY

CHAPTER SUMMARY

Among vertebrates, fish show by far the greatest variability of alternative reproductive tactics (ARTs). Usually, males attempting to monopolize access to females or fertilizations are parasitized by conspecific male competitors (Taborsky 1997). This is so common in fish (Mank and Avise 2006) that it appears to be the rule rather than the exception: in fish with external fertilization, 170 species belonging to 32 families have been described to show ARTs (Table 10.1). Apart from being common, ARTs in fish are also exceptionally variable. Parasitic males exploiting the effort of conspecific competitors may do so by surreptitious participation in spawning; they may mimic females in appearance and behavior to reach their goal; intercept approaching mates or steal eggs from neighbors to attract mates to their nest; force copulations in viviparous species; gain access to mates by cooperating with their competitors; or oust a territory owner aggressively to spawn in his nest before letting him care for their brood. Sometimes, three or more alternative tactics may exist within a species (Taborsky 1994, 2001, Avise et al. 2002). Our understanding of sexual selection mechanisms and the concepts underlying conventional classifications of mating patterns largely ignore the existence and importance of ARTs (Emlen and Oring 1977, Wittenberger 1979, Davies 1991, Andersson 1994, 2005). One could argue that the way reproductive behaviour in animals is viewed and categorized today would be different if it had been developed on the basis of fish reproduction instead of bird mating systems. The existence and form of ARTs in fish is important also for population ecology, conservation, and speciation (the latter because of hybridization events caused by ARTs: Taborsky 1994, Wirtz 1999). In this chapter, I discuss why alternative reproductive behaviors are so frequent in fish compared to other taxa; how this relates to sperm competition and how males of different types cope with it; to what extent ARTs

result from phenotypic plasticity or fixed life-history patterns; our understanding of the origin of alternative reproductive phenotypes and the importance of genes and environment; how alternative mating patterns in fish may be maintained in a population and why cooperation between reproductive competitors may be involved; what role females play for male alternative behaviors; and what forms of ARTs we find in female fish. Finally, I shall discuss important areas for future research of alternative reproductive phenotypes in fish.

10.1 WHY ARE ARTs SO PROMINENT IN FISH?

There are four potential reasons why alternative reproductive tactics (ARTs) are more frequent and more variable in male fish than in males of other vertebrate taxa (see Table 10.1).

(1) *Fertilization mechanism.* The vast majority of fish taxa show external fertilization of eggs (Breder and Rosen 1966). This has two important consequences. First, it is difficult for males to monopolize access to partners or fertilizable eggs. Mate guarding is not really an option when eggs are fertilized outside of the female body and potential competitors can access these eggs in a three-dimensional space. Second, external fertilization selects for large numbers of sperm, which in turn is a precondition for a successful role in sperm competition. In contrast, males fertilizing eggs inside the female will be selected to economize in gametic expenditure (Parker 1984).

The variability of ARTs in fish also relates to their diverse spawning patterns. In fish with external fertilization eggs may be released in the water column (pelagic spawning), on the ground, or on/in a substrate (demersal or benthic spawning). With pelagic spawning,

Alternative Reproductive Tactics, ed. Rui F. Oliveira, Michael Taborsky, and H. Jane Brockmann. Published by Cambridge University Press.
© Cambridge University Press 2008.

Table 10.1. *Fish species with external fertilization exhibiting alternative reproductive tactics. This table lists 46 examples of teleosts with external fertilization and ARTs that had not been included in a previous review of the subject (Taborsky 1994). In total (Table 1 from Taborsky 1994 and this table) there are 170 species from 32 families included*

Family	Species	References
Acipenseridae	*Acipenser fulvescens*	Bruch and Binkowski 2002
Salmonidae	*Thymallus thymallus*	Darchambeau and Poncin 1997
	Oncorhynchus masou	Nikolsky 1963, Koseki and Maekawa 2000, Yamamoto and Edo 2002, Kano *et al.* 2006
	O. tshawytscha[a]	Taylor 1989, Beckman and Larsen 2005
	O. mykiss	Liley *et al.* 2002, Seamons *et al.* 2004
	Salvelinus leucomaenis	Maekawa *et al.* 1994
	S. confluentus	James and Sexauer 1997
Esocidae	*Esox lucius*[b]	Fabricius and Gustafson 1958
Cyprinidae	*Clinostomus elongates*[b]	Koster 1939
	Rhinichthys atratulus[b]	Raney 1940
	Rhodeus ocellatus[b]	Kanoh 1996
	Rutilus rutilus[b]	Wedekind 1996
	Pseudorasbora parva[b]	Maekawa *et al.* 1996
Gasterosteidae	*Spinachia spinachia*	Jones *et al.* 1998
	Hypoptychus dybowskii	Akagawa and Okiyama 1993, Narimatsu and Munehara 2001
Oryziidae	*Oryzias latipes*[b]	Grant *et al.* 1995
Gadidae	*Gadus morhua*	Hutchings *et al.* 1999, Bekkevold *et al.* 2002
Serranidae	*Serranus tabacarius*[b]	Petersen 1995
Centrarchidae	*Lepomis punctatus*	DeWoody *et al.* 2000
	L. marginatus	Mackiewicz *et al.* 2002
Percidae	*Perca fluviatilis*[b]	Treasurer 1981, Wirtz and Steinmann 2006
Cichlidae	*Ctenochromis horei*[b]	Ochi 1993
	Cyathopharynx furcifer	Rossiter and Yamagishi 1997
	Telmatochromis temporalis	Mboko and Kohda 1999, Katoh *et al.* 2005
	T. vittatus	Ota and Kohda 2006
	Julidochromis ornatus	Awata *et al.* 2005, 2006
Pomacentridae	*Stegastes nigricans*[b]	Karino and Nakazono 1993
	Abudefduf abdominalis[b]	Tyler 1995
	Chromis chromis	Picciulin *et al.* 2004
Labridae	*Thalassoma duperrey*	Hourigan *et al.* 1991
	Halichoeres marginatus[b]	Shibuno *et al.* 1993
Scaridae	*Sparisoma cretense*	De Girolamo *et al.* 1999
Pinguipedidae	*Parapercis snyderi*	Ohnishi *et al.* 1997
Hexagrammidae	*Hexagrammos otakii*	Munehara and Takenaka 2000
Gobiidae	*Gobio gobio*	Poncin *et al.* 1997
	Zosterisessor ophiocephalus	Mazzoldi *et al.* 2000
	Gobius niger	Mazzoldi and Rasotto 2002, Immler *et al.* 2004
	Bathygobius fuscus	Taru *et al.* 2002
	Rhinogobius sp.[a]	Okuda *et al.* 2003
	Lythrypnus dalli[a]	Drilling and Grober 2005

Table 10.1. (*Cont.*)

Family	Species	References
Hypoptychidae	*Hypoptychus dybowskii*[b]	Akagawa and Okiyama 1993
Blenniidae	*Salaria pavo*[b]	Ruchon *et al.* 1995, Gonçalves *et al.* 1996, 2003
	Scartella cristata	Neat *et al.* 2003a, Mackiewicz *et al.* 2005
	Salaria fluviatilis	Neat *et al.* 2003b
Tripterygiidae	*Axoclinus nigricaudus*	Neat 2001
Monacanthidae	*Rudarius ercodes*[b]	Akagawa and Okiyama 1995, Kawase and Nakazono 1995

[a] Existence of ARTs as derived from precocious maturation of a proportion of males.
[b] Examples included in the compilation of the number of species per family with simultaneous parasitic spawning in Taborsky 1998 (Table 1), but without species name and references.

several males may release sperm at a time causing scramble competition among sperm (Bekkevold *et al.* 2002). However, also in such seemingly egalitarian spawning assemblages there may be differences in the degree of investment of males in privileged access to eggs and some sort of pair spawning (Brawn 1961, Hutchings *et al.* 1999), with females benefiting from optimal mate selection (Rudolfsen *et al.* 2005). A well-known case is the spawning stations defended by males in many reef fishes (Warner and Robertson 1978, Warner and Hoffmann 1980, Thresher 1984). With demersal spawning, there is enormous variation (a) in the way eggs are fertilized, including catfish where females drink sperm and fertilize eggs externally after intestinal sperm passage (Kohda *et al.* 1995), or sculpin in which copulation may be combined with external fertilization (Munehara 1988, Munahara *et al.* 1989, 1991); and (b) in the participation of male competitors in spawning, as well as in their alternative reproductive behavior (see Taborsky 1994 for review); males may either attempt to monopolize the spawning site (bourgeois tactic: Taborsky 1997) or sneak secretly towards it, streak rapidly into a nest or enter it boldly in female attire (female mimicry: Barlow 1967, Dominey 1980, Goncalves *et al.* 1996) or as a despotic usurper (piracy: van den Berghe 1988). Males in unfavorable positions may even fan their sperm towards a spawning pair (Gronell 1989).

ARTs and sperm competition do occur also in connection with internal fertilization in fish, however (Winge 1937, Hildemann and Wagner 1954, Darling *et al.* 1980, Farr 1980a, b, Constantz 1984, Munehara

et al. 1990, Bisazza 1993, Evans *et al.* 2003a, Marcias-Garcia and Saborio 2004), which shows that fertilization mode is clearly not the only reason for the prevalence of variable reproductive behaviors in fish. When fertilization is internal, alternative tactics often involve a courting type displaying to females before mating and a coercive type forcing copulations surreptitiously by gonopodial thrusts (Farr 1980a, b, Zimmerer 1982, Ryan and Causey 1989). Conditions for ARTs and adaptations of males differ at various levels between fish species showing internal and external fertilization (Taborsky 1998), but to my knowledge a comparative study clarifying the importance of the fertilization mode for the existence and evolution of ARTs in fish with external and internal fertilization is still missing. A group of particular interest is the seahorses and pipefish (Syngnathidae), in which there is often internal fertilization in the *male* sex (Fiedler 1954, Vincent *et al.* 1992, Jones and Avise 1997a, b, 2001, Avise *et al.* 2002). In species with sex role reversal we should expect that ARTs develop primarily in the female sex, as females compete for males more than the other way round (Berglund *et al.* 1989, 2005, Rosenqvist 1990, Berglund and Rosenqvist 1993, Jones *et al.* 2000, 2001, 2005). However, due to their high initial investment in gametes females gain less from employing bourgeois and parasitic reproductive behaviors than males (see below), so it is perhaps not surprising that female ARTs appear not to exist in syngnathids (Vincent *et al.* 1992, Jones and Avise 2001).

(2) *Indeterminate growth.* In contrast to homeotherm vertebrates, the vast majority of fishes do not stop to

grow after maturation. This has important effects on reproduction. In the context of ARTs, the most significant consequence is that there are often enormous intrasexual size differences within species (Taborsky 1999). Perhaps the most impressive example is Atlantic salmon, where anadromous males after foraging at sea are more than 600 times heavier when reproducing than males remaining in their natal rivers ("mature parr": Jones and King 1952, Jones 1959, Gage et al. 1995, Fleming 1996, 1998, Esteve 2005). No wonder that these males cannot compete with each other on equal terms for the fertilization of eggs. The intrasexual size dimorphism (ISD) is much smaller in animals with determinate growth. In 490 species of passeriform birds listed by Dunning (1992), for example, the largest males of a species are on average only 19% (or 1.19 times) heavier than their smallest male conspecifics (median: Taborsky 1999). Apparently, no specialized parasitic reproductive tactics occur in passerine birds. On the contrary, in eight species of fish with ARTs from which such data were available, comprising salmon, sunfish, cichlids, wrasses, and the triplefin blenny, the largest males of a species were on average 18 times heavier than their smallest male conspecifics (range 7 to 625: Taborsky 1999). This small sample may not be representative, but it helps to illustrate the point. A large ISD will select for divergence in optimal reproductive tactics of competitors, with large individuals taking advantage of monopolizing mate access (bourgeois tactic) and small ones using surreptitious alternatives (parasitic tactics). Of course a large ISD in a species may not only be a cause for the evolution of ARTs, but it will be affected by this process in turn. A proper comparative analysis testing for the potential influence of indeterminate growth and ISD on the evolution of ARTs should proceed in two steps. First, major taxa with and without indeterminate growth should be compared with regard to ISD levels and the occurrence of ARTs using appropriate comparative techniques (Harvey and Pagel 1991). Second, among a taxon with indeterminate growth, the occurrence of ARTs should be related to the magnitude of ISD. Even though this analysis cannot separate completely between cause and effect (because once ARTs have evolved this will feed back on the evolution of ISD levels), it may give a clear hint on a potential functional relationship.

(3) *Parental roles.* The mode of brood care may influence the evolution of ARTs for two reasons. First, paternal investment is particularly frequent in fish (Blumer 1979, 1982, Clutton-Brock 1991). This opens a potential for male competitors to exploit such investment by the performance of ARTs, similar to the adoption of egg dumping by females, where parental investment can be exploited in a similar way (Andersson 1984, Tallamy 2005). To test whether this exploitation potential may be an important evolutionary cause of alternative reproductive phenotypes in fish I checked brood-care patterns of the two fish families from which the most species are known to have specialized ARTs, cichlids and wrasses (Taborsky 1999). Of 14 cichlid species with specialized ARTs, nine have female-only care and five are biparental; of 25 wrasses, 17 show no parental care and eight have paternal care only. These data do not suggest a strong influence of paternal care prevalence on the prevalence of ARTs in fish by creating an exploitation potential for parasitic tactics, but a proper comparative analysis is still missing. Second, fish are much more diverse in their brood-care behavior than other classes of vertebrates, which may be an important cause of the great variability of ARTs in this taxon. While in amphibians and reptiles there is either no brood care or female care only (with very few exceptions), in birds biparental care prevails and in mammals we find female care only, with some male participation in a few taxa. Fish exhibit all possible patterns from no care to uniparental (male or female care only), biparental, multiparental, and alloparental care (Breder and Rosen 1966, Blumer 1982, Taborsky 1994), which may in turn influence variability in mating patterns and ARTs. In fish we even find the unlikely combination of internal fertilization with paternal care (Ragland and Fischer 1987), which may also involve ARTs (Munehara et al. 1990).

(4) *Sex determination.* The flexible sex determination of fishes may be another peculiarity adding to the variability of reproductive phenotypes found in this taxon, though it may not be responsible for the prevalence and evolution of ARTs in this group in general. No other vertebrate class disposes of so many different genetic and environmental mechanisms of sex determination, which is reflected in the existence of gonochorism, simultaneous and sequential hermaphroditism, and the latter with either males (protandry) or females (protogyny) preceding each other (Demski 1987, Shapiro 1987, Ross 1989). Sometimes, gonochorism and sex change coexist

within species (Robertson *et al.* 1982). This flexibility potential allows for adaptive responses of sex allocation to environmental conditions (Warner *et al.* 1975, Sunobe and Nakazono 1993, Kuwamura *et al.* 1994, Devlin and Nagahama 2002, Godwin *et al.* 2003, Munoz and Warner 2003, 2004, Munday *et al.* 2006). The high variability of sex determination and differentiation mechanism may also facilitate the evolution of alternative tactics in fish (see Chapter 7). ARTs are widespread in hermaphroditic species, both with simultaneous (Reinboth 1962, Fischer 1980, 1986, Petersen and Fischer 1986, 1996, Petersen 1987, 1990, 1995, Cheek *et al.* 2000) and sequential hermaphrodites (Robertson and Choat 1974, Warner *et al.* 1975, Robertson and Warner 1978, Warner and Robertson 1978, Warner and Hoffman 1980, Lejeune 1987, Colin and Bell 1991, Shibuno *et al.* 1993, Ohnishi *et al.* 1997, Munoz and Warner 2004, Drilling and Grober 2005).

In comparison to invertebrates, ARTs in fish are characterized mainly by the dichotomy between bourgeois monopolization of reproduction and parasitic exploitation, i.e., the producer/scrounger paradigm (Barnard 1984), whereas in insects, for example, the alternatives often involve "staying" versus "roving" (or "dispersing"), as mating sites may differ for individuals performing divergent tactics (see Chapter 8). No example comes to my mind showing a similar ecological separation of spawning sites in fish, together with the respective divergence in reproductive tactics employed by alternative phenotypes. This does not mean, however, that such cases do not exist; we have sufficient information about reproductive patterns from only a minute proportion of the more than 25 000 species of teleosts (<5% of described species). The occurrence of intraspecific morph divergence resulting from other than sexual selection processes (e.g., trophic niche specialization) may provide the conditions required for the evolution of ARTs that are not based on inter-tactic competition (see below).

10.2 ADAPTATIONS TO SPERM COMPETITION

The high incidence of simultaneous parasitic spawning creates an enormous potential for sperm competition in fish (Taborsky 1994, 1998, Petersen and Warner 1998). As outlined above, the prevalence of external fertilization in an aquatic environment is a precondition for this pattern. It could be, however, that the predominant presence of sperm

competition in externally as opposed to internally fertilizing fish is more apparent than real. In external fertilization, the existence of sperm competition can be much more easily observed than with internal fertilization, where matings are temporally separated (Petersen and Warner 1998). In a comparison of multiple mate participation between some fish species with external and internal fertilization (the latter including mouthbrooding cichlids), DeWoody and Avise (2001) found evidence for greater numbers of mates in external fertilizers, which meets our expectation. I predict that sperm competition and multiple parentage is more common with external fertilization in general, due to (a) the intrinsic problem of mating monopolization (Taborsky 1994, 2001), (b) the superior possibilities of females to influence fertilization when it is internal (Pitcher *et al.* 2003, Pilastro *et al.* 2004), which limits the success potential of non-preferred males, and (c) the selective advantage of sperm economy when fertilization is internal as opposed to external (Parker 1984).

The potential of sperm competition selects for optimal allocation of resources in response to the risk and intensity of sperm competition with regard to sperm production (Parker 1990a, b, 1993, Parker and Begon 1993, Parker and Ball 2005) and release (Shapiro *et al.* 1994, Parker *et al.* 1996, Pilastro *et al.* 2002, Evans *et al.* 2003a, Zbinden *et al.* 2003, 2004); sperm morphology and physiology (Parker 1984, 1993, Parker and Begon 1993, Gage *et al.* 1995, 2004, Ball and Parker 1996, Stockley *et al.* 1997, Leach and Montgomerie 2000, Balshine *et al.* 2001a, Vladic and Järvi 2001, Vladic *et al.* 2002, Burness *et al.* 2004, Snook 2005); male morphology and physiology (de Jonge *et al.* 1989, Bass 1992, Brantley *et al.* 1993a,b, Scaggiante *et al.* 1999; Taborsky 1999, Oliveira *et al.* 2001a-f, 2003, 2005); and male behavior (see Taborsky 1994, 1999 for review). Table 10.2 summarizes traits influenced by sperm competition at different levels. In general, adaptations to monopolization (bourgeois tactic) are caused by intrasexual selection, intersexual selection, or by both processes, whereas adaptations to reproductive parasitism result usually from intrasexual selection only, as females typically either ignore or attempt to avoid male reproductive parasites (Taborsky 1999). As resources and reserves are usually limited, allocation decisions in both tactics inevitably involve trade-offs (Snook 2005), causing for example a negative correlation between behavioral and gonadal effort. This is probably one reason why testes per body mass (measured by the gonadosomatic index, GSI) are usually larger in parasitic than in bourgeois males (Taborsky 1994, 1998), although this relationship may be influenced also by different levels of sperm competition faced by these males (Parker 1990b, Petersen and Warner 1998) or by different options at sperm release (distance, timing: Parker 1990a, Foote *et al.* 1997, Blanchfield and Ridgway 1999, Hutchings *et al.* 1999, Liley *et al.* 2002, Blanchfield *et al.* 2003, Hoysak *et al.* 2004, Stoltz and Neff 2006) and effects of allometric organ growth (Tomkins and Simmons 2002, Stoltz *et al.* 2005).

Ejaculates competing for fertilization may either do so on equal terms like in a lottery ("fair raffle"), or sperm from one or another ejaculate may have superior fertilization potential ("loaded raffle": Parker 1990a, Neff and Wahl 2004). If more than one male is participating in spawning, the fertilization potential varies with potential differences in sperm (e.g., swimming speed: Kazakov 1981, Gage *et al.* 1995, 2004), ejaculates (ejaculate size and density often differs between tactics: Pilastro and Bisazza 1999, Schärer and Robertson 1999, Alonzo and Warner 2000a, Leach and Montgomerie 2000, Neff *et al.* 2003; and it may be adjusted to sperm competition intensity and potential female fecundity: Shapiro *et al.* 1994, Evans *et al.* 2003a, Aspbury and Gabor 2004, Zbinden *et al.* 2004), or male behavior (e.g., timing and proximity at sperm release: Gile and Ferguson 1995, Hutchings *et al.* 1999, Hoysak and Liley 2001, Stoltz and Neff 2006). In internal fertilization, mating sequence may be an important determinant of fertilization potential (Evans and Magurran 2001, Neff and Wahl 2004; but see Marcias-Garcia and Saborio 2004), and post-copulatory choice of females may add another source of nonrandom variation (Evans *et al.* 2003b, Pitcher *et al.* 2003). It is almost inevitable that differences in fertilization potential exist on one or the other of these levels, even in broadcast spawners (e.g., Brawn 1961, Wedekind 1996, Hutchings *et al.* 1999; see Marshall *et al.* 2004 for an example in marine invertebrates); therefore, the "fair raffle" is probably nonexistent in the fertilization of fish eggs. This has been suggested also by empirical studies including both internal (Evans and Magurran 2001, Neff and Wahl 2004) and external fertilizers (Vladic and Järvi 2001, Neff *et al.* 2003, Hoysak *et al.* 2004).

It should be stressed that some of the alleged functions of apparent adaptations to sperm competition (Table 10.2) still need further clarification, as the functional significance of important sperm characteristics is not well understood. Flagellum length was assumed to relate to sperm swimming speed (Gomendio and Roldan 1991, Ball and Parker 1996, Stockley *et al.* 1997, Balshine *et al.* 2001a), for example, but in the most intensively studied species, Atlantic salmon, this is not supported (Gage *et al.* 1998, 2002), and adaptive differences in sperm dimensions between bourgeois and

parasitic males were not found (Vladic *et al.* 2002). Evidence for such differences between males displaying ARTs in bluegill sunfish is controversial (Leach and Montgomerie 2000, Schulte-Hostedde and Burness 2005), and in sockeye salmon sperm from alternative male phenotypes did not differ in motility or fertilization propensity (Hoysak and Liley 2001). Also, a trade-off was assumed between sperm swimming speed and sperm longevity (Stockley *et al.* 1997), but again this could not be confirmed in a study of Atlantic salmon (Gage *et al.* 2002), where a trade-off between sperm size and numbers was not confirmed either (Gage *et al.* 1998). In roach there is even a positive correlation between sperm velocity and longevity (Kortet *et al.* 2004b). The functional importance of sperm head dimensions is unclear, except that a larger midpiece may relate to greater energy reserves and high energy production (Jamieson 1991, Lahnsteiner 2003). Sperm number has been shown to be important for fertilization probability in bluegill sunfish (Neff *et al.* 2003), but not so in Atlantic salmon (Vladic *et al.* 2002, Gage *et al.* 2004). Sperm number is a function of both ejaculate size and density (Gage *et al.* 1995, Schärer and Robertson 1999, Vladic and Järvi 2001, Liley *et al.* 2002, Neff *et al.* 2003), which may however correlate with each other either negatively or positively (Leach and Montgomerie 2000, Wirtz and Steinmann 2006). In direct ejaculate competition a numerical disadvantage of sperm of parasitic males may be compensated for by superior sperm quality in Atlantic salmon (Vladic and Järvi 2001, Vladic *et al.* 2002, Gage *et al.* 2004), whereas a similar relationship in bluegill sunfish appears to exist with opposite signs: here bourgeois males can make up for lower ejaculate sperm densities by their greater sperm longevity (Neff *et al.* 2003, Schulte-Hostedde and Burness 2005), even though initial sperm swimming speed is higher in parasite male sperm (Burness *et al.* 2004, Schulte-Hostedde and Burness 2005). Often, stripped volume of sperm is used as an approximation to natural ejaculate size (Gage *et al.* 1995, Stockley *et al.* 1997, Leach and Montgomerie 2000, Liley *et al.* 2002), calibration is usually missing. Overall, it seems that we need a better understanding of the fertilization mechanisms in fish and improved experimental scrutiny to study predictions from sperm competition theory at the level of spermatozoa (see Engqvist and Reinhold 2005).

10.3   MAINTENANCE OF ARTs IN A POPULATION

A fundamental question in the study of ARTs is how discontinuous phenotypic variation of conspecific, same-sex

Table 10.2. *Levels of adaptation to sperm competition*

Level	Trait	Functional significance	Evidence (examples)	References
Sperm				
Morphology	Sperm length	Swimming speed	Interspecific comparison	Balshine *et al.* 2001a
				Gage *et al.* 2002, Stockley *et al.* 1997
		Longevity (neg.)	*Salmo salar*, interspecific comparison	Snook 2005 (review)
	Sperm head	Midpiece size (energy reserves and production)	Internal vs. external fertilizers	Jamieson 1991 (review) (cf. Lahnsteiner 2003)
Physiology	Metabolism	Swimming speed	*Salmo salar*	Kazakov 1981, Gage *et al.* 1995, 2004
			Salvelinus alpinus	Rudolfsen *et al.* 2006
			Gasterosteus aculeatus	De Fraipont *et al.* 1993
			Oncorhynchus mykiss	Lahnsteiner *et al.* 1998
			Lepomis macrochirus	Burness *et al.* 2004, 2005, Schulte–Hostedde and Burness 2005
		Longevity	*Symphodus melops*	Uglem *et al.* 2001
			Gasterosteus aculeatus	De Fraipont *et al.* 1993
			Salmo salar	Gage *et al.* 1995
			Lepomis macrochirus	Neff *et al.* 2003
			Symphodus melops	Uglem *et al.* 2002
			Oncorhynchus mykiss	Lahnsteiner 1998
		Metabolic biochemistry	*Salmo salar*	Vladic and Järvi 2001
Ejaculate				
Spermatozoa	Ejaculate size	Sperm number	*Salmo salar*	Kazakov 1981, Gage *et al.* 1995
			S. gairdneri	Linhart 1984
			Gasterosteus aculeatus	De Fraipont *et al.* 1993
			Gambusia holbrooki	Evans *et al.* 2003a
			Thalassoma bifasciatum	Shapiro *et al.* 1994
			Symphodus ocellatus	Alonzo and Warner 2000a
			Hexagrammos otakii	Munehara and Takenaka 2000
			Interspecific meta-analysis	Stockley *et al.* 1997

Table 10.2. (cont.)

Level	Trait	Functional significance	Evidence (examples)	References
	Ejaculate density	Sperm concentration	*Oncorhynchus mykiss* *Salvelinus alpinus*	Liley *et al.* 2002 Lijedal and Folstad 2003, Rudolfsen *et al.* 2006
			Lepomis macrochirus	Leach and Montgomerie 2000, Neff *et al.* 2003
			Thalassoma bifasciatum	Schärer and Robertson 1999
Accessory substances	Ejaculate composition	Biochemistry of seminal plasma	Interspecific comparison	Piironen and Hyvärinen 1983
		Mucins	*Zosterisessor ophiocephalus, Gobius niger, Knipowitschia panizzae*	Marconato *et al.* 1996
			Zosterisessor ophiocephalus	Ota *et al.* 1996, Scaggiante *et al.* 1999, Mazzoldi *et al.* 2000
			Gobius niger	Rasotto and Mazzoldi 2002
Male Morphology	Body size	Access to mates	General pattern	Taborsky 1999 (review)
			Oncorhynchus kisutch	Healey and Prince 1998
			Lamprologus callipterus	Taborsky 2001, Sato *et al.* 2004
	Testis size	Ejaculate production	Many species	Billard 1986 (review), Taborsky 1994 (review), Petersen and Warner 1998 (review)
			Oncorhynchus masou	Koseki and Maekawa 2002
			Neolamprologus pulcher	Fitzpatrick *et al.* 2006
			Telmatochromis vittatus	Ota and Kohda 2006a
			Thalassoma duperrey	Hourigan *et al.* 1991
			Symphodus melops	Uglem *et al.* 2000, 2001, 2002
			Zosterisessor ophiocephalus	Scaggiante *et al.* 2004
			Salaria pavo	Oliveira *et al.* 2001f
			Axoclinus nigricaudus	Neat 2001
			Porichthys notatus	Brantley and Bass 1994

Trait	Function	Species / context	References
Testis structure		Interspecific meta-analysis	Stockley et al. 1997
Accessory glands	Ejaculate production	*Tripterygion tripteronotus*	De Jonge et al. 1989
		Zosterisessor ophiocephalus	Billard 1986 (review), Lahnsteiner et al. 1993, Scaggiante et al. 1999
		Gobius niger	Rasotto and Mazzoldi 2002, Immler et al. 2004
	Sperm survival	*Salaria pavo*	Lahnsteiner et al. 1990, Ruchon et al. 1995
		Scartella cristata	Neat et al. 2003a
		Tripterygion tripteronotus	De Jonge et al. 1989
		Porichthys notatus	Barni et al. 2001
		Interspecific comparison (gobies)	Mazzoldi et al. 2005
		Gobiidae	Miller 1984 (review), Fishelson 1991 (review)
		Blennioidei	Rasotto 1995 (review), Richtarski and Patzner 2000 (review)
	Signaling	*Salaria pavo*	Laumen et al. 1974, Lahnsteiner et al. 1993
		Scartella cristata	Neat et al. 2003a
		Axoclinus nigricaudus, A. carminalis	Neat 2001
Secondary sex traits	Signaling	Color: examples from 10 families	Taborsky 1994 (review)
		Gasterosteus aculeatus (color)	Frischknecht 1993, Wedekind et al. 1998
		Salaria pavo (nuptial crest)	Eggert 1932, Ruchon et al. 1995, Gonçalves et al. 1996
		Parablennius sanguinolentus (anal gland)	Oliveira et al. 2001c
		Rutilus rutilus (skin tubercles)	Wedekind 1992, Kortet et al. 2003, 2004a
		Porichthys notatus (sound organ, neural modifications)	Bass and Andersen 1991, Bass 1992, Brantley et al. 1993a,

Table 10.2. (cont.)

Level	Trait	Functional significance	Evidence (examples)	References
		Weapons	Zosterisessor ophiocephalus (sound-production)	Brantley and Bass 1994, Bass et al. 1996
			Oncorhynchus kisutch	Malavasi et al. 2003
			Salmo salar (hooknose)	Davidson 1935
				Tchernavin 1938, Jones 1959
	Male dimorphism	Concealment (female mimicry)	>30 species (10 different families)	Taborsky 1994 (review)
			Oncorhynchus masou	Kano et al. 2006
			Rhinogobius sp.	Okuda et al. 2003
			Salaria pavo	Gonçalves et al. 1996, 2005
Physiology	Endocrine regulation	Control of organs and behavior		Moore 1991 (review), Brantley et al.1993b (review), Oliveira 2006 (review)
			Salmo salar, Salaria pavo, Parablennius parvicornis	Stuart-Kregor et al. 1981, Oliveira et al. 2001b, c, d, 2005
			Oreochromis mossambicus	Oliveira et al. 2005
			Zosterisessor ophiocephalus	Scaggiante et al. 2004, 2006
			Tripterygion tripteronotus	De Jonge et al. 1989
			Interspecific comparison	Seiwald and Patzner 1989
	Pheromones	Communication		Liley and Stacey 1983 (review), Stacey 2003 (review), Stacey et al. 2003 (review)
			Petromyzon marinus	Li et al. 2002
			Salmo salar	Moore and Scott 1991
			Salvelinus alpinus	Sveinsson and Hara 1995
			Gobius niger	Colombo et al. 1980, Locatello et al. 2002
	Energy expenditure	Signaling	Gasterosteus aculeatus	Frischknecht 1993
		provisioning	Neolamprologus pulcher	Grantner and Taborsky 1998

260

Growth	Strategic size adjustment	*N. pulcher*	Taborsky 1984, Heg *et al.* 2004
Behavior	Tactic applied	Access type	147 species of 29 families	Taborsky 1994, 1998, 1999 (reviews)
	Timing of sperm release	Synchronization with egg laying	*Salmo salar*	Mjolnerod *et al.* 1998
			Oncorhynchus nerka	Foote *et al.* 1997, Hoysak *et al.* 2004
			Rhodeus sericeus	Reichard *et al.* 2004a
			Lepomis macrochirus	Stoltz and Neff 2006
			Hexagrammos otakii	Munehara and Takenaka 2000
	Proximity at sperm release	Position in sperm race	*Gadus morhua*	Hutchings *et al.* 1999
			Oncorhynchus keta	Schröder 1981, 1982
			O. kisutch	Gross 1985
			O. nerka	Hanson and Smith 1967, Foote *et al.* 1997
			Salvelinus fontinalis	Blanchfield *et al.* 2003
			Lepomis macrochirus	Fu *et al.* 2001, Stoltz and Neff 2006

reproductive competitors is maintained in a population (Gross 1996; see Chapters 1 and 2). This question may be dealt with at three different levels (Austad 1984, Taborsky 1998):

10.3.1 Pattern

Variation in reproductive phenotypes may result from opportunistic responses to specific conditions, from a switch between tactics at some stage in life, or from the occurrence of divergent fixed patterns persisting for life (Henson and Warner 1997, Taborsky 1998, Brockmann 2001, Rhen and Crews 2002). All three possibilities are frequently found in fish. The first, most flexible, pattern is represented, for instance, by territorial neighbors taking advantage of nearby spawning events through attempted fertilizations of eggs by parasitic intrusions. This has been described for sticklebacks (e.g., van den Assem 1967, Li and Owings 1978, Rowland 1979, Wootton 1984), but it occurs also in many other fish families including suckers, sunfish, cichlids, damselfish, parrotfish, surgeonfish, and trypterygiids (see Taborsky 1994 for review). This opportunistic and fully conditional reproductive parasitism is reminiscent of extra-pair fertilizations in birds, where territorial neighbors also perform both roles, bourgeois and parasitic, simultaneously and depending on momentary circumstances (Petrie and Kempenaers 1998; see Chapter 13). In contrast to such "simultaneous" ARTs, the most widespread pattern in fish is to make use of large body size by monopolizing resources and access to mates, whereas small size provides benefits for surreptitious reproductive tactics (Gross 1984, Taborsky 1994, 1999; but see Koseki and Maekawa 2000 and Lee and Bass 2004 for examples showing that large reproductive parasites outcompete smaller ones). Males of some fish species switch repeatedly back and forth between tactics (e.g., Morris 1952, Barlow 1967, Foote and Larkin 1988; "reversible" ARTs: see Chapter 1, Figure 1.1). In wrasses and cichlids the largest males in a population, which usually perform the bourgeois tactic, were found to take over nests of conspecifics temporarily for spawning, while leaving the respective nest owners with nest tending and brood care thereafter (termed "piracy": van den Berghe 1988, Mboko and Kohda 1999, Ota and Kohda 2006a, b). In general, conditional choice of tactics in fish is usually dependent on relative body size (Farr et al. 1986, De Jonge and Videler 1989, Sigurjonsdottir and Gunnarson 1989, Foote 1990, Blanchfield et al. 2003), but body condition, prior residence, sex ratio, and population density may also take effect (reviewed in Taborsky 1994; see also Lee 2005).

Alternatively to a purely opportunistic tactic choice, many fish species exhibit sequential tactics changing from parasitic to bourgeois male behavior at some stage in life (e.g., Wirtz 1978, Warner 1982, 1984a, b, Warner and Hoffman 1980, Magnhagen 1992, De Fraipont et al. 1993). This ontogenetic switch pattern has a strong causal relationship to indeterminate growth as exhibited by most fishes (Taborsky 1999; see above), which generates age-related size variation of reproductive competitors within each sex. A sequential choice of tactics is not confined to species with indeterminate growth, however, even though size dependence is still of paramount importance also in species halting growth at maturity (Constantz 1975, Farr 1980b). A similar mechanism of size dependent tactic choice can be observed in species with ontogenetic sex change. Here, the switch is not from one to another reproductive tactic within a sex, but from one sex to another (Warner et al. 1975, Shapiro 1987, Ross 1989). The direction of switch depends on environmental conditions determining which sex can benefit more from large size (Warner et al. 1975, Munoz and Warner 2004). In sequential ART choice the direction of change is always from parasitic to bourgeois because the latter tactic inevitably benefits more from large size. The piracy tactic mentioned above may appear to be an exception, but in the cases described it appears to be an opportunistic, reversible ART rather than a case of truly sequential tactic performance (see Ota and Kohda [2006a] for a potential exception).

In a number of fish species tactics are fixed for life (Dominey 1980, Gross 1982, 1984, Jennings and Philipp 1992, Martin and Taborsky 1997, DeWoody et al. 2000b). Here, conditions early in life can determine which tactic to choose, or tactic variation depends on a genetic polymorphism (Taborsky 1998; see below). In the first case, a threshold mechanism depending on growth conditions may be responsible for the tactic expressed by an individual. A partial influence of that sort has been suggested in Atlantic salmon (Hutchings and Myers 1994), where thresholds may apparently vary between habitats within a population (Aubin-Horth and Dodson 2004). In temperate species exposed to strong seasonal variation of reproductive conditions, date of birth might have a strong influence on tactic expression ("birthdate effect": Taborsky 1998), because in short-lived species individuals born at different times in the season will face different growth periods before maturation (Cargnelli and Gross 1996, Pastres et al. 2002). It is hitherto unknown, however, whether such birthdate effect indeed determines fixed tactics, but in the Mediterranean wrasse Symphodus ocellatus this seems likely (Alonzo et al.

Box 10.1 *Lamprologus callipterus*, a cichlid fish with fixed and flexible ARTs

Reproduction of the endemic Lake Tanganyika cichlid *Lamprologus callipterus* is a paradigm for the astounding complexity potential of ARTs in fishes. The reproductive biology and behavior of this species is characterized by adaptations to their peculiar breeding substrate and pronounced conflict within and between the sexes.

(1) *Breeding in shells*. *Lamprologus callipterus* is an obligatory "snail brooder," which means that females spawn exclusively in empty snail shells, preferably of the species *Neothauma tanganicense*. For spawning, females enter a shell completely where they attach about 70 eggs to the inner wall of the shell. After spawning they stay in the shell for 10–14 days to guard and oxygenate the brood by frequent fanning. They leave the shell when the fry are free-swimming, which disperse independently from the female. Females lose a lot of weight during brood care and need to recover and gain weight before spawning again at a different nest after several weeks (Sato 1994).

(2) *Three ARTs within a species*. Males may use three alternative tactics to reproduce. (a) The largest males ("nest males") in the population defend nests consisting of empty snail shells, which they collect in the nest vicinity and from neighboring nests or which they have inherited from a previous nest

owner. (b) Males of all sizes may parasitize the reproductive effort of nest owners by sneaking or darting into the nest when a female is spawning and releasing sperm into the shell opening ("sneakers"). Obviously, this is not an easy task because nest owners attempt to keep reproductive competitors at bay. (c) "Dwarf males," which are on average only 2.5% the size (mass) of nest males, enter shells with spawning females and try to wriggle past them into the inner whorl of the shell. They remain at this safe and privileged position and attempt to fertilize eggs in competition with the nest owner until the female has finished spawning (Taborsky 1998, 2001, Sato *et al.* 2004) (Figure 10.1).

(3) *Extreme sexual size dimorphism*. Nest males are on average more than 12 times bigger (in mass) than females, which is the most extreme sexual size dimorphism known among animals with males exceeding females in size. They need to pass a threshold size to be able to carry shells into their nest, and shell-carrying efficiency increases with body size. Females, on the other hand, must be small enough to fit into shells. Their body size is strongly dependent on the availability of shell sizes, so the size distribution of females is apparently triggered by shell size distribution in a population. Intersexual selection (female choice) is not involved in the enormous sexual size dimorphism of this species, but large nest males benefit in competition with other bourgeois males. Gastropod shells being the only breeding substrate used by this species determine this extraordinary sexual size dimorphism (Schütz and Taborsky 2000, 2005, Schütz *et al.* 2006).

(4) *Mating pattern, reproductive tactic, and sexual conflict affected by breeding substrate*. The nests of this species are a unique extended phenotype that alters the environment considerably. This also allows other species to settle and breed in these accumulations of shells, hence often several species share and use the same shell. *Lamprologus callipterus* nest owners are extremely haremic, with harem sizes and mating patterns depending on shell distribution (mean harem sizes in different populations = 2.4–5.5 females, maximum = 18). When stealing shells containing a female from another nest, nest males attempt to oust her, which usually results in infanticide. At locations where shells are virtually unlimited ("shell beds"),

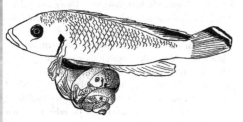

Figure 10.1 A bourgeois (i.e., nest) male positioned with his genital papilla over the opening of a snail shell during sperm release. A female is depositing eggs inside the shell, while a parasitic (i.e., dwarf) male competing for egg fertilization is contained further inside the shell. These parasites benefit from their privileged position close to the egg deposition site; they succeed in sperm competition despite their absolutely smaller testes sizes and sperm numbers. (Drawing by Barbara Taborsky.)

Box 10.1   (Cont.)

males do not need to carry shells, hence they start to defend territories at a much smaller body size and they do not pass the size threshold for shell carrying (Sato 1994, Sato and Gashagaza 1997, Taborsky 2001, Maan and Taborsky 2007).

(5) *Fixed and conditional ARTs.* Reproductive tactics in this species are both conditional and fixed. Males follow one of two principal life-history pathways. (a) They may become bourgeois nest owners that attempt to monopolize access to females by investing in territory defense, nest building, courtship, and defense of females and their broods. A proportion of these males matures at a much smaller size than required for nesting and act opportunistically as sneakers, attempting to fertilize eggs when a nest male is spawning. These males express a conditional ART. (b) The second life-history pathway is exhibited by dwarf males, which start reproducing by their specialized wriggling tactic early in life. They halt growth soon after maturation and remain small for life. Their lifespan is much shorter than that of bourgeois males. This tactic is fixed and genetically determined by a Mendelian polymorphism (Taborsky 2001; N. Rijneveld, M. Taborsky, and S. Wirtz, unpublished data).

(6) *Strong sperm competition.* Sperm competition is intense in this species, with very diverging options of males employing different tactics. Spawning lasts on average between 6 and 7 hours, and DNA analyses revealed that nest males monopolize most spawnings successfully. However, when a dwarf male succeeds in entering a shell and passing the female, he usually fertilizes the majority of eggs. The relative size of the female and shell determines whether a dwarf male is able to obtain his privileged position in the tip of the shell. Dwarf males invest relatively more into gonads than nest males do. They spend considerable time feeding in between their reproductive attempts and their condition factor is much higher than that of size-matched males developing into the bourgeois male morph. Nest males release sperm several hundred times per spawning, and both the rate of sperm release and the number of sperm released per bout decline towards the end of a spawning. This suggests sperm limitation, which should take effect particularly when a male spawns with several females concurrently. Nest males do not take up food during their territory tenure, which usually lasts 45 days. During this period they lose a lot of weight, which may ultimately urge them to take a prolonged break to recover from their investment before starting a new nest (Meidl 1999, Pachler 2001, Sato et al. 2004; N. Rijneveld, T. Sato, M. Taborsky, and S. Wirtz, unpublished data).

2000). The second potential mechanism causing fixed ARTs, a genetic polymorphism, has been described for live-bearing poeciliid fishes (Zimmerer and Kallmann 1989, Ryan et al. 1990, 1992, Erbelding-Denk et al. 1994). I should like to stress that genetic and environmental influences on tactic expression do not exclude each other also in fixed ARTs, which will be discussed below.

Both fixed and flexible tactics may co-occur within a species (Box 10.1). In a West African cichlid, *Pelvicachromis pulcher*, males come in two color morphs, yellow and red. Red morph males always reproduce as bourgeois territory owners, whereas yellow morph males may switch from a parasitic "satellite" tactic employed within the territory boundaries of a red morph harem-owner to a bourgeois tactic if conditions allow (Martin and Taborsky 1997). In the Mediterranean wrasse *Symphodus ocellatus*, otolith growth data revealed that three different life-history pathways coexist within this species (Alonzo et al. 2000). Males

with little growth during their first year of life reproduce as reproductive parasites throughout life. Males growing faster during the first year may either start as reproductive parasites when 1 year old and switch to a bourgeois nest builder tactic 1 year later, or they may refrain from reproduction when 1 year old (which speeds up their growth: Taborsky 1998) and only reproduce as bourgeois nest builders when 2 years of age (Alonzo et al. 2000). It is likely that a threshold mechanism based on size or growth speed is responsible for the decision of males to choose one or the other tactic, but this needs to be clarified.

10.3.2   Mechanism

To understand the origin of ARTs in a species we need to address both ultimate (evolutionary) and proximate (mechanistic) levels (see Chapter 1 for discussion). As with most biological traits studied to date, proximate causes of discontinuous ARTs involve genetic and environmental

Box 10.2   Salmon life history and reproductive patterns

Atlantic salmon (*Salmo salar*) females dig nests in a gravel area of a stream bed (called "redd") where they deposit a clutch of eggs, usually during October and November. After spawning females bury their eggs with gravel before moving on to dig another nest. Males do not participate in nest digging, nor do they participate in any form of guarding or brood care. Eggs develop slowly over the winter and hatch in spring. The hatchlings (called "alevins") remain buried in the gravel for up to 5 weeks while they absorb their large yolk sacs, before emerging at a size of about 2.5 cm. Young salmon (called "parr" before seaward migration) defend feeding territories in shallow areas in the river, but in the winter they usually stay under rocks on the bottom of a stream. Male parr may mature at 1– 4 years of age and will then attempt to participate in spawnings as sneakers. This holds particularly for fast-growing males, whereas males growing slowly and all females do not mature but instead migrate to sea in this age range (then called "smolts"). These anadromous males and females spend between 1 and 3 years at sea before returning to the freshwater system where they were born.

By then, anadromous males have reached between 1 and 10 kg of weight and develop a hooked lower jaw ("kype"), which they use in aggression mainly against other anadromous males and mature male parr. They may grab the latter and vigorously shake them between their jaws, which sometimes kills them (Hutchings and Myers 1987). On the spawning grounds anadromous males search and fight for mates which they court, with body size being an important determinant of their success. In contrast, parr sneakers are small and inconspicuous in appearance and behavior (they may weigh less than 10 g, being 1000 times lighter than bourgeois males in extreme cases). They establish a dominance hierarchy downstream of a courting anadromous male. From this position they sneak under the egg-laying female or dart in to shed sperm close to her, which results in their sperm competing for fertilization with sperm from the bourgeois anadromous male (Figure 10.2). Despite investing proportionally more in testes and sperm (Vladic and Järvi 2001) male parr are more likely than anadromous males to breed again, either prior to or following a migration to sea. Single male parr fertilize on average about 5% of eggs in a nest, and the proportion of eggs fertilized by male parr increases with their relative frequency, exceeding 20% at a parr : anadromous male ratio of 20 : 1 (Hutchings and Myers 1988). The relative frequency of mature parr males varies between populations, sometimes exceeding anadromous male numbers by 20 : 1.

A combination of additive genetic effects, parental life history, and habitat quality ultimately shapes juvenile growth rate, which together with thresholds for tactic choice are the main determinants of male life-history tactics in salmon (Garant *et al.* 2003, Aubin-Horth and Dodson 2004). Parr males attain successful reproduction more likely than anadromous males due to their younger age at maturation, but their fertilization success and number of females they spawn with is smaller than in anadromous males (Thomaz *et al.* 1997, Garant *et al.* 2002). The pattern described here is representative for other salmon species as well, with differences mainly found in semelparous species (Koseki and Fleming 2006). For additional information on salmon reproduction see Orton *et al.* 1938, Jones and Orton 1940, Jones 1959, Fleming and Gross 1994, Fleming 1996, 1998, Hendry and Stearns 2003, and references provided in the main text).

Figure 10.2 Sketch of sperm competition in spawning salmon, including an egg-depositing female (center), a large bourgeois male, and three small parasitic males. (After McCart 1970.)

components. In fish, origin of variation has been studied extensively in salmonids (Box 10.2). In the Atlantic salmon *Salmo salar* parasitic males (mature parr) grow quicker during some period of development than their bourgeois male conspecifics (Alm 1959, Thorpe and Morgan 1980,

Rowe and Thorpe 1990). This growth difference is already apparent before the young start feeding, i.e., in the first 20 days after hatching (Aubin-Horth and Dodson 2004), which can be due to genetic differences or maternal effects (e.g., Reznick 1981, 1982, Travis 1981, Wright *et al.* 2004); the only energy source of larvae is yolk during this period. A paternal (i.e., genetic) influence on growth differences and tactic expression in Atlantic salmon males has been demonstrated repeatedly (Thorpe *et al.* 1983, Glebe and Saunders 1986, Garant *et al.* 2002, 2003), but it is unclear whether a maternal effect is additionally present (see Bailey *et al.* 1980). This early growth and size difference in males translates into size differences between males in the next spring (i.e., shortly before parasitic males mature).

However, environment has a significant effect on juvenile Atlantic salmon growth and maturation as well (Metcalfe *et al.* 1988, 1989, Baum *et al.* 2004), which may override effects of genetic differences (Garant *et al.* 2003, Aubin-Horth and Dodson 2004). Brain gene-expression profiles in Atlantic salmon differ between the bourgeois male tactic and both females and parasitic males, all measured at about 1 year of age (Aubin-Horth *et al.* 2005a). This supports the hypothesis that the parasitic-type males follow the default developmental pathway, whereas bourgeois-type males failing to reach a threshold size will suppress maturation (Thorpe *et al.* 1998). Size thresholds responsible for tactic expression may differ between individuals within a population, and they can be affected by environmental conditions (Aubin-Horth and Dodson 2004, Baum *et al.* 2004). Overall, these results of Atlantic salmon demonstrate the importance of gene–environment interactions in the determination of male reproductive tactics (Aubin-Horth *et al.* 2005b; see Heath *et al.* 2002 for another salmon example, *Oncorhynchus tshawytscha*). On the ultimate level, this allows for local adaptations to different habitats (Riddell *et al.* 1981, Verspoor and Jordan 1989, Landry and Bernatchez 2001; see Taylor 1991 for review), which seems of particular importance in the river systems in which salmon live, with their high degree of habitat heterogeneity.

In other fish taxa, both environmental and genetic effects on tactic choice have also been demonstrated, but in comparison to salmon information is limited. Growth and size were found to be genetically determined in some live-bearing poeciliids (Zimmerer and Kallmann 1989, Ryan *et al.* 1990, 1992, Erbelding-Denk *et al.* 1994), which also decide on the choice of tactic: large males court females, while small males reproduce by coerced copulations (Farr 1980a, b, Constantz 1984, Bisazza 1993). Evidence for a genetic component of tactic choice exists also from sunfish (Dominey 1980, Gross 1982, Neff 2004) and cichlids (Martin and Taborsky 1997). Environmental influence on tactic expression has been found in many species: most often relative competitive ability (Farr *et al.* 1986, Sigurjonsdottir and

Gunnarson 1989, Foote 1990) and population parameters influencing the intensity of intrasexual competition (Kodric-Brown 1981, 1986) are of paramount importance, but ecological cues like the risk of predation may affect tactic choice as well (Godin 1995, Dill *et al.* 1999, Hamilton *et al.* 2006). As in Atlantic salmon, both genetic and environmental effects have been demonstrated to influence ART expression within a species (Constantz 1975, Zimmerer and Kallmann 1989, Martin and Taborsky 1997) and I expect that interaction between these effects is the rule rather than the exception.

In all these cases it is unknown, however, whether and how threshold mechanisms decide about tactic expression, even though it is very likely that size- and growth-related threshold mechanisms are of fundamental importance. Size variation of mature fish is usually large because of indeterminate growth (see above), and arguably, size is the most important determinant of the relative success of alternative tactics in reproductive competition (Taborsky 1999). If tactic expression is based on threshold mechanisms, the evolution of traits and reaction norms in different tactics is partly uncoupled, which greatly facilitates divergence in ART evolution (West-Eberhard 1989, 2003; see also Chapter 5).

10.3.3 Evolution

On the ultimate level, we can distinguish between ARTs resulting from diverging conditions to individuals in a population (e.g., superior versus "best of a bad job" tactics; Dawkins 1980; see Hazel *et al.* 1990), and those in which individuals are not constrained by quality differences but are selected to take one or the other alternative, dependent on crucial extrinsic variables like prevailing tactic frequency in the population, or environmental and social parameters (Shuster and Wade 2003). In this latter case frequency dependent selection will cause equal average pay-offs of tactics (Gadgil 1972, Shuster and Wade 1991, Ryan *et al.* 1992, Hutchings and Myers 1994; see Chapter 2).

As outlined above, the usual pattern of fish ARTs is exploitation of reproductive effort of bourgeois males by same-sex competitors that are either inferior in competitive ability (sneakers, satellites) or using favorable conditions opportunistically (territorial neighbors, pirates). In any case, they save the costs of prezygotic (courtship, nest building, competitor exclusion) and postzygotic (guarding, brood care) investment shown by bourgeois males (Taborsky 1994). Usually, the reproductive success of parasitic males has been estimated to be inferior to that of bourgeois conspecifics, despite the frequently expected fitness equality. Often, however, these estimates were based on inadequate measures like numbers of spawnings divided by the number of competitors in cases where more than one male was observed to participate in a spawning, which does not tell us too much about actual fertilizations achieved because of the important effects of position, timing, ejaculate size, and sperm quality (see Taborsky 1994 for discussion). Male wrasses, for example, were found to adjust their ejaculate size to the numbers of males participating in spawning (Shapiro et al. 1994). Sperm quality has been found to differ between males performing different tactics (De Fraipont et al. 1993, Gage et al. 1995, 2004, Uglem et al. 2001, Vladic and Järvi 2001, Neff et al. 2003), and the same is true for ejaculate size and density (Schärer and Robertson 1999, Alonzo and Warner 2000a, Leach and Montgomerie 2000, Vladic and Järvi 2001, Vladic et al. 2002, Neff et al. 2003).

A better measure of relative male success can be obtained by genotyping young together with potential sires. The problem here is that genetic information should exist from all participants at a spawning to obtain a reliable estimate. This is rarely the case, as reproductive parasites are usually furtive, quick, and hard to get hold of. Therefore, often genetic information exists only from young and the bourgeois male guarding them. Nevertheless, parentage reconstruction on the basis of DNA markers provides very valuable estimates on relative success rates of males. Reviewing the literature, DeWoody and Avise (2001) and Avise et al. (2002) reported that in 11 such cases including sunfish, largemouth bass, sticklebacks, sand goby, darters, and mottled sculpin, between 2% and 21% of offspring (mean 11.8%) were not sired by the nest-tending male. It is not known, however, how much of these proportions were due to reproductive parasitism by simultaneous parasitic spawning (SPS), and how much was due to other causes of paternity variation, such as nest take-overs and egg stealing (Constantz 1985, Sargent 1989, Rico et al. 1992, Jones et al. 1998). For a sample of 13 species (same as above plus Atlantic salmon and brown trout) genetic evidence suggests a spawning participation of reproductive parasites in almost one-third of nests (mean = 31.1%, range 0–100%; calculated from tables in DeWoody and Avise 2001, Avise et al. 2002).

Detailed knowledge about the relative reproductive success of bourgeois and parasitic males due to the application of genetic markers exists in bluegill sunfish Lepomis macrochirus, where individual parasitic sneakers were found to fertilize on average 89% of eggs and parasitic satellite males 67% of eggs when participating in spawning (Fu et al.

2001). However, SPS occurs only in 10.3% of spawnings, therefore the bourgeois nest males still appear to have the highest fertilization success (79% on average: Neff 2001). Satellites have probably a higher overall reproductive success than sneakers, because they have access to many more matings (Gross 1982). In Atlantic salmon, relative male success under sperm competition was found to vary with (experimental) conditions (Mjolnerod et al. 1998, Taggart et al. 2001, Jones and Hutchings 2002). Individual parasitic males fertilized between 1.2% and 26% of eggs in a nest (Hutchings and Myers 1988, Thomaz et al. 1997, Jones and Hutchings 2002). Estimates derived from a seminatural experimental set-up with competition between four bourgeois and 20 parasitic males suggests that the former are about 13 times more successful (Jones and Hutchings 2002). It has been assumed that this difference is even more pronounced in nature, because more parasitic males are often participating, and individual success of parasites declines with their number. The total success of parasitic spawnings at a nest was found to vary between 5% and 90% in eight studies of Atlantic salmon, depending on the number of competitors and environmental conditions (see Jordan and Youngson 1992, Avise et al. 2002, Jones and Hutchings 2002).

Specific adaptations exist in reproductive parasites to cope with the high levels of sperm competition they face (see above). In Atlantic salmon this includes sperm flagellum structure, ATP content (Vladic and Järvi 2001, Vladic et al. 2002), and sperm performance (especially swimming speed: Gage et al. 2004; also in bluegill sunfish: Burness et al. 2004). An artificial fertilization experiment with standardized ejaculate sizes suggested a higher fertilization efficiency of sneaker sperm than bourgeois male sperm (Vladic et al. 2002), and in another salmonid, Arctic charr Salvelinus alpinus, sperm of subordinate males produced more offspring than that of dominant males, probably due also to a higher fertilization success (Figenschou et al. 2007). In bluegill sunfish young sired by parasitic males showed higher growth rates and survival probabilities when exposed to their major predator (Neff 2004), suggesting that fitness differences between offspring may additionally compensate for differences in fertilization success.

These apparent adaptations of parasitic males on the gametic level may somewhat compensate for their inferior role at spawning, which is caused by the behavioral monopolization of bourgeois competitors. But how does this affect the balance of lifetime reproductive success to be expected by males performing different tactics? Clearly, snapshot appraisals of relative parentage at competitive spawnings will not yield sensible estimates of relative lifetime fitnesses. Using age-specific survival data from the field and strategy-specific fertilization data from the laboratory, Hutchings and Myers (1994) estimated the fitness associated with parasitic and bourgeois male tactics in Atlantic salmon. They proposed a mechanism involving polygenic thresholds of age at maturity, which is a largely environmentally controlled trait, to generate evolutionarily stable equilibrium frequencies between ARTs. In their threshold model they found that the equilibrium point between parasitic and bourgeois male frequencies depended strongly on age at maturity and hence growth rates, leading to higher expected equilibrium frequencies of parasitic males the lower the age at maturity (i.e., the faster they grow) (Figure 10.3). Bourgeois male success varied little with changing frequencies of parasitic males, while the latter's success depended strongly on their relative frequencies, as predicted. In general, parasitic males achieve large fitness gains by maturing early and thereby reducing pre-maturity mortality, despite fertilizing relatively few eggs. The model predicted that frequency-dependent selection maintains an evolutionary stable continuum of tactic frequencies through selection on growth rate thresholds for parasitic male maturation, with the incidence of parasitic male maturity and body size fitting a normal cumulative density function (see also Hazel et al. 1990). This has been confirmed in a New-foundland population of Atlantic salmon (Hutchings and Myers 1994). This informed model of Atlantic salmon tactics makes a number of predictions that have not yet been tested experimentally. A straightforward test would be to check for the influence of experimentally varied growth rates on relative tactic frequencies, for example, which could test the predictions on a quantitative level.

The search for equilibrium frequencies of ARTs is difficult particularly because (1) the probability of reaching maturity, (2) the average reproductive rate, and (3) the length of the reproductive lifespan of males performing alternative tactics are usually all hard to measure. In live-bearing swordtails Xiphophorus nigrensis Ryan et al. (1992) tested the prediction of equal fitnesses by estimating mating success and survival of parasitic and courting males. In this species, a genetic polymorphism with three alleles located at a single Y-chromosome locus are held responsible for the expression of male reproductive tactics. Therefore, the coexistence of ARTs should involve equivalent fitness expectations of different male types that are stabilized by frequency dependent selection. Indeed, the fitnesses did not differ significantly between male types, but the assumption of frequency-dependent selection has not yet been tested.

The relative fitness expectations of specialized reproductive competitors are influenced by a variety of traits reflecting an "arms race" of adaptations and counter-adaptations between bourgeois and parasitic males, resem-bling the situation in interspecific host–parasite interactions (Davies 1989). These traits are not confined to the level of competitive behavior before and during spawning, or to the morphology, physiology, and performance of gonads, ejaculates, and spermatozoa. Neff (2003) found that bour-geois males in bluegill sunfish differentiate between eggs fertilized by them or by parasitic males. They reduce brood care at the egg and larval stages with increasing proportions of parasitized fertilizations, based on the visual presence of parasitic male during spawning and distinguishing between own and foreign offspring by chemical cues (Neff and Sherman 2005). A similar influence of paternity on the pro-pensity of nest males to protect a brood was demonstrated experimentally in the Lake Tanganyika cichlid Lamprologus callipterus (M. E. Maan and M. Taborsky, unpublished data), suggesting that such discrimination of brood-caring males might be widespread. This will again shift the balance between the pay-offs of parasitic and bourgeois tactics and may therefore influence the equilibrium point of tactic frequencies, which clearly illustrates the complexity involved when attempting to obtain reliable fitness meas-ures of ARTs in fish.

In addition to the conventional bourgeois–parasite paradigm as extensively discussed here other divergent reproductive niches may also exist for conspecific, same-sex competitors in fish reproduction. For example, when reproductive habitats differ discontinuously (see Chapter 2), or when competitors differ in important features due to natural selection on other or more general traits (e.g., body size: Pigeon et al. 1997, Lu and Bernatchez 1999, Jonsson and Jonsson 2001, Trudel et al. 2001, Snorrason and Skúlason 2004). Trophic morphs have been described in several fish species (e.g., Sage and Selander 1975, Turner and Grosse 1980, Meyer 1987, Robinson and Wilson 1994), but it is yet unclear how such morph polymorphisms affect reproductive tactics and how in turn reproduction feeds back on phenotype variation (Parker et al. 2001). Alternative tactics may also exist in the context of mate choice (Kawase and Nakazono 1996). When female peacock wrasse (Symphodus tinca), for example, cannot spawn with preferred nest males, they spawn with males that do not provide paternal care outside of territories, where egg

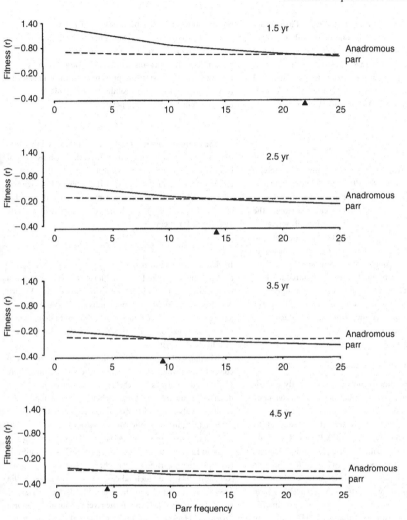

Figure 10.3 Jeffrey Hutchings and Ransom Myers modeled the lifetime fitness expectancies for bourgeois (i.e., anadromous; solid lines) and parasitic (i.e., mature parr; dashed lines) Atlantic salmon males based on data collected in Newfoundland. They found that the fitness of both male types depended strongly on the assumed age at maturity (note the heights of both curves dropping from the top to the bottom figures; these figures assume decision points increasing from 1.5 to 4.5 years of age). In addition, the Darwinian fitness of the parasitic male type was strongly affected by their frequency, while the fitness of bourgeois males was largely unaffected. Equilibrium frequencies of parasitic males (i.e., where fitness lines cross, denoted by black triangles) declined with "decision age at maturity." This probably causes parasitic males to start to reproduce after only 1 year and bourgeois males, which migrate and spend at least a year at sea, to start to reproduce only 3 years after that point. (From Hutchings and Myers 1994, with kind permission of Springer Science and Business Media.)

survival is much lower (Warner *et al.* 1995). Throughout the season female preference for one or the other type of spawning varies, which may also result from intrasexual competition. In many other cases of choice variation it is unclear, however, whether and to what extent the observed variation depends on intrasexual competition (Warner 1987, Henson and Warner 1997, Alonzo and Warner 2000c; see Chapter 18).

10.4   COOPERATION AS AN ART

In many fish species reproductive competitors cooperate to obtain fertilizations (Taborsky 1987, 1994, 2001). This may seem strange at first glance, but if the fitness benefits of investment in cooperating with a competitor outweigh the costs, such effort can be favored by natural selection. Cooperative behavior offered by reproductive parasites to bourgeois competitors, for example, can be a means to reduce the costs of competition, in order to provide incentives for tolerance by the bourgeois resource holders. The reproductive access, effort, and success of cooperating competitors may be skewed to various degrees (Taborsky 1999, 2001), depending on the stage in reproduction at which collaboration occurs.

10.4.1   Skew in access to matings

Sometimes male competitors may not differ greatly in their abilities to obtain fertilizations, or they may need to cooperate obligatorily to reproduce. Striking examples of this pattern include the North American suckers (Catostomidae) by default spawning in trios (Reighard 1920, Page and Johnston 1990, Taborsky 1994). Two males bearing breeding tubercles that roughen their body surface adjoin a female on either side and press against her flanks, which leads to simultaneous release of gametes of all three partners. Six hypotheses have been proposed to explain this apparently symmetric cooperative spawning pattern (Taborsky 1994), but up to now they remain untested. Joint spawning of two or more males without apparent asymmetries and aggression between them has been observed also in other freshwater fishes such as lake trout (*Salvelinus namaycush*: Royce 1951) and yellowfin shiner (*Notropis lutipinnis*: Wallin 1989).

Usually, however, there are asymmetries between competitors for fertilizations, which cause systematic differences in access: some individuals may have privileged access due to differences in quality or conditions, while others may need to find alternative ways to compete for

reproduction. This has been described extensively above, where bourgeois and parasitic tactics were discussed. One way to overcome competitive shortcomings for males is to cooperate with bourgeois partners, if these accept such a relationship or at least do not prevent it. For this to happen, there are two principle possibilities: bourgeois males may either benefit from such cooperation:

$$b_{coop} > c_{coop},$$

i.e., their fitness benefits of cooperation (b) exceed the costs (c), as is the case in mutualistic relationships; or, if cooperation does not pay for bourgeois males, they may be either physically unable to expel the satellite (Yanagisawa 1987), or the costs of expulsion may surpass its benefits (Kodric-Brown 1977). So bourgeois males will tolerate satellites if for them

$$b_{coop} < c_{coop}; \qquad c_{exp}/b_{exp} > 1.$$

In this case, not both partners benefit from such apparent cooperation, so on the functional level the relationship can be regarded as parasitic rather than mutualistic. A third possibility would be that one individual would force a partner to cooperate (Tebbich *et al.* 1996), but I do not know of any example from fish reproduction where this seems to be a likely explanation.

Examples where bourgeois males tolerate the presence of competitors, which are usually called "satellites," are widespread in fishes. In the literature, cases have been described from at least 22 species belonging to ten different families (Taborsky 1994). Satellites may join in territory defence (e.g., in the cichlid *Sarotherodon alcalicus*: Albrecht 1968; and in the wrasses *Symphodus ocellatus*, *S. roissali*, and *S. tinca*: Lejeune 1985, Taborsky *et al.* 1987), nest building (e.g., in the cyprinids river chub *Nocomis micropogon*: Reighard 1943; and bluehead chub *N. leptocephalus*: Wallin 1989), courtship (e.g., in the greenside darter *Etheostoma blennioides*: Fahy 1954; and in the river redhorse *Mosostoma carinatum*: Hackney *et al.* 1967), and brood care (e.g., in the cichlids *Pelvicachromis pulcher*: Martin and Taborsky 1997; and *Julidochromis ornatus*: Heg and Bachar 2006). Satellite males often behave submissively towards bourgeois males (e.g., in the ostraciid *Lactoria fornasini*: Moyer 1979; and in the wrasse *Symphodus ocellatus*: Taborsky *et al.* 1987), and the latter may even court satellites (e.g., in the wrasse *Coris julis*: Lejeune 1985), which suggests that their presence is beneficial to the territory owners.

Satellite and bourgeois males may or may not have different life-history trajectories (Martin and Taborsky 1997,

Alonzo et al. 2000, Uglem et al. 2000, Taborsky 2001). In several cichlids with long-term pair bonds, males (*Pelvicachromis pulcher*: Martin and Taborsky 1997) or members of both sexes may be tolerated and share in reproduction with the territory owners (*Neolamprologus brichardi/N. pulcher*: Taborsky 1984, 1985, Dierkes et al. 1999, Heg et al. 2006; *N. multifasciatus*: Kohler 1998, Taborsky 2001; *N. savoryi*: Heg et al. 2005; *Julidochromis marlieri*: Yamagishi 1988, Taborsky 1994; *J. ornatus*: Awata et al. 2005, Heg and Bachar 2006). Such relationships may persist for much longer than for the production of one brood, and they may be characterized by a very high degree of philopatry of subordinate "helpers" (Taborsky and Limberger 1981, Awata et al. 2005, Dierkes et al. 2005), even if a switch of the latter between groups may occur (Stiver et al. 2004, Bergmüller et al. 2005a, Heg and Bachar 2006). Relatedness between dominant breeders and subordinate helpers that compete for reproduction in such groups has been shown to be typically low (Kohler 1998, Awata et al. 2005, Dierkes et al. 2005).

10.4.2 Skew in behavioral effort

When competitors do not differ strongly in their competitive abilities, they may share courtship, defense and nest building fairly equally, as in lake trout (*Salvelinus namaycush*: Royce 1951), greenside darter (*Etheostoma blennioides*: Fahy 1954), bluehead chub (*Nocomis leptocephalus*: Wallin 1989), or the river redhorse (*Moxostoma carinatum*: Hackney et al. 1967, Page and Johnston 1990). More often, bourgeois males spend much more effort than satellites or helpers, and the latter engage only in certain duties such as defense against other males (e.g., in the wrasse *Halichoeres maculipinna*: Thresher 1979; and in the blenny *Parablennius sanguinolentus*: Santos 1985, Oliveira et al. 2002a). In some species, however, the defense effort of satellite males or helpers may exceed that of territory owners (*Neolamprologus brichardi/N. pulcher*: Taborsky et al. 1986; *Symphodus ocellatus*: Taborsky et al. 1987; *Pelvicachromis pulcher*: Martin and Taborsky 1997). When territory maintenance and brood care is shown by helpers as well (Taborsky and Limberger 1981, Taborsky 1984), it may exceed the effort of dominant male breeders (Kohler 1998, Taborsky and Grantner 1998), but relatedness between group members may influence the effort of helpers (Stiver et al. 2005).

In the Lake Tanganyika cichlid *Neolamprologus pulcher* it has been shown experimentally that subordinate group members of both sexes pay to be allowed to stay in the territory (Balshine-Earn et al. 1998, Bergmüller and Taborsky

2005, Bergmüller et al. 2005b). Male helpers in this species often, and females sometimes, participate in spawning (Taborsky 1985, Dierkes et al. 1999), i.e., they are reproductive competitors of the dominant territory owners. The situation may be similar in *N. multifasciatus* (Kohler 1998) and *Julidochromis ornatus* (Awata et al. 2005, 2006, Heg and Bachar 2006), but pay-to-stay has not been tested in these two species.

10.4.3 Skew in reproductive success

At the level of genetic fitness, behavior that may appear to be of mutual benefit may turn out to be neutral or even harmful to one of the cooperating partners. For instance, interactions of satellite males with females that are ready to spawn in the wrasse *Symphodus ocellatus* look like "herding in," but actually reduce the chances that a female will visit a nest (Taborsky 1994). A positive relationship between the presence of satellites and bourgeois male success was found both in Mediterranean wrasses (*Symphodus ocellatus*: Taborsky 2001) and in Azorean rock-pool blennies (*Parablennius sanguinolentus parvicornis*: Oliveira et al. 2002a), which are two cases of temporary cooperation during spawning between nonrelatives. This may indicate that bourgeois males benefit from the presence and activities of satellites, but alternatively it might simply reflect the greater attractiveness of successful nests to potential satellites. Only experiments can distinguish between these two possibilities.

In more stable associations between cooperators, greater reproductive success of breeders with helpers than without has been demonstrated in cichlids. In *Neolamprologus pulcher*, females with helpers lay more eggs (Taborsky 1984) because of energy saving (Taborsky and Grantner 1998, Balshine et al. 2001b), and young survive better when helpers are present, as shown by field experiments (Brouwer et al. 2005). In the West African cichlid *Pelvicachromis pulcher*, harem males with helpers sired and produced more offspring with their male helpers than pair males lacking them (Martin and Taborsky 1997). This is apparently not the case in *Julidochromis ornatus*, where a tendency was found that more young are produced when a helper (mostly male) is present, but nearly half of the young produced are sired by male helpers (Awata et al. 2005).

When comparing reproductive success between male tactics, a similar picture emerges in cooperative relationships as in purely parasitic ones. In general, it seems that the males showing more reproductive effort, i.e., the bourgeois competitors, are more successful in reproductive competition than males performing alternative tactics. This does not mean, however, that their average lifetime fitness will exceed that of satellite or helper males. Let us first consider cooperative breeders, where performing the helper tactic is a transient stage during life; both roles are performed by the same type of males, so there is no point in searching for differences in lifetime reproductive success (Taborsky 1984, 1985, 1994). At any given instance of competition for fertilizations, however, bourgeois males apparently get the greater share, as is predicted by reproductive skew theory assuming dominance inequalities between group members (Vehrencamp 1983, Keller and Reeve 1994, Clutton-Brock 1998, Reeve et al. 1998, Cant and Johnstone 2000, Johnstone 2000, Cant 2006). In *Neolamprologus pulcher*, individual male helpers may fertilize up to 36% of eggs of a clutch, but their average contribution is much lower (Dierkes et al. 1999, Heg et al. 2006). In *Pelvicachromis pulcher*, dominant male satellites obtained 29% of the fertilizations of clutches spawned with the harem males they were supporting, while second and third satellites in the hierarchy obtained only 14% and 5% of fertilizations of a clutch, respectively (Martin and Taborsky 1997). In the cooperatively breeding cichlid *Julidochromis ornatus*, male helpers sired on average 44% of offspring, which was not significantly different from the fertilization success of bourgeois territory owners (Awata et al. 2005).

The second possibility is that cooperation occurs between males performing different tactics throughout their lifetimes. This has been observed in the Mediterranean wrasse *Symphodus ocellatus*, where satellites may either belong to a male type with retarded growth that remains parasitic for life, or to a type that is large enough to behave as satellites already in their first year, but switch to the bourgeois tactic in their second reproductive season (Alonzo et al. 2000). Here the rate of fertilization attempts of individual satellites is only about 20% of that of bourgeois nest males (Taborsky et al. 1987), but the fertilization success and hence reproductive success still needs to be clarified with the help of genetic markers (N. Basieux and M. Taborsky, unpublished data). Due to this lack of information lifetime reproductive success cannot be compared between both types of males at the present stage.

10.5   THE ROLE OF FEMALE BEHAVIOR IN MALE ARTs

In species with external fertilization, the possibilities for females to exercise a mating preference are somewhat limited because of the omnipresent risk that nonpreferred males may participate in spawning. Still, females can

influence the probability that their eggs will be fertilized by a particular mate or by males of a particular type in two ways: (1) they may approach preferred partners for spawning and avoid nonpreferred ones, and (2) they may decide to spawn with a preferred partner even though there is the risk that nonpreferred males may participate in a spawning. In general, females prefer to spawn with bourgeois rather than parasitic males (e.g., *Salvelinus alpinus*: Sigurjonsdottir and Gunnarsson 1989; *Hypoptychus dybowskii*: Akagawa and Okiyama 1993; see Taborsky 1994 for examples of 12 more species belonging to eight teleost families), and a similar preference seems to prevail also in fish with internal fertilization (Farr 1980a, b, Bisazza 1993). As will be outlined below, such preference may depend on selection for genetic quality ("good-genes" benefits: Taborsky 1999) or on discriminating brood care behavior of bourgeois males, and it may be influenced by fertilization probabilities in case of sperm limitation, by assessment costs and abilities of females, and by intersexual conflict and the costs entailed on females through interactions with males. It is important to stress here that female choice patterns can strongly influence the evolution and equilibrium frequencies of male ARTs (Alonzo and Warner 2000b, Luttbeg 2004), even if choice is not tactic-specific but focused on some other parameter, such as male size or age (Henson and Warner 1997). Also, females may accept multiple matings because of various benefits (Evans and Magurran 2000).

In some species females refuse to spawn even with their preferred type of males, if males of the nonpreferred type are around. In the Mediterranean wrasse *Symphodus ocellatus*, the same courtship behavior shown by different types of males has opposite effects on the spawning propensity of females. When bourgeois nest males approach females in the nest vicinity and show "contact following," the contacted females enter the nest with greater likelihood, which often leads to spawning; but they are more likely to leave the nest area when the same behavior is performed by satellite or sneaker males, so in these males this behavior has a repelling rather than attracting function (Taborsky 1994). In this species and in the congener *S. tinca*, removal of parasitic males around focus nests increased spawning rates of females in these nests 3–8 times (van den Berghe et al. 1989, Wernerus 1989, Alonzo and Warner 2000c). This is particularly interesting as assumed fertilization rates, egg mortalities, and the quality of subsequent paternal care did not differ whether or not parasitic males participated in spawning (van den Berghe et al. 1989). Bourgeois and sneaker males in this species follow different life-history trajectories (Alonzo *et al.* 2000), however, which may relate to quality differences between them (Taborsky 1998).

In contrast to this general pattern of female preference for fertilization by bourgeois males, in particular cases females seem to be indifferent (Gross 1991a) or may even show a spawning preference for bourgeois males that are likely to be parasitized by other males at spawning. Female European bitterling *Rodeus sericeus* deposit eggs onto the gills of freshwater mussels, where they are well aerated during embryonic development (Reynolds *et al.* 1997). Bourgeois males guard territories around living mussels and court females. Parasitic spawning occurs frequently, both by nonterritorial floaters and bourgeois neighbors (Smith *et al.* 2004). Experiments revealed that females spawn preferentially at territories when parasitic males are nearby (Candolin and Reynolds 2002, Smith and Reichard 2005). This is apparently an adaptive female decision, because the proportion of eggs fertilized increases with the number of ejaculations performed shortly before and after egg deposition, and more ejaculations occur when more than one male is participating in spawning (Smith and Reichard 2005). This hints at sperm limitation for females which might be a result of a modulation of ejaculate investment by males (Wedell *et al.* 2002). Interestingly, in bitterling there are two forces acting against female preference to spawn when participation of parasitic males is likely. First, territorial males lead females preferentially to mussels where no parasites have released sperm before, even if the quality of these mussels is low (Smith *et al.* 2002). Second, higher male intrusion rates cause more territorial aggression by bourgeois owners, which in turn reduces the female propensity to spawn in these territories (Candolin and Reynolds 2002). In an experiment, the number of unsuccessful spawning attempts significantly increased with higher local male density, with females ceasing to spawn as males interrupted courtship to engage in aggressive attacks against rivals (Reichard *et al.* 2004b). At very high male densities territoriality collapses and many males participate in each spawning, with the effect that females are able to spawn with fewer courtship interruptions (Reichard *et al.* 2004b).

A similar female preference for spawning when parasitic males participate in fertilizations was suggested for bluegill sunfish, *Lepomis macrochirus*. Here, females were found to release about three times more eggs when parasitic males participated in spawning than when they were spawning with the territory owner alone (Fu *et al.* 2001). This study did not allow one to distinguish, however, whether this relationship was due to females adjusting egg deposition to parasite participation, or whether causality worked in the opposite direction, i.e., more parasitic males joined in spawning when females increased egg-deposition frequency. More recent results suggest that female choice in this species is probably more complex than initially thought. The quality of offspring sired by parasitic males is apparently greater than that of bourgeois males, as revealed by a comparison of maternal half-siblings sired in vitro: parasite male offspring grew faster and to a larger size than bourgeois male offspring while feeding endogenously on their yolk sac (Neff 2004). Nest males exhibit increased levels of brood care, however, when the proportion of parasite offspring in their nests is lower, as assessed by the visual presence of parasitic males during spawning and olfactory cues released by newly hatched eggs (Neff 2003, Neff and Sherman 2005). Probably for the latter reason, females classified as being of higher quality (based on a combined measure of their condition index, parasite load, and a fluctuating asymmetry score) spawned preferably in nests with lower parasite male paternity (see Chapter 17). There was a positive relationship between female quality and the paternal care their offspring received, and larger females preferred to spawn with higher-quality bourgeois males. So in effect, the mate choice pattern of bluegill sunfish females seems to follow the conventional paradigm of bourgeois male preference. This seems to run against female interests at the level of offspring quality, but it is a concession to the bourgeois male brood care response to the level of parasitized fertilizations. Hence it may be viewed as a consequence of an intersexual conflict where the bourgeois male's fitness interests prevail.

In salmon offspring of parasitic males also grow faster than those of bourgeois conspecifics (Alm 1959, Thorpe and Morgan 1980, Garant *et al.* 2002, Aubin-Horth and Dodson 2004). Faster growth leads to earlier maturation, which often results in a reduced mortality risk up to the start of reproduction (Hager and Noble 1976, Clarke and Blackburn 1994, Garant *et al.* 2002, Aubin-Horth *et al.* 2005c; see Sogard 1997 for review). This may suggest that salmon females should spawn preferentially with parasitic males, especially when bourgeois males do not contribute to nest preparation or brood care. This hypothesis was tested in Pacific coho salmon, *Oncorhynchus kisutch* (Watters 2005). Indeed, females held the oviposition posture longer when parasitic males were present at spawning, and they tended to show more nest digging when only a parasitic male was nearby than when bourgeois males attended as well (however, the sample size was only five here and the statistical analysis used was based on assumptions that cannot be tested); both results may be regarded as evidence for female choice of parasitic males (Watters 2005). There are alternative possibilities to explain these results, however. The longer spawning duration might have been caused by disturbance during spawning when parasitic males interfere. It is unknown how this prolonged oviposition posture relates to the number of deposited eggs, as egg number was not determined. The alleged preference to dig when only a parasitic male is present – if confirmed with a larger sample size – cannot be explained easily if alternative behaviors shown by females and their total time budget are unknown. In contrast to the interpretation of female preference towards parasitic males, there is indeed evidence for female preference of bourgeois males; (1) only one female mated with a lone parasitic male, whereas six females mated when only bourgeois males were present (and eight females mated in attendance of both types of males); (2) a high rate of chases by bourgeois males towards parasites was a good predictor of female spawning; (3) female spawning increased when crossing-over and nudging occurred, which are courtship behaviors mainly shown by bourgeois males (Watters 2005).

In conclusion, hitherto existing evidence in this species again suggests that females rather adopt the conventional preference of bourgeois males, if there is any tactic-specific preference at all. This has also been shown for Atlantic salmon (de Gaudemar et al. 2000).

When aiming to predict optimal choice in whichever behavioral context, the costs of choosing are an important parameter (Dawkins and Guilford 1991, Guilford and Dawkins 1991, Warner et al. 1995). Costs of assessment determine also the accuracy with which quality differences of targets can be gauged. In the context of mate choice, if females are able to assess the quality of potential partners accurately, low-quality males should not attempt to compete with high-quality contenders but rather employ alternative tactics that circumvent female choice. By using a dynamic state variable game model, Luttbeg (2004) showed that assessment capability of females can be an important determinant of the relative pay-offs of male tactics and hence of the evolution of ARTs. The better the assessment abilities of females, the higher the relative benefits of lower-quality males to opt out of the rat race for soaring preference scores, and hence the more likely it becomes that ARTs will evolve. In addition, due to female choice the variation in quality of bourgeois males may influence whether parasitic males can invade a population (Hamilton et al. 2006). When

there is a large difference in quality among bourgeois males, parasitic males are more likely to persist. Also, Hamilton and colleagues (2006) showed in one of their game-theoretic models that more females should be expected to visit poor bourgeois males when the cost to females of parasitic males is high. This hints at an intricate relationship between the equilibrium frequencies of different male tactics and female choice patterns (see Alonzo and Warner 2000b, and Chapter 18), which in turn relate to the quality variation among males, assessment costs and capabilities of females, and the costs entailed to them by interactions with males.

10.6   ARTs IN FEMALE FISH

In general, female ARTs are far less understood than male ARTs (see Chapter 18); this clearly applies also to fish. Intraspecific brood parasitism would be the equivalent of parasitic male tactics, and it seems to be rare in fish (Taborsky 1994, Wisenden 1999). It is as yet unclear, however, to what extent this rarity is apparent rather than real, because intraspecific brood parasitism is hard to detect (Andersson 1984, MacWhirter 1989). Some instances of intraspecific egg or offspring dumping in fish have been described. In the weakly electric fish *Eigenmannia virescens* females defend spawning territories around clumps of floating vegetation. During spawning, neighboring females sneak into the plant thicket and release eggs at the spawning site (Hagedorn and Heiligenberg 1985). Apparently, there is no brood care in this species, so brood parasitism brings about little costs to the parasitized territory holder, which reflects conditions in many insects known to show intraspecific brood parasitism (Tallamy 2005). It is unknown whether the functional cause for egg dumping in *Eigenmannia* is a shortage of males, or pronounced differences in the quality of spawning sites. Another probable case of egg dumping has been observed in the biparental largemouth bass (*Micropterus salmoides*: DeWoody et al. 2000a).

In mouthbrooding cichlids of the great lakes of East Africa, mixed maternity was detected in the mouths of egg-brooding females (Kellogg et al. 1998), but it is unclear whether this was the consequence of egg dumping. This is also true for other examples of maternal mouthbrooders from Lake Malawi (Ribbink et al. 1980). In several species of biparental Lake Tanganyika cichlids showing a combination of mouthbrooding of the early offspring stages and subsequent guarding, mixed broods of different conspecific females are commonly found, which is probably mainly due to "farming out" of young (Yanagisawa 1985, 1986,

Ochi *et al.* 1995, Ochi and Yanagisawa 2005). During the guarding stage, free-swimming offspring are collected in the mouth of one of the parents, who then moves to another territory where breeders guard a batch of more or less similar-sized young to release their own young into this guarded school (also called "young dumping": Taborsky 1994). In the best-studied species, *Perissodus microlepis*, however, farming out is almost exclusively performed by male guarders (Ochi and Yanagisawa 2005). In the South American convict cichlids, *Archocentrus nigrofasciatus* (formerly *Cichlasoma nigrofasciatum*), a biparental demersal spawner, transfers of free-swimming young between conspecific broods occured in 42% of 232 monitored natural broods (Wisenden and Keenleyside 1992). When males were experimentally removed, females often transferred young to neighboring guarders, which adopted them usually when they were not bigger than their own young. Such intraspecific adoptions have been reported frequently in the literature (see Taborsky [1994] referring to examples from 16 species, mainly cichlids), but in most cases it is unclear whether egg or young dumping is responsible for this alloparental care, and if the latter occurs, whether females are responsible. Intraspecific brood mixing may result from other mechanisms (reviews: Taborsky 1994, Wisenden 1999) even including kidnapping of neighboring young (McKaye and McKaye 1977, Mrowka 1987), which may be regarded as the antipode of offspring dumping.

In comparison to other taxa showing extensive care of eggs or young, e.g., birds and insects, the prevalence of female reproductive parasitism in fish is still low, especially when compared to the ubiquitous distribution of *male* ARTs in fish. The reason for this discrepancy is twofold. Due to higher prezygotic investment in females in general, males usually compete for access to female gametes much more than vice versa (Bateman 1948, Trivers 1972, Andersson 1994). Therefore, males can parasitize each other's effort to obtain access to fertilizable gametes, whereas females cannot usually make use of exploiting each other's effort (Taborsky 1994). However, females may parasitize postzygotic investment of conspecifics if it exists. This leads to the second reason why female reproductive parasitism is relatively rare in fishes: male brood care is more common than female-only care or biparental care (Breder and Rosen 1966, Baylis 1981), hence males may more easily exploit each other also at this level. This contrasts with conditions in birds, where female postzygotic reproductive investment is usually high and – as expected – also often exploited by other females (Yom-Tov 1980, 2001, Andersson 1984).

A peculiar situation where intraspecific brood parasitism may pay is given in cooperative breeders. Here, subordinate mature females may participate in reproduction, which can lead to increased levels of aggression of breeder males towards their partners, and even a reversal of the helping function: breeder females caring for the brood of their helper (Taborsky 1985). Even though this appears to be rare in the highly social African cichlid *Neolamprologus brichardi/pulcher*, recent experiments revealed that if potential breeding substrate in the territory is outspaced, the probability of female helper reproduction increases substantially (D. Heg, unpublished data). The importance of reproductive parasitism by female helpers in this species is corroborated by the fact that male and female helpers receive similar aggression levels by breeders, and female helpers are as likely to be expelled from the territory as male helpers are (Taborsky 1985; male helpers act as reproductive parasites as well: Dierkes *et al.* 1999, see above), despite the fact that female helpers are more related to breeders than male helpers are (Dierkes *et al.* 2005). In *Julidochromis ornatus*, female helpers sired a large proportion of young in the field, as revealed by DNA microsatellite analyses (Awata *et al.* 2005). While in this species the helpers and breeders are usually unrelated (seven of nine genotyped mature female helpers: Awata *et al.* 2005), cooperatively breeding groups of *Neolamprologus multifasciatus* consist usually of relatives (Kohler 1998), with a female bias among helpers (Taborsky 2001). Here again subordinate female group members regularly share in reproduction (10 of 18 analyzed groups: Kohler 1998). In *N. savoryi*, breeding groups consist usually of one haremic male and several breeding females forming subgroups in the territory, each with own helpers many of which are sexually mature and including female helpers (Heg *et al.* 2005). It is very likely that several females in the group breed, possibly including also female helpers within subgroups. This still needs to be checked by genetic analyses, however.

It is worth noting here that in group-breeding species reproductive competition between males and females have different fitness effects on dominant breeders (Taborsky 2001). Male and female breeders may suffer from offspring production of subordinate group members due to space limitations, additional effort (e.g., in brood care and social interactions), and competition for help from other group members. In addition, male breeders suffer from reduced production of their own offspring, whereas reproductive parasitism of female helpers does not necessarily impede the reproductive output of female breeders. Therefore, one should expect that reproductive competition, aggression, and dominance interactions should be more intense between male than between female group members. This was not confirmed, however, for *N. brichardi/pulcher* (Taborsky 1985), a species in which helpers of both sexes were found to pay for being allowed to stay (Balshine-Earn *et al.* 1998, Bergmüller and Taborsky 2005, Bergmüller *et al.* 2005b). In such a case, unrelated helpers are not expected to compensate fully for the costs they pose on breeders (Hamilton and Taborsky 2005).

In contrast to intraspecific reproductive parasitism of females, interspecific brood parasitism has been described in many fishes (reviews: Taborsky 1994, Wisenden 1999). Here, females dump eggs into nests of other species, or either sex dumps offspring into guarded schools of young (McKaye and Oliver 1980, McKaye *et al.* 1992, Ochi and Yanagisawa 1996, Ochi *et al.* 2001). I will not consider this further here, as interspecific brood parasitism is not an ART (see Chapter 1).

10.7 FUTURE RESEARCH

The study of ARTs in fish is a prospering field. Our understanding of evolutionary mechanisms underlying ARTs and reproductive behavior in general has greatly advanced by this research (see examples given above). As with any significant progress in science, the more we know, the more questions are turning up, and the questions gain in general interest, depth, and importance. Here I mention a few areas of research that I think deserve special attention in the study of fish ARTs. This list is by no means exhaustive.

(1) *Fertilization.* We need to better understand the mechanisms involved in fertilization, especially if it is external as is the case in most fishes. The performance of sperm in water and its effect on fertilization probability is hardly understood (Petersen *et al.* 2001). Water is usually a very turbid medium (see Neat *et al.* 2003b) at spawning that poses enormous osmotic stress and may strongly limit the intrinsic movement abilities of spermatozoa. Few data exist on functional properties of sperm and accessory secretions, although there are notable exceptions (e.g., Lahnsteiner *et al.* 1998, Vladic *et al.* 2002, Mansour *et al.* 2004, Burness *et al.* 2005). A comparative analysis of the prevalence and variability of ARTs in fishes with external and internal fertilization could reveal the significance of the fertilization mode for the evolution of alternative mating behaviors.

(2) *Sperm competition.* For our understanding of behavioral, morphological, and physiological trade-offs of ARTs it is of paramount importance to unravel the causes of success and failure during sperm competition in fish species with SPS. In species with male ARTs, characteristics need to be determined of males, their spawning behavior, timing of sperm release, use of space during spawning, reproductive morphology, physiological performance, ejaculate features, sperm number and quality, the nature and effect of accessory substances, and the performance of sperm in ovarian fluid (see Gage *et al.* 1995, Marconato *et al.* 1996, Scaggiante *et al.* 1999, Leach and Montgomerie 2000, Hoysak and Liley 2001, Vladic and Järvi 2001, Liley *et al.* 2002, Turner and Montgomerie 2002, Neff *et al.* 2003, Burness *et al.* 2004, Casselman and Montgomerie 2004, Urbach *et al.* 2005, Rudolfsen *et al.* 2006). Sperm competition experiments under controlled conditions can provide essential information in this context (Neff *et al.* 2003, Gage *et al.* 2004, Linhart *et al.* 2005, Schulte-Hostedde and Burness 2005). Unsolved riddles include tactic-specific adaptations in sperm turnover, the functional significance of sperm swimming speed and its alleged trade-off with sperm longevity, and the tactic-specific decisions made within the trade-offs involving different functional responses to sperm competition (e.g., ejaculate size and quality vs. behavioral effort: Alonzo and Warner 2000a; see also Snook 2005, Rudolfsen *et al.* 2006).

(3) *Causal factors.* What are the intrinsic and environmental criteria deciding whether ARTs in a species occur, and whether they will be fixed or flexible? In particular, the threshold criteria that trigger tactic expression are barely understood, despite strong theoretical expectations and ample evidence that specific factors are particularly important, such as size and growth related threshold mechanisms. Fish ARTs provide an excellent paradigm for the study of gene–environment interactions in the manifestation of life-history strategies at large (Parker *et al.* 2001). Promising progress has been made here recently mainly by research on Atlantic salmon (e.g., Garant *et al.* 2002, 2003, Aubin-Horth and Dodson 2004, Baum *et al.* 2004, Aubin-Horth *et al.* 2005b).

(4) *Frequency dependence.* Polymorphisms can originate and be maintained by density- and frequency-dependent disruptive selection (Maynard Smith 1982, Wilson 1989, Gross 1991b, Pfennig 1992, Lucas and Howard

1995, Brockmann 2001; see Chapter 2). However, despite a firm theoretical foundation, the empirical basis is weak in fish and other taxa for our belief that frequency dependence of tactics is an important component of the evolution of ARTs (Gross 1996). Supporting evidence (e.g., Ryan *et al.* 1992, Hutchings and Myers 1994) needs additional scrutiny, but experimental studies of frequency dependence of fish ARTs are lacking. Measurement of lifetime fitnesses of ARTs must account for age at maturity and the probability to reach it, the reproductive rate (i.e., sired offspring), and the length of the reproductive lifespan; these data are difficult to obtain in the wild, especially in a sperm competition scenario and when individuals use different ecologies throughout life (Hutchings and Jones 1998, Quinn and Myers 2004) and are subjected to important effects of spatial and temporal dynamics (e.g., Metcalfe and Thorpe 1990, Letcher and Gries 2003, Baum *et al.* 2004, Aubin-Horth *et al.* 2005b).

(5) *Physiological mechanisms.* By their unparalleled variation in the modes of reproduction, teleost fishes are prime candidates for the study of proximate causes of sexual plasticity, which is a major component of the expression of ARTs (Taborsky 1994). For example, size, growth, and maturation, which are major determinants of tactic choice and expression, turned out to be significantly influenced by social conditions, sometimes in an intricate manner (Hofmann *et al.* 1999, Heg *et al.* 2004, Hobbs *et al.* 2004). The endocrine mediation of ARTs in vertebrates is a particularly active field of research (Knapp 2003, Oliveira *et al.* 2005, Bender *et al.* 2006), which revealed that sex steroids and neuropeptides are prime agents for the expression of different reproductive tactics within a sex (Moore *et al.* 1998, Foran and Bass 1999, Goodson and Bass 2001). However, a causal link between steroids and sperm quality is less clear (Miura *et al.* 1992, Borg 1994, Uglem *et al.* 2002). In general, studies of regulatory mechanisms of reproductive physiology in fish exhibiting ARTs have challenged the conventional view that androgens are the sole or major agents controlling male reproductive characters. While this may hold for morphological features, where in fish 11-ketotestosterone triggers the expression of morphological traits of bourgeois males, it does not seem to generally apply to immediate regulation of male behavior, where regulatory processes may be more complex and tactic specific (Oliveira *et al.* 2001d,

2005). The study of the role and importance of target tissues and receptors of endocrine agents is an eminently important aim to further our understanding of the regulation of reproductive traits in fishes performing different tactics (Todo *et al.* 1999, Kim *et al.* 2002, Olsson *et al.* 2005, Chen and Fernald 2006; see also Douard *et al.* 2003). Variation in receptor densities in specific organs, for example, allow the fine-tuning effects of circulating hormone levels and thereby a compartmentalization of androgen effects (Ketterson and Nolan 1999), which may be a powerful mechanism in regulating the expression of ARTs (Oliveira 2006; see also Chapter 7). Feedback mechanisms between behavior, the social environment, and hormonal and neural physiology will need special attention in our study of proximate mechanisms in ARTs (Oliveira *et al.* 2002b, Fernald 2003, Oliveira 2004). A new and promising area to further our understanding of proximate mechanisms is the application of functional genomics to the study of the expression of alternative phenotypes (Hofmann 2003; cf. Robinson *et al.* 2005), which already showed that gene expression in salmon males is tactic specific (Aubin-Horth *et al.* 2005a).

(6) *Cooperative competitors.* Reciprocity between competitors appear to be a widespread mechanism in fish reproduction (see above). In most cases, it is still a riddle, however, how such cooperation between unrelated individuals is stabilized in evolution (Trivers 1971, Dugatkin and Mesterton-Gibbons 1996, Dugatkin 1997, Nowak *et al.* 2004). How do cooperators avoid being exploited? In symmetrical relationships, like in the spawning associations between male suckers (Page and Johnston 1990), are cooperators closely related (kin selection), do they gain mutual benefits (mutualism) or reciprocate by helping each other in turn (direct reciprocity), does one partner force the other one to participate (forced cooperation), or is it actually a parasitic relationship where the participants cannot effectively avoid being parasitized because prevention costs more than acceptance (Taborsky 1994)? In the more widespread case, where relations between cooperators are asymmetrical due to differences in dominance status or reproductive investment, similar mechanisms may apply. At present it is obscure in most cases why the investing bourgeois partner accepts a potential defector, and how it avoids being exploited. Evidence exists only in few cases that

bourgeois males accrue fitness benefits from relationships with unrelated subordinates (Martin and Taborsky 1997, Taborsky 2001, Oliveira *et al.* 2002a, Brouwer *et al.* 2005, Stiver *et al.* 2005), and a stage-structured pay-to-stay model has shown that unrelated subordinate cooperators are not expected to fully compensate for the costs they pose on dominants, at least in a group-living scenario (Hamilton and Taborsky 2005). It is usually less difficult to explain the behavior of the subordinate partners, because they often have no better option to reproduce due to their competitive inferiority (Taborsky 1987, 1999). It is currently unclear whether fish when performing alternative reproductive tactics may employ direct, indirect, or generalized reciprocity, as demonstrated experimentally in jays, monkeys, and rats (de Waal and Berger 2000, Stephens *et al.* 2002, Hauser *et al.* 2003, Spahni 2005, Rutte and Taborsky 2007 a, b). Outside of ARTs, cooperation in fish has been attributed to direct reciprocity in the context of predator inspection behaviour (Milinski 1987, Milinski *et al.* 1990a, b, Dugatkin 1991), albeit the evidence is controversial (Huntingford *et al.* 1994, Connor 1996, Stephens *et al.* 1997; see also Dugatkin and Mesterton-Gibbons 1996, Noe 2006). In addition, the interspecific interaction between cleaner fish and their clients shows elements of indirect reciprocity (Bshary 2002, Bshary and Grutter 2006). In the context of fish ARTs, interactions between unrelated, cooperating reproductive competitors are a highly suitable paradigm for future studies of reciprocity under natural conditions.

(7) *Sexual conflict.* The role of females in the evolution of male ARTs is still poorly understood. Female choice patterns can significantly influence the relative frequencies of male tactics in a population (Henson and Warner 1997, Alonzo and Warner 2000b, Luttbeg 2004; see Chapters 17 and 18). There is ample evidence for a female preference of bourgeois males, even though female choice of male types should be balanced or females should be indifferent to male types if these are maintained by frequency dependence, where lifetime reproductive success of different male tactics is similar at equilibrium. However, if direct fitness benefits exist from choice of a particular male type or mating situation, females should indeed be selective (Henson and Warner 1997; see also Garant *et al.* 2005). Recent evidence suggests that paternal care decisions depending on parasitic male spawning

participation, fertilization insurance, and relative genetic quality of males are major factors influencing female choice among males displaying different tactics (e.g., Neff 2003, Smith and Reichard 2005; see Chapter 17). The intersexual conflict is of particular interest here because in contrast to more conventional forms of male–female conflict (Arnqvist and Rowe 2005) three or more "partners" are involved in this game. The assessment ability of females is an additional riddle, because costs and sensory limitations may alter the optimal preference of females considerably, with respective effects on mating success of males and hence on the evolution of ARTs (Luttbeg 2004). This promising field of research is still at an early stage, and a theoretical and empirical research focus is required to unravel the co-evolution of male and female reproductive tactics and characters (see also Stockley *et al.* 1996, Stockley 1997, and Chapter 18).

In conclusion, alternative reproductive tactics in fish are a demanding and fascinating research field providing unique opportunities to clarify general mechanisms, both at the ultimate and proximate levels. Fish are ideal model systems for the study of ARTs because of their enormous variation of reproductive tactics and because of methodological feasibility, including experimental approaches in the laboratory and field. I predict that this very active research field is still going to grow exponentially over the next decade or two.

Acknowledgments
I am very grateful for stimulating discussions and constructive comments of Jane Brockmann and Rui Oliveira throughout compiling this work, and for helpful comments by an anonymous referee.

References
Akagawa, I. and Okiyama, M. 1993. Alternative male mating tactics in *Hypoptychus dybowskii* (Gasterosteiformes): territoriality, body size and nuptial coloration. *Japanese Journal of Ichthyology* 40, 343–350.

Akagawa, I. and Okiyama, M. 1995. Reproductive behaviour of the filefish *Rudarius ercodes*: male spawning parades and female choice. *Environmental Biology of Fishes* 43, 51–62.

Albrecht, H. 1968. Freiwasserbeobachtungen an Tilapien (Pisces, Cichlidae) in Ostafrika. *Zeitschrift für Tierpsychologie* 25, 377–394.

Alm, G. 1959. Connection between maturity, size and age in fishes. *Report of the Institute of Freshwater Research Drottningholm* 40, 5–145.

Alonzo, S. H. and Warner, R. R. 2000a. Allocation to mate guarding or increased sperm production in a Mediterranean wrasse. *American Naturalist* 156, 266–275.

Alonzo, S. H. and Warner, R. R. 2000b. Female choice, conflict between the sexes and the evolution of male alternative reproductive behaviours. *Evolutionary Ecology Research* 2, 149–170.

Alonzo, S. H. and Warner, R. R. 2000c. Dynamic games and field experiments examining intra- and intersexual conflict: explaining counterintuitive mating behavior in a Mediterranean wrasse, *Symphodus ocellatus*. *Behavioral Ecology* 11, 56–70.

Alonzo, S. H., Taborsky, M., and Wirtz, P. 2000. Male alternative reproductive behaviors in a Mediterranean wrasse, *Symphodus ocellatus*: evidence from otoliths for multiple life-history pathways. *Evolutionary Ecology Research* 2, 1–11.

Andersson, M. 1984. Brood parasitism within species. In C. J. Barnard (ed.) *Producers and Scroungers: Strategies of Exploitation and Parasitism*, pp. 195–228. Beckenham, UK: Croom Helm.

Andersson, M. 1994. *Sexual Selection*. Princeton, NJ: Princeton University Press.

Andersson, M. 2005. Evolution of classical polyandry: three steps to female emancipation. *Ethology* 111, 1–23.

Arnqvist, G. and Rowe, L. 2005. *Sexual Conflict*. Princeton, NJ: Princeton University Press.

Aspbury, A. S. and Gabor, C. R. 2004. Differential sperm priming by male sailfin mollies (*Poecilia latipinna*): effects of female and male size. *Ethology* 110, 193–202.

Aubin-Horth, N. and Dodson, J. J. 2004. Influence of individual body size and variable thresholds on the incidence of a sneaker male reproductive tactic in Atlantic salmon. *Evolution* 58, 136–144.

Aubin-Horth, N., Landry, C. R., Letcher, B. H., and Hofmann, H. A. 2005a. Alternative life histories shape brain gene expression profiles in males of the same population. *Proceedings of the Royal Society of London B* 272, 1655–1662.

Aubin-Horth, N., Letcher, B. H., and Hofmann, H. A. 2005b. Interaction of rearing environment and reproductive tactic on gene expression profiles in Atlantic salmon. *Journal of Heredity* 96, 261–278.

Aubin-Horth, N., Ryan, D. A. J., Good, S. P., and Dodson, J. J. 2005c. Balancing selection on size: effects on the incidence of an alternative reproductive tactic. *Evolutionary Ecology Research* 7, 1171–1182.

Austad, S. N. 1984. A classifiaction of alternative reproductive behaviors and methods of field-testing ESS models. *American Zoologist* 24, 309–319.

Avise, J. C., Jones, A. G., Walker, D., and DeWoody, J. A. 2002. Genetic mating systems and reproductive natural histories of fishes: lessons for ecology and evolution. *Annual Review of Genetics* 36, 19–45.

Awata, S., Munehara, H., and Kohda, M. 2005. Social system and reproduction of helpers in a cooperatively breeding cichlid fish (*Julidochromis ornatus*) in Lake Tanganyika: field observations and parentage analyses. *Behavioral Ecology and Sociobiology* 58, 506–516.

Awata, S., Heg, D., Munehara, H., and Kohda, M. 2006. Testis size depends on social status and the presence of male helpers in the cooperatively breeding cichlid *Julidochromis ornatus*. *Behavioral Ecology* 17, 372–379.

Bailey, J. K., Saunders, R. L., and Buzeta, M. I. 1980. Influence of parental smolt age and sea age on growth and smolting of hatchery reared Atlantic salmon (*Salmo salar*). *Canadian Journal of Fisheries and Aquatic Sciences* 37, 1379–1386.

Ball, M. A., and Parker, G. A. 1996. Sperm competition games: external fertilization and "adaptive" infertility. *Journal of Theoretical Biology* 180, 141–150.

Balshine-Earn, S., Neat, F. C., Reid, H., and Taborsky, M. 1998. Paying to stay or paying to breed? Field evidence for direct benefits of helping behavior in a cooperatively breeding fish. *Behavioral Ecology* 9, 432–438.

Balshine, S., Leach, B. J., Neat, F., Werner, N. Y., and Montgomerie, R. 2001a. Sperm size of African cichlids in relation to sperm competition. *Behavioral Ecology* 12, 726–731.

Balshine, S., Leach, B., Neat, F., et al. 2001b. Correlates of group size in a cooperatively breeding cichlid fish. *Behavioral Ecology and Sociobiology* 50, 134–150.

Barlow, G. W. 1967. Social behavior of a South American leaf fish, *Polycentrus schomburgkii*, with an account of recurring pseudofemale behavior. *American Midland Naturalist* 78, 215–234.

Barnard, C. J. (ed.) 1984. *Producers and Scroungers: Strategies of Exploitation and Parasitism*. Beckenham, UK: Croom Helm.

Barni, A., Mazzoldi, C., and Rasotto, M. B. 2001. Reproductive apparatus and male accessory structures in two batrachoid species (Teleostei, Batrachoididae). *Journal of Fish Biology* 58, 1557–1569.

Bass, A. H. 1992. Dimorphic male brains and alternative reproductive tactics in a vocalizing fish. *Trends in Neurosciences* 15, 139–145.

Bass, A. H. and Andersen, K. 1991. Intersexual and intrasexual dimorphisms in the vocal control system of a teleost fish: motor axon number and size. *Brain, Behavior, and Evolution* 37, 204–214.

Bass, A. H., Horvath, B. J., and Brothers, E. B. 1996. Nonsequential developmental trajectories lead to dimorphic vocal circuitry for males with alternative reproductive tactics. *Journal of Neurobiology* 30, 493–504.

Bateman, A. J. 1948. Intrasexual selection in *Drosophila*. *Heredity* 2, 349–368.

Baum, D., Laughton, R., Armstrong, J. D., and Metcalfe, N. B. 2004. Altitudinal variation in the relationship between growth and maturation rate in salmon parr. *Journal of Animal Ecology* 73, 253–260.

Baylis, J. R. 1981. The evolution of parental care in fishes, with reference to Darwin's rule of male sexual selection. *Environmental Biology of Fishes* 6, 223–251.

Beall, E. and de Gaudemar, B. 1999. Plasticity of reproductive behaviour in Atlantic salmon *Salmo salar* (Salmonidae) in relation to environmental factors. *Cybium* 23, 9–28.

Beckman, B. R. and Larsen, D. A. 2005. Upstream migration of minijack (age-2) Chinook salmon in the Columbia River: behavior, abundance, distribution, and origin. *Transactions of the American Fisheries Society* 134, 1520–1541.

Bekkevold, D., Hansen, M. M., and Loeschcke, V. 2002. Male reproductive competition in spawning aggregations of cod (*Gadus morhua*, L.). *Molecular Ecology* 11, 91–102.

Bender, N., Heg, D., Hamilton, I. M., *et al.* 2006. The relationship between social status, behaviour, growth and steroids in male helpers and breeders of a cooperatively breeding cichlid. *Hormones and Behavior* 50, 173–182.

Berglund, A. and Rosenqvist, G. 1993. Selective males and ardent femals in pipefishes. *Behavioral Ecology and Sociobiology* 32, 331–336.

Berglund, A., Rosenqvist, G., and Svensson, I. 1989. Reproductive success of females limited by males in two pipefish species. *American Naturalist* 133, 506–516.

Berglund, A., Widemo, M. S., and Rosenqvist, G. 2005. Sex-role reversal revisited: choosy females and ornamented, competitive males in a pipefish. *Behavioral Ecology* 16, 649–655.

Bergmüller, R. and Taborsky, M. 2005. Experimental manipulation of helping in a cooperative breeder: helpers "pay to stay" by pre-emptive appeasement. *Animal Behaviour* 69, 19–28.

Bergmüller, R., Heg, D., Peer, K., and Taborsky, M. 2005a. Extended safe havens and between-group dispersal of helpers in a cooperatively breeding cichlid. *Behaviour* 142, 1643–1667.

Bergmüller, R., Heg, D., and Taborsky, M. 2005b. Helpers in a cooperatively breeding cichlid stay and pay or disperse and breed, depending on ecological constraints. *Proceedings of the Royal Society of London B* 272, 325–331.

Billard, R. 1986. Spermatogenesis and spermatology of some teleost fish species. *Reproduction Nutrition Development* 26, 877–920.

Bisazza, A. 1993. Male competition, female mate choice and sexual size dimorphism in poeciliid fishes. *Marine Behaviour and Physiology* 23, 257–286.

Blanchfield, P. J. and Ridgway, M. S. 1999. The cost of peripheral males in a brook trout mating system. *Animal Behaviour* 57, 537–544.

Blanchfield, P. J., Ridgway, M. S., and Wilson, C. C. 2003. Breeding success of male brook trout (*Salvelinus fontinalis*) in the wild. *Molecular Ecology* 12, 2417–2428.

Blumer, L. S. 1979. Male parental care in bony fishes. *Quarterly Review of Biology* 54, 149–161.

Blumer, L. S. 1982. A bibliography and categorization of bony fishes exhibiting parental care. *Zoological Journal of the Linnean Society* 76, 1–22.

Borg, B. 1994. Androgens in teleost fishes. *Comparative Biochemistry and Physiology C* 109, 219–245.

Brantley, R. K. and Bass, A. H. 1994. Alternative male spawning tactics and acoustic signaling in the plainfin midshipman fish, *Porichthys notatus*. *Ethology* 96, 213–232.

Brantley, R. K., Tseng, J., and Bass, A. H. 1993a. The ontogeny of intersexual and intrasexual vocal muscle dimorphisms in a sound producing fish. *Brain, Behavior, and Evolution* 42, 336–349.

Brantley, R. K., Wingfield, J. C., and Bass, A. H. 1993b. Sex steroid levels in *Porichthys notatus*, a fish with alternative reproductive tactics, and a review of the hormonal bases for male dimorphism among teleost fishes. *Hormones and Behavior* 27, 332–347.

Brawn, V. M. 1961. Reproductive behaviour of the cod (*Gadus callarias* L.). *Behaviour* 18, 177–198.

Breder, C. M. and Rosen, D. E. 1966. *Modes of Reproduction in Fishes*. Garden City, NY: Natural History Press.

Brockmann, H. J. 2001. The evolution of alternative strategies and tactics. *Advances in the Study of Behavior* 30, 1–51.

Brouwer, L., Heg, D., and Taborsky, M. 2005. Experimental evidence for helper effects in a cooperatively breeding cichlid. *Behavioral Ecology* 16, 667–673.

Bruch, R. M. and Binkowski, F. P. 2002. Spawning behavior of lake sturgeon (*Acipenser fulvescens*). *Journal of Applied Ichthyology* 18, 570–579.

Bshary, R. 2002. Biting cleaner fish use altruism to deceive image-scoring client reef fish. *Proceedings of the Royal Society of London B* 269, 2087–2093.

Bshary, R. and Grutter, A. S. 2006. Image scoring and cooperation in a cleaner fish mutualism. *Nature* 441, 975–978.

Burness, G., Casselman, S. J., Schulte-Hostedde, A. I., Moyes, C. D., and Montgomerie, R. 2004. Sperm swimming speed and energetics vary with sperm competition risk in bluegill (*Lepomis macrochirus*). *Behavioral Ecology and Sociobiology* 56, 65–70.

Burness, G., Moyes, C. D., and Montgomerie, R. 2005. Motility, ATP levels and metabolic enzyme activity of sperm from bluegill (*Lepomis macrochirus*). *Comparative Biochemistry and Physiology A* 140, 11–17.

Candolin, U. and Reynolds, J. D. 2002a. Adjustments of ejaculation rates in response to risk of sperm competition in a fish, the bitterling (*Rhodeus sericeus*). *Proceedings of the Royal Society of London B* 269, 1549–1553.

Candolin, U. and Reynolds, J. D. 2002b. Why do males tolerate sneakers? Tests with the European bitterling, *Rhodeus sericeus*. *Behavioral Ecology and Sociobiology* 51, 146–152.

Cant, M. A. 2006. A tale of two theories: parent–offspring conflict and reproductive skew. *Animal Behaviour* 71, 255–263.

Cant, M. A. and Johnstone, R. A. 2000. Power struggles, dominance testing, and reproductive skew. *American Naturalist* 155, 406–417.

Cargnelli, L. M. and Gross, M. R. 1996. The temporal dimension in fish recruitment: birth date, body size, and size-dependent survival in a sunfish (bluegill: *Lepomis macrochirus*). *Canadian Journal of Fisheries and Aquatic Sciences* 53, 360–367.

Casselman, S. J. and Montgomerie, R. 2004. Sperm traits in relation to male quality in colonial spawning bluegill. *Journal of Fish Biology* 64, 1700–1711.

Cheek, A. O., Thomas, P., and Sullivan, C. V. 2000. Sex steroids relative to alternative mating behaviors in the simultaneous hermaphrodite *Serranus subligarius* (Perciformes: Serranidae). *Hormones and Behavior* 37, 198–211.

Chen, C. C. and Fernald, R. D. 2006. Distributions of two gonadotropin-releasing hormone receptor types in a cichlid fish suggest functional specialization. *Journal of Comparative Neurology* 495, 314–323.

Clarke, W. C. and Blackburn, J. 1994. Effect of growth on early sexual maturation in stream-type chinook salmon (*Oncorhynchus tshawytscha*). *Aquaculture* 121, 95–103.

Clutton-Brock, T. H. 1991. *The Evolution of Parental Care*. Princeton, NJ: Princeton University Press.

Clutton-Brock, T. H. 1998. Reproductive skew, concessions and limited control. *Trends in Ecology and Evolution* 13, 288–292.

Colin, P. L. and Bell, L. J. 1991. Aspects of the spawning of labrid and scarid fishes (Pisces: Labroidei) at Enewetak Atoll, Marshall Islands with notes on other families. *Environmental Biology of Fishes* 32, 229–260.

Colombo, L., Marconato, A., Belvedere, P. C., and Frisco, C. 1980. Endocrinology of teleost reproduction: a testicular steroid pheromone in the black goby, *Gobius jozo* L. *Bolletin Zoologico* 47, 355–364.

Connor, R. C. 1996. Partner preferences in by-product mutualisms and the case of predator inspection in fish. *Animal Behaviour* 51, 451–454.

Constantz, G. D. 1975. Behavioural ecology of mating in the male Gila topminnow, *Poeciliopsis occidentalis*. *Ecology* 56, 966–973.

Constantz, G. D. 1984. Sperm competition in poeciliid fishes. In R. L. Smith (ed.) *Sperm Competition and the Evolution of Animal Mating Systems*, pp. 465–485. Orlando, FL: Academic Press.

Constantz, G. D. 1985. Alloparental care in the tessellated darter, *Etheostoma olmstedi* (Pisces: Percidae). *Environmental Biology of Fishes* 14, 175–183.

Darchambeau, F. and Poncin, P. 1997. Field observations of the spawning behaviour of European grayling. *Journal of Fish Biology* 51, 1066–1068.

Darling, J. D. S., Noble, M. L., and Shaw, E. 1980. Reproductive strategies in the surfperches. 1. Multiple insemination in natural populations of the shiner perch, *Cymatogaster aggregata*. *Evolution* 34, 271–277.

Davidson, F. A. 1935. The development of the secondary sexual characters in the pink salmon (*Oncorhynchus kisutch*). *Journal of Morphology* 57, 169–183.

Davies, N. B. 1991. Mating systems. In J. R. Krebs and N. B. Davies (eds.) *Behavioural Ecology: An Evolutionary Approach*, Oxford, UK: Blackwell Scientific.

Davies, N. B., Bourke, F. G., and Brooke, M. D. 1989. Cuckoos and parasitic ants: interspecific brood parasitism as an evolutionary arms race. *Trends in Ecology and Evolution* 4, 274–278.

Dawkins, M. S. and Guilford, T. 1991. The corruption of honest signaling. *Animal Behaviour* 41, 865–873.

Dawkins, R. 1980. Good strategy or evolutionarily stable strategy? In G. W. Barlow and J. Silverberg (eds.) *Sociobiology: Beyond Nature/Nurture?*, pp. 331–367. Boulder, CO: Westview Press.

De Gaudemar, B., Bonzom, J. M., and Beall, E. 2000. Effects of courtship and relative mate size on sexual motivation in Atlantic salmon. *Journal of Fish Biology* 57, 502–515.

De Girolamo, M., Scaggiante, M., and Rasotto, M. B. 1999. Social organization and sexual pattern in the Mediterranean parrotfish *Sparisoma cretense* (Teleostei : Scaridae). *Marine Biology* 135, 353–360.

De Fraipont, M., FitzGerald, G. J., and Guderley, H. 1993. Age-related differences in reproductive tactics in the three-spined stickleback, *Gasterosteus aculeatus*. *Animal Behaviour* 46, 961–968.

De Jonge, J. and Videler, J. J. 1989. Differences between the reproductive biologies of *Tripterygion tripteronotus* and *Tripterygion delaisi* (Pisces, Perciformes, Tripterygiidae): the adaptive significance of an alternative mating strategy and a red instead of a yellow nuptial color. *Marine Biology* 100, 431–437.

De Jonge, J., de Ruiter, A. J. H., and van den Hurk, R. 1989. Testis–testicular gland complex of two *Tripterygion* species (Blennioidei, Teleostei): differences between territorial and non-territorial males. *Journal of Fish Biology* 35, 497–508.

Demski, L. S. 1987. Diversity in reproductive patterns and behavior in teleost fishes. In D. Crenos (ed.) *Psychobiology of Reproductive Behavior: An Evolutionary Perspective*, pp. 1–27. Englewood Cliffs, NJ: Prentice Hall.

Devlin, R. H. and Nagahama, Y. 2002. Sex determination and sex differentiation in fish: an overview of genetic, physiological, and environmental influences. *Aquaculture* 208, 191–364.

De Waal, F. B. M. and Berger, M. L. 2000. Payment for labour in monkeys. *Nature* 404, 563.

DeWoody, J. A. and Avise, J. C. 2001. Genetic perspectives on the natural history of fish mating systems. *Journal of Heredity* 92, 167–172.

DeWoody, J. A., Fletcher, D. E., Wilkins, S. D., Nelson, W. S., and Avise, J. C. 2000a. Genetic monogamy and biparental care in an externally fertilizing fish, the largemouth bass (*Micropterus salmoides*). *Proceedings of the Royal Society of London B* 267, 2431–2437.

DeWoody, J. A., Fletcher, D. E., Mackiewicz, M., Wilkins, S. D., and Avise, J. C. 2000b. The genetic mating system of spotted sunfish (*Lepomis punctatus*): mate numbers and the influence of male reproductive parasites. *Molecular Ecology* 9, 2119–2128.

Dierkes, P., Taborsky, M., and Kohler, U. 1999. Reproductive parasitism of broodcare helpers in a cooperatively breeding fish. *Behavioral Ecology* 10, 510–515.

Dierkes, P., Heg, D., Taborsky, M., Skubic, E., and Achmann, R. 2005. Genetic relatedness in groups is sex-specific and declines with age of helpers in a cooperatively breeding cichlid. *Ecology Letters* 8, 968–975.

Dill, L. M., Hedrick, A. V., and Fraser, A. 1999. Male mating strategies under predation risk: do females call the shots? *Behavioral Ecology* 10, 452–461.

Dominey, W. J. 1980. Female mimicry in male bluegill sunfish: a genetic polymorphism? *Nature* 284, 546–548.

Douard, V., Robinson-Rechavi, M., Laudet, V., and Guiguen, Y. 2003. Molecular evolution of angrogen receptors in fish. *Fish Physiology and Biochemistry* 28, 207–208.

Drilling, C. C. and Grober, M. S. 2005. An initial description of alternative male reproductive phenotypes in the bluebanded goby, *Lythrypnus dalli* (Teleostei, Gobiidae). *Environmental Biology of Fishes* 72, 361–372.

Dugatkin, L. A. 1991. Dynamics of the tit-for-tat strategy during predator inspection in the guppy (*Poecilia reticulata*). *Behavioral Ecology and Sociobiology* 29, 127–132.

Dugatkin, L. A. 1997. *Cooperation among Animals: An Evolutionary Perspective*. Oxford, UK: Oxford University Press.

Dugatkin, L. A. and Mesterton-Gibbons, M. 1996. Cooperation among unrelated individuals: reciprocal altruism, by-product mutualism and group selection in fishes. *Biosystems* 37, 19–30.

Dunning, J. B. 1992. *CRC Handbook of Avian Body Masses*. Boca Raton, FL: CRC Press.

Eggert, B. 1932. Zur Kenntnis der Biologie, der sekundären Geschlechtsmerkmale und des Eies von *Blennius pavo*. *Zeitschrift für Morphologie und Ökologie der Tiere* 24, 682–703.

Emlen, S. T. and Oring, L. W. 1977. Ecology, sexual selection, and the evolution of mating systems. *Science* 197, 215–223.

Engqvist, L. and Reinhold, K. 2005. Pitfalls in experiments testing predictions from sperm competition theory. *Journal of Evolutionary Biology* 18, 116–123.

Erbelding-Denk, C., Schröder, J. H., Schartl, M., *et al.* 1994. Male polymorphism in *Limia perugiae* (Pisces: Poeciliidae). *Behavioral Genetics* 24, 95–101.

Esteve, M. 2005. Observations of spawning behaviour in Salmoninae: *Salmo, Oncorhynchus* and *Salvelinus*. *Reviews in Fish Biology and Fisheries* 15, 1–21.

Evans, J. P. and Magurran, A. E. 2000. Multiple benefits of multiple mating in guppies. *Proceedings of the National Academy of Sciences of the United States of America* 97, 10074–10076.

Evans, J. P. and Magurran, A. E. 2001. Patterns of sperm precedence and predictors of paternity in the Trinidadian guppy. *Proceedings of the Royal Society of London B* 268, 719–724.

Evans, J. P., Pierotti, M., and Pilastro, A. 2003a. Male mating behavior and ejaculate expenditure under sperm competition risk in the eastern mosquitofish. *Behavioral Ecology* 14, 268–273.

Evans, J. P., Zane, L., Francescato, S., and Pilastro, A. 2003b. Directional postcopulatory sexual selection revealed by artificial insemination. *Nature* 421, 360–363.

Fabricius, E. and Gustafson, K. J. 1958. Some new observations on the spawning behaviour of the pike, *Esox lucius* L. In *Report No. 39*, pp. 23–54. Drottningholm, Sweden: Fishery Board of Sweden.

Fahy, W. E. 1954. The life history of the northern greenside darter, *Etheostoma blennioides blennioides* Rafinesque. *Journal of the Elisha Mitchell Scientific Society* 70, 139–205.

Farr, J. A. 1980a. The effects of sexual experience and female receptivity on courtship–rape decisions in male guppies, *Poecilia reticulata* (Pisces: Poeciliidae). *Animal Behaviour* 28, 1195–1201.

Farr, J. A. 1980b. Social behavior patterns as determinants of reproductive success in the guppy, *Poecilia reticulata* Peters (Pisces, Poeciliidae): an experimental study of the effects of intermale competition, female choice, and sexual selection. *Behaviour* 74, 38–91.

Farr, J. A., Travis, J., and Trexler, J. C. 1986. Behavioral allometry and interdemic variation in sexual behavior of the sailfin molly, Poecilia latipinna (Pisces, Poeciliidae). *Animal Behaviour* 34, 497–509.

Fernald, R. D. 2003. How does behavior change the brain? Multiple methods to answer old questions. *Integrative and Comparative Biology* 43, 771–779.

Fiedler, K. 1954. Vergleichende Verhaltensstudien an Seenadeln, Schlangennadeln und Seepferdchen. *Zeitschrift für Tierpsychologie* 11, 358–416.

Fischer, E. A. 1980. The relationship between mating system and simultaneous hermaphroditism in the coral-reef fish, *Hypoplectrus nigricans* (Serranidae). *Animal Behaviour* 28, 620–633.

Fischer, E. A. 1986. Mating systems of simultaneously hermaphroditic serranid fishes. *Proceedings of the 2nd International Conference of Indo-Pacific Fishes.*

Fishelson, L. 1991. Comparative cytology and morphology of seminal vesicles in male gobiid fishes. *Japanese Journal of Ichthyology* 38, 17–30.

Fitzpatrick, J. L., Desjardins, J. K., Stiver, K. A., Montgomerie, R., and Balshine, S. 2006. Male reproductive suppression in the cooperatively breeding fish *Neolamprologus pulcher*. *Behavioral Ecology* 17, 25–33.

Fleming, I. A. 1996. Reproductive strategies of Atlantic salmon: ecology and evolution. *Reviews in Fish Biology and Fisheries* 6, 379–416.

Fleming, I. A. 1998. Pattern and variability in the breeding system of Atlantic salmon (*Salmo salar*), with comparisons to other salmonids. *Canadian Journal of Fisheries and Aquatic Sciences* 55, 59–76.

Fleming, I. A. and Gross, M. R. 1994. Breeding competition in a pacific salmon (coho, *Oncorhynchus kisutch*): measures of natural and sexual selection. *Evolution* 48, 637–657.

Foote, C. J. 1990. An experimental comparison of male and female spawning territoriality in a pacific salmon. *Behaviour* 115, 283–314.

Foote, C. J. and Larkin, P. A. 1988. The role of male choice in the assortative mating of anadromous and non-anadromous sockeye salmon (*Oncorhynchus nerka*). *Behaviour* 106, 43–62.

Foote, C. J., Brown, G. S., and Wood, C. C. 1997. Spawning success of males using alternative mating tactics in sockeye salmon, *Oncorhynchus nerka*. *Canadian Journal of Fisheries and Aquatic Sciences* 54, 1785–1795.

Foran, C. M. and Bass, A. H. 1999. Preoptic GnRH and AVT: axes for sexual plasticity in teleost fish. *General and Comparative Endocrinology* 116, 141–152.

Frischknecht, M. 1993. The breeding coloration of male three-spined sticklebacks (*Gasterosteus aculeatus*) as an indicator of energy investment and vigour. *Evolutionary Ecology* 7, 439–450.

Fu, P., Neff, B. D., and Gross, M. R. 2001. Tactic-specific success in sperm competition. *Proceedings of the Royal Society of London B* **268**, 1105–1112.

Gadgil, M. 1972. Male dimorphism as a consequence of sexual selection. *American Naturalist* **106**, 574–580.

Gage, M. J. G., Stockley, P., and Parker, G. A. 1995. Effects of alternative male mating strategies on characteristics of sperm production in the Atlantic salmon (*Salmo salar*): theoretical and empirical investigations. *Philosophical Transactions of the Royal Society of London B* **350**, 391–399.

Gage, M. J. G., Stockley, P., and Parker, G. A. 1998. Sperm morphometry in the Atlantic salmon. *Journal of Fish Biology* **53**, 835–840.

Gage, M. J. G., MacFarlane, C., Yeates, S., Shackleton, R., and Parker, G. A. 2002. Relationships between sperm morphometry and sperm motility in the Atlantic salmon. *Journal of Fish Biology* **61**, 1528–1539.

Gage, M. J. G., Macfarlane, C. P., Yeates, S., et al. 2004. Spermatozoal traits and sperm competition in Atlantic salmon: relative sperm velocity is the primary determinant of fertilization success. *Current Biology* **14**, 44–47.

Garant, D., Fontaine, P. M., Good, S. P., Dodson, J. J., and Bernatchez, L. 2002. The influence of male parental identity on growth and survival of offspring in Atlantic salmon (*Salmo salar*). *Evolutionary Ecology Research* **4**, 537–549.

Garant, D., Dodson, J. J., and Bernatchez, L. 2003. Differential reproductive success and heritability of alternative reproductive tactics in wild Atlantic salmon (*Salmo salar* L.). *Evolution* **57**, 1133–1141.

Garant, D., Dodson, J. J., and Bernatchez, L. 2005. Offspring genetic diversity increases fitness of female Atlantic salmon (*Salmo salar*). *Behavioral Ecology and Sociobiology* **57**, 240–244.

Gile, S. R. and Ferguson, M. M. 1995. Factors affecting male potency in pooled gamete crosses of rainbow trout, *Oncorhynchus mykiss*. *Environmental Biology of Fishes* **42**, 267–275.

Glebe, B. D. and Saunders, L. 1986. Genetic factors in sexual maturity of cultured Atlantic salmon (*Salmo salar* L.) parr and adults reared in sea cages. In D. J. Meerburg (ed.) *Salmonid Age at Maturity*, pp. 24–29.

Godin, J.-G. J. 1995. Predation risk and alternative mating tactics in male Trinidadian guppies (*Poecilia reticulata*). *Oecologia* **103**, 224–229.

Godwin, J., Luckenbach, J. A., and Borski, R. J. 2003. Ecology meets endocrinology: environmental sex determination in fishes. *Evolution and Development* **5**, 40–49.

Gomendio, M. and Roldan, E. M. S. 1991. Sperm competition influences sperm size in mammals. *Proceedings of the Royal Society of London B* **243**, 181–185.

Gonçalves, D., Fagundes, T., and Oliveira, R. F. 2003. Reproductive behaviour of sneaker males of the peacock blenny. *Journal of Fish Biology* **63**, 528–532.

Gonçalves, D., Matos, R., Fagundes, T., and Oliveira, R. F. 2005. Bourgeois males of the peacock blenny, *Salaria pavo*, discriminate female mimics from females? *Ethology* **111**, 559–572.

Gonçalves, E. J., Almada, V. C., Oliveira, R. F., and Santos, A. J. 1996. Female mimicry as a mating tactic in males of the blenniid fish *Salaria pavo*. *Journal of the Marine Biological Association of the United Kingdom* **76**, 529–538.

Goodson, J. L. and Bass, A. H. 2001. Social behavior functions and related anatomical characteristics of vasotocin/vasopressin systems in vertebrates. *Brain Research Reviews* **35**, 246–265.

Grant, J. W. A., Bryant, M. J., and Soos, C. E. 1995. Operational sex ratio, mediated by synchrony of female arrival, alters the variance of male mating success in Japanese medaka. *Animal Behaviour* **49**, 367–375.

Grantner, A. and Taborsky, M. 1998. The metabolic rates associated with resting, and with the performance of agonistic, submissive and digging behaviours in the cichlid fish *Neolamprologus pulcher* (Pisces: Cichlidae). *Journal of Comparative Physiology B* **168**, 427–433.

Gronell, A. M. 1989. Visiting behavior by females of the sexually dichromatic damselfish, *Chrysiptera cyanea* (Teleostei, Pomecentridae): a probable method of assessing male quality. *Ethology* **81**, 89–122.

Gross, M. R. 1982. Sneakers, satellites and parentals: polymorphic mating strategies in North American sunfishes. *Zeitschrift für Tierpsychologie* **60**, 1–26.

Gross, M. R. 1984. Sunfish, salmon, and the evolution of alternative reproductive strategies and tactics in fishes. In G. Potts and R. J. Wootton (eds.) *Fish Reproduction: Strategies and Tactics*, pp. 55–75. London: Academic Press.

Gross, M. R. 1985. Disruptive selection for alternative life histories in salmon. *Nature* **313**, 47–48.

Gross, M. R. 1991a. Salmon breeding behavior and life history evolution in changing environments. *Ecology* **72**, 1180–1186.

Gross, M. R. 1991b. Evolution of alternative reproductive strategies: frequency-dependent sexual selection in male bluegill. *Philosophical Transactions of the Royal Society of London B* **332**, 59–66.

Gross, M. R. 1996. Alternative reproductive strategies and tactics: diversity within sexes. *Trends in Ecology and Evolution* **11**, 92–98.

Guilford, T. and Dawkins, M. S. 1991. Receiver psychology and the evolution of animal signals. *Animal Behaviour* **42**, 1–14.

Hackney, P. A., Tatum, W. M., and Spencer, S. L. 1967. Life history study of the river redhorse, *Moxostoma carinatum* (Cope), in the Cahaba River, Alabama, with notes on the management of the species as a sport fish. *Proceedings of the 21st Annual Conference S. E. Association for Game and Fisheries Commerce*, pp. 324–332.

Hagedorn, M. and Heiligenberg, W. 1985. Court and spark: electric signals in the courtship and mating of gymnotoid fish. *Animal Behaviour* **33**, 254–265.

Hager, R. C. and Noble, R. E. 1976. Relation of size at release of hatchery-reared coho salmon to age, size and sex composition of returning adults. *Progressive Fish Culturist* **38**, 144–147.

Hamilton, I. M. and Taborsky, M. 2005. Unrelated helpers will not fully compensate for costs imposed on breeders when they pay to stay. *Proceedings of the Royal Society of London B* **272**, 445–454.

Hamilton, I. M., Haesler, M. P., and Taborsky, M. 2006. Predators, reproductive parasites, and the persistence of poor males on leks. *Behavioral Ecology* **17**, 97–107.

Hanson, A. J. and Smith, H. D. 1967. Mate selection in a population of sockeye salmon (*Oncorhynchus nerka*) of mixed age-groups. *Journal of the Fisheries Research Board of Canada* **24**, 1955–1977.

Harvey, P. H. and Pagel, M. D. 1991. *The Comparative Method in Evolutionary Biology*. Oxford, UK: Oxford University Press.

Hauser, M. D., Chen, M. K., Chen, F., Chuang, E., and Chuang, E. 2003. Give unto others: genetically unrelated cotton-top tamarin monkeys preferentially give food to those who altruistically give food back. *Proceedings of the Royal Society of London B* **270**, 2363–2370.

Hazel, W. N., Smock, R., and Johnson, M. D. 1990. A polygenic model for the evolution and maintenance of conditional strategies. *Proceedings of the Royal Society of London B* **242**, 181–187.

Healey, M. C. and Prince, A. 1998. Alternative tactics in the breeding behaviour of male coho salmon. *Behaviour* **135**, 1099–1124.

Heath, D. D., Rankin, L., Bryden, C. A., Heath, J. W., and Shrimpton, J. M. 2002. Heritability and Y-chromosome influence in the jack male life history of chinook salmon (*Oncorhynchus tshawytscha*). *Heredity* **89**, 311–317.

Heg, D. and Bachar, Z. 2006. Cooperative breeding in the Lake Tanganyika cichlid *Julidochromis ornatus*. *Environmental Biology of Fishes* **76**, 265–281.

Heg, D., Bender, N., and Hamilton, I. 2004. Strategic growth decisions in helper cichlids. *Proceedings of the Royal Society of London B* **271**, S505–S508.

Heg, D., Bachar, Z., and Taborsky, M. 2005. Cooperative breeding and group structure in the Lake Tanganyika cichlid *Neolamprologus savoryi*. *Ethology* **111**, 1017–1043.

Heg, D., Bergmüller, R., Bonfils, D., *et al.* 2006. Cichlids do not adjust reproductive skew to the availability of independent breeding options. *Behavioral Ecology* **17**, 419–429.

Hendry, A. P. and Stearns, S. C. (eds.) 2003. *Evolution Illuminated: Salmon and Their Relatives*. Oxford, UK: Oxford University Press.

Henson, S. A. and Warner, R. R. 1997. Male and female alternative reproductive behaviors in fishes: a new approach using intersexual dynamics. *Annual Review of Ecology and Systematics* **28**, 571–592.

Hildemann, W. H. and Wagner, E. D. 1954. Intraspecific sperm competition in *Lebistes reticulatus*. *American Naturalist* **88**, 87–91.

Hobbs, J. P. A., Munday, P. L., and Jones, G. P. 2004. Social induction of maturation and sex determination in a coral reef fish. *Proceedings of the Royal Society of London B* **271**, 2109–2114.

Hofmann, H. A. 2003. Functional genomics of neural and behavioral plasticity. *Journal of Neurobiology* **54**, 272–282.

Hofmann, H. A., Benson, M. E., and Fernald, R. D. 1999. Social status regulates growth rate: consequences for life-history strategies. *Proceedings of the National Academy of Sciences of the United States of America* **96**, 14171–14176.

Hourigan, T. F., Nakamura, M., Nagahama, Y., Yamauchi, K., and Grau, E. G. 1991. Histology, ultrastructure, and in vitro steroidogenesis of the testes of two male phenotypes of the protogynous fish *Thalassoma duperrey* (Labridae). *General and Comparative Endocrinology* **83**, 193–217.

Hoysak, D. J. and Liley, N. R. 2001. Fertilization dynamics in sockeye salmon and a comparison of sperm from alternative male phenotypes. *Journal of Fish Biology* **58**, 1286–1300.

Hoysak, D. J., Liley, N. R., and Taylor, E. B. 2004. Raffles, roles, and the outcome of sperm competition in sockeye salmon. *Canadian Journal of Zoology/Revue Canadienne de Zoologie* **82**, 1017–1026.

Huntingford, F. A., Lazarus, J., Barrie, B. D., and Webb, S. 1994. A dynamic analysis of cooperative predator inspection in sticklebacks. *Animal Behaviour* **47**, 413–423.

Hutchings, J. A. and Jones, M. E. B. 1998. Life history variation and growth rate thresholds for maturity in Atlantic salmon, *Salmo salar*. *Canadian Journal of Fisheries and Aquatic Sciences* **55**, 22–47.

Hutchings, J. A. and Myers, R. A. 1987. Escalation of an asymmetric contest: mortality resulting from mate competition in Atlantic salmon, *Salmo salar*. *Canadian Journal of Zoology/Revue Canadienne de Zoologie* **65**, 766–768.

Hutchings, J. A. and Myers, R. A. 1988. Mating success of alternative maturation phenotypes in male Atlantic salmon, *Salmo salar*. *Oecologia* **75**, 169–174.

Hutchings, J. A. and Myers, R. A. 1994. The evolution of alternative mating strategies in variable environments. *Evolutionary Ecology* **8**, 256–268.

Hutchings, J. A., Bishop, T. D., and McGregor-Shaw, C. R. 1999. Spawning behaviour of Atlantic cod, *Gadus morhua*: evidence of mate competition and mate choice in a broadcast spawner. *Canadian Journal of Fisheries and Aquatic Sciences* **56**, 97–104.

Immler, S., Mazzoldi, C., and Rasotto, M. B. 2004. From sneaker to parental male: change of reproductive traits in the black goby, *Gobius niger* (Teleostei: Gobiidae). *Journal of Experimental Zoology* **301A**, 177–185.

James, P. and Sexauer, H. 1997. Spawning behaviour, spawning habitat and alternative mating strategies in an adfluvial population of bull trout. In W. Mackay, M. Brewin, and M. Monita (eds.) *Friends of the Bull Trout Conference Proceedings*, pp. 325–329.

Jamieson, B. G. M. 1991. *Fish Evolution and Systematics: Evidence from Spermatozoa.* Cambridge, UK: Cambridge University Press.

Jennings, M. J. and Philipp, D. P. 1992. Reproductive investment and somatic growth rates in longear sunfish. *Environmental Biology of Fishes* **35**, 257–271.

Johnstone, R. A. 2000. Models of reproductive skew: a review and synthesis. *Ethology* **106**, 5–26.

Jones, A. G. and Avise, J. C. 1997a. Microsatellite analysis of maternity and the mating system in the Gulf pipefish *Syngnathus scovelli*, a species with male pregnancy and sex-role reversal. *Molecular Ecology* **6**, 203–213.

Jones, A. G. and Avise, J. C. 1997b. Polygynandry in the dusky pipefish *Syngnathus floridae* revealed by microsatellite DNA markers. *Evolution* **51**, 1611–1622.

Jones, A. G. and Avise, J. C. 2001. Mating systems and sexual selection in male-pregnant pipefishes and seahorses: insights from microsatellite-based studies of maternity. *Journal of Heredity* **92**, 150–158.

Jones, A. G., Ostlund-Nilsson, S., and Avise, J. C. 1998. A microsatellite assessment of sneaked fertilizations and egg thievery in the fifteenspine stickleback. *Evolution* **52**, 848–858.

Jones, A. G., Rosenqvist, G., Berglund, A., Arnold, S. J., and Avise, J. C. 2000. The Bateman gradient and the cause of sexual selection in a sex-role-reversed pipefish. *Proceedings of the Royal Society of London Series B* **267**, 677–680.

Jones, A. G., Walker, D., and Avise, J. C. 2001. Genetic evidence for extreme polyandry and extraordinary sex-role reversal in a pipefish. *Proceedings of the Royal Society of London B* **268**, 2531–2535.

Jones, A. G., Rosenqvist, G., Berglund, A., and Avise, J. C. 2005. The measurement of sexual selection using Bateman's principles: an experimental test in the sex-role-reversed pipefish *Syngnathus typhle*. *Integrative and Comparative Biology* **45**, 874–884.

Jones, J. W. 1959. *The Salmon.* London: Collins.

Jones, J. W. and King, G. M. 1952. The spawning of the male salmon parr (*Salmo salar* Limn. juv). *Proceedings of the Zoological Society of London* **122**, 615–619.

Jones, J. W. and Orton, J. H. 1940. The paedogenetic male cycle in *Salmo salar* L. *Proceedings of the Royal Society of London B* **128**, 485–499.

Jones, M. W. and Hutchings, J. A. 2002. Individual variation in Atlantic salmon fertilization success: implications for effective population size. *Ecological Applications* **12**, 184–193.

Jonsson, B. and Jonsson, N. 2001. Polymorphism and speciation in Arctic charr. *Journal of Fish Biology* **58**, 605–638.

Jordan, W. C. and Youngson, A. F. 1992. The use of genetic marking to assess the reproductive success of mature male Atlantic salmon parr (*Salmo salar*, L.) under natural spawning conditions. *Journal of Fish Biology* **41**, 613–618.

Kano, Y., Shimizu, Y., and Kondou, K. 2006. Status-dependent female mimicry in landlocked red-spotted masu salmon. *Journal of Ethology* **24**, 1–7.

Kanoh, Y. 1996. Pre-oviposition ejaculation in externally fertilizing fish: how sneaker male rose bitterlings contrive to mate. *Ethology* **102**, 883–899.

Karino, K. and Kobayashi, M. 2005. Male alternative mating behaviour depending on tail length of the guppy, *Poecilia reticulata*. *Behaviour* 142, 191–202.

Karino, K. and Nakazono, A. 1993. Reproductive behavior of the territorial herbivore *Stegastes nigricans* (Pisces: Pomacentridae) in relation to colony formation. *Journal of Ethology* 11, 99–110.

Katoh, R., Munehara, H., and Kohda, M. 2005. Alternative male mating tactics of the substrate brooding cichlid *Telmatochromis temporalis* in Lake Tanganyika. *Zoological Science* 22, 555–561.

Kawase, H. and Nakazono, A. 1995. Predominant maternal egg care and promiscuous mating system in the Japanese filefish, *Rudarius ercodes* (Monacanthidae). *Environmental Biology of Fishes* 43, 241–254.

Kawase, H. and Nakazono, A. 1996. Two alternative female tactics in the polygynous mating system of the threadsail filefish, *Stephanolepis cirrhifer* (Monacanthidae). *Ichthyological Research* 43, 315–323.

Kazakov, R. V. 1981. Peculiarities of sperm production by anadromous and parr Atlantic salmon (*Salmo salar* L.) and fish cultural characteristics of such sperm. *Journal of Fish Biology* 18, 1–8.

Keller, L. and Reeve, H. K. 1994. Partitioning of reproduction in animal societies. *Trends in Ecology and Evolution* 9, 98–103.

Kellogg, K. A., Markert, J. A., Stauffer, J. R., and Kocher, T. D. 1998. Intraspecific brood mixing and reduced polyandry in a maternal mouth-brooding cichlid. *Behavioral Ecology* 9, 309–312.

Ketterson, E. D. and Nolan, V. 1999. Adaptation, exaptation, and constraint: a hormonal perspective. *American Naturalist* 154, S4–S25.

Kim, S. J., Ogasawara, K., Park, J. G., Takemura, A., and Nakamura, M. 2002. Sequence and expression of androgen receptor and estrogen receptor gene in the sex types of protogynous wrasse, *Halichoeres trimaculatus*. *General and Comparative Endocrinology* 127, 165–173.

Knapp, R. 2003. Endocrine mediation of vertebrate male alternative reproductive tactics: the next generation of studies. *Integrative and Comparative Biology* 43, 658–668.

Kodric-Brown, A. 1977. Reproductive success and the evolution of breeding territories in pupfish (*Cyprinodon*). *Evolution* 31, 750–766.

Kodric-Brown, A. 1981. Variable breeding systems in pupfishes (Genus *Cyprinodon*): adaptions to changing environments. In R. Naiman and D. L. Soltz (eds.) *Fishes in North American Deserts*, pp. 205–235.

Kodric-Brown, A. 1986. Satellites and sneakers: opportunistic male breeding tactics in pupfish (*Cyprinodon pecosensis*). *Behavioral Ecology and Sociobiology* 19, 425–432.

Kohda, M., Tanimura, M., Kikue-Nakamura, M., and Yamagishi, S. 1995. Sperm drinking by female catfishes: a novel mode of insemination. *Environmental Biology of Fishes* 42, 1–6.

Kohler, U. 1998. *Zur Struktur und Evolution des Sozialsystems von Neolamprologus multifasciatus (Cichlidae, Pisces), Dem Kleinsten Schneckenbuntbarsch des Tanganjikasees*. Aachen, Germany: Shaker-Verlag.

Kortet, R., Vainikka, A., Rantala, M. J., Jokinen, I., and Taskinen, J. 2003. Sexual ornamentation, androgens and papillomatosis in male roach (*Rutilus rutilus*). *Evolutionary Ecology Research* 5, 411–419.

Kortet, R., Taskinen, J., Vainikka, A., and Ylonen, H. 2004a. Breeding tubercles, papillomatosis and dominance behaviour of male roach (*Rutilus rutilus*) during the spawning period. *Ethology* 110, 591–601.

Kortet, R., Vainikka, A., Rantala, M. J., and Taskinen, J. 2004b. Sperm quality, secondary sexual characters and parasitism in roach (*Rutilus rutilus* L.). *Biological Journal of the Linnean Society* 81, 111–117.

Koseki, Y. and Fleming, I. A. 2006. Spatio-temporal dynamics of alternative male phenotypes in coho salmon populations in response to ocean environment. *Journal of Animal Ecology* 75, 445–455.

Koseki, Y. and Maekawa, K. 2000. Sexual selection on mature male parr of masu salmon (*Oncorhynchus masou*): does sneaking behavior favor small body size and less-developed sexual characters? *Behavioral Ecology and Sociobiology* 48, 211–217.

Koseki, Y. and Maekawa, K. 2002. Differential energy allocation of alternative male tactics in masu salmon (*Oncorhynchus masou*). *Canadian Journal of Fisheries and Aquatic Sciences* 59, 1717–1723.

Koster, W. J. 1939. Some phases of the life history and relationships of the cyprinid, *Clinostomus elongatus* (Kirtland). *Copeia*, 201–208.

Kuwamura, T., Nakashima, Y., and Yogo, Y. 1994. Sex change in either direction by growth rate advantage in the monogamous coral goby, *Paragobiodon echinocephalus*. *Behavioral Ecology* 5, 434–438.

Lahnsteiner, F. 2003. The spermatozoa and eggs of the cardinal fish. *Journal of Fish Biology* 62, 115–128.

Lahnsteiner, F., Richtarski, U., and Patzner, R.A. 1990. Functions of the testicular gland in two blenniid fishes, *Salaria* (=*Blennius*) *pavo* and *lipophrys* (=*Bleennius*) *delmatinus* (Blenniidae, Teleosei) as revealed by electron microscopy and enzyme histochemistry. *Journal of Fish Biology* 37, 85–97.

Lahnsteiner, F., Nussbaumer, B., and Patzner, R.A. 1993. Unusual testicular accessory organs, the testicular blind pouches of blennies (Teleostei, Blenniidae): fine structure, (enzyme) histochemistry and possible functions. *Journal of Fish Biology* 42, 227–241.

Lahnsteiner, F., Berger, B., Weismann, T., and Patzner, R.A. 1998. Determination of semen quality of the rainbow trout, *Oncorhynchus mykiss*, by sperm motility, seminal plasma parameters, and spermatozoal metabolism. *Aquaculture* 163, 163–181.

Landry, C. and Bernatchez, L. 2001. Comparative analysis of population structure across environments and geographical scales at major histocompatibility complex and microsatellite loci in Atlantic salmon (*Salmo salar*). *Molecular Ecology* 10, 2525–2539.

Laumen, J., Pern, U., and Blum, V. 1974. Investigations on function and hormonal regulation of anal appendices in *Blennius pavo* (Risso). *Journal of Experimental Zoology* 190, 47–56.

Leach, B. and Montgomerie, R. 2000. Sperm characteristics associated with different male reproductive tactics in bluegills (*Lepomis macrochirus*). *Behavioral Ecology and Sociobiology* 49, 31–37.

Lee, J.S.F. 2005. Alternative reproductive tactics and status-dependent selection. *Behavioral Ecology* 16, 566–570.

Lee, J.S.F. and Bass, A.H. 2004. Does exaggerated morphology preclude plasticity to cuckoldry in the midshipman fish (*Porichthys notatus*)? *Naturwissenschaften* 91, 338–341.

Lejeune, P. 1985. Etude écoéthologique des comportements reproducteurs et sociaux des *Labridae* méditerranéens des genres *Symphodus* Rafinesque, 1810 et *Coris* Lacepède, 1802. *Cahiers d'Ethologie Appliquée* 5, 1–208.

Lejeune, P. 1987. The effect of local stock density on social-behavior and sex change in the mediterranean labrid *Coris julis*. *Environmental Biology of Fishes* 18, 135–141.

Letcher, B.H. and Gries, G. 2003. Effects of life history variation on size and growth in stream-dwelling Atlantic salmon. *Journal of Fish Biology* 62, 97–114.

Li, S.K. and Owings, D.H. 1978. Sexual selection in the three-spined stickleback. 2. Nest raiding during the courtship phase. *Behaviour* 64, 298–304.

Li, W.M., Scott, A.P., Siefkes, M.J., et al. 2002. Bile acid secreted by male sea lamprey that acts acts as a sex pheromone. *Science* 296, 138–141.

Liley, N.R. and Stacey, N.E. 1983. Hormones, pheromones, and reproductive behavior in fish. In W.S. Hoar, D.J. Randall, and E.M. Donaldson (eds.) *Fish Physiology*, pp. 1–63. New York: Academic Press.

Liley, N.R., Tamkee, P., Tsai, R., and Hoysak, D.J. 2002. Fertilization dynamics in rainbow trout (*Oncorhynchus mykiss*): effect of male age, social experience, and sperm concentration and motility on in vitro fertilization. *Canadian Journal of Fisheries and Aquatic Sciences* 59, 144–152.

Liljedal, S. and Folstad, I. 2003. Milt quality, parasites, and immune function in dominant and subordinate Arctic charr. *Canadian Journal of Zoology/Revue Canadienne de Zoologie* 81, 221–227.

Linhart, O. 1984. Evaluation of sperm in some salmonids. *Bulletin VURH Vodnany* 20, 20–34.

Linhart, O., Rodina, M., Gela, D., Kocour, M., and Vandeputte, M. 2005. Spermatozoal competition in common carp (*Cyprinus carpio*): what is the primary determinant of competition success? *Reproduction* 130, 705–711.

Locatello, L., Mazzoldi, C., and Rasotto, M.B. 2002. Ejaculate of sneaker males is pheromonally inconspicuous in the black goby, *Gobius niger* (Teleostei, Gobiidae). *Journal of Experimental Zoology* 293, 601–605.

Lu, G.Q. and Bernatchez, L. 1999. Correlated trophic specialization and genetic divergence in sympatric lake whitefish ecotypes (*Coregonus clupeaformis*): support for the ecological speciation hypothesis. *Evolution* 53, 1491–1505.

Lucas, J.R. and Howard, R.D. 1995. On alternative reproductive tactics in anurans: dynamic games with density and frequency dependence. *American Naturalist* 146, 365–397.

Luttbeg, B. 2004. Female mate assessment and choice behavior affect the frequency of alternative male mating tactics. *Behavioral Ecology* 15, 239–247.

Mann, M.E. and Taborsky, M. 2007. Sexual conflict over breeding substrate causes female expulsion and offspring loss in a cichlid fish. *Behavioral Ecology*, in press.

Macias-Garcia, C. and Saborio, E. 2004. Sperm competition in a viviparous fish. *Environmental Biology of Fishes* 70, 211–217.

Mackiewicz, M., Fletcher, D. E., Wilkins, S. D., DeWoody, J. A., and Avise, J. C. 2002. A genetic assessment of parentage in a natural population of dollar sunfish (*Lepomis marginatus*) based on microsatellite markers. *Molecular Ecology* 11, 1877–1883.

Mackiewicz, M., Porter, B. A., Dakin, E. E., and Avise, J. C. 2005. Cuckoldry rates in the Molly Miller (*Scartella cristata*; Blenniidae), a hole-nesting marine fish with alternative reproductive tactics. *Marine Biology* 148, 213–221.

MacWhirter, R. B. 1989. On the rarity of intraspecific brood parasitism. *Condor* 91, 485–492.

Maekawa, K., Nakano, S., and Yamamoto, S. 1994. Spawning behavior and size assortative mating of Japanese charr in an artificial lake-inlet stream system. *Environmental Biology of Fishes* 39, 109–117.

Maekawa, K., Iguchi, K., and Katano, O. 1996. Reproductive success in male Japanese minnows, *Pseudorasbora parva*: observations under experimental conditions. *Ichthyological Research* 43, 257–266.

Magnhagen, C. 1992. Alternative reproductive behaviour in the common goby, *Pomatoschistus microps*: an ontogenetic gradient? *Animal Behaviour* 44, 182–184.

Malavasi, S., Torricelli, P., Lugli, M., Pranovi, F., and Mainardi, D. 2003. Male courtship sounds in a teleost with alternative reproductive tactics, the grass goby, *Zosterisessor ophiocephalus*. *Environmental Biology of Fishes* 66, 231–236.

Mank, J. E. and Avise, J. C. 2006. Comparative phylogenetic analysis of male alternative reproductive tactics in ray-finned fishes. *Evolution* 60, 1311–1316.

Mansour, N., Lahnsteiner, F., and Patzner, R. A. 2004. Seminal vesicle secretion of African catfish, its composition, its behaviour in water and saline solutions and its influence on gamete fertilizability. *Journal of Experimental Zoology A* 301A, 745–755.

Marconato, A., Rasotto, M. B., and Mazzoldi, C. 1996. On the mechanism of sperm release in three gobiid fishes (Teleostei: Gobiidae). *Environmental Biology of Fishes* 46, 321–327.

Marshall, D. J., Steinberg, P. D., and Evans, J. P. 2004. The early sperm gets the good egg: mating order effects in free spawners. *Proceedings of the Royal Society of London B* 271, 1585–1589.

Martin, E. and Taborsky, M. 1997. Alternative male mating tactics in a cichlid, *Pelvicachromis pulcher*: a comparison of reproductive effort and success. *Behavioral Ecology and Sociobiology* 41, 311–319.

Maynard Smith, J. 1982. *Evolution and the Theory of Games*. Cambridge, UK: Cambridge University Press.

Mazzoldi, C. and Rasotto, M. B. 2002. Alternative male mating tactics in *Gobius niger*. *Journal of Fish Biology* 61, 157–172.

Mazzoldi, C., Scaggiante, M., Ambrosin, E., and Rasotto, M. B. 2000. Mating system and alternative male mating tactics in the grass goby *Zosterisessor ophiocephalus* (Teleostei: Gobiidae). *Marine Biology* 137, 1041–1048.

Mazzoldi, C., Petersen, C. W., and Rasotto, M. B. 2005. The influence of mating system on seminal vesicle variability among gobies (Teleostei, Gobiidae). *Journal of Zoological Systematics and Evolutionary Research* 43, 307–314.

Mboko, S. K. and Kohda, M. 1999. Piracy mating by large males in a monogamous substrate-breeding cichlid in Lake Tanganyika. *Journal of Ethology* 17, 51–55.

McCart, P. 1970. A polymorphic population of *Oncorhynchus nerka* in Babine Lake, British Columbia. Ph.D. thesis, University of British Columbia, Vancouver, Canada.

McKaye, K. R. and McKaye, N. M. 1977. Communal care and kidnapping of young by parental cichlids. *Evolution* 31, 674–681.

McKaye, K. R. and Oliver, M. K. 1980. Geometry of a selfish school: defense of cichlid young by bagrid catfish in Lake Malawi, Africa. *Animal Behaviour* 28, 1287–1292.

McKaye, K. R., Mughogho, D. E., and Lovullo, T. J. 1992. Formation of the selfish school. *Environmental Biology of Fishes* 35, 213–218.

Meidl, P. 1999. Microsatellite analysis of alternative mating tactics in *Lamprologus callipterus*. Diploma thesis, University of Vienna, Austria.

Metcalfe, N. B. and Thorpe, J. E. 1990. Determinants of geographical variation in the age of seaward migrating salmon, *Salmo salar*. *Journal of Animal Ecology* 59, 135–145.

Metcalfe, N. B., Huntingford, F. A., and Thorpe, J. E. 1988. Feeding intensity, growth rates, and the establishment of life history patterns in juvenile Atlantic salmon *Salmo salar*. *Journal of Animal Ecology* 57, 463–474.

Metcalfe, N. B., Huntingford, F. A., Graham, W. D., and Thorpe, J. E. 1989. Early social status and the develoment of life history strategies in Atlantic salmon. *Proceedings of the Royal Society of London B* 236, 7–19.

Meyer, A. 1987. Phenotypic plasticity and heterochrony in *Cichlasoma managuense* (Pisces, Cichlidae) and their implications for speciation in cichlid fishes. *Evolution* 41, 1357–1369.

Milinski, M. 1987. Tit for tat in sticklebacks and the evolution of cooperation. *Nature* 325, 433–435.

Milinski, M., Külling, D., and Kettler, R. 1990a. Tit for tat: sticklebacks (*Gasterosteus aculeatus*) "trusting" a cooperating partner. *Behavioral Ecology* 1, 7–11.

Milinski, M., Pfluger, D., and Küelling, D. K. 1990b. Do sticklebacks cooperate repeatedly in reciprocal pairs? *Behavioral Ecology and Sociobiology* 27, 17–21.

Miller, P. J. 1984. The tokology of gobiid fishes. In G. W. Potts and R. J. Wootton (eds.) *Fish Reproduction: Strategies and Tactics*, pp. 119–154. New York: Academic Press.

Miura, T., Yamauchi, K., Takahashi, H., and Nagahama, Y. 1992. The role of hormones in the acquisition of sperm motility in salmonid fish. *Journal of Experimental Zoology* 261, 359–363.

Mjolnerod, I. B., Fleming, I. A., Refseth, U. H., and Hindar, K. 1998. Mate and sperm competition during multiple-male spawnings of Atlantic salmon. *Canadian Journal of Zoology/Revue Canadienne de Zoologie* 76, 70–75.

Moore, A. and Scott, A. P. 1991. Testosterone is a potent odorant in precocious male Atlantic Salmon (*salmo salar* L.) parr. *Philosophical Transactions of the Royal Society of London B* 332, 241–244.

Moore, M. C. 1991. Application of organization-activation theory to alternative male reproductive strategies : a review. *Hormones and Behavior* 25, 154–179.

Moore, M. C., Hews, D. K., and Knapp, R. 1998. Hormonal control and evolution of alternative male phenotypes: generalizations of models for sexual differentiation. *American Zoologist* 38, 133–151.

Morris, D. 1952. Homosexuality in the ten-spined stickleback (*Pygosteus pungitius* L.). *Behaviour* 4, 233–261.

Moyer, J. T. 1979. Mating strategies and reproductive behavior of ostraciid fishes at Miyake-Jima, Japan. *Japanese Journal of Ichthyology* 26, 148–160.

Mrowka, W. 1987. Egg stealing in a mouthbrooding cichlid fish. *Animal Behaviour* 35, 923–925.

Munday, P. L., Buston, P. M., and Warner, R. R. 2006. Diversity and flexibility of sex-change strategies in animals. *Trends in Ecology and Evolution* 21, 89–95.

Munehara, H. 1988. Spawning and subsequent copulating behavior of the elkhorn sculpin *Alcichthys alcicornis* in an aquarium. *Japanese Journal of Ichthyology* 35, 358–364.

Munehara, H. and Takenaka, O. 2000. Microsatellite markers and multiple paternity in a paternal care fish, *Hexagrammus otakii*. *Journal of Ethology* 18, 101–104.

Munehara, H., Takano, K., and Koya, Y. 1989. Internal gametic association and external fertilization in the elkhorn sculpin, *Alcichthys alcicornis*. *Copeia*, 673–678.

Munehara, H., Okamoto, H., and Shimazaki, K. 1990. Paternity estimated by isozyme variation in the marine sculpin *Alcichthys alcicornis* (Pisces, Cottidae) exhibiting copulation and paternal care. *Journal of Ethology* 8, 21–24.

Munehara, H., Takano, K., and Koya, Y. 1991. The little dragon sculpin *Blespias cirrhosus*: another case of internal gametic association and external fertilization. *Japanese Journal of Ichthyology* 37, 391–394.

Munoz, R. C. and Warner, R. R. 2003. Alternative contexts of sex change with social control in the bucktooth parrotfish, *Sparisoma radians*. *Environmental Biology of Fishes* 68, 307–319.

Munoz, R. C. and Warner, R. R. 2004. Testing a new version of the size-advantage hypothesis for sex change: sperm competition and size-skew effects in the bucktooth parrotfish, *Sparisoma radians*. *Behavioral Ecology* 15, 129–136.

Narimatsu, Y. and Munehara, H. 2001. Territoriality egg desertion and mating success of a paternal care fish, *Hypoptychus dybowskii* (Gasterosteiformes). *Behaviour* 138, 85–96.

Neat, F. C. 2001. Male parasitic spawning in two species of triplefin blenny (Tripterigiidae): contrasts in demography, behaviour and gonadal characteristics. *Environmental Biology of Fishes* 61, 57–64.

Neat, F. C., Locatello, L., and Rasotto, M. B. 2003a. Reproductive morphology in relation to alternative male reproductive tactics in *Scartella cristata*. *Journal of Fish Biology* 62, 1381–1391.

Neat, F. C., Lengkeek, W., Westerbeek, E. P., Laarhoven, B., and Videler, J. J. 2003b. Behavioural and morphological differences between lake and river populations of *Salaria fluviatilis*. *Journal of Fish Biology* 63, 374–387.

Neff, B. D. 2001. Genetic paternity analysis and breeding success in bluegill sunfish (*Lepomis macrochirus*). *Journal of Heredity* 92, 111–119.

Neff, B. D. 2003. Decisions about parental care in response to perceived paternity. *Nature* 422, 716–719.

Neff, B. D. 2004. Increased performance of offspring sired by parasitic males in bluegill sunfish. *Behavioral Ecology* 15, 327–331.

Neff, B. D. and Gross, M. R. 2001. Dynamic adjustment of parental care in response to perceived paternity. *Proceedings of the Royal Society of London B* 268, 1559–1565.

Neff, B. D. and Sherman, P. W. 2005. In vitro fertilization reveals offspring recognition via self-referencing in a fish with paternal care and cuckoldry. *Ethology* 111, 425–438.

Neff, B. D. and Wahl, L. M. 2004. Mechanisms of sperm competition: testing the fair raffle. *Evolution* 58, 1846–1851.

Neff, B. D., Fu, P., and Gross, M. R. 2003. Sperm investment and alternative mating tactics in bluegill sunfish (*Lepomis macrochirus*). *Behavioral Ecology* 14, 634–641.

Nikolsky, G. V. 1963. *The Ecology of Fishes*. London: Academic Press.

Njiwa, J. R. K., Muller, P., and Klein, R. 2004. Variations of sperm release in three batches of zebrafish. *Journal of Fish Biology* 64, 475–482.

Noe, R. 2006. Cooperation experiments: coordination through communication versus acting apart together. *Animal Behaviour* 71, 1–18.

Nowak, M. A., Sasaki, A., Taylor, C., and Fudenberg, D. 2004. Emergence of cooperation and evolutionary stability in finite populations. *Nature* 428, 646–650.

Ochi, H. 1993. Mate monopolization by a dominant male in a multi-male social group of a mouthbrooding cichlid *Ctenochromis horei. Japanese Journal of Ichthyology* 40, 209–218.

Ochi, H. and Yanagisawa, Y. 1996. Interspecific brood-mixing in Tanganyikan cichlids. *Environmental Biology of Fishes* 45, 141–149.

Ochi, H. and Yanagisawa, Y. 2005. Farming-out of offspring is a predominantly male tactics in a biparental mouthbrooding cichlid *Perrisodus microlepis. Environmental Biology of Fishes* 73, 335–340.

Ochi, H., Yanagisawa, Y., and Omori, K. 1995. Intraspecific brood mixing of the cichlid fish *Perissodus microlepis* in Lake Tanganyika. *Environmental Biology of Fishes* 43, 201–206.

Ochi, H., Onchi, T., and Yanagisawa, Y. 2001. Alloparental care between catfishes in Lake Tanganyika. *Journal of Fish Biology* 59, 1279–1286.

Ohnishi, N., Yanagisawa, Y., and Kohda, M. 1997. Sneaking by harem masters of the sandperch, *Parapercis snyderi. Environmental Biology of Fishes* 50, 217–223.

Okuda, N., Ito, S., and Iwao, H. 2003. Female mimicry in a freshwater goby *Rhinogobius* sp. OR. *Ichthyological Research* 50, 198–200.

Oliveira, R. F. 2004. *Social Modulation of Androgens in Vertebrates: Mechanisms and Function.*

Oliveira, R. F. 2006. Neuroendocrine mechanisms of alternative reproductive tactics in fish. In K. A. Sloman, R. W. Wilson, and S. Balshine (eds.) *Behaviour: Interactions with Physiology*, New York: Elsevier.

Oliveira, R. F., Almada, V. C., Gonçalves, E. J., Forsgren, E., and Canario, A. V. M. 2001a. Androgen levels and social interactions in breeding males of the peacock blenny. *Journal of Fish Biology* 58, 897–908.

Oliveira, R. F., Canario, A. V. M., and Grober, M. S. 2001b. Male sexual polymorphism, alternative reproductive tactics, and androgens in combtooth blennies (Pisces: Blenniidae). *Hormones and Behavior* 40, 266–275.

Oliveira, R. F., Canario, A. V. M., Grober, M. S., and Santos, R. S. 2001c. Endocrine correlates of male polymorphism and alternative reproductive tactics in the Azorean rock-pool blenny, *Parablennius sanguinolentus parvicornis. General and Comparative Endocrinology* 121, 278–288.

Oliveira, R. F., Carneiro, L. A., Gonçalves, D. M., Canario, A. V. M., and Grober, M. S. 2001d. 11-ketotestosterone inhibits the alternative mating tactic in sneaker males of the peacock blenny, *Salaria pavo. Brain Behavior and Evolution* 58, 28–37.

Oliveira, R. F., Carneiro, L. A., Canario, A. V. M., and Grober, M. S. 2001e. Effects of androgens on social behavior and morphology of alternative reproductive males of the azorean rock-pool blenny. *Hormones and Behavior* 39, 157–166.

Oliveira, R. F., Gonçalves, E. J., and Santos, R. S. 2001f. Gonadal investment of young males in two blenniid fishes with alternative mating tactics. *Journal of Fish Biology* 59, 459–462.

Oliveira, R. F., Carvalho, N., Miranda, J., *et al.* 2002a. The relationship between the presence of satellite males and nest-holders' mating success in the Azorean rock-pool blenny *Parablennius sanguinolentus parvicornis. Ethology* 108, 223–235.

Oliveira, R. F., Hirschenhauser, K., Carneiro, L. A., and Canario, A. V. M. 2002b. Social modulation of androgen levels in male teleost fish. *Comparative Biochemistry and Physiology B* 132, 203–215.

Oliveira, R. F., Hirschenhauser, K., Canario, A. V. M., and Taborsky, M. 2003. Androgen levels of reproductive competitors in a co-operatively breeding cichlid. *Journal of Fish Biology* 63, 1615–1620.

Oliveira, R. F., Ros, A. F. H., and Gonçalves, D. M. 2005. Intra-sexual variation in male reproduction in teleost fish: a comparative approach. *Hormones and Behavior* 48, 430–439.

Olsson, P. E., Berg, A. H., von Hofsten, J., *et al.* 2005. Molecular cloning and characterization of a nuclear androgen receptor activated by

11-ketotestosterone. *Reproductive Biology and Endocrinology* **3**, 00–00.

Orton, J. H., Jones, J. W., and King, G. M. 1938. The male sexual stage in salmon parr (*Salmo salar* L. juv.). *Proceedings of the Royal Society of London B* **125**, 103–114.

Ota, D., Marchesan, M., and Ferrero, E. A. 1996. Sperm release behaviour and fertilization in the grass goby. *Journal of Fish Biology* **49**, 246–256.

Ota, K. and Kohda, M. 2006a. Description of alternative male reproductive tactics in a shell-brooding cichlid, *Telmatochromis vittatus* in Lake Tanganyika. *Journal of Ethology* **24**, 9–15.

Ota, K. and Kohda, M. 2006b. Nest use by territorial males in a shell-brooding cichlid: the effect of reproductive parasitism. *Journal of Ethology* **24**, 91–95.

Pachler, G. 2001. Growth, spawning behavior, and ejaculate characteristics: adaptations to sperm competition in *Lamprologus callipterus*. Diploma thesis, University of Vienna, Austria.

Page, L. M. and Johnston, C. E. 1990. Spawning in the creek chubsucker, *Erimyzon oblongus*, with a review of spawning behavior in suckers (*Catostomidae*). *Environmental Biology of Fishes* **27**, 265–272.

Parker, G. A. 1984. Sperm competition and the evolution of animal mating strategies. In R. L. Smith (ed.) *Sperm Competition and the Evolution of Animal Mating Systems*, pp. 1–60. Orlando, FL: Academic Press.

Parker, G. A. 1990a. Sperm competition games: raffles and roles. *Proceedings of the Royal Society of London B* **242**, 120–126.

Parker, G. A. 1990b. Sperm competition games: sneaks and extra-pair copulations. *Proceedings of the Royal Society of London B* **242**, 127–133.

Parker, G. A. 1993. Sperm competition games: sperm size and sperm number under adult control. *Proceedings of the Royal Society of London B* **253**, 245–254.

Parker, G. A. and Ball, M. A. 2005. Sperm competition, mating rate and the evolution of testis and ejaculate sizes: a population model. *Biology Letters* **1**, 235–238.

Parker, G. A. and Begon, M. E. 1993. Sperm competition games: sperm size and sperm number under gametic control. *Proceedings of the Royal Society of London B* **253**, 255–262.

Parker, G. A., Ball, M. A., Stockley, P., and Gage, M. J. G. 1996. Sperm competition games: individual assessment of sperm competition intensity by group spawners. *Proceedings of the Royal Society of London B* **263**, 1291–1297.

Parker, H. H., Noonburg, E. G., and Nisbet, R. M. 2001. Models of alternative life-history strategies, population structure and potential speciation in salmonid fish stocks. *Journal of Animal Ecology* **70**, 260–272.

Pastres, R., Pranovi, F., Libralato, S., Malavasi, S., and Torricelli, P. 2002. "Birthday effect" on the adoption of alternative mating tactics in *Zosterisessor ophiocephalus*: evidence from a growth model. *Journal of the Marine Biological Association of the United Kingdom* **82**, 333–337.

Petersen, C. W. 1987. Reproductive behaviour and gender allocation in *Serranus fasciatus*, a hermaphroditic reef fish. *Animal Behavior* **35**, 1601–1614.

Petersen, C. W. 1990. The relationship among population density, individual size, mating tactics and reproductive success in a hermaphroditic fish, *Serranus fasciatus*. *Behaviour* **113**, 57–80.

Petersen, C. W. 1995. Reproductive behavior, egg trading, and correlates of male mating success in the simultaneous hermaphrodite, *Serranus tabacarius*. *Environmental Biology of Fishes* **43**, 351–361.

Petersen, C. W. and Fischer, E. A. 1986. Mating system of the hermaphroditic coral-reef fish, *Serranus baldwini*. *Behavioral Ecology and Sociobiology* **19**, 171–178.

Petersen, C. W. and Fischer, E. A. 1996. Intraspecific variation in sex allocation in a simultaneous hermaphrodite: the effect of individual size. *Evolution* **50**, 636–645.

Petersen, C. W. and Warner, R. R. 1998. Sperm competition in fishes. In T. R. Birkhead and A. P. Moller (eds.) *Sperm Competition and Sexual Selection*, pp. 435–463. San Diego, CA: Academic Press.

Petersen, C. W., Warner, R. R., Shapiro, D. Y., and Marconato, A. 2001. Components of fertilization success in the bluehead wrasse, *Thalassoma bifasciatum*. *Behavioral Ecology* **12**, 237–245.

Petrie, M. and Kempenaers, B. 1998. Extra-pair paternity in birds: explaining variation between species and populations. *Trends in Ecology and Evolution* **13**, 52–58.

Pfennig, D. W. 1992. Polyphenism in spadefoot toad tadpoles as a locally adjusted evolutionarily stable strategy. *Evolution* **46**, 1408–1420.

Picciulin, M., Verginella, L., Spoto, M., and Ferrero, E. A. 2004. Colonial nesting and the importance of the brood size in male parasitic reproduction of the Mediterranean damselfish *Chromis chromis* (Pisces: Pomacentridae). *Environmental Biology of Fishes* **70**, 23–30.

Pigeon, D., Chouinard, A., and Bernatchez, L. 1997. Multiple modes of speciation involved in the parallel evolution of sympatric morphotypes of lake whitefish (*Coregonus clupeaformis*, Salmonidae). *Evolution* 51, 196–205.

'iironen, J. and Hyvärinen, H. 1983. Composition of the milt of some teleost fishes. *Journal of Fish Biology* 22, 351–361.

'ilastro, A. and Bisazza, A. 1999. Insemination efficiency of two alternative male mating tactics in the guppy (*Poecilia reticulata*). *Proceedings of the Royal Society of London B* 266, 1887–1891.

'ilastro, A., Scaggiante, M., and Rasotto, M. B. 2002. Individual adjustment of sperm expenditure accords with sperm competition theory. *Proceedings of the National Academy of Sciences of the United States of America* 99, 9913–9915.

'ilastro, A., Simonato, M., Bisazza, A., and Evans, J. P. 2004. Cryptic female preference for colorful males in guppies. *Evolution* 58, 665–669.

'itcher, T. E., Neff, B. D., Rodd, F. H., and Rowe, L. 2003. Multiple mating and sequential mate choice in guppies: females trade up. *Proceedings of the Royal Society of London B* 270, 1623–1629.

'oncin, P., Jeandarme, J., Rinchard, J., and Kestemont, P. 1997. Preliminary results on the spawning behaviour of the gudgeon, *Gobio gobio*, in aquarium. *Bulletin Français de la Pêche et de la Pisciculture*, 547–555.

Quinn, T. P. and Myers, K. W. 2004. Anadromy and the marine migrations of Pacific salmon and trout: Rounsefell revisited. *Reviews in Fish Biology and Fisheries* 14, 421–442.

.agland, H. C. and Fischer, E. A. 1987. Internal fertilization and male parental care in the scalyhead sculpin, *Artedius harringtoni*. *Copeia*, 1059–1062.

.aney, E. C. 1940. Comparison of the breeding habits of two subspecies of black-nosed dace, *Rhinichthys atratulus* (Hermann). *American Midland Naturalist* 23, 399–403.

.asotto, M. B. and Mazzoldi, C. 2002. Male traits associated with alternative reproductive tactics in *Gobius niger*. *Journal of Fish Biology* 61, 173–184.

.eeve, H. K., Emlen, S. T., and Keller, L. 1998. Reproductive sharing in animal societies: reproductive incentives or incomplete control by dominant breeders? *Behavioral Ecology* 9, 267–278.

.eichard, M., Smith, C., and Jordan, W. C. 2004a. Genetic evidence reveals density-dependent mediated success of alternative mating behaviours in the European bitterling (*Rhodeus sericeus*). *Molecular Ecology* 13, 1569–1578.

.eichard, M., Jurajda, P., and Smith, C. 2004b. Male–male interference competition decreases spawning rate in the

European bitterling (*Rhodeus sericeus*). *Behavioral Ecology and Sociobiology* 56, 34–41.

Reighard, J. 1920. The breeding behaviors of the suckers and minnows. 1. The suckers. *Biological Bulletin* 38, 1–32.

Reighard, J. 1943. The breeding habits of the river chub, *Nocomis micropogon* (Cope). *Papers of the Michigan Academy of Sciences, Arts and Letters* 28, 397–423.

Reinboth, R. 1962. Morphologische und funktionelle Zweigeschlechtlichkeit bei marinen Teleostern (Serranidae, Sparidae, Centracanthidae, Labridae). *Zoologisches Jahrbuch (Phyisologie)* 69, 405–480.

Reynolds, J. D., Debuse, V. J., and Aldridge, D. C. 1997. Host specialisation in an unusual symbiosis: European bitterlings spawning in freshwater mussels. *Oikos* 78, 539–545.

Reznick, D. 1981. Grandfather effects: the genetics of interpopulation differences in offspring size in the mosquito fish. *Evolution* 35, 941–953.

Reznick, D. 1982. Genetic determination of offspring size in the guppy (*Poecilia reticulata*). *American Naturalist* 120, 181–188.

Rhen, T. and Crews, D. 2002. Variation in reproductive behaviour within a sex: neural systems and endocrine activation. *Journal of Neuroendocrinology* 14, 517–531.

Ribbink, A. J., Marsh, A. C., Marsh, B., and Sharp, B. J. 1980. Parental behavior and mixed broods among cichlid fish of Lake Malawi. *South African Journal of Zoology* 15, 1–6.

Richtarski, U. and Patzner, R. A. 2000. Comparative morphology of male reproductive systems in Mediterranean blennies (Blenniidae). *Journal of Fish Biology* 56, 22–36.

Rico, C., Kuhnlein, U., and FitzGerald, G. J. 1992. Male reproductive tactics in the threespine stickleback: an evaluation by DNA fingerprinting. *Molecular Ecology* 1, 79–87.

Riddell, B. E., Leggett, W. C., and Saunders, R. L. 1981. Evidence of adaptive polygenic variation between two populations of Atlantic salmon (*Salmo salar*) native to tributaries of the SW Miramichi River, NB. *Canadian Journal of Fisheries and Aquatic Sciences* 38, 321–333.

Rios-Cardenas, O. and Webster, A. S. 2005. Paternity and paternal effort in the pumpkinseed sunfish. *Behavioral Ecology* 16, 914–921.

Robertson, D. R. and Choat, J. H. 1974. Protogynous hermaphroditism and social systems in labrid fish. *Proceedings of the 2nd International Coral Reef Symposium* 1, 217–225.

Robertson, D. R. and Warner, R. R. 1978. Sexual patterns in the labrid fishes of the Western Caribbean. 2. The parrotfishes (Scaridae). *Smithsonian Contributions to Zoology* 255, 1–26.

Robertson, D. R., Reinboth, R., and Bruce, R. W. 1982. Gonochorism, protogyneous sex-change and spawning in three sparisomatinine parrotfishes from the western Indian Ocean. *Bulletin of Marine Science* 32, 868–879.

Robinson, B. W. and Wilson, D. S. 1994. Character release and displacement in fishes: a neglected literature. *American Naturalist* 144, 596–627.

Robinson, G. E., Grozinger, C. M., and Whitfield, C. W. 2005. Sociogenomics: social life in molecular terms. *Nature Reviews Genetics* 6, 257–316.

Ros, A. F. H., Bouton, N., Santos, R. S., and Oliveira, R. F. 2006. Alternative male reproductive tactics and the immunocompetence handicap in the Azorean rock-pool blenny, *Parablennius parvicornis*. *Proceedings of the Royal Society of London B* 273, 901–909.

Rosenqvist, G. 1990. Male mate choice and female–female competition for mates in the pipefish *Nerophis ophidion*. *Animal Behaviour* 39, 1110–1115.

Ross, R. M. 1989. The evolution of sex-change mechanisms in fishes. *Environmental Biology of Fishes* 19, 81–93.

Rossiter, A. and Yamagishi, S. 1997. Intraspecific plasticity in the social system and mating behaviour of a lek-breeding cichlid fish. In H. Kawanabe, M. Hori, and M. Nagoshi (eds.) *Fish Communities in Lake Tanganyika*, pp. 193–217. Kyoto, Japan: Kyoto University Press.

Rowe, D. K. and Thorpe, J. E. 1990. Differences in growth between maturing and non-maturing male Atlantic salmon, *Salmo salar* L., parr. *Journal of Fish Biology* 36, 643–658.

Rowland, W. J. 1979. Stealing fertilizations in the fourspine stickleback, *Apeltes quadracus*. *American Naturalist* 114, 602–604.

Royce, W. F. 1951. Breeding habits of lake trout in New York. *Fishery Bulletin* 59, 59–76.

Ruchon, F., Laugier, T. and Quignard, J. P. 1995. Alternative male reproductive strategies in the peacock blenny. *Journal of Fish Biology* 47, 826–840.

Rudolfsen, G., Figenschou, L., Folstad, I., Nordeide, J. T., and Soreng, E. 2005. Potential fitness benefits from mate selection in the Atlantic cod (*Gadus morhua*). *Journal of Evolutionary Biology* 18, 172–179.

Rudolfsen, G., Figenschou, L., Folstad, I., Tveiten, H., and Figenschou, M. 2006. Rapid adjustments of sperm characteristics in relation to social status. *Proceedings of the Royal Society of London B* 273, 325–332.

Rutte, C. and Taborsky, M. 2007a. Generalized reciprocity in rats. *PLoS Biology* 5, 1421–1425.

Rutte, C. and Taborsky, M. 2007b. The influence of social experience on cooperative behaviour of rats (*Rattus norvegicus*): direct vs. generalized reciprocity. *Behavioral Ecology and Sociobiology*, DOI: 10.1007/s00265-007-0474-3.

Ryan, M. J. and Causey, B. 1989. Alternative mating behaviour in the sword tails *Xiphophorus pygmaeus* (Pisces: Poeciliidae). *Behavioral Ecology and Sociobiology* 24, 341–348.

Ryan, M. J., Hews, D. K., and Wagner, W. E. 1990. Sexual selection on alleles that determine body size in the swordtail *Xiphophorus nigrensis*. *Behavioral Ecology and Sociobiology* 26, 231–237.

Ryan, M. J., Pease, C. M., and Morris, M. R. 1992. A genetic polymorphism in the swordtail, *Xiphophorus nigrensis*: testing the prediction of equal fitnesses. *American Naturalist* 139, 21–31.

Sage, R. D. and Selander, R. K. 1975. Trophic radiation through polymorphism in cichlid fishes. *Proceedings of the National Academy of Sciences of the United States of America* 72, 4669–4673.

Santos, R. S. 1985. Parentais e satélites: tácticas alternativas de acasalamento nos machos de *Blennius sanguinolentus* pallas (Pisces: Blenniidae). *Arquipélago: Life and marine Sciences* 6, 119–146.

Sargent, R. C. 1989. Allopaternal care in the fathead minnow, *Pimephales promelas*: stepfathers discriminate against their adopted eggs. *Behavioral Ecology and Sociobiology* 25, 379–386.

Sato, T. 1994. Active accumulation of spawning substrate: a determinant of extreme polygyny in a shell-brooding cichlid fish. *Animal Behaviour* 48, 669–678.

Sato, T. and Gashagaza, M. M. 1997. shell-brooding cichlid fishes of Lake Tanganyika: their habitats and mating systems. In H. kawanabe, M. Hori, and M. Nagoshi eds. *Fish communities in Lake Tanganyika* pp. 221–240. Kyoto, Japan: Kyoto University Press.

Scaggiante, M., Mazzoldi, C., Petersen, C. W., and Rasotto, M. B. 1999. Sperm competition and mode of fertilization in the grass goby *Zosterisessor ophiocephalus* (Teleostei: Gobiidae). *Journal of Experimental Zoology* 283, 81–90.

Scaggiante, M., Grober, M. S., Lorenzi, V., and Rasotto, M. B. 2004. Changes along the male reproductive axis in response to social context in a gonochoristic gobiid, *Zosterisessor ophiocephalus* (Teleostei, Gobiidae), with alternative mating tactics. *Hormones and Behavior* 46, 607–617.

Scaggiante, M., Grober, M. S., Lorenzi, V., and Rasotto, M. B. 2006. Variability of GnRH secretion in two goby species with socially controlled alternative male mating tactics. *Hormones and Behavior* 50, 107–117.

Schärer, L. and Robertson, D. R. 1999. Sperm and milt characteristics and male v. female gametic investment in the Caribbean reef fish, *Thalassoma bifasciatum*. *Journal of Fish Biology* 55, 329–343.

Schröder, S. L. 1981. *The Role of Sexual Selection in Determining Overall Mating Patterns and Mate Choice in Chum Salmon*. Seattle, WA: University of Washington Press.

Schröder, S. L. 1982. The influence of intrasexual competition on the distribution of chum salmon in an experimental stream. (Ed. by E. L. Brannon & E. O. Salo), pp. 275–285. Seattle, School of Fisheries, University of Washington.

Schulte-Hostedde, A. I. and Burness, G. 2005. Fertilization dynamics of sperm from different male mating tactics in bluegill (*Lepomis macrochirus*). *Canadian Journal of Zoology/ Revue Canadienne de Zoologie* 83, 1638–1642.

Schütz, D. and Taborsky, M. 2000. Giant males or dwarf females: what determines the extreme sexual size dimorphism in *Lamprologus callipterus*? *Journal of Fish Biology* 57, 1254–1265.

Schütz, D. and Taborsky, M. 2005. The influence of sexual selection and ecological constraints on an extreme sexual size dimorphism in a cichlid. *Animal Behaviour* 70, 539–549.

Schütz, D., Parker, G. A., Taborsky, M., and Sato, T. 2006. An optimality approach to male and female body sizes in an extremely size-dimorphic cichlid fish. *Evolutionary Ecology Research* 8, 1393–1408.

Seamons, T. R., Bentzen, P., and Quinn, T. P. 2004. The mating system of steelhead, *Oncorhynchus mykiss*, inferred by molecular analysis of parents and progeny. *Environmental Biology of Fishes* 69, 333–344.

Shapiro, D. Y. 1987. Differentiation and evolution of sex change in fishes. *BioScience* 37, 490–497.

Shapiro, D. Y., Marconato, A., and Yoshikawa, T. 1994. Sperm economy in a coral reef fish, *Thalassoma bifasciatum*. *Ecology* 75, 1334–1344.

Shibuno, T., Chiba, I., Gushima, K., Kakuda, S., and Hashimoto, H. 1993. Reproductive behavior of the wrasse, *Halichoeres marginatus*, at Kuchierabu-jima. *Japanese Journal of Ichthyology* 40, 351–359.

Shuster, S. M. and Wade, M. J. 1991. Female copying and sexual selection in a marine isopod crustacean. *Animal Behaviour* 42, 1071–1078.

Shuster, S. M. and Wade, M. J. 2003. *Mating Systems and Strategies*. Princeton, NJ: Princeton University Press.

Sigurjonsdottir, H. and Gunnarsson, K. 1989. Alternative mating tactics of arctic charr, *Salvelinus alpinus*, in Thingvallavatn, Iceland. *Environmental Biology of Fishes* 26, 159–176.

Smith, C. and Reichard, M. 2005. Females solicit sneakers to improve fertilization success in the bitterling fish (*Rhodeus sericeus*). *Proceedings of the Royal Society of London B* 272, 1683–1688.

Smith, C., Douglas, A., and Jurajda, P. 2002. Sexual conflict, sexual selection and sperm competition in the spawning decisions of bitterling, *Rhodeus sericeus*. *Behavioral Ecology and Sociobiology* 51, 433–439.

Smith, C., Reichard, M., Jurajda, P., and Przybylski, M. 2004. The reproductive ecology of the European bitterling (*Rhodeus sericeus*). *Journal of Zoology* 262, 107–124.

Snook, R. R. 2005. Sperm in competition: not playing by the numbers. *Trends in Ecology and Evolution* 20, 46–53.

Snorrason, S. S. and Skúlason, S. 2004. Adaptive speciation in northern freshwater fishes: patterns and processes. In U. Dieckmann, H. Metz, M. Doebeli, and D. Tautz (eds.) *Adaptive Speciation*, pp. 210–228. Cambridge, UK: Cambridge University Press.

Sogard, S. M. 1997. Size-selective mortality in the juvenile stage of teleost fishes: a review. *Bulletin of Marine Science* 60, 1129–1157.

Spahni, C. 2005. Indirect reciprocity in rats. Master's thesis, University of Berne, Switzerland.

Stacey, N. 2003. Hormones, pheromones and reproductive behavior. *Fish Physiology and Biochemistry* 28, 229–235.

Stacey, N., Chojnacki, A., Narayanan, A., Cole, T., and Murphy, C. 2003. Hormonally derived sex pheromones in fish: exogenous cues and signals from gonad to brain. *Canadian Journal of Physiology and Pharmacology* 81, 329–341.

Stephens, D. W., Anderson, J., and Benson, K. E. 1997. On the spurious occurrence of tit for tat in pairs of predator-approaching fish. *Animal Behaviour* 53, 111–131.

Stephens, D. W., McLinn, C. M., and Stevens, J. R. 2002. Discounting and reciprocity in an iterated prisoner's dilemma. *Science* 298, 2216–2218.

Stiver, K. A., Dierkes, P., Taborsky, M., and Balshine, S. 2004. Dispersal patterns and status change in a co-operatively breeding cichlid, *Neolamprologus pulcher*: evidence from microsatellite analyses and behavioural observations. *Journal of Fish Biology* 65, 91–105.

Stiver, K. A., Dierkes, P., Taborsky, M., Gibbs, H. L., and Balshine, S. 2005. Relatedness and helping in fish: examining the theoretical predictions. *Proceedings of the Royal Society of London B* 272, 1593–1599.

Stockley, P. 1997. Sexual conflict resulting from adaptations to sperm competition. *Trends in Ecology and Evolution* 12, 154–159.

Stockley, P., Gage, M. J. G., Parker, G. A., and Moller, A. P. 1996. Female reproductive biology and the coevolution of ejaculate characteristics in fish. *Proceedings of the Royal Society of London B* 263, 451–458.

Stockley, P., Gage, M. J. G., Parker, G. A., and Moller, A. P. 1997. Sperm competition in fishes: the evolution of testis size and ejaculate characteristics. *American Naturalist* 149, 933–954.

Stoltz, J. A. and Neff, B. D. 2006. Male size and mating tactic influence proximity to females during sperm competition in bluegill sunfish. *Behavioral Ecology and Sociobiology* 59, 811–818.

Stoltz, J. A., Neff, B. D., and Olden, J. D. 2005. Allometric growth and sperm competition in fishes. *Journal of Fish Biology* 67, 470–480.

Stuart-kregor, P. A. C., Sumpter, J. P., and Dodd, J. M. 1981. The involvement of gonodatropin and sex steroids in the control of reproduction in the parr and adults of the Atlantic Salmon, *salmo salar*. *Journal of Fish Biology* 18, 59–72.

Sunobe, T. and Nakazono, A. 1993. Sex-change in both directions by alteration of social dominance in *Trimma okinawae* (Pisces, Gobiidae). *Ethology* 94, 339–345.

Sveinsson, T. and Hara, T. J. 1995. Mature males of Arctic charr, *Salvelinus alpinus*, release F-type prostaglandins to attract conspecific mature females and stimulate their spawning behaviour. *Environmental Biology of Fishes* 42, 253–266.

Taborsky, M. 1984. Broodcare helpers in the cichlid fish *Lamprologus brichardi*: their costs and benefits. *Animal Behaviour* 32, 1236–1252.

Taborsky, M. 1985. Breeder–helper conflict in a cichlid fish with broodcare helpers: an experimental analysis. *Behaviour* 95, 45–75.

Taborsky, M. 1987. Cooperative behaviour in fish: coalitions, kin groups and reciprocity. In Y. Ito, J. L. Brown, and J. Kikkawa (eds.) *Animal Societies: Theories and Facts*, pp. 229–237. Tokyo: Japanese Science Society Press.

Taborsky, M. 1994. Sneakers, satellites, and helpers: parasitic and cooperative behavior in fish reproduction. *Advances in the Study of Behavior* 23, 1–100.

Taborsky, M. 1997. Bourgeois and parasitic tactics: do we need collective, functional terms for alternative reproductive behaviours? *Behavioral Ecology and Sociobiology* 41, 361–362.

Taborsky, M. 1998. Sperm competition in fish: bourgeois males and parasitic spawning. *Trends in Ecology and Evolution* 13, 222–227.

Taborsky, M. 1999. Conflict or cooperation: what determines optimal solutions to competition in fish reproduction? In R. F. Oliveira, V. Almada, and E. Gonçalves (eds.) *Behaviour and Conservation of Littoral Fishes*, pp. 301–349. Lisbon: Instituto Superior de Psicologia Aplicada.

Taborsky, M. 2001. The evolution of parasitic and cooperative reproductive behaviors in fishes. *Journal of Heredity* 92, 100–110.

Taborsky, M. and Grantner, A. 1998. Behavioural time-energy budgets of cooperatively breeding *Neolamprologus pulcher* (Pisces: Cichlidae). *Animal Behaviour* 56, 1375–1382.

Taborsky, M. and Limberger, D. 1981. Helpers in fish. *Behavioral Ecology and Sociobiology* 8, 143–145.

Taborsky, M., Hert, E., Siemens, M., and Stoerig, P. 1986. Social behaviour of *Lamprologus* species: functions and mechanisms. *Annales du Musée Royal d'Afrique Centrale (Zoologie)* 251, 7–11.

Taborsky, M., Hudde, B., and Wirtz, P. 1987. Reproductive behaviour and ecology of *Symphodus (Crenilabrus) ocellatus*, a European wrasse with four types of male behaviour. *Behaviour* 102, 82–118.

Taggart, J. B., McLaren, I. S., Hay, D. W., Webb, J. H., and Youngson, A. F. 2001. Spawning success in Atlantic salmon (*Salmo salar* L.): a long-term DNA profiling-based study conducted in a natural stream. *Molecular Ecology* 10, 1047–1060.

Tallamy, D. W. 2005. Egg dumping in insects. *Annual Review of Entomology* 50, 347–370.

Taru, M., Kanda, T., and Sunobe, T. 2002. Alternative mating tactics of the gobiid fish *Bathygobius fuscus*. *Journal of Ethology* 20, 9–12.

Taylor, E. B. 1989. Precocial male maturation in laboratory-reared populations of chinook salmon, *Oncorhynchus tshawytscha*. *Canadian Journal of Zoology/Revue Canadienne de Zoologie* 67, 1665–1669.

Taylor, E. B. 1991. A review of local adaptation in Salmonidae, with particular reference to Pacific and Atlantic salmon. *Aquaculture* 98, 185–207.

Tchernavin, V. 1938. Changes in the salmon skull. *Transactions of the Zoological Society of London* 24, 104–184.

Tebbich, S., Taborsky, M., and Winkler, H. 1996. Social manipulation causes cooperation in keas. *Animal Behaviour* 52, 1–10.

Thomaz, D., Beall, E., and Burke, T. 1997. Alternative reproductive tactics in Atlantic salmon: factors affecting mature parr success. *Proceedings of the Royal Society of London B* 264, 219–226.

Thorpe, J. E. and Morgan, R. I. G. 1980. Growth rate and smolting rate of progeny of male Atlantic salmon parr (*Salmo salar* L.). *Journal of Fish Biology* 17, 451–459.

Thorpe, J. E., Morgan, R. I. G., Talbot, L., and Miles, M. S. 1983. Inheritance of developmental rates in Atlantic salmon (*Salmo salar*). *Aquaculture* 33, 119–128.

Thorpe, J. E., Mangel, M., Metcalfe, N. B., and Huntingford, F. A. 1998. Modelling the proximate basis of salmonid life-history variation, with application to Atlantic salmon, *Salmo salar* L. *Evolutionary Ecology* 12, 581–599.

Thresher, R. E. 1979. Social behavior and ecology of two sympatric wrasses (Labridae: *Halichoeres* spp.) off the coast of Florida. *Marine Biology* 53, 161–172.

Thresher, R. E. 1984. *Reproduction in Reef Fishes*. Neptune City, NJ: TFH Publications.

Todo, T., Ikeuchi, T., Kobayashi, T., and Nagahama, Y. 1999. Fish androgen receptor: cDNA cloning, steroid activation of transcription in transfected mammalian cells, and tissue mRNA levels. *Biochemical and Biophysical Research Communications* 254, 378–383.

Tomkins, J. L. and Simmons, L. W. 2002. Measuring relative investment: a case study of testes investment in species with alternative male reproductive tactics. *Animal Behaviour* 63, 1009–1016.

Travis, J. 1981. Control of larval growth variation in a population of *Pseudacris triseriata* (Anura, Hylidae). *Evolution* 35, 423–432.

Treasurer, J. W. 1981. Some aspects of the reproductive biology of perch, *Perca fluviatilis* L.: fecundity, maturation and spawning behaviour. *Journal of Fish Biology* 18, 729–740.

Trivers, R. L. 1971. Evolution of reciprocal altruism. *Quarterly Review of Biology* 46, 35–48.

Trivers, R. L. 1972. Parental investment and sexual selection. In B. Campbell (ed.) *Sexual Selection and the Descent of Man*, pp. 139–179. Chicago, IL: Aldine Press.

Trudel, M., Tremblay, A., Schetagne, R., and Rasmussen, J. B. 2001. Why are dwarf fish so small? An energetic analysis of polymorphism in lake whitefish (*Coregonus clupeaformis*). *Canadian Journal of Fisheries and Aquatic Sciences* 58, 394–405.

Turner, B. J. and Grosse, D. J. 1980. Trophic differentiation in *Ilyodon*, a genus of stream-dwelling goodeid fishes: speciation versus ecological polymorphism. *Evolution* 34, 259–270.

Turner, E. and Montgomerie, R. 2002. Ovarian fluid enhances sperm movement in Arctic charr. *Journal of Fish Biology* 60, 1570–1579.

Tyler, W. A. III 1995. The adaptive significance of colonial nesting in a coral-reef fish. *Animal Behaviour* 49, 949–966.

Uglem, I., Rosenqvist, G., and Wasslavik, H. S. 2000. Phenotypic variation between dimorphic males in corkwing wrasse. *Journal of Fish Biology* 57, 1–14.

Uglem, I., Galloway, T. F., Rosenqvist, G., and Folstad, I. 2001. Male dimorphism, sperm traits and immunology in the corkwing wrasse (*Symphodus melops* L.). *Behavioral Ecology and Sociobiology* 50, 511–518.

Uglem, I., Mayer, I., and Rosenqvist, G. 2002. Variation in plasma steroids and reproductive traits in dimorphic males of corkwing wrasse (*Symphodus melops* L.). *Hormones and Behavior* 41, 396–404.

Urbach, D., Folstad, I., and Rudolfsen, G. 2005. Effects of ovarian fluid on sperm velocity in Arctic charr (*Salvelinus alpinus*). *Behavioral Ecology and Sociobiology* 57, 438–444.

van den Assem, J. 1967. Territory in the three-spined stickleback *Gasterosteus aculeatus* L. *Behaviour* (Suppl. 16), 164.

van den Berghe, E. P. 1988. Piracy: a new alterantive male reproductive tactic. *Nature* 334, 697–698.

van den Berghe, E. P., Wernerus, F., and Warner, R. R. 1989. Female choice and the mating cost of satellite males: evidence of choice for good genes? *Animal Behaviour* 38, 875–884.

Vandenhurk, R. and Resink, J. W. 1992. Male reproductive system as sex pheromone producer in teleost fish. *Journal of Experimental Zoology* 261, 204–213.

Vehrencamp, S. L. 1983. Optimal degree of skew in cooperative societies. *American Zoologist* 23, 327–335.

Verspoor, E. and Jordan, W. C. 1989. Genetic variation at the *ME-2* locus in the Atlantic salmon within and between rivers: evidence for its selective maintenance. *Journal of Fish Biology* 35, 205–213.

Vincent, A., Ahnesjö, I., Berglund, A., and Rosenqvist, G. 1992. Pipefishes and seahorses: are they all sex role reversed? *Trends in Ecology and Evolution* 7, 237–241.

Vives, S. P. 1990. Nesting ecology and behavior of hornyhead chub *Nocomis biguttatus*, a keystone species in Allequash Creek, Wisconsin. *American Midland Naturalist* 124, 46–56.

Vladic, T. V. and Järvi, T. 2001. Sperm quality in the alternative reproductive tactics of Atlantic salmon: the importance of the loaded raffle mechanism. *Proceedings of the Royal Society of London B* 268, 2375–2381.

Vladic, T. V., Afzelius, B. A., and Bronnikov, G. E. 2002. Sperm quality as reflected through morphology in salmon alternative life histories. *Biology of Reproduction* 66, 98–105.

Wallin, J. E. 1989. Bluehead chub (*Nocomis leptocephalus*) nests used by yellowfin shiners (*Notropis lutipinnis*). *Copeia*, 1077–1080.

Warner, R. R. 1982. Mating systems, sex change and sexual demography in the rainbow wrasse, *Thalassoma lucasanum*. *Copeia*, 653–661.

Warner, R. R. 1984a. Deferred reproduction as a response to sexual selection in a coral reef fish: a test of the life historical consequences. *Evolution* 38, 148–162.

Warner, R. R. 1984b. Mating behavior and hermaphroditism in coral-reef fishes. *American Scientist* 72, 128–136.

Warner, R. R. 1987. Female choice of sites versus mates in a coral reef fish, *Thalassoma bifasciatum*. *Animal Behaviour* 35, 1470–1478.

Warner, R. R. and Hoffman, S. G. 1980. Local-population size as a determinant of mating system and sexual composition in two tropical marine fishes (*Thalassoma* spp.). *Evolution* 34, 508–518.

Warner, R. R. and Robertson, D. R. 1978. Sexual patterns in the labrid fishes of the Western Caribbean. 1. The wrasses (Labridae). *Smithsonion Contributions to Zoology* 254, 27–00.

Warner, R. R., Robertson, D. R., and Leigh, E. G. J. 1975. Sex change and sexual selection. *Science* 190, 633–638.

Warner, R. R., Lejeune, P., and Van den Berghe, E. 1995. Dynamics of female choice for parental care in a fish species where care is facultative. *Behavioral Ecology* 6, 73–81.

Watters, J. V. 2005. Can the alternative male tactics "fighter" and "sneaker" be considered "coercer" and "cooperator" in coho salmon? *Animal Behaviour* 70, 1055–1062.

Wedekind, C. 1992. Detailed information about parasites revealed by sexual ornamentation. *Proceedings of the Royal Society of London B* 247, 169–174.

Wedekind, C. 1996. Lek-like spawning behaviour and different female mate preferences in roach (*Rutilus rutilus*). *Behaviour* 133, 681–695.

Wedekind, C., Meyer, P., Frischknecht, M., Niggli, U. A., and Pfander, H. 1998. Different carotenoids and potential information content of red coloration of male three-spined stickleback. *Journal of Chemical Ecology* 24, 787–801.

Wedell, N., Gage, M. J. G., and Parker, G. A. 2002. Sperm competition, male prudence and sperm-limited females. *Trends in Ecology and Evolution* 17, 313–320.

Wernerus, F. 1989. Etude des mécanismes sous-tendant les systémes d'appariement de quatre espéces de poissons labridés méditerranéens des genres *Symphodus* Rafinesque, 1810, et *Thalassoma* Linné 1758. *Cahiers d'Ethologie appliquée* 9, 117–320.

West-Eberhard, M. J. 1989. Phenotypic plasticity and the origins of diversity. *Annual Review of Ecology and Systematics* 20, 249–278.

West-Eberhard, M. J. 2003. *Developmental Plasticity and Evolution*. Oxford, UK: Oxford University Press.

Wilson, D. S. 1989. The diversification of single gene pools by density- and frequency-dependent selection. In D. Otte and J. A. Endler (eds.) *Speciation and Its Consequences*, pp. 366–385. Sunderland, MA: Sinauer Associates.

Winge, O. 1937. Succession of broods in Lebistes. *Nature* 140, 467.

Wirtz, P. 1978. The behaviour of the mediterranean *Tripterygion* species (Pisces, Blennioidei). *Zeitschrift für Tierpsychologie* 48, 142–174.

Wirtz, P. 1999. Mother species–father species: unidirectional hybridization in animals with female choice. *Animal Behaviour* 58, 1–12.

Wirtz, S. and Steinmann, P. 2006. Sperm characteristics in perch *Perca fluviatilis* L. *Journal of Fish Biology* 68, 1896–1902.

Wisenden, B. D. 1999. Alloparental care in fishes. *Reviews in Fish Biology and Fisheries* 9, 45–70.

Wisenden, B. D. and Keenleyside, M. H. A. 1992. Intraspecific brood adoption in convict cichlids: mutual benefit. *Behavioral Ecology and Sociobiology* 31, 263–269.

Wittenberger, J. F. 1979. The evolution of mating system in birds and mammals. In P. Mailer and A. Vandenburgh (eds.) *Social Behavior and Communication*, pp. 271–349. New York: Plenum Press.

Wootton, R. J. 1984. *A Functional Biology of Sticklebacks*. Beckenham, UK: Croom Helm.

Wright, H. A., Wootton, R. J., and Barber, I. 2004. Interpopulation variation in early growth of threespine sticklebacks (*Gasterosteus aculeatus*) under laboratory conditions. *Canadian Journal of Fisheries and Aquatic Sciences* 61, 1832–1838.

Yamagishi, S. 1988. Polyandry and helper in a cichlid fish *Julidochromis marlieri*. In H. Kawanabe and M. K. Kwetuenda (eds.) *Ecological and Limnological Study on Lake Tanganyika and Its Adjacent Regions*, vol. 5, pp. 21–22. Kyoto, Japan: Kyoto University Press.

Yamamoto, T. and Edo, K. 2002. Reproductive behaviors related to life history forms in male masu salmon, *Oncorhynchus masou* Breboort, in Lake Toya, Japan. *Journal of Freshwater Ecology* 17, 275–281.

Yanagisawa, Y. 1985. Parental strategy of the cichlid fish *Perissodus microlepis*, with particular reference to intraspecific brood "farming out." *Environmental Biology of Fishes* 12, 241–249.

Yanagisawa, Y. 1986. Parental care in a monogamous mouthbreeding cichlid *Xenotilapia flavipinnis* in Lake Tanganyika. *Japanese Journal of Ichthyology* 33, 249–261.

Yanagisawa, Y. 1987. Social organization of a polygyneous cichlid *Lamprologus furcifer* in Lake Tanganyika. *Japanese Journal of Ichthyology* 34, 82–90.

Yom-Tov, Y. 1980. Intraspecific nest parasitism in birds. *Biological Review* 55, 93–108.

Yom-Tov, Y. 2001. An updated list and some comments on the occurrence of intraspecific nest parasitism in birds. *Ibis* 143, 133–143.

Zbinden, M., Mazzi, D., Kunzler, R., Largiader, C. R., and Bakker, T. C. M. 2003. Courting virtual rivals increase ejaculate size in sticklebacks (*Gasterosteus aculeatus*). *Behavioral Ecology and Sociobiology* 54, 205–209.

Zbinden, M., Largiader, C. R., and Bakker, T. C. M. 2004. Body size of virtual rivals affects ejaculate size in sticklebacks. *Behavioral Ecology* 15, 137–140.

Zimmerer, E. J. 1982. Size related courtship strategies in the pygmy swordtail, *Xiphophorus nigrensis*. *American Zoologist* 22, 910–919.

Zimmerer, E. J. and Kallmann, K. D. 1989. Genetic basis for alternative reproductive tactics in the pygmy swordtail, *Xiphophorus nigrensis*. *Evolution* 43, 1298–1307.

11 · Alternative reproductive tactics in amphibians

KELLY R. ZAMUDIO AND LAUREN M. CHAN

CHAPTER SUMMARY

Frogs and salamanders, the two most diverse lineages of amphibians, differ significantly in reproductive mode, morphology, and behavior. We review reproductive tactics in these lineages and consider their distribution among taxa in light of phylogeny, ecology, and organismal traits. Together these groups show a surprising diversity of alternative reproductive phenotypes that can roughly be divided into two classes: those that increase an individual's chance of mate acquisition (e.g., satellite or intercepting males) and those that directly increase fertilization success (e.g., spermatophore capping, clutch piracy, or multimale spawning). Our survey underscores the fact that mode of fertilization (internal or external) and operational sex ratios at breeding aggregations have important implications for the frequency and nature of alternative reproductive tactics. However, our understanding of the evolution of amphibian alternative reproductive tactics is hampered by a lack of detailed information for many species. For example, most alternative tactics have been described in temperate amphibians despite the fact that tropical species account for most of the taxonomic, morphological, behavioral, and ecological diversity in this group. A challenge for future studies will be to further describe and categorize the diversity of reproductive tactics in amphibians to uncover general patterns within and across lineages in the factors that modulate polymorphism in reproductive phenotypes.

11.1 INTRODUCTION

In the last few decades, organismal evolutionary biologists have become increasingly aware of inter-individual differences in mate acquisition abilities that may underlie the evolution of alternative strategies or tactics for reproductive success (Shuster and Wade 2003). In many species or populations adults differ in mating phenotypes and this presumably affects their fitness when in direct competition with distinct rivals; both theoretical and empirical studies have made progress in explaining how such polymorphism can evolve and be maintained within a population when alternate states have different fitness outcomes (Parker 1984, Lucas and Howard 1995, Gross 1996). One generalization from these studies is that interactions among competing individuals within an immediate breeding group, the "neighborhood" or social context for mating, influences both the intensity and direction of sexual selection (Zamudio and Sinervo 2000, 2003) and ultimately the fitness of alternative reproductive phenotypes (Emlen and Oring 1977, West-Eberhard 1991). The importance of social environment is especially evident in amphibian breeding systems. Amphibian breeding groups are extremely dynamic in that the temporal and spatial distributions of individuals that define the social context for breeding are highly variable (Wells 1977a, Douglas 1979) and this certainly influences the diversity and maintenance of reproductive phenotypes and tactics.

The three living lineages of amphibians – salamanders, frogs, and caecilians – diverged about 300 million years ago (Trueb and Cloutier 1991, Laurin and Reiz 1997), although the phylogenetic relationship among these three lineages is still debated (Feller and Hedges 1998, Zardoya and Meier 2000). The oldest fossils from each group are from the Jurassic, 213–144 mya (Wake 1997), thus, each lineage has had a long independent evolutionary history. Differences among amphibian lineages include behavior, morphology, and physiology (Wake 1993, Wake and Dickie 1998, Houck and Arnold 2003, Lehtinen and Nussbaum 2003, Scheltinga and Jamieson 2003a, b, Tyler 2003). External versus internal fertilization, vocalization versus chemosensation for mate attraction, and aquatic versus direct-developing larvae are just a few examples of broad differences among these groups that have implications for the nature of sexual selection and the evolution of reproductive strategies in each group.

Alternative Reproductive Tactics, ed. Rui F. Oliveira, Michael Taborsky, and H. Jane Brockmann. Published by Cambridge University Press.
© Cambridge University Press 2008.

Heritable alternative strategies have been identified in a number of animal taxa (e.g., Zimmerer and Kallman 1989, Shuster and Wade 1991, Ryan *et al.* 1992, Lank *et al.* 1995, Sinervo and Zamudio 2001); however, none of the alternative reproductive phenotypes described in amphibians is known to have a genetic basis, therefore the reproductive diversity in this lineage results from conditional strategies that allow individuals to assess their fitness potential and express tactics that maximize their fitness in any particular context (Gross 1996). A fair number of alternative reproductive tactics have been identified in frogs and salamanders (no data exist for caecilians); most of these are variations in male courting behavior (Verrell 1989, Halliday and Tejedo 1995). In amphibians, tactics adopted by courting individuals typically change over a relatively short temporal scale, ranging from minutes in some cases of short-term context-dependent behavioral choices to seasons in tactics that are dependent on body size, condition, or social status. In general, the ability for opportunistic capitalization, the temporal and spatial distribution of resources and mates, and the condition or social status of courting individuals are "criteria" used in adopting alternative mating tactics. While the average fitness between tactics is unequal, individuals must vary sufficiently in reproductive success that the evolution of alternative phenotypes is favored (Lucas and Howard 1995, Gross 1996).

In this chapter we review examples of reproductive tactics described in amphibians and consider their distribution among taxa with respect to phylogeny, ecology, and organismal traits. Because of the largely different tactics of salamanders and frogs, we first describe our findings in each of these groups separately, and then comment on the similarities and differences in behavioral diversity in each lineage. Rather than provide an exhaustive list of reproductive tactics in amphibians we chose to exemplify all known categories of tactics and highlight those that have been best studied. We define an alternative reproductive tactic as any behavioral or morphological variant that is bi- or multimodal in distribution within a population, and can be classified into discrete alternatives directly involved in mate acquisition and/or mating success. Therefore, we do not consider as tactics behaviors that are present in all individuals but vary continuously in degree or intensity (such as territoriality, degree of aggressiveness, or intensity of coloration or signal). Many of the phenotypes we classify as tactics may well be the result of context-dependent behaviors and it remains to be studied whether individuals are more likely to adopt one behavior over the other, or

whether these behaviors are dependent only on the immediate social context of individual males. Regardless of the proximal determinants and the longevity of a particular behavior in a courting male, the tactics described here are likely the product of selection for phenotypic diversity as part of a conditional strategy. Thus, presumably each of these behaviors offers fitness advantages that would not have accrued in their absence.

We are only beginning to appreciate the diversity in amphibian behavioral polymorphism in natural populations. Our survey reveals a bias toward detailed studies in temperate species, despite the fact that all three amphibian lineages have radiations in the tropics, and that tropical lineages exhibit surprising diversity in reproductive modes and behaviors (Duellman 2003, Haddad and Prado 2005). This regional (and hence partly phylogenetic) bias in behavioral studies limits our ability to detect macroevolutionary behavioral patterns within and among lineages and test their generality. Herpetologists have exciting work ahead to characterize mating tactics and strategies in many of the groups that remain unexplored. We end our chapter with suggestions for future research that will substantially contribute to our understanding of the evolution and maintenance of alternative behavioral phenotypes in amphibians.

11.2 ALTERNATIVE MATING TACTICS IN FROGS

Frogs and toads comprise the most speciose group of amphibians with more than 5000 species globally in 33 families (Frost 2004). Despite the tremendous diversity in modes of reproduction and parental care (Duellman 2003, Lehtinen and Nussbaum 2003), the number of alternative reproductive behaviors known in frogs is relatively low; all are male courtship tactics described from only a few families. No genetically based strategies have been identified; therefore, these behaviors are most likely alternative reproductive tactics within condition-dependent strategies (Gross 1996). In cases where we have adequate data, the average fitness of the alternative tactic is less than that of the main tactic (e.g., Sullivan 1982, Krupa 1989, Haddad 1991). Many aspects of anuran reproductive biology have undoubtedly influenced the evolution of tactics in this group as they limit the ways in which alternative behavioral phenotypes can be reproductively successful. While some anurans have internal fertilization (e.g., *Ascaphus*, *Necto-phrynoides*) the widespread condition is external fertilization accomplished during amplexus (Duellman and Trueb

1994). Male reproductive success requires mate acquisition, monopolization, and fertilization success. Not surprisingly, we see the evolution among frogs of alternative tactics aimed at increasing a male's success at all of these stages. A second trait common to many anurans is vocal communication and it plays a central role in social interactions including mate attraction, courtship behaviors, and defensive behaviors (e.g., Fellers 1979, Arak 1983a, Smith and Roberts 2003). This characteristic of anurans, and the apparent ubiquitous importance of this male cue for mate attraction and female choice, has favored the evolution of tactics that capitalize on proximity to calling individuals to increase access to females.

Reproductive tactics in anurans can be partitioned into two groups: those in which males attempt to increase their chances of amplexing females and fertilizing clutches and those in which males attempt to gain access to paired females and fertilize a portion of the clutch during oviposition. This dichotomy has been discussed as "satellite" versus "sneaker" tactics (Fukuyama 1991); however, in this review we define them more broadly as tactics concerned with mate acquisition and monopolization versus those concerned with partial fertilization of the clutch of a potentially polyandrous female (Table 11.1).

11.2.1 Tactics related to mate acquisition

Nonvocalizing male alternative tactics have been described for a number of species where dominant males vocalize and/or defend sites or females; most of these occur in mating systems where the operational sex ratio (OSR: Emlen and Oring 1977) is male-biased. In many instances, phenotypes are bimodal with callers and satellite males; however, in some taxa active searching, interference, or displacement occurs in addition to satellite behavior resulting in a trimodal conditional strategy.

CALLER–SATELLITES

The most widely reported alternative mating tactic among anurans is caller–satellite associations in which satellite males sit silently near vocalizing males at choruses (Wells 1977a). Individuals may maintain tactics over the course of several nights, but switching between vocalizing and parasitizing often takes place within a single night (e.g., Sullivan 1982, Perrill and Magier 1988). The frequency of satellites per calling male varies across species: most calling males are parasitized in *Bufo cognatus* (95%: Sullivan 1982) and relatively few in hylid frogs (e.g., 6.9% in *Hyla picta*: Roble

1985; 2–14% in *Pseudacris crucifer*: Forester and Lykens 1986). Wells (1977b) suggested that satellites assuming these tactics are either attempting to intercept females attracted to callers, or are waiting to take over calling sites or territories vacated by the resident male. Interception of females by satellite males has been observed in a number of caller–satellite systems and the relative success of this alternative tactic varies widely (Table 11.1). In treefrogs, such as *H. cinerea* and *H. minuta*, callers are aggressive towards satellites and satellite males successfully intercept gravid females attracted to callers with some success (43% and 38% of females were in amplexus with satellite males in *H. cinerea* and *H. minuta*, respectively: Perrill et al. 1978, Haddad 1991). In other cases, interception rates are much lower and it appears that satellite behavior must be finely tuned to be successful. For example, females of *B. cognatus* and *Scinax ruber* (previously *Ololygon rubra*) initiate amplexus by touching their chosen mate; satellites are successful at maintaining amplexus only if they intercept females at the exact moment she makes her choice (Sullivan 1982, Bourne 1993). Females are capable of actively dislodging males in many taxa; thus, it may be that females only tolerate amplexus by satellites because they are misled about the identity of the clasping male. In species with intercepting tactics, some satellite males will often begin vocalizing if resident callers are removed (e.g., *H. cinerea*: Perrill et al. 1982; *Acris crepitans*: Perrill and Magier 1988; *H. minuta*: Haddad 1991; *Eleutherodactylus johnstonei*: Ovaska and Hunte 1992). This is in contrast to species such as *Uperoleia rugosa* and *Rana clamitans* in which satellite males do not attempt interception and only attempt to take over vacated territories (Wells 1977b, Robertson 1986a). Robertson (1986b) found that maintaining a territory was costly for *U. rugosa*, but males that assumed satellite tactics gained weight and sometimes took over a calling site later in the breeding season. This suggests that satellite behavior may be the best tactic for males in poor condition to maximize reproductive success over the course of the breeding season. Similarly, large resident male *R. clamitans* aggressively defend oviposition sites; smaller males are unable to monopolize calling sites and do not attempt interception, instead they benefit by waiting for a vacancy (Wells 1977b).

Condition dependence is a common theme among anuran satellite tactics; however, the switch point for tactic choice may depend on body condition and/or assessment of potential attractiveness given the social context of the breeding aggregation (Arak 1988, Lucas and Howard 1995,

Table 11.1. Alternative reproductive tactics in anurans. For each species we report the primary versus the alternative tactic(s), the success rate (percentage of total females in amplexus that are paired with the alternative tactic), and the context for the occurrence of alternative phenotypes. We also gather data from the literature on four parameters that may play important roles in the evolution of reproductive diversity: the longevity of a particular tactic in an individual ("Time"), whether the species usually breeds under a skewed operational sex ratio ("OSR"), whether these tactics are associated with differences in body size ("BS"), and the length of the breeding season ("BR"). Dashes indicate parameters that have not been studied or could not be inferred from literature reports

Species	Tactic	Success rate	Context[a]	Time[b]	OSR[c]	BS[d]	BR[e]	Reference
Bufonidae								
Bufo americanus	Active searchers within pond vs. peripheral males ("gauntlet")	~40%	CD	S, L	M	S	P	Forester and Thompson 1998
Bufo americanus	Callers vs. joiners	–	CD	–	M	S	E	Kaminsky 1997
Bufo bufo	Active searchers (terrestrial) vs. interfering males	39%	–	S	M	–	E	Davies and Halliday 1979
Bufo bufo	Early-arriving aggressive males vs. late-arriving peripheral males	74%	CD	S, L	M	S	E	Loman and Madsen 1986
Bufo calamita	Switchers vs. stayers	switchers > stayers	–	L	F	no	P	Denton and Beebee 1993
Bufo calamita	Callers vs. intercepting satellites	20%	CD	S	M	S	P	Arak 1988
Bufo cognatus	Callers vs. intercepting satellites	8%	CD	S	M	S	E	Krupa 1989, Leary et al. 2004
Bufo cognatus	Callers vs. intercepting satellites	23%	DD	S	M	–	E	Sullivan 1982
Bufo woodhousii	Callers vs. active searchers and intercepting satellites	0%, 9%	DD	S	M	no	E	Sullivan 1989
Bufo woodhousii	Callers vs. peripheral males	–	CD	S	M	S	P	Leary et al. 2004
Hylidae								
Acris crepitans	Callers vs. intercepting satellites	40%	–	S	–	no	P	Perrill and Magier 1988
Hyla chrysoscelis	Callers vs. intercepting satellites	–	–	S	–	no	–	Roble 1985
Hyla cinerea	Callers vs. intercepting satellites	43%	CD	S	–	no	P	Perrill et al. 1978, Perrill et al. 1982

303

Table 11.1. (cont.)

Species	Tactic	Success rate	Context[a]	Time[b]	OSR[c]	BS[d]	BR[e]	Reference
Hyla ebraccata	Callers vs. intercepting satellites	38%	–	–	–	–	PE	Miyamoto and Cane 1980
Hyla minuta	Callers vs. intercepting satellites	38%	CD	S	–	S	P	Haddad 1991
Hyla picta	Callers vs. satellites	–	–	S	–	no	P	Roble 1985
Hyla versicolor	Callers vs. satellites	–	RD	S	–	no	P	Fellers 1979
Pseudacris crucifer	Callers vs. intercepting satellites	–	CD	S	–	S	P	Forester and Lykens 1986
Pseudacris regilla	Callers vs. intercepting satellites	–	–	S	–	no	P	Fellers 1979, Perrill 1984
Pseudacris triseriata	Callers vs. satellites	–	–	S	–	no	–	Roble 1985
Scinax ruber	Callers vs. intercepting satellites and interfering males	7%, 78%	CD, OPP	S	M	S, L	PE	Bourne 1992, Bourne 1993
Leptodactylidae								
Eleutherodactylus johnstonei	Callers vs. intercepting satellites and interfering males	–	CD, OPP	S	–	S, no	P	Ovaska and Hunte 1992
Leptodactylus chaquensis	Callers vs. joiners	–	CD	–	–	S	PE	Prado et al. 2000, Prado and Haddad 2003
Leptodactylus podicipinus	Callers vs. sneakers	–	RD/CD	–	–	S	C	Prado et al. 2000, Prado and Haddad 2003
Physalaemus signifer	Callers vs. intercepting satellites	17%	P	S	M	no	E	Wogel et al. 2002
Myobatrachidae								
Crinia georgiana	Callers vs. joiners	50% paternity	OPP	S	M	L dorsally	P	Byrne and Roberts 1999, Roberts et al. 1999, Smith and Roberts 2003, Byrne and Roberts 2004

Crinia georgiana	Callers vs. satellites	CD	24–80%	S	M	S	P	Byrne and Roberts 2004
Uperoleia rugosa [a]	Callers vs. satellites	CD	–	S	M	S	P	Robertson 1986a
Ranidae								
Rana catesbeiana	Callers vs. intercepting satellites	CD	2.7%	S	M	S	P	Emlen 1968, Howard 1978
Rana clamitans	Callers vs. satellites	RD	0–low	S, L	–	S	P	Wells 1977b
Rana temporaria	Amplexus vs. clutch piracy	OPP	43.8% multiple paternity	S	M	no	E	Vieites et al. 2004
Rana virgatipes	Callers vs. satellites	RD	–	S	–	S	P	Given 1988
Rhacophoridae								
Chiromantis rufescens	Amplexus vs. pair joiners	OPP	–	S	–	no	P	Coe 1974
Chiromantis xerampelina	Amplexus vs. pair joiners	OPP	–	S	M	no	P	Coe 1974, Jennions et al. 1992
Rhacophorus arboreus	Amplexus vs. pair joiners	OPP	–	S	M	no	P	Toda 1989, Kasuya et al. 1996
Rhacophorus schlegelii	Amplexus vs. pair joiners	OPP	–	S	–	no	P	Kato 1956, Fukuyama 1991

[a] CD, condition dependent; DD, density dependent; RD, resource acquisition or defense; OPP, opportunistic behavior; many of the condition dependent tactics are also density dependent, thus we use DD only in cases where there was no discernible differences among phenotypes.

[b] S, short term, individual can switch tactics in minutes or days; L, long term, individual uses the same tactic for at least one season.

[c] M, male biased; F, female biased; E, equal.

[d] S, smaller; L, larger.

[e] P, prolonged; E, explosive; PE, explosive bouts within a prolonged season; C, continuous, *sensu* Wells 1977a.

Lucas *et al.* 1996). Body size difference among satellites and callers are common (Table 11.1) and may be an important determinant of mating tactics either because of aggressive abilities or due to call attractiveness. Smaller males are often unable to usurp territories or successfully defend calling sites or females (e.g., Emlen 1968, Wells 1977b, Howard 1978, Robertson 1986b); in addition, given the high cost of vocalizations, satellite individuals may be unable to produce attractive or more intense calls (Wells 2001, Leary *et al.* 2004). Body size is not the only determinant; in *Physalaemus signifer*, a highly explosive tropical breeder, first males to arrive at the breeding site vocalize, while later ones assume satellite roles. Although arrival time may be condition dependent, Wogel *et al.* (2002) did not find any differences in body size among individuals assuming alternative tactics.

PERIPHERAL SATELLITES/ACTIVE SEARCHERS

Some nonvocalizing males adopt positions peripheral to a group of calling males actively searching and attempting to intercept females attracted to the calling group. These tactics have been reported in some anurans including *Bufo woodhousii* (Sullivan 1989, Leary *et al.* 2004), *B. americanus* (Fairchild 1984, Forester and Thompson 1998), and *B. bufo* (Davies and Halliday 1979, Loman and Madsen 1986). Smaller males patrol the periphery of a calling group in what has been referred to as "gauntlet behavior" (Forester and Thompson 1998) and attempt to intercept incoming females before they arrive at the group. It is unclear whether these peripheral satellites represent a distinct alternative tactic or are males employing active searching near the calling group. Regardless, this behavior is similar to satellite behavior in that it relies on interception for mating success but peripheral satellites are not associated with specific calling males (Leary *et al.* 2004).

Active searching can be common in breeding aggregations; in *Bufo bufo*, terrestrial amplexus appears to be the primary tactic with 84.4% of females arriving at the pond already paired (Davies and Halliday 1979). Unpaired males search at spawning sites or along the periphery of the pond for either single females or pairs in amplexus, which they attempt to disrupt. The success of this tactic can be high with unpaired males obtaining 38.6% of the females. In a different population of *B. bufo*, Loman and Madsen (1986) found two distinct size-based tactics. Large males arrived early and attempted to displace or dislodge males in amplexus while smaller males arrived later and instead searched for unpaired females. Interestingly, in

B. bufo calling is reduced and the mating system is characterized as "scramble competition" (Arak 1983b); in this species, terrestrial amplexus as well as interference tactics may be associated with increased role of aggressive male–male interactions versus female choice.

SIZE-BASED TRIMODAL TACTICS

Some species exhibit a third tactic in addition to callers and satellites. In *Scinax ruber* females prefer males of intermediate size. Smaller males assume silent satellite roles whereas larger males attempt to dislodge intermediate-sized males in amplexus; larger males are successful at amplexing 78% of females (Bourne 1992, 1993). While assortative mating increases fertilization success of males in one size category, male aggression results in greatest mating success for larger males that interfere with mating pairs. *Eleutherodactylus johnstonei* provides a second example of trimodal behavioral phenotypes; this species exhibits the typical condition-dependent caller–satellite polymorphism (Ovaska and Hunte 1992), but some vocalizing males employ an opportunistic strategy aggregating around pairs and interfering with courtship. Fitness of each tactic in this system is not known, although both satellite and opportunistic males may be maintained in the species even in the face of low success because of the low cost of the tactics.

11.2.2. Tactics related to fertilization success

MULTIMALE SPAWNING

Instances of multiple males clustering around a single female (in some cases referred to as mating balls: Verrell and McCabe 1986) have been described in a number of families including Bufonidae (Davies and Halliday 1979, Forester and Thompson 1998), Hylidae (Pyburn 1970, Roberts 1994), Leptodactylidae (Prado and Haddad 2003), Myobatrachidae (Roberts *et al.* 1999), and Rhacophoridae (Coe 1974, Feng and Narins 1991, Kasuya *et al.* 1996). In some cases, it is unclear whether individuals are attempting to dislodge males in amplexus or attempting to steal fertilization through deliberate sneaking or joining behaviors. Although there are many described instances of multiple male aggregations, we have surprisingly little data on the fertilization or fitness success of primary versus joining males. Here we limit our discussion to those species where the intent of the alternate male seems to be attempted fertilization rather than merely physically interfering with the male in amplexus.

One of the best-characterized mating systems with multimale spawning is found in a myobatrachid frog, *Crinia georgiana*; this species is a prolonged breeder and males adopt a caller–satellite system. Females are attracted to males with particular call characteristics (Smith and Roberts 2003) and initiate amplexus (Byrne and Roberts 1999). Satellites have variable success obtaining mates depending on density and the intensity of male–male competition (Byrne and Roberts 2004), but one field study found that half of all matings included more than one male (Roberts *et al.* 1999). Based on focal observations of mated pairs, researchers describe that the courting males held females in inguinal amplexus and satellite males clasped the pair ventrally such that the cloacae of the two males were opposite one another. Subsequent males did not assume any particular position and paternity tests on several clutches found that fertilization is achieved primarily, if not completely, by the first and second males (Roberts *et al.* 1999). Thus, in this system priority has consequences for fertilization and adopting both satellite and peripheral tactics may be ways of obtaining favorable positions in spawning groups.

Some species of Rhacophoridae exhibit multimale spawning and construct a foam nest for egg deposition (e.g., *Chiromantis, Rhacophorus*: Jennions *et al.* 1992, Kasuya *et al.* 1996). In most of these cases, female choice for particular males does not seem to play a large role; females approach chorusing groups and a male grasps her in amplexus, peripheral males then join the group and clasp the pair during nest formation. In *C. xerampelina* for example, the pair hangs from a branch and the female beats an arboreal foam nest as the primary male releases sperm moving his egs down the female's dorsum. At the same time, males hanging on either side of the pair position their cloacae next to the females for 1–3-second bouts presumably depositing sperm. Jennions *et al.* (1992) found that peripheral males further from the pair did not commonly attempt to obtain cloacal proximity to the female. In *C. xerampelina*, 66% of females spawned in this way with more than one male compared to 81.4% in *R. arboreus* (4.5 and 3.5 males per female respectively: Toda 1989, Jennions *et al.* 1992). Peripheral versus amplexing males did not differ significantly in body size and over half of the males adopted both tactics, suggesting it is a highly opportunistic behavior. Fertilization success of individual males is not known for any of these multimale spawns in Old World treefrogs; however, it seems likely that males adopting different tactics may have partial fertilization success. Multiple paternity has

been documented in the red-eyed treefrog, *Agalychnis callidryas* (D'Orgeix and Turner 1995), a species with similar group spawning at arboreal egg-laying sites (Pyburn 1970). Although no data are available, multiple paternity may be present in many other species for which group spawning has been observed (e.g., *Bufo bufo*: Verrell and McCabe 1986; *Polypedates leucomystax*: Feng and Narins 1991; *Agalychnis saltator*: Roberts 1994).

Group spawning has also been described in species that build cavity or basin nests for egg deposition. In *Rhacophorus schlegelii* paired females dig backward into the soil creating a cavity in which they beat a foam nest (Fukuyama 1991). After the pair creates the nest sneaker males dig into the nest and join the spawning pair. Fukuyama (1991) found that in 10 of 12 laboratory observations one to four males would sneak into the nest presumably releasing sperm. A similar sneaking situation occurs in a leptodactylid frog in the Brazilian Pantanal, *Leptodactlyus podicipinus*, where males build and defend nest sites from which they call to attract females. Prado and Haddad (2003) observed a female pairing with a calling male at a nest site. As the frogs began spawning a single smaller male was observed in between the resident male and female, and both males churned the foam nest as eggs were deposited, likely fertilized by both males. Multimale spawning was also observed in the sympatric *L. chaquensis* which breeds in shallow ponds; males aggregate around small puddles and vocalize (Prado *et al.* 2000, Prado and Haddad 2003). The largest male in the center of the nest was louder and aggressive towards other callers in the nest area; as a female entered the nest males attempted to clasp her but the largest male obtained amplexus and spawning began immediately with the pair churning the foam in the nest (Prado and Haddad 2003) (Figure 11.1). Without touching the pair, additional males (up to seven) entered the nest and began churning the foam with their hind limbs while facing away from the pair and presumably releasing sperm. In addition to these nest-building species, a single anecdote for multimale spawning exists for *Bufo americanus*, a free-spawning explosive aggregate breeder. Kaminsky (1997) observed two small satellite males as they followed amplexed pairs and aligned their cloacae with females as they moved around the pond. Unfortunately there is no description of egg laying, thus it is unclear whether these sneaking males achieved fertilization. These last three examples are described from few field observations, yet they are suggestive of condition-dependent "sneaking" of fertilizations. While we do not have paternity data to assess the reproductive success of each behavior,

Figure 11.1 Multimale spawning in *Leptodactylus chaquensis*: an amplecting pair with seven "joiner" males. (From Prado and Haddad 2003, with permission.)

these examples underscore the potential for undiscovered diversity in anuran mating tactics.

Multimale spawning in anuran species is phylogenetically widespread and sperm competition may be an important determinant of male reproductive fitness in these cases (Jennions and Passmore 1993). Testis mass (adjusted for body size) in many multimale spawning species is significantly larger than that of pair-spawning relatives (Kusano *et al.* 1991, Prado and Haddad 2003) suggesting the evolution of increased ejaculate expenditure. For example, testis mass in *Chiromantis xerampelina* is 7.79% of body mass compared to 0.25–1.11% in other rhacophorids; testes of *Leptodactylus chaquensis* weigh 4.13% of body mass compared to 0.04–0.12% in other *Leptodactylus* species. Testis mass and the probability of group spawning are correlated among Australian myobatrachids (see Byrne *et al.* 2002 and references therein). The evolution of enlarged testes in leptodactylid, myobatrachid, and rhacophorid frogs suggests that group spawning is not merely the result of heightened aggression or unusually high population densities, rather it likely occurs often such that the male "joining tactics" are important ways to increase individual reproductive success.

CLUTCH PIRACY

Perhaps the most unusual multiple-male tactic is clutch piracy in the common frog, *Rana temporaria* (Vieites *et al.* 2004) (Figure 11.2). In this explosively breeding species courting males amplex females in vernal pools and sneaker males (referred to as "pirates") follow the pair to the site of oviposition. After the egg mass has been deposited and the pair has left, the sneaking male or males clasp the egg clutch, in some instances crawling into it and moving among the eggs, while releasing sperm. More than 84% of clutches are attended by pirates, and a mean of 24.1% of the embryos (*n* = 16 clutches) are fathered by pirate males. The reproductive success of sneakers is highly variable with 5% to 100% of a clutch fathered by a pirate and with four of seven clutches fathered only by the parental male and the first pirate to arrive at the clutch. Individuals switch between parental and pirate tactics, although the frequencies and factors influencing tactic choice remain to be determined (Vieites *et al.* 2004).

SWITCHING VERSUS STAYING

We found only one tactic that was stable across breeding seasons and this was in an isolated population of *Bufo calamita*, which is also unusual in having female-biased sex ratios (1.77–3.38 females : males post-breeding season; Denton and Beebee 1993). Approximately half of the breeding males are "stayers" at a single pond, and often at a

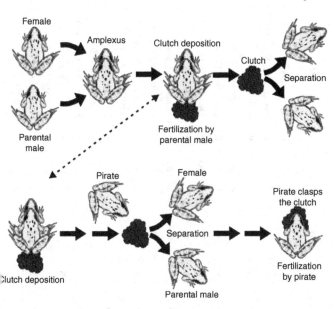

Figure 11.2 Clutch piracy in *Rana temporaria*. Mating sequence with single male adopting the primary tactic of courtship and amplexus (top) and with egg clasping by a "pirate" male (bottom). (From Vieites *et al.* 2004, with permission.)

ingle calling site across breeding nights whereas the other males are "switchers" moving among breeding ponds throughout the season. Only four of 45 individuals adopted oth tactics across 3 years suggesting that it is not genetically polymorphic despite being fairly static. Switchers chieved higher mating success than stayers by mating with multiple females; however, predation risks or possibly high energetic costs associated with moving among ponds might ompensate for the lower mating success of the staying actic (Denton and Beebee 1993).

1.3   ALTERNATIVE MATING TACTICS IN SALAMANDERS

'he caudate amphibians (including salamanders and newts, nembers of the Urodela) are unique among living amphiians in a number of derived reproductive characters. These pecializations are not surprising given the antiquity of the neages within Amphibia. For example, in sharp contrast) anurans, most salamanders have internal fertilization Salthe 1967, Wake and Dickie 1998). The exceptions are ryptobranchidae, Hynobiidae, and Sirenidae; at least the rst two, and possibly all three, are relatively basal lineages in

the salamander phylogeny (Larson *et al.* 2003 and references therein). Therefore, the largest diversification of salamanders occurred after the evolution of internal fertilization. Gamete transfer in internally fertilizing salamanders is indirect (via deposition of a spermatophore by courting males) and therefore does not involve copulation. Both indirect gamete transfer and internal fertilization of eggs have important implications for the evolution of alternative tactics; a large number of mating tactics described in salamanders exploit the period of time when male gametes are in the environment, before being transferred to the females (Verrell 1989). In contrast to anurans, salamanders do not usually communicate with sound in the context of reproduction, thus the signals used in mate recognition and choice are primarily olfactory, visual, or tactile (Arnold 1977, Arnold and Houck 1982), requiring closer proximity of individuals in mated pairs than is necessary for mate choice in systems with broadcast advertisement. Not surprisingly, alternative behaviors that have evolved in salamanders are often mediated by the immediate and close presence of mates and competing males.

Three main categories of tactics are evident from examining the summary table of diversity in salamander behavioral phenotypes (Table 11.2): the first two categories

Table 11.2. Alternative reproductive tactics in salamanders. For each species we report the reproductive tactic(s) and their success rate(s) (quantified as the percentage of times a male adopting each tactic successfully inseminated a female, in laboratory or field experiments). We also gather from the literature data on six parameters that may play important roles in the evolution of reproductive diversity: the context for the use of these alternative phenotypes, the longevity of a particular tactic in an individual ("Time"), whether the species usually breeds under a skewed operational sex ratio ("OSR"), whether these tactics are associated with distinct phenotypes, such as different secondary sexual characteristics ("SSC") or body size ("BS"), and the length of the breeding season ("BR"). Dashes indicate parameters that have not been studied or could not be inferred from literature reports

Species	Tactic	Success rate	Context[a]	Time[b]	OSR[c]	SSC	BS	BR[d]	Reference
Salamandridae									
Cynops ensicauda	Male courtship vs. mimicry of female behavior	–	OPP	S	–	–	–	P	Sparreboom 1994
Notophthalmus viridescens	Amplexus vs. female mimicry	64/31 lab	OPP	S	M	–	–	P	Verrell 1982, Verrell 1983
		6.1/6.2 field	OPP	S	M	no	no	P	Massey 1988
	Amplexus vs. display	64/29	OPP	S	M	–	–	P	Verrell 1982, Verrell 1983
	Same-sex courtship	–	OPP	M	M	–	–	P	Massey 1988
Taricha granulosa	Pairing vs. non-pairing	–	CD	L	M	yes	yes	P	Janzen and Brodie 1989
	Amplexus vs. intruding	–	OPP	S	M	no	no	P	Janzen and Brodie 1989
Triturus alpestris	Facultative paedomorphosis	1.7/1.7 lab	CD	L	M	yes	yes	P	Denoël 2002, Denoël 2003
	Courting vs. interference courting	–/7.5	OPP	S	M	–	–	P	Verrell 1988
	Courting (luring) vs. waiting	30/6 lab	OPP	S	M	–	no	P	Denoël et al. 2001
Triturus boscai	Mimicry of female behavior	–	OPP	S	M	–	–	P	Faria 1993, Faria 1995
	Courting (luring) vs. waiting	43/0 field	OPP	S	M	–	–	P	Rafinski and Pecio 1992, Faria 1993, Faria 1995
Triturus cristatus	Mimicry of female behavior	–	OPP	S	M	–	–	P	Zuiderwijk and Sparreboom 1986
Triturus italicus	Mimicry of female behavior	–	OPP	S	M	–	–	P	Giacoma and Crusco 1987
Triturus marmoratus	Mimicry of female behavior	–	OPP	S	M	–	–	P	Zuiderwijk and Sparreboom 1986
Triturus vulgaris	Mimicry of female behavior	–/22 lab	OPP	S	M	–	–	P	Verrell 1984a, Verrell and McCabe 1988

Ambystomatidae

Ambystoma annulatum	Spermatophore capping	OPP	S	M	–	E	Spotila and Beumer 1970
Ambystoma barbouri	Spermatophore capping	OPP	S	M	–	E	Petranka 1982
Ambystoma gracile	Spermatophore capping	OPP	S	M	–	E	Arnold 1977
Ambystoma jeffersonianum	Spermatophore capping	OPP	S	M	–	E	Uzzell 1969
Ambystoma laterale	Spermatophore capping	OPP	S	M	–	E	Arnold 1977
Ambystoma macrodactylum	Spermatophore capping	OPP	S	M	–	E	Arnold 1977
Ambystoma maculatum	Mimicry of female behavior	OPP	S	M	–	E	Verrell and Pelton 1996
Ambystoma maculatum	Spermatophore capping	OPP	S	M	–	E	Arnold 1976
Ambystoma mexicanum	Spermatophore capping	OPP	S	M	–	E	Arnold 1977
Ambystoma opacum	Spermatophore capping	OPP	S	M	–	E	Noble and Brady 1933, Arnold 1977
Ambystoma talpoideum	Facultative paedomorphosis	CD	L	M	yes	E	Krenz and Sever 1995, Semlitsch 1985, Ryan and Plague 2004
	Mimicry of female behavior	OPP	S	M	–	E	Verrell and Krenz 1998
	Courting vs. blocking	OPP	S	M	–	E	Verrell and Krenz 1998
Ambystoma texanum	Spermatophore capping	OPP	S	M	–	E	Arnold 1977, McWilliams 1992
Ambystoma tigrinum	Mimicry of female behavior	OPP	S	M	no	E	Arnold 1976, Howard et al. 1997
	Spermatophore capping	OPP	S	M	–	E	Arnold 1976

Plethodontidae

Desmognathus ochrophaeus	Mimicry of female behavior	OPP	S	E	–	P	Arnold 1977
Ensatina eschscholtzii	Mimicry of female behavior	OPP	S	E	–	P	Arnold 1977
Eurycea cirrigera	Mimicry of female behavior	OPP	S	E	–	P	Thomas 1989
Plethodon glutinosus	Mimicry of female behavior	OPP	S	E	–	P	Arnold 1972
Plethodon jordani	Mimicry of female behavior	OPP	S	E	–	P	Organ 1958, Arnold 1976

Table 11.2. (cont.)

Species	Tactic	Success rate	Context[a]	Time[b]	OSR[c]	SSC	BS	BR[d]	Reference
Plethodon ouachitae	Mimicry of female behavior	–	OPP	S	E	–	–	P	Arnold 1972
Plethodon yonahlossee	Mimicry of female behavior	–	OPP	S	E	–	–	P	Arnold 1977
Pseudotriton ruber	Mimicry of female behavior	–	OPP	S	E	–	–	P	Organ and Organ 1968
Hynobiidae									
Hynobius leechii	Monopolist vs. scrambler	100/16.7 lab	RD	S	M	–	yes	E	Park *et al.* 1996, Park and Park 2000
Hynobius nigrescens	Monopolist vs. scrambler	100/57.8 lab	RD	S	M	yes	no	E	Usuda 1993, Hasumi 1994, Hasumi 2001
	Territorial vs. opportunists	–	RD	S	M	–	no	E	Usuda 1997
Hynobius retardatus	Monopolist vs. scrambler	100/59.2 lab	RD	S	M	–	–	E	Sasaki 1924, Sato 1992
Cryptobranchidae									
Andrias japonicus	Den master vs. satellite	–	RD	L	E	–	yes	P/E	Kawamichi and Ueda 1998
Cryptobranchus alleganiensis	Territorial vs. satellite	–	RD	L	M/E	–	–	P/E	Smith 1907, Bishop 1943, Nickerson and Mays 1973a

[a] CD, condition dependent; RD, resource acquisition or defense; OPP, opportunistic behavior.
[b] S, short term, individual can switch tactics in minutes or days; L, long term, individual uses the same tactic for at least one season.
[c] M, male biased; F, female biased; E, equal.
[d] P, prolonged; E, explosive; C, continuous, or a combination if population variation exists.

nique to salamanders, are those that depend on paedo-morphosis, the retention of larval characteristics in sexually nature adults, and those that include some form of sexual nterference via female mimicry. The third category ncludes alternative tactics that occur in the context of efense of a resource necessary for reproduction (such as est, or egg-deposition site). These patterns are in sharp ontrast to those in anurans, where most tactics include nales competing for calling or egg-deposition sites; inter-rence never includes female mimicry, and paedomorph-sis is known in two species of anuran (Bokerman 1974, Iaddad and Prado 2005).

1.3.1 Tactics involving facultative paedomorphosis

notable exception to the pattern of short-term behavioral lasticity in salamander tactics can be found in species that mploy alternate reproductive tactics based on facultative aedomorphosis (Table 11.2). In these cases, the alternative henotypes are fixed for the reproductive season, although 1ey may change in the course of an individual's reproductive fetime (Winne and Ryan 2001). A number of proximate 1echanisms have been proposed for the expression of Iternative life cycles in facultatively paedomorphic sala-1anders (Harris 1987, Whiteman 1994, Denoël and Poncin 001), including larval growth rate, density, food availability, 1d genetics (Semlitsch and Gibbons 1985, Harris et al. 990). Regardless of the proximal cues, these alternate life 'stories obviously have important implications for present 1d future reproductive success, thus, if both paedomorphic 1d metamorphosed individuals exist in the same population 1d compete for mates, we consider them here as alternative 2productive tactics despite the complexity of ultimate and 1oximal determinants (Semlitsch 1985, Krenz and Sever 995). Paedomorphosis is facultative in a number of sala-1anders (Whiteman 1994) but only in a few of these cases do e find metamorphosed and paedomorphic reproductive 1ults in the same populations (Semlitsch 1985, 1987, Denoël 002, Ryan and Plague 2004). Despite the large differences 1 body size and development of secondary sexual charac-ristics that are common between metamorphs and paedo-1orphs (Kalezic et al. 1996, Denoël et al. 2001), there seems 1 be surprisingly little effect of metamorphic state on the 1urtship behaviors and tactics employed by these males or 1 their success in mate acquisition (Denoël et al. 2001, 1enoël 2002). The most likely advantage for this alternate 2productive tactic seems to be a temporal one, in that

paedomorphic adults breed earlier, and thus benefit from higher survival rates of their hatchlings (Krenz and Sever 1995, Ryan and Plague 2004); however, this pattern has not yet been demonstrated in all species that exhibit this reproductive tactic.

11.3.2 Sexual interference and female mimicry

Sexual interference is defined as a behavior on the part of a male that alters the probability that a competing male will be successful in the courtship, transfer of gametes, or fertilization of a female (Arnold 1976, Verrell 1982). In salamanders, the most common form of interference is dis-ruption of a mating pair through physical displacement of courting males, wrestling, or overt forms of aggression such as biting or chasing (Verrell 1989, Halliday and Tejedo 1995). Oftentimes the winner in these aggressive interactions is the largest male (Houck 1988); thus, although this behavior may provide benefits to interrupting males, it is not a tactic with alternative states, as males are either more aggressive (or larger) and successfully chase away potential rivals, or they are the losers in these physical combats. A second form of sexual interference, in which the alternative tactics are more clearly delimited, involves behaviors that prevent the suc-cessful transfer of male gametes to the female. This form of alternative tactics is unique because of the indirect mode of gamete transfer. These alternative tactics have evolved in the Ambystomatidae, Plethodontidae and Salamandridae (Table 11.2) and can take two forms. The first of these is spermatophore capping (Figure 11.3). This tactic is most prevalent in the Ambystomatidae, where males in breeding aggregations cover spermatophores of other males with their own (Arnold 1976, 1977) and make the sperm cap of the original male inaccessible. The incidence of this behavior varies among species of Ambystoma; in laboratory studies 7.5% of the spermatophores produced by A. texanum were multiple structures (McWilliams 1992), compared to 57% and 63% for A. maculatum and A. tigrinum respectively (Arnold 1976). Within Ambystoma, spermatophore capping is most common in species with extremely short and explosive breeding systems (Arnold 1977); males deposit numerous spermatophores on the substrate but deposition is not part of a courtship series that requires close associations with females. Rather, Ambystoma spermatophores are placed in large numbers in patches along the bottom of breeding ponds, and females retrieve only some of them, thus the courtship investment per spermatophore is low, but so is the

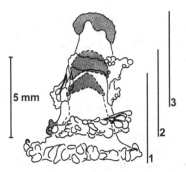

Figure 11.3 Sexual interference in *Ambystoma maculatum* occurs via spermatophore capping. Only the sperm cap from the topmost spermatophore (3) is accessible to the female. (From Arnold 1976, with permission.)

success rate of each one that is deposited (Arnold 1977). Despite its occurrence in this relatively well-studied genus, we still know surprisingly little about the exact fitness costs and benefits of spermatophore capping, primarily because we lack basic comparative information about the cost of gamete production, and the frequency and social context for this behavior in species of different lineages. For example, it is not known whether males can identify a spermatophore as their own, or if they merely cover a proportion of the spermatophores they encounter. Likewise, we know very little about female choice of single or capped spermatophores, or about female selectivity of gametes in general in salamanders. Thus, although we have relatively complete information on spermatophore deposition rates and probabilities of spermatophore encounter for some species (Arnold 1976, 1977, McWilliams 1992), we are lacking information on the probability of sperm cap retrieval by females. Perhaps trade-offs between cost of gamete production, cost of courtship behaviors, and spermatophore success rate may explain some of the variation in breeding behavior and alternative tactics within *Ambystoma*, in much the same way that differences in reproductive investment have been used to explain differences between plethodontids and ambystomatids (Arnold 1977).

The second behavior used to disrupt gamete transfer is female mimicry which occurs when a male interrupts a pair that is already in courtship (Verrell 1989). Female mimicry occurs most frequently (albeit not exclusively) in plethodontids and salamandrids, two families in which males perform elaborate and species-specific displays in the process of sperm transfer, and where females use specific behaviors to indicate receptivity (Halliday 1974, Arnold 1976, Houck and Arnold 2003). Plethodontid salamanders deposit very few spermatophores (between two and ten) in a reproductive bout but invest heavily in courtship displays. Male plethodontid salamanders have evolved elaborate behavioral repertoires to entice females to reciprocate in courtship and pick up their spermatophores (e.g., Verrell 1982, Halliday 1990). The cost to the males of these complex courtship behaviors is poorly known (but see Bennet and Houck 1983); however they must not be trivial, as many salamanders invest substantial time and effort in mating displays (Stebbins 1949, Organ 1960). In plethodontids, males court receptive females in a stereotyped "tail-straddling walk," in which the male leads a receptive female until she is behind him and aligned with the long axis of his body; he then deposits a spermatophore and guides her over it until she retrieves the sperm cap with her cloaca. This process includes tactile stimulation on the part of the receptive female, usually through nudges or pressing at the base of the male's tail, to indicate receptivity. An interrupting male approaches a courting pair and mimics the female by nudging the male's tail, thus prompting early spermatophore deposition. The interrupting male then begins his own courtship and deposits his own spermatophore for the receptive female. Plethodontid males also use female mimicry when not in the immediate presence of females. In a number of species, male–male elicitation of a tail-straddle walk has been observed (Arnold 1977); if successful, the leading male will be duped in to depositing a spermatophore with no chances of fertilization. Given that once a male deposits a spermatophore he may not be able to do so again for some time, these male–male courtships are likely an adaptive form of sexual interference rather than errors in sex identification (Arnold 1976). Therefore, female mimicry (in the form of male–male courtship) in this family occurs as a means of exhausting the future reproductive potential of rival males.

Courtship in the Salamandridae also includes fairly complex interactions between males and females, culminating in spermatophore deposition and transfer (Halliday 1974, 1975, Verrell 1982, 1984a). The salamandrids for which we have the most complete data on reproductive behavior are European newts in the genus *Triturus* and the North American red-spotted newt, *Notophthalmus viridescens* (Halliday 1974, 1975, 1977, Verrell 1982, 1984a, 1984b). Species of *Triturus* have relatively prolonged breeding seasons of 2–4 months; however, OSR can change ignificantly during the course of the mating season (Verrell nd Halliday 1985, Verrell and McCabe 1988) such that emales are in relatively short supply during some periods. The smooth newt (*T. vulgaris*) has a fairly stereotyped equence of male courting behaviors, and males perform a 'creep" behavior and slowly move away from the female just efore spermatophore deposition; female tail-touching is sually the stimulus used by the male for spermatophore eposition and males resume movement with a "creep on" nd "brake," to lead the female's cloaca just above the ocation of the spermatophore (Halliday 1977) (Figure 11.4). t is at the spermatophore transfer phase that most sexual nterference happens in this species; a male newt that ncounters a female already engaged in courtship will wait ntil the primary male is in the "creeping" phase, and mimic he female's tail-touching, causing the first male to deposit is spermatophore while the usurper leads the female away nd begins his own courtship and "creep" (Verrell 1984a) Figure 11.4). Species of *Triturus* vary in the frequency of emale mimicry. For example, interfering male Bosca's newts *T. boscai*) rarely tail-touch courting males (Faria 1995); nstead, an approaching male adopts a "waiting" tactic and lowly fan his tail (presumably sending olfactory stimuli to he female) while waiting for the primary male to flick his tail t the female, at which point the intruding male engages in push-tail" behavior and disrupts the mating pair. Faria 1995) did not observe a successful sperm transfer in any of he case of sexual interference in *T. boscai*, suggesting that hese behaviors drastically lower male reproductive success. .ikewise, in the alpine newt (*T. alpestris*), interfering males pproach a courting pair and attempt an "interference ourtship" in the vicinity of the female. Some of these side isplays result in successful transfer of spermatophores, ut at a relatively low rate (approximately 7.5% of inter-ctions: Verrell 1988). Female mimicry in the form of male-nale courtship has been noted in *Triturus* (Halliday 1974, ireen 1989) but at very low frequencies; therefore, if it ccurs as a form of sexual interference it must not be used ery often.

The New World red-spotted newt, *Notophthalmus iridescens*, exhibits a form of female mimicry similar to its)ld World relatives. Males of this species, when courting a onresponsive female, engage in long amplexus, clasping female around the neck with their hindlimbs and sing repetitive head movements to apply genial (cheek) land secretions on the female's nares (Verrell 1982) Figure 11.5A and B). If a female is quickly receptive, the nale will bypass amplexus and use a lateral "hula" display

before depositing a spermatophore (Verrell 1982). A third tactic is to not amplex or display, but interfere with a clasped pair at the moment when a male dismounts to deposit their spermatophore for the female (Figure 11.5C). Male red-spotted newts use female mimicry and touch the tail of the first male to trick the courting male into depositing a spermatophore; they then deposit their own spermatophore for the female (Verrell 1982, 1983, Massey 1988). Interestingly, this alternative tactic yields approximately equal fitness opportunities (Massey 1988). Amplexing and interfering males do not differ in insemination success, despite the presumed advantage offered by the prolonged amplexus, and the possible cost to amplecting males (Verrell 1985). *Notophthalmus* are sexually dimorphic; males have deeper and longer tails (Able 1999, Gabor *et al.* 2000). Field and laboratory studies have demonstrated that male tail depth and size are positively correlated with variation in male courtship success (Able 1999, Gabor *et al.* 2000), but these studies did not distinguish between the two forms of courtship (amplexus versus hula display); nonetheless, if deeper-tailed males are more successful in either form of court-ship, then shallow-tailed males may be the ones primarily using interference tactics. A fourth tactic, referred to as "pseudofemale behavior," was described in red-spotted newts (Massey 1988). In this case, males that are searching for females sometimes clasp other males; amplexing males will court their same-sex partners for the same amount of time typical of male–female amplexus. When the amplexing male dismounts, the courted male may display the typical female cloacal nudging behavior and in this way elicit spermatophore deposition from the clasping male; Massey (1988) reported that 50% of the male–male courtship encounters successfully resulted in spermato-phore deposition. This case of sexual interference is similar to the same-sex courtship observed in *Plethodon jordani* (Arnold 1976), in that the interfering male does not gain any direct insemination benefits, but does cause competing males to produce spermatophores that are not utilized, thus costing the competing male reproductive potential.

Reproductive dynamics differ in the large-bodied North American newts in the genus *Taricha*, where interference does not involve female mimicry, but rather a condition-dependent switch among tactics. Courtship has been best studied in the rough-skinned newt, *T. granulosa* (Arnold 1972, Janzen and Brodie 1989, Propper 1991). Male *Taricha* amplex females and continue grasping them as other males

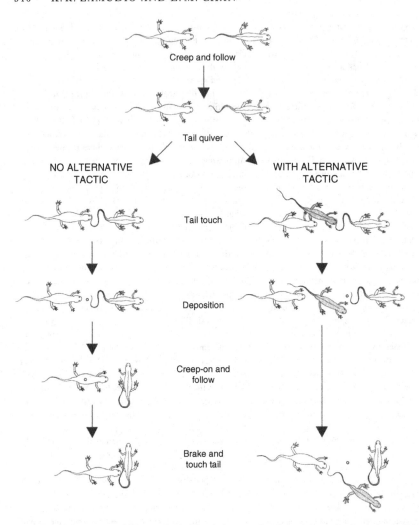

Figure 11.4 Mating sequence of *Triturus vulgaris* with a single courting male (left) and with an additional male adopting the alternative tactic (right). Males lead female in a stereotyped creeping sequence before depositing spermatophores.

Interfering males use female mimicry to elicit spermatophore deposition by the courting male. (Modified from an original drawing by T. Halliday and from Halliday 1977 and Verrell 1998, with permission.)

attempt to interfere, forming a mating ball composed of a paired male and female, and multiple intruders. In the only field observations of a mating aggregation of *T. granulosa*, males could be categorized in hierarchical groups with different mating opportunities and tactics. First, males at

the aggregation either engaged in amplexus (or attempted amplexus), or they were nonpaired (Janzen and Brodie 1989). However, it is unclear whether the nonpaired males were adopting alternate tactics, or whether they were simply excluded from reproduction due to the preponderance

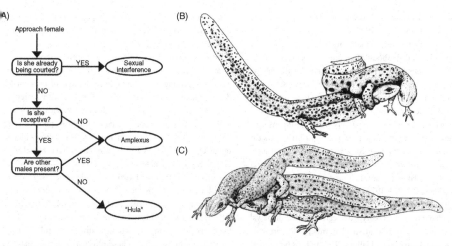

(A)

Approach female

Is she already being courted? — YES → Sexual interference

NO

Is she receptive? — NO

YES

Are other males present? — YES → Amplexus

NO

"Hula"

(B)

(C)

igure 11.5 Alternative behaviors for courtship in *Notophthalmus iridescens* (A), a male and female in amplexus (B), and an amplecting pair with an interfering male (C). ((A) and (B) modified from Halliday 1990, with permission; (C) courtesy of T. Halliday.)

f larger males. Unpaired males were significantly smaller in ody size, tail length, and tail height than those in amplexus. mong the males that were attempting amplexus or successfully amplexing a female, the correlation with body size isappeared. Males that were either in single pairs, in a mating ball, or attempting to intrude in a mating ball, were 1e same size and had the same degree of development of econdary sexual characteristics. In this system there is pparently condition dependence for pairing opportunity, ut once a threshold for mating has been met, the tactics tilized by males are opportunistic. Both pre- and post-asemination amplexus is common and amplexus can be ery prolonged (median 7 hours, range 40 minutes to 2 days: ropper 1991). After post-insemination amplexus the male ismounts, deposits a spermatophore, and clasps the female gain. Given the ubiquity of female mimicry in other sala-andrids, it might seem surprising that *Taricha* have not volved similar forms of interference; this difference may be xplained by the difference in investment in mate-guarding 1 New World newts (Halliday 1977), although more etailed observations of courtship success in *Taricha* are eeded.

Ambystomatidae also employ female mimicry, but the use f this behavior varies among species depending on their articular mating system and investment in courtship Arnold 1976, 1977). Within *Ambystoma*, there is substantial among-species variation in the number of individuals that participate in a typical breeding aggregation, the time individual males spend courting females, and the number of spermatophores males deposit relative to time spent courting (Arnold 1977). For example, highly explosive species such as *A. maculatum*, *A. opacum*, and *A. texanum* deposit many spermatophores in one mating bout and do not orient toward females or directly interact with females during the spermatophore transfer phase of courtship (Noble and Brady 1933, Arnold 1976, McWilliams 1992). *Ambystoma tigrinum*, *A. dumerilii*, *A. mexicanum*, *A. talpoideum*, and *A. laterale* insert a tail-nudging walk between spermatophore depositions, form temporary associations with females in courting pairs, and therefore deposit fewer spermatophores per unit time, but presumably increase the probability of gamete transfer to females (Arnold 1977, Verrell and Krenz 1998). Female mimicry in *Ambystoma* is seen mostly in these cases where males and females engage in pairwise courtship displays. In *A. tigrinum*, interfering males insert themselves between the courting male and female and nudge the leading male's cloaca. When the primary male deposits a spermatophore, the intruder covers it with his own for fertilization of the female. In metamorphosed breeding individuals of *A. talpoideum*, males creep in front of the female before spermatophore deposition and undulate their tails; the interfering males block access of the female to the original males tail, and also elicit spermatophore deposition with tail nudges (Verrell and Krenz 1998).

The prevalence of female mimicry behavior in Salamandridae and Plethodontidae and its occurrence only in those *Ambystoma* that have more prolonged male–female courtship sequences suggests that this form of sexual interference is most efficient in cases where courtship involves a close behavioral interplay between males and females mating in pairs or in relatively small mating groups. Although individuals in these groups exhibit little physical contact during mating (when compared to species that have prolonged amplexus), the species-specific behavioral sequences are crucial in stimulating females into responsiveness (Halliday 1974, 1977, Moore *et al.* 1979, Teyssedre and Halliday 1986). Both families have prolonged breeding seasons, elaborate and stereotyped species-specific displays' and in general, deposition of small numbers of spermatophores per breeding bout (Halliday 1990). This behavioral complexity and high investment in courtship is evident to an extreme degree in terrestrially breeding plethodontids, where we find the highest occurrence of long-term pair bonds, territoriality, and even social monogamy among males and females in mated pairs (Jaeger and Forester 1993, Gillette *et al.* 2000). Combined, these reproductive characteristics favor the evolution of tactics such as female mimicry that exploit behaviors crucial to spermatophore transfer during courtship. In contrast, spermatophore capping is most prevalent in species with aggregate or explosive mating systems, and the behavior is generally exhibited in large multimale courting groups. Not surprisingly, these species also exhibit relatively high levels of multiple mating and multiple paternity (Tennessen and Zamudio 2003, Myers and Zamudio 2004). In female mimicry and spermatophore capping, we do not know whether individual males preferentially adopt these tactics based on some inherent characteristic (e.g., body size, strength) that may be used by females for mate choice, or whether males are opportunistic depending on the immediate availability of mates or presence of competitors. Despite many unknowns, it is obvious that the evolution of indirect internal fertilization has opened important opportunities for the evolution of alternative tactics in this lineage.

In addition to sexual interference, salamanders also employ short-term tactics according to the receptivity of females they are courting (Verrell 1982, Denoël *et al.* 2001). In these cases, it is clear that the alternate behaviors are employed by the same individual sometimes in short succession, however, we do not know whether different males

employ these tactics to different degrees. In *Triturus alpestris*, female receptivity seems to be the main determinant of male choice between a "waiting" tactic, where males wait for a female response before depositing spermatophores, and a "luring" tactic, where males use quivering and distal lures before depositing a spermatophore in front of the female. The frequency of successful insemination varies significantly among males adopting different tactics (64% for males who wait for female to come into a responsive state compared to 8% for luring males) and there is no evidence of correlation with male body size or condition (Denoël *et al.* 2001). Similarly, male red-spotted newts (*Notophthalmus viridescens*) adopt one of two behaviors depending on the initial response of the female (Verrell 1982, 1983). If a female is responsive, males perform a lateral display (hula display); on the other hand, if females are not responsive the male clasps her in a prolonged amplexus, followed by spermatophore transfer. These alternative tactics may be elicited by a female stimulus (receptivity), but in red-spotted newts, experimental manipulations of densities show that the males favor amplexus in the increased presence of competitors, presumably as a defense against sexual interference (Verrell 1983). This defense against sexual interference comes at the cost of time, energy, and reduced encounter rate with other females.

11.3.3   Tactics associated with resource defense

The final pattern evident from our survey is that, in contrast to the pattern in anurans, very few alternative tactics in salamanders are related to defense of a resource necessary for reproduction (such as nest, or egg-deposition site). In salamanders, the cases that do involve defense of a resource are found exclusively in lineages with external fertilization. *Hynobius* exhibit behavioral tactics that most closely approximate the typical anuran caller–satellite system. In *H. nigrescens*, females deposit bilobed egg sacs in the water attached by adhesive tips to a stick or branch. Males fertilize the eggs once they are in the water. Just before oviposition, a female finds an appropriate site and rubs her cloaca against the substrate to attach her egg mass. In *H. nigrescens*, males defend sticks or twigs before females arrive (Usuda 1997). Egg-sac deposition is not random, likely because females select optimal sites for oviposition; in one study, 20 of 22 eggs sacs were deposited on five twigs that were in close proximity (Usuda 1997). Therefore, this behavior is a form of short-term territorial defense, and nonterritorial males emain near the optimal sites as opportunists. Interestingly, the territorial males did not differ from opportunists in size and mass (Usuda 1997). In other species of *Hynobius* males are attracted to ovipositing females, and embrace or amplex the egg mass during oviposition (Sasaki 1924, Thorn 1962, Sato 1992); In some cases, a male's efforts to embrace the egg sac accelerate the process, because repeated clasping motions pull the eggs from her cloaca; this phenomenon has been termed "midwifing" (Usuda 1993). In all species of *Hynobius* that have been studied, the first male to reach an ovipositing female becomes the "monopolist" and remains clasped tightly to the egg sac while fertilizing the eggs in an attempt to prevent other males from doing the same. Secondary males join as "scramblers" and attempt to fertilize some of the eggs by releasing gametes around the monopolizing male (Sato 1992, Usuda 1993, Hasumi 1994, Park *et al.* 1996, Park and Park 2000). In *Hynobius*, the resource in short supply is female eggs. Because fertilization is external, behavior of the monopolists is similar to amplexus and mate defense that is seen in many species of frogs, except that male attention is directed to the gametes, instead of the female.

Two species of Cryptobranchidae, an ancient lineage of salamanders that includes only three species distributed in Asia and North America, also exhibit tactics that involve defense of a limited resource. All cryptobranchids are external fertilizers, and females select and mate with males that defend nest sites. In *Andrias japonicus*, the Japanese giant salamander, alternative male tactics are driven by limited availability of appropriate egg-deposition sites. Males occupy long burrows with underwater openings in the banks of streams, and defend these aggressively against intruders. The largest males in the population have been called "den masters" because they are the only nest-site holders and tend the egg masses of several females. The remaining small males in the population use satellite or "sneaker" tactics; they are active in the vicinity of dens and enter burrows occupied by mating pairs. In a detailed observational study of two nests, Kawamichi and Ueda (1998) observed a nest defended by a den master and visited by ten sneaker males during the spawning season. These ten males made 47 intrusions, each immediately after females entered the nest. It is not known whether satellite males benefit directly from fertilization of some of the female's eggs, although this is certainly a possibility. Data for *Cryptobranchus alleganiensis*, the single North American species of this lineage, are less conclusive, but also suggestive of alternative male mating behaviors. Early studies

of reproductive behavior (Smith 1907, Bishop 1941, Nickerson and Mays 1973a, b) suggest that males defend excavated nest sites from intruders (Nickerson and Tohulka 1986), and that once a female selects a nest for spawning, other males approach the pair and release sperm in the vicinity of the spawning pair (Smith 1907). It is not known whether territory holders in this species are larger males, or whether the secondary males at spawning sites are simply those who have not successfully obtained a mating site; however, it seems likely that this system is similar to that observed in *Andrias*, and that resource defense (in this case an excavated nest site where eggs can be guarded) underlies differences among males in reproductive tactics.

Plethodontid salamanders also defend territories that are used for reproduction; however, territories are defended year round and may be primarily a means of sequestering foraging sites and appropriate microhabitat patches (Mathis 1989, Gabor 1995, Jaeger *et al.* 1995). Nonetheless, some interesting patterns in the spatial distribution of males and females has led researchers to suggest that males may be adopting different tactics in their territorial "decisions" (Jaeger *et al.* 2000). Adult *Plethodon cinereus* adopt different spatial arrangements, and these naturally have consequences for their mating strategies: some co-defend territories with an adult of the opposite sex, others defend territories singly, but overlap in part with territories of members of opposite sex, and finally, some defend territories singly with no apparent overlap (Mathis 1991a, Jaeger *et al.* 1995). If these spatial arrangements reflect mate choice and availability, those pairs that live together in the same place would favor monogamy while overlapping territories would favor polygyny, polyandry, or polygynandry. We are just beginning to unravel some of the intricacies of social interactions among salamanders (Jaeger and Forester 1993) and the variation that exists among species and individuals, thus it remains to be seen whether these territorial patterns are modulated by social interactions in the context of reproduction and mating strategies, or if they are driven primarily by ecological or spatial constraints. Given the degree of communication, discrimination, and social interactions in plethodontids that has been detected in field and laboratory studies (Mathis 1991b, Jaeger *et al.* 1995, 2002, Guffey *et al.* 1998), it will not be surprising if terrestrial salamanders exhibit substantially more behavioral complexity than we currently recognize. (Box 11.1)

The approximately 500 species of salamanders are included in ten families, yet the majority of alternative behavioral tactics have been described in the families

---

Box 11.1   Female reproductive tactics in amphibians

A surprising result of our survey was the apparent lack of alternative female reproductive phenotypes in amphibians. This may result from male-biased OSR that is typical of most amphibian breeding groups, such that there is little competition among females for males or their gametes, and thus less selection for reproductive tactics. Accordingly, one of the few examples of a female alternative tactic was reported in breeding populations of *Triturus vulgaris* with equal or female-biased sex ratios (Waights 1996). Early in the season, prior to ovulation, the limiting resource in breeding aggregations is sperm (because of a physiological limit on spermatophore production rates by the few males present). Under these conditions, females interrupt and attempt to displace courting females instead of engaging in normal courtship behaviors. These interrupting females either touch the tail of the male to elicit spermatophore deposition, or they wait until the courting female does so and "dart" in between them to pick up the spermatophore. Once ovulation commences in the population and the OSR shifts to male-biased, the limiting resource becomes female gametes, and males display the tactics typical for this species. Likewise, in a study of mate selection in *Bufo bufo*, Davies and Halliday (1977) report on context-dependent female behaviors that are also suggestive of alternative tactics. Females already in amplexus can influence the reproductive success of males in male-biased breeding aggregations. In their study, females in amplexus with suitable males moved away from lone males to avoid dislodgement. However, when females were paired with unsuitable males they instead moved toward lone males of appropriate size to induce male–male competition and obtain a better mate.

Variation in post-fertilization behaviors among females has been observed in salamanders; although these are not behavioral reproductive tactics associated with increasing fertilization success, they are alternative reproductive behaviors with important fitness consequences. The best studied of these occurs in females of the plethodontid *Hemidactylium scutatum*, that show tactics related to nest site selection and parental care. Females adopt one of three

tactics (Blanchard 1934, Harris *et al.* 1995): they may lay eggs in solitary nests and brood them, they may lay their eggs in a joint nest with other females and brood them (lay-and-stay females), or they may leave them to be brooded by another female (lay-and-leave females). This species has been the focus of substantial study by Harris and colleagues; they studied a population where 30–52% of the nests were communal, up to 45–70% of females were lay-and-leave females, and nests included clutches from two to seven females, but as a rule joint nests had only one attendant (Harris and Gill 1980; Harris *et al.* 1995). Brooding in this species has clear advantages for embryonic survival and hatching success rate (Harris and Gill 1980), possibly due to decreased fungal infections from antimicrobial components in the skin of brooding females (Harris and Gill 1980, R. N. Harris, pers. comm.). Lay-and-stay females were in significantly better condition as measured by weight/length ratio (Harris and Ludwig 2004) and tended to lose more weight during the reproductive season than deserting females (Harris *et al.* 1995). On the other hand, lay-and-leave females were significantly lower in body condition, suggesting that females in poor condition will stay with, and perhaps be attracted to, females that are in good condition for joint nesting (Harris and Ludwig 2004).

The selective advantage for lay-and-stay females is not yet clear, especially since the second female to join a nest is equally likely to be the lay-and-stay female as the founding female; the brooding female may benefit from the presence of additional eggs in her nest, possibly due to a "dilution" effect of the probability of predation (Harris *et al.* 1995). The benefits for females that abandon eggs are clearer, they do not incur the cost of brooding and can return to foraging soon after oviposition, thus potentially increasing their chance of reproduction the following year (Harris and Ludwig 2004). This population study underscores the importance and value of long-term studies of organisms in their environment. Through mark-recapture studies and multiple years of observation, Harris and colleagues have shown that females switch tactics between years, and that female size and condition, and population density play a role in the decisions of females in any one season (Harris *et al.* 1995, Harris and Ludwig 2004).

Ambystomatidae, Salamandridae, and Plethodontidae. This distribution begs the question: why do we not know of more diversity in reproductive behaviors, especially in other families? One possible answer is that behavioral

phenotypes do not exist in other families; however, given the diversity of reproductive modes and courtship behaviors in the less-known lineages of salamanders, this explanation seems unlikely. The answer may lie in the fact that we ave very few data on courtship and reproductive behaviors or many salamander species. In part this may be due to the naccessibility of their breeding sites and crypticity of their ehaviors, but we also lack basic descriptions and studies f reproductive patterns and behaviors in entire families. A recent survey of salamander courtship and mating ehavior by Houck and Arnold (2003) underscores the aucity of data. Reproduction in the remaining salamander amilies is poorly understood: courtship has been studied in hree species of Hynobiidae and one Proteidae; mating ehavior and courtship sequences are poorly described or ompletely unknown in Cryptobranchidae, Sirenidae, Dicamptodontidae, Amphiumidae, and Rhyacotritonidae Houck and Arnold 2003). Collectively, the diversity in nest ites, mode of fertilization, length of breeding period, and nating densities in these poorly studied groups rivals that ound in the better-known families; therefore we can be ertain that further efforts in basic description and obser- ation of salamander reproductive biology and behavior vill reveal more incidences of alternative behavioral henotypes.

1.4   COMPARATIVE AND EVOLUTIONARY THEMES

1.4.1   Operational sex ratio and amphibian alternative tactics

Dur survey underscores the importance of density of reeding aggregations and OSR in modulating the fre- uency and nature of alternative reproductive tactics in mphibians. Many authors before us have noted this pattern Arnold 1976, 1977, Wells 1977a); it is not surprising that lternative tactics have evolved repeatedly in species with ggregate breeding. Although OSRs in breeding ponds can ary over short time periods as a function of density, fre- uency, and individual decisions on the part of breeding emales and males (Lucas and Howard 1995, Lucas et al. 996, McCauley et al. 2000), alternative mating tactics in rogs and salamanders occur, with few exceptions, in reeding aggregations that have a male-biased OSR (Tables 1.1 and 11.2). The single example of a female-biased OSR n anurans was in an isolated population of Bufo calamita Denton and Beebee 1993) that exhibited the unique switching" and "staying" breeding tactics in contrast to the aller-satellite system in populations with male-biased OSR Arak 1988). Although it is unclear how often female-biased SR occurs in this species, this example underscores how

population-level differences in demographic structure can mediate the social context and the direction of sexual selection on short temporal and spatial scales. In laboratory studies of salamander mating, evidence for the role of density or OSR in modulating behavioral tactics is some- what equivocal; in some species males increase their courtship efforts to overcome increased competition in the presence of rival males (McWilliams 1992), whereas in others, males reduce their efforts, abandon them altogether (Uzzell 1969, Uzendoski et al. 1993, Verrell and Krenz 1998), or are indifferent to the presence of additional males (Arnold 1976, Petranka 1982). Experimental manipulations are not available for most amphibian taxa, and in those taxa where it is feasible, it would be informative to measure the frequency and outcome of breeding tactics in these different contexts.

In general, the distribution of tactics among amphibian lineages suggests that higher densities and skewed sex ratios result in increased mate competition and influence tactic diversity and prevalence. This pattern is undoubtedly related to the context dependence and relatively short lon- gevity of amphibian reproductive tactics. We found few examples of long-term tactics (those with which individuals adopt the same tactic for a season or more) in either sala- manders or frogs. This may be related to the overall dependence of male tactics on the density of competing males; short-term changes in densities or OSR (sometimes as short as among days within season: Douglas 1979) should select for the ability to change tactics on relatively short temporal scales. Flexibility or plasticity over short time- frames allows individuals to fine-tune their behavioral responses and maximize fitness depending on the immedi- ate social context. This seems to be the norm for the con- ditional strategies of amphibians.

One clear exception to the usual pattern of tactic pre- ponderance and density of competitors is found in pletho- dontid salamanders. Plethodontids are unique among salamanders in that they do not breed in large groups; this is especially true of the terrestrial forms. Breeding adults defend territories, and engage in complex social inter- actions, including pair bonding. Therefore, although sex ratios in the population may not be equal, most interactions among breeding adults happen in pairs, with little oppor- tunity for the evolution of satellite or sneaking behavior. In fact, the most frequent tactic in this group is pre-emptive sexual interference through female mimicry; plethodontids are limited in the number of spermatophores they produce, thus this alternative tactic exhausts the rivaling male's fertilization potential (Arnold 1977). This behavior may gain no immediate fitness (in terms of fertilization opportunity) for the female mimic; however, if interactions among males are limited to a small group that are competing in adjacent territories, the tactic likely results in overall success given that the duped male may not be able to fertilize the next receptive female. This category of tactic, pre-emptive rather than offering immediate reproductive success, has been reported only in plethodontids and in a few species of salamandrids.

The reproductive modes in anurans may also have consequences for the likelihood of direct competition among mating males. Thirty-nine reproductive modes have been described in anurans (Haddad and Prado 2005) and they determine the context of sexual selection and thus influence the diversity of behavioral strategies that are successful. For example, in *Hyla leucopygia* eggs are deposited in hidden nests constructed by males such that the potential for the evolution of particular alternative tactics such as satellite males, multimale spawning, and sneaking behaviors does not exist (Haddad and Sawaya 2000). Males of some species in the *Leptodactylus fuscus* group also construct subterranean chambers where females oviposit. In these species, the nests are hidden and have small openings that the male can physically obstruct after entrance of the female, thus eliminating the possibility of multimale spawning (Martins 1988). Prado and Haddad (2003) found that *fuscus* group members have small testes relative to body size suggesting that this reproductive mode prevents multimale spawning and hence the selective pressures for enlarged testes.

Given the importance of density in the evolution of amphibian alternative mating tactics, length of the breeding season will likely also play an important role in behavioral diversity within species (Wells 1977a). In general, populations with explosive breeding patterns tend to exhibit higher densities, with an overall greater number of competing males, but with a less strongly skewed OSR than species with extended or continuous breeding patterns (Wells 1977a, Arak 1983b). We would expect that amphibians with explosive breeding seasons might have different frequencies or types of reproductive tactics when compared to species that have prolonged breeding seasons (Arnold 1977, Halliday and Tejedo 1995, Verrell and Krenz 1998). This pattern does not seem to be a general one across all amphibians. For instance, among anurans, caller–satellite systems occur in many species, both explosive and prolonged breeders (Table 11.1). In salamanders, breeding-season length varies systematically with phylogenetic

lineage; therefore, differences in behavioral tactics cannot be attributed to breeding season length alone. Given the nonlinear relationship between length of breeding season and OSR, and species-specific changes in OSR during the course of the breeding season, we will need more detailed data to test for a link between temporal reproductive patterns and diversity in behavioral tactics.

The expression of alternative reproductive behaviors presumably increases individual fitness, and in fact, in some amphibian species we have been able to quantify the relative benefits of alternative tactics. However, we know far less about the potential costs. Depending on circumstances, the behaviors themselves may result in fitness disadvantages. For example, in the genus *Bufo*, active searching by males is common and larger males displace smaller amplecting rivals. This tactic bypasses courtship behaviors involved in species recognition, resulting in a higher probability of erroneous amplexus and hybridization with heterospecific females (Haddad *et al.* 1990). Another potentially negative consequence is increased likelihood of predation. In large *Bufo* aggregations, actively searching males will cause disturbance and noise while attempting to intercept females, possibly increasing attraction of predators (Haddad and Bastos 1997).

11.4.2 Phylogeny, fertilization mode, and amphibian alternative tactics

Our survey reveals macroevolutionary patterns that transcend the deep phylogenetic divergence between frogs and salamanders. One of these patterns is the convergence of tactic types within lineages that have similar fertilization modes. Internal fertilization is found in most salamanders, a few frogs, and all caecilians, while external fertilization is found in nearly all frogs, and only two families of salamanders. In salamanders with internal fertilization, there are evident phylogenetic patterns in three well-studied families that each consistently use a distinct tactic category. Ambystomatids use spermatophore capping as a means of sexual interference, with evidence of female mimicry in a few species. Plethodontids use female mimicry (in the form of male–male courtship), to reduce the reproductive potential of rival males. Salamandrids also use female mimicry, but rarely in male–male courtship; it is most often used to interrupt the courtship of a courting pair. Reproductive dynamics of the few internally fertilizing frogs and the cryptic caecilians are poorly understood. Although internal fertilization has evolved very differently in each of these amphibian lineages (Wake and Dickie 1998), it will be teresting to compare the diversity in reproductive henotypes in these groups and how they differ from those und in taxa with external fertilization.

Phylogenetic patterns of tactics used by externally fer- lizing amphibians are less consistent and tactics are often hylogenetically widespread. For instance, satellite and gauntlet" males occur in five of the six frog families ncluded in this review as well as the two externally fertili- ing salamander families. In frogs, satellites occur most ommonly among bufonids and hylids (Table 11.1); mem- ers of both families generally lay eggs away from calling tes such that mate acquisition and mate defense is of pri- ary importance to reproductive success. In contrast, ranids, ynobiids, and cryptobranchids usually involve attempted nonopolization of a resource such as a nest site (in crypto- ranchids and ranids), or a clutch of eggs (in hynobiids), with atellite or sneaker behaviors on the part of additional males. n these cases heightened aggression might have evolved as exual defense to deter satellite males from intercepting emales. External fertilizing frogs, primarily of the families hacophoridae and Leptodactylidae, can also participate in aultimale spawning behaviors where mated pairs are tied to visible and obvious nest and hence are unable to avoid ining males. Despite these general patterns, instances of atellite tactics and multimale spawning in other frog families uggest that reproductive strategies are especially flexible in nurans. The differences in tactics employed by internally nd externally fertilizing lineages are likely a result of the volution of a "breeding phenotype," which includes the oncerted evolution of density and temporal aspects of the nating system, as well as the relative investment in gametes nd courtship in each lineage.

1.5 FUTURE RESEARCH

he study of the origin, diversification, and maintenance of ehaviors is most complete if we consider both process and attern at hierarchical scales of analysis, encompassing nultiple scientific fields and approaches. Tinbergen (1963) et forth a conceptual framework for ethology in the form of ur aims or goals for the comprehensive study of behavior. hese aims are to study the control (or causative mechan- ms) of behaviors, their ontogeny (genetics and develop- nent during the life of an individual), their evolution istorical change over time or across taxa), and their func- on (or adaptive value). Tinbergen (1951) acknowledged verlap among fields that address these questions, and ressed that a coherent and complete study of behavior

should adequately address each aim and integrate among them. Other authors have recently clarified or amended the goals in ethology, but also reasserted the value of this framework (Dewsbury 1992, Burghardt 1997). The inter- vening years have seen multiple shifts in the emphasis placed on these aims, as well as the increased separation and inde- pendence of the conceptual fields that address the four types of questions. Nonetheless, any organismal biologist studying behavior can appreciate the depth of understanding one would obtain if in fact this research agenda were complete for their focal species; therefore, this framework remains an important landmark and a continuing challenge for studies of behavior. It is clear from our amphibian survey as well as the other taxonomic summaries in this book, that the study of alternative behaviors has focused primarily on the function or adaptive value of various tactics. It is not surprising, given that alternative tactics evolve specifically for increases in reproductive success, that fitness is often the first attribute of behavioral tactics examined in most systems. In amphibians, we have made less progress in the study of causation or control factors (internal or external) that underlie behaviors, the study of evolutionary patterns of behavioral change in diversifying lineages, and the study of individual ontogenies. Future research on alternative tactics should involve increased efforts in those particular arenas.

Our understanding of the evolution of alternative reproductive tactics in amphibians is greatly hampered by a lack of detailed information. Therefore, a major challenge for future studies of alternative reproductive phenotypes will be to further describe and categorize the diversity of reproductive tactics in these groups. Most of the approxi- mately 5000 species of frogs, 500 species of salamanders, and 160 species of caecilians have some element of crypticity to their reproduction due to inaccessibility of breeding sites, relatively short periods of reproductive activity, low dens- ities, or a combination of these factors. Most alternative tactics have been described in temperate amphibians despite the fact that most amphibian taxonomic diversity occurs in the tropics, and tropical species account for much of the morphological, taxonomic, behavioral, and ecological diversity in this group. Certainly, the overall diversity of reproductive tactics is much greater than summarized here and we have much more to learn about alternative repro- ductive phenotypes in this diverse lineage.

Our understanding of the causation or control mecha- nisms underlying the expression of behavioral tactics is improving, and our next challenge will be to uncover the general patterns within and across lineages in the factors that modulate behavioral polymorphism. We mentioned the importance of density and OSR and the possible roles they play in sexual selection, but we do not yet have sufficient detail on the proximal mechanisms that result from these demographic conditions, and how they shape the expression and incidence of alternate behavioral phenotypes in natural populations. For example, in high-density breeding aggregations, what role, if any, does relatedness or kinship among breeding adults play in tactics adopted by males? What are the costs of calling, producing sperm, courting, and amplexing, and how do those trade-off with the benefits of various tactics? Does previous experience play a role in choice of tactics? Answering these questions will involve extensive and detailed behavioral observations, experimental manipulations, comparative studies, and the incorporation of genetic techniques that quantify the fitness consequences of alternative tactics. For tactics that are condition dependent, as in the case of smaller satellite males, proximate explanations will most likely be found in individual characters related to physical condition. For alternative phenotypes that are more variable over short time periods, and not directly correlated with condition, it will be important to know whether these behaviors can be changed at any moment depending on the immediate social context, or if hormone levels, aggression, previous mating history, or social experiences might influence the tactics available to a courting individual. In that case, one might expect variation among males in the propensity to switch tactics even when exposed to the same context and stimuli.

Our search for general patterns in the evolution of behavioral phenotypes has thus far focused on interspecific or interlineage comparisons. These macroevolutionary comparisons are an excellent place to start; however, they can be nicely complemented with intraspecific or population-level comparisons that are most fruitful for uncovering the environmental or social determinants of reproductive behaviors. Populations of the same species in different regions often experience drastically different local densities, OSRs, and length of breeding seasons; not surprisingly we also see substantial variation in the occurrence and frequency of tactics among populations, seasons, and even breeding nights (e.g., Semlitsch 1985, Arak 1988, Sullivan 1989, Byrne 2002). These variable species are ideal candidates for studies of factors modulating or controlling the expression of behavioral tactics.

Over the last few decades new empirical and analytical techniques have become available for analyses of paternity and relatedness in animal taxa, and we are just beginning

to reap the benefits of those results (Roberts et al. 1999, Gabor et al. 2000, Jones et al. 2002, Vieites et al. 2004). Paternity assays can alert us to the presence of alternative behaviors that may have gone unnoticed in field or laboratory observations. For example, multiple paternity was detected genetically in *Rana temporaria* before the observation of clutch piracy (Laurila and Seppä 1998, Vieites et al. 2004). Similarly, multiple paternity has been found in *R. dalmatina* (Lodé and Lesbarrères 2004) although the behavioral mechanism is still unknown. Given the difficulty of observing many amphibian reproductive behaviors, it is likely that these discoveries will happen repeatedly as we survey more species, and will lead us to species for which detailed observational studies might be particularly fruitful. More importantly, paternity studies will allow us to more accurately document the fitness of individuals adopting each tactic and thus their adaptive value. Until now, success in mate acquisition has been used as a proxy for fitness, with the implicit assumption that processes such as sperm competition, mating precedence, and cryptic female choice will have minimal effects on relative parental contributions to the clutch. We know from recent discoveries in other taxa that this assumption is false (Eberhard 1996, Danielsson 2001, Cordoba et al. 2003), resulting in an oversimplification of the parameters involved in sexual selection. Current analytical techniques allow us to measure the realized outcome of alternative tactics, even in systems with multiple mating, multiple paternity, and large clutches (Fiumera et al. 2001, Blouin 2003). Given that the number of offspring produced by courting individuals is not the result of a simple formula, quantifying the relative success of individuals adopting different tactics will be a large step toward understanding the evolution and maintenance of behavioral diversity in amphibians.

Acknowledgments

We thank the editors for the opportunity to contribute to this volume, and Harry W. Greene, Jeanne M. Robertson, Rulon W. Clark, Kurt E. Galbreath, and Christine E. Voyer, for useful discussions and comments on earlier versions of the manuscript. We thank Célio Haddad and Cynthia Prado for calling our attention to literature and for helpful discussions. We especially acknowledge all those who collected the data summarized in this review; a synthetic survey of behavioral patterns in a large and diverse group such as amphibians is only possible because many dedicated biologists and behaviorists spent numerous hours studying and describing the biology of their organisms.

References

Able, D. J. 1999. Scramble competition selects for greater tailfin size in male red-spotted newts (Amphibia: Salamandridae). *Behavioral Ecology and Sociobiology* 46, 423–428.

Arak, A. 1983a. Sexual selection by male–male competition in natterjack toad choruses. *Nature* 306, 261–262.

Arak, A. 1983b. Male–male competition and mate choice in anuran amphibians. In P. Bateson (ed.) *Mate Choice*, pp. 181–210. Cambridge, UK: Cambridge University Press.

Arak, A. 1988. Callers and satellites in the natterjack toad: evolutionary stable decision rules. *Animal Behaviour* 36, 416–422.

Arnold, S. J. 1972. The evolution of courtship behaviors in salamanders. Ph.D. dissertation, University of Michigan, Ann Arbor, MI.

Arnold, S. J. 1976. Sexual behavior, sexual interference, and sexual defense in the salamanders *Ambystoma maculatum*, *Ambystoma tigrinum*, and *Plethodon jordani*. *Zeitschrift für Tierpsychologie* 41, 247–300.

Arnold, S. J. 1977. The evolution of courtship behavior in New World salamanders with some comments on Old World salamandrids. In D. H. Taylor and S. I. Guttman (eds.) *The Reproductive Biology of Amphibians*, pp. 141–183. New York: Plenum Press.

Arnold, S. J. and Houck, L. D. 1982. Courtship pheromones: evolution by natural and sexual selection. In M. Nitecki (ed.) *Biochemical Aspects of Evolutionary Biology*, pp. 173–211. Chicago, IL: University of Chicago Press.

Bennet, A. F. and Houck, L. D. 1983. The energetic cost of courtship and aggression in a plethodontid salamander. *Ecology* 64, 979–985.

Bishop, S. C. 1941. The salamanders of New York. *New York State Museum Bulletin* 324, 1–365.

Bishop, S. C. 1943. *Handbook of Salamanders: The Salamanders of the United States, of Canada, and of Lower California*. Ithaca, NY: Comstock.

Blanchard, F. N. 1934. The relation of the female four-toed salamander to her nest. *Copeia*, 137–138.

Blouin, M. S. 2003. DNA-based methods for pedigree reconstruction and kinship analysis in natural populations. *Trends in Ecology and Evolution* 18, 503–511.

Bokerman, W. C. A. 1974. Observações sobre desenvolvimento precoce em *Sphaenorhynchus bromelicola* Bok. 1966 (Anura, Hylidae). *Revista Brasileira de Biologia* 34, 35–41.

Bourne, G. R. 1992. Lekking behavior in the neotropical frog *Ololygon rubra*. *Behavioral Ecology and Sociobiology* 31, 173–180.

Bourne, G. R. 1993. Proximate costs and benefits of mate acquisition at leks of the frog *Ololygon rubra*. *Animal Behaviour* 45, 1051–1059.

Burghardt, G. M. 1997. Amending Tinbergen: a fifth aim for ethology. In R.W. Mitchell, N.S. Thompson, and H.L. Miles (eds.) *Anthropmorphism, Anectclotes, and Animals*. Albany, NY: State University of New York Press.

Byrne, P. G. 2002. Climatic correlates of breeding, simultaneous polyandry and potential for sperm competition in the frog *Crinia georgiana*. *Journal of Herpetology* 36, 124–129.

Byrne, P. G. and Roberts, J. D. 1999. Simultaneous mating with multiple males reduces fertilization success in the myobatrichid frog *Crinia georgiana*. *Proceedings of the Royal Society of London B* 266, 717–721.

Byrne, P. G. and Roberts, J. D. 2004. Intrasexual selection and group spawning in quacking frogs (*Crinia georgiana*). *Behavioral Ecology* 15, 872–882.

Byrne, P. G., Roberts, J. D., and Simmons, L. W. 2002. Sperm competition selects for increased testis mass in Australian frogs. *Journal of Evolutionary Biology* 15, 347–355.

Coe, M. 1974. Observations on the ecology and breeding biology of the genus *Chiromantis* (Amphibia: Rhacophoridae). *Journal of Zoology* 172, 13–34.

Cordoba, A. A., Uhia, E., and Rivera, A. C. 2003. Sperm competition in Odonata (Insecta): the evolution of female sperm storage and rivals' sperm displacement. *Journal of Zoology (London)* 261, 381–398.

Danielsson, I. 2001. Antagonistic pre- and post-copulatory sexual selection on male body size in a water strider (*Gerris lacustris*). *Proceedings of the Royal Society of London B* 268, 77–81.

Davies, N. B. and Halliday, T. R. 1977. Optimal mate selection in the toad *Bufo bufo*. *Nature* 269, 56–58.

Davies, N. B. and Halliday, T. R. 1979. Competitive mate searching in male common toads, *Bufo bufo*. *Animal Behaviour* 27, 1253–1267.

Denoël, M. 2002. Paedomorphosis in the alpine newt (*Triturus alpestris*): decoupling behavioural and morphological change. *Behavioral Ecology and Sociobiology* 52, 394–399.

Denoël, M. 2003. Effect of rival males on the courtship of paedomorphic and metamorphic *Triturus alpestris* (Amphibia: Salamandridae). *Copeia*, 618–623.

Denoël, M. and Poncin, P. 2001. The effect of food on growth and metamorphosis of paedomorphs in *Triturus alpestris apuanus*. *Archiv für Hydrobiologie* 152, 661–670.

Denoël, M., Poncin, P., and Ruwet, J.-C. 2001. Alternative mating tactics in the alpine newt, *Triturus alpestris alpestris*. *Journal of Herpetology* 35, 62–67.

Denton, J. S. and Beebee, T. J. C. 1993. Reproductive strategies in a female-biased population of natterjack toads, *Bufo calamita*. *Animal Behaviour* 46, 1169–1175.

Dewsbury, D. A. 1992. On the problems studied in ethology, comparative psychology, and animal behavior. *Ethology* 92, 89–107.

D'Orgeix, C. A. and Turner, B. J. 1995. Multiple paternity in the red-eyed treefrog *Agalychnis callidryas* (Cope). *Molecular Ecology* 4, 505–508.

Douglas, M. E. 1979. Migration and sexual selection in *Ambystoma jeffersonianum*. *Canadian Journal of Zoology* 57, 2303–2310.

Duellman, W. E. 2003. An overview of anuran phylogeny, classification, and reproductive modes. In B. G. M. Jamieson (ed.) *Reproductive Biology and Phylogeny of Anura*, pp. 1–18. Enfield, UK: Science Publishers.

Duellman, W. E. and Trueb, L. 1994. *Biology of Amphibians*. Baltimore, MD: Johns Hopkins University Press.

Eberhard, W. G. 1996. *Female Control: Sexual Selection by Cryptic Female Choice*. Princeton, NJ: Princeton University Press.

Emlen, S. T. 1968. Territoriality in the bullfrog, *Rana catesbeiana*. *Copeia*, 240–243.

Emlen, S. T. and Oring, L. W. 1977. Ecology, sexual selection, and the evolution of mating systems. *Science* 197, 215–223.

Fairchild, L. 1984. Male reproductive tactics in an explosive breeding toad population. *American Zoologist* 24, 407–418.

Faria, M. M. 1993. Sexual behavior of Bosca's newt, *Triturus boscai*. *Amphibia–Reptilia* 14, 169–185.

Faria, M. M. 1995. A field study of reproductive interactions in Bosca's newt, *Triturus boscai*. *Amphibia–Reptilia* 16, 357–374.

Feller, A. E. and Hedges, S. B. 1998. Molecular evidence for the early history of living amphibians. *Molecular Phylogenetics and Evolution* 9, 509–516.

Fellers, G. M. 1979. Aggression, territoriality, and mating behaviour in North American treefrogs. *Animal Behaviour* 27, 107–119.

Feng, A. S. and Narins, P. M. 1991. Unusual mating behavior of Malaysian treefrogs, *Polypedates leucomystax*. *Naturwissenschaften* 78, 364–365.

Fiumera, A. C., DeWoody, Y. D., DeWoody, J. A., Asmussen, M. A., and Avise, J. C. 2001. Accuracy and precision of methods to estimate the number of parents contributing to a half-sib progeny array. *Journal of Heredity* 92, 120–126.

Forester, D. C. and Lykens, D. V. 1986. Significance of satellite males in a population of spring peepers (*Hyla crucifer*). *Copeia*, 719–724.

Forester, D. C. and Thompson, K. J. 1998. Gauntlet behaviour as a male sexual tactic in the American toad (Amphibia: Bufonidae). *Behavior* 135, 99–119.

Frost, D. R. 2004. *Amphibian Species of the World: An Online Reference*, version 3.0 (22 Aug. 2004). Available online at http://research.amnh.org/herpetology/amphibia/index.html.

Fukuyama, K. 1991. Spawning behaviour and male mating tactics of a foam-nesting treefrog, *Rhacophorus schlegelii*. *Animal Behaviour* 42, 193–199.

Gabor, C. R. 1995. Resource quality affects the agonistic behaviour of terrestrial salamanders. *Animal Behaviour* 49, 71–79.

Gabor, C. R., Krenz, J. D., and Jaeger, R. G. 2000. Female choice, male interference, and sperm precedence in the red-spotted newt. *Behavioral Ecology* 11, 115–124.

Giacoma, C. and Crusco, N. 1987. Courtship and male interference in the Italian newt: a field study. *Monitore Zoologico Italiano* 21, 190–191.

Gillette, J. R., Jaeger, R. G., and Peterson, M. G. 2000. Social monogamy in a territorial salamander. *Animal Behaviour* 59, 1241–1250.

Given, M. F. 1988. Territoriality and aggressive interactions of male carpenter frogs, *Rana virgatipes*. *Copeia*, 411–421.

Green, A. J. 1989. The sexual behavior of the great crested newt, *Triturus cristatus* (Amphibia: Urodela: Salamandridae). *Ethology* 83, 129–153.

Gross, M. R. 1996. Alternative reproductive strategies and tactics: diversity within sexes. *Trends in Ecology and Evolution* 11, 92–98.

Guffey, C. A., MaKinster, J. G., and Jaeger, R. G. 1998. Familiarity affects interactions between potentially courting territorial salamanders. *Copeia*, 205–208.

Haddad, C. F. B. 1991. Satellite behavior in the neotropical treefrog *Hyla minuta*. *Journal of Herpetology* 25, 226–229.

Haddad, C. F. B. and Bastos, R. P. 1997. Predation on the toad *Bufo crucifer* during reproduction (Anura: Bufonidae). *Amphibia Reptilia* 18, 295–298.

Haddad, C. F. B. and Prado, C. P. A. 2005. Reproductive modes in frogs and their unexpected diversity in the Atlantic forest of Brazil. *BioScience* 55, 207–217.

Haddad, C. F. B. and Sawaya, R. J. 2000. Reproductive modes of Atlantic forest hylid frogs: a general overview and the description of a new mode. *Biotropica* 32, 862–871.

Haddad, C. F. B. *et al.* 1990. Natural hybridization between *Bufo ictericus* and *Bufo crucifer*. *Revista Brasileira de Biologia* 50, 739–744.

Halliday, T. R. 1974. Sexual behavior of the smooth newt, *Triturus vulgaris* (Urodela: Salamandridae). *Journal of Herpetology* 8, 277–292.

Halliday, T. R. 1975. An observational and experimental study of sexual behaviour in the smooth newt, *Triturus vulgaris* (Amphibia, Salamandridae). *Animal Behaviour* 23, 291–322.

Halliday, T. R. 1977. The courtship of European newts: an evolutionary perspective. In D. H. Taylor and S. I. Guttman (eds.) *The Reproductive Biology of Amphibians*, pp. 185–232. New York: Plenum Press.

Halliday, T. R. 1990. The evolution of courtship behavior in newts and salamanders. *Advances in the Study of Behavior* 19, 137–169.

Halliday, T. R. and Tejedo, M. 1995. Intrasexual selection and alternative mating behaviour. In H. Heatwole and B. K. Sullivan (eds.) *Amphibian Biology*, vol. 2, *Social Behaviour*, pp. 419–468. Chipping Norton, NSW: Surrey Beatty and Sons.

Harris, R. N. 1987. Density-dependent paedomorphosis in the salamander *Notophthalmus viridescens dorsalis*. *Ecology* 68, 705–712.

Harris R. N. and Gill, D. E. 1980. Communal nesting, brooding behavior, and embryonic survival of the four-toed salamander *Hemidactylium scutatum*. *Herpetologica* 36, 141–144.

Harris, R. N. and Ludwig, P. M. 2004. Resource level and reproductive frequency in female four-toed salamanders, *Hemidactylium scutatum* (Caudata: Plethodontidae). *Ecology* 85, 1585–1590.

Harris, R. N., Semlitsch, R. D., Wilbur, H. M., and Fauth, J. E. 1990. Local variation in the genetic basis of paedomorphosis in the salamander *Ambystoma talpoideum*. *Evolution* 44, 1588–1603.

Harris, R. N., Hames, W. W., Knight, I. T., Carreno, C. A., and Vess T. J. 1995. An experimental analysis of joint nesting in the salamander *Hemidactylium scutatum* (Caudata: Plethodontidae): the effects of population density. *Animal Behaviour* 50, 1309–1316.

Hasumi, M. 1994. Reproductive behavior of the salamander *Hynobius nigrescens*: monopoly of egg sacs during scramble competition. *Journal of Herpetology* 28, 264–267.

Hasumi, M. 2001. Sexual behavior in female-biased operational sex ratios in the salamander *Hynobius nigrescens*. *Herpetologica* 57, 396–406.

Houck, L. D. 1988. The effect of body size on male courtship success in a plethodontid salamander. *Animal Behaviour* 36, 837–842.

Houck, L. D. and Arnold, S. J. 2003. Courtship and mating behavior. In B. G. M. Jamieson (ed.) *Reproductive Biology and Phylogeny of Urodela*, pp. 383–424. Enfield, UK: Science Publishers.

Howard, R. D. 1978. The evolution of mating strategies in bullfrogs, *Rana catesbeiana*. *Evolution* 32, 850–871.

Howard, R. D., Moorman, R. S., and Whiteman, H. H. 1997. Differential effects of mate competition and mate choice on eastern tiger salamanders. *Animal Behaviour* 53, 1345–1356.

Jaeger, R. G. and Forester, D. C. 1993. Social behavior of plethodontid salamanders. *Herpetologica* 49, 163–175.

Jaeger, R. G., Wicknick, J. A., Griffis, M. R., and Anthony, C. D. 1995. Socioecology of a terrestrial salamander: juveniles enter adult territories during stressful foraging periods. *Ecology* 76, 533–543.

Jaeger, R. G., Peterson, M. G., and Gillette, J. R. 2000. A model of alternative mating strategies in the redback salamander, *Plethodon cinereus*. In R. C. Bruce, R. G. Jaeger, and L. D. Houck (eds.) *The Biology of Plethodontid Salamanders*, pp. 441–450. New York: Plenum Press.

Jaeger, R. G., Gillette, J. R., and Cooper, R. C. 2002. Sexual coercion in a territorial salamander: males punish socially polyandrous female partners. *Animal Behaviour* 63, 871–877.

Janzen, F. J. and Brodie, E. D. III. 1989. Tall tails and sexy males: sexual behavior of the rough-skinned newts (*Taricha granulosa*) in a natural breeding pond. *Copeia*, 1068–1071.

Jennions, M. D. and Passmore, N. I. 1993. Sperm competition in frogs: testis size and a "sterile male" experiment on *Chiromantis xerampelina* (Rhacophoridae). *Biological Journal of the Linnean Society* 50, 211–220.

Jennions, M. D., Backwell, P. R. Y., and Passmore, N. I. 1992. Breeding behaviour of the African frog, *Chiromantis xerampelina*: multiple spawning and polyandry. *Animal Behaviour* 44, 1091–1100.

Jones, A. G., Adams, E. M., and Arnold, S. J. 2002. Topping off: a mechanism of first-male sperm precedence in a vertebrate. *Proceedings of the National Academy of Sciences of the United States of America* 99, 2078–2081.

Kalezic, M. L., Cvetkovic, D., Djorovic, A., and Dzukic, G. 1996. Alternative life-history pathways: paedomorphosis and adult fitness in European newts (*Triturus vulgaris* and *T. alpestris*). *Journal of Zoological Systematics and Evolutionary Research* 34, 1–7.

Kaminsky, S. K. 1997. *Bufo americanus* reproduction. *Herpetological Review* 28, 84.

Kasuya, E., Hirota, M., and Shigehara, H. 1996. Reproductive behavior of the Japanese treefrog, *Rhacophorus arboreus* (Anura: Rhacophoridae). *Researches on Population Ecology* 38, 1–10.

Kato, K. 1956. Ecological notes on the green frogs during the breeding season. 2. Breeding habit and others. *Japanese Journal of Ecology* 6, 57–61.

Kawamichi, T. and Ueda, H. 1998. Spawning at nests of extra-large males in the giant salamander *Andrias japonicus*. *Journal of Herpetology* 32, 133–136.

Krenz, J. D. and Sever, D. M. 1995. Mating and oviposition in paedomorphic *Ambystoma talpoideum* precedes the arrival of terrestrial males. *Herpetologica* 51, 387–393.

Krupa, J. L. 1989. Alternative mating tactics in the Great Plains toad. *Animal Behaviour* 37, 1035–1043.

Kusano, T., Toda, M., and Fukuyama, K. 1991. Testes size and breeding systems in Japanese anurans with special reference to large testes in the treefrog, *Rhacophorus arboreus* (Amphibia: Rhacophoridae). *Behavioral Ecology and Sociobiology* 29, 27–31.

Lank, D. B., Smith, C. M., Hanotte, O., Burke, T., and Cooke, F. 1995. Genetic polymorphism for alternative mating behaviour in lekking male ruff, *Philomachus pugnax*. *Nature* 378, 59–62.

Larson, A., Weisrock, D. W., and Kozak, K. 2003. Phylogenetic systematics of salamanders (Amphibia: Urodela): a review. In B. G. M. Jamieson (ed.) *Reproductive Biology and Phylogeny of Urodela*, pp. 31–108. Enfield, UK: Science Publishers.

Laurila, A. and Seppä, P. 1998. Multiple paternity in the common frog (*Rana temporaria*): genetic evidence from tadpole kin groups. *Biological Journal of the Linnean Society* 63, 221–232.

Laurin, M. and Reisz, R. 1997. A new perspective on tetrapod phylogeny. In S. S. Suemida and K. L. Martin (eds.) *Amniote Origins*, pp. 9–59. New York: Academic Press.

Leary, C. J., Jessop, T. S., Garcia, A. M., and Knapp, R. 2004. Steroid hormone profiles and relative body condition of calling and satellite toads: implications for proximate

regulation of behavior in anurans. *Behavioral Ecology* 15, 313–320.

Lehtinen, R. M. and Nussbaum, R. A. 2003. Parental care: a phylogenetic perspective. In B. G. M. Jamieson (ed.) *Reproductive Biology and Phylogeny of Anura*, pp. 343–386. Enfield, UK: Science Publishers.

Lodé, T. and Lesbarrères, D. 2004. Multiple paternity in *Rana dalmatina*, a monogamous territorial breeding anuran. *Naturwissenschaften* 91, 44–47.

Loman, J. and Madsen, T. 1986. Reproductive tactics of large and small *Bufo bufo*. *Oikos* 46, 57–61.

Lucas, J. R. and Howard, R. D. 1995. On alternative reproductive tactics in anurans: dynamic games with density and frequency dependence. *American Naturalist* 146, 365–397.

Lucas, J. R., Howard, R. D., and Palmer, J. G. 1996. Caller and satellites: chorus behaviour in anurans as a stochastic dynamic game. *Animal Behaviour* 51, 501–518.

Martins, M. 1988. Reproductive biology of *Leptodactylus fuscus* in Boa Vista, Roraima (Amphibia: Anura). *Revista Brasileira de Biologia* 48, 969–977.

Massey, A. 1988. Sexual interactions in red-spotted newt populations. *Animal Behaviour* 36, 205–210.

Mathis, A. 1989. Do seasonal spatial distributions in a tewestrial salamander reflect reproductive behavior or territoriality? *Copeia*, 788–791.

Mathis, A. 1991a. Territories of male and female terrestrial salamanders: costs, benefits, and intersexual spatial associations. *Oecologia* 86, 433–440.

Mathis, A. 1991b. Large male advantage for access to females: evidence of male–male competition and female discrimination in a terrestrial salamander. *Behavioral Ecology and Sociobiology* 29, 133–138.

McCauley, S. J., Bouchard, S. S., Farina, B. J., *et al.* 2000. Energetic dynamics and anuran breeding phenology: insights from a dynamic game. *Behavioral Ecology* 11, 429–436.

McWilliams, S. R. 1992. Courtship behavior of the small-mouthed salamander (*Ambystoma texanum*): the effects of conspecific males on male mating tactics. *Behavior* 121, 1–19.

Miyamoto, M. M. and Cane, J. H. 1980. Behavioral observations of noncalling males in Costa Rican *Hyla ebraccata*. *Biotropica* 12, 225–227.

Moore, F. L., McCormack, C., and Swanson, L. 1979. Induced ovulation: effects of sexual behavior and insemination on ovulation and progesterone levels in

Taricha granulosa. General and Comparative Endocrinology **39**, 262–269.

Myers, E. M. and Zamudio, K. R. 2004. Multiple paternity in an aggregate breeding amphibian: the effect of reproductive skew on estimates of male reproductive success. *Molecular Ecology* **13**, 1951–1963.

Nickerson, M. A. and Mays, C. E. 1973a. The hellbenders: North American "Giant Salamanders." *Milwaukee Public Museum Publications in Biology and Geology* **1**, 1–106.

Nickerson, M. A. and Mays, C. E. 1973b. A study of the Ozark hellbender *Cryptobranchus alleganiensis bishopi* Grobman. *Ecology* **54**, 1164–1165.

Nickerson, M. A. and Tohulka, M. D. 1986. The nests and nest site selection by Ozark hellbenders, *Cryptobranchus alleganiensis bishopi* Grobman. *Transactions of the Kansas Academy of Science* **89**, 66–69.

Noble, G. K. and Brady, M. K. 1933. Observations on the life history of the marbled salamander, *Ambystoma opacum* Gravenhorst. *Zoologica* **11**, 89–132.

Organ, J. A. 1958. Courtship and spermatophore of *Plethodon jordani metcalfi*. *Copeia* **1958**, 251–259.

Organ, J. A. 1960. The courtship and spermatophore of the salamander *Plethodon glutinosus*. *Copeia* **1960**, 34–40.

Organ, J. A. and Organ, D. J. 1968. Courtship behavior of the red salamander *Pseudotriton ruber*. *Copeia*, 217–223.

Ovaska, K. and Hunte, W. 1992. Male mating behavior of the frog *Eleutherodactylus johnstonei* (Leptodactylidae) in Barbados, West Indies. *Herpetologica* **48**, 40–49.

Park, D. and Park, S. R. 2000. Multiple insemination and reproductive biology of *Hynobius leechii*. *Journal of Herpetology* **34**, 594–598.

Park, S. R., Park, D. S., and Yang, S. Y. 1996. Courtship, fighting behaviors, and sexual dimorphism of the salamander, *Hynobius leechi*. *Korean Journal of Zoology* **39**, 437–446.

Parker, G. A. 1984. Evolutionary stable strategies. In J. R. Krebs and N. B. Davies (eds.) *Behavioral Ecology: An Evolutionary Approach*, pp. 30–61. Sunderland, MA: Sinauer Associates.

Perrill, S. A. 1984. Male mating behavior in *Hyla regilla*. *Copeia* **1984**, 727–732.

Perrill, S. A. and Magier, M. 1988. Male mating behavior in *Acris crepitans*. *Copeia* **1988**, 245–248.

Perrill, S. A., Gerhardt, H. C., and Daniel, R. E. 1978. Sexual parasitism in the green tree frog (*Hyla cinerea*). *Science* **200**, 1179–1180.

Perrill, S. A., Gerhardt, H. C., and Daniel, R. E. 1982. Mating strategy shifts in male green treefrogs (*Hyla cinerea*): an experimental study. *Animal Behaviour* **30**, 43–48.

Petranka, J. W. 1982. Courtship behavior of the small-mouthed salamander (*Ambystoma texanum*) in central Kentucky. *Herpetologica* **38**, 333–336.

Prado, C. P. A. and Haddad, C. F. B. 2003. Testes size in leptodactylid frogs and occurrence of multimale spawning in the genus *Leptodactylus* in Brazil. *Journal of Herpetology* **37**, 354–362.

Prado, C. P. A., Uetanabaro, M., and Lopes, F. S. 2000. Reproductive strategies of *Leptodactylus chaquensis* and *L. podicipinus* in the Pantanal, Brazil. *Journal of Herpetology* **34**, 135–139.

Propper, C. R. 1991. Courtship in the rough-skinned newt *Taricha granulosa*. *Animal Behaviour* **41**, 547–557.

Pyburn, W. F. 1970. Breeding behavior of the leaf-frogs *Phyllomedusa callidryas* and *Phyllomedusa dacnicolor* in Mexico. *Copeia*, 209–218.

Rafinski, J. and Pecio, A. 1992. The courtship behaviour of the Bosca's newt, *Triturus boscai* (Amphibia: Salamandridae). *Folia Biologica* **40**, 155–165.

Roberts, J. D., Standish, R. J., Byrne, P. G., and Doughty, P. 1999. Synchronous polyandry and multiple paternity in the frog *Crinia georgiana* (Anura: Myobatrachidae). *Animal Behaviour* **57**, 721–726.

Roberts, W. E. 1994. Explosive breeding aggregations and parachuting in a neotropical frog, *Agalychnis saltator* (Hylidae). *Journal of Herpetology* **28**, 193–199.

Robertson, J. G. M. 1986a. Female choice, mating strategies and the role of vocalizations in the Australian frog *Uperoleia rugosa*. *Animal Behaviour* **34**, 773–784.

Robertson, J. G. M. 1986b. Male territoriality, fighting and assessment of fighting ability in the Australian frog *Uperoleia rugosa*. *Animal Behaviour* **34**, 763–772.

Roble, S. M. 1985. Observations on satellite males in *Hyla chrysoscelis*, *Hyla picta*, and *Pseudacris triseriata*. *Journal of Herpetology* **19**, 432–436.

Ryan, M. J., Pease, C. M., and Morris, M. R. 1992. A genetic polymorphism in the swordtail *Xiphiphorus nigrensis*: testing the prediction of equal fitness. *American Naturalist* **139**, 21–31.

Ryan, T. J. and Plague, G. R. 2004. Hatching asynchrony, survival, and the fitness of alternative adult morphs in *Ambystoma talpoideum*. *Oecologia* **140**, 46–51.

Salthe, S. E. 1967. Courtship patterns and the phylogeny of urodeles. *Copeia*, 100–117.

330   K. R. ZAMUDIO AND L. M. CHAN

Sasaki, M. 1924. On a Japanese salamander in Lake Kuttarush which propagates like the axolotl. *Journal of the College of Agriculture Hokkaido Imperial University* 15, 12–23.

Sato, T. 1992. Reproductive behavior in the Japanese salamander *Hynobius retardatus*. *Japanese Journal of Herpetology* 14, 184–190.

Scheltinga, D. M. and Jamieson, B. G. M. 2003a. The mature spermatozoon. In B. G. M. Jamieson (ed.) *Reproductive Biology and Phylogeny of Urodela*, pp. 203–274. Enfield, UK: Science Publishers.

Scheltinga, D. M. and Jamieson, B. G. M. 2003b. Spermatogenesis and the mature spermatozoon: form, function, and phylogenetic implications. In B. G. M. Jamieson (ed.) *Reproductive Biology and Phylogeny of Anura*, pp. 343–386. Enfield: Science Publishers.

Semlitsch, R. D. 1985. Reproductive strategy of a facultatively paedomorphic salamander *Ambystoma talpoideum*. *Oecologia* 65, 305–313.

Semlitsch, R. D. 1987. Paedomorphosis in *Ambystoma talpoideum*: effects of density, food and pond-drying. *Ecology* 68, 994–1002.

Semlitsch, R. D. and Gibbons, J. W. 1985. Phenotypic variation in metamorphosis and paedomorphosis in the salamander *Ambystoma talpoideum*. *Ecology* 66, 1123–1130.

Shuster, S. M. and Wade, M. J. 1991. Equal mating success among male reproductive strategies in a marine isopod. *Nature* 350, 608–610.

Shuster, S. M. and Wade, M. J. 2003. *Mating Systems and Strategies*. Princeton, NJ: Princeton University Press.

Sinervo, B. and Zamudio, K. R. 2001. The evolution of alternative reproductive strategies: fitness differential, heritability, and genetic correlation between the sexes. *Journal of Heredity* 92, 198–205.

Smith, B. G. 1907. The life history and habits of *Cryptobranchus alleghaniensis*. *Biological Bulletin* 13, 5–39.

Smith, M. J. and Roberts, J. D. 2003. Call structure may affect male mating success in the quacking frog *Crinia georgiana* (Anura: Myobatrachidae). *Behavioral Ecology and Sociobiology* 53, 221–226.

Sparreboom, M. 1994. On the sexual behavior of the sword-tailed newt, *Cynops ensicauda* (Hallowell, 1860). *Abhandlungen und Berichte für Naturkunde* 17, 151–161.

Spotila, J. R. and Beumer, R. J. 1970. The breeding habits of the ringed salamander, *Ambystoma annulatum* (Cope), in northeastern Arkansas. *American Midland Naturalist* 84, 77–89.

Stebbins, R. C. 1949. Courtship of the plethodontid salamander *Ensatina eschscholtzii*. *Copeia*, 274–281.

Sullivan, B. K. 1982. Male mating behaviour in the Great Plains toad (*Bufo cognatus*). *Animal Behaviour* 30, 939–940.

Sullivan, B. K. 1989. Mating system variation in Woodhouse's toad (*Bufo woodhousii*). *Ethology* 83, 60–68.

Tennessen, J. A. and Zamudio, K. R. 2003. Early male reproductive advantage, multiple paternity and sperm storage in an amphibian aggregate breeder. *Molecular Ecology* 12, 1567–1576.

Teyssedre, C. and Halliday, T. R. 1986. Cumulative effect of male's display in the sexual behavior of the smooth newt, *Triturus vulgaris* (Urodela, Salamandridae). *Ethology* 71, 89–102.

Thomas, J. S. 1989. Courtship, male–male competition, and male aggressive behavior of the salamander *Eurycea bislineata*. M.S. thesis, University of Southwestern Louisiana, Lafayette, LA.

Thorn, R. 1962. Protection of a brood by a male salamander, *Hynobius nebulosus*. *Copeia*, 638–640.

Tinbergen, N. 1951. *The Study of Instinct*. Oxford, UK: Clarendon Press.

Tinbergen, N. 1963. On aims and methods of ethology. *Zeitschrift für Tierpsychologie* 20, 410–433.

Toda, M. 1989. The life history of Japanese forest frog, *Rhacophorus arboreus* in the Kanazawa Castle. M.S. thesis, Kanazawa University, Japan.

Trueb, L. and Cloutier, R. 1991. A phylogenetic investigation of the inter- and intrarelationships of the Lissamphibia (Amphibia: Temnospondyli). In H. P. Schultze and L. Trueb (eds.) *Origins of the Major Groups of Tetrapods: Controversies and Consensus*, pp. 223–313. Ithaca, NY: Cornell University Press.

Tyler, M. J. 2003. The gross anatomy of the reproductive system. In B. G. M. Jamieson (ed.) *Reproductive Biology and Phylogeny of Anura*, pp. 19–26. Enfield, UK: Science Publishers.

Usuda, H. 1993. Reproductive behavior of *Hynobius nigrescens*, with special reference to male midwife behavior. *Japanese Journal of Herpetology* 15, 86–90.

Usuda, H. 1997. Individual relationship of male aggressive behavior during the reproductive season of *Hynobius nigrescens*. *Japanese Journal of Herpetology* 17, 53–61.

Uzendoski, K., Maksymovitch, E., and Verrell, P. 1993. Do the risks of predation and intermale competition affect courtship behavior in the salamander *Desmognathus ochrophaeus*? *Behavioral Ecology and Sociobiology* 32, 421–427.

Jzzell, T. M. 1969. Notes on spermatophore production by salamanders of the *Ambystoma jeffersonianum* complex (Amphibia, Caudata). *Copeia*, 602–612.

'errell, P. A. 1982. The sexual behavior of the red-spotted newt, *Notophthalmus viridescens* (Amphibia: Urodela: Salamandridae). *Animal Behaviour* 30, 1224–1236.

'errell, P. A. 1983. The influence of the ambient sex ratio and intermale competition on the sexual behavior of the red spotted newt, *Notophthalmus viridescens* (Amphibia: Urodela: Salamandridae). *Behavioral Ecology and Sociobiology* 13, 307–313.

'errell, P. A. 1984a. Sexual interference and sexual defense in the smooth newt, *Triturus vulgaris* (Amphibia: Urodela: Salamandridae). *Zeitschrift für Tierpsychologie* 66, 242–254.

'errell, P. A. 1984b. Responses to different densities of males in the smooth newt, *Triturus vulgaris*: "One at a time, please." *Journal of Herpetology* 18, 482–484.

'errell, P. A. 1985. Is there an energetic cost to sex? Activity courtship mode and breathing in the red-spotted newt *Notophthalmus viridescens*. *Monitore Zoologico Italiano* 19, 121–128.

'errell, P. A. 1988. Sexual interference in the alpine newt, *Triturus alpestris* (Amphibia: Urodela: Salamandridae). *Zoological Science* 5, 159–164.

'errell, P. A. 1989. The sexual strategies of natural populations of newts and salamanders. *Herpetologica* 45, 265–282.

'errell, P. A. and Halliday, T. R. 1985. Reproductive dynamics of a population of smooth newts, *Triturus vulgaris*, in southern England. *Herpetologica* 41, 386–395.

'errell, P. A. and Krenz, J. D. 1998. Competition for mates in the mole salamander, *Ambystoma talpoideum*: tactics that may maximize male mating success. *Behaviour* 135, 121–138.

'errell, P. A. and McCabe, N. R. 1986. Mating balls in the common toad, *Bufo bufo*. *British Herpetological Society Bulletin* 16, 28–29.

errell, P. A. and McCabe, N. R. 1988. Field observations of the sexual behavior of the smooth newt, *Triturus vulgaris vulgaris* (Amphibia, Salamandridae). *Journal of Zoology (London)* 214, 533–545.

errell, P. A. and Pelton, J. 1996. The sexual strategy of the central long-toed salamander, *Ambystoma macrodactylum colombianum*, in south-eastern Washington. *Journal of Zoology (London)* 240, 37–50.

ieites, D. R., Nieto-Román, S., Barluenga, M., *et al.* 2004. Post-mating clutch piracy in an amphibian. *Nature* 431, 305–308.

/aights, V. 1996. Female sexual interference in the smooth newt, *Triturus vulgaris vulgaris*. *Ethology* 102, 736–747.

Wake, M. H. 1993. Evolution of oviductal gestation in amphibians. *Journal of Experimental Zoology* 266, 394–413.

Wake, M. H. 1997. Amphibian locomotion in evolutionary time. *Zoology* 100, 141–151.

Wake, M. H. and Dickie, R. 1998. Oviduct structure and function and reproductive modes in amphibians. *Journal of Experimental Zoology* 282, 477–506.

Wells, K. D. 1977a. The social behaviour of anuran amphibians. *Animal Behaviour* 25, 666–693.

Wells, K. D. 1977b. Territoriality and male mating success in the green frog (*Rana clamitans*). *Ecology* 58, 750–762.

Wells, K. D. 2001. The energetics of calling in frogs. In M. J. Ryan (ed.) *Anuran Communication*, pp. 45–60. Washington, DC: Smithsonian Institution Press.

West-Eberhard, M. J. 1991. Sexual selection and social behavior. In M. H. Robinson and L. Tiger (eds.) *Man and Beast Revisited*, pp. 159–172. Washington, DC: Smithsonian Institution Press.

Whiteman, H. H. 1994. Evolution of facultative paedomorphosis in salamanders. *Quarterly Review of Biology* 69, 205–221.

Winne, C. T. and Ryan, T. J. 2001. Aspects of sex-differences in the expression of an alternative life cycle in the salamander *Ambystoma talpoideum*. *Copeia*, 143–149.

Wogel, H., Abrunhosa, P. A., and Pombal Jr., J. P. 2002. Atividade reprodutiva de *Physalaemus signifer* (Anura, Leptodactylidae) em ambiente temporário. *Inheringia* 92, 57–70.

Zamudio, K. R. and Sinervo, B. 2000. Polygyny, mate guarding, and posthumous fertilizations as alternative male mating strategies. *Proceedings of the National Academy of Sciences of the United States of America* 97, 14427–14432.

Zamudio, K. R. and Sinervo, B. 2003. Ecological and social contexts for the evolution of alternative mating strategies. In S. F. Fox, J. K. McCoy, and T. A. Baird (eds.) *Lizard Social Behavior*, pp. 83–106. Baltimore, MD: Johns Hopkins University Press.

Zardoya, R. and Meyer, A. 2000. Mitochondrial evidence on the phylogenetic position of caecilians (Amphibia: Gymnophiona). *Genetics* 155, 765–775.

Zimmerer, E. J. and Kallman, K. D. 1989. Genetic basis for alternative reproductive tactics in the pygmy swordtail, *Xiphiphorus nigrensis*. *Evolution* 43, 1298–1307.

Zuiderwijk, A. and Sparreboom, M. 1986. Territorial behavior in crested newt *Triturus cristatus* and marbled newt *T. marmoratus* (Amphibia, Urodela). *Bijdragen tot de Dierkunde* 56, 205–213.

12 · Alternative reproductive tactics in reptiles

RYAN CALSBEEK AND BARRY SINERVO

CHAPTER SUMMARY

In this chapter we explore the diversity of alternative reproductive tactics (ARTs) exhibited by reptiles. There is a rich literature on ARTs in a broad diversity of reptile lineages, and our contribution is therefore not an exhaustive one. Rather, we attempt to cover topics of general significance to many fields of study, including differences in male and female reproductive behavior, sex ratio adjustment and progeny gender manipulation, and the role of parthenogenesis in mating systems. Our goal is to provide a representative portrait of the diversity of tactics displayed among reptilian lineages, but we often illustrate more elaborate points using data from side-blotched lizards, *Uta stansburiana*, a system that we have been working on together for a decade. Throughout the chapter we emphasize the distinction between the proximate and the ultimate mechanisms that underlie the evolution of alternative tactics. We conclude with a brief discussion of potentially exciting future research directions in reptilian systems.

12.1 INTRODUCTION

Exploring alternative reproductive tactics (ARTs) in reptiles presents a great challenge given the diverse nature of these taxa. Modern reptilian lineages are paraphyletic with ancient histories. Some extinct reptilian groups such as the dinosaurs undoubtedly exhibited alternative reproductive tactics, given the documented dichotomy between precocial (Geist and Jones 1996, Varricchio *et al.* 1997) and altricial young (Horner 2000) in various dinosaur lineages. This dichotomy in the developmental tactics of young is strongly associated with the mating systems in the surviving descendants of dinosaurs, the birds. In addition, dinosaurs exhibited many social behavior patterns such as herding (Lockley *et al.* 2002) and communal nesting (Horner and Makela 1979). Finally, dinosaurs also exhibited elaborate

sexually dimorphic ornaments (e.g., hadrosaurs: Horner 2000) indicating the potential for strong sexual selection. The effects of social selection, mating system, sexually selected ornaments, and life-history tactics comprise the basic selective attributes that are conducive to the evolution of alternative reproductive tactics in males and females.

Modern reptile lineages include a tremendous diversity of reproductive tactics that span the entire gambit of tactics employed by most other vertebrate taxa. Here, we review the evolution of alternative reproductive tactics of males and females for the extant reptilian lineages (squamates, turtles, and crocodilians) and compare and contrast these patterns with other groups (Table 12.1). We also show how the interaction between alternative male and female reproductive tactics generates a variety of cryptic and active choice mechanisms in females and may select for adaptive sex ratio adjustment. Finally, we point toward important future directions for studies of alternative reproductive tactics in reptiles.

12.2 ALTERNATIVE MALE TACTICS

True monogamy is extremely rare among reptiles and throughout the animal kingdom in general. Molecular evidence from sleepy lizards, *Tiliqua rugosa*, has demonstrated high rates of mate fidelity (Gardner *et al.* 2002). Field studies of sleepy lizards in Australia have indicated that 75% of field-collected clutches had single sires, and no males were observed to have fertilized multiple females. However, males of most reptile species tend towards polygyny, and polygynous mating systems are far more conducive to the evolution of alternative mating strategies than are monogamous systems (though monogamous species are still subject to sexual selection) (Darwin 1871, Fisher 1958). This is because polygyny tends to increase the variance in reproductive success among males, and to cope with this

Alternative Reproductive Tactics, ed. Rui F. Oliveira, Michael Taborsky, and H. Jane Brockmann. Published by Cambridge University Press.
© Cambridge University Press 2008.

Table 12.1. *Taxonomic summary of the tactics described in this chapter*

Species	Tactic	Sex[a]	References[b]
Tiliqua rugosa	Monogamy	M, F	Gardner *et al.* 2002, How and Bull 2002
Urosaurus ornatus	Nomadic vs. territorial	M	Thompson and Moore 1991a, b
Uta stansburiana	Female mimicry and territoriality	M	Sinervo and Lively 1996, Sinervo and Calsbeek 2003, Sinervo and Clobert 2003, Calsbeek and Sinervo 2004
Thamnophis sirtalis	Female mimicry	M	Shine and Mason 2001, Shine *et al.* 2003
Crocodylus acutus, C. palustris	Parental care	F and rarely M	Lang *et al.* 1986, Platt and Thorbjarnarson 2000
Uta stansburiana	*r*- vs. *K*- life histories	F	Sinervo *et al.* 2000
Amphibolurus muricatus, Chelydra serpentina, Chrysemys picta, Uta stansburiana	Progeny gender manipulation	F	Packard *et al.* 1987, Harlow and Taylor 2000, Morjan and Janzen 2003, Calsbeek and Sinervo 2004
Lacerta vivipara	Oviparity vs. viviparity	F	De Fraipont *et al.* 1996
Cnemidophorus spp., *Heteronotia binoei, Lacerta* spp., *Ramphotyphlops braminus, Python molurus*	Parthenogenesis vs. sexuality	F	Moore and Crews 1986, Crews and Young 1991, Godwin *et al.* 1996, Rocha *et al.* 1997

[a] Indicates whether the tactic is employed by males, females, or both.
[b] Although our citation list is not exhaustive, it covers the majority of different reproductive tactics present among reptiles.

variance, males often adopt alternative solutions to the problem of obtaining mates (Andersson 1994). The degree of polygyny exhibited by different males is often related to their social status. For example, throat color is a reliable indicator of status in tree lizards (Thompson and Moore 1991a, Thompson *et al.* 1993) and is correlated with alternative reproductive behavior. Alternative throat colors serve as a badge (Thompson and Moore 1991b) to indicate relative behavioral differences among males. It is these behavioral differences among males that we focus on below in describing the evolution of alternative male reproductive tactics.

Perhaps the simplest set of alternative reproductive tactics among males is related to variation in dominance or aggression. The evolution of discrete variation in dominance behavior among males may manifest itself as female mimicry and sneaky behavior by males that are otherwise unable

to compete with larger, older males or males of higher status. Although the proximate mechanisms leading to the evolution of these alternatives are beyond the scope of this chapter, we refer interested readers to recent discussions of these mechanisms by Sinervo and Calsbeek (2003). Briefly, perturbations of endocrine pathways that mediate aggressive behavior and dominance may result in the build-up of correlations between the genes for hormone expression and the genes for morphology and behavior. For example, genetic combinations of bright colors and aggressive behavior become coupled in one set of individuals (i.e., morphs) while at the same time, combinations of cryptic color patterns and furtive behavior become coupled in others. When correlations become more obvious and fixed as discrete alternative strategies, selection will continually refine these alternative suites of complementary traits through the action of frequency-dependent selection (Sinervo *et al.* 2000).

Female mimicry by male reptiles may take a variety of forms. For example, morphological and behavioral attributes of sneaky, male side-blotched lizards, *Uta stansburiana*, are similar in many respects to those of female *Uta* (Sinervo and Lively 1996). Sneaky males are often smaller than the two other territorial morphs of males in this group. Sneaky males tend to be more cryptic, are more likely to freeze than flee when threatened, and have yellow throat colors similar to female lizards. Finally, sneaky male *Uta* often have the distended abdomen typical of gravid females burdened with a clutch of eggs. These males tend not to defend territories but choose rather to float around the territories of larger, more dominant males and steal copulations by subterfuge (Zamudio and Sinervo 2000). Yellow males sneak onto the territories of orange males and copulate with the females residing there.

A yellow-throated sneaker male can transform into a territorial blue-throated male during a single reproductive season, and the proximate control is related to elevated levels of the gonadal steroid testosterone (Sinervo *et al.* 2000). However, the ability of sneaker males to transform from the yellow- to blue-throated male strategy is restricted to males with a *by* genotype at the OBY locus (i.e., carrying both *b* and *y* color alleles). The OBY locus of side-blotched lizards is named for the distinctive orange, blue, or yellow colors on the throats of both males and females (Sinervo *et al.* 2000, Sinervo and Clobert 2003). In contrast to the developmental plasticity exhibited in *by* male genotypes, sneaker males with the *yy* genotype are fixed for the sneaker strategy throughout life. The environmental cue for transformation from yellow to blue in *by* males is related to the presence versus the absence of orange neighbors (Sinervo 2001). When orange males (i.e., with genotypes *oo*, *bo*, and *yo*) are present as nearest neighbors, *by* males remain as sneaker, but if orange males are absent in the social neighborhood, *by* males transform to the territorial, blue-throated males (B. Sinervo, L. Hazard, D. Costa, and K. Nagy, unpublished data). Such a strategy is highly adaptive (Sinervo 2001) as *by* sneaker males can cuckold orange males if they are present (Zamudio and Sinervo 2000), but *by* males exhibiting the territorial strategy of a blue male can gain fitness against other sneaker males by mutually defending territory with other blue males (Sinervo and Clobert 2003; and see below).

In a closely related group, tree lizards, *Urosaurus ornatus*, adopt either satellite or nomadic strategies depending on different environmental conditions (Moore *et al.* 1998). Satellite males are sedentary and reside around the territory boundaries of other males. However, under stressful conditions (e.g., drought years), satellite males become nomadic wanderers. The behavioral switch appears to be mediated by differential sensitivity to progesterone (Moore *et al.* 1998) and is thus at least partially under the influence of plastic hormonal regulation.

Thus, while nonterritorial male strategies exist in both *Uta* and *Urosaurus*, *Uta* has both a fixed genetic alternative tactic (*yy* genotype) and a condition-dependent alternative tactic (*by* genotype), while in *Urosaurus* the sneaker strategy is conditionally determined based on environmental (stress) conditions (Knapp *et al.* 2003). Curiously, morphs using similar tactics in *Uta* and *Urosaurus* have different body sizes and throat colors (Sinervo *et al.* 2000). Nonterritorial *Urosaurus* morphs have orange throat fans and large body size (Hews *et al.* 1994), whereas the large-bodied, orange-throated male phenotype is characteristic of the most aggressive territorial strategy in *Uta* (Calsbeek *et al.* 2002). Thus, while the proximate mechanisms for hormonal regulation and color expression are conserved across reptile lineages, it appears that different associations between specific colors (e.g., sneaker yellow in *Uta* and sneaker orange in *Urosaurus*) and alternative behavioral strategies (e.g., aggressive orange in *Uta* and aggressive blue-orange in *Urosaurus*: Thompson and Moore 1991a, b) are possible in different taxa. This implies that color evolution is not constrained by the evolution of hormonal and behavioral strategies.

Other forms of female mimicry may be chemical rather than morphological. Alternative snake reproductive tactics include a chemical signal that mimics that of sexually receptive females (Shine and Mason 2001, Shine *et al.* 2003). Males intertwined in the confines of high-density hibernacula are lured to female-mimicking males by the chemical guise. The female mimics have newly emerged from overwintering sites and thus have a body temperature of only a few degrees above zero Celsius. The mimicry is thought to confer an advantage in natural selection not sexual selection. The active and warm-bodied males are attracted to the cold, female-mimicking males and thereby unintentionally warm them with their advances, allowing the male mimics an opportunity to gain a thermal advantage.

Alternative behavioral strategies in snakes are not limited to cases of female mimicry. Some male garter snakes (*Thamnophis sirtalis parietalis*) leave a gelatinous plug in the female's cloaca following insemination (Shine *et al.* 2000). Plugs serve to prevent other males from copulating with the female by physically blocking intromission. The cloacal plugs are effective blocks to repeated copulation by the female for only a few days, but this appears to be long enough for the female's sexual receptivity to abate, consequently enhancing the inseminating male's fitness (Shine et al. 2000).

Sneaky mating tactics aside, alternative territorial strategies may also result in differential access to females. Examples of alternative strategies for territoriality are numerous and form the backbone of a rich history of studies in behavioral ecology. Reptiles provide examples of resource defense polygyny (e.g., *Uta palmeri*: Hews 1988), lekking behavior patterns (e.g., marine iguanas: Wikelski et al. 1996), and alternative usurping and defending behavior patterns (e.g., *Uta stansburiana*: Calsbeek et al. 2002). Although not all forms of territoriality are associated with alternative reproductive tactics within a species, the despotic nature of territoriality, and indeed resource defense in general, by definition can lead to alternative fitness strategies. For example, our recent work has uncovered the existence of two alternative strategies for territoriality: usurp and defend (Calsbeek et al. 2002). Usurper males tend to be large, aggressive, and in the case of *Uta stansburiana*, tend to possess an orange throat. Usurper males target high-quality territories (e.g., high female densities) and attempt to take them over from neighbors. Defender males are, on average, smaller, less aggressive than usurpers, and in the case of *U. stansburiana*, possess a blue throat. Defenders usually defend territories near their natal site containing only a few females. Orange-throated usurpers often have large harems of up to seven females and typically obtain access to many of these females by defeating neighboring, blue-throated defenders in contests over territory. Orange-throated males, however, often have a higher mortality rate (Sinervo and Lively 1996) and, in some situations, may sire no progeny despite having potentially very high siring success in some situations (Calsbeek et al. 2002). Thus, while the usurper strategy has a potentially high reproductive pay-off through siring success, the mortality risks from male–male competition that are involved in territory takeover result in a highly variable outcome. By contrast, the more monogamous strategy, defend, has a lower average reproductive pay-off but is evolutionarily stable given its significantly lower variance in outcome.

Within the defender territorial strategy of blue-throated, male (*bb* genotype) side-blotched lizards, we have uncovered a further distinction that allows some males to disperse and settle beside a genetically similar neighbor, cooperatively defend their adjoining territories from sneaker males, and thereby achieve high fitness. Based on detailed maternal and paternal pedigrees, it is clear that cooperative *bb* males are not kin but are nonetheless genetically similar. Mutual attraction and cooperation between such blue defenders arise in part from the throat color locus (only males with a *bb* genotype cooperate). The mutual attraction and high fitness through cooperation exhibited by *bb* genotypes that are genetically similar are complemented by *oo*, *bo*, and *yo* genotypes (i.e., O strategy) that have low fitness when they settle next to genetically similar males (Sinervo and Clobert 2003). Thus, O phenotypes are hyperdispersed with respect to genetic similarity in contrast to genetically similar *bb* genotypes that are aggregated.

The added requirement of genetic similarity at loci that are unlinked to the OBY locus of side-blotched lizards implies that recognition is also due to many factors across the genome other than throat color alone that are responsible for repulsion and attraction behavior of males. Furthermore, because male fitness is contingent upon the presence or absence of a genetically similar but unrelated neighbor, the *bb* genotype either reflects a greenbeard allele or a case of group selection (Sinervo and Clobert 2003). Pedigree information indicates that none of the cooperative blue males that have high genetic similarity are kin; thus, kin selection cannot explain the evolution of cooperation between blue neighbors. The presence of *b* alleles in both males, along with the high pay-off from social cooperation, satisfies the two conditions for greenbeard alleles. Dawkins (1976) hypothesized that alternative tactics of social behavior might evolve from genic selection if such greenbeard alleles confer both recognition of individuals carrying the same alleles (e.g., the attraction of *bb* male *Uta* at settlement) and enhancement of the likelihood of cooperation (e.g., *bb* males cooperate to defend their females against sneaker yellow males). Cooperation among *bb* males serves to further reduce the variance in their reproductive success and contributes to their stability with usurper males.

12.3 VARIANCE IN REPRODUCTIVE SUCCESS WITHIN STRATEGIES

So what are the consequences of the difference in the variance in reproductive success within male reproductive tactics of *Uta*? Bateman (1948) explained why females were generally the choosier sex by showing that the variance in male reproductive success was much greater than that of females. Males in nature are generally limited by access to large numbers of females, while females are more likely to be limited by the numbers of eggs which they can produce; ergo the adage that "sperm is cheap" (but see Olsson et al. 1997). In the case of reptiles, however, alternative reproductive tactics generate intrasexual variance in reproductive success that cannot be explained by Bateman's principle alone. In the case of usurper/defender and cooperation dynamics, the resulting intrasexual variance in reproductive success may lead to stronger sexual selection on some male tactics compared with others. For example, orange-throated males are more strongly sexually selected in the sense that there is higher variation in reproductive success within the orange morph. In contrast, blue-throated males have additional components of variation in reproductive success that arise from "social selection" (West-Eberhard 1983). For example, bb males that find a genetically similar partner have much higher fitness than bb males that are unable to find a genetically similar neighbor (Sinervo and Clobert 2003).

Whether similar intrasexual variation in fitness exists within females has yet to be convincingly demonstrated, but such studies could provide a fruitful direction for future research. Certainly distinguishing between the sexual (e.g., male–male competition, female choice) versus social (e.g., cooperation) agents of selection that generate variation in reproductive success in alternative male strategies may provide a key to understanding the environmental circumstances under which discrete alternative tactics evolve.

12.4   ALTERNATIVE FEMALE TACTICS

Female mating behavior patterns can be as diverse as those seen in males. The same tactical distinction between single and multiple mates can be made for female reptiles. Indeed, polyandry may be the most common mating system for reptiles. However, having one versus multiple partners in both sexes poses a different problem in reptiles compared with studies of other taxa such as mammals (Ciszek 2000) and birds (e.g., dunnocks: Davies and Lundberg 1984, Davies 1985). The classical explanation for polyandry in these taxa is that it is a socially mediated switch that is often related to the amount of parental care given by males. For example, male dunnocks provide extensive care of young at the nest, freeing females to alternatively visit the nest sites of several males. However, few extant reptiles exhibit paternal care or elaborate levels of maternal care after oviposition or birth with a few minor exceptions. The exceptions include crocodilians, in which there is often female (Platt and

Thorbjarnarson 2000) and occasionally male (Lang et al. 1986) protection until after hatching; skinks, where some nest guarding has been observed (Duffield and Bull 2002); and live-bearing crotalids, where Greene (1988) has reported guarding of progeny. Alternative reproductive tactics in female reptiles are generally related to variation in yolk-provisioning strategies (Sinervo 1993, 1994), egg retention, or variation in the degree of viviparity (see below). However, explaining the evolution of polyandry as an alternative mating tactic in reptiles requires explanations that go beyond differences in paternal versus maternal care.

Given the generally low levels of parental care among reptiles, there are many other factors that could potentially drive the evolution of polyandry in reptiles besides levels of paternal versus maternal investment. The first factor related to the existence of alternative reproductive tactics in females, which relaxes the strength of Bateman's principle relative to other social pressures. Polyandry in reptiles, which is related to alternative female reproductive tactics, is driven by the phenomenon of social selection, which leads to the evolution of bright ornaments in females. Most female side-blotched lizards, for example, display one of two alternative throat colors, orange or yellow. The alternative throat colors are each associated with alternative reproductive tactics that are expressed as different life-history strategies. Females compete for territory space and females with bright orange throats secure the best territories but only when their strategy is both rare and at low density. Orange-throated females lay large clutches of small eggs (i.e., orange females are K strategists); yellow-throated females lay small clutches of large eggs (i.e., r strategists). These two alternative life-history strategies are each favored every other year owing to oscillations in population density, that increase the recruitment of orange female progeny in low-density years and of yellow female progeny in high-density years (Sinervo et al. 2000).

Alternative, female life-history strategies should also have associated preferences for alternative male genotypes (Alonzo and Sinervo 2001), due to the interplay between alleles for male and female throat color, population density, and progeny sex. Because only one female morph has high fitness, depending on whether the population is at low density (e.g., orange) or high density (e.g., yellow), the variance in reproductive success of females may be large – similar to that normally observed in males. It is possible that alternative female morphs will therefore also be under selection for different levels of polyandry. This is likely to be the case for Uta females, and it could be a potentially important aspect of sexual selection in other reptile mating systems.

Given the potential for active female choice in reptiles, the importance of alternative female preferences as agents of sexual selection cannot be overstated. There is abundant evidence that female preferences have important consequences for reptile evolution. Monogamy is notoriously rare among reptiles but, as noted above, appears generally to be the case for sleepy lizards (How and Bull 2002). Most female lizards mate with multiple males and the evolutionary explanations for this phenomenon are complex. Polyandry can increase genetic diversity within clutches of eggs, which may be especially important in small populations (Yasui 2001). Female *Uta* are among the most promiscuous amniote vertebrates ever studied. Estimates from field-caught gravid females indicate that at least 80% of female side-blotched lizards mate with multiple males. This is necessarily an underestimate, since many females will have mated with several males but may have subsequently chosen to use the sperm from a single sire to fertilize her eggs. In extreme cases, a female may lay five eggs with five unique sires (Zamudio and Sinervo 2000).

High levels of polyandry in side-blotched lizards may be explained either by the intense competition among males for mate acquisition (Sinervo and Lively 1996, Sinervo et al. 2000) or by an adaptive female preference for multiple mates. Mating with males of different throat colors gives females an opportunity to produce a diverse array of progeny genotypes to face a variety of social situations in a very dynamic social environment (Sinervo et al. 2001). In many cases, females can use the sperm from males long after the actual act of copulation. For side-blotched lizards, this temporal separation between copulation and fertilization has been observed over periods of several months (Zamudio and Sinervo 2000); in some snakes and turtles, sperm may remain viable for several years (Seigel and Ford 1987).

2.5 ADAPTIVE GENDER ADJUSTMENT

Mating with multiple sires also affords females the opportunity to engage in cryptic choice for different sperm (Eberhard 1996). Cryptic reproductive tactics in reptiles are known to include selection to avoid inbreeding in snakes (Madsen et al. 1992) or to alleviate the costs of ontogenetic conflict between the sexes (Calsbeek and Sinervo 2002, Calsbeek and Sinervo 2004). For example, when certain genes have alternative fitness optima in males and females,

natural and sexual selection at those loci will generate ontogenetic conflict between the sexes (Rice and Chippindale 2001, Rice and Holland 1997). This form of genetic conflict is typified by *Drosophila*, in which genetic variation for male and female life-history traits has opposing fitness optima. Ontogenetic conflict can be viewed as life-history trade-offs between sexes or between juvenile and adult phases (Sinervo and Calsbeek 2003). Conflict can be alleviated by sex-limited expression of traits that are under conflicting selection regimes (Chippindale et al. 2001) For example, the sex specificity of male versus female development can evolve via sex steroids specific to male (e.g., testosterone) versus female traits (e.g., estrogen) (Sinervo and Calsbeek 2003). Likewise, ontogenetic conflict between juvenile and adult traits is alleviated by the steroid control over the development of secondary sexual traits at maturity.

Another way to alleviate ontogenetic conflict would be for females to sort male benefit/female detriment genes into sons and female benefit/male detriment genes into daughters. Side-blotched lizards employ just such a strategy to produce both high-quality sons and daughters (Calsbeek et al. 2002, Calsbeek and Sinervo 2004). Many reptiles are sexually dimorphic in body size, and in territorial species, large male body size would confer an advantage in male–male conflict. Female side-blotched lizards sort sperm from large sires to produce sons of high quality, and at the same time, they sort sperm from small males into daughters of high quality (e.g., survival to maturity). The ability to sort sperm implies that females recognize sperm genotypes on the basis of X- and Y-chromosomes. Sperm recognition also appears to be possible in mice (Vacquier 1995), and some other reptiles have also shown evidence for adaptive sperm-sorting behavior (Madsen et al. 1992, Olsson et al. 1996). However, the generality of sperm sorting and subsequent adaptive sex-ratio distortion still needs to be explored in other reptile groups.

Sex-ratio adjustment can also occur via environmentally induced maternal effects with adaptive consequences. Progeny gender is under environmental influence in numerous species of reptiles (Harlow 2000, Harlow and Taylor 2000, Elf et al. 2002, Milnes et al. 2002, Shine et al. 2002) and is thought to have an adaptive explanation (Shine 1999). For example, turtles are able to manipulate progeny sex by varying the depth at which eggs are buried in nests (Packard et al. 1987, Morjan and Janzen 2003). The adaptive significance of environmental sex determination has been a subject of much debate (reviewed in Shine 1999), but it is generally thought that females can maximize their fitness by giving progeny an opportunity to develop into the gender that will perform best given the environmental conditions. One problem with this argument is that temperature-dependent sex determination will necessarily lead to a confounding of environmental effects and progeny gender effects. Shine (1999) points out, and we agree, that hormonal manipulations that override temperature effects will be an important next step towards understanding the fitness effects of sex-ratio adjustment in these taxa.

We also suggest that a female might manipulate the sex ratio of her clutch to ameliorate the potential ontogenetic conflict arising from both her genotype and the genotype of her sire. For example, in turtles, females are often the larger sex and thus a small-bodied female (i.e., due to genetic causes) might oviposit in soil with a temperature that will generate all-male broods. Conversely, a female that mated with a large male might oviposit in soil with a temperature regime that will generate all-female broods.

Incubation regimes have broad evolutionary significance to reptilian lineages that go beyond the sex determination of progeny. The evolution of alternative reproductive modes like viviparity and oviparity (Shine 1985) has occurred repeatedly within species of reptiles (De Fraipont et al. 1996). Within Lacerta vivipara, for example, viviparity has evolved independently four times (Surget-Groba et al. 2001). Often, shifts in reproductive mode have been linked to changes in thermal regime – specifically to latitudinal or elevational gradients in temperature that select for different optimal gestation times in viviparous forms or optimal egg retention in oviparous forms. At higher latitudes or elevations, egg-retention times may increase to facilitate embryonic development (Guillette et al. 1980, Guillette 1981), and egg morphology may similarly be under selection to increase nutrient exchange between mother and offspring (Heulin et al. 2002). A possible alternative tactic to increased retention time is egg guarding (Shine and Guillette 1988), a strategy that may represent a transitional stage from oviparity to viviparity or that may be part of a polymorphic set of alternatives to deal with the problem of juvenile development (De Fraipont et al. 1996).

12.6 BREAKING DOWN THE GENDER BARRIER IN ALTERNATIVE REPRODUCTIVE TACTICS

Several reptile lineages initially composed of both sexually active males and sexually active females undergo a sexual transition to clonal or parthenogenetic populations of females. Perhaps the best-known examples of parthenogenetic lizards are the whiptails, Cnemidophorus, found in the desert southwest of the United States and now also known to occur in South America (Rocha et al. 1997). In all-female populations of Cnemidophorus, alternative female reproductive behavior patterns arise during different stages of the ovulatory cycle. Pre-ovulatory females with high levels of plasma estrogen (Moore and Crews 1986) exhibit sexually receptive behavior directed towards other females (Godwin et al. 1996). In contrast, post-ovulatory females with high levels of progesterone will often mount pre-ovulatory females. Thus, pairs of females alternately engage in pseudo-copulatory behavior patterns that actually stimulate reproduction (Crews and Young 1991), which is thought to enhance clutch size in the wild.

Because sexual behavior in such parthenogenetic lizards might have an adaptive function (e.g., enhancing clutch size), such genes for sexual behavior are not expected to accumulate mutations as is often observed in other parthenogenetic species (e.g., Drosophila: Nuzhdin and Petrov 2003). If sexual behavior patterns were extinguished owing to a deleterious mutation in a given parthenogenetic clone of Cnemidophorus, such a clone would have a fitness disadvantage compared to a clone that had retained intact the genes for sexual behavior. The clones engaging in pseudo-copulation would have the attendant fitness benefits of enhanced clutch size through pseudo-copulatory stimulation of the endocrine system.

An interesting evolutionary interaction with sexual species arises from the retention of sexual behavior patterns. Parthenogenetic females should be capable of copulating with males of sexual lineages. The females of such asexual lineages already have a twofold advantage compared to sexual lineages because sexual females produce males that do not contribute to population growth. If parthenogenetic forms also engage in pseudo-copulation with sexual males, the males may become sperm depleted and thus sexual females may be sperm limited. Thus, the fitness of females from the sexual lineage would be reduced even further relative to asexual clones if males became limiting.

Such a three-sex system is, in fact, a highly unstable evolutionary game that makes the asexual lineages even more prone to extinction without such sexual interactions. Maynard Smith (1978) formulated such a three-sex system to address the question of why there are only two sexes. Any asymmetry in fitness between two of the sexes would serve to eliminate one of the sexes; thus, only two-sex systems or one-sex systems are evolutionarily stable. *Cnemidophorus* provides an interesting evolutionary example of this idea. It would be very interesting to know whether sperm becomes limiting in male *Cnemidophorus* in areas where they contact parthenogenetic lineages.

Asexual reproduction is also prevalent in other lizards (Moritz 1983, Murphy *et al.* 2000) and snakes (Nussbaum 1980, Groot *et al.* 2003). Why should asexuality be such a prevalent reproductive tactic? The fitness benefits of clonal reproduction are clear. Clonal organisms that produce only daughters can reproduce at double the rate of their sexually reproducing counterparts. However, because clonal lineages cannot purge mutations via recombination, asexual species are doomed to an evolutionary dead end. The predictable outcome of all asexual lineages is extinction (Lynch and Gabriel 1990, Howard and Lively 1998), either from mutation accumulation or by parasites that target the common and genetically homogeneous asexual clones. Whether effects as basic as mutation accumulation (e.g., reduced fertility, etc.) or susceptibility to parasitic attack are features of reptilian parthenogenetic forms is currently unknown and would be very interesting avenues for investigation.

12.7 CONCLUSIONS

In sum, the phenomenal diversity exhibited by reptilian mating systems should make them a primary target for future studies aimed at understanding the evolution of reproductive tactics. Reptiles represent a continuum of alternative reproductive tactics. The group is ideal for studying mating behavior in natural environments since many taxa are highly tractable owing to, for example, high degrees of territoriality (e.g., Stamps 1994, Stamps and Krishnan 1994a, b, 1995) or, more generally, low dispersal distances (Sinervo and Clobert 2003). Fitness components can be assessed with relative ease because both male and female reproductive success is often readily measured both in the field and laboratory (Shine and Greer 1991, Olsson *et al.* 1994, 1996, Zamudio and Sinervo 2000). Together, these traits should facilitate many major advances on the subject of the evolution of alternative reproductive tactics. We suggest that two of the most fruitful avenues for immediate research should be to understand the differential selection pressures generated by intrasexual variance in reproductive success among tactics and to explore the potential impact of cryptic choice (Eberhard 2000) under genetic and environmental sex determination.

References

Alonzo, S. H. and Sinervo, B. 2001. Mate choice games, context-dependent good genes, and genetic cycles in the side-blotched lizard, *Uta stansburiana*. *Behavioral Ecology and Sociobiology* 49, 176–186.

Andersson, M. 1994. *Sexual Selection*. Princeton, NJ: Princeton University Press.

Bateman, A. J. 1948. Intrasexual selection in *Drosophila*. *Heredity* 2, 349–368.

Calsbeek, R. and Sinervo, B. 2002. Uncoupling direct and indirect components of female choice in the wild. *Proceedings of the National Academy of Sciences of the United States of America* 99, 14897–14902.

Calsbeek, R. and Sinervo, B. 2004. Progeny sex is determined by relative male body size within polyandrous females' clutches: cryptic mate choice in the wild. *Journal of Evolutionary Biology* 17, 464–470.

Calsbeek, R., Alonzo, S. H., Zamudio, K., and Sinervo, B. 2002. Sexual selection and alternative strategies generate demographic stochasticity in small populations. *Proceedings of the Royal Society of London B* 269, 157–164.

Chippindale, A. K., Gibson, J. R., and Rice, W. R. 2001. Negative genetic correlation for adult fitness between sexes reveals ontogenetic conflict in *Drosophila*. *Proceedings of the National Academy of Sciences of the United States of America* 98, 1671–1675.

Ciszek, D. 2000. New colony formation in the "highly inbred" eusocial naked mole-rat: outbreeding is preferred. *Behavioral Ecology* 11, 1–6.

Crews, D. and Young, L. J. 1991. Pseudocopulation in nature in a unisexual whiptail lizard. *Animal Behaviour* 42, 512–514.

Darwin, C. 1871. *The Descent of Man and Selection in Relation to Sex*. London: John Murray.

Davies, N. B. 1985. Cooperation and conflict among dunnocks, *Prunella modularis*, in a variable mating system. *Animal Behaviour* 33, 628–648.

Davies, N. B. and Lundberg, A. 1984. Food distribution and a variable mating system in the dunnock, *Prunella modularis*. *Journal of Animal Ecology* 53, 895–912.

Dawkins, R. 1976. *The Selfish Gene*. Oxford, UK: Oxford University Press.

De Fraipont, M., Clobert, J., and Barbault, R. 1996. The evolution of oviparity with egg guarding and viviparity in lizards and snakes: a phylogenetic analysis. *Evolution* 50, 391–400.

Duffield, G. A. and Bull, C. M. 2002. Stable social aggregations in an Australian lizard, *Egernia stokesfi*. *Naturwissenschaften* **89**, 424–427.

Eberhard, W. G. 1996. *Female Control: Sexual Selection by Cryptic Female Choice*. Princeton, NJ: Princeton University Press.

Eberhard, W. G. 2000. Criteria for demonstrating postcopulatory female choice. *Evolution* **54**, 1047–1050.

Elf, P. K., Lang, J. W., and Fivizzani, A. J. 2002. Dynamics of yolk steroid hormones during development in a reptile with temperature-dependent sex determination. *General and Comparative Endocrinology* **127**, 34–39.

Fisher, R. A. 1958. *The Genetical Theory of Natural Selection*. New York: Dover.

Gardner, M. G., Bull, C. M., and Cooper, S. J. B. 2002. High levels of genetic monogamy in the group-living Australian lizard *Egernia stokesii*. *Molecular Ecology* **11**, 1787–1794.

Geist, N. R. and Jones, T. D. 1996. Juvenile skeletal structure and the reproductive habits of dinosaurs. *Science* **272**, 712–714.

Godwin, J., Hartman, V., Grammer, M., and Crews, D. 1996. Progesterone inhibits female-typical receptive behavior and decreases hypothalamic estrogen and progesterone receptor messenger ribonucleic acid levels in whiptail lizards (Genus *Cnemidophorus*). *Hormones and Behavior* **30**, 138–144.

Greene, H. W. 1988. Antipredator mechanisms in reptiles. In C. Gans and R. B. Huey (eds.) *Biology of the Reptilia*, vol. 16, pp. 1–152. New York: Wiley-Liss.

Groot, T. V. M., Bruins, E., and Breeuwer, J. A. J. 2003. Molecular genetic evidence for parthenogenesis in the Burmese python, *Python molurus bivittatus*. *Heredity* **90**, 130–135.

Guillette, L. J. 1981. On the occurence of oviparous and viviparous forms of the Mexican lizard *Sceloporu aeneus*. *Herpetologica* **37**, 11–15.

Guillette, L. J., Jones, K., Fitzgerald, T., and Smith, H. M. 1980. Evolution of viviparity in the lizard genus *Sceloporus*. *Herpetologica* **36**, 201–215.

Harlow, P. S. 2000. Incubation temperature determines hatchling sex in Australian rock dragons (Agamidae: Genus *Ctenophorus*). *Copeia*, 958–964.

Harlow, P. S. and Taylor, J. E. 2000. Reproductive ecology of the jacky dragon (*Amphibolurus muricatus*): an agamid lizard with temperature-dependent sex determination. *Austral Ecology* **25**, 640–652.

Heulin, B., Ghielmi, S., Vogrin, N., Surget-Groba, Y., and Guillaume, C. P. 2002. Variation in eggshell characteristics and in intrauterine egg retention between two oviparous clades of the lizard *Lacerta vivipara*: insight into the oviparity–viviparity continuum in squamates. *Journal of Morphology* **252**, 255–262.

Hews, D. K. 1988. Resource defense and sexual selection on male head size in the lizard *Uta palmeri*. *American Zoologist* **28**, A52.

Hews, D. K., Knapp, R., and Moore, M. C. 1994. Early exposure to androgens affects adult expression of alternative male types in tree lizards. *Hormones and Behavior* **28**, 96–115.

Horner, J. R. 2000. Dinosaur reproduction and parenting. *Annual Review of Earth and Planetary Sciences* **28**, 19–45.

Horner, J. R. and Makela, R. 1979. Nest of juveniles provides evidence of family structure among dinosaurs. *Nature* **282**, 296–298.

How, T. L. and Bull, C. M. 2002. Reunion vigour: an experimental test of the mate guarding hypothesis in the monogamous sleepy lizard (*Tiliqua rugosa*). *Journal of Zoology (London)* **257**, 333–338.

Howard, R. S. and Lively, C. M. 1998. The maintenance of sex by parasitism and mutation accumulation under epistatic fitness functions. *Evolution* **52**, 604–610.

Knapp, R., Hews, D. K., Thompson, C. W., Ray, L. E., and Moore, M. C. 2003. Environmental and endocrine correlates of tactic switching by nonterritorial male tree lizards (*Urosaurus ornatus*). *Hormones and Behavior* **43**, 83–92.

Lang, J. W., Whitaker, R., and Andrews, H. 1986. Male parental care in mugger crocodiles. *National Geographic Research* **2**, 519–525.

Lockley, M., Schulp, A. S., Meyer, C. A., Leonardi, G., and Mamani, D. K. 2002. Titanosaurid trackways from the Upper Cretaceous of Bolivia: evidence for large manus, wide-gauge locomotion and gregarious behaviour. *Cretaceous Research* **23**, 383–400.

Lynch, M. and Gabriel, W. 1990. Mutation load and the survival of small populations. *Evolution* **44**, 1725–1737.

Madsen, T., Shine, R., Loman, J., and Hakansson, T. 1992. Why do female adders copulate so frequently? *Nature* **355**, 440–442.

Maynard Smith, J. 1978. *The Evolution of Sex*. Cambridge, UK: Cambridge University Press

Milnes, M. R., Roberts, R. N., and Guillette, L. J. 2002. Effects of incubation temperature and estrogen exposure on aromatase activity in the brain and gonads of embryonic alligators. *Environmental Health Perspectives* **110**, 393–396.

Moore, M. C. and Crews, D. 1986. Sex steroid hormones in natural populations of a sexual whiptail lizard, *Cnemidophorus inornatus*, a direct evolutionary ancestor of a unisexual, parthenogenic lizard. *General and Comparative Endorcinology* 63, 424–430.

Moore, M. C., Hews, D. K., and Knapp, R. 1998. Hormonal control and evolution of alternative male phenotypes: generalizations of models for sexual differentiation. *American Zoologist* 38, 133–151.

Moritz, C. 1983. Parthenogenesis in the endemic Australian lizard *Heteronotia binoei* (Gekkonidae). *Science* 220, 735–737.

Morjan, C. L. and Janzen, F. J. 2003. Nest temperature is not related to egg size in a turtle with temperature-dependent sex determination. *Copeia*, 366–372.

Murphy, R. W., Fu, J. Z., Macculloch, R. D., Darevsky, I. S., and Kupriyanova, L. A. 2000. A fine line between sex and unisexuality: the phylogenetic constraints on parthenogenesis in lacertid lizards. *Zoological Journal of the Linnean Society* 130, 527–549.

Nussbaum, R. A. 1980. The Brahminy blind snake (*Ramphotyphlops braminus*) in the Seychelles Archipelago: distribution, variation, and further evidence for parthenogenesis. *Herpetologica* 36, 215–221.

Nuzhdin, S. V. and Petrov, D. A. 2003. Transposable elements in clonal lineages: lethal hangover from sex. *Biological Journal of the Linnean Society* 79, 33–41.

Olsson, M., Madsen, T., Shine, R., Gullberg, A., and Tegelstrom, H. 1994. Rewards of promiscuity. *Nature* 372, 230.

Olsson, M., Shine, R., Gullberg, A., Madsen, T., and Tegelström, H. 1996. Female lizards control paternity of their offspring by selective use of sperm. *Nature* 383, 585.

Olsson, M., Madsen, T., and Shine, R. 1997. Is sperm really so cheap? Costs of reproduction in male adders, *Vipera berus*. *Proceedings of the Royal Society of London B* 264, 455–459.

Packard, G. C., Packard, M. J., Miller, K., and Boardman, T. J. 1987. Influence of moisture, temperature, and substrate on snapping turtle eggs and embryos. *Ecology* (*Tempe*) 68, 983–993.

Platt, S. G. and Thorbjarnarson, J. B. 2000. Nesting ecology of the American crocodile in the coastal zone of Belize. *Copeia*, 869–873.

Rice, W. R. and Chippindale, A. K. 2001. Intersexual ontogenetic conflict. *Journal of Evolutionary Biology* 14, 685–693.

Rice, W. R. and Holland, B. 1997. The enemies within: intergenomic conflict, interlocus contest evolution (ICE),

and the intraspecific Red Queen. *Behavioral Ecology and Sociobiology* 41, 1–10.

Rocha, C. F. D., Bergallo, H. G., and Peccinini Seale, D. 1997. Evidence of a unisexual population of the Brazilian whiptail lizard genus *Cnemidophorus* (Teiidae), with description of a new species. *Herpetologica* 53, 374–382.

Seigel, R. A. and Ford, N. B. 1987. Reproductive ecology. In R. A. Seigel, J. T. Collins, and S. S. Novak (eds.) *Snakes: Ecology and Evolutionary Biology*, pp. 210–252. New York: Macmillan.

Shine, R. 1985. The evolution of viviparity in reptiles: an ecological analysis. In C. Gans and F. Billett (eds.) *Biology of the Reptilia*, vol. 0, pp. 00–00. New York: Wiley-Liss.

Shine, R. 1999. Why is sex determined by nest temperature in many reptiles? *Trends in Ecology and Evolution* 14, 186–189.

Shine, R. and Greer, A. E. 1991. Why are clutch sizes more variable in some species than in others? *Evolution* 45, 1696–1706.

Shine, R. and Guillette, L. J. 1988. The evolution of viviparity in reptiles: a physiological model and its ecological consequences. *Journal of Theoretical Biology* 132, 43–50.

Shine, R. and Mason, R. T. 2001. Courting male garter snakes (*Thamnophis sirtalis parietalis*) use multiple cues to identify potential mates. *Behavioral Ecology and Sociobiology* 49, 465–473.

Shine, R., Olsson, M. M., and Mason, R. T. 2000. Chastity belts in gartersnakes: the functional significance of mating plugs. *Biological Journal of the Linnean Society* 70, 377–390.

Shine, R., Elphick, M. J., and Donnellan, S. 2002. Co-occurrence of multiple, supposedly incompatible modes of sex determination in a lizard population. *Ecology Letters* 5, 486–489.

Shine, R., Phillips, B., Waye, H., LeMaster, M., and Mason, R. T. 2003. Chemosensory cues allow courting male garter snakes to assess body length and body condition of potential mates. *Behavioral Ecology and Sociobiology* 54, 162–166.

Sinervo, B. 1993. The effect of offspring size on physiology and life history: manipulation of size using allometric engineering. *BioScience* 43, 210–218.

Sinervo, B. 1994. Experimental tests of reproductive allocation paradigms. In L. J. Vitt and E. R. Pianka (eds.) *Lizard Ecology: Historical and Experimental Perspectives*, pp. 73–93. Princeton, NJ: Princeton University Press.

Sinervo, B. 2001. Selection in local neighborhoods, graininess of social environments, and the ecology of alternative strategies. In Editor (ed.) *Model Systems in Behavioral*

Ecology, pp. 191–226. Princeton, NJ: Princeton University Press.

Sinervo, B. and Calsbeek, R. 2003. Physiological epistasis, ontogenetic conflict and natural selection on physiology and life history. *Integrative and Comparative Biology* 43, 419–430.

Sinervo, B. and Clobert, J. 2003. Morphs, dispersal behavior, genetic similarity, and the evolution of cooperation. *Science* 300, 1949–1951.

Sinervo, B. and Lively, C. M. 1996. The rock–paper–scissors game and the evolution of alternative male reproductive strategies. *Nature* 380, 240–243.

Sinervo, B., Svensson, E., and Comendant, T. 2000. Density cycles and an offspring quantity and quality game driven by natural selection. *Nature* 406, 985–988.

Sinervo, B., Bleay, C., and Adamopoulou, C. 2001. Social causes of correlational selection and the resolution of a heritable throat color polymorphism in a lizard. *Evolution* 55, 2040–2052.

Stamps, J. 1994. *Territorial Behavior: Testing the Assumptions.* New York: Academic Press.

Stamps, J. A. and Krishnan, V. V. 1994a. Territory acquisition in lizards. 1. First encounters. *Animal Behaviour* 47, 1375–1385.

Stamps, J. A. and Krishnan, V. V. 1994b. Territory acquisition in lizards. 2. Establishing social and spatial relationships. *Animal Behaviour* 47, 1387–1400.

Stamps, J. A. and Krishnan, V. V. 1995. Territory acquisition in lizards. 3. Competing for space. *Animal Behaviour* 49, 679–693.

Surget-Groba, Y., Heulin, B., Guillaume, G. P., *et al.* 2001. Intraspecific phylogeography of *Lacerta vivipara* and the

evolution of viviparity. *Molecular Phylogenetics and Evolution* 18, 449–459.

Thompson, C. W. and Moore, M. C. 1991a. Syntopic occurence of multiple dewlap color morphs in male tree lizards. *Copeia* 1991, 493–503.

Thompson, C. W. and Moore, M. C. 1991b. Throat colour reliably signals status in male tree lizards, *Urosaurus ornatus*. *Animal Behaviour* 42, 745–754.

Thompson, C. W., Moore, I. T., and Moore, M. C. 1993. Social, environment and genetic factors in the ontogeny of phenotypic differentiation in a lizard with alternative male reproductive strategies. *Behavioral Ecology and Sociobiology* 33, 137–146.

Vacquier, V. D. 1995. Evolution of gamete recognition. *Science* 281, 1995–1998.

Varricchio, D. J., Jackson, F., Borkowski, J. J., and Horner, J. R. 1997. Nest and egg clutches of the dinosaur *Troodon formosus* and the evolution of avian reproductive traits. *Nature* 385, 247–250.

West-Eberhard, M. J. 1983. Sexual selection, social competition, and speciation. *Quarterly Review of Biology* 58, 155–183.

Wikelski, M., Carbone, C., and Trillmich, F. 1996. Lekking in marine iguanas: female grouping and male reproductive strategies. *Animal Behaviour* 52, 581–596.

Yasui, Y. 2001. Female multiple mating as a genetic bet-hedging strategy when mate choice criteria are unreliable. *Ecological Research* 16, 605–616.

Zamudio, K. and Sinervo, B. 2000. Polygyny, mate-guarding, and posthumous fertilizations as alternative male strategies. *Proceedings of the National Academy of Sciences of the United States of America* 97, 14427–14432.

13 · Alternative reproductive tactics in birds

OLIVER KRÜGER

CHAPTER SUMMARY

Birds, as one of the most-studied taxa among all organisms, have provided some of the best examples of alternative reproductive strategies and tactics. In this chapter, I first review the few cases where a genetic polymorphism has either been documented or is the most likely underlying cause for the alternative strategies observed. They range from the classic ruff example, where males either defend a territory as independents or try to obtain matings as nonterritorial satellites, to the different egg morphs described in the common cuckoo and plumage polymorphisms associated with different life-history strategies. In birds, commonly employed conditional strategies that increase fitness are intraspecific brood parasitism in females and extra-pair copulation behavior in males. Finally I discuss the interactions between the sexes, the influence these interactions have on strategies followed by one sex, and the importance of incorporating these interactions in future models. I also suggest reasons to explain why certain strategies, such as parasitic male reproductive behavior, are so rare in birds compared to other taxa.

13.1 INTRODUCTION

The evolution of alternative reproductive phenotypes is often believed to be associated with sexual selection (Neff 2001). Since the study of sexual selection and mate choice has figured prominently in ornithology over the last two decades (Andersson 1994), it comes as no surprise that a number of examples have been discovered in this taxon.

Gross's (1996) classification divides alternative reproductive phenotypes in birds into two groups. Within a population of a species, *alternative strategies* represent two or more genetic types having equal average fitness and an evolutionarily stable state frequency that is selected for by negative frequency-dependent selection. As in other taxa, alternative strategies are relatively rare among birds. *Conditional strategies*, by contrast, are commonly found and may be thought of as representing a single genotype. The two different phenotypes are referred to as alternative reproductive tactics that have unequal average fitness but equal fitness at an evolutionarily stable strategy switch point. This switch point may be the product of selection. Status-dependent selection is the mechanism by which an individual adopts a particular tactic depending on its own status relative to population variation in condition (Badyaev and Hill 2002). This classification has been criticized for relying on mechanism rather than behavior and for ignoring that, in many examples, genetic differences and environmental influences are not mutually exclusive (Brockmann 2001). Hence, it can be considered somewhat artificial, but I still find it the most useful for bringing some order to the examples that will follow while explicitly acknowledging its problems.

In this review, I consider alternative reproductive phenotypes to be a bimodal or trimodal distribution of phenotypes in terms of behavior, morphology, or life history. I do not consider continuously varying phenotypes, such as the size of a secondary sexual character like badge size (Qvarnström 1999), to be true alternative reproductive phenotypes.

Although alternative reproductive strategies and tactics are concerned with intrasexual variation (Gross 1996), the variety of alternative phenotypes encountered is, of course, quite different in females and males. Many alternative phenotypes in males have evolved to give their bearers better access to mates, whether it is through mimicking a female in plumage and behavior (Slagsvold and Sætre 1991) or through a completely different mating strategy (Lank *et al.* 1995). In contrast, females often use alternative reproductive tactics to boost their fertility (Brown and Brown 1998) or to optimize a given reproductive event in terms of probability of success (Gibbs *et al.* 2000) or costs

Alternative Reproductive Tactics, ed. Rui F. Oliveira, Michael Taborsky, and H. Jane Brockmann. Published by Cambridge University Press.
© Cambridge University Press 2008.

(Sasvári and Hegyi 2000). These differences between the sexes are a result of anisogamy (Parker *et al.* 1972); hence, they are not taxon specific. As in other taxa, such as fish and reptiles (Gross 1996, Sinervo and Lively 1996), dimorphism or polymorphism in bird coloration is often associated with alternative reproductive phenotypes.

13.2 ALTERNATIVE STRATEGIES

Gross's (1996) influential review of alternative reproductive phenotypes listed one avian example of alternative strategies. Since then, evidence for the existence of others has mounted. I consider six additional species to exhibit alternative strategies (Table 13.1).

The classic example remains male behavior in the lek mating system of the ruff (*Philomachus pugnax*) (see Box 13.1), for which the widely adopted term "satellite" was coined (Hogan-Warburg 1966). There are two types of males: *independent* males, which develop a dark plumage nape in the breeding season and defend a territory on the lek and *satellite* males, which possess a light plumage nape and move between independent males' territories. Lank *et al.* (1995) have shown that the inheritance of these two behavior patterns is indeed the result of a genetic polymorphism, comprising two alleles at a single locus that fixes the strategy for life. However, there is a further complexity in the form of a conditional strategy within the *independent* males: the majority of males are referred to as *marginals*, outnumbering the territorial independents by as much as 4 : 1 (Hogan-Warburg 1966, Widemo 1998). Lank *et al.* (1995) did not report any differences in growth rate between the two strategies; Bachman and Widemo (1999) found that satellite males had shorter tarsi and wings and were lighter than independent males. Satellite males are also rarer compared to independent males, comprising around 15–20% of the male population (Lank *et al.* 1995, Widemo 1998). In a 7-year study in Sweden, Widemo (1998) reported that satellite males achieved 9% of the observed copulations, lower than expected by their frequency in the population. This indicates either an unstable frequency of satellites in this population or, as the author states, that satellites obtain matings away from the lek or have a longer life expectancy. However, other studies reported that satellite males performed as many copulations as expected by their frequency in the population (Hogan-Warburg 1966, Rhijn 1991, Thuman 2003). It might be that the satellite strategy has lower benefits, but also lower costs (Widemo 1998). The maintenance of the two strategies appears to be

stabilized by female choice (Hugie and Lank 1997, Widemo 1998, Lank *et al.* 2002), but the lower frequency of satellite males consistently reported indicates unequal fitness if both strategies are equally common in a population. This consistent difference in the proportion of the two strategies within a population might indicate that negative frequency-dependent selection operates until the evolutionarily stable state frequency is reached where both strategies have equal average fitness. Why the equilibrium frequency sits at about 15% remains unclear. However, Hoglund *et al.* (1993) report differences in resident and satellite success as a function of lek size. Thus, the distribution of lek sizes may be involved in determining the equilibrium frequency.

An example of alternative strategies in females has recently been found in an obligate brood parasite, the common cuckoo (*Cuculus canorus*). Cuckoo females lay distinct egg types that match the eggs of some of their hosts that reject a nonmimetic egg (Davies 2000). On the basis of these egg types, the cuckoo is divided into gentes, and Gibbs *et al.* (2000) showed that these gentes are restricted to female lineages. Differentiation between gentes occurs in maternally inherited mitochondrial DNA, but not in nuclear DNA. Cuckoo females specialize on a host species (Marchetti *et al.* 1998) and the matching egg pattern is most likely determined by genes on the female-specific sex chromosome W (in birds, females are the heterogametic sex). Speciation as a consequence of this differentiation and specialization on a host species seems to be prevented by the cross-mating by males. The cuckoo also provides the best circumstantial evidence for frequency-dependent selection in any avian system with alternative strategies. Its fitness is strongly influenced by the host response to brood parasitism and there is good evidence that as parasitism rate in a population increases, so do host defenses and vice versa (Davies *et al.* 1996, Brooke *et al.* 1998).

An example where both sexes show alternative life history strategies has been described in the black-bellied seedcracker (*Pyrenestes ostrinus*). There are two distinct bill types in this species, a large-billed morph and a small-billed morph. The two morphs have different ecological niches; hence, there is less intraspecific competition (Smith 1987). The large-billed morph is adapted to feed on large, hard seeds while the small-billed morph feeds on smaller, soft seeds. Because the food of the two morphs is differently affected by precipitation, they follow alternative strategies in the initiation of nesting (Smith 1990): large-billed morphs nest early and small-billed morphs nest late in a season. This difference in nesting time shifts this otherwise trophic polymorphism to an alternative reproductive

Table 13.1. *Classification of alternative reproductive phenotypes in birds listing some conditional strategies*

Alternative strategies					Conditional strategy (alternative tactics)				
Species	Phenotypes	Sex[a]	Fitness[b]	References[c]	Species	Phenotypes	Sex[a]	Fitness[b]	References[c]
Ruff, *Philomachus pugnax*	Independent/satellite	M	=	Lank et al. 1995	ca. 240 species	Nest/nest and parasitize	F	≠	Eadie et al. 1998, Yom-Tov 2001
Common cuckoo, *Cuculus canorus*	Host gentes	F	?	Gibbs et al. 2000	Many species	Nest/nest and extra-pair copulation	M	≠	Petrie and Kempenaers 1998
Black-bellied seedcracker, *Pyrenestes ostrinus*	Large bill and early breeding/Small bill and late breeding	M, F	=	Smith 1987, Smith 1990	Pied flycatcher, *Ficedula hypoleuca*	Dominant/female mimic	M	≠	Slagsvold and Sætre 1991
White-throated sparrow, *Zonotrichia albicollis*	White stripe/tan stripe Different life histories	M, F	=	Tuttle 2003	Lazuli bunting, *Passerina amoena*	Dull/intermediate/bright	M	≠	Greene et al. 2000
Paradise flycatcher, *Terpsiphone mutata*	Rufous/white Different life histories	M	?	Mulder et al. 2002	House finch, *Carpodacus mexicanus*	Red and no parental care/yellow and parental care	M	≠	Badyaev and Hill 2002
Common buzzard, *Buteo buteo*	Dark/intermediate/light Different life histories	M, F	≠	Krüger et al. 2001	Tree sparrow, *Passer montanus*	Solitary breeding/colony	M	≠	Sasvári and Hegyi 2000
Arctic skua, *Stercorarius parasiticus*	Dark/intermediate/light Different life histories	M, F	=	O'Donald 1983			F		

[a] M and F denote alternative reproductive phenotypes in males and females, respectively.
[b] The symbols indicate whether the average fitness of the two alternative reproductive phenotypes is equal (=) or unequal (≠), or whether there is insufficient information (?).
[c] Numerals indicate key references where most of the data are provided.

345

strategy (see also Chapter 2). Apart from this seasonal effect, the morphs did not differ in clutch size, brood size, or predation risk (Smith 1990). Females of both bill morphs mate at random, and pairs consisting of both morphs also produce offspring with both morphs. This cross-mating prevents speciation occurring. The random mating behavior might be selected for by an unpredictable environment where neither morph enjoys a higher fitness over a lifetime.

Box 13.1   The mating strategies of the ruff

The unrivalled "star" among all bird species of alternative reproductive strategies is the ruff (*Philomachus pugnax*), a lekking sandpiper breeding across the Palearctic. Males arrive on leks before females and mating activity peaks in early May. Males follow two genetic strategies: they are either *independent* males that can defend a territory or *satellite* males that do not defend a territory. Lank *et al.* (1995) have provided evidence that this dichotomy in behavior is consistent with a single-locus, two-allele autosomal genetic polymorphism (Table 13.2). By rearing chicks from fathers following different mating strategies, they showed that the observed frequencies of the two strategies are in line with such a simple genetic model. Nested within the genetically fixed independent male strategy is a conditional strategy where males are either *residents* (holding a territory) or *marginals* (nonterritorial birds). Marginals try to become residents by ousting other residents from their territories or by setting up new ones, hence there are transitions between residents and marginals within and between breeding seasons. In the most detailed study so far, Widemo (1998) found that satellite males obtained only around 10% of the copulations even though they commonly comprise around 20% of the male population (Table 13.3). Interestingly, the differences in mating strategy are also reflected in morphological differences: independent males are larger than satellite males (Lank *et al.* 1995, Bachman and Widemo 1999) (Figure 13.1) and there are also differences in time budgets for foraging and aggression. Whereas independents invest in body mass and fat to increase their endurance capabilities as territory holders (either present or future), satellite males invest less in body mass and fat reserve, maybe to reduce the costs of flight (Bachman and Widemo 1999). Further studies are needed on this fascinating system to determine costs and benefits of the different strategies and tactics more precisely.

The white-throated sparrow (*Zonotrichia albicollis*) is another example where both sexes show alternative reproductive strategies (Knapton and Falls 1983, Houtman and Falls 1994, Tuttle 2003). As in ruffs, behavioral differences are correlated with a plumage marker, which is either a white or a tan stripe over the eye. Differences in behavior and plumage are controlled by a karyotypically visible chromosomal inversion (Thorneycroft 1975), making this the only case where a genetic marker for alternative behavioral types is known. In both sexes, the white morph is much more aggressive (Watt *et al.* 1984, Piper and Wiley 1989). White males show a lower level of parental care, seek more extra-pair copulations than tan males. White males settle in high-density areas to increase their chances of obtaining extra-pair copulations; tan males settle in low-density areas to minimize the risk of losing paternity to neighbors (Formica *et al.* 2004). These two territory settlement patterns do not result in differences in territory size or structure between white and tan males. White females pursue conspecific brood parasitism at a much higher rate and solicit copulations more often than tan females. Both morphs seem to have equal fitness because, although white males gain fitness through extra-pair copulations, they lose fitness through extra-pair copulations solicited by their own female and their own poor parental care. However, the crucial mechanism maintaining the polymorphism is that the two morphs show negative assortative mating, so far a unique mating pattern among birds (Lank 2002). This ensures that roughly equal frequencies of both morphs are produced in each breeding attempt. Assortative mating is strongly selected against because a double inversion resulting from a white-white pair could be lethal (Thorneycroft 1975), and trade-offs between parental care and extra-pair copulations would be unbalanced in assortatively mating pairs.

The Madagascar paradise flycatcher (*Terpsiphone mutata*) is another recent example of alternative strategies associated with a plumage dimorphism. Male paradise flycatchers occur as either a rufous or a white morph. Recent work by Mulder *et al.* (2002) has shown that these morphs are not due to seasonal dimorphism or delayed plumage maturation, but seem to be two consistent phenotypes and most likely genotypes. Data from chicks followed until plumage maturation and from trapped adults show that the white morph comprises about two-thirds of the males in this population. The two morphs seem to follow different life-history strategies – the white morph adopts adult breeding plumage by the age of 3 years; the rufous morph between

Table 13.2. *Autosomal models for the inheritance of male mating strategies in ruff*

Father	Maternal grandfather[b]	n	Proportion of independent male offspring[a]		
			Expected (Ss)	Expected (Ii)	Observed
I	I	6	0.96	0.92	0.83
I	U	19	0.92	0.89	1.00
I	S	4	0.70	0.80	0.75
S	I + U	19	0.44	0.61	0.62
S	S	11	0.33	0.30	0.36

[a] Satellite dominance (Ss), independent dominance (Ii).
[b] I, independent; U, unknown; S, satellite.
Source: After Lank *et al.* (1995) with permission.

Table 13.3. *Percentage of male ruffs of different strategies/tactics and their corresponding percentage of copulations*

Year	Residents	Resident copulations	Marginals	Marginal copulations	Satellites	Satellite copulations
1990	33.3	91.7	47.0	0.0	19.7	8.3
1991	25.2	83.3	56.3	3.5	18.5	13.2
1992	28.6	90.7	52.4	1.2	19.0	8.1
Mean	29.0	88.6	51.9	1.6	19.1	9.2

Source: After Widemo (1998) with permission.

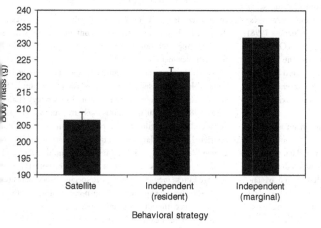

Figure 13.1 Mean body mass of the three different male ruff
mating strategies/tactics (± Se). (Drawn from data provided in
Bachman and Widemo 1999.)

Table 13.4. *Offspring morph percentages in relation to parental combination in the common buzzard, based on 7 years of data (1998–2004), with 422 broods and 724 chicks*

Parental combination	n	Chicks (%)		
		Dark	Intermediate	Light
Dark × Dark	2	100	0	0
Dark × Intermediate	88	46.6	53.4	0
Dark × Light	10	0	100	0
Intermediate × Intermediate	362	17.1	66.0	16.9
Intermediate × Light	212	0.9	50.0	49.1
Light × Light	50	0	0	100

the ages of 3 and 6 years. Apart from this difference, no detailed information on other parts of the life history is available, but it seems that white males also sing more than their rufous counterparts (Lank 2002).

A classic case of a plumage polymorphism resulting in alternative reproductive strategies was discovered by O'Donald (1983) in the Arctic skua (*Stercorarius parasiticus*). These skuas occur in pale, intermediate, or dark ventral plumage morphs and pedigree data indicate that the morphs are genetically determined. In O'Donald's study, females seemed to prefer to mate with dark males, producing a sexual-selection pressure favoring the dark morph, but pale birds nested at a younger age, resulting in a natural-selection pressure favoring the pale morph. According to these results, the plumage morphs thus seem to be associated with different life-history strategies. However, a more recent study (Phillips and Furness 1998) could not detect many of the selection pressures O'Donald reported and it remains unclear whether the morphs indeed have different life-history strategies and whether there is an evolutionarily stable state frequency. The alternative hypothesis is that fitness clines exist, with forms mixing in overlap zones.

A similar example of a plumage polymorphism associated with different life-history strategies has been documented in the common buzzard (*Buteo buteo*). Like the skua, there are three morphs: dark, intermediate, and light. Pedigree data indicate that plumage morph might be controlled by one locus with two alleles, intermediates being heterozygotes (Krüger *et al.* 2001) (Table 13.4), with pairs mating assortatively. More surprisingly, the three morphs differ greatly in their life history and fitness. Intermediate birds achieve a higher annual breeding success, have a much longer lifespan, and occupy higher-quality territories, so this system might be

one of the few examples of a heterozygous advantage. These differences result in intermediates having twice as many offspring during their lifetime as dark and light birds (Krüger and Lindström 2001) (Figure 13.2). Because the three morphs differ in fitness, with no frequency-dependent selection, there is no evolutionarily stable state frequency; hence the polymorphism should not be stable. However, intermediate pairs continuously produce dark and light offspring because of the inheritance system, so the population could only evolve towards monomorphism if a genetic modifier evolved (Lank 2002).

13.3 CONDITIONAL STRATEGY WITH ALTERNATIVE TACTICS

In contrast to the few examples of alternative strategies described above, the literature contains many studies documenting conditional strategies in birds (see Table 13.1 for examples mentioned below).

By far the most common conditional strategy pursued by female birds is conspecific brood parasitism. Around 185 species have been documented as using this strategy (Eadie *et al.* 1998, Davies 2000) and it is likely that many more will be added. Conspecific brood parasitism in birds is not restricted to particular taxa and life-history strategies, but is generally associated with high fertility (Arnold and Owens 2002). There are also numerous examples of both types of average fitness inequalities: in some species, brood parasites achieve lower average fitness than nonparasites; in others, they enhance their reproductive success by both laying some eggs parasitically and nesting on their own, and have a higher average fitness than nonparasites.

Well-documented examples of species in which females use brood parasitism to achieve at least some fitness include

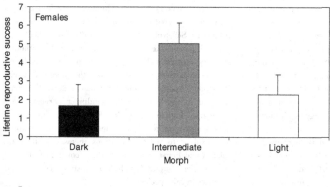

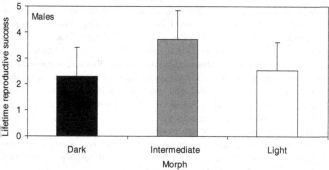

Figure 13.2 Mean (+ Se) lifetime reproductive success in relation to plumage morph in the common buzzard. (Modified from Krüger and Lindström 2001.)

starlings (*Sturnus vulgaris*), eastern bluebirds (*Sialis sialis*), white-fronted bee-eaters (*Merops bullockoides*), and lesser snow geese (*Chen caerulescens*) (Davies 2000). In these cases, some females in the population are unable to secure a nest site and resort to parasitic laying. In addition, females that have lost some or all of their eggs due to predation or disturbance also salvage reproductive success by becoming brood parasites. In these examples, it is commonly low-status individuals that have to use this tactic.

In sharp contrast, there are a number of species in which high-status females use brood parasitism as a tactic to enhance their fitness above that of nonparasitic females. In a long-term study on cliff swallows (*Petrochelidon pyrrhonota*), Brown and Brown (1998) found that parasitic females were high-quality individuals and had a much higher survival rate than nonparasitic ones, resulting in higher lifetime reproductive success for parasitic females.

Similarly, Åhlund and Andersson (2001) showed that goldeneye ducks (*Bucephala clangula*) were able to double their reproduction by engaging in brood parasitism. In this example, brood parasitism is used by high-quality females where their limiting factor is not fertility but parental care. However, brood parasitism as a tactic used by high-quality females might well be constrained, because it is costly to the host (see Sorenson 1997). In a classic simulation paper, Eadie and Fryxell (1992) showed that population demography has profound implications on the frequency of brood parasitism. Focusing on another species of goldeneye duck (*Bucephala islandica*), they predicted an evolutionarily stable strategy switch point of 23% parasitic eggs, which was very close to the 17% observed in the population. Incorporating density dependence, they showed that brood parasitism is a "best-of-a-bad-job" tactic at low population densities, regardless of the parasitism frequency. However, at high population densities, parasitism follows negative frequency-dependent selection so that when parasitism is very common, nesting females achieve higher reproduction. This study neatly illustrates the complexities involved when considering optimal tactics and switch points in conditional strategies.

A mixture of salvaging reproductive success and fitness augmentation can also be seen in American coots (*Fulica americana*: studied by Lyon 1993, 1998, 2003) and moorhens (*Gallinula chloropus*: studied by McRae 1997, 1998). In both species, some females laid parasitic eggs because they could not obtain a territory or because their own nest was depredated; other females laid parasitic eggs in addition to their own successful nest.

The corresponding male conditional strategy to augment fitness is extra-pair copulation behavior, which has been documented in many species (Birkhead and Møller 1992, Petrie and Kempenaers 1998). Because there are excellent recent reviews of extra-pair copulation behavior in birds (Griffith *et al.* 2002, Magrath and Komdeur 2003, Westneat and Stewart 2003), there is no need to discuss it at length here. Extra-pair paternity ranges from 0% to around 80% among bird species (Petrie and Kempenaers 1998) and leads to a much higher variance in reproductive success among males compared to females in socially monogamous species. It is thought that females commonly control the success of any copulation attempt (Birkhead and Møller 1993) and so the literature has mainly focused on costs and benefits of extra-pair copulations to females. Besides potential direct (paternal care, etc.) and indirect (genetic) benefits, there are substantial costs, such as the loss of paternal care from the social father, search and assessment costs, and the risk of sexually transmitted diseases (Petrie and Kempenaers 1998). Recently, Arnqvist and Kirkpatrick (2005) concluded that there is negative selection on extra-pair copulation behavior in females: the widespread occurrence of this behavior might best be explained as selection on male behavior where the costs of resistance in females are larger than the benefits. An additional trade-off limiting male EPC behavior might be paternal care (Magrath and Komdeur 2003), which might also be essential in explaining why male parasitic reproductive tactics are noticeably rarer in birds (see below).

One type of conditional strategy commonly found in fish (Taborsky 1994), but only rarely documented in birds, is female mimicry by males. In the pied flycatcher (*Ficedula hypoleuca*), first-year males either mimic females in plumage or develop full male breeding plumage. Slagsvold and Sætre (1991) showed that territorial males accepted female-like males as territory holders in their vicinity, but not males in full breeding plumage. However, the advantage of female-like males in gaining a territory incurs a cost in the form of increased female aggression. Moreover, female-like males arrive later on territories in spring and seem to be in worse condition. Slagsvold and Sætre (1991) concluded that, although female-like males do better than males without a territory, they are making the best of a bad job, considering their comparatively low status and the increased female aggression. Another example of female mimicry has been reported in the buff-breasted sandpiper (*Tryngites subruficollis*) by Lanctot *et al.* (1997). Here, males mimic female behavior patterns to disrupt copulation by the territorial male and/or to sneak copulations themselves. Recently Jukema and Piersma (2004) reported female-like males in ruff, further complicating the mating strategies employed by males in this species.

An intriguing example of female mimicry has recently been described in the lazuli bunting (*Passerina amoena*). Yearling males occur as dull, female-like males, as intermediates, and as bright males. Greene *et al.* (2000) showed that this polymorphism is under disruptive sexual selection. Dull yearling males were allowed to settle in higher-quality territories then they would normally be able to defend because neighboring bright males tolerated them. Bright yearling males were also successful in obtaining medium- or high-quality territories further from older males, due to their competitive abilities. However, intermediate males were not of high enough quality to outcompete other males and they were not tolerated by bright males, so they had a very low pairing success. Female preference for high-quality territories was the driving force for the disruptive selection. In addition, Greene *et al.* (2000) showed that by allowing dull yearlings to occupy neighboring territories, bright males reduced their chance of being cuckolded and increased their likelihood of obtaining extra-pair copulations as a consequence of having less attractive neighbors.

Another example of a conditional strategy based on a sexual ornamentation has been described in the house finch (*Carpodacus mexicanus*). Males in this species develop a carotenoid-based breast coloration during the breeding season and Badyaev and Hill (2002) found a correlation between the different reproductive tactics of males and the level of ornamentation. Brightly ornamented red males paired with females initiating breeding early in the season, but they provided little provisioning for the female and the brood. In contrast, less brightly ornamented yellow males paired with females that nested later in the season, but they fed the female and the brood much more. The third morph, intermediate between red and yellow in its ornamentation, also followed an intermediate reproductive tactic. Interestingly, the three

levels of plumage ornamentation achieve different reproductive success depending on age and partner. Among first-
time breeders, red males achieved the greatest reproductive
success because of their mate's quality. However, among
older breeders, intermediate males achieved the highest
reproduction because of the combined effects of relatively
early breeding and higher levels of male parental care.

Another example of an evolutionarily stable strategy
switch point with regard to age has been documented in the
tree sparrow (*Passer montanus*). Both sexes in this species
either breed solitarily or in colonies. Sasvári and Hegyi
(2000) found that for females, the highest fitness was
obtained by breeding in a colony in the first year of life and
then breeding solitarily in all subsequent years. In contrast,
males obtained the highest fitness by breeding colonially
throughout their life. For females, the large costs of solitarily breeding in the first year of life seem to be so large that
colonial breeding is favored, although productivity was
lower. This example points towards the complex interactions not only between alternative reproductive tactics,
but also between the sexes. Males always favor colonial
breeding but females favor colonial breeding only in their
first year of life. The mix of alternative reproductive tactics
might be stalemates of this conflict, as in the dynamic
mating system of dunnocks (Davies 1992).

One conditional strategy that is common in fish and
insects (Gross 1996) but not in birds is sneaking versus
fighting for a territory as true alternative tactics. One
potential example involves the ruff again, where around 1%
of males do not develop a nuptial plumage and are referred
to as "naked-nape" males (Hogan-Warburg 1966). They try
to sneak copulations on the territories of resident males by
mimicking females, but it is not known whether these males
are using a conditional tactic or whether they are indeed a
third alternative strategy in this species (Jukema and
Piersma 2004).

Considering the efforts devoted to studying avian mating systems, it is interesting that no conditional strategy of
sneaking has been definitely reported, although extra-pair
copulations by males are often referred to as sneaking in the
literature. Why is a true parasitic reproductive strategy in
males not more common in birds? In comparison to fish,
female choice is strong in birds, which limits male opportunities to sneak copulations in the same way as fish can,
because of internal fertilization and the lack of intromittent
organs. Hence males can rarely force a copulation onto a
female, with some notable exceptions, such as in species of
waterfowl (Briskie 1998). Another contributing factor is

that biparental care enhances breeding success significantly
in many bird species (Clutton-Brock 1991), which greatly
reduces the pay-off for a parasitic male strategy. The relatively small clutch sizes of birds compared to other taxa
further reduces the potential benefits of a parasitic male
reproductive strategy for birds when compared with other
taxa such as fish and insects.

13.4 INTERACTIONS BETWEEN THE SEXES

Many alternative male tactics are thought to have evolved
under the theme of salvaging at least some reproductive
success and, although they increase the variance in male
reproductive success, they also increase the opportunity for
female choice. The examples of alternative strategies and
conditional strategies highlighted above should, at least in
some cases, lead to mate-choice decisions by males and
females (female choice has been largely ignored but see
Alonzo and Warner 1999, 2000 for an example in fish).

Consider the ruff with its two alternative strategies. If
Widemo (1998) is correct in postulating that the satellite
strategy is low cost and low benefit, with equal average
fitness compared to the resident strategy, a female should
make mate-choice decisions based on her condition. It is
well established that early conditions can have a profound
effect on lifetime fitness in birds (Lindström 1999), so even
in precocial birds, female condition at laying might have
long-lasting effects. Females in bad condition should
therefore mate with a satellite male because producing low-
quality offspring of the resident strategy will yield lower
fitness, since not all residents manage to secure a territory.
These potential effects have so far been largely overlooked
in ruff (but see Lank 2002). The house finch example
provided by Badyaev and Hill (2002) also highlights the
complexities of female choice when alternative reproductive
phenotypes are involved. Status-dependent selection gives
rise to the alternative tactics in males and it should also
affect female choice. If males provide a mixture of direct
(parental care) and indirect benefits (male condition as
expressed in the ornamentation) to females, then females
should choose depending on their own condition. If a large
male ornament signals good condition but low parental care,
and this has a heritable component, then females in good
condition should choose a highly ornamented male because
the indirect benefits will exceed the costs of low parental
care, for which a high-quality female can compensate
(Kokko 1998, Badyaev and Hill 2002).

The cliff swallow provides another example of the complex interactions between the sexes when alternative reproductive phenotypes are involved. Brown and Brown (1996, 1998) not only found that parasitic females had higher average fitness than nonparasitic females, they also reported that males engaging in extra-pair copulations had significantly lower survival rates compared to males that did not. They concluded that parasitic females are high-quality individuals, whereas males engaging in extra-pair copulations are low-quality individuals and females should resist them. Females should therefore preferentially pair with males that do not engage in extra-pair copulations because these males will be of higher quality and will show a higher presence at the nest and guard their female more intensively, lowering the female's risk of forced copulations by inferior males. Males should always try to pair up with parasitic females because these females have a higher fitness thus minimizing the male's risk of being cuckolded.

In addition to mate choice decisions discussed above, selection on one sex can constrain the evolution of alternative reproductive phenotypes in the other sex, as exemplified by the common cuckoo. In this species, males may even compromise the fitness of females. Host behavior strongly influences fitness in common cuckoos and the different gentes with their host-specific eggs have presumably evolved as an adaptation during the host–parasite arms race (Davies 2000). Host-specific chick begging behavior would be a distinct advantage but, apart from call rate, Butchart et al. (2003) did not find a difference in call note or structure. The problem here is that host-specific begging behavior cannot be encoded in genes on the female W-chromosome, otherwise only female offspring would benefit from it. Due to the cross-mating by male cuckoos, females are actually constrained in their alternative strategies. As long as polygyny yields more benefits to males than the better adapted host-specific behavior of the chick, mating across gentes by males will be favored.

Where to next? The amount and quality of work being conducted on birds will continue to produce important insights into the evolution and maintenance of alternative reproductive phenotypes. Theoretical studies need to incorporate the idea that alternative reproductive phenotypes in one sex might be affected not only by fitness pay-offs but also by the behavior of the other sex. Gross (1996) stated that theoretical studies should reconcile frequency-dependent selection with status-dependent selection. There might also be a need to incorporate the strategies of the two sexes explicitly rather than just through fitness pay-offs (see Hugie and Lank 1997, Alonzo and Warner 1999). Empirical studies might be able to test whether status-dependent mate choice exists in females of species where alternative phenotypes exist. Another potentially rewarding area of future research might be to use the wealth of information in birds for comparative studies, which might further elucidate life-history or ecological correlates of alternative reproductive phenotypes. Arnold and Owens (2002) have demonstrated that for conspecific brood parasitism, for example, most of the variation between species can be explained by differences in fertility due to phylogenetic inertia rather than ecology.

Acknowledgments

I am indebted to the editors for the invitation to contribute to this book and to Mike Brooke, Nick Davies, Camilla Hinde, David Lank, Nick MacGregor, Joah Madden, Raoul Mulder, Andrew Radford, Nat Seddon, Claire Spottiswoode, Michael Taborsky, and Fritz Trillmich for many stimulating discussions and helpful suggestions that greatly improved this chapter.

References

Åhlund, M. and Andersson, M. 2001. Female ducks can double their reproduction. *Nature* **414**, 600–601.

Alonzo, S. H. and Warner, R. R. 1999. A trade-off generated by sexual conflict: Mediterranean wrasse males refuse present mates to increase future success. *Behavioral Ecology* **10**, 105–111.

Alonzo, S. H. and Warner, R. R. 2000. Female choice, conflict between the sexes and the evolution of male alternative reproductive behaviours. *Evolutionary Ecology Research* **2**, 149–170.

Andersson, M. 1994. *Sexual Selection*. Princeton, NJ: Princeton University Press.

Arnold, K. E. and Owens, I. P. F. 2002. Extra-pair paternity and egg dumping in birds: life history, parental care and the risk of retaliation. *Proceedings of the Royal Society of London B* **269**, 1263–1269.

Arnqvist, G. and Kirkpatrick, M. 2005. The evolution of infidelity in socially monogamous passerines: the strength of direct and indirect selection on extrapair copulation behavior in females. *American Naturalist* **165**, S26–S37.

Bachman, G. and Widemo, F. 1999. Relationships between body composition, body size and alternative reproductive tactics in a lekking sandpiper, the ruff (*Philomachus pugnax*). *Functional Ecology* **13**, 411–416.

Badyaev, A. V. and Hill, G. E. 2002. Paternal care as a conditional strategy: distinct reproductive tactics associated with elaboration of plumage ornamentation in the house finch. *Behavioral Ecology* 13, 591–597.

Birkhead, T. R. and Møller, A. P. 1992. *Sperm Competition in Birds: Evolutionary Causes and Consequences.* London: Academic Press.

Birkhead, T. R. and Møller, A. P. 1993. Female control of paternity. *Trends in Ecology and Evolution* 8, 100–104.

Briskie, J. V. 1998. Avian genitalia. *Auk* 115, 826–828.

Brockmann, H. J. 2001. The evolution of alternative strategies and tactics. *Advances in the Study of Behavior* 30, 1–51.

Brooke, M. de L., Davies, N. B., and Noble, D. G. 1998. Rapid decline of host defences in response to reduced cuckoo parasitism: behavioural flexibility of reed warblers in a changing world. *Proceedings of the Royal Society of London B* 265, 1277–1282.

Brown, C. R. and Brown, M. B. 1996. *Coloniality in the Cliff Swallow: The Effect of Group Size on Social Behaviour.* Chicago, IL: University of Chicago Press.

Brown, C. R. and Brown, M. B. 1998. Fitness components associated with alternative reproductive tactics in cliff swallows. *Behavioral Ecology* 9, 158–171.

Butchart, S. H. M., Kilner, R. M., Fuisz, T., and Davies, N. B. 2003. Differences in the nestling begging calls of hosts and host-races of the common cuckoo, *Cuculus canorus. Animal Behaviour* 65, 345–354.

Clutton-Brock, T. H. 1991. *The Evolution of Parental Care.* Princeton, NJ: Princeton University Press.

Davies, N. B. 1992. *Dunnock Behaviour and Social Evolution.* Oxford, UK: Oxford University Press.

Davies, N. B. 2000. *Cuckoos, Cowbirds and Other Cheats.* London: T. and A. D. Poyser.

Davies, N. B., Brooke, M. de L., and Kacelnik, A. 1996. Recognition errors and probability of parasitism determine whether reed warblers should accept or reject mimetic cuckoo eggs. *Proceedings of the Royal Society of London B* 263, 925–931.

Eadie, J. M. and Fryxell, J. M. 1992. Density dependence, frequency dependence, and alternative nesting strategies in goldeneyes. *American Naturalist* 140, 621–641.

Eadie, J. M., Sherman, P., and Semel, B. 1998. Conspecific brood parasitism, population dynamics, and the conservation of cavity-nesting birds. In T. Caro (ed.) *Behavioural Ecology and Conservation Biology*, pp. 306–340. New York: Oxford University Press.

Formica, V. A., Gonser, R. A., Ramsay, S., and Tuttle, E. M. 2004. Spatial dynamics of alternative reproductive strategies: the role of neighbors. *Ecology* 85, 1125–1136.

Gibbs, H. L., Sorenson, M. D., Marchetti, K., *et al.* 2000. Genetic evidence for female host-specific races of the common cuckoo. *Nature* 407, 183–186.

Greene, E., Lyon, B. E., Muehter, V. E., *et al.* 2000. Disruptive sexual selection for plumage coloration in a passerine bird. *Nature* 407, 1000–1003.

Griffith, S. C., Owens, I. P. F., and Thuman, K. A. 2002. Extrapair paternity in birds: a review of interspecific variation and adaptive function. *Molecular Ecology* 11, 2195–2212.

Gross, M. R. 1996. Alternative reproductive strategies and tactics: diversity within sexes. *Trends in Ecology and Evolution* 11, 92–98.

Hogan-Warburg, A. J. 1966. Social behaviour of the ruff, *Philomachus pugnax. Ardea* 54, 108–229.

Hoglund, J., Montgomerie, R., and Widemo, F. 1993. Costs and consequences of variation in the size of ruff leks. *Behavioral Ecology and Sociobiology* 32, 31–39.

Houtman, A. M. and Falls, J. B. 1994. Negative assortative mating in the white-throated sparrow, *Zonotrichia albicollis*: the role of mate choice and intra-sexual competition. *Animal Behaviour* 48, 377–383.

Hugie, D. M. and Lank, D. B. 1997. The resident's dilemma: a female-choice model for the evolution of alternative male reproductive strategies in lekking male ruffs (*Philomachus pugnax*). *Behavioral Ecology* 8, 218–225.

Jukema, J. and Piersma, T. 2004. Kleine mannelijke Kemphanen met vrouwelijk broedkleed: bestaat er een derde voortplantingsstrategie de faar? *Limosa* 77, 1–10.

Knapton, R. W. and Falls, J. B. 1983. Differences in parental contribution among pair types in the polymorphic white-throated sparrow. *Canadian Journal of Zoology* 61, 1288–1292.

Kokko, H, 1998. Should advertising parental care be honest? *Proceedings of the Royal Society of London B* 265, 1871–1878.

Krüger, O. and Lindström, J. 2001. Lifetime reproductive success in common buzzard *Buteo buteo*: from individual variation to population demography. *Oikos* 93, 260–273.

Krüger, O., Lindström, J., and Amos, W. 2001. Maladaptive mate choice maintained by heterozygote advantage. *Evolution* 55, 1207–1214.

Lanctot, R. B., Scribner, K. T., Kempenaers, B., and Weatherhead, P. J. 1997. Lekking without a paradox in the buff-breasted sandpiper. *American Naturalist* 149, 1051–1070.

Lank, D. B. 2002. Diverse processes maintain plumage polymorphisms in birds. *Journal of Avian Biology* 33, 327–330.

Lank, D. B. and Smith, C. M. 1987. Conditional lekking in ruff (*Philomachus pugnax*). *Behavioral Ecology and Sociobiology* 20, 137–145.

Lank, D. B., Smith, C. M., Hanotte, O., Burke, T. A., and Cooke, F. 1995. Genetic polymorphism for alternative mating behaviour in lekking male ruff, *Philomachus pugnax*. *Nature* 378, 59–62.

Lank, D. B., Smith, C. M., Hanotte, O., et al. 2002. High frequency of polyandry in a lek mating system. *Behavioral Ecology* 13, 209–215.

Lindström, J. 1999. Early development and fitness in birds and mammals. *Trends in Ecology and Evolution* 14, 243–248.

Lyon, B. E. 1993. Brood parasitism as a flexible female reproductive tactic in American coots. *Animal Behaviour* 46, 911–928.

Lyon, B. E. 1998. Optimal clutch size and conspecific brood parasitism. *Nature* 392, 380–383.

Lyon, B. E. 2003. Ecological and social constraints on conspecific brood parasitism by nesting female American coots (*Fulica americana*). *Journal of Animal Ecology* 72, 47–60.

Magrath, M. J. L. and Komdeur, J. 2003. Is male care compromised by additional mating opportunity? *Trends in Ecology and Evolution* 18, 424–430.

Marchetti, K., Nakamura, H., and Gibbs, H. L. 1998. Host-race formation in the common cuckoo. *Science* 282, 471–472.

McRae, S. B. 1997. A rise in nest predation enhances the frequency of intraspecific brood parasitism in a moorhen population. *Journal of Animal Ecology* 66, 143–153.

McRae, S. B. 1998. Relative reproductive success of female moorhens using conditional strategies of brood parasitism and parental care. *Behavioral Ecology* 9, 93–100.

Mulder, R. A., Ramiarison, R., and Emahalala, R. E. 2002. Ontogeny of male plumage dichromatism in Madagascar paradise flycatchers *Terpsiphone mutata*. *Journal of Avian Biology* 33, 342–348.

Neff, B. D. 2001. Alternative reproductive tactics and sexual selection. *Trends in Ecology and Evolution* 16, 669–682.

O'Donald, P. 1983. *The Arctic Skua: A Study of the Ecology and Evolution of a Seabird*. Cambridge, UK: Cambridge University Press.

Parker, G. A., Baker, R. R., and Smith, V. C. F. 1972. The origin and evolution of gamete dimorphism and the male–female phenomenon. *Journal of Theoretical Biology* 36, 529–553.

Petrie, M. and Kempenaers, B. 1998. Extra-pair paternity in birds: explaining variation between species and populations. *Trends in Ecology and Evolution* 13, 52–58.

Phillips, R. A. and Furness, R. W. 1998. Polymorphism, mating preferences and sexual selection in the Arctic skua. *Journal of Zoology (London)* 245, 245–252.

Piper, W. H. and Wiley, R. H. 1989. Distinguishing morphs of the white-throated sparrow in basic plumage. *Journal of Field Ornithology* 60, 73–83.

Qvarnström, A. 1999. Different reproductive tactics in male collared flycatchers signalled by size of secondary character. *Proceedings of the Royal Society of London B* 266, 2089–2093.

Rhijn, J. G. 1991. *The Ruff*. London: T. and A. D. Poyser.

Sasvári, L. and Hegyi, Z. 2000. Mate fidelity, divorce and sex-related differences in productivity of colonial and solitary breeding tree sparrows. *Ethology Ecology and Evolution* 12, 1–12.

Sinervo, B. and Lively, C. M. 1996. The rock–paper–scissors game and the evolution of alternative male strategies. *Nature* 380, 240–243.

Slagsvold, T. and Sætre, G. P. 1991. Evolution of plumage color in male pied flycatchers (*Ficedula hypoleuca*): evidence for female mimicry. *Evolution* 45, 910–917.

Smith, T. B. 1987. Bill size polymorphism and intraspecific niche utilization in an African finch. *Nature* 329, 717–719.

Smith, T. B. 1990. Comparative breeding biology of the two bill morphs of the black-bellied seedcracker. *Auk* 107, 153–160.

Sorenson, M. D. 1997. Effects of intra- and interspecific brood parasitism on a precocial host, the canvasback, *Aythya valisineria*. *Behavioral Ecology* 8, 153–161.

Taborsky, M. 1994. Sneakers, satellites and helpers: parasitic and cooperative behavior in fish reproduction. *Advances in the Study of Behavior* 23, 1–100.

Thorneycroft, H. B. 1975. A cytogenetic study of the white-throated sparrow, *Zonotrichia albicollis* (Gmelin). *Evolution* 29, 611–621.

Thuman, K. E. 2003. Female reproductive strategies in the ruff (*Philomachus pugnax*). Ph.D. thesis, Uppsala University, Sweden.

Tuttle, E. M. 2003. Alternative reproductive strategies in the white-throated sparrow: behavioural and genetic evidence. *Behavioral Ecology* 14, 425–432.

Watt, D. J., Ralph, C. J., and Atkinson, C. T. 1984. The role of plumage polymorphism in dominance relationships of the white-throated sparrow. *Auk* 101, 110–120.

Westneat, D. F. and Stewart, I. R. K. 2003. Extra-pair paternity in birds: causes, correlates, and conflict. *Annual Review of Ecology and Systematics* 34, 365–396.

Widemo, F. 1998. Alternative reproductive strategies in the ruff, *Philomachus pugnax*: a mixed ESS? *Animal Behaviour* 56, 329–336.

Yom-Tov, Y. 2001. An updated list and some comments on the occurrence of intraspecific nest parasitism in birds. *Ibis* 143, 133–143.

14 · Alternative reproductive tactics in nonprimate male mammals

JERRY O. WOLFF

CHAPTER SUMMARY

Alternative reproductive tactics in male mammals fall into two categories: unequal pay-offs, in which a younger or subordinate individual assumes a lower fitness tactic in response to the frequency and competitive ability of other males in the population, and equal pay-offs, in which ecological or environmental factors dictate alternative tactics based on distribution of resources, population density, and demographic conditions. The tactic(s) used in both situations are conditional and are based on social and ecological environments, respectively, and on the relative social status of competitors. The decision-making rules for adopting a particular tactic for any individual at any one time in its life are based on its age, status, competitive ability, or current environmental conditions; these rules will ultimately maximize the individual's lifetime reproductive success. The reproductive tactic used by an individual may vary throughout its lifetime or seasonally. Herein I provide examples of selective forces and relative fitness pay-offs for alternative reproductive tactics used by a variety of male mammals under different social and ecological conditions.

14.1 INTRODUCTION

Mating systems are viewed as an outcome of individual, reproductive strategies that have been subjected to natural selection and have become characteristics of given species (Gross 1996). A reproductive strategy may have one or more alternative tactics, each conditionally dependent on factors such as social status, frequency of alternatives, population density, and environmental factors (Dominey 1984, Taborsky 1994, Gross 1996, Brockmann 2001). Following the logic of Gross (1996), I view alternative reproductive tactics (actually phenotypes) from two

perspectives. Within any given species, there is one tactic that works "best"; that is, it results in the greatest number of matings and reproductive success for the individual using that tactic. This tactic is typically used by the dominant or most competitive individuals in the population and may be equated with the "bourgeois" tactic as described by Taborsky (2001). Other (subordinate) individuals in the population that are not as competitive using this tactic choose an alternative one that, while providing them with the greatest success they can achieve at the time, considering their current condition, is usually less successful than the dominant tactic. Among the reproductive tactics available within a population, the secondary tactic is conditional on the current social and environmental milieu and can be thought of as an alternative to the best or bourgeois tactic. At the individual level, however, alternative reproductive tactics should be evolutionarily stable if they maximize the fitness of individuals within the constraints of the immediate social and ecological pressures. Thus, individuals must assess and make decisions based on the current situation and prospects for future reproductive success. The pay-offs to the alternative tactics are not equal when assessed at the population level; rather one tactic is more successful than an alternative (Repka and Gross 1995). That said, alternative tactics are still more successful than not breeding; therefore, less competitive individuals adopt a tactic that is the best they can achieve given their relative ranking compared to reproductive competitors. Little, if any, evidence exists to show that the various tactics used by wild mammals result from a genetic polymorphism and thus are not mixed or alternative strategies (sensu Gross). In fact, the various tactics are used by the same individuals throughout their lifetime or even within the same breeding season, depending on competition with other males and environmental, demographic, or social conditions. The optimum

Alternative Reproductive Tactics, ed. Rui F. Oliveira, Michael Taborsky, and H. Jane Brockmann. Published by Cambridge University Press.
© Cambridge University Press 2008.

reproductive tactic for an individual at any one time is likely conditional and is based on the frequency of alternatives being used in the population and the associated trade-offs of these alternatives (e.g., Repka and Gross 1995, Brockmann 2001, Taborsky 2001).

The objectives of this chapter are to provide examples of alternative reproductive tactics in nonprimate male mammals and to explain their apparent adaptive significance with respect to selection pressures from social, demographic, and ecological environments (summarized in Table 14.1). I will describe alternative reproductive tactics in which one "best" tactic is used by the dominant and most competitive individuals resulting in the most mating opportunities and a second alternative tactic is used by subordinate or less competitive males resulting in fewer mating opportunities. In this case, the alternative is conditional and the pay-offs unequal. However, there are some alternative reproductive tactics that can have equal pay-offs and that are more dependent on ecological or demographic constraints than on social competition.

14.2 ALTERNATIVE REPRODUCTIVE TACTICS IN MALES: UNEQUAL PAY-OFFS

14.2.1 Harem defense polygyny

One of the most common reproductive tactics among mammals is harem-defense polygyny in which a dominant male defends access to several females. Harems can be defended seasonally or throughout the year depending on the species and length of the breeding season. Alternative tactics employed by subordinate males involve some variation of a sneak behavior in which subordinate males "sneak" a copulation from females being guarded by a harem-master when he is not looking or when he cannot guard all of his females at once. These two tactics may be similar to the bourgeois and parasitic tactics described for fish (Taborsky 2001). Examples of harem-defense polygyny include northern (*Mirounga angustirostris*: Le Boeuf 1974) and southern (*M. leonina*: Hoelzel *et al.* 1999) elephant seals, plains zebras (*Equus burchelli*: Rubenstein 1986), feral horses (*E. caballus*: Berger 1986), red deer (*Cervus elaphus*: Clutton-Brock *et al.* 1982), and other ungulates (Table 14.1). Within both species of elephant seals, males that are of similar size to females mimic them behaviorally and occasionally sneak copulations when the harem-master is not looking. Additionally, some males wait until late in the season when females return to the water

and are no longer guarded by dominant bulls to obtain copulations. However, these late copulations are probably not successful because females most likely have mated previously (Le Boeuf 1974, Hoelzel *et al.* 1999).

Alternative tactics for male plains zebras and horses include wandering and searching for females or waiting to depose harem-master stallions (Berger 1986, Rubenstein 1986, Asa 1999). Berger (1986) described three tactics used by feral horses to acquire mates. Forty-five percent of females in harems were acquired by wandering bachelors that encountered them by chance, 48% were acquired by deposing resident stallions through aggression and fighting (see Figure 14.1), and 6% were acquired through cooperative coalitions of unrelated males (Figure 14.2; discussed below). Dominant stallions acquire harems, whereas subordinate males are either relegated to bachelorhood or, in some cases, are accepted in subservient roles within a harem where they may sire some offspring (Berger 1986, Asa 1999, Linklater and Cameron 2000). Harem-masters sire the most offspring, whereas the relative reproductive success of subordinates and bachelors is considerably lower and probably comparable. Among Grevy's zebras (*Equus grevyi*), territorial stallions typically obtain the majority of matings. Ginsberg and Rubenstein (1990) reported that bachelor males obtained 9% of copulations using a sneak or follower tactic.

Attempted copulations by subordinate, usually younger, males are common in other species of mammals. In red deer, 7- to 12-year-old stags hold harems. Occasionally younger stags run through a harem scattering females and causing a disruption that requires the harem-master to concentrate on rounding up the scattered females (Clutton-Brock *et al.* 1982). During this time, younger stags are able to sneak copulations. Whether sneak copulations are with fertile females and whether they produce young is not known. Similar patterns of sneak copulations occur in waterbuck (*Kobus ellipsiprymnus*: Wirtz 1981), oribi (*Ourebia ourebi*: Arcese 1999), and other ungulates (Gosling 1986, Rubenstein 1986). Although these alternative reproductive tactics are qualitatively regarded as more or less competitive, the absolute reproductive success of each has been rarely quantified.

14.2.2 Resource-defense polygyny

A common reproductive tactic for male mammals is to acquire mating opportunities by defending an essential resource that is used by females. Males that are not

Table 14.1. Dominant (tactic 1) and subordinate (tactic 2) tactics of representative species of nonprimate mammals and the selective pressures for the alternative reproductive tactic

Tactic 1	Tactic 2	Species	Selective force	References
Harem–defense polygyny	Female mimicry, sneak, mate late in season	Northern elephant seals (*Mirounga angustirostris*), southern elephant seals (*M. leonina*)	Body size and age (dominance)	Le Boeuf (1974), Hoelzel *et al.* (1999)
	Sneaks, satellites (wandering), steal females, harem takeovers	Plains zebras (*Equus burchelli*), feral horses (*E. caballus*), red deer (*Cervus elaphus*)	Age and dominance	Clutton-Brock *et al.* (1982)
	Coalition of subordinate on dominant male's territory or solitary bachelor	Feral horses (*Equus caballus*), waterbuck (*Kobus ellipsiprymnus*), oribi (*Ourebia ourebi*), Grevy's zebra (*Equus grevyi*)	Frequency of dominant males, density	Wirtz (1981, 1982), Berger (1986), Ginsberg and Rubenstein (1990), Arcese (1999), Asa (1999), Feh (1999), Linklater and Cameron (2000)
Resource–defense polygyny	Satellite and intercept females, or sneaks	White-lined bat (*Saccopteryx bilineata*)	Frequency of dominant males	Heckel and Helversen (2002)
	Subordinate on dominant's territory	Waterbuck (*Kobus ellipsiprymnus*)	Frequency of dominant males, density	Wirtz (1981, 1982)
	Following	Fallow deer (*Dama dama*), pronghorn (*Antilocapra americana*), chamois (*Rupricapra pyrenaica*), humpback whale (*Megaptera novaeangliae*), marmots (*Marmota marmota*)	Habitat quality, age (dominance), density	Lovari and Locati (1991), Moore *et al.* (1995), Byers (1997), Goossens *et al.* (1998), Clapham (2000)
	Lek	Topi (*Damaliscus lunatus*)	Age structure, dominance	Gosling and Petrie (1990), Bro-Jorgensen and Durant (2003)
Lek	Resource territories, satellites interrupt females	Fallow deer (*Dama dama*), lechwe (*Kobus leche*), blackbuck (*Antilope cervicapra*), Uganda kob (*Kobus kob*), topi (*Damaliscus lunatus*)	Frequency of dominant males, density	Leuthold (1966), Gosling and Petrie (1990), Balmford and Blakeman (1991), Clutton-Brock *et al.* (1992, 1993), Nefdt and Thirgood (1997), Isvaran and Jhala (2000), Bro-Jorgensen and Durant (2003)

		Species	Dependency	References
Territorial	Wandering	Ground squirrels (*Spermophilus* sp.), prairie vole (*Microtus ochrogaster*), alpine marmot (*Marmota marmota*), black-tailed prairie dog (*Cynomys ludovicianus*), spear-nosed bat (*Phyllostomus hastatus*), bearded seals (*Erignathus barbatus*)	Unknown, probably frequency dependent; Frequency of dominant males, density	McGracken and Bradbury (1981), Getz et al. (1993), Boellstorff et al. (1994), Travis et al. (1996), Groossens et al. (1998), Lacey and Wieczorek (2001), Solomon and Jacquot (2002), Parijs et al. (2003)
Tending	Coursing	Bighorn sheep (*Ovis canadensis*), Soay sheep (*Ovis aries*)	Frequency of dominant males, density; spiral or scurred horn phenotypes	Hogg (1984), Hogg and Forbes (1997), Clutton-Brock et al. (2004), Pemberton et al. (2004), Stevenson et al. (2004)
	Challenging, breeding late in the season	Bison (*Bison bison*), Africa elephant (*Loxodonta africana*), Soay sheep (*Ovis aries*)	Frequency of dominant males, age-dependent status	Poole (1989), Wolff (1998), Preston et al. (2001)
	Roaming	Raccoons (*Procyon lotor*)	Frequency of dominant males, age-dependent status	Gehrt and Fritzell (1999)
	Satellite	Gray squirrels (*Sciurus carolinensis*), fox squirrels (*S. niger*), ground squirrels (*Spermophilus* spp.), bridled nailtail wallaby (*Onychogalea fraenata*)	Frequency of dominant males, age-dependent status; number and spacing of females	Schwagmeyer and Parker (1987), Sherman (1989), Koprowski (1993a, b), Waterman (1998), Fisher and Lara (1999)
Mate guarding	Searching	Idaho ground squirrel (*Spermophilus brunneus*)	Frequency of dominant males and dispersion of females	Sherman (1989)
Searching	Mate guarding	Belding's ground squirrel (*Spermophilus beldingi*)	Frequency of males and dispersion of females	Sherman (1989)
Dominance	Alliances and coalitions	Dolphins (*Tursiops* spp.), cheetah (*Acinonyx jubatus*), African lion (*Panthera leo*), feral horses (*Equus caballus*)	Frequency of dominant males; male age structure	Packer et al. (1991), Caro (1994), Feh (1999), Connor et al. (2000)
	Sexual coercion	Northern elephant seals (*Mirounga angustirostris*), sea otters (*Enhydra lutris*), fallow deer (*Dama dama*), feral horses (*Equus caballus*)	Frequency of dominant males and dispersion of females	Le Boeuf (1974), Berger (1986), Clutton-Brock et al. (1992), Mestal (1994)

Figure 14.1 Acquiring a harem by fighting and deposing harem-masters is the most successful reproductive tactic in wild horses. (Photograph by Heidi L. Hopkins.)

competitive enough to defend a resource-based territory typically adopt a satellite or peripheral mating tactic. Dominant male white-lined bats (*Saccopteryx bilineata*) defend roosting areas that are limited and variable in quality. Subordinate males use an alternative peripheral tactic in which they compete directly for mating opportunities when females are active and away from the nesting site (Heckel and Helversen 2002). Both tactics are effective, but the territorial tactic results in nearly twice as many matings as the peripheral tactic. Peripheral males are younger than territorial males and will switch to territorial behavior if a roosting site becomes available or if they are able to displace a territorial male or both. Once territorial males are ousted, they do not switch to a peripheral tactic. Nonterritorial or usurper males prefer to hang around and take over territories with large rather than small harems (Voigt and Streigh 2003). The number of males adopting the peripheral tactic appears to be a function of the number of high-quality resource territories occupied by dominant males (Heckel and Helversen 2002).

Fallow deer (*Dama dama*) characteristically use a lek mating system (described below); however, considerable variation occurs within and among populations depending on a series of behavioral and ecological variables. In a nonlekking population, males use four different tactics with varying degrees of success (Alvarez *et al.* 1990,

Moore *et al.* 1995). The most successful tactic is low-fidelity territoriality in which 5- to 7-year-old dominant bucks defend territories, but also follow doe herds for part of the day (56.2% of matings). Followers obtain 19.5% of the matings. High-fidelity territorial bucks have smaller territories than low-fidelity bucks but remain on their territories all day and obtain 17.5% of matings (Figure 14.2). Satellite males that position themselves around the territories of dominant bucks obtain 6.9% of matings. The tactics employed are largely a function of age with older males using the more successful tactic. Some bucks that are territorial during one year can become followers the next when food resources on their territories are diminished.

Among African antelope, dominant males exhibit harem- or resource-defense polygyny, but commonly tolerate subordinate males on their territories or in their social group. These subordinate males may eventually inherit the territory or harem and have higher lifetime reproductive success than they would have by emigrating and attempting to form a new social unit. Wirtz (1981, 1982) reported that at any one time only 7% of male waterbucks were territorial and about half of these tolerated from one to three younger satellite males on their territories; the remaining 84% were bachelors. Satellite males switched back and forth between being a satellite and joining the bachelor herd, but many of them eventually inherited a territory (Wirtz 1981). Apparently, the territorial males benefited in some way by having the satellite males on their territories.

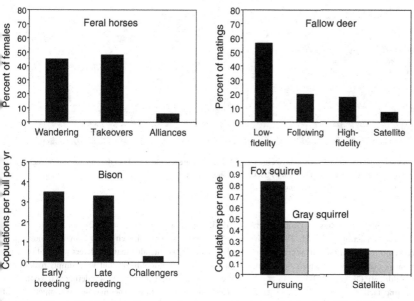

Figure 14.2 Percent of females acquired by males using alternative reproductive tactics by feral horses (Berger 1986); percent of matings by male fallow deer using four tactics (Moore et al. 1995); percent of matings by bull bison using three tactics (Wolff 1998; unpublished data); and number of copulations/male/bout in fox squirrels and gray squirrels (Koprowski 1993a, b).

Resource-defense polygyny is also the predominant mating tactic used by pronghorn (Antilocapra americana: Byers 1997), Apennine chamois (Rupricapra pyrenaica: Lovari and Locati 1991), humpback whales (Megaptera novaeangliae: Clapham 2000), and alpine marmots (Marmota marmota: Goossens et al. 1998). Among pronghorns and chamois, the oldest (>4 years) bucks defend large, resource-based territories, exclude younger males, and mate with females that enter the territory (Figure 14.3). Subordinate males adopt a following tactic; however, females resist advances by young follower males and the majority of copulations occur on territories. In large expanses of the ocean, male humpback whales space themselves out in prime foraging areas and produce songs that attract females and repel other males (Clapham 2000). As an alternative, some nonterritorial males intercept and follow females before they get to the singing males (Clapham 2000). It is not known which tactic is most effective, but it appears that the same individuals can switch from one tactic to the other at different times during their lives (Connor et al. 2000). Switching between territoriality and intercepting may be an example of frequency-dependent mating tactics.

14.2.3 Leks, territories, and satellites

Leks are an unusual breeding system in which females congregate on small, clustered breeding sites defended by individual males. Leks occur in a number of species of mammals (Clutton-Brock et al. 1993) but have been studied most extensively in ungulates, particularly fallow deer. Lekking fallow deer bucks use three mating tactics with differing degrees of success. The most productive tactic for males is to defend a breeding site within clusters of males at a lek; the second most productive tactic is to defend single, resource-based territories; and the third is to intercept and mate with females as they move onto territories or breeding sites at leks (Alvarez et al. 1990, Clutton-Brock et al. 1992) (Table 14.1). Similar tactics have been documented in lechwe antelope (Kobus leche: Nefdt and Thirgood 1997) and blackbuck (Antilope cervicapra: Isvaran and Jhala 2000). As the breeding season approaches, females leave the herd and congregate on leks where some males obtain high reproductive success. Males on territories have moderate success because female density and frequency of visits are less than on leks affording these males fewer mating opportunities. Females are harassed

Figure 14.3 Dominant pronghorn bucks >4 years old typically hold resource-based territories and acquire harems. Younger bucks follow individuals or groups of females and are only successful when there are few older bucks in the population. (Photographs by John Byers.)

the most while they are in the herd, less on resource territories, and least on leks when densities are high. The relative reproductive success of males varies and depends on a combination of resource availability and female movements and aggregations (Nefdt and Thirgood 1997).

These three mating tactics also occur in the Uganda kob (*Kobus kob*: Leuthold 1966), apparently with the same variance in reproductive success as in fallow deer (Rubenstein 1986). In the topi (*Damaliscus lunatus*), the variance in reproductive success of males on leks is high with few males getting the majority of matings (Gosling and Petrie 1990, Bro-Jørgensen and Durant 2003). Males on leks are larger and older than those on resource-based territories and have greater immediate and long-term reproductive success than those defending resource-based territories (see also Balmford and Blakeman 1991). However, at any given moment, each individual likely assesses its options and adopts the tactic that gives the best net reward.

14.2.4 Territoriality versus wandering

A relatively common set of male alternative reproductive tactics in rodents is to be either territorial or nonterritorial and wander in search of mates. Prairie voles (*Microtus ochrogaster*) use both of these tactics. About 55% of males in a population are territorial and form a bond with one female. They mutually occupy and defend the space and mate monogamously (Getz et al. 1993). The remaining 45% of males are wanderers with large home ranges and no established bond with any one female. Males can switch from one tactic to the other. The two tactics do not appear to be based on physical condition, density, or male "quality" (Getz et al.

1993, Solomon and Jacquot 2002) nor is the relative reproductive success of each tactic known. Solomon and Jacquot (2002) concluded that wanderers were not of lower quality than territorial males, but were simply making the best of a bad job. Alternatively, wandering may be the best tactic: resident territorial males may be less competitive and thus mate guard and stay with just one female. To distinguish between these possibilities we would need to know the relative competitive abilities and reproductive success of males using the two tactics.

Switching between territoriality and wandering occurs in a number of other species. In arctic ground squirrels (*Spermophilus parryi*) territorial males have priority access to females and sire the most offspring; however, some wandering males are able to dominate agonistic interactions on the day of a female's estrus and achieve some reproductive success (Lacey and Wieczorek 2001). A similar pattern of territorial defense and extra-pair copulations with wandering, satellite, or neighboring males occurs in alpine marmots (Goossens et al. 1998), black-tailed prairie dogs (*Cynomys ludovicianus*: Travis et al. 1996), ground squirrels (*Spermophilus* sp.: e.g., Boellstorff et al. 1994), and spear-nosed bats (*Phyllostomus hastatus*: McCracken and Bradbury 1981) (Table 14.1). In bearded seals (*Erignathus barbatus*), dominant males defend aquatic territories and subordinate males wander or "roam" over large areas that overlap each other and several territories. Territorial males have a longer trill to their calls than roaming males, which may be an indicator of male quality (Parijs et al. 2003). Territoriality and wandering appear to be conditional reproductive tactics that occur in a variety of taxa and ecological conditions. Unfortunately little is known regarding the relative fitness benefits of each tactic.

14.2.5   Tending, challenging, and coursing

Among nonterritorial ungulates, the most common mating tactic is the tending of individual females by males, though alternative tactics do exist. Rocky Mountain bighorn rams (*Ovis canadensis*) use three distinct tactics in competition for mates (Hogg 1984). Two tactics, tending and blocking (males guard and block movements of individual females prior to estrus and until mating occurs), feature defense and cooperative mating over a prolonged consort period of up to 3 days. Coursing is the attack on dominant rams by subordinates during which subordinates are able to copulate with females for a very brief, perhaps a few seconds, contact time. Coursing is costly because the long chases over rough terrain may result in injury to either males or females. The coursing tactic can be equally successful to the cooperative tactics: Hogg and Forbes (1997) found that 43% and 47% of lambs were sired by males using the coursing and cooperative tactics, respectively

Three reproductive tactics are used by the American bison (*Bison bison*) to acquire matings (Wolff 1998) (Figure 14.4). Females and their young live in large fluid herds throughout the year and males are either solitary or live in bachelor groups except during the late July–August rut. The most successful bulls enter the herd early during the rut and tend females for up to 3 days before mating with them. During tending one male stands alongside a female and guards her until after copulation. A second alternative is for satellite males to challenge the tending bull and take the female away following defeat of the tending bull in aggressive threats or fights (Figure 14.5). This tactic is similar to that of the coursing rams in which subordinate males incite females to run away from the tending males. A third tactic is for less competitive males to enter herds late in the season when the dominant males, exhausted and weakened, have left the herd or are less competitive (Figure 14.4). All three tactics lead to reproductive success, though tending early and late in the season are the most successful (Wolff 1998) (Figure 14.2). The mating success of bulls late in the season is due, in part, to the lower number of bulls in the herd (35 of 38 bulls in early season and 14 of 46 in late season obtained at least one copulation). It is not known, however, if the reproductive success of late matings is comparable to that of early matings because calf survival might be lower in the former. A comparable tactic occurs in elephants (*Loxodonta africana*) in which musth males over 35 years of age dominate the tending and breeding during mid-estrus and younger 25- to 35-year-old males obtain

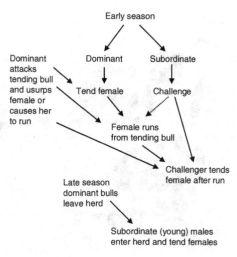

Figure 14.4 Alternative reproductive tactics used by bull bison and their relative success are a function of competitive ability and season (from Wolff 1998). A similar pattern of behavior is observed in bighorn sheep (Hogg 1984) and Soay sheep (Pemberton *et al.* 2004) (see text).

some matings early and late in the season (Poole 1989). It is not clear how successful these early and late matings are or whether or not they produce offspring.

Raccoons (*Procyon lotor*) exhibit a reproductive pattern similar to that of many ungulates. Gehrt and Fritzell (1999) described a pattern in which several males aggregate around permanent water sources (feeding areas) and then individual, dominant males develop consortships with individual females (a form of tending). An alternative tactic is for some males, considered subordinate, to roam and encounter females away from groups of dominant males. Gehrt and Fritzell (1999) found that promiscuity was common and that 38% of the females consorted with more than one male.

14.2.6   Pursuit or satellite

In some species in which males do not defend territories, several, often large numbers, of males aggregate and pursue estrous females. Male eastern gray squirrels (*Sciurus carolinensis*: Koprowski 1993a) and fox squirrels (*S. niger*: Koprowski 1993b) congregate around females during their single day of estrus and use two tactics to achieve matings. The most successful tactic is active pursuit

Figure 14.5 Bison bull tending a cow and being challenged by three satellite bulls. (Photograph by J. O. Wolff.)

and it is used by dominant males >2.75 years old that are able to defend proximity to females. Satellite males are subordinate and remain dispersed in the female's home range and obtain some matings when females escape from dominant, pursuing males. Although dominant males achieve most of the matings, once a female escapes from a dominant male, matings are evenly distributed between pursuer and satellite males (Figure 14.2). The active pursuit and the satellite tactics appear to be conditional tactics in which young, subordinate males are making the best of a bad job (Koprowski 1993a, b). Similar tactics occur in thirteen-lined ground squirrels (*Spermophilus tridecemlineatus*: Schwagmeyer and Parker 1987) and Belding's ground squirrels (*S. beldingi*: Sherman 1989).

Sherman (1989) made a comparison of reproductive tactics used by Idaho ground squirrels (*Spermophilus brunneus*) and Belding's ground squirrels. In Idaho ground squirrels, mate guarding and defending an insemination post-copulation are more successful than abandoning the female and searching for additional females. In contrast, in Belding's ground squirrels, searching after a copulation is more successful than mate guarding. In Idaho ground squirrels, females are widely spaced, searching is costly, and unguarded females mate multiply. In addition, the last male to mate sires most of the offspring. In Belding's ground squirrels, additional females are readily available and

accessible, and the first male to mate sires most of the offspring. Thus each species uses a different tactic, each having a different pay-off for the male depending on the distribution of females and the probability of multiple inseminations; mate guarding is best for Idaho ground squirrels and searching is best for Belding's ground squirrels (Sherman 1989).

In Cape ground squirrels (*Xerus inauris*), females can breed throughout the year but the number of females in estrus at any one time is small resulting in an operational sex ratio at any given time of about 10 males to 1 estrous female (Waterman 1998). Males roam in an amicable hierarchical band in search of estrous females. Dominant males are the most successful at finding estrous females and they obtain the majority of copulations, but females mate with several males and consequently subordinate males also obtain mating opportunities. Although aggression is minimal among band members, disruptions during copulations do occur, which in turn provide more opportunities for mating by subordinates. Living in a hierarchical group provides an increased level of protection for males in the group through improved vigilance. Hierarchical groups provide some immediate mating opportunities and the potential to move up the hierarchy with age and status (Waterman 1998).

In bridled nailtail wallabies (*Onychogalea fraenata*), males have home ranges that overlap those of two or more females (Fisher and Lara 1999). Occasionally, groups of large, dominant males will aggregate in areas of high densities of females. Alternatively, when individual females are in estrus, as many as six males may follow them and vie for opportunities to mate. The largest and presumed oldest males that have the largest home ranges and encounter the most females obtain the most matings. Females have relatively long estrous periods and roam widely during estrus, inciting male–male competition. This behavior attracts the most dominant males who sire most of the offspring; however, satellite males that have small home ranges can find lone females and account for up to 10% of the matings. This mating pattern is similar to that described for most other large terrestrial macropods (reviewed in Fisher and Lara 1999).

14.2.7 Alliances and coalitions

In many species of social mammals, single, dominant males attempt to sequester a female or are chosen by females for exclusive mating. However, an alternative tactic in these species is for two or more individuals to join forces against conspecifics, especially dominant individuals that are in a long-term alliance or coalition (Harcourt and de Waal 1992). Male alliances are generally rare in mammals (Clutton-Brock 1989) but have been documented in dolphins, primates, ungulates, and carnivores (e.g., Harcourt and de Waal 1992, Caro 1994, van Hoof and van Schaik 1994, Feh 1999, Connor et al. 2000) and may even be the dominant reproductive tactic where it occurs.

In dolphins (e.g., Tursiops aduncus and T. truncatus), alliances of two or three males cooperate to control the movements of a female to monopolize her during the period when she is fertile (Connor et al. 2000). Alliances or consorts of males seem to be more successful in controlling reproductive access to females than single individuals (Connor et al. 1996, 2000). Some alliances seem to switch often, but many are long-term and may last for several years or a lifetime. The breeding season is relatively long for dolphins and females may have extended periods of receptivity that last several months. The long period of receptivity may provide opportunities for several males to mate, thus reducing competition within a coalition (Connor et al. 1996).

Alliances consisting of two to three males, often brothers, cooperatively defend territories against other males in cheetahs (Acinonyx jubatus: Caro 1994) and lions (Panthera leo: Packer et al. 1991). Single males, on the other hand, wander over large areas and avoid territorial males while attempting to sequester lone females. Cooperation and territoriality is the most adaptive tactic for cheetahs. Caro (1994) observed that single males exhibit more stress and are less likely to acquire mates than territorial males. In cheetahs, pairs of males often scent mark their territories and perch on lookout posts, whereas single males are less conspicuous and avoid detection (Figure 14.6). Similarly lone male lions are not able to acquire or defend a pride of females or food resources. So, as in dolphins, single male cheetahs and lions do not appear to be successful in defending reproductive females and consequently alliances become the more successful tactic.

Figure 14.6 Coalitions of two male cheetahs are territorial and spend a significant amount of time on vantage points scouring for intruding single males. Dominant males scent mark their territories, whereas single intruder males avoid detection. (Photograph by Tim Caro.)

The mating system of feral horses is characterized by a single-male, harem-defense polygyny (see above); however, Feh (1999) described a situation in which low-ranking sons of low-ranking mares developed alliances with the dominant stallion that could last a lifetime. Both stallions would confront intruders or the dominant stallion would tend the females while the subordinate displayed toward an advancing rival. The subordinate of the pair sired about 25% of the foals born into the alliance harem, which is significantly more foals than low-ranking stallions sired by adopting a "sneak" mating tactic. The alliance appears to be based on a mutualism from which both the dominant and subordinate benefit (Feh 1999). A similar behavioral pattern was observed in oribi where dominant territorial males accepted subordinates on their territories. Socially dominant males that shared territories with subordinates were replaced by rivals less often than males that defended territories without auxiliary males. Sixty-three percent of the auxiliary males were likely offspring of territorial males, but 37% were not related. Auxiliary males obtained some direct and indirect (in the case of fathers and sons) fitness while occupying joint territories and eventually obtained territories of their own (Arcese 1999). Thus cooperation can be an adaptive competitive tactic between dominants and subordinates (see also Taborsky 2001).

Alliances are almost always limited to two or three males. Three is the maximum number of males for nonkin alliances in chimpanzees (*Pan troglodytes*), dolphins, cheetahs, and lions (Packer *et al.* 1988, 1991, Caro 1994, Watts 1998, Connor *et al.* 2000). Three may be the optimum number for maximizing individual reproductive success of males. A fourth male may have limited opportunity against three companion males if there are a limited number of females in estrus at any given time and therefore may be better off solitarily or forming another alliance of two or three (Connor *et al.* 2000).

14.2.8   Sexual coercion

In multimale societies or those in which males cannot always mate-guard or protect a female, an alternative tactic for males is to coerce females into mating with them and to avoid any sexual association with competitor males (Smuts and Smuts 1993, Clutton-Brock and Parker 1995). Males may coerce females by forced copulations, harassment (repeated attempts to copulate with a female), intimidation (males punish females that refuse to mate with them), and

violence if a female mates with another male (Clutton-Brock and Parker 1995). Sexual coercion as an alternative mating tactic is most common among primates (Smuts and Smuts 1993; and see Chapters 15 and 18), but has also been reported in elephant seals (Le Boeuf 1974), sea otters (*Enhydra lutris*: Mestal 1994), fallow deer (Clutton-Brock *et al.* 1992), and horses (Berger 1983).

14.2.9   An example of genetic polymorphism

Soay sheep (*Ovis aries*) are unusual in having a genetic polymorphism for horn size and shape (Doney *et al.* 1974). Approximately 85% of the males grow spiral horns and 15% have small, deformed horns referred to as "scurred" (Clutton-Brock *et al.* 2004). Horns are used in conflict among males with spiral-horned males dominating scurred males. Consequently, the two morphs use different reproductive tactics. The tactic for the dominant male is to tend or form consorts with individual estrous females (Pemberton *et al.* 2004). Small, young, and scurred males follow a more opportunistic coursing tactic similar to that used by bighorn sheep. Scurred and young males harass consort males causing females to run away from their tending males, which results in a chase by subordinate males. Subordinate males often obtain matings during these chases before the dominant males can catch and defend their females. Forced copulations by a succession of males often occur during these disruptive chases. The probability of siring at least one young in a given year is density dependent and ranges from <0.1 to 0.4 for yearlings and 0.2 to 0.55 for adults. Younger males will eventually become dominant and switch to the consort tactic, whereas scurred males will use the coursing tactic throughout their lifetimes. In some cases, the subordinate coursing tactic may account for the entire lifetime reproductive success of some rams (Pemberton *et al.* 2004). How the scurred morph is maintained in the population is not known, but reproductive success of this phenotype may be frequency dependent or attributable to its longer life expectancy (Stevenson *et al.* 2004). Alternatively, the genetic polymorphism for horn size may be a function of domestication and not an adaptation per se as it is not known to occur in wild sheep or other ungulates.

Multimale mating and sperm competition commonly occur in Soay sheep (Preston *et al.* 2001). Dominant rams usually tend females and achieve most of the matings and sire most of the offspring early during the rutting season. However as the season progresses, these males experience sperm depletion and their siring success declines as a result of sperm competition with subordinate males. Thus, by waiting until later in the rutting season, subordinate rams can achieve relatively high reproductive success.

14.2.10 Summary of alternative reproductive tactics: unequal pay-offs

In the section above, I provide examples that illustrate how some species have one tactic that results in the most mating opportunities and alternative tactics that are conditionally dependent on the frequency of dominant (tactic 1) individuals competing for females. In this sense, subordinate and often younger individuals make the best of the current situation by adopting an alternative tactic that provides some immediate reproductive success or makes them more competitive for future reproductive success. These alternative tactics may be optimal based on the current situation and competitive ability of the individuals involved, but one tactic has a higher immediate fitness advantage than the other. The switch point at which pay-offs are equal is a function of an individual's condition and the frequency of alternatives in the population (*sensu* Repka and Gross 1995). Differential pay-offs of alternative tactics must be considered at two levels. In the short term, pay-offs differ in reproductive success; however, over the lifetime of an individual, a combination of condition-dependent tactics may be the optimum strategy for maximizing lifetime reproductive success. In some situations, however, alternative tactics are based on environmental or phenotypic features rather than social status and can have equal pay-offs. Below I provide examples of alternative reproductive tactics that result in equal pay-offs (or at least are cases in which individuals are not making the best of a bad job).

14.3 ALTERNATIVE REPRODUCTIVE TACTICS IN MALES: EQUAL PAY-OFFS

14.3.1 Size and development

Successful reproduction within a cohort of males can vary seasonally based on body size and growth rate. In yellow-pine chipmunks (*Tamias amoenus*), Schulte-Hostedde and Millar (2002) found that small males were more aggressive and able to dominate larger males in one-on-one

encounters, but larger males were more successful in mating chases for direct access to estrous females. Schulte-Hostedde and Millar were not able to quantify reproductive success of the two phenotypes but concluded they were probably comparable. A similar relationship occurs in laboratory mice (*Mus musculus*) in which small males maintain dominance over larger individuals through increased scent marking, but at a cost of reduced growth and body size. As a result, small males become more vulnerable to dominance reversals later in life (Gosling et al. 2000). Trade-offs in dominance with respect to body size and growth rates purportedly result in comparable levels of fitness. Common shrews (*Sorex araneus*) use either a territorial or a wandering tactic based on their body size and rate of growth. In the spring, larger type B males have large territories that overlap several females, whereas smaller type A males use a wandering tactic (Stockley et al. 1994, 1996). By fall, the type A males achieve comparable size to type B males and are more successful later in the season resulting in relatively equal mating success for the two types. Stockley et al. concluded that mate-searching tactics might be conditional upon the timing of sexual maturation. There is no evidence in these three examples that differences in size and growth rates result from a genetic polymorphism, but the relative success of each phenotype could be subject to seasonal, frequency-dependent mating success.

14.3.2 Distribution and abundance of food

Reproductive tactics of males are commonly dependent on the dispersion of females, which in turn is a function of the distribution and abundance of resources, primarily food (e.g., Maher and Lott 2000) (Figure 14.7). Territoriality with resource-defense polygyny is the most common mating system of African antelope and other species of ungulates (Gosling 1986). Resource-defense polygyny requires a dependable food resource such that females predictably congregate on the territory. Males in several species of ungulates defend territories that are centered on high-quality, clumped resources. In many cases, males defend these territories long before the breeding season and thus claim residence as females move onto the sites during the rut (e.g., pronghorn: Byers 1997; alpine chamois: Hardenburg et al. 2000; red deer: Carranza et al. 1995; impala: Jarman 1979; feral asses [*Equus asinus*]: Woodward 1979; hartebeest [*Alcephalus buselaphus*], wildebeest [*Connochaetes taurinus*], topi [*Damaliscus korrigum*], and others, reviewed in Gosling

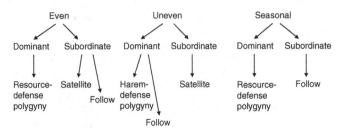

Figure 14.7 The dispersion and seasonal distribution of food resources dictates alternative reproductive tactics of male African ungulates in response to movement of females.

1986, Thirgood *et al.* 1999, and Maher and Lott 2000). An alternative to the sit-and-wait, resource-defense polygyny in these and other species of ungulates when resources become diminished is for males to switch to a wandering, following, or harem-defense tactic (Figure 14.7). These tactics are most likely to occur when the chance of a receptive female being in a given location is low. These conditions often include an unpredictable food resource such that females are always moving, and/or exhibit a low degree of breeding synchrony in response to unpredictable environmental change (Gosling 1986). A resource-defense polygyny tactic can switch to a following, harem-defense polygyny when there are changes in the distribution and predictability of resources. In this sense, each tactic may be equally successful for males and dictated by ecological rather than social environments.

The distribution of females, likely in response to food availability, also affected the breeding behavior of male red foxes (*Vulpes vulpes*) and harbor seals (*Phoca vitulina*). When food resources were abundant, red foxes on Round Island, Alaska were polygynous with higher reproductive success for males and females than individuals mating monogamously (Zabel and Taggart 1989). Nesting failure of seabirds following an El Niño event reduced food resources and foxes switched from polygyny to monogamy. Male harbor seals encounter females by distributing themselves on land where females bask in the sun and in the water where they go to forage (Coltman *et al.* 1999). The latter tactic appears to result in fewer matings but is probably relatively successful depending on the frequency of the former. The distribution of resting and foraging areas provides an opportunity for males to adopt alternative reproductive tactics.

14.3.3 Density

Because male tactics are a function of the dispersion of females, changes in density and spatial distribution of females often result in trade-offs in the relative success of different reproductive tactics. For instance, Langbein and Thirgood (1989) described as many as seven alternative tactics used by fallow deer including following, harems, dominance groups, stands (small territories), temporary stands, multiple stands, and leks. The ecological variables that exerted the greatest influence on mating systems were buck density and doe density. Reproductive success and the frequency of the various tactics used by males varied considerably depending on the dispersion of females and local mate competition resulting in conditional and equally successful mating tactics.

Musk oxen (*Ovibos moschatus*) males can maximize the number of matings per unit time by using two mating tactics with different pay-offs depending on demography. Increases in female group density and operational sex ratio favor "roving" among female groups, whereas low group density, low operational sex ratio, and increased time in nonmating activities favor a "staying" tactic (Forchhammer and Boomsma 1998). Under this scenario, males should spend the most time with females early and late in the rut and the least time during peak rut, which appears to be the common tactic used in this and other species of non-territorial ungulates (Geist 1971, Clutton-Brock *et al.* 1982, Prins 1995, Wolff 1998).

In white-footed mice (*Peromyscus leucopus*), males typically are territorial and overlap the territories of one to three females. However, when densities are low and females widely spaced, males use a wandering tactic and move from female to female after their current mate is pregnant (Wolff and Cicirello 1990). As densities increase, males switch back to a resource-defense territorial tactic. Switching from territoriality to wandering appears to be conditionally dependent on the density and spacing of females.

In all of the above cases, alternative reproductive tactics require optimal decision-making based on current

onditions. The tactics used differ from making the best of a
ad job because each tactic is equally successful and is based
n ecological or demographic factors rather than on status-
dependent selective pressures. Additional examples of
ariation in mating behavior associated with habitat, dens-
ty, food, and other socioecological factors are described in
ott (1984).

4.5 CONCLUSIONS

Alternative reproductive tactics in male mammals fall into
two categories based on their fitness potential. One tactic
used by the dominant and most competitive males is
deemed "best" when it results in the greatest number of
breeding opportunities. Subordinate individuals are rele-
ated to choosing alternative tactics that result in lower
immediate reproductive success under the current social
nd ecological constraints. In some cases, alliances between
dominants and subordinates can be an adaptive, competitive
actic. These alternative tactics result from social pressures
nd are status-based; subordinate individuals adopt inferior
actics, but those that still provide some immediate or future
reproductive success. Alternative tactics that produce few
ffspring can still be evolutionarily stable because males use
n evolved decision rule to adopt whichever tactic is optimal
onsidering their current status and the frequency of other
males using the dominant tactic. In this scenario, alternative
actics are conditional and based on an individual's status
nd the frequency of alternatives within its social environ-
nent. In the second category, variance in ecological par-
meters, such as the temporal and spatial distribution of
ood and the density and dispersion of females, can favor
witching among alternative tactics that have equal pay-offs.
Thus, the best tactic under one set of ecological conditions
nay not be the best under a different ecological setting. In
his latter case, selective pressures for switching among
lternative tactics result from changes in ecological and
emographic parameters rather than social pressures.

Two areas of uncertainty regarding alternative repro-
uctive tactics in mammals are the relative fitness pay-offs
or each tactic and the genetic variation for competitive
bility. The currency currently used to estimate fitness pay-
ffs is the number of copulations for each tactic; however,
ew studies have been able to quantify accurately the
umber of young sired by males using each tactic. Also, we
now very little about the lifetime reproductive success of
ge- or status-specific reproductive tactics and when in an
ndividual's life it switches to an alternative. Second, we

know very little regarding genetic variation for competitive
ability. Little evidence exists for a genetic polymorphism for
alternative reproductive tactics for male mammals with the
exception of horn phenotype in Soay sheep. This remains a
fruitful area for future research. Future research should
focus on accurate assessments of reproductive success and
on the genetic variation that contributes to these fitness pay-
offs for alternative reproductive tactics in male mammals.

References

Alvarez, F., Braza, F., and San Jose, C. 1990. Coexistence of
   territoriality and harem defense in a rutting fallow deer
   population. *Journal of Mammalogy* 71, 692–695.

Arcese, P. 1999. Effect of auxiliary males on territory
   ownership in the oribi and the attributes of multimale
   groups. *Animal Behaviour* 57, 61–71.

Asa, C. S. 1999. Male reproductive success in free-ranging
   feral horses. *Behavioral Ecology and Sociobiology* 47,
   89–93.

Balmford, A. and Blakeman, S. 1991. Horn and body
   measurements of topi in relation to a variable mating
   system. *African Journal of Ecology* 29, 37–42.

Berger, J. 1983. Induced abortion and social factors in wild
   horses. *Nature* 303, 59–61.

Berger, J. 1986. *Wild Horses of the Great Basin*. Chicago,
   IL: University of Chicago Press.

Boellstorff, D. E., Owings, D. H., Penedo, M. C. T., and
   Hersek, M. J. 1994. Reproductive behaviour and multiple
   paternity of California ground squirrels. *Animal Behaviour*
   47, 1057–1064.

Brockmann, H. J. 2001. The evolution of alternative
   reproductive strategies and tactics. *Advances in the Study of
   Behavior* 30, 1–51.

Bro-Jørgensen, J. and Durant, S. M. 2003. Mating strategies
   of topi bulls: getting in the centre of attention. *Animal
   Behaviour* 65, 585–594.

Byers, J. A. 1997. *American Pronghorn*. Chicago, IL:
   University of Chicago Press.

Caro, T. M. 1994. *Cheetahs of the Serengeti Plains*. Chicago,
   IL: University of Chicago Press.

Carranza, J., Garcia-Munoz, A. J., and de Dios Vargas, J.
   1995. Experimental shifting from harem defence to
   territoriality in rutting red deer. *Animal Behaviour* 49,
   551–554.

Clapham, P. J. 2000. The humpback whale: seasonal feeding
   and breeding in a baleen whale. In J. Mann, R. C. Connor,
   P. L. Tyack, and H. Whitehead (eds.) *Cetacean Societies*,
   pp. 173–196. Chicago, IL: University of Chicago Press.

Clutton-Brock, T. H. 1989. Mammalian mating systems. *Proceedings of the Royal Society of London B* 236, 339–372.

Clutton-Brock, T. H. and Parker, G. A. 1995. Sexual coercion in animal societies. *Animal Behaviour* 49, 1345–1365.

Clutton-Brock, T. H., Guiness, F. E., and Albon, S. D. 1982. *Red Deer: Behavior and Ecology of Two Sexes.* Chicago, IL: University of Chicago Press.

Clutton-Brock, T. H., Price, O. F., and MacColl, A. D. C. 1992. Mate retention, harassment, and the evolution of ungulate leks. *Behavioral Ecology* 3, 234–242.

Clutton-Brock, T. H., Deutsch, J. C., and Nefdt, R. J. C. 1993. The evolution of ungulate leks. *Animal Behaviour* 46, 1121–1138.

Clutton-Brock, T. H., Pemberton, J., Coulson, T., Stevenson, I. R., and MacColl, A. D. C. 2004. The sheep of St. Kilda. In T. H. Clutton-Brock and J. Pemberton (eds.) *Soay Sheep: Dynamics and Selection in an Island Population,* pp. 17–51. Cambridge, UK: Cambridge University Press.

Coltman, D. W., Bowen, W. D., and Wright, J. M. 1999. A multivariate analysis of phenotype and paternity in male harbor seals, *Phoca vitulina,* at Sable Island, Nova Scotia. *Behavioral Ecology* 10, 169–177.

Connor, R. C., Richards, A. F., Smolker, R. A., and Mann, J. 1996. Patterns of female attractiveness in Indian Ocean bottlenose dolphins. *Behaviour* 133, 37–69.

Connor, R. C., Read, A. J., and Wrangham, R. 2000. Male reproductive strategies and social bonds. In J. Mann, R. C. Connor, P. L. Tyack, and H. Whitehead (eds.) *Cetacean Societies,* pp. 247–269. Chicago, IL: University of Chicago Press.

Dominey, W. J. 1984. Alternative mating tactics and evolutionary stable strategies. *American Zoologist* 24, 385–396.

Doney, J. M., Rider, M. L., Gunn, R. G., and Grubb, P. 1974. Colour, conformation, affinities, fleece and patterns of inheritance of the Soay sheep. In P. A. Jewell, C. Milner, and J. M. Boyd (eds.) *The Ecology of the Soay Sheep of St. Kilda,* pp. 88–125. London: Athlone Press.

Feh, C. 1999. Alliances and reproductive success in Camargue stallions. *Animal Behaviour* 57, 705–713.

Fisher, D. O. and Lara, M. C. 1999. Effects of body size and home range on access to mates and paternity in male bridled nailtail wallabies. *Animal Behaviour* 58, 121–130.

Forchhammer, M. C. and Boomsma, J. J. 1998. Optimal mating strategies in nonterritorial ungulates: a general model tested on muskoxen. *Behaviour* 9, 136–143.

Gehrt, S. D. and Fritzell, E. K. 1999. Behavioural aspects of the raccoon mating system: determinants of consortship success. *Animal Behaviour* 57, 593–601.

Geist, V. 1971. *Mountain Sheep: A Study in Behavior and Evolution.* Chicago, IL: University of Chicago Press.

Getz, L. L., McGuire, B., Pizzuto, T., Hofmann, J. E., and Frase, B. 1993. Social organization of the prairie vole (*Microtus ochrogaster*). *Journal of Mammalogy* 74, 44–51.

Ginsberg, J. R. and Rubenstein, D. I. 1990. Sperm competition and variation in zebra mating behavior. *Behavioral Ecology and Sociobiology* 26, 427–434.

Goossens, B., Graziani, L., Waits, L. P., *et al.* 1998. Extra-pair paternity in the monogamous Alpine marmot revealed by nuclear DNA microsatellite analysis. *Behavioral Ecology and Sociobiology* 43, 281–288.

Gosling, L. M. 1986. The evolution of mating strategies in male antelope. In D. I. Rubenstein and R. W. Wrangham (eds.) *Ecological Aspects of Social Evolution,* pp. 244–281. Princeton, NJ: Princeton University Press.

Gosling, L. M. and Petrie, M. 1990. Lekking in topi: a consequence of satellite behaviour by small males at hotspots. *Animal Behaviour* 40, 272–287.

Gosling, L. M., Roberts, S. C., Thornton, E. A., and Andrews, M. J. 2000. Life history costs of olfactory status signaling in mice. *Behavioral Ecology and Sociobiology* 48, 328–332.

Gross, M. R. 1996. Alternative reproductive strategies and tactics: diversity within sexes. *Trends in Ecology and Evolution* 11, 92–98.

Harcourt, A. H. and de Waal, F. B. M. 1992. *Coalitions and Alliances in Humans and Other Primates.* Oxford, UK: Oxford University Press.

Hardenberg, von A., Bassano, B., Peracino, A., and Lovari, S. 2000. Male alpine chamois territories at hotspots before the mating season. *Ethology* 106, 617–630.

Heckel, G. and von Helversen, O. 2002. Male tactics and reproductive success in the harem polygynous bat *Saccopteryx bilineata. Behavioral Ecology* 13, 750–756.

Hoelzel, A. R., Le Boeuf, B. J., Reiter, J., and Campagna, C. 1999. Alpha-male paternity in elephant seals. *Behavioral Ecology and Sociobiology* 46, 298–306.

Hogg, J. T. 1984. Intrasexual competition in bighorn sheep: multiple creative male strategies. *Science* 225, 526–529.

Hogg, J. T. and Forbes, S. H. 1997. Mating in bighorn sheep: frequent male reproduction via a high-risk "unconventional" tactic. *Behavioral Ecology and Sociobiology* 41, 33–48.

svaran, K. and Jhala, Y. V. 2000. Variation in lekking costs in blackbuck (*Antilope cervicapra*): relationship to lek-territory location and female mating patterns. *Behaviour* 137, 547–563.

arman, M. V. 1979. Impala social behaviour: territory, hierarchy, mating, and use of space. *Advances in Ethology* 21, 1–92.

.oprowski, J. L. 1993a. Alternative reproductive tactics in male eastern gray squirrels: "making the best of a bad job." *Behavioral Ecology* 4, 165–171.

.oprowski, J. L. 1993b. Behavioral tactics, dominance, and copulatory success among male fox squirrels. *Ethology Ecology and Evolution* 5, 169–176.

.acey, E. A. and Wieczorek, J. R. 2001. Territoriality and male reproductive success in arctic ground squirrels. *Behavioral Ecology* 12, 626–632.

.angbein, J. and Thirgood, S. J. 1989. Variation in mating systems of fallow deer (*Dama dama*) in relation to ecology. *Ethology* 83, 195–214.

.e Boeuf, B. J. 1974. Male–male competition and reproductive success in elephant seals. *American Zoologist* 14, 163–176.

.euthold, W. 1966. Variations in territorial behavior of Uganda kob *Adenota kob thomasi* (Neumann 1896). *Behaviour* 27, 214–257.

.inklater, W. L. and Cameron, E. Z. 2000. Tests for cooperative behaviour between stallions. *Animal Behaviour* 60, 731–734.

.ott, D. F. 1984. Intraspecific variation in social systems of wild vertebrates. *Behaviour* 28, 266–325.

.ovari, S. and Locati, M. 1991. Temporal relationships, transitions and structure of the behavioural repertoire in male Apennine chamois during the rut. *Behaviour* 119, 77–103.

.aher, C. R. and Lott, D. F. 2000. A review of ecological determinants of territoriality within vertebrate species. *American Midland Naturalist* 143, 1–29.

.cCracken, G. F. and Bradbury, J. W. 1981. Social organization and kinship in the polygynous bat *Phyllostomus hastatus*. *Behavioral Ecology and Sociobiology* 8, 11–34.

.estal, R. 1994. The seamy side of sea otter life. *New Scientist* 1913, 5–6.

.oore, N. P., Kelly, P. F., Cahill, J. P., and Hayden, T. J. 1995. Mating strategies and mating success of fallow (*Dama dama*) bucks in a non-lekking population. *Behavioral Ecology and Sociobiology* 36, 91–100.

Nefdt, R. J. C. and Thirgood, S. J. 1997. Lekking, resource defense, and harassment in two subspecies of lechwe antelope. *Behavioral Ecology* 8, 1–9.

Packer, C., Herbst, L., Pusey, A. E., *et al.* 1988. Reproductive success in lions. In T. H. Clutton-Brock (ed.) *Reproductive Success*, pp. 363–383. Chicago, IL: University of Chicago Press.

Packer, C., Gilbert, D. A., Pusey, A. E., and O'Brien, S. J. 1991. A molecular genetic analysis of kinship and cooperation in African lions. *Nature* 351, 562–565.

Parijs, S. M. V., Lydersen, C., and Kovacs, K. M. 2003. Vocalizations and movements suggest alternative mating tactics in male bearded seals. *Animal Behaviour* 65, 273–283.

Pemberton, J., Coltman, D. W., Smith, J. A., and Bancroft, D. R. 2004. Mating patterns and male breeding success. In T. H. Clutton-Brock and J. Pemberton (eds.) *Soay Sheep: Dynamics and Selection in an Island Population*, pp. 166–189. Cambridge, UK: Cambridge University Press.

Poole, J. H. 1989. Mate guarding, reproductive success and female choice in African elephants. *Animal Behaviour* 37, 842–849.

Preston, B. T., Stevenson, I. R., Pemberton, J. M., and Wilson, K. 2001. Dominant rams lose out by sperm depletion. *Nature* 409, 681–682.

Prins, H. H. T. 1995. *Ecology and Behaviour of the African Buffalo*. London: Chapman and Hall.

Repka, J. and Gross, M. R. 1995. The evolutionarily stable strategy under individual condition and tactic frequency. *Journal of Theoretical Biology* 176, 27–31.

Rubenstein, D. I. 1986. Ecology and sociality in horses and zebras. In D. I. Rubenstein and R. W. Wrangham (eds.) *Ecological Aspects of Social Evolution*, pp 282–302. Princeton, NJ: Princeton University Press.

Schulte-Hostedde, A. I. and Millar, J. S. 2002. "Little chipmunk" syndrome? Male body size and dominance in captive yellow-pine chipmunks (*Tamias amoenus*). *Ethology* 108, 127–137.

Schwagmeyer, P. L. and Parker, G. A. 1987. Queuing for mates in thirteen-lined ground squirrels. *Animal Behaviour* 35, 1015–1025.

Sherman, P. W. 1989. Mate guarding as paternity insurance in Idaho ground squirrels. *Nature* 338, 418–420.

Smuts, B. B. and Smuts, R. 1993. Male aggression and sexual coercion of females in nonhuman primates and other mammals: evidence and theoretical implications. *Advances in the Study of Behavior* 22, 1–63.

Solomon, N. G. and Jacquot, J. J. 2002. Characteristics of resident and wandering prairie voles, *Microtus ochrogaster*. *Canadian Journal of Zoology* 8, 951–955.

Stevenson, I. R., Pemberton, J., Preston, B. T., Pemberton, J. R., and Wilson K. 2004. Adaptive reproductive strategies. In T. H. Clutton-Brock and J. Pemberton (eds.) *Soay Sheep: Dynamics and Selection in an Island Population*, pp. 243–275. Cambridge, UK: Cambridge University Press.

Stockley, P., Searle, J. B., Macdonald, D. W., and Jones, C. S. 1994. Alternative reproductive tactics in male common shrews: relationships between mate-searching behaviour, sperm production, and reproductive success as revealed by DNA fingerprinting. *Behavioral Ecology and Sociobiology* 34, 71–78.

Stockley, P., Searle, J. B., Macdonald, D. W., and Jones, C. S. 1996. Correlates of reproductive success within alternative mating tactics of the common shrew. *Behavioral Ecology* 7, 334–340.

Taborsky, M. 1994. Sneakers, satellites, and helpers: parasitic and cooperative behavior in fish reproduction. *Advances in the Study of Behavior* 23, 1–100.

Taborsky, M. 2001. The evolution of bourgeois, parasitic, and cooperative reproductive behaviors in fish. *Journal of Heredity* 92, 100–110.

Thirgood, S., Langbein, J., and Putman, R. J. 1999. Intraspecific variation in ungulate mating strategies: the case of the fallow deer. *Advances in the Study of Behavior* 28, 333–361.

Travis, S. E., Slobodchikoff, C. N., and Keim, P. 1996. Social assemblages and mating relationships in prairie dogs: a DNA analysis. *Behavioral Ecology* 7, 95–100.

van Hoof, J. A. R. A. M. and van Schaik, C. P. 1994. Male bonds: affiliative relationships among nonhuman primate males. *Behaviour* 130, 309–337.

Voigt, C. C. and Streigh, W. J. 2003. Queuing for harem access in colonies of the greater sac-winged bat. *Animal Behaviour* 65, 149–156.

Waterman, J. M. 1998. Mating tactics of male Cape ground squirrels, *Xerus inauris*: consequences of year-round breeding. *Animal Behaviour* 56, 459–466.

Watts, D. P. 1998. Coalitionary mate guarding by male chimpanzees at Ngogo, Kibale National Park, Uganda. *Behavioral Ecology and Sociobiology* 44, 43–55.

Wirtz, P. 1981. Territorial defence and territorial take-over by satellite males in the waterbuck *Kobus ellipsiprymnus* (Bovidae). *Behavioral Ecology and Sociobiology* 8, 161–162.

Wirtz, P. 1982. Territory holders, satellite males, and bachelor males in a high density population of waterbuck (*Kobus ellipsiprymnus*) and their association with conspecifics. *Zeitschrift für Tierpsychologie* 58, 277–300.

Wolff, J. O. 1998. Breeding strategies, mate choice, and reproductive success in the American bison. *Oikos* 84, 529–544.

Wolff, J. O. and Cicirello, D. M. 1990. Mobility versus territoriality: alternative reproductive strategies in a polygynous rodent. *Animal Behaviour* 39, 1222–1224.

Woodward, S. L. 1979. The social system of feral asses (*Equus asinus*). *Zeitschrift für Tierpsychologie* 49, 304–316.

Zabel, C. J. and Taggart, S. J. 1989. Shift in red fox, *Vulpes vulpes*, mating system associated with El Niño in the Bering Sea. *Animal Behaviour* 38, 830–838.

15 · Alternative reproductive tactics in primates

JOANNA M. SETCHELL

CHAPTER SUMMARY

A wide diversity of reproductive strategies and alternative reproductive tactics (ARTs) have evolved to promote the reproductive success of individual male and female primates. Intraspecific variation in male mating strategies has received far more attention than flexibility of reproductive behavior in female primates. However, female primates may also employ ARTs, with important implications for lifetime reproductive success. ARTs in primates tend to be limited to behavior, gonads, and physiology and are rarely associated with dramatic alternative morphologies, although striking exceptions to this rule exist. This is likely due to the advantages of plasticity and the lower costs of adjustment according to changing characteristics of the individual and social conditions. Most ARTs in primates appear to be "best-of-a-bad-job" phenotypes, whereby inferior individuals, or those in a suboptimal situation, make the most of any opportunity available to gain reproductive success. With the exception of female reproductive suppression in common marmosets, relatively little is known about the life-history pathways underlying ARTs and the factors that determine their expression. Finally, male and female reproductive strategies are intricately linked in primates, and interactions between the sexes play an important role in the evolution of primate ARTs.

5.1 INTRODUCTION

The adaptive adjustment of individuals to differences in their social and ecological environment is expected to lead to intraspecific variation in reproductive tactics (Rubenstein 1980, Dunbar 1982, Clutton-Brock 1989, Davies 1991, Lott 1991). Where consistent and discrete variation occurs in the reproductive behavior of one sex within a population, and the tactics serve the same functional end, they are referred to as alternative reproductive tactics (ARTs) (Brockman *et al.*

1979, Rubenstein 1980, Dominey 1984). ARTs have been demonstrated for a variety of taxa, including insects, Crustacea, fish, amphibians, reptiles, birds, and mammals (Henson and Warner 1997, Brockmann 2001, Schuster and Wade 2003; and see relevent chapters in this volume). Alternative behavioral tactics may occur with no associated morphological differences, but often co-occur with specific morphological, physiological, and life-history differences. For example, "resident" male ruffs (*Philomachus pugnax*) have dark plumage and defend courts on a lek; "satellite" males are white and share courts with resident males (van Rhijn 1973). Similarly big-horned adult male scarab beetles (*Onthophagus* spp.) fight for access to females and males with no or tiny horns mate sneakily (Cooke 1990, Emlen 1994).

Studies of primates can contribute to our general understanding of the evolution of mating systems, reproductive strategies, and ARTs for several reasons. First, primates exhibit complex social behavior that is likely to be reflected in their reproductive tactics (Kappeler and van Schaik 2002, Setchell and Kappeler 2003). Second, primates exhibit a diversity of ecological (Sussman 1999, 2000, in press) and life-history strategies (Kappeler and Pereira 2003), as well as a variety of social organizations, social structures, and mating systems that is rivaled by few other mammalian orders (Kappeler and van Schaik 2002). We may therefore expect equally diverse reproductive strategies and tactics (Setchell and Kappeler 2003). Finally, a great deal of detailed knowledge exists concerning primate behavior and ecology in comparison to other mammals (Smuts *et al.* 1987, Lee 1999, Kappeler 2000b, Kappeler and Pereira 2003), making the order a rich source of comparative data for the investigation of mammalian reproductive strategies.

In this chapter I review the current state of knowledge of ARTs in primates. I begin with a brief introduction to ARTs, before examining types and patterns of variability in reproductive tactics employed by individual male and

Alternative Reproductive Tactics, ed. Rui F. Oliveira, Michael Taborsky, and H. Jane Brockmann. Published by Cambridge University Press.
© Cambridge University Press 2008.

female primates, and the circumstances under which alternatives have evolved. I then examine briefly the role that interactions between the sexes plays in the expression and evolution of reproductive strategies and end with some general conclusions regarding the study of ARTs in primates.

15.2   ASSESSING ARTs

ARTs can be viewed as representing strategies for the allocation of time and/or resources to different activities to maximize individual fitness (Brockmann 2001). A complete understanding of ARTs therefore requires an integrated investigative approach combining patterns of expression, behavior, morphology, development, physiology, life-history pathways, relative fitness pay-offs, and the genetic basis of alternative phenotypes (Brockmann 2001). However, such knowledge is available for relatively few taxa (Henson and Warner 1997).

Taborsky (1998) has suggested that the evolution and maintenance of ARTs in a population can be assessed at three separate levels: determination, plasticity, and selection. Determination refers to whether reproductive phenotypes are genetically or environmentally determined (West-Eberhard 1979). Genetically based phenotypes are fixed, meaning that an individual can only display the phenotype determined by its genes, as in swordtails (*Xiphophorus nigrensis* and *X. pygmaeus*: Ryan and Causey 1989), ruffs (*Philomachus pugnax*: Lank et al. 1995), and the marine isopod *Paracerceis sculpta* (Schuster and Wade 1991). By contrast, where phenotypes are environmentally determined (facultative), each individual has the potential to display more than one phenotype. However, Taborsky (1998) and several other authors have pointed out that the distinction between genetic and environmental determination represents an artificial dichotomy, and that conditions (environmental, social, and individual) are likely to affect the expression of all ARTs (West-Eberhard 1979, Caro and Bateson 1986, Gross 1996, Brockmann 2001).

Plasticity refers to the underlying mechanisms that regulate alternative phenotypes (Taborsky 1998). Reproductive phenotypes may be fixed for life over an individual's lifetime (irreversible), or have the potential to change over lifetime (reversible). Reversible phenotypes can be simultaneous, where an individual is able to change back and forth between different phenotypes, or sequential, where a one-time switch exists (Caro and Bateson 1986, Gross 1996, Brockmann 2001). Moore (1991) has proposed differing

proximate hormonal influences for developmentally fixed and plastic phenotypes. In irreversible cases, hormones play an organizational role prior to adulthood, but hormonal levels do not vary among adult phenotypes. In reversible phenotypes, however, hormones play an activational role during adulthood, and variation occurs in hormonal characters between adult phenotypes (Moore 1991). The relative costs and benefits of ARTs for an individual change as a function of age, relative size, body condition, future reproductive opportunity, the intensity of intrasexual competition, and prior residence or environmental conditions such as predation risk (Taborsky 1998).

Finally, ARTs can be assessed in terms of why and how selection favors more than one tactic, and why variants are maintained in a population (Taborsky 1998). ARTs have traditionally been viewed as being maintained via frequency-dependent selection (Rubenstein 1980, Hutchings and Myers 1994, Gross 1996), or by differences in the quality of individuals (Dawkins 1980, Davies 1982, Dunbar 1982, Hazel et al. 1990). In the former situation, the relative fitnesses of different ARTs depend on their frequency in the population, resulting in a stable mixture of phenotypes (evolutionarily stable strategy [ESS]: Maynard Smith 1982). In the latter, the fitness benefits of different ARTs need not be equal and the relative costs and benefits of ARTs differ between individuals due to differences in status, such as ontogenetic stage, age, body condition, and experience (Gross 1996). Lower-quality individuals maximize their lifetime reproductive success by adopting alternative tactics and making the "best of a bad job" (Dawkins 1980, Davies 1982, Dunbar 1982), rather than by attempting to monopolize access to females, even if the ARTs employed do not provide similar fitness pay-offs. Similarly, young, competitively inferior males may employ opportunistic "side-payment tactics" (Dunbar 1982), making the best of their current situation while investing in growth at the expense of high immediate levels of reproduction. ARTs thus represent an optimal response to the particular situation in which an individual finds itself.

Brockmann (2001) has shown that two (or more) ARTs are maintained in a population where the fitness curves cross, and that the factor most likely to cause this to occur is a switch in behavior based on the relationship between fitness and individual status (status-dependent selection: Gross 1996). Frequency-dependent selection thus may play a role in all ARTs, including those that are condition dependent, with the ESS switch point determining both the condition at which an individual changes and the stable frequency of ARTs determining both the condition at which an individual changes, and the stable frequency of ARTs in the population (Repka and Gross 1995, Gross and Repka 1998, Brockmann 2001). Finally, Shuster and Wade 2003) suggest that current theories of status-dependent selection (e.g., Gross 1996) are limited by their use of the average fitness of males, ignoring the variance between individual males. Whereas the status-dependent selection model suggests that no heritable variation exists for ARTs, Shuster and Wade (2003) argue that genetic polymorphism in male mating behavior may be more common than presently recognized.

5.3   ARTs IN MALE PRIMATES

As in other taxa (Henson and Warner 1997), ARTs arise in male primates as a consequence of "winner-take-all" situations, in which reproductive success is skewed to a few dominant males in a population, who employ "bourgeois" mate acquisition tactics (Taborsky 1994, 1997). The resulting reproductive skew favors the evolution of any alternative, parasitic" behavior that allows inferior-quality males to obtain at least some reproductive success, while avoiding the risks of attempting to gain high rank (Taborsky 1994, 1997). The degree of reproductive skew in male primates is determined by the monopolizability of females, which in turn is determined by their spatial distribution, the degree of synchrony of female receptive periods, and the absolute number of females in group-living species (Mitani et al. 1996a, b, Nunn 1999, Kappeler 2000a). Most primates live in bisexual groups, where male dominance rank reflects relative power in excluding other males from resources, including estrous females (van Noordwijk and van Schaik 2004). Skewed reproductive success may also occur where females are dispersed, if dominant males are able to defend home ranges that encompass those of several females to the exclusion of rival males. Reproductive skew may also occur under less obvious circumstances. For example, extra-pair fertilizations (EPFs) are known to occur in pair-living primates (Fietz et al. 2000), meaning that some males may be able to increase their relative reproductive success at the expense of other males. However, currently available paternity data are insufficient to determine whether skew occurs in such situations.

Competing primate males have evolved a variety of ARTs by which to overcome the monopolization of females by primary access males (Table 15.1). These can be divided into pre-mating, mating, and post-mating ARTs (Taborsky 1999, 2001). Pre-mating ARTs are concerned with obtaining access to mates, mating ARTs concern actual mating behavior, and post-mating ARTs are concerned with the degree of male investment in parental care. In this section I describe the different ARTs known for male primates, then use Taborsky's (1998) three levels of assessment (determination, plasticity, and selection) to examine the evolution and maintenance of these ARTs in primate populations.

15.3.1   Pre-mating ARTs in male primates

Male primates can increase their likelihood of future mating access to receptive females in various ways, depending on the strategies of females and of other males and on their degree of engagement in male–male competition. In a dispersed social system, males may opt to defend a territory or to be nomadic. In group-living species, males live with a group, alone, or in an all-male band. Possibilities for ARTs thus include dispersal, transfer, and group residency decisions, as well as the tactics used to obtain a group of females (in a one-male, multifemale social system), attract females, and increase the number of females available for fertilization. Finally, male behavioral strategies may be accompanied by morphological differences: while dominant males show maximal development of secondary sexual characters, subordinate males may suppress the development of such characters, reducing inter-male competition and investment in reproduction and facilitating the use of behavioral ARTs.

TERRITORIAL VERSUS NOMADIC MALES
The best-studied example of territorial versus nomadic male ARTs in primates occurs in orang-utans (Box 15.1). A similar situation appears to occur in some galagos and pottos where dominant, territorial A males have full adult body weight, and subordinate B or vagabond males occupy peripheral ranges (Table 15.1). In these cases, the strategy employed by a male is likely to be dependent on age and condition, and nomadic or B males may be younger males waiting for the opportunity to establish a territory (Dixson 1998). Little is known about the reproductive pay-offs of the different tactics, although larger male Galago moholi have higher mating success than smaller males (Pullen et al. 2000).

GROUP MEMBERSHIP VERSUS INCURSIONS
In group-living species, a prime adult male can achieve high reproductive success as a residential male in a bisexual group, particularly if he is top-ranking (Ohsawa et al. 1993, Borries 2000, Takahashi 2001). However, young or post-prime males with lower competitive ability have lower

Table 15.1. *Incidence of ARTs in primates by family*

Family[a]	Species	Common name	Sex	Type[b]	ART	References
Cheirogaleidae	*Microcebus murinus*	Lesser mouse lemur	M	Pre-	Alternative morphologies	Perret 1992
Lemuridae	*Lemur catta*	Ring-tail lemur	M	Pre-	Transfer decisions	Sussman 1992
Megaladapidae	—					
Indridae	*Propithecus verreauxi*	Sifaka	M	Pre-	Alternative morphologies	Kraus et al. 1999
Daubentoniidae	—					
Loridae	*Perodicticus potto*	Potto	M	Pre-	Territorial vs. nomadic	Charles-Dominique 1977, Bearder 1987
Galagonidae	*Galago moholi*	Mohol's galago	M	Pre-	Territorial vs. nomadic	Charles-Dominique 1977, Bearder 1987, Pullen et al. 2000
	Galagoides demidoff	Demidoff's galago	M	Pre-	Territorial vs. nomadic	Charles-Dominique 1977, Bearder 1987
Tarsiidae						
Cebidae	*Alouatta seniculus*	Red howler monkey	M	Pre-	Dispersal decisions	Pope 1990, 1998
			M	Pre-	Obtaining a harem	Crockett and Pope 1993
			F	Pre-	Dispersal decisions	Crockett and Pope 1993, Pope 2000
	Callithrix jacchus	Common marmoset	F	Pre-	Reproductive suppression	Abbott et al. 1990, 1998, Digby 1995, Lazaro-Perea et al. 2000, Nievergelt et al. 2000
	Cebus apella	Brown capuchin	F	Mating	Harassment and avoidance	Janson 1984
	Leontopithecus rosalia	Golden lion tamarin	M	Pre-	Dispersal decisions	Baker et al. 1993
	Saguinus fuscicollis	Saddle-back tamarin	M	Pre-	Dispersal decisions	Goldizen et al. 1996
	Saguinus mystax	Moustached tamarin	M	Pre-	Dispersal decisions	Garber et al. 1993
			M	Pre-	Alternative morphologies	Garber et al. 1996
	Saimiri oerstedii	Costa Rican squirrel monkey	M	Pre-	Dispersal decisions	Boinski and Mitchell 1994
Aotidae	—					
Pithecidae	—					
Atelidae	—					
Cercopithecidae	*Cercopithecus ascanius*	Red-tail monkey	M	Pre-	Group membership vs. incursions	Tsingalia and Rowell 1984, Cords 1987, 1988, 2000, Struhsaker 1988
			F	Mating	Pseudo-estrus	Cords 1984, 1987

Cercopithecus mitis	Blue monkey	M	Pre-	Group membership vs. incursions	Tsingalia and Rowell 1984, Cords 1987, 1988, 2000
		F	Mating	Pseudo-estrus	Fairgrieve 1995
	Samango monkey (*C. m. albogularis*)	M	Pre-	Group membership vs. incursions	Henzi and Lawes 1988
Erythrocebus patas	Patas monkey	M	Pre-	Group membership vs. incursions	Ohsawa *et al.* 1993
		M	Mating	Surreptitious mating	Ohsawa *et al.* 1993
		F	Mating	Harassment and avoidance	Loy and Loy 1977
		F	Mating	Pseudo-estrus	Loy 1975
Macaca fuscata	Japanese macaque	M	Pre-	Group membership vs. incursions	Huffman 1991, Sprague *et al.* 1998, Soltis *et al.* 2001, Takahashi 2001
		M	Pre-	Transfer decisions	Suzuki *et al.* 1998
		M	Pre-	"Friendships"	Takahata 1982
		M	Mating	Surreptitious mating	Manson 1996, Soltis *et al.* 2001
		F	Mating	Surreptitious mating	Soltis *et al.* 2001
		F	Mating	Pseudo-estrous	Okayusu 1992
Macaca fascicularis	Long-tail macaque	M	Post-	Protection of infants	van Noordwijk and van Schaik 1988
Macaca mulatta	Rhesus macaque	M	Pre-	Group membership vs. incursions	Berard *et al.* 1993
		M	Pre-	Transfer decisions	Lindburg 1969, Drickamer and Vessey 1973
		M	Pre-	"Friendships"	Manson 1994
		M	Mating	Surreptitious mating	Berard *et al.* 1994
		M	Mating	Mate selectivity	Chapais 1983
		F	Mating	Harassment and avoidance	Loy 1971
Macaca sinica	Toque macaque	M	Pre-	Group membership vs. incursions	Keane *et al.* 1997
Mandrillus sphinx	Mandrill	M	Pre-	Alternative morphologies	Wickings and Dixson 1992, Setchell and Dixson 2001a, b
		M	Pre-	Group membership vs. incursions	Setchell 1999
		M	Mating	Surreptitious mating	Setchell 1999
		M	Mating	Mate selectivity	Setchell 1999

Table 15.1. (cont.)

Family[a]	Species	Common name	Sex	Type[b]	ART	References
			F	Mating	Surreptitious mating	Author's unpublished observations
	Papio anubis	Olive or anubis baboon	M	Pre-	Transfer decisions	Packer 1979a
			M	Pre-	"Friendships"	Smuts 1985, Bercovitch 1991
			M	Mating	Coalitions	Packer 1977, Bercovitch 1988
			M	Mating	Surreptitious mating	Smuts 1985
			M	Mating	Mate selectivity	Scott 1984, Smuts 1985
	Papio cynocephalus	Yellow baboon	M	Mating	Coalitions	Noë and Sluijter 1990, Alberts et al. 2003
			M	Post-	Protection of infants	Buchan et al. 2003
	Papio ursinus	Chacma baboon	M	Pre-	"Friendships"	Palombit et al. 1997, 2001
			M	Post-	Protection of infants	Palombit et al. 1997, 2000
	Papio hamadryas	Hamadryas baboon	M	Pre-	Obtaining a harem	Kummer 1968, 1995, Sigg et al. 1982, Abegglen 1984
	Presbytis thomasi	Thomas' langur	M	Pre-	Dispersal decisions	Steenbeek et al. 2000
			M	Pre-	Obtaining a harem	Steenbeek et al. 2000
			F	Pre-	Transfer decisions	Sterck 1997, Steenbeek et al. 2000
	Theropithecus gelada	Gelada	M	Pre-	Obtaining a harem	Dunbar and Dunbar 1975, Dunbar 1982, 1984
			M	Mating	Surreptitious mating	Dunbar 1984
			F	Mating	Harassment and avoidance	Dunbar 1984
			F	Mating	Pseudo-estrus	Mori 1979
	Semnopithecus entellus	Hanuman langur	M	Pre-	Group membership vs. incursions	Rajpurohit and Mohnot 1988, Borries 2000, Launhardt et al. 2001
			M	Pre-	Obtaining a harem	Rajpurohit et al. 1995
			F	Mating	Harassment and avoidance	Sommer and Rajpurohit 1989, Sommer et al. 1992
			F	Mating	Pseudo-estrus	Hrdy 1977, Sommer 1994

378

Hylobatidae —

Hominidae

Family	Species	Sex	Timing[b]	ART	References
Gorilla gorilla	Mountain gorilla (G. g. beringei)	M	Pre-	Dispersal decisions	Harcourt and Stewart 1981, Robbins 1995, 1999, Watts 2000
		M	Pre-	Obtaining a harem	Harcourt 1978, Harcourt and Stewart 1981, Watts 1990
		F	Pre-	Transfer decisions	Stewart and Harcourt 1987, Watts 1989, 1990
		F	Mating	Surreptitious mating	Watts 2000
	Western lowland gorilla (G. g. gorilla)	F	Pre-	Transfer decisions	Stokes et al. 2003
Pan troglodytes	Chimpanzee	M	Mating	Coalitions	Tutin 1979, Hasegawa and Hiraiwa-Hasegawa 1983, Watts 1998
		M	Mating	Surreptitious mating	Tutin 1979
		M	Mating	Mate selectivity	Tutin 1979
		F	Mating	Surreptitious mating	Tutin 1979
Pongo spp.	Orang-utan	M	Pre-	Territorial vs. nomadic	MacKinnon 1974, Galdikas 1981, 1985a, Mitani 1985, Schürmann and van Hoof 1986
		M	Pre-	Alternative morphologies	MacKinnon 1974, Graham and Nadler 1990, Maggioncalda et al. 1999, 2000, Utami 2000
		M	Mating	Coercion vs. female choice	Galdikas 1981, 1985b, Mitani 1985, Utami 2000, Utami et al. 2002

[a] ARTs have been reported for 8 of 15 primate families (taxonomy after Groves 2001). Families for which ARTs have not been reported are marked —.

[b] ARTs occur pre-mating, mating, or post-mating.

Box 15.1 ARTs in male orang-utans

Orang-utans are semisolitary, living in overlapping home ranges in the rain forests of Sumatra and Borneo (Delgado and van Schaik 2000). Reproductive males may hold a territory and attempt to monopolize matings with sympatric females (bourgeois males), or they may be nomadic, avoiding contact with territorial males and mating with females in other males' ranges (parasitic males) (Table 15.1). Territorial males show full secondary sexual development, including fatty facial flanges, long thick hair, and a throat sac, and make a characteristic male "long call" (MacKinnon 1974) (Figure 15.1). By contrast, nomadic males do not show flange, hair, and throat sac development, and do not make long calls (MacKinnon 1974), although they are sexually mature and fertile (Dixson et al. 1982). These two types of males are discrete alternatives and employ different mating ARTs. Flanged male orang-utans tend to consort with reproductive females, unflanged males tend to use force to copulate with females outside consortships (Galdikas 1981, 1985b, Mitani 1985). However, both flanged and unflanged males have been observed to use both mating tactics (Utami 2000) with the mating tactic employed likely dependent on the exact situation in which a male finds himself, as well as on individual social relationships.

Development from "unflanged" to "flanged" status can occur at any time after puberty, and males may remain in a state of "arrested development" for more than 20 years in the wild (Utami 2000). Male tactics depend on the social

environment: unflanged males are subordinate to flanged males and develop adult secondary sexual characteristics when dominant males are removed or leave the area (Graham and Nadler 1990, Utami 2000). Endocrine studies in captive orang-utans have shown that arrested males have significantly lower levels of circulating testosterone, dihydrotestosterone (DHT), and luteinizing hormone (LH) than adolescent males that are developing adult secondary sexual characteristics (Maggioncalda et al. 1999). This suggests that arrested males lack the hormone levels necessary for secondary sexual development, although they have sufficient testicular steroids, LH, and follicle-stimulating hormone (FSH) to be fertile. Levels of growth hormone are also reduced in arrested males by comparison with developing males, which may explain their smaller body mass (Maggioncalda et al. 2000).

Dominant male orang-utans benefit from suppressing development in rival males by reducing male–male competition for access to females. However, facultative suppression of secondary sexual development may also be adaptive in subordinate males by acting to ameliorate intermale competition, to reduce investment in secondary sexual traits, and to facilitate alternative behavioral mating tactics (Dixson 1998, Setchell 2003). Indeed, Utami et al. (2002) have shown that both morphs of male sire offspring in the wild, suggesting that unflanged males are successful at obtaining fertilizations, and do not delay reproduction until they are sufficiently dominant to develop full secondary sexual characteristics.

reproductive success as a group-associated male and may increase their reproductive success by living alone or in an all-male group and visiting one or more groups during the mating season to mate opportunistically (Japanese macaques, mandrills, samango monkeys, Hanuman langurs) (Table 15.1). Bourgeois, resident males appear to have a reproductive advantage over intruders, but parasitic extra-group males avoid the costs of dominance and have been shown to sire offspring (Berard et al. 1993, Keane et al. 1997, Launhardt et al. 2001, Soltis et al. 2001). A similar situation occurs in some one-male, multifemale species, where all-male bands enter a bisexual group during the annual mating season and mate with females (blue monkeys, red-tail monkeys, and patas monkeys) (Table 15.1).

The success of intruder male tactics will depend on the ability of top-ranking group males to exclude newcomers, thereby reducing their likelihood of gaining access to fertile

females. Thus multimale influxes in blue monkeys are more likely where there are more estrous females available and when there are many days with multiple estrous females available (Cords 2000). The reaction of females to mating attempts from intruding males is also likely to play a role in male reproductive success.

DISPERSAL DECISIONS

In many group-living, nonhuman primate species, males leave their natal group whereas females remain (Pusey and Packer 1987). Male dispersal decisions begin with whether to disperse or to remain in the natal group, and the tactics employed depend on social conditions and demography. Alternative dispersal decisions are not strictly ARTs, but they heavily influence the availability of mating partners and thus lead directly to ARTs. For example, mature male Thomas' langurs either remain in their natal

Figure 15.1 Flanged male orang-utan (*Pongo abelii*). Unflanged males lack the fatty facial flanges and throat sac, have shorter hair and do not make the characteristic male "long call." (Photograph by Benoit Goossens.)

group as subordinate males, forming an age-graded group (Eisenberg *et al.* 1972), or disperse as juveniles and join an all-male band. Remaining males do not obtain any reproductive success and have not been observed to eventually take over their natal group. However, they increase the length of their father's tenure thereby increasing their own inclusive fitness. They also gain in experience and delay joining an all-male band, where costs are higher than life in a bisexual group (Steenbeek *et al.* 2000b). Male tactics depend on social conditions: age-graded groups develop where male tenure is long enough for male infants to mature in their father's group. However, if a takeover occurs, the new resident male expels any juvenile males (Steenbeek *et al.* 2000b).

Further examples of alternative male dispersal decisions include Costa Rican squirrel monkeys, red howler monkeys, mountain gorillas, and callitrichids (Table 15.1). Male Costa Rican squirrel monkeys either remain in their natal group to breed or disperse and cooperate with age-mates to invade another established group and expel the resident males. The latter strategy occurs more rarely and appears to represent the "best of a bad job," where males are forced out of their natal group due to intrasexual competition (Boinski and Mitchell 1994). Male red howler monkeys either remain in their natal group and aid their father in interactions with extra-group males or disperse. Here male tactics depend on the likelihood of successfully taking over another group, which in turn depends on mean group size and population density (Pope 1990, 1998). In mountain gorillas, the majority of maturing males remain in their natal group, while approximately one-third (36%) emigrate (Robbins 1999). The former tactic appears to be more effective. Additionally, remaining males also gain indirect fitness benefits by protecting infants born in their natal group. Dispersing males are less successful and may never gain females. The tactic pursued depends on the within-group sex ratio, the age of the current dominant male, and the breeding queue length, and may also depend on the male's relationship to the current male, as males that remain in their father's group inherit his females (Harcourt and Stewart 1981, Robbins 1995, Watts 2000). Finally, male callitrichids either stay in their natal group as subordinate "helpers" to wait for an opportunity to breed, disperse to found a new breeding group, or join another established group as either a breeding male or a subordinate. Opportunities for dispersing males are dependent on population demography and density; in a saturated habitat males may do better waiting for an opportunity to breed in their natal group, as dispersing males have high mortality and poor chances of finding breeding opportunities elsewhere (golden lion tamarins, moustached tamarins, saddle-back tamarins) (Table 15.1).

TRANSFER DECISIONS

In addition to natal dispersal, males of multimale, multifemale group-living species may subsequently transfer from group to group (secondary transfer: Pusey and Packer 1987). For example, adult male Japanese macaques tend to join troops with few or no males (Suzuki et al. 1998), male rhesus macaques transfer to groups with higher male: female ratios (Drickamer and Vessey 1973), and olive baboons and ring-tail lemurs transfer into groups containing more available cycling females (Packer 1979a, Sussman

1992). Males may also transfer between groups briefly during the mating season but return to their long-term group afterwards (rhesus macaques: Lindburg 1969).

The optimal transfer strategy will differ among individual males according to characteristics of the male, such as relative age and competitive ability, status, tenure, and social relationships in the current group, and whether he will leave behind offspring vulnerable to infanticide. The demography of groups available for transfer may also be a factor, such as the number and relative competitive ability of males and the number of available females (Altmann 2000, van Noordwijk and van Schaik 2004). A male should transfer if the benefits, in terms of improved mating access to females, outweigh the costs, which are made up of the cost of any transition period (e.g., risk of predation and starvation: Alberts and Altmann 1995), and the cost of immigration (e.g., injury risk: Cheney and Seyfarth 1983, Zhao 1996). Determining the long-term reproductive payoffs of male transfer decisions over a male primate's career is not easy, due to the long lifespan of primates and the difficulty of following the fate of dispersing males (Alberts and Altmann 1995). However, using comparative data, van Noordwijk and van Schaik (2004) have shown that male transfer decisions are strongly affected by the degree of reproductive skew in favor of the top-ranking males in local groups, and that decisions vary predictably with age.

OBTAINING A HAREM

In one-male, multifemale group-living species, a male can obtain a group of females in several ways. A male may aggressively take over a breeding group, in a high-risk, high-benefit strategy (geladas, Thomas' langurs, Hanuman langurs, mountain gorillas, red howler monkeys); remain in his natal group or join another group as a subordinate "follower" and wait in a breeding queue (hamadryas baboons) or obtain females by group fission (geladas); or he may acquire females one or two at a time from other groups (mountain gorillas, hamadryas baboons, Thomas' langurs) (Table 15.1). The tactics employed by a male will be dependent on many factors but are known to be correlated with population demography and density (Watts 2000). Male reproductive success in one-male, multifemale groups is dependent on tenure length and the size and stability of groups (hamadryas baboons: F. Colmenares, unpublished data, cited in Watts 2000). In Thomas' langurs, male tactics depend on female group size (groups with more adult females are more likely to be taken over) and the tenure of group males (females are more likely to leave a long-tenure male) (Steenbeek *et al.* 2000b). In geladas, population growth and increased group size lead to more group fission and thus should lead to more males employing a follower strategy relative to takeover males (Dunbar 1984).

As with transfer decisions, little is known about the relative long-term fitness consequences of these ARTs. However, study of geladas has shown that a male that effects a forceful takeover begins his reproductive career with more females than a "follower" male. A takeover male must wait until he is competitive enough to succeed in male–male competition, and his reproductive career thus begins later than that of a "follower," although they are likely to achieve sexual maturity at the same age. "Followers" have higher chances of success in obtaining females, start reproducing earlier, and have a longer reproductive life, but begin with fewer females. Over a lifetime, these ARTs appear to yield very similar reproductive pay-offs (Dunbar and Dunbar 1975, Dunbar 1982, 1984).

"FRIENDSHIPS"

Subordinate males may obtain sexual access to at least one female at little risk of aggression from dominant males by forming special relationships, or "friendships," with particular females (Japanese macaques, olive baboons, chacma baboons, rhesus macaques) (Table 15.1). Such associations also act to increase female fitness, as male friends protect their infants from infanticidal males (Palombit *et al.* 1997).

ALTERNATIVE MORPHOLOGIES

Although dominant males are generally in their prime, and are therefore likely to be the largest males and in the best physical condition, behavioral ARTs in male primates are not generally paired with dramatic alternative morphologies. However, we have seen that behavioral ARTs in male orangutans are accompanied by striking differences in appearance and physiology (Box 15.1). Further examples of morphological differences between males include reduced body mass and condition, reduced testicular volume, decreased levels of circulating testosterone, reduced development of secondary sexual traits, and smaller and less active scent glands in subordinate males (lesser mouse lemurs, moustached tamarins, sifaka, mandrills) (Table 15.1). These differences may be due to physiological suppression by the dominant male, or may be mediated by olfactory cues from dominants (Schilling *et al.* 1984, Perret and Schilling 1987a, b) or possibly by visual and/or auditory signals (Maggioncalda *et al.* 1999, Setchell 2003). However, suppression is nonpermanent. For example, even the lowest-ranking adult male in a mandrill group

can develop the impressive red facial coloration and other secondary sexual traits of dominant males if given the opportunity (Setchell and Dixson 2001b) (Figure 15.2).

Suppression of rival males confers a reproductive advantage on the dominant by lessening reproductive competition by reducing sperm competition if suppressed males manage to mate (they have smaller testes), and by reducing their attractiveness to females, if females prefer to mate with males showing full secondary sexual development (orang-utans: Schürmann 1982, Utami 2000; mandrills: Setchell 2005). However, as in "unflanged" male orangutans (Box 15.1), facultative suppression may also be adaptive in subordinate individuals, by reducing inter-male competition and investment in reproduction and facilitating the use of behavioral ARTs (Dixson 1998, Setchell 2003). "Arrested" males thus possess the most adaptive traits for the way in which they seek fertilizations.

15.3.2 Mating ARTs in male primates

The majority of mating ARTs in male primates exist as alternatives to monopolization of females by a dominant male. Such parasitic ARTs include coalition formation and sneak mating and exploit the investment of bourgeois males while avoiding some of the costs of male–male competition and the constraints imposed by mate guarding on foraging activity (Packer 1979b, Bercovitch 1983, Alberts *et al.* 1996). Mate choice by both sexes will also influence male mating tactics: males may coerce females that are unwilling to mate.

COALITIONS

Males in multimale, multifemale group-living species may form coalitions to force a dominant male to give up a receptive female, leading to mating access to the female (baboons, chimpanzees) (Table 15.1), a cooperative behavior similar to that found in reproductive competition in a number of fish species (Taborsky, Chapter 10, this volume). Mate guarding by dominant males may become less effective when many males are present (Watts 1998) and the expression of coalitions may also depend on the age structure of males in a group, their tenure, and their social relationships. For example, Alberts *et al.* (2003) conclude that baboon coalitions are more likely when more and older males are present in the group.

SURREPTITIOUS MATING

Low-ranking and extra-group males of many primate species use opportunistic and/or surreptitious mating tactics

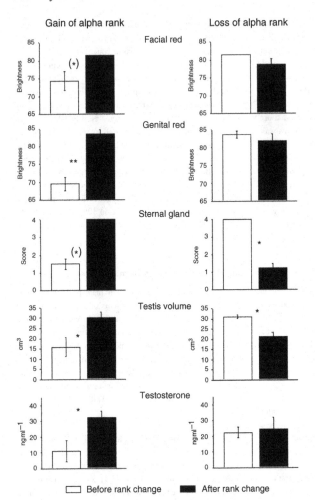

Gain of alpha rank   Loss of alpha rank

Facial red

Genital red

Sternal gland

Testis volume

Testosterone

☐ Before rank change   ■ After rank change

Figure 15.2 Red coloration on the face and genitalia, sternal gland activity, testis size, and testosterone levels in male mandrills before and after gain and loss of alpha status (based on a correlational study). Significance levels: (*) $P = 0.08$, *$P = 0.05$, **$P = 0.01$ results of paired tests ($n = 4$). (From Setchell and Dixson 2001, with permission.)

(chimpanzees, mandrills, baboons, geladas, patas monkeys, rhesus macaques, Japanese macaques) (Table 15.1, Figure 15.3), and sneak copulations that have been shown to result in fertilizations (Ohsawa *et al.* 1993, Berard *et al.* 1994, Manson 1996, Launhardt *et al.* 2001). The opportunity for, and siring success of, sneak copulations will depend on the ability of high-ranking males to monopolize

females, and therefore on the number of males and females in a group, the number of extra-group males, and the synchrony of female receptivity.

COERCION VERSUS FEMALE CHOICE

Males may use sexual coercion to force females to mate with them against female preference, particularly in species

Figure 15.3 An adolescent male mandrill (aged 7 years, mass approximately 20 kg) mates sneakily with a female showing a large sexual swelling, but who was unlikely to be ovulating at the time of the mating. Males attain adult size (approximately 31 kg) and appearance at 9–10 years (Setchell and Dixson 2002). (Photograph of the mandrill colony at the Centre International de Recherches Médicales, Franceville, Gabon, by Joanna M. Setchell.)

where adult males are larger than adult females (Smuts and Smuts 1993). For example, flanged male orang-utans tend to consort with reproductive females, while unflanged males tend to use force to copulate with females outside consortships (see Box 15.1).

MATE SELECTIVITY

Males may differ in mate selectivity, depending on their dominance rank. A dominant male that has free choice but is unable to monopolize all available females should concentrate his mating effort on the females who are most likely to conceive and raise his offspring to maturity. Dominant males may thus show less sexual interest in nulliparous females, which are typically less fertile and less adequate mothers than older, experienced females (Altmann 1980), by comparison with females who have already produced at least one infant (baboons, mandrills) (Table 15.1). Dominant males may also prefer to mate with high-ranking females (review in Berenstain and Wade 1983), which may be more fertile and able to invest more in resulting offspring (van Noordwijk and van Schaik 1999, Setchell et al. 2002). Lower-ranking males, on the other hand, for whom choice of mates is limited by male–male competition, should be more likely to mate when the chance arises, even if the

chances of fertilization are lower. Thus, while dominant males mate selectively, concentrating their mating attempts when a female is most likely to ovulate, subordinate males mate opportunistically with any female and at times when the female may be less likely to ovulate (rhesus macaques, baboons, chimpanzees, mandrills) (Table 15.1).

15.3.3   Post-mating ARTs in male primates

Once a new bourgeois male has obtained the position of breeding male in a bisexual group, he may kill the offspring of previous males to increase the number of females available for fertilization because death of an unweaned infant accelerates the resumption of ovarian cycles in females (Hrdy 1979; see van Schaik and Janson 2000 for a recent review of evidence for infanticide in primates). Use of this tactic depends on a male's rank and his previous reproductive history within the group. DNA analyses of wild Hanuman langurs have confirmed that male attackers were not related to their infant victims and that presumed killers were the likely sire of subsequent infants (Borries et al. 1999). Further, Palombit et al. (2000) have shown that male chacma baboons express infanticidal behavior facultatively, depending on attributes of the alpha male or conditions of male tenure. The expression of infanticide is also likely to depend on the number of other males in a group, as the presence of other males increases the costs of infanticide, while mating competition decreases the benefits by reducing the ability of the infanticidal male to monopolize subsequent fertilizations (Palombit et al. 2000).

In the majority of primate species males contribute little to the survival of offspring post conception. However, males may protect infants that are likely to be their offspring from other, infanticidal males (long-tail macaques, baboons) (Table 15.1). This expression of protective behavior depends on the likelihood of paternity (Buchan et al. 2003).

In an exception to the general primate rule, male callitrichids show extensive infant care (Goldizen 1987, Heymann 2000), and infant survival correlates with the number of adult males present in a group (Garber et al. 1984, Koenig 1995). Extreme reproductive skew in females of these species reduces reproductive opportunities for males, which either breed themselves or help raise the offspring of other males (Goldizen 1987). Male tactics depend on social status. Mating and paternity are concentrated in the behaviorally dominant resident male (Baker et al. 1993, Nievergelt et al. 2000), while subordinates appear to employ a waiting strategy in the hope of obtaining a breeding position in the future. Subordinates may also gain inclusive fitness, if they are related to the breeding male, but the only genetic study of relatedness available suggests that this is not the case (Nievergelt et al. 2000).

15.3.4 Determination, plasticity, and selection of ARTs in male primates

DETERMINATION

No evidence currently exists for the genetic determination of male ARTs in primates, although it has been suggested that a genetic polymorphism might underlie flange development in orang-utans (van Hooff and Utami 2004). Environmental modification, however, appears ubiquitous. ARTs are expressed conditionally, although it is possible that genetic variation occurs between individuals in the position of the switch point at which they change from one tactic to another. Male reproductive decisions are dependent on asymmetries in competitive ability, which influence the costs and benefits of each tactic for the individual concerned.

PLASTICITY

Plasticity in male primate ARTs occurs at both the population and the individual level. With the exception of flange development in male orang-utans, which is sequential with a one-time switch point, all reproductive phenotypes appear to be reversible and individuals are capable of changing back and forth between different phenotypes if the opportunity occurs. For example, while former, overthrown harem owners or dominant males are unlikely to attempt to attain top rank a second time, they will do so if a suitable opportunity occurs, such as a lack of rival males (e.g., mandrills: author's observations). This flexibility is likely to underlie the predominance of behavioral, rather than morphological ARTs, giving males the possibility to facultatively adjust to changing conditions. Reversibility implies hormonal differences during adult life (Moore 1991), and studies have demonstrated that male ARTs are indeed associated with hormonal differences (e.g., Wickings and Dixson 1992, Maggioncalda et al. 1999, 2000, Setchell and Dixson 2001a, b).

SELECTION

The pay-off of the monopolization of reproductive females by dominant males, and conversely the success of male ARTs, can be examined by investigating the degree of male reproductive skew. Cowlishaw and Dunbar (1991, 1992)

have shown that as group size increases, high-ranking males lose their ability to monopolize access to females, meaning that parasitic male ARTs are more successful. At the population level, Alberts et al. (2003) investigated the relationship between male rank and mating success in yellow baboons, using 32 group–years of data. As expected, dominance rank was an important predictor of male mating success: males who spent extended periods at high rank experienced an overall reproductive advantage over males that did not do so. However, parasitic ARTs were more successful when there were many adult males in the group, when group males differed greatly in age, and when the highest ranking male maintained his rank for only short periods.

By their very nature, some male mating tactics (e.g., mate guarding) are easier to observe than others are (e.g., sneak copulations). Thus, although studies of mating success are useful, paternity determination is necessary to truly investigate the impact of ARTs on male reproductive success. In a review of paternity studies in group-living primates, van Noordwijk and van Schaik (2004) have recently shown that an increase in the number of adult males or females in a group is significantly correlated with a decrease in the percentage paternity concentration in the top-ranking male (and therefore a corresponding increase in the success of male parasitic ARTs). As predicted, seasonality of reproduction also had an effect on the concentration of paternity in the top-ranking male, independent of the number of males (van Noordwijk and van Schaik 2004).

A career perspective (van Noordwijk and van Schaik 2001, 2004) and knowledge of life-history pathways are necessary when considering the costs and benefits of ARTs (Caro and Bateson 1986). For example, we cannot draw conclusions concerning the lifetime success of a male from a 1- or 2-year study, when a male's career may last far longer and involve periods of low, mid, and high rank, and accordingly different reproductive tactics and varying pay-offs. However, particularly little is known concerning the lifetime reproductive success of males following different strategies, or potential associated differences in reproductive lifespan. From the available information, the majority of ARTs in male primates appear to be "best-of-a-bad-job" situations, where lower-quality males do the best they can to achieve at least some reproductive success. However, frequency-dependent pay-offs to reproductive competitors displaying different ARTs do appear to occur in gelada (Dunbar 1982, 1984) and may also occur in orang-utans (Utami et al. 2002).

5.4   ARTs IN FEMALE PRIMATES

n general, far less is known about the sexual strategies of emale primates than those of males (Setchell and Kappeler 003), and attention has generally focused on male ARTs, as 1 other taxa (Henson and Warner 1997). However, female rimates show ARTs at both pre-mating and mating levels, lthough there is as yet no evidence for post-mating ARTs n female primates.

5.4.1   Pre-mating ARTs in female primates

ISPERSAL DECISIONS
emales of most group-living primate species remain in their atal group to breed (Pusey and Packer 1987, Pope 2000). Iowever in some species, females may disperse, giving rise to cope for alternative dispersal decisions based on social onditions and breeding opportunities. For example, female ed howler monkeys either remain in their natal group to reed or disperse and form new groups with other dispersing emales (Crockett and Pope 1993, Pope 2000). The latter is he most common tactic but is more costly, and many dis-ersing females die without reproducing. Dispersers suffer nore injuries, have a nutrient-deficient diet by comparison to roup members, and have delayed age at first breeding (Pope 000). Dispersal tactics in females of this species depend on he number of resident reproductive females in a group and he presence of the mother. Females always disperse where here are already four or more resident females, and the nother is always present when maturing females remain Crockett and Pope 1993, Pope 2000). As in males, alternative emale dispersal decisions do not represent ARTs per se, but hey clearly lead to differences in reproduction between emales.

RANSFER DECISIONS
emales of some primate species may transfer from one roup to another during their adult life (Pusey and Packer 987). Transfer tactics are likely to change with female age nd future reproductive potential (Dunbar 1979) and will epend on the social situation. For example, females may nigrate to avoid potentially infanticidal males (Thomas' ngurs, mountain gorillas, western lowland gorillas) Table 15.1). Female Thomas' langurs transfer when the urrent resident male is no longer able to protect offspring rom other males, and females are thus more likely to leave a te-tenure male than a short-tenure male (Steenbeek et al. 000a). In mountain gorillas, females prefer to transfer into groups with more than one adult male and fewer females (Watts 2000), in which females enjoy lower risk of infanti-cide (Robbins 1995, Watts 2000) and significantly shorter inter-birth intervals than in single-male groups (Gerard-Steklis and Steklis 2001). Female western lowland gorilla transfer tactics are also related to social and group demo-graphic factors. In this species, females do not have the option of transferring into multimale groups, which do not occur, but female immigration rates are negatively related to group size and emigration rates are positively related to group size (Stokes et al. 2003).

REPRODUCTIVE SUPPRESSION
As in males, ARTs can be expected to evolve in females when there is intense intrasexual competition and high repro-ductive skew. A prime example of such conditions occurs in cooperatively breeding marmosets. High-ranking female marmosets interrupt the copulations of other females, and reproductive function is suppressed in subordinate females (Abbott et al. 1990). Suppression is reversible (Abbott et al. 1998) and some subordinate females do reproduce in the wild (Nievergelt et al. 2000), although any offspring produced are likely to be killed (Digby 1995, Lazaro-Perea et al. 2000). Clearly, the best tactic for an adult female is to be dominant, but subordination with reproductive suppression may rep-resent an alternative tactic whereby females avoid wasting reproductive effort while they wait to obtain a breeding position.

Social stress, due to harassment or aggression from high-ranking females, may also act to lower the reproductive success of low-ranking rivals in groups where multiple females breed (Dunbar 1980, 1988). This may represent a tactic by high-ranking females to reduce future competition for nutritional resources from the offspring of subordinates, while low-ranking rivals make the best of a bad job.

15.4.2   Mating ARTs in female primates

Like males, females mate both within consortships and sneakily (Japanese macaques, chimpanzees, mountain gor-illas, mandrills) (Table 15.1). The mating mode employed depends on the risk of "punishment" by males (sexual coercion); females mate surreptitiously with subordinate or extra-group males, but consort openly with dominant group males. Other possibilities for mating ARTs in female primates include mate choice, competition between females for matings, and the timing of mating behavior. However, it is not yet clear whether discrete differences in reproductive tactics occur between females, and whether these are adaptive.

MATE CHOICE

Considerable evidence exists that female primates show mate choice (see Paul 2002 for a recent review). However, while the male chosen may differ between females (e.g., choice for genetic compatibility: Sauermann et al. 2001), the rule remains the same (choose the "best" male), thus there are no known mate-choice ARTs for female primates.

HARASSMENT AND AVOIDANCE

In one-male, multifemale groups, high-ranking females attempt to prevent low-ranking females from mating through aggression and harassment (patas monkeys, geladas, Hanuman langurs) (Table 15.1). This may be because the sperm of the dominant male is limited (Dewsbury 1982, Marson et al. 1989, Wedell et al. 2002, Preston et al. 2003), or because dominant females try to reduce competition of their own offspring with potential offspring of subordinates by preventing the latter from reproducing. Dominant females in multimale, multifemale groups may also aggressively interrupt matings involving subordinate females (rhesus macaques: Loy 1971), and the simple presence of dominant females may inhibit subordinate females from interacting with males (brown capuchins: Janson 1984). These ARTs are expressed according to social environment: dominant females harass subordinates, while low-ranking females avoid harassment at the cost of reproductive opportunities, a tactic that represents the "best of a bad job."

TIMING OF MATING BEHAVIOR

Female primates show situation-dependent flexibility in mating behavior that may act to reduce the risk of sexually selected infanticide by nonsire males (Hrdy 1979, Hrdy and Whitten 1987, Smuts and Smuts 1993, van Schaik et al. 1999). For example, where group takeover by a new male leads to a high risk of infanticide for infants sired by a previous male, females that are already pregnant solicit the new male for mating ("pseudo-estrus") (Table 15.1), resulting in paternity confusion and reduction in the risk of infanticide.

15.4.3   Post-mating ARTs in female primates

Investment in an individual offspring enhances that offspring's chance of survival, while at the same time diminishing a female's ability to invest in future reproduction by reducing her fertility or chances of survival (Fisher 1930, Trivers 1972). Potential post-mating ARTs in female primates may therefore include facultative adjustment of investment according to aspects of female condition, such as age, rank, or body condition (Trivers and Willard 1973) or of offspring quality, such as the identity of the sire or the sex of the infant (Qvanstrom and Price 2001). Evidence exists that female primates invest in infant growth and survival according to their own age, growth status, rank, and physical condition (rhesus macaques: Simpson et al. 1981, Gomendio 1990, Bercovitch et al. 1998; mandrills: Setchell et al. 2001, 2002; baboons: Altmann 1980, Johnson 2003; chimpanzees: Pusey et al. 1997). Females may also terminate investment in a developing fetus when a newly dominant male is likely to commit infanticide. The benefits from mating with a new dominant male outweigh the costs of terminating current investment in the offspring of another male (baboons: Pereira 1983; humans: Forbes 1997; Hanuman langurs: Lhota et al. 2001). Finally, many studies have examined whether female primates manipulate birth sex ratios according to the social environment or adjust their parental investment according to the sex of an infant. However, these questions have proved difficult to resolve (van Schaik and Hrdy 1991, Hiraiwa-Hasegawa 1993, Silk et al. 1993, Brown 2001, Bercovitch 2002, Brown and Silk 2002), and although different sex allocation in offspring would represent alternative allocation phenotypes, it would not represent ARTs. The basic female tactic appears to be the same in all cases: maximize the benefits and minimize the costs of that investment in each offspring in terms of reproductive fitness. Female primates do not appear to employ post-mating ARTs, although the possibility exists that they may employ post-mating ARTs in terms of fertilization control by physiological mechanisms (cryptic female choice: Eberhard 1985, 1996).

15.4.4   Determination, plasticity, and selection of ARTs in female primates

As with male primates, all ARTs identified for female primates are influenced by the social conditions in which a female finds herself. Female ARTs are not fixed and are not accompanied by alternative morphotypes. Instead they represent an adaptive response to current conditions and characteristics of the individual female, which determine the relative costs of the tactics. Tactics can thus change if conditions change. The reproductive pay-offs are generally unequal, and female ARTs represent the "best of a bad job" y subordinate females, or females that have fewer resources to invest in reproduction.

5.5 INTERACTIONS BETWEEN MALE AND FEMALE STRATEGIES

Interactions between the sexes play an important role in the expression and evolution of primate reproductive strategies (Setchell and Kappeler 2003), and of ARTs in general (Henson and Warner 1997). Female reproductive strategies may alter the costs and benefits of male ARTs. For example, males may compete for access to reproductive females, but females can increase reproductive skew among males by showing mate preference for dominant males (orang-utans: Utami 2000; mandrills: Setchell 2002), creating a situation that favors the evolution of male ARTs. Alternatively, female choice for sneak matings with subordinate or extra-group males may act to reduce the dominant male's potential to monopolize females and increase the success of male ARTs (Soltis et al. 2001). Male strategies may also limit or determine female strategies. For example, male sexual coercion may prevent females from mating with preferred males, and male infanticide will terminate a female's investment in an infant. Females therefore adjust their mating tactics according to the risk of coercion. The extent to which the strategies of males or females determine the sire of an infant differs with circumstance, and differences may occur between two studies of the same species (Soltis et al. 1997a, b, 2001).

5.6 CONCLUSIONS

This review allows us to draw some general conclusions concerning primate ARTs. First, a wide diversity of reproductive strategies and ARTs has evolved to promote the reproductive success of individual primates. Underlying this diversity is the bourgeois/parasite paradigm (Taborsky 1994, 1997) and the extent to which individual males are able to monopolize access to mates and the resources available to females for investment in infant growth and survival (Trivers 1972, Emlen and Oring 1977). Perhaps because the effects of male–male competition can be dramatic, intraspecific variation in male mating strategies has received far more attention than the flexibility of reproductive behavior in female primates. However, females may also employ ARTs, with important implications for lifetime reproductive success.

Second, the types of ARTs employed by the two sexes show both similarities and differences. Pre-mating

ARTs in both males and females involve dispersal and transfer decisions, involvement in intrasexual competition, and physiological suppression. However, whereas male–male competition is generally related to access to females, female–female competition is more often related to access to other resources. Both sexes may mate sneakily or with a consort partner, and both may employ mate-choice tactics. However, male tactics may involve coercion of females, while females respond to the risk of sexual coercion by manipulating their own mating tactics. Finally, post-mating investment tactics are much more important in females, as females are responsible for the majority of parental investment in most primate species, but these do not appear to involve ARTs.

Third, ARTs in primates occur within, as well as between, individuals. They tend to be limited to behavior, gonads, and physiology and are rarely associated with dramatic alternative morphologies. This is likely due to the advantages of plasticity and the lower costs of adjustment according to changes in the characteristics of the individual (age, ontogenetic stage) and in social conditions (e.g., rank).

Fourth, most ARTs in primates appear to be "best-of-a-bad-job" phenotypes, whereby inferior individuals, or those in a suboptimal situation, make the most of any opportunity available to gain reproductive success.

Fifth, in contrast to some other taxa (Henson and Warner 1997, Alonzo et al. 2000), and with the exception of female reproductive suppression in common marmosets (Abbott et al. 1998), relatively little is known about the life-history pathways underlying ARTs and the factors that determine their expression. This is due to the difficulty of studying long-lived species, and the fact that primates are not as easy to manipulate experimentally as invertebrates or fish. However, much remains to be learned from analyses of the entire careers of wild individuals (e.g., van Noordwijk and van Schaik 1999, 2001), which would allow us to determine the relative pay-offs of alternative tactics, the influence of development on lifetime strategies, and how ARTs may change over a lifetime (Setchell and Lee 2004).

Finally, male and female reproductive strategies are intricately linked in primates (Setchell and Kappeler 2003, van Schaik et al. 2004), and interactions between the sexes play an important role in the evolution of primate ARTs.

Acknowledgments

I am grateful to the editors for their invitation to contribute to this volume on alternative reproductive tactics, to Jane Brockmann, Michael Taborsky, and anonymous reviewers for helpful comments on the manuscript, and to Peter

Kappeler for extensive discussion of sexual selection in primates.

References

Abbott, D. H., George, L. M., Barrett, J., et al. 1990. Social control of ovulation in marmoset monkeys: a neuroendocrine basis for the study of infertility. In T. E. Zeigler and F. B. Bercovitch (eds.) *Socioendocrinology of Primate Reproduction*, pp. 135–158. New York: Wiley-Liss.

Abbott, D. H., Saltzman, W., SchultzDarken, N. J., and Tannenbaum, P. L. 1998. Adaptations to subordinate status in female marmoset monkeys. *Endocrinology* 119, 261–274.

Abegglen, J. J. 1984. *On Socialization in Hamadryas Baboons*. Cranbury, NJ: Associated University Presses.

Alberts, S. C. and Altmann, J. 1995. Balancing costs and opportunities: dispersal in male baboons. *American Naturalist* 145, 279–306.

Alberts, S. C., Altmann, J., and Wilson, M. L. 1996. Mate guarding constrains foraging activity of male baboons. *Animal Behaviour*, 51, 1269–1277.

Alberts, S. C., Watts, H., and Altmann, J. 2003. Queuing and queue jumping: long-term patterns of dominance rank and mating success in male savannah baboons. *Animal Behaviour*, 65, 821–840.

Alonzo, S. H., Taborsky, M., and Wirtz, P. 2000. Male alternative reproductive behaviours in a Mediterranean wrasse, *Symphodus ocellatus*: evidence from otolioths for multiple life-history pathways. *Evolutionary Ecology Research* 2, 997–1007.

Altmann, J. 1980. *Baboon Mothers and Infants*. Chicago, IL: University of Chicago Press.

Altmann, J. 2000. Models of outcome and process: predicting the number of males in primate groups. In P. M. Kappeler (ed.) *Primate Males: Causes and Consequences of Variation in Group Composition*, pp. 236–247. Cambridge, UK: Cambridge University Press.

Baker, A. J., Dietz, J. M., and Kleiman, D. G. 1993. Behavioural evidence for monopolization of paternity in multi-male groups of golden lion tamarins. *Animal Behaviour* 46, 1091–1103.

Bearder, S. K. 1987. Lorises, bushbabies and tarsiers: diverse societies in solitary foragers. In B. B. Smuts, D. L. Cheney, R. M. Seyfarth, R. W. Wrangham, and T. T. Struhsaker (eds.) *Primate Societies*, pp. 11–24. Chicago, IL: University of Chicago Press.

Berard, J. D., Nurnberg, P., Epplen, J. T., and Schmidtke, J. 1993. Male rank, reproductive behavior, and reproductive

success in free-ranging rhesus macaques. *Primates* 34, 481–489.

Berard, J. D., Nurnberg, P., Epplen, J. T., and Schmitdke, J. 1994. Alternative reproductive tactics and reproductive success in male rhesus macaques. *Behaviour* 129, 177–201.

Bercovitch, F. B. 1983. Time budgets and consortships in olive baboons (*Papio anubis*). *Folia Primatologica* 41, 180–190.

Bercovitch, F. B. 1988. Coalitions, cooperation, and reproductive tactics among adult male olive baboons. *Animal Behaviour*, 36, 1198–1209.

Bercovitch, F. B. 1991. Mate selection, consortship formation and reproductive tactics in adult female savanna baboons. *Primates* 32, 437–452.

Bercovitch, F. B. 2002. Sex-biased parental investment in primates. *International Journal of Primatology* 23, 905–921.

Bercovitch, F. B., Lebron, M. R., Martinez, H. S., and Kessler, M. J. 1998. Primigravidity, body weight, and costs of rearing first offspring in rhesus macaques. *American Journal of Primatology* 46, 135–144.

Berenstain, L. and Wade, T. D. 1983. Intrasexual selection and male mating strategies in baboons and macaques. *International Journal of Primatology* 4, 201–235.

Boinski, S. and Mitchell, C. L. 1994. Male residence and association patterns in Costa Rican squirrel monkeys (*Saimiri oerstedi*). *American Journal of Primatology* 34, 157–169.

Borries, C. 2000. Male dispersal and mating season influxes in Hanuman langurs living in multi-male groups. In P. M. Kappeler (ed.) *Primate Males: Causes and Consequences of Variation in Group Composition*, pp. 46–58. Cambridge, UK: Cambridge University Press.

Borries, C., Launhardt, K., Epplen, C., Epplen, J. T., and Winkler, P. 1999. DNA analyses support the hypothesis that infanticide is adaptive in langur monkeys. *Proceedings of the Royal Society of London B* 266, 901–904.

Brockman, D. K., Grafen, A., and Dawkins, R. 1979. Evolutionary stable nesting strategy in a digger wasp. *Journal of Theoretical Biology* 7, 473–496.

Brockmann, H. J. 2001. The evolution of alternative strategies and tactics. *Advances in the Study of Behavior* 30, 1–51.

Brown, G. R. 2001. Sex-biased investment in non-human primates: can Trivers' and Willard's theory be tested? *Animal Behaviour* 61, 683–694.

Brown, G. R. and Silk, J. B. 2002. Reconsidering the null hypothesis: is maternal rank associated with birth sex ratios in primate groups? *Proceedings of the National Academy of Sciences of the United States of America* 99, 11252–11255.

Buchan, J. C., Alberts, S. C., Silk, J. B., and Altmann, J. 2003. True paternal care in a multi-male primate society. *Nature* **425**, 179–180.

Caro, T. M. and Bateson, P. P. G. 1986. Organization and ontogeny of alternative tactics. *Animal Behaviour* **34**, 1483–1499.

Chapais, B. 1983. Matriline membership and male rhesus reaching high ranks in their natal troops. In R. A. Hinde (ed.) *Primate Social Relationships: An Integrated Approach*, pp. 171–175. Oxford, UK: Blackwell Scientific.

Charles-Dominique, P. 1977. *Ecology and Behaviour of Nocturnal Primates*. London: Duckworth.

Cheney, D. L. and Seyfarth, R. M. 1983. Non-random dispersal in free ranging vervet monkeys: social and genetic consequences. *American Naturalist* **122**, 392–412.

Clutton-Brock, T. H. 1989. Mammalian mating systems. *Proceedings of the Royal Society of London B* **236**, 339–372.

Cooke, D. F. 1990. Differences in courtship, mating and post-copulatory behaviour between male morphs of the dung beetle *Onthophagus binodis* Thunberg (Coleoptera: Scarabaeidae). *Animal Behaviour* **40**, 428–436.

Cords, M. 1984. Mating patterns and social structure in red-tail monkeys (*Cercopithecus ascanius*). *Zeitschrift für Tierpsychologie* **64**, 313–329.

Cords, M. 1987. Forest guenons and patas monkeys: male–male competition in one-male groups. In B. B. Smuts, D. L. Cheney, R. M. Seyfarth, R. W. Wrangham, and T. T. Struhsaker (eds.) *Primate Societies*, pp. 98–111. Chicago, IL: University of Chicago Press.

Cords, M. 1988. Mating systems of forest guenons: a preliminary review. In A. Gautier-Hion, F. Bourlière, J. P. Gautier, and J. Kingdon (eds.) *A Primate Radiation: Evolutionary Biology of the African Guenons*, pp. 323–339. Cambridge, UK: Cambridge University Press.

Cords, M. 2000. The number of males in guenon groups. In P. M. Kappeler (ed.) *Primate Males: Causes and Consequences of Variation in Group Composition*, pp. 84–96. Cambridge, UK: Cambridge University Press.

Cowlishaw, G. and Dunbar, R. I. M. 1991. Dominance ranking and mating success in male primates. *Animal Behaviour* **41**, 1045–1056.

Cowlishaw, G. and Dunbar, R. I. M. 1992. Dominance and mating success: a reply to Barton and Simpson. *Animal Behaviour* **44**, 1162–1163.

Crockett, C. M. and Pope, T. R. 1993. Consequences of sex differences in dispersal for juvenile red howler monkeys. In M. E. Pereira and L. A. Fairbanks (eds.) *Juvenile Primates: Life History, Development and Behaviour*, pp. 104–118. Oxford, UK: Oxford University Press.

Davies, N. B. 1982. Behaviour and competition for scarce resources. In King's College Sociobiology Group (ed.) *Current Problems in Sociobiology*, pp. 363–380. Cambridge, UK: Cambridge University Press.

Davies, N. B. 1991. Mating systems. In J. B. Krebs and N. B. Davies (eds.) *Behavioural Ecology*, pp. 263–294. Oxford, UK: Blackwell Scientific.

Dawkins, R. 1980. Good strategy or evolutionary stable strategy? In G. W. Barlow and J. Silverberg (eds.) *Sociobiology: Beyond Nature/Nuture*, pp. 331–367. Boulder, CO: Westview Press.

Delgado, R. and van Schaik, C. P. 2000. The behavioural ecology and conservation of the orangutan (*Pongo pygmaeus*): a tale of two islands. *Evolutionary Anthropology* **9**, 201–208.

Dewsbury, D. A. 1982. Ejaculate cost and male choice. *American Naturalist* **119**, 601–610.

Digby, L. J. 1995. Infant care, infanticide, and female reproductive strategies in polygynous groups of common marmosets (*Callithrix jacchus*). *Behavioral Ecology and Sociobiology* **37**, 51–61.

Dixson, A. F. 1998. *Primate Sexuality: Comparative Studies of the Prosimians, Monkeys, Apes and Human Beings*. Oxford, UK: Oxford University Press.

Dixson, A. F., Knight, J., Moore, H. D., and Carman, M. 1982. Observations on sexual development in male orangutans, *Pongo pygmaeus*. *International Zoo Yearbook* **22**, 222–227.

Dominey, W. J. 1984. Alternative mating tactics and evolutionarily stable strategies. *American Zoologist* **24**, 385–396.

Drickamer, L. C. and Vessey, S. 1973. Group changing in free ranging male rhesus monkeys. *Primates* **14**, 359–368.

Dunbar, R. I. M. 1979. Age-dependent changes in sexual skin colour and associated phenomena in female gelada baboons. *Journal of Human Evolution* **6**, 667–672.

Dunbar, R. I. M. 1980. Determinants and evolutionary consequences of dominance among female gelada baboons. *Behavioral Ecology and Sociobiology* **7**, 253–265.

Dunbar, R. I. M. 1982. Intraspecific variations in mating strategy. In P. P. G. Bateson (ed.) *Perspectives in Ethology*, vol. 5, pp. 385–431. New York: Plenum Press.

Dunbar, R. I. M. 1984. *Reproductive Decisions: An Economic Analysis of Gelada Baboon Social Strategies*. Princeton, NJ: Princeton University Press.

Dunbar, R. I. M. 1988. *Primate Social Systems*. Ithaca, NY: Comstock Press.

Dunbar, R. I. M. and Dunbar, E. P. 1975. *Social Dynamics of Gelada Baboons*. Basel, Switzerland: Karger.

Eberhard, W. G. 1985. *Sexual Selection and Animal Genitalia*. Cambridge, MA: Harvard University Press.

Eberhard, W. G. 1996. *Female Control: Sexual Selection by Cryptic Female Choice*. Princeton, NJ: Princeton University Press.

Eisenberg, J. F., Muckenhirn, N. A., and Rudran, R. 1972. The relation between ecology and social structure in primates. *Science* 176, 863–874.

Emlen, D. J. 1994. Environmental control of horn length dimorphism in the beetle *Onthophagus acuminatus* (Coleoptera: Scarabaeidae). *Proceedings of the Royal Society of London B* 256, 131–136.

Emlen, S. T. and Oring, L. W. 1977. Ecology, sexual selection, and the evolution of mating systems. *Science* 197, 215–223.

Fairgrieve, C. 1995. Infanticide and infant eating in the blue monkey (*Cercopithecus mitis stuhlmanni*) in the Budongo forest reserve, Uganda. *Folia Primatologica* 64, 69–72.

Fietz, J., Zischler, H., Schwiegk, C., *et al.* 2000. High rates of extra-pair young in the pair-living fat-tailed dwarf lemur, *Cheirogaleus medius*. *Behavioral Ecology and Sociobiology* 49, 8–17.

Fisher, R. A. 1930. *The Genetical Theory of Natural Selection*. Oxford, UK: Oxford University Press.

Forbes, L. 1997. The evolutionary biology of spontaneous abortion in humans. *Trends in Ecology and Evolution* 12, 446–450.

Galdikas, B. 1981. Orangutan reproduction in the wild. In C. Graham (ed.) *Reproductive Biology of the Great Apes*, pp. 281–300. New York: Academic Press.

Galdikas, B. 1985a. Adult male sociality and reproductive tactics among orangutans at Tanjung Putting. *Folia Primatologica* 45, 9–24.

Galdikas, B. 1985b. Subadult male orangutan sociality and reproductive behavior at Tanjung Putting. *American Journal of Primatology* 8, 87–99.

Garber, P. A., Moya, L., and Malaga, C. 1984. A preliminary field study of the moustached tamarin monkey (*Saguinus mystax*) in northeastern Peru: questions concerned with the evolution of a communal breeding system. *Folia Primatologica* 42, 17–32.

Garber, P. A., Moya, L., Pruetz, J. D., and Ique, C. 1996. Social and seasonal influences on reproductive biology in male moustached tamarins (*Saguinus mystax*). *American Journal of Primatology* 38, 29–46.

Garber, P. A., Encarnacion, F., Moya, L., and Pruetz, J. D. 1993. Demographic and reproductive patterns in moustached tamarin monkeys (*Saguinus mystax*): implications for reconstructing platyrrhine mating systems. *American Journal of Primatology* 29, 235–254.

Gerard-Steklis, N. and Steklis, H. D. 2001. Reproductive benefits for female mountain gorillas in multi-male groups. *American Journal of Primatology* 54, 60–61.

Goldizen, A. W. 1987. Tamarins and marmosets: communal care of offspring. In B. B. Smuts, D. L. Cheney, R. M. Seyfarth, R. W. Wrangham, and T. T. Struhsaker (eds.) *Primate Societies*, pp. 34–43. Chicago, IL: University of Chicago Press.

Goldizen, A. W., Mendelson, J., van Vlaardingen, M., and Terborgh, J. 1996. Saddle-back tamarin (*Saguinus fuscicollis*) reproductive strategies: evidence from a thirteen-year study of a marked population. *American Journal of Primatology* 38, 57–83.

Gomendio, M. 1990. The influence of maternal rank and infant sex on maternal investment trends in rhesus macaques: birth sex-ratios, inter-birth intervals and suckling patterns. *Behavioral Ecology and Sociobiology* 27, 365–375.

Graham, C. and Nadler, R. 1990. Socioendocrine interactions in great ape reproduction. In T. E. Zeigler and F. B. Bercovitch (eds.) *Socioendocrinology of Primate Reproduction*, pp. 35–58. New York: Wiley-Liss.

Gross, M. R. 1996. Alternative reproductive strategies and tactics: diversity between the sexes. *Trends in Ecology and Evolution* 11, 92–98.

Gross, M. R. and Repka, J. 1998. Stability with inheritance in the conditional strategy. *Journal of Theoretical Biology* 192, 445–453.

Groves, C. 2001. *Primate Taxonomy*. Washington, DC: Smithsonian Institution Press.

Harcourt, A. H. 1978. Strategies of emigration and transfer by primates, with particular reference to gorillas. *Zeitschrift für Tierpsychologie* 48, 401–420.

Harcourt, A. H. and Stewart, K. J. 1981. Gorilla male relationships: can differences during immaturity lead to contrasting reproductive tactics in adulthood? *Animal Behaviour* 29, 206–210.

Hasegawa, T. and Hiraiwa-Hasegawa, M. 1983. Opportunistic and restrictive matings among wild chimpanzees in the Mahale mountain, Tanzania. *Journal of Ethology* 1, 75–85.

Hazel, W. N., Smock, R., and Johnson, M. D. 1990. A polygenic model for the evolution and maintenance of conditional strategies. *Proceedings of the Royal Society of London B* **242**, 181–187.

Henson, S. A. and Warner, R. R. 1997. Male and female alternative reproductive behaviors in fishes: a new approach using intersexual dynamics. *Annual Review of Ecology and Systematics* **28**, 571–592.

Henzi, S. P. and Lawes, M. 1988. Strategic responses of male samago monkeys *Cercopithecus mitis* to a reduction in the availability of receptive females. *International Journal of Primatology* **9**, 479–495.

Heymann, E. W. 2000. The number of males in callitrichine groups and its implications for callitrichine social evolution. In P. M. Kappeler (ed.) *Primate Males: Causes and Consequences of Variation in Group Composition*, pp. 64–71. Cambridge, UK: Cambridge University Press.

Hiraiwa-Hasegawa, M. 1993. Skewed birth sex ratios in primates: should high-ranking mothers have daughters or sons? *Trends in Ecology and Evolution* **8**, 395–400.

Hrdy, S. B. 1977. *The Langurs of Abu*. Cambridge, MA: Harvard University Press.

Hrdy, S. B. 1979. Infanticide among animals: a review, classification, and examination of the implications for the reproductive strategies of females. *Ethology and Sociobiology* **1**, 13–40.

Hrdy, S. B. and Whitten, P. L. 1987. Patterning of sexual activity. In B. B. Smuts, D. L. Cheney, R. M. Seyfarth, R. W. Wrangham, and T. T. Struhsaker (eds.) *Primate Societies*, pp. 370–384. Chicago, IL: University of Chicago Press.

Huffman, M. A. 1991. History of the Arashiyama Japanese macaques in Kyoto. In L. M. Fedigan and P. J. Asquith (eds.) *The Monkeys of Arashiyama: Thirty-Five Years of Research in Japan and the West*, pp. 21–53. Albany, NY: State University of New York Press.

Hutchings, J. A. and Myers, R. A. 1994. The evolution of alternative mating strategies in variable environments. *Evolutionary Ecology* **8**, 256–268.

Janson, C. 1984. Female choice and the mating system of the brown capuchin monkey, *Cebus apella* (Primates, Cebidae). *Zeitschrift für Tierpsychologie* **65**, 177–200.

Johnson, S. E. 2003. Life history and the competitive environment: trajectories of growth, maturation, and reproductive output among chacma baboons. *American Journal of Physical Anthropology* **120**, 83–98.

Kappeler, P. M. 2000a. Causes and consequences of unusual sex ratios among lemurs. In P. M. Kappeler (ed.) *Primate Males: Causes and Consequences of Variation in Group Composition*, pp. 55–63. Cambridge, UK: Cambridge University Press.

Kappeler, P. M. (ed.) 2000b. *Primate Males: Causes and Consequences of Variation in Group Composition*. Cambridge, UK: Cambridge University Press.

Kappeler, P. M. and Pereira, M. E. (eds.) 2003. *Primate Life Histories and Socioecology*. Chicago, IL: University of Chicago Press.

Kappeler, P. M. and van Schaik, C. P. 2002. The evolution of primate social systems. *International Journal of Primatology* **23**, 707–740.

Keane, B., Dittus, W. P. J., and Melnick, D. J. 1997. Paternity assessment in wild groups of toque macaques *Macaca sinica* at Polonnaruwa, Sri Lanka using molecular markers. *Molecular Ecology* **6**, 267–282.

Koenig, A. 1995. Group size, composition, and reproductive success in wild common marmosets (*Callithrix jacchus*). *American Journal of Primatology* **35**, 311–317.

Kraus, C., Heistermann, M., and Kappeler, P. M. 1999. Physiological suppression of sexual function of subordinate males: a subtle form of intrasexual competition among male sifakas (*Propithecus verreauxi*)? *Physiology and Behavior* **66**, 855–861.

Kummer, H. 1968. *Social Organization of Hamadryas Baboons: A Field Study*. Chicago, IL: University of Chicago Press.

Kummer, H. 1995. *In Quest of the Sacred Baboon: A Scientist's Journey*. Princeton, NJ: Princeton University Press.

Lank, D. B., Smith, C. M., Hanotte, O., Burke, T., and Cooke, F. 1995. Genetic polymorphism for alternative mating behaviour in lekking male ruff (*Philomachus pugnax*). *Nature* **368**, 59–62.

Launhardt, K., Borries, C., Hardt, C., Epplen, J. T., and Winkler, P. 2001. Paternity analysis of alternative male reproductive routes among the langurs (*Semnopithecus entellus*) of Ramnagar. *Animal Behaviour* **61**, 53–64.

Lazaro-Perea, C., Castro, C. S. S., Harrison, R., et al. 2000. Behavioral and demographic changes following the loss of the breeding female in cooperatively breeding marmosets. *Behavioral Ecology and Sociobiology* **48**, 137–146.

Lee, P. C. (ed.) 1999. *Comparative Primate Socioecology*. Cambridge, UK: Cambridge University Press.

Lhota, S., Havlicek, J., and Bartos, L. 2001. Abortions in Hanuman langurs (*Semnopithecus entellus*): a female reproductive strategy? *Primate Report* **60–1**, 27–28.

Lindburg, D. G. 1969. Rhesus monkeys: mating season mobility of adult males. *Science* **166**, 1176–1178.

Lott, D. F. 1991. *Intraspecific Variation in the Social Systems of Wild Vertebrates*. Cambridge, UK: Cambridge University Press.

Loy, J. 1971. Estrous behaviour of free-ranging rhesus monkeys (*Macaca mulatta*). *Primates* 12, 1–31.

Loy, J. 1975. The copulatory behaviour of adult male patas monkeys, *Erythrocebus patas*. *Journal of Reproduction and Fertility* 45, 193–195.

Loy, J. and Loy, K. 1977. Sexual harassment among captive patas monkeys *Erythrocebus patas*. *Primates* 18, 691–699.

MacKinnon, J. R. 1974. The ecology and behaviour of wild orang-utans (*Pongo pygmaeus*). *Animal Behaviour* 22, 3–74.

Maggioncalda, A. N., Sapolsky, R. M., and Czekala, N. M. 1999. Reproductive hormone profiles in captive male orangutans: implications for understanding developmental arrest. *American Journal of Physical Anthropology* 109, 19–32.

Maggioncalda, A. N., Czekala, N. M., and Sapolsky, R. M. 2000. Growth hormone and thyroid stimulating hormone concentrations in captive male orangutans: implications for understanding developmental arrest. *American Journal of Primatology* 50, 67–76.

Manson, J. H. 1994. Mating patterns, mate choice, and birth season heterosexual relationships in free-ranging rhesus macaques. *Primates* 35, 417–433.

Manson, J. H. 1996. Male dominance and mount series duration in Cayo Santiago rhesus macaques. *Animal Behaviour* 51, 1219–1231.

Marson, J., Gervais, D., Meuris, S., Cooper, R. W., and Jouannet, P. 1989. Influence of ejaculation frequency on semen characteristics in chimpanzees (*Pan troglodytes*). *Journal of Reproduction and Fertility* 85, 43–50.

Maynard Smith, J. 1982. *Evolution and the Theory of Games*. Cambridge, UK: Cambridge University Press.

Mitani, J. C. 1985. Mating behavior of male orangutans in the Kutai Game Reserver, Indonesia. *Animal Behaviour* 33, 392–402.

Mitani, J. C., Gros-Louis, J., and Manson, J. H. 1996a. Number of males in primate groups: comparative tests of competing hypotheses. *American Journal of Primatology* 38, 315–332.

Mitani, J. C., Gros-Louis, J., and Richards, A. F. 1996b. Sexual dimorphism, the operational sex ratio, and the intensity of male competition in polygynous primates. *American Naturalist* 147, 966–980.

Moore, M. C. 1991. Application of organization-activation theory to alternative male reproductive strategies: a review. *Hormones and Behavior* 25, 154–179.

Mori, A. 1979. Unit formation and the emergence of a new leader. In M. Kawai (ed.) *Contributions to Primatology, Ecological and Sociological Studies of Gelada Baboons*, vol. 16, pp. 155–181. Basel, Switzerland: Karger.

Nievergelt, C. M., Digby, L. J., Ramakrishnan, U., and Woodruff, D. S. 2000. Genetic analysis of group composition and breeding system in a wild common marmoset (*Callithrix jacchus*) population. *International Journal of Primatology* 21, 1–20.

Noë, R. and Sluijter, A. A. 1990. Reproductive tactics of male savanna baboons. *Behaviour* 113, 117–170.

Nunn, C. L. 1999. The number of males in primate social groups: a comparative test of the socioecological model. *Behavioral Ecology* 46, 1–13.

Ohsawa, H., Inoue, M., and Takenata, O. 1993. Mating strategy and reproductive success of male patas monkeys (*Erythrocebus patas*). *Primates* 35, 533–544.

Okayusu, N. 1992. Prolonged estrous in female Japanese macaques (*Macaca fuscata yakui*) and the social influence on estrus: with special reference to male intertroop movement. In J. E. Fa and D. G. Lindburg (eds.) *Evolution and Ecology of Macaque Societies*, pp. 342–368. Cambridge, UK: Cambridge University Press.

Packer, C. 1977. Reciprocal altruism in *Papio anubis*. *Nature* 265, 441–443.

Packer, C. 1979a. Inter-troop transfer and inbreeding avoidance in *Papio anubis*. *Animal Behaviour* 27, 1–36.

Packer, C. 1979b. Male dominance and reproductive activity in *Papio anubis*. *Animal Behaviour* 27, 37–45.

Palombit, R. A., Seyfarth, R. M., and Cheney, D. L. 1997. The adaptive value of "friendships" to female baboons: experimental and observational evidence. *Animal Behaviour* 54, 599–614.

Palombit, R. A., Cheney, D., Fischer, J., et al. 2000. Male infanticide and defense of infants in chacma baboons. In C. P. van Schaik and C. H. Janson (eds.) *Infanticide by Males and its Implications*, pp. 123–152. Cambridge, UK: Cambridge University Press.

Palombit, R. A., Cheney, D. L., and Seyfarth, R. M. 2001. Female–female competition for male "friends" in wild chacma baboons, *Papio cynocephalus ursinus*. *Animal Behaviour* 61, 1159–1171.

Paul, A. 2002. Sexual selection and mate choice. *International Journal of Primatology* 23, 877–904.

Pereira, M. E. 1983. Abortion following the immigration of an adult male baboon (*Papio cynocephalus*). *American Journal of Primatology* 4, 93–98.

Perret, M. E. 1992. Environmental and social determinants of sexual function in the male lesser mouse lemur (*Microcebus murinus*). *Folia Primatologica* 59, 1–25.

Perret, M. E. and Schilling, A. 1987a. Intermale sexual effect elicited by volatile urinary ether extract in *Microcebus murinus* (Prosimian, Primates). *Journal of Chemical Ecology* 13, 495–507.

Perret, M. E. and Schilling, A. 1987b. Role of prolactin in a pheromone-like sexual inhibition in the male lesser mouse lemur. *Journal of Endocrinology* 114, 279–287.

Pope, T. R. 1990. The reproductive consequences of male cooperation in the red howler monkey: paternity exclusion in multi-male and single-male troops using genetic markers. *Behavioral Ecology and Sociobiology* 27, 439–446.

Pope, T. R. 1998. Effects of demographic change on group kin structure and gene dynamics of populations of red howling monkeys. *Journal of Mammalogy* 79, 692–712.

Pope, T. R. 2000. The evolution of male philopatry in neotropical monkeys. In P. M. Kappeler (ed.) *Primate Males: Causes and Consequences of Variation in Group Composition*, pp. 219–235. Cambridge, UK: Cambridge University Press.

Preston, B. R., Stevenson, I. R., and Wilson, K. 2003. Soay rams target reproductive activity towards promiscuous females' optimal insemination period. *Proceedings of the Royal Society of London B* 270, 2073–2078.

Pullen, S. L., Bearder, S. K., and Dixson, A. F. 2000. Preliminary observations on sexual behavior and the mating system in free-ranging lesser galagos (*Galago moholi*). *American Journal of Primatology* 51, 179–188.

Pusey, A. E. and Packer, C. 1987. Dispersal and philopatry. In B. B. Smuts, D. L. Cheney, R. M. Seyfarth, R. W. Wrangham, and T. T. Struhsaker (eds.) *Primate Societies*, pp. 250–266. Chicago, IL: University of Chicago Press.

Pusey, A. E., Williams, J., and Goodall, J. 1997. The influence of dominance rank on the reproductive success of female chimpanzees. *Science* 277: M828–M831.

Qvanstrom, A. and Price, T. D. 2001. Maternal effects, paternal effects and sexual selection. *Trends in Ecology and Evolution* 16, 95–100.

Rajpurohit, L. S. and Mohnot, S. M. 1988. Fate of ousted male residents of one-male bisexual troops of Hanuman langurs *Presbytis entellus* at Jodhpur, Rajasthan, India. *Human Evolution* 3, 309–318.

Rajpurohit, L. S., Sommer, V., and Mohnot, S. M. 1995. Wanderers between harems and bachelor bands: male

Hanuman langurs (*Presbytis entellus*) at Jodhpur in Rajasthan. *Behaviour* 132, 255–299.

Repka, J. and Gross, M. R. 1995. The evolutionarily stable strategy under individual condition and tactic frequency. *Journal of Theoretical Biology* 176, 27–31.

Robbins, M. M. 1995. A demographic analysis of male life history and social structure of mountain gorillas. *Behaviour* 132, 21–47.

Robbins, M. M. 1999. Male mating patterns in wild multimale mountain gorilla groups. *Animal Behaviour* 57, 1013–1020.

Rubenstein, D. 1980. On the evolution of alternative mating strategies. In J. E. R. Staddon (ed.) *Limits to Action: The Allocation of Individual Behaviour*, pp. 65–100. New York: Academic Press.

Ryan, M. J. and Causey, B. A. 1989. Alternative mating behavior in the swordtails *Xiphophorus nigrensis* and *Xiphophorus pygmaeus* (Pisces: Poeciliidae). *Behavioral Ecology and Sociobiology* 24, 341–348.

Sauermann, U., Nürnberg, P., Bercovitch, F. B., *et al.* 2001. Increased reproductive success of MHC class II heterozygous males among free-ranging rhesus macaques. *Human Genetics* 108, 249–254.

Schilling, A., Perret, M., and Predine, J. 1984. Sexual inhibition in a prosimian primate: a pheromone-like effect. *Journal of Endocrinology* 102, 143–151.

Schürmann, C. L. 1982. Mating behaviour of wild orangutans. In L. de Boer (ed.) *The Orang-Utan: Its Biology and Conservation*, pp. 271–286. The Hague, Netherlands: Junk Publishers.

Schürmann, C. L. and van Hoof, J. A. R. A. M. 1986. Reproductive strategies of the orang-utan: new data and a reconsideration of existing social models. *International Journal of Primatology* 7, 265–287.

Schuster, S. M. and Wade, M. J. 1991. Equal mating success among alternative reproductive strategies in a marine isopod. *Nature* 350, 608–610.

Schuster, S. M. and Wade, M. J. 2003. *Mating Systems and Strategies*. Princeton, NJ: Princeton University Press.

Schwagmeyer, P. L. 1979. The Bruce effect: an evaluation of male/female advantages. *American Naturalist* 114, 932–939.

Scott, L. M. 1984. Reproductive behavior of adolescent female baboons (*Papio anubis*) in Kenya. In M. F. Small (ed.) *Female Primates: Studies by Women Primatologists*, pp. 77–100. New York: Alan R. Liss.

Setchell, J. M. 1999. Socio-sexual development in the male mandrill (*Mandrillus sphinx*). Ph.D. thesis, University of Cambridge, UK.

Setchell, J. M. 2003. The evolution of alternative reproductive morphs in male primates. In C. B. Jones (ed.) *Sexual Selection and Reproductive Competition in Primates: New Perspectives and Directions*, pp. 413–435. American Society of Primatologists.

Setchell, J. M. 2005. Do female mandrills (*Mandrillus sphinx*) prefer brightly coloured males? *International Journal of Primatology* 26, 713–732.

Setchell, J. M. and Dixson, A. F. 2001a. Arrested development of secondary sexual adornments in subordinate adult male mandrills (*Mandrillus sphinx*). *American Journal of Physical Anthropology* 115, 245–252.

Setchell, J. M. and Dixson, A. F. 2001b. Changes in the secondary sexual adornments of male mandrills (*Mandrillus sphinx*) are associated with gain and loss of alpha status. *American Journal of Primatology* 39, 177–184.

Setchell, J. M. and Kappeler, P. M. 2003. Selection in relation to sex in primates. *Advances in the Study of Behavior* 33, 87–173.

Setchell, J. M. and Lee, P. C. 2004. Development and sexual selection in primates. In P. M. Kappeler and C. P. van Schaik (eds.) *Sexual Selection in Primates: New and Comparative Perspectives*, pp. 175–195. Cambridge, UK: Cambridge University Press.

Setchell, J. M., Lee, P. C., Wickings, E. J., and Dixson, A. F. 2001. Growth and ontogeny of sexual size dimorphism in the mandrill (*Mandrillus sphinx*). *American Journal of Physical Anthropology* 115, 349–360.

Setchell, J. M., Lee, P. C., Wickings, E. J., and Dixson, A. F. 2002. Reproductive parameters and maternal investment in mandrills (*Mandrillus sphinx*). *International Journal of Primatology* 23, 51–68.

Sigg, H., Stolba, A., Abegglen, J. J., and Dasser, V. 1982. Life history of hamadryas baboons: physical development, infant mortality, reproductive parameters, and family relationships. *Primates* 23, 473–487.

Silk, J. B., Shmt, J., Roberts, J., and Kusnitz, J. 1993. Gestation length in rhesus macaques. *International Journal of Primatology* 14, 95–104.

Simpson, M. J. A., Simpson, A. E., Hooley, J., and Zunz, M. 1981. Infant-related influences on birth interval in rhesus monkeys. *Nature* 290, 49–51.

Smuts, B. B. 1985. *Sex and Friendships in Baboons*. Hawthorne, NY: Aldine Press.

Smuts, B. B. and Smuts, R. W. 1993. Male aggression and sexual coercion of females in nonhuman primates and other mammals: evidence and theoretical implications. *Advances in the Study of Behavior* 22, 1–63.

Smuts, B. B., Cheney, D. L., Seyfarth, R. E., Wrangham, R. W., and Struhsaker, T. T. (eds.) 1987. *Primate Societies*. Chicago, IL: University of Chicago Press.

Soltis, J., Mitsunaga, F., Shimizu, K., *et al.* 1997a. Sexual selection in Japanese macaques. 2. Female mate choice and male–male competition. *Animal Behaviour* 54, 737–746.

Soltis, J., Mitsunaga, F., Shimizu, K., Yanagihara, Y., and Nozaki, M. 1997b. Sexual selection in Japanese macaques. 1. Female mate choice or male sexual coercion? *Animal Behaviour* 54, 725–736.

Soltis, J., Thomsen, R., and Takenaka, O. 2001. The interaction of male and female reproductive strategies and paternity in wild Japanese macaques, *Macaca fuscata*. *Animal Behaviour* 62, 485–494.

Sommer, V. 1994. Infanticide among the langurs of Jodhpur: testing the sexual selection hypothesis with a long-term record. In S. Parmigiani and F. S. von Saal (eds.) *Infanticide and Parental Care*, pp. 155–168. Chur, Switzerland: Harwoon Academic.

Sommer, V. and Rajpurohit, L. S. 1989. Male reproductive success in harem troops of Hanuman langurs (*Presbytis entellus*). *International Journal of Primatology* 10, 293–317.

Sommer, V., Srivastava, A., and Borries, C. 1992. Cycles, sexuality, and conception in free-ranging langurs (*Presbytis entellus*). *American Journal of Primatology* 28, 1–27.

Sprague, D. S., Suzuki, S., Takahashi, H., and Sato, S. 1998. Male life history in natural populations of Japanese macaques: migration, dominance rank, and troop participation of males in two habitats. *Primates* 39, 351–363.

Steenbeek, R., Sterck, E. H. M., de Vries, H., and van Hooff, J. A. R. A. M. 2000. Costs and benefits of the one-male, age-graded and all-male phases in wild Thomas' langur groups. In P. M. Kappeler (ed.) *Primate Males: Causes and Consequences of Variation in Group Composition*, pp. 130–145. Cambridge, UK: Cambridge University Press.

Sterck, E. H. M. 1997. Determinants of female dispersal in Thomas' langurs. *American Journal of Primatology* 42, 179–198.

Stewart, K. J. and Harcourt, A. H. 1987. Gorillas: variation in female relationships. In B. B. Smuts, D. L. Cheney, R. M. Seyfarth, R. W. Wrangham, and T. T. Struhsaker (eds.) *Primate Societies*, pp. 165–177. Chicago, IL: University of Chicago Press.

Stokes, E. J., Parnell, R. J., and Olejniczak, C. 2003. Female dispersal and reproductive success in wild western lowland gorillas (*Gorilla gorilla gorilla*). *Behavioral Ecology and Sociobiology* 54, 329–339.

truhsaker, T. T. 1988. Male tenure, multi-male influxes and reproductive success in redtail monkeys (*Cercopithecus ascanius*). In A. Gautier-Hion, F. Boulière, J. P. Gautier, and J. Kingdon (eds.) *A Primate Radiation: Evolutionary Biology of the African Guenons*, pp. 340–363. Cambridge, UK: Cambridge University Press.

ussman, R. W. 1992. Male life-history and intergroup mobility among ringtailed lemurs (*Lemur catta*). *International Journal of Primatology* 13, 395–413.

ussman, R. W. 1999. *Primate Ecology and Social Structure*, vol. 1, *Lorises, Lemurs and Tarsiers*. Needham Heights, MA: Pearson Custom.

ussman, R. W. 2000. *Primate Ecology and Social Structure*, vol. 2, *New World Monkeys*. Needham Heights, MA: Pearson Custom.

ussman, R. W. 2006. *Primate Ecology and Social Structure*, vol. 3, *Old World Monkeys and Apes*. Needham Heights, MA: Pearson Custom.

uzuki, S., Hill, D. A., and Sprague, D. S. 1998. Intertroop transfer and dominance rank structure of non-natal male Japanese macaques in Yakushima, Japan. *International Journal of Primatology* 19, 703–722.

aborsky, M. 1994. Sneakers, satellites, and helpers: parasitic and cooperative behaviour in fish reproduction. *Advances in the Study of Behavior* 23, 1–100.

aborsky, M. 1997. Bourgeois and parasitic tactics: do we need collective, functional terms for alternative reproductive behaviours? *Behavioral Ecology and Sociobiology* 41, 361–362.

aborsky, M. 1998. Sperm competition in fish: "bourgeois" males and parasitic spawning. *Trends in Ecology and Evolution* 13, 222–227.

aborsky, M. 1999. Conflict or cooperation: what determines optimal solutions to competition in fish reproduction? In V. C. Almada, R. F. Oliveira, and E. J. Gonçalves (eds.) *Behaviour and Conservation of Littoral Fishes*, pp. 301–349. Lisbon: Instituto Superior de Psicologia Aplicada.

aborsky, M. 2001. The evolution of bourgeois, parasitic, and cooperative reproductive behaviors in fishes. *Journal of Heredity* 92, 100–110.

akahashi, H. 2001. Influence of fluctuation in the operational sex ratio to mating of troop and non-troop male Japanese macaques for four years on Kinkazan island, Japan. *Primates* 42, 183–191.

akahata, Y. 1982. Social relations between adult males and females of Japanese monkeys in the Arashiyama B troop. *Primates* 23, 1–23.

Trivers, R. L. 1972. Parental investment and sexual selection. In B. Campbell (ed.) *Sexual Selection and the Descent of Man*, pp. 136–179. Chicago, IL: Aldine Press.

Trivers, R. L. and Willard, D. 1973. Natural selection of parental ability to vary the sex ratio of offspring. *Science* 179, 90–92.

Tsingalia, H. M. and Rowell, T. E. 1984. The behaviour of adult male blue monkeys. *Zeitschrift für Tierpsychologie* 64, 253–268.

Tutin, C. E. G. 1979. Mating patterns and reproductive strategies in a community of wild chimpanzees (*Pan troglodytes*). *Behavioral Ecology and Sociobiology* 6, 29–38.

Utami, S. S. 2000. Bimaturism in orang-utan males: reproductive and ecological strategies. Ph.D. thesis, University of Utrecht, the Netherlands.

Utami, S. S., Goossens, B., Bruford, M. W., de Ruiter, J., and van Hooff, J. A. R. A. M. 2002. Male bimaturism and reproductive success in Sumatran orang-utans. *Behavioral Ecology* 13, 643–652.

van Hooff, J. A. R. A. M. and Utami, S. S. 2004. Male bimaturism and sexual selection in orangutans. In P. K. Kappeler and C. P. van Schaik (eds.) *Sexual Selection in Primates: New and Comparative Perspectives*, pp. 196–207. Cambridge, UK: Cambridge University Press.

van Noordwijk, M. A. and van Schaik, C. P. 1999. The effects of dominance rank and group size on female lifetime reproductive success in wild long-tailed macaques, *Macaca fascicularis*. *Primates* 40, 105–130.

van Noordwijk, M. A. and van Schaik, C. P. 2001. Career moves: transfer and rank challenge decisions by male long-tailed macaques. *Behaviour* 138, 359–395.

van Noordwijk, M. A. and van Schaik, C. P. 2004 Sexual selection and the careers of primate males: paternity concentration, dominance acquisition tactics and transfer decisions. In P. K. Kappeler and C. P. van Schaik (eds.) *Sexual Selection in Primates: New and Comparative Perspectives*, pp. 208–229. Cambridge, UK: Cambridge University Press.

van Rhijn, J. G. 1973. Behavioural dimorphism in male ruffs, *Philomachus pugnax* (L.). *Behaviour* 47, 153–229.

van Schaik, C. P. and Hrdy, S. B. 1991. Intensity of local resource competition shapes the relationship between maternal rank and sex ratios at birth in cercopithecine primates. *American Naturalist* 138, 1555–1562.

van Schaik, C. P. and Janson, C. H. (eds.) 2000. *Infanticide by Males and Its Implications*. Cambridge, UK: Cambridge University Press.

van Schaik, C. P., van Noordwijk, M. A., and Nunn, C. L. 1999. Sex and social evolution in primates. In P. C. Lee (ed.) *Comparative Primate Socioecology*, pp. 204–240. Cambridge, UK: Cambridge University Press.

van Schaik, C. P., Pradhahn, G. R., and van Noordwijk, M. A. 2004. Mating conflict in primates: infanticide, sexual harassment and female sexuality. In P. M. Kappeler and C. P. van Schaik (eds.) *Sexual Selection in Primates: New and Comparative Perspectives*, pp. 131–150. Cambridge, UK: Cambridge University Press.

Watts, D. P. 1989. Infanticide in mountain gorillas: new cases and a reconsideration of the evidence. *Ethology* **81**, 1–18.

Watts, D. P. 1990. Ecology of gorillas and its relation to female transfer in mountain gorillas. *International Journal of Primatology* **11**, 21–45.

Watts, D. P. 1998. Coalitionary mate guarding by male chimpanzees at Ngogo, Kibale National Park, Uganda. *Behavioral Ecology and Sociobiology* **44**, 43–55.

Watts, D. P. 2000. Causes and consequences of variation in male mountain gorilla life histories and group membership. In P. M. Kappeler (ed.) *Primate Males: Causes and Consequences of Variation in Group Composition*, pp. 169–180. Cambridge, UK: Cambridge University Press.

Wedell, N., Gage, M. J. G., and Parker, G. A. 2002. Sperm competition, male prudence and sperm-limited females. *Trends in Ecology and Evolution* **17**, 313–320.

West-Eberhard, M. J. 1979. Sexual selection, social competition, and evolution. *Proceedings of the American Philosophical Society* **123**, 222–234.

Wickings, E. J. and Dixson, A. F. 1992. Testicular function, secondary sexual development and social status in male mandrills (*Mandrillus sphinx*). *Physiology and Behavior* **52**, 909–916.

Zhao, Q.-K. 1996. Etho-ecology of Tibetan macaques at Mount Emei, China. In J. E. Fa and D. G. Lindburg (eds.) *Evolution and Ecology of Macaques Societies*, pp. 263–289. Cambridge, UK: Cambridge University Press.

Part IV
Emerging perspectives on alternative reproductive tactics

16 · Communication and the evolution of alternative reproductive tactics

DAVID M. GONÇALVES, RUI F. OLIVEIRA, AND PETER K. McGREGOR

CHAPTER SUMMARY

In this chapter, concepts derived from communication network theory are applied to the understanding of the evolution of signals in species with alternative reproductive tactics (ARTs). These species are particularly interesting to consider from the perspective of communicating in a network because the signaling and receiving behavior of different reproductive phenotypes can be expected to be subject to diverse selection pressures. We begin by briefly introducing ARTs and communication networks. Then the consequences of communicating in a network are considered from the perspective of the several reproductive phenotypes occurring in species with ARTs, both as signalers and receivers. Finally, the evolutionary outcome of conflict and cooperation between these reproductive phenotypes is predicted in an integrative approach, and new directions are proposed to test some of the hypotheses derived.

6.1 INTRODUCTION

Alternative reproductive tactics (ARTs) is the term used to refer to variation in mating behavior found within a species. As the topic is the subject of this book, we will only briefly introduce ARTs in relation to signaling. More detailed information on ARTs can be found in several chapters in this book and recent reviews (e.g., Brockmann 2001, Shuster and Wade 2003).

For simplicity, we have only considered male ARTs. This choice reflects the facts that male ARTs are more common than female ARTs (but see Alonzo, Chapter 18, this volume) and that many more examples of male ARTs have been described. Nevertheless, the ideas presented here extend directly to female ARTs. The bias towards fish examples in this chapter reflects the abundant literature on fish ARTs.

16.1.1 Bourgeois, sneaker, female-mimicking, and cooperative males

Males may reproduce by investing primarily in direct access to, and defense of, reproductive resources ("bourgeois males"). Other males may access these resources either by a quick and inconspicuous approach ("sneaker males"), by mimicking females ("female mimics") or by cooperating with bourgeois males ("cooperative males") (Taborsky 1994, 1997, 1998, 1999).

Sneakers and female mimics are expected to decrease the bourgeois male's success. For example, in the beetle *Onthophagus taurus*, the bourgeois male's share of paternity declines with increasing sneaking pressure (Hunt and Simmons 2002). Contrarily, cooperative males are subordinates who overall increase the bourgeois male's reproductive success by investing in female attraction, territory defense, or parental care. As an example, in the cooperatively breeding fish *Neolamprologus pulcher*, subordinate males increase the reproductive success of the bourgeois male by helping with parental duties and territory defense (Brouwer *et al.* 2005) and these cooperative males benefit by siring some of the offspring (Balshine-Earn *et al.* 1998, Dierkes *et al.* 1999). Sharing of reproductive resources is usually explained by two types of models: optimal skew models assume that bourgeois males control the access to reproductive resources and allow cooperative males to access resources in exchange for their cooperative efforts, and incomplete control models assume that cooperative males forcibly gain access to those resources due to incomplete control by the bourgeois male, thus also reproducing parasitically (e.g., Emlen *et al.* 1998, Reeve *et al.* 1998, Johnstone and Cant 1999, Kokko 2003). In both cases, however, conflict between bourgeois and cooperative males occurs on the level of access (allowed or forced) to reproductive resources.

Alternative Reproductive Tactics, ed. Rui F. Oliveira, Michael Taborsky, and H. Jane Brockmann. Published by Cambridge University Press. © Cambridge University Press 2008.

16.2   COMMUNICATION NETWORKS

Signals produced by animals are often detected by more than one receiver simultaneously. As a result, most animals communicate in a network with several individuals occurring within communication distance (McGregor 1993, McGregor and Dabelsteen 1996). However, although conflict and cooperation between senders and receivers have long been recognized as selection pressures shaping the nature and design of signaling and receiving systems (e.g., Dawkins and Krebs 1978, Krebs and Dawkins 1984), only recently has the role of other parties in a communication network (e.g., eavesdroppers or audiences; see below) been considered when studying the evolution of communication (e.g., Johnstone 2000, 2001).

16.2.1   Eavesdropping

A consequence of animals communicating within a network is that information produced by a signaler is more widely available than the signaler–receiver dyad that is usually considered. An important class of such extra receivers has been termed "eavesdroppers" (McGregor 1993, McGregor and Dabelsteen 1996). Recently, Peake (2005) has clarified the definition of eavesdropping in the context of animal communication as "the use of information in signals by individuals other than the primary target." We will use this

definition throughout. One reason for doing so is that it specifically avoids the effects of the presence of eavesdroppers on communication, and this is important because eavesdroppers can confer benefits as well as impose more obvious costs. Peake (2005) has also made a distinction between two types of eavesdropping. *Interceptive eavesdropping* refers to the use of information contained in a signal intended (in the evolutionary sense) for another individual, as, for example, when a bat locates a male frog based on the calls produced to attract female frogs (Figure 16.1A). Interceptive eavesdroppers usually use broadcast signals as the source of information, are usually heterospecifics, and generally produce a negative or zero pay-off to signalers (Peake 2005). *Social eavesdropping* refers to the gathering of information from signaling interactions between conspecifics in which the eavesdropper plays no part. For example, in the fighting fish *Betta splendens*, males pay more attention to a pair of interacting than noninteracting males and are more reluctant to approach and display towards a male that they have observed winning an interaction than towards a loser but there is no such difference in response to males that have won and lost interactions out of sight of the subject (Oliveira *et al.* 1998) (Figure 16.1B). Social eavesdroppers thus extract and may use detailed information from social interactions, and this may result in a negative, neutral, or positive pay-off to signalers (Peake 2005).

(A) Interceptive eavesdropping

(B) Social eavesdropping

Figure 16.1 Two distinct types of eavesdropping. (A) In interceptive eavesdropping information contained in a signal intended for another animal is used. In the example, frog-eating bats locate prey by intercepting their mating calls. (B) In social eavesdropping animals use information gathered during signaling interactions. For example, eavesdropping males of the fighting fish *Betta splendens* are less likely to initiate a fight with a male observed winning an interaction than with a loser male. (After Peake 2005.)

16.2.2   Audience effects

During a social interaction, signalers may also adjust their behavior according to the presence and nature of animals other than those directly involved in the interaction. This has been termed the "audience effect," and it has been demonstrated in a number of species (e.g., Evans and Marler 1984, Gyger et al. 1986, Hector et al. 1989, Marler and Evans 1996, Doutrelant et al. 2001, Matos and McGregor 2002, Matos et al. 2003, Dzieweczynski et al. 2005; reviewed by McGregor and Peake 2000, Matos and Schlupp 2005). For example, male fighting fish change the nature of their aggressive displays during male–male agonistic interactions depending on the gender of the audience (Doutrelant et al. 2001, Matos and McGregor 2002, Dzieweczynski et al. 2005). In nature, audiences are also likely to act as eavesdroppers on most occasions. For example, during a male–female sexual interaction, the presence of another female may create an audience effect i.e., influence the displays of the sexual pair), and at the same time, she may be a social eavesdropper (i.e., collect and use information from the interaction between the pair). For implicity, we will assume that all audiences are possible eavesdroppers and are considered as such by signalers (for a discussion of the distinction between apparent and evolutionary audiences, see Matos and Schlupp 2005).

16.2.3   Fitness consequences of eavesdropping

The effect of eavesdropping on the general design of signaling and receiving systems will depend on its fitness consequences to both signalers and receivers. It is probably reasonable to assume that if an animal eavesdrops it has, on average, benefited from the behavior in the past. It is less straightforward to make generalizations about the animals that are eavesdropped upon, particularly those involved in a signaling interaction where several combinations of fitness consequences are possible, including different consequences for each individual. The examples below illustrate the range of outcomes expected from the occurrence of eavesdropping.

EAVESDROPPERS HAVE FITNESS COSTS

If eavesdropping is common and has a fitness cost for both signalers and receivers, eavesdropping pressure should promote inconspicuous, cheap, and directional signals i.e., "conspiratorial whispers": Dawkins and Krebs 1978, Maynard Smith 1991, Johnstone and Grafen 1992, Johnstone 2000). Examples of animals decreasing signal intensity with increasing eavesdropping pressure are common: several species of petrels stop producing mating calls when playback simulates the presence of a predator (Mougeot and Bretagnolle 2000), and pipefish *Sygnathus typhle* decrease courtship display frequency and take longer to court females with increasing eavesdropping pressure from predators (Fuller and Berglund 1996).

EAVESDROPPERS HAVE FITNESS BENEFITS

If eavesdropping benefits the signaler and is positive or neutral for receivers, signals should contain features that enhance information transfer to eavesdroppers. For example, during sexual interactions females may copy the mate choice of other females and prefer to associate with males previously observed in the company of females (e.g., Dugatkin and Godin 1992). A successfully courting male is likely to gain fitness benefits (e.g., more matings) if other females eavesdrop upon its interaction with the primary female. The primary female may not suffer any cost from eavesdropping females; indeed, it may even gain benefits as in species where the probability of nest abandonment by males decreases with increasing numbers of eggs or young in the nest (e.g., Taborsky et al. 1987). There is abundant empirical evidence that successful males use more conspicuous displays during sexual interactions than less successful males, but whether this aims, at least partially, to enhance information transfer to eavesdropping females is unclear.

EAVESDROPPERS HAVE BOTH FITNESS COSTS AND BENEFITS

Eavesdropping may have opposite fitness outcomes on interacting individuals. For instance, a proposed function of long-range copulation calls by females in birds and mammals is to attract not only the pair male but also extra-pair males in order to promote male–male competition and possibly gain both direct and genetic benefits (e.g., Cox and La Boeuf 1977, Birkhead and Møller 1992). The paired male may pay a cost if eavesdropping occurs (e.g., lost fertilizations), and females may gain from eavesdropping (e.g., the eavesdropping male may be of superior quality). In these cases signals will result from a compromise between costs and benefits for signalers and receivers. In this example, paired males may be unresponsive to female signals above a certain threshold or may punish females observed signaling to extra-pair males (e.g., Valera et al. 2003). Females should signal at a level where benefits of extra-pair male attraction compensate the costs of retaliation by the paired male.

16.3   COMMUNICATION AND ARTs

The adaptive significance of morph-specific traits that are such distinctive features of ARTs has been thoroughly investigated, but little attention has been devoted to the role of conflict and cooperation between signalers, receivers, and eavesdroppers in shaping the evolution of communication traits in the context of ARTs. In other words, some of the differences observed between alternative reproductive morphs may relate to their different roles in the communication network environment. For example, sneakers or female mimics have opposite fitness consequences for bourgeois males. Therefore, we might expect cooperative males to signal their tactic to bourgeois males while sneakers and female mimics should not. Such differences will result in different selection pressures acting on the signaling and receiving systems of the various alternative reproductive phenotypes, leading to differences in their sensory and receiving systems. In this section, we explore in detail the influence of intraspecific interactions on the signaling and receiving behavior of bourgeois males, females, and parasitic or cooperative males.

16.3.1   The bourgeois male perspective

SIGNALING BEHAVIOR IN RELATION TO
EAVESDROPPING PRESSURE
Bourgeois male sexual signals should attract females while minimizing the likelihood of sexual parasitism by other males. These are conflicting interests as signals produced by bourgeois males for female attraction may be subject to eavesdropping by other males seeking access to the bourgeois males' reproductive resources. Thus, in species with ARTs, bourgeois male sexual signals generally represent a trade-off between female attraction and attracting unwanted male competitors (Table 16.1, Figure 16.2). If eavesdropping by other males decreases the bourgeois male's reproductive success, this should promote a decrease in the conspicuousness of sexual signals produced by the bourgeois male (e.g., intensity, frequency) as eavesdropping pressure increases.

Many examples have been described in support of this prediction, probably because in several species both bourgeois males and females do not benefit from advertising mating events to eavesdropping males. One such example is the Mediterranean wrasse *Symphodus ocellatus*. In this species bourgeois males actively defend a nest and court females while smaller sneaker males stay close to actively spawning nests and try to achieve parasitic fertilizations of

eggs (Taborsky *et al.* 1987, Taborsky 1994). The reproductive success of both bourgeois males and females decreases with increasing parasitic pressure (Alonzo and Warner 1999, 2000). Field experiments have shown that bourgeois males dynamically adjust their signaling behavior according to parasitic pressure. As predicted, when the number of sneakers in the vicinity of nests was experimentally decreased, a larger number of bourgeois males courted females (Figure 16.3A), and the reverse was true when there was an increase in the number of sneakers (Figure 16.3B) (Alonzo and Warner 1999, 2000; see also van den Berghe *et al.* 1989). The male's unresponsiveness to females leads to a decrease in the number of sneakers in the vicinity of the nest that potentially increases the bourgeois male's future reproductive success by decreasing parasitic fertilizations of eggs (Alonzo and Warner 1999, 2000). Similar results were found for the three-spined stickleback *Gasterosteus aculeatus*, where bourgeois males reduce their courtship rate towards females in the presence of potential sneakers (Le Comber *et al.* 2003).

Besides decreasing the conspicuousness, frequency, or duration of signals in the presence of eavesdroppers, bourgeois males may also include signaling components that diminish the probability of eavesdropping. This has been suggested for an Australian bushcricket of the genus *Caecidia*. In this species calling males add a loud chirping sound, not used in female attraction, to the end of their female-calling song. Females respond with a short click soon after the male call. Hammond and Bailey (2003) suggest that the chirping component of the male call masks the female response so that eavesdropping males are unable to intercept the female based on her response. The authors also suggest that the calling male is likely to be able to hear the female response shortly before or during pauses in the syllables of the mask while an eavesdropping male will not.

However, bourgeois males will have higher reproductive success if cooperative males are attracted by their signals (this is still eavesdropping by Peake's [2005] definition, because the primary targets are females). In this scenario, bourgeois male sexual signals should become more conspicuous when the benefits of attracting other males (e.g., an increase in female attraction) outweigh its costs (e.g., lost fertilizations: Figure 16.2C). The hypothesis that male sexual signals directed to females may incorporate conspicuous features to promote interception by eavesdropping cooperative males has not been investigated. Bourgeois males may also signal directly to other males in order to promote their cooperation. For example, in the lek-breeding ruff.

Table 16.1. *A hypothetical example of variation in female and parasitic male attraction in relation to the intensity of a sexual signal produced by a bourgeois male*

Signal intensity	Number of females attracted F	Total number of eggs laid[a] $E = F \times 10$	Number of parasitic males attracted P	Total number of eggs fertilized by parasitic males[b] $L = (P \times E)/10$	Proportion of eggs fertilized by the bourgeois male $M = (E - L)/E$	Total number of eggs fertilized by the bourgeois male $S = E - L$
1	10	1	1	0.9	9
2	20	2	4	0.8	16
3	30	3	9	0.7	21
4	40	4	16	0.6	24
5	50	5	25	0.5	25
6	60	6	36	0.4	24
7	70	7	49	0.3	21
8	80	8	64	0.2	16
9	90	9	81	0.1	9

[a] Assuming each female lays 10 eggs.
[b] Assuming each parasitic male fertilizes 10% of the eggs.

Box 16.1   Lek breeding: the ruff

In the lek-breeding ruff, *Philomachus pugnax*, "independent" males defend territories inside leks where many males aggregate to perform sexual displays towards females. Nonterritorial "satellite" males move between territories, displaying in the independents' courts and trying to copulate with females when they enter the territory (Hogan-Warburg 1966, van Rhijn 1991, Höglund and Alatalo 1995, Lank *et al.* 1995, Hugie and Lank 1997). Independent and satellite alternative strategies are genetically determined, and independents have darker plumage than satellites (Lank *et al.* 1995). Breeding plumage is highly variable between individuals but highly stable within the same animal (Lank *et al.* 1995), and territorial males can presumably individually identify satellite males by their plumage (Lank and Dale 2001). Larger leks are preferred by females that seem to be attracted to territories with both types of males (van Rhijn 1973, Lank and Smith 1992, Höglund and Alatalo 1995, Höglund *et al.*

1993, Widemo and Owens 1995, Widemo 1998). Independents try to recruit satellites into their territory by directing signals to satellites similar to those produced during courtship sequences (Figure 16.4A). Independent male's reproductive success is predicted to be maximum in intermediate-sized leks, as a decrease in the proportion of copulations attained by the territorial male in larger leks offsets the increase in female visits (Widemo and Owens 1995, 1999) (Figure 16.4B). As lek size increases, the control of the territorial male over the reproduction of satellites in its court decreases. Territorial males do not evict satellites from their territory but try to prevent them from mating with the female by placing their bill over the satellite's head in a "mutual squat" that seems to prevent satellites from leaving to mate with females (Hogan-Warburg 1966, van Rhijn 1991, Höglund *et al.* 1993, Hugie and Lank 1997) (Figure 16.4C). If a satellite male is nevertheless seen trying to mate with a female, the territorial male may attack and expel that male from the lek (Hogan-Warburg 1993).

Philomachus pugnax, females prefer territories with both territorial and nonterritorial ("satellite") males, and territorial males actively recruit satellites to their territories by directing displays toward satellites similar to the ones

performed toward females during courtship sequences (Box 16.1). Thus, under some conditions, an increase in female attraction or offspring survival due to the presence of other males seems to overcome the costs of lost fertilizations.

(A)

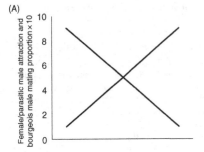

(B)

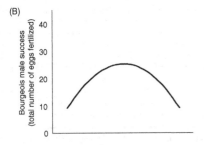

(C)

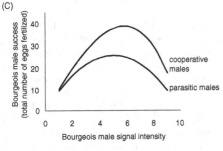

Figure 16.2 (A) Female attraction may increase with the intensity of the bourgeois male sexual signals. Parasitic male attraction may also increase with signal intensity, leading to a decrease in the proportion of fertilizations achieved by the bourgeois male. (B) Bourgeois male sexual signaling intensity should reflect the trade-off between female attraction and mating opportunities lost to parasitic males. (C) The presence of cooperative males may increase the bourgeois male fitness, for example, if the benefits of an increase in female attraction outweigh the costs of lost fertilizations. In these conditions, sexual signals produced by the bourgeois male are expected to increase in conspicuousness. (A) and (B) data from Table 16.1; (C) data from Table 16.2.

Females may also eavesdrop on the bourgeois male sexual signals. Females have been shown to copy the choice of other females in order to select males. Males observed by eavesdropping females being preferred by other females increase in attractiveness (e.g., Dugatkin and Godin 1992); males observed being rejected by females decrease in attractiveness (Witte and Ueding 2003). For example, females of the sailfin molly *Poecilia latipinna* prefer males that they have seen in the presence of other females (e.g., Witte and Ryan 2002) and males observed being rejected by females decrease in attractiveness (Witte and Ueding 2003). It can thus be predicted that bourgeois males should avoid having their signals eavesdropped upon by females when the probability of female rejection is high, but if the probability of female rejection is low, signals should increase in conspicuousness in the presence of female eavesdroppers (Figure 16.5). Bourgeois male sexual signals will therefore be partially shaped by the fitness consequences that eavesdropping imposes on the male, which in turn depends on the nature of the eavesdroppers.

RECEIVING BEHAVIOR IN RELATION TO EAVESDROPPING PRESSURE

Eavesdropping is not only expected to influence signal production by bourgeois males but also the way bourgeois males receive and interpret signals. Again, the nature of the eavesdroppers (i.e., females, parasitic males, or cooperative males) will impose different selection pressures on bourgeois males' receiving systems.

The receiving systems of bourgeois males should be selected to detect males using parasitic tactics (e.g., sneakers and female mimics) in order to minimize costs of parasitism. This is likely to be a difficult task as parasitic males are expected to evolve behavioral and morphological adaptations that make such detection difficult. Sneaker males use inconspicuous or darting behavior to avoid detection by bourgeois males. Female mimics imitate female morphology and behavior for the same reason. Thus, an evolutionary arms race between the bourgeois male's detection and discriminatory abilities and the parasitic male's signaling system and reproductive behavior is expected. In other words, the occurrence of eavesdropping by parasitic male in species with ARTs will likely be one of the selection pressures shaping the nature of sensory and perceptive systems of bourgeois males.

Bourgeois males' receiving systems should also be adjusted to detect cooperative males. In the context of ARTs, cooperative males usually pay some price to stay

Table 16.2. *A hypothetical example of variation in female and cooperative male attraction in relation to the intensity of a sexual signal produced by a bourgeois male. Cooperative males that intercept the male signal and move to the bourgeois male territory are assumed to further increase female attraction but also to reproduce parasitically within the territory*

Signal intensity	No. cooperative males attracted C	No. females attracted by the bourgeois male's signal (F') + by the presence of cooperative males (F'')[a] $F'+F''=F$	Total number of eggs laid[b] $E = F \times 10$	Total number of eggs fertilized by cooperative males[c] $L = (C \times E)/10$	Proportion of eggs fertilized by the bourgeois male $M = (E-L)/E$	Total number of eggs fertilized by the bourgeois male $S = E - L$
	1	1+0.1=1.1	11	1.1	0.9	9.9
	2	2+0.4=2.4	24	4.8	0.8	19.2
	3	3+0.9=3.9	39	11.7	0.7	27.3
	4	4+1.6=5.6	56	22.4	0.6	33.6
	5	5+2.5=7.5	75	37.5	0.5	37.5
	6	6+3.6=9.6	96	57.6	0.4	38.4
	7	7+4.9=11.9	119	83.3	0.3	35.7
	8	8+6.4=14.4	144	115.2	0.2	28.8
	9	9+8.1=17.1	171	153.9	0.1	17.1

Assuming females are attracted by cooperative males by $F'' = C^2/10$.
Assuming each female lays 10 eggs.
Assuming each cooperative male fertilizes 10% of the eggs.

in the male's territory and to have privileged access to reproductive resources (e.g., Martin and Taborsky 1997, Balshine-Earn *et al.* 1998, Bergmüller *et al.* 2005, Bergmüller and Taborsky 2005). This suggests that bourgeois males are able to individually recognize cooperative males and that cooperative males benefit from this recognition (but see Pfeiffer *et al.* 2005). However, although individual recognition has been demonstrated in several taxa (e.g., invertebrates: Karavanich and Atema 1998; fish: Höjesjö *et al.* 1998; reptiles: Olsson 1994; birds: Whitfield 1987; mammals: Sayigh *et al.* 1999), empirical evidence for direct reciprocity in the context of ARTs is lacking, and examples of reproductive concessions of bourgeois males to cooperative males are rare (Clutton-Brock *et al.* 2001). More likely, in most of these systems, bourgeois males are unable fully to control the reproduction of cooperative males that, once in the territory, may use inconspicuous approaches to access females. As a consequence, most cooperative males will reproduce parasitically. Identifying a cooperative male as such should be an easy task for a bourgeois male as cooperative males should signal their cooperative nature, but detecting parasitic events by these cooperative males is likely to be more difficult. In the ruff, territorial males

adopt a specific behavior that tries to prevent satellite males from accessing females, and satellite males may be expelled from the lek if seen mating (Box 16.1). In the cichlid *Neolamprologus pulcher*, cooperative males detected parasitizing fertilizations were expelled from the group by the breeding pair (Dierkes *et al.* 1999). Thus, the receiving systems of bourgeois males should be selected to detect parasitic events by both cooperative and parasitic males. Hypothetically, this detection may be easier in the case of cooperative males because cooperative males need to advertise their cooperative nature. In many species both cooperative and truly parasitic males occur, providing a good model to test this hypothesis. Also, a comparison of the properties of the bourgeois males' receiving systems in populations with different degrees of prevalence of parasitic males may reveal adaptations to eavesdropping pressure, but no comparative analyses on this issue in the context of ARTs have been conducted to date.

Finally, bourgeois males' receiving systems may also be tuned to eavesdrop on signals produced by other bourgeois males. For example, in the cricket frog *Acris crepitans*, males may either call to attract females or wait in the proximity of calling males and try to intercept approaching females

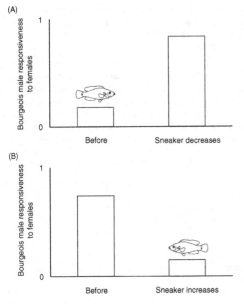

Figure 16.3 In the ocellated wrasse *Symphodus ocellatus*, bourgeois males' success decreases as the number of sneakers around the nest increases. (A) Males increase their courtship displays to females when the number of sneakers is experimentally decreased and (B) decrease their responsiveness to females when sneakers are experimentally increased. (Data from Alonzo and Warner 1999.)

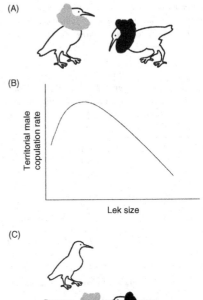

Figure 16.4 In the ruff *Philomachus pugnax*, territorial males (dark neck plumage) seem to actively recruit satellite males (light neck plumage) to their territories (A), displaying the same courtship displays towards satellites as towards females. The territorial male's reproductive success (B) is predicted to be maximum in medium-sized leks. When a female enters the territory (C), the male tries to control the satellite by placing its bill over the satellite's head and preventing access to the female. (Adapted from Hugie and Lank 1997, Widemo and Owens 1999.)

("satellite males"). Small, calling males, presented with playbacks of low-frequency calls typical of large males, may stop calling and switch into the satellite tactic within minutes (Wagner 1992) (Figure 16.6). The proportion of males switching into the satellite tactic correlates with the decrease in the frequency of the call, that is, with an increase in apparent male size. Thus, bourgeois males may also gain by increasing the probability of detection of signals from other bourgeois males in order to adjust their own signal production and even to switch between reproductive tactics.

16.3.2   The parasitic and cooperative male perspectives

SIGNALING TO BOURGEOIS MALES AND FEMALES
Cooperative males should signal their cooperative intentions to bourgeois males in order to access reproductive

resources. Accordingly, cooperative males usually look distinct both from females and from bourgeois males. For example, satellite males of the ruff have a light plumage distinct from the darker plumage of territorial males (Lank and Dale 2001) (Box 16.1). In another example, in the cooperatively breeding cichlid *Pelvicachromis pulcher*, cooperative males have a yellow coloration whereas bourgeois males have a red coloration (Martin and Taborsky 1997).

Cooperative behavior patterns like nest building or territory defense should be performed within visual range of the bourgeois male because cooperative males not seen helping

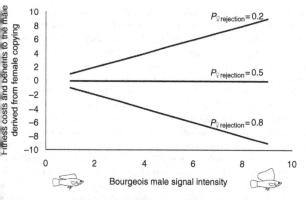

Figure 16.5 Males of the sailfin molly *Poecilia latipinna* observed by eavesdropping females mating with other females increase in attractiveness while rejected males decrease in attractiveness to the eavesdropping females. The intensity of the bourgeois male sexual signals is thus likely to depend on the probability of female rejection ($P_{♀rejection}$). Successful males that are usually not rejected by females (i.e., $P_{♀rejection} < 0.5$) are expected to produce conspicuous signals to benefit from female copying. Unsuccessful males (i.e., $P_{♀rejection} > 0.5$) should display less conspicuous signals to avoid having their signals intercepted by eavesdropping females.

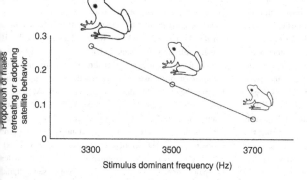

Figure 16.6 Calling males of the cricket frog *Acris crepitans* may switch to a satellite tactic when presented with low-frequency playbacks, typical of large males. More males switch into the satellite tactic as the frequency of the calls decreases. (Data from Wagner 1992.)

are often attacked by territorial owners (e.g., Balshine-Earn et al. 1998). In the cooperatively breeding cichlids *Neolamprologus brichardi* and *N. pulcher* from Lake Tanyika, helpers compete among themselves for access to positions closer to the brood chamber and helpers close to the nest help more (Werner et al. 2003). This investment should be matched by increased access to reproductive resources, either through reproductive concessions by the bourgeois male or forced access, with more efficient cooperators achieving proportionally higher gains. In these two cichlid species, individuals who help more spend more time inside the brood chamber (Werner et al. 2003), potentially having more fertilization opportunities. However, empirical evidence is lacking that helpers in cooperative species with ARTs gain reproductive opportunities in relation to their helping effort. Indeed in *P. pulcher* dominant helpers gain more parasitic fertilizations than subordinate helpers but their helping rate does not differ (Martin and Taborsky 1997) and in long-tailed tits *Aegithalos caudatus* helping does not relate to shared paternity (Hatchwell *et al.* 2002). This suggests that, at least in some species, cooperative males access the bourgeois male's reproductive resources not by concessions from the bourgeois male but by competing among themselves for access to the best positions for parasitic fertilizations and escaping the bourgeois male's control (Hogan-Warburg 1993, Martin and Taborsky 1997, Werner *et al.* 2003). Cooperating with bourgeois males may allow them to stay in the territory (the "pay-to-stay" hypothesis: e.g., Kokko *et al.* 2002, Bergmüller *et al.* 2005), and once in the territory competition among cooperative males for access to parasitic fertilizations occurs, independently of helping effort. Thus, cooperative males are expected, on the one hand, to develop signals to facilitate transmission of their cooperative intentions to bourgeois males and, on the other hand, to stop signaling and assume sneaking or darting behavior during parasitic events.

Selection should favor parasitic males with adaptations that increase their ability to access reproductive resources without being identified as parasites by bourgeois males and thus they are not expected to advertise their tactic. Parasitic males may rely on small size and speed to quickly access females or nests, or they may mimic females or other bourgeois males. For example, in the shell-brooding cichlid *Lamprologus callipterus*, bourgeois males gather and place shells at the nest entrance that are used by females to lay eggs. During spawning events "dwarf" males make use of their small size (approximately 2.5% the weight of bourgeois males: Sato 1994, Sato *et al.* 2004) to dart quickly inside the shell where a female is spawning and parasitically fertilize eggs from inside the shell (Sato *et al.* 2004). Bourgeois males are too large to enter the shell and evict dwarf males. Thus, dwarf males' success depends on an inconspicuous and fast approach, and both their morphology and behavior are adapted to avoid detection by the bourgeois male.

Males that mimic females in order to participate in mating events are particularly interesting to consider under the framework of communication network theory as they rely on deception to reproduce. Although qualitative observations in several species have suggested that female mimics are indistinguishable from females to the eyes of bourgeois males (e.g., bluegill sunfish *Lepomis macrochirus*: Gross and Charnov 1980, Dominey 1981), only in the peacock blenny *Salaria pavo* has this hypothesis been tested (Gonçalves *et al.* 2005). Bourgeois males of *S. pavo*

sequentially presented with females and female-mimicking males matched for size attacked and courted females and female mimics equally, suggesting that female mimics were able to deceive bourgeois males. However, not all female mimics were equally efficient and larger female mimics were attacked more and courted less by bourgeois males (Gonçalves *et al.* 2005) (Box 16.2). These results may suggest an evolutionary arms race between the female-mimicking signaling mechanisms and the bourgeois males' discrimination systems, with larger female mimics being more easily discriminated by bourgeois males. More generally, the occurrence of female mimics in a population is likely to complicate a bourgeois male's decision to accept or reject a courting conspecific into his nest or territory as it may be a parasitic male instead of a female. Female rejection by bourgeois males is thus likely to increase with the frequency of female mimics in the population.

Female mimicry is also interesting to consider under the scenario of nonindependent mechanisms of choice. Females may eavesdrop on a male–female interaction and copy the choice of other females, for example, to decrease mate-searching costs (e.g., Dugatkin and Godin 1992). When female mimics occur (assuming they also deceive females), females may be observing either a female or a female-like parasitic male courting a bourgeois male. If females still copy the choice of other females in such a system, female mimics may signal to males in the presence of females in order to manipulate female eavesdroppers and increase the probability that mating events will take place in that nest. Interestingly, in the peacock blenny female mimics will perform conspicuous female-like courtship behavior to a bourgeois male even if no spawning event is taking place (D. Gonçalves, personal observations). Whether this is to incite potential eavesdropping females to spawn remains to be tested. In conclusion, signal manipulation by female mimics will certainly be a selection pressure shaping the way other parties communicate in a network.

RECEIVING SIGNALS FROM BOURGEOIS MALES AND FEMALES

Parasitic and cooperative males need to locate potential reproductive opportunities. In general, this will be achieved in two steps. First, these males need to locate reproductive areas with high potential for parasitic reproduction and should try to gain a privileged position within those areas. Second, once in reproductive areas these males need to identify and participate in mating events. Locating

Box 16.2   Sexual dimorphism and courtship

Peacock blennies have pronounced sexual dimorphism with bourgeois males being much larger than females and having a set of well-developed secondary sexual characters, such as a conspicuous head crest (Fishelson 1963, Patzner *et al.* 1986) (Figure 16.7). In a sex-role-reversed population in southern Portugal, females court males using a complex courtship display involving beating the pectoral fins and opening and closing the mouth in synchrony while displaying a typical nuptial coloration (Almada *et al.* 1995). Small males mimic female morphology (Figure 16.7) and complex courtship behavior in order to approach the nest of bourgeois males and release sperm during spawning events (Gonçalves *et al.* 1996). These sneaker males compete for access to the best spawning locations, and successful bourgeois males have more and larger sneakers in the vicinity of their nests (Gonçalves *et al.* 2003a).

Sneakers seem to use both independent and non-independent (i.e., eavesdropping) mechanisms to choose successful males. When given a choice, sneakers prefer to associate with larger nesting males (Gonçalves *et al.* 2003b). Larger males are more frequently courted by females (T. Fagundes, D. Gonçalves, and R. F. Oliveira, unpublished data) and have higher reproductive success (Gonçalves *et al.* 2002); therefore, by associating with large males, sneakers are probably increasing their probability of participating in spawning events. The importance for *S. pavo* sneakers of eavesdropping on sexual interactions to choose successful males was evident in two experiments. Using a copying paradigm, sneakers were shown to prefer to associate with bourgeois males previously seen in the company of females (Gonçalves *et al.* 2003b), and in a second experiment sneakers increased their female-like courtship frequency when observing a female courting a male (R. J. Matos, D. Gonçalves, R. F. Oliveira, and P. K. McGregor, unpublished data). Thus, sneakers are probably using the female's presence and courtship displays as indicators of male quality, and this is likely to correlate with potential opportunities for future parasitic reproduction. It seems plausible that in other systems with ARTs, parasitic males increase their reproductive opportunities both by independent mechanisms of choice and by eavesdropping on the choice of females or even of other parasitic males.

In *S. pavo*, sneakers rely on female mimicry to reproduce. However, bourgeois males should respond by developing good discrimination mechanisms and females by developing a divergent morphology and behavior to signal to nesting males that they are females and not sneakers. The efficiency of female mimicry in *S. pavo* was tested at two levels: a visual model was constructed to estimate how similar sneaker color patterns appear to both females and males and behavioral tests were performed to assess the bourgeois males' behavior towards sneakers and females.

The visual model incorporated visual pigment absorbance and lens transmission data (from White *et al.* 2005), reflectance patterns from several body parts of the three morphs, and ambient light measurements (M. Cummings, D. Gonçalves, and R.F. Oliveira, unpublished data). The model estimated that, for bourgeois males, the color patterns of sneakers and females are much more similar than the color patterns of sneakers and bourgeois males, suggesting that sneakers mimic female colors efficiently. This idea was further tested in a laboratory experiment. Nesting males were sequentially presented with a sneaker and a female matched for size and their aggressive and courtship behavior recorded. Small female mimics were apparently able to deceive bourgeois males, as there was no difference in the amount of courtship and agonistic displays directed by bourgeois males towards small parasitic males or matched-for-size females (Gonçalves *et al.* 2005) (Figure 16.7). However, as body size increased, female mimicry efficiency apparently decreased and sneakers were attacked more and courted less by bourgeois males (Gonçalves *et al.* 2005) (Figure 16.7). An increase in body size may potentially facilitate discrimination by bourgeois males. If this is the case, a large courting female should be more easily correctly identified by the bourgeois male than a smaller one. Larger females were courted more and attacked less than smaller females, although there are alternative explanations for this observation (Gonçalves *et al.* 2005). These results are likely to reflect the conflicts in *S. pavo* derived from the existence of female mimicry. Interestingly, there are differences in the visual sensitivity of sneakers and bourgeois males of *S. pavo* (White *et al.* 2004), raising the possibility that these relate to the different visual tasks alternative reproducing males need to perform.

eproductive areas may not depend on eavesdropping. Parasitic males may, for example, choose to associate with males or reproductive sites that have been previously

preferred by females or to follow reproductively active females until they mate. In the peacock blenny, for instance, both females and sneakers prefer to associate with a large

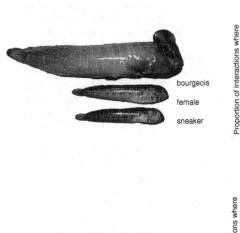

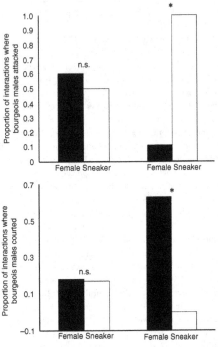

Figure 16.7 In the peacock blenny *Salaria pavo*, small sneaker males mimic the female's morphology and behavior to approach the nests of bourgeois males. In the field bourgeois males attack and court small sneakers in the same proportion as small females. Larger sneakers are more attacked and less courted than larger females. Results marked * are significant at the 0% level; n.s., not significant. (Data from Gonçalves *et al.* 2005.)

male and larger males are more successful in the field (Gonçalves *et al.* 2002, 2003b) (Box 16.2). In another example, male crickets *Acheta domesticus* show phonotaxis for male calls, with small males responding more strongly to playbacks of male calls that are also preferred by females. These small males avoid contact with the speaker, suggesting they eavesdrop on male signals to identify and approach calling males, probably to increase the probability of intercepting females, but avoid direct contact with the larger calling males (Kiflawi and Gray 2000) (Figure 16.8).

Nevertheless, eavesdropping on sexual signals is probably also used widely by parasitic males to locate breeding areas. For instance, sneaker males of *P. notatus* approach a speaker playing back a bourgeois male sexual call, suggesting this signal is used to locate nests (Brantley and Bass 1994, McKibben and Bass 1998, Bass and McKibben 2003). If eavesdropping males locate breeding areas based on bourgeois male signals, their receiving systems should be well tuned to these signals.

Although there are abundant examples of female sensory systems matching the properties of male calls (e.g., Sisneros and Bass 2003, Sisneros *et al.* 2004), evidence that this is also the case for parasitic and cooperative males is scarce. Nevertheless, sensory differences between male morphs have been identified (e.g., White *et al.* 2004), and this may relate to the different tasks these males need to accomplish. Again, cooperative males would be expected to have good sensory matching to bourgeois male signals as such mechanisms increase the success of the bourgeois male. Bourgeois males may signal directly to cooperative males, as happens in the ruff (van Rhijn 1973, Widemo 1998) (Box 16.1) or include components that facilitate their detection by the receiving system of cooperative males. There are no such advantages to parasitic males; therefore

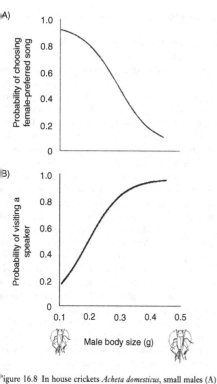

Figure 16.8 In house crickets *Acheta domesticus*, small males (A) show phonotaxis to playback calls of other males preferred by females and (B) avoid contact with the speaker. The results suggest that small males may eavesdrop on calling males' sexual signals to adopt a satellite tactic and intercept females. (Data from Kiflawi and Gray 2000.)

ss of a match is expected between the signals of bourgeois males and the receiving systems of parasitic males.

Once in reproductive areas, both cooperative and parasitic males will, on most occasions, try to escape control of the bourgeois male and reproduce parasitically. Eavesdropping on sexual interactions between males and females seems a crucial task for these males as fertilization must, on most occasions, occur during a limited period of time. There is evidence that eavesdropping males pay more attention to sexual signals as a function of parasitic opportunities. For example, in the Mediterranean wrasse the number of sneakers around a nest increases when the bourgeois male courts females more frequently and decreases when male responsiveness to females decreases (Alonzo and

Warner 1999). In another example, small noncalling males of the grasshopper *Bullacris membracioides* do not respond to playbacks of bourgeois male sexual calls but move towards a speaker playing back a female response call (Donelson and van Staaden 2005). This suggests that these parasitic males are trying to locate and intercept females based on their sexual response calls (Donelson and van Staaden 2005), and the neurophysiology of their auditory system well is adjusted to this task (van Staaden *et al.* 2003). Thus, eavesdropping on sexual signals seems to be ubiquitous in species with ARTs and crucial for the success of both cooperative and parasitic males. Undoubtedly eavesdropping will influence the design of parasitic and cooperative males' receiving systems.

16.3.3   The female perspective

SIGNALING TO BOURGEOIS, COOPERATIVE, AND
PARASITIC MALES
Potentially, females may gain, lose, or suffer no effect by mating with eavesdropping males. The direction of these effects will influence female signaling behavior. Although females are the choosier sex and their sexual signals are less elaborated than in males (except in sex-role-reversed species), females also need to signal their reproductive condition to males and to compete for access to high-quality males. Thus, female signals are also subject to eavesdropping. Whether females will promote or avoid eavesdropping depends on the fitness consequences for the female.

When females benefit from being fertilized by both bourgeois and satellite or parasitic males, they should actively seek multiple-male reproductive situations. For instance, in bluegill sunfish females allow parasitic males to participate in spawning (Gross 1991), in the ruff females seem to prefer to mate in courts co-occupied by satellite and territorial males (Lank and Smith 1992), and in the bluehead wrasse smaller females seek group spawning (Warner 1987, 1990). However, it is unclear in these examples if females actively promote aggregations of males by, for example, signaling both towards bourgeois and parasitic males. Sex-role-reversed species with male ARTs offer a good opportunity to test female preference for distinct male morphs as female courtship behavior is more conspicuous than in species with standard sex roles. In the poly-gynandrous dunnock *Prunella modularis*, females solicit copulations equally from dominant alpha males that attempt to guard females and from subordinate beta males (Davies *et al.* 1996). Females increase their solicitation rates towards males who had fewer opportunities to mate (Figure 16.9),

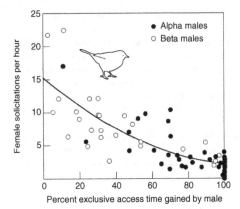

Figure 16.9 In polygynandrous dunnocks *Prunella modularis*, female solicitation rates towards alpha and beta males do not differ and decline with access time gained by the male. (After Davies *et al.* 1996.)

and males invest in parental care in proportion to mating success (Hartley and Davies 1994, Davies *et al.* 1996). Thus, females seem to be maximizing their own reproductive success by manipulating, through sexual signaling, the proportion of shared mating and thus of parental help with alpha and beta males.

In another example, females of the European bitterling fish *Rhodeus sericeus* increase the frequency of conspicuous behavior patterns in the presence of sneaker males prior to spawning, and the participation of sneakers in spawning increases the fertilization success of the eggs (Smith and Reichard 2005). Thus, when females have a net gain by having some of their eggs fertilized by parasitic males, they are expected either to signal directly to these males or to include conspicuous components in their signals to bourgeois males in order to increase the probability of signal interception by eavesdropping parasitic males.

Several alternative explanations have been proposed to explain why females are expected to mate with males of both morphs (e.g., production of offspring from both morphs at the evolutionarily stable strategy frequency if the reproductive strategy is heritable: Henson and Warner 1997, Hugie and Lank 1997, Alonzo and Warner 2000), but a discussion of the reproductive advantages of this and other female mating tactics is beyond the scope of this chapter.

In other species females have been shown to prefer to mate as a pair with bourgeois males. Several hypotheses

have been advanced to explain why females may prefer to mate with bourgeois males; these include gaining both direct benefits (e.g., better paternal care of the eggs) and indirect benefits (e.g., more fit offspring). This seems to be the case in the Mediterranean wrasse *S. ocellatus*, where females apparently choose sneaker-free opportunities to spawn (van den Berghe *et al.* 1989, Alonzo and Warner 2000). In this species, when sneakers are experimentally removed, females increase their spawning rate fourfold, and the nest success may increase threefold (van den Berghe *et al.* 1989, Alonzo and Warner 2000). In these systems, females should avoid parasitic males, and female signals directed to bourgeois males are expected to be conspiratorial. When female mimics occur and impose costs on females, an evolutionary arms race is expected: females should signal to bourgeois males that they are females and parasitic males should counteract with improved female mimicry. This may lead to the evolution of more complex female displays or behavior that are increasingly difficult to mimic. As an example, in the sex-role-reversed population of the peacock blenny described above, females produce a complex courtship display that sneakers imitate. Small sneakers are better at deceiving bourgeois males, presumably because an increase in target area facilitates discrimination (Gonçalves *et al.* 2005) (Box 16.2). These results are likely to reflect an evolutionary arms race where females try to advertise their sex to bourgeois males, female mimics try to deceive bourgeois males, and bourgeois males try to discriminate females from female mimics.

In species where females would prefer to mate with parasitic rather than bourgeois males, female signals should be conspiratorial and directed to parasitic males. In some species females may gain genetic benefits from mating with parasitic males. For example, both in bluegill sunfish and in the Atlantic salmon *Salmo salar* fry from eggs fertilized by sneakers grow faster when compared with fry from eggs fertilized by bourgeois males (Garant *et al.* 2002, Neff 2004). In coho salmon *Oncorhynchus kisutch*, there is some evidence that females prefer to mate with parasitic males and only mate with larger bourgeois males to avoid the costs of coercion (Watters 2005). Thus, the traditional view that females, when given a choice, prefer to mate with bourgeois rather than parasitic males may prove incorrect for some species. In the context of communication networks, this means another level needs to be considered, with possible cooperation during sexual interactions between females and parasitic males and conflict between bourgeois males and both females and parasitic males.

RECEIVING SIGNALS FROM BOURGEOIS, COOPERATIVE, AND PARASITIC MALES

Since selection favors females that maximize their long-term reproductive success in mate-choice decisions, females should be tuned to male signals and selected to evaluate male quality from the properties of the signal. For instance, female gray treefrogs *Hyla versicolor* prefer male calls of longer duration (Klump and Gerhardt 1987), and the progeny of "long-callers" are more fit, thus providing females with indirect fitness benefits (Welch *et al.* 1998). In species with ARTs, cooperative and parasitic males eavesdrop on male–female interactions; therefore, a female's decision to respond to appropriate bourgeois male signals should incorporate the reproductive consequences for her of mating with these eavesdropping males. Females' receiving systems should thus be tuned not only for the detection and evaluation of bourgeois males' signals but also for the detection of potential eavesdropping males. Again, identification of cooperative males should be easier than detection of parasitic males as the former usually have distinctive traits to signal their cooperative nature. Discrimination of parasitic males by females' receiving systems should also be facilitated when females gain from their presence. In this scenario, females and parasitic males may signal to each other in order to increase the probability of mating, and bourgeois males may eavesdrop on this interaction, reducing the conspicuousness of the signal. This hypothesis raises again the interesting possibility that the receiving systems of reproductive morphs within the same population may differ according to their position in the communication network. In this example, females may be better tuned to parasitic male signals than bourgeois males.

In species where females pay a fitness cost by mating with parasitic males, their receiving system should also be selected to identify these males in order to avoid parasitic fertilizations. In this scenario, however, parasitic males are expected to counteract with strategies that decrease their detection by females.

In both situations, females are expected to gain an advantage by discriminating parasitic or cooperative males from bourgeois males. In the first case, discrimination would allow females to select multiple male mating scenarios or to choose parasitic males, and in the second case it would allow females to avoid these males. Female discrimination of alternative morphs seems evident in many species. Female swordtails *Xiphophorus nigrensis*, for example, clearly avoid small males, preferring to mate with larger courting males (for a review see Ryan and Rosenthal 2001). Parasitic males

may counteract, reproducing by forced copulations (e.g., swordtails), fast access to the female (e.g., ruff), or female mimicry (e.g., peacock blenny). It is unclear if female mimics are also able to deceive females.

16.4   INTEGRATIVE APPROACH AND FUTURE DIRECTIONS

Game-theoretical models have shown that eavesdropping will influence the way animals communicate (e.g., Johnstone 2000, 2001). Species with ARTs are no exception and, as explained in the previous section, the properties of signaling and receiving systems of bourgeois males, females, parasitic and cooperative males will be influenced by the complex communication network in which animals live. Functional approaches to the study of animal communication in these species will thus need to consider the costs and benefits imposed by eavesdropping on each member of a communication network.

Alternative morphs play different roles in their communication network, and thus different evolutionary pressures act on their signaling and receiving systems. The often dramatic difference in traits between males reproducing using alternative tactics is an obvious consequence of these pressures. The hypothesis that alternative morphs also show differences in their receiving and signaling systems in relation to their particular mode of reproduction has been less explored. For instance, while bourgeois males need to detect females, parasitic males may reproduce by intercepting bourgeois males' signals. Differences in the sensory and receiving apparatus of alternative morphotypes relating to the distinct roles they play in the communication network are likely to be widespread.

Surprisingly, although research has revealed a plasticity in signal production in relation to eavesdropping pressure, as shown by some of the above examples, a demonstration that eavesdropping influences the evolution of signals in species with ARTs is still lacking. For this, inter- and intraspecific comparative approaches may prove particularly successful, as has been the case in other communication systems. For example, John Endler's work with guppies *Poecilia reticulata* has demonstrated that the male courtship coloration correlates negatively with predation pressure (Endler 1977, 1978, 1980). Furthermore, in populations with stronger predation pressure, male guppies show a lower frequency of sigmoid displays towards females and a higher frequency of forced copulation attempts, presumably because sigmoid displays are more conspicuous to predators relying on interceptive eavesdropping (Luyten and Liley 1985, Endler 1987, Magurran and Seghers 1990). Similar comparative approaches could be carried out in species with ARTs where the degree of eavesdropping pressure varies between populations. For example, in the peacock blenny several populations with ARTs have been described. In two populations, a scarcity of nest sites leading to a strong male–male competition for nests is likely to explain the presence of sneakers (Ruchon et al. 1995, Gonçalves et al. 1996). In other populations, nest-site availability is higher (e.g., in the Adriatic) and the frequency of sneakers in the population is lower (J. Saraiva and R. F. Oliveira, unpublished data). Qualitative observations suggest that in this population male courtship signals are more conspicuous, although it is unclear if this is a consequence of a lower eavesdropping pressure (J. Saraiva and R. F. Oliveira, unpublished data).

Interspecific comparative analyses on the signaling properties of species with ARTs under different eavesdropping pressure may also prove rewarding. By including species with and without ARTs while controlling for phylogeny, one could test whether the properties of signals would change in a predictable way with eavesdropping pressure. For example, when eavesdroppers impose a cost on bourgeois males, one of the predictions would be that bourgeois male sexual signals should include shorter-range components when eavesdropping pressures increase. When female mimics occur, the rate of female rejection by bourgeois males may increase with the frequency of female mimics in the population.

Finally, understanding the output of the complex interaction between and within the sexes in species with ARTs is only likely to be possible with a combination of experimental and observational work aided by mathematical models. A full understanding of animal communication in these systems will necessarily include measuring the fitness consequences of signal production and reception for bourgeois males, females, parasitic and cooperative males and combining these results in holistic models.

Acknowledgments

The research on the peacock blenny conducted in our laboratory and described in this chapter was supported by several grants from Fundação para a Ciência e a Tecnologia (FCT). The writing of this chapter was funded by the plurianual program of FCT (UIandD 331/2001) and by a postdoctoral grant to D.G. (SFRH/BPD/7188/2001). We are also grateful to one anonymous reviewer and Jane Brockmann for comments on a previous version of the chapter.

References

Almada, V. C., Gonçalves, E. J., Oliveira, R. F., and Santos, E. J. 1995. Courting females: ecological constraints affect sex roles in a natural population of the blenniid fish *Salaria pavo. Animal Behaviour* 49, 1125–1127.

Alonzo, S. H. and Warner, R. R. 1999. A trade-off generated by sexual conflict: Mediterranean wrasse males refuse present mates to increase future success. *Behavioral Ecology* 10, 105–111.

Alonzo, S. H. and Warner, R. R. 2000. Dynamic games and field experiments examining intra- and intersexual conflict: explaining counterintuitive mating behavior in a Mediterranean wrasse, *Symphodus ocellatus. Behavioral Ecology* 11, 56–70.

Balshine-Earn, S., Neat, F., Reid, H., and Taborsky, M. 1998. Paying to stay or paying to breed? Field evidence for direct benefits of helping in a cooperatively breeding fish. *Behavioral Ecology* 9, 432–438.

Bass, A. H. and McKibben, J. R. 2003. Neural mechanisms and behaviors for acoustic communication in teleost fish. *Progress in Neurobiology* 69, 1–26.

Bergmüller, R. and Taborsky, M. 2005. Experimental manipulation of helping in a cooperative breeder: helpers "pay to stay" by pre-emptive appeasement. *Animal Behaviour* 69, 19–28.

Bergmüller, R., Heg, D., and Taborsky, M. 2005. Helpers in a cooperatively breeding cichlid stay and pay or disperse and breed, depending on ecological constraints. *Proceedings of the Royal Society of London B* 272, 325–331.

Birkhead, T. R. and Möller, A. P. 1992. *Sperm Competition in Birds: Evolutionary Causes and Consequences.* London: Academic Press.

Brantley, R. K. and Bass, A. H. 1994. Alternative male spawning tactics and acoustic signals in the plainfin midshipman fish *Porichthys notatus* Girard (Teleostei, Batrachoididae). *Ethology* 96, 213–232.

Brockmann, H. J. 2001. The evolution of alternative strategies and tactics. *Advances in the Study of Behavior* 30, 1–51.

Brouwer, L., Heg, D., and Taborsky, M. 2005. Experimental evidence for helper effects in a cooperatively breeding cichlid. *Behavioral Ecology* 16, 667–673.

Clutton-Brock, T. H., Brotherton, P. N. M., Russell, A. F., et al. 2001. Cooperation, control, and concession in meerkat groups. *Science* 291, 478–481.

Cox, C. R. and Le Boeuf, J. B. 1977. Female incitation of male–male competition: a mechanism in sexual selection. *American Naturalist* 111, 317–335.

Davies, N. B., Hartley, I. R., Hatchwell, B. J., and Langmore, N. E. 1996. Female control of copulations to maximize male help: a comparison of polygynandrous alpine accentors, *Prunella collaris*, and dunnocks, *P. modularis*. *Animal Behaviour* 51, 27–47.

Dawkins, R. and Krebs, J. R. 1978. Animal signals: information or manipulation? In J. R. Krebs and N. B. Davies (eds.) *Behavioural Ecology*, pp. 282–304. Oxford, UK: Blackwell Scientific.

Dierkes, P., Taborsky, M., and Kohler, U. 1999. Reproductive parasitism of broodcare helpers in a cooperatively breeding fish. *Behavioral Ecology* 10, 510–515.

Dominey, W. J. 1981. Maintenance of female mimicry as a reproductive strategy in bluegill sunfish (*Lepomis macrochirus*). *Environmental Biology of Fishes* 6, 59–64.

Donelson, N. C. and van Staaden, M. J. 2005. Alternate tactics in male bladder grasshoppers *Bullacris membracioides* (Orthoptera: Pneumoridae). *Behaviour* 142, 761–778.

Doutrelant, C., McGregor, P. K., and Oliveira, R. F. 2001. The effect of an audience on intrasexual communication in male Siamese fighting fish, *Betta splendens*. *Behavioral Ecology* 12, 283–286.

Dugatkin, L. A. and Godin, J. G. 1992. Reversal of female mate choice by copying in the guppy (*Poecilia reticulata*). *Proceedings of the Royal Society of London B* 249, 179–184.

Dzieweczynski, T. L., Earley, R. L., Green, T. M., and Rowland, W. J. 2005. Audience effect is context dependent in Siamese fighting fish, *Betta splendens*. *Behavioral Ecology* 16, 1025–1030.

Emlen, S. T., Reeve, H. K., and Keller, L. 1998. Reproductive skew: disentangling concessions from control. *Trends in Ecology and Evolution* 13, 458–459.

Endler, J. A. 1977. Natural selection on color patterns in *Poecilia reticulata*. *Evolution* 34, 76–94.

Endler, J. A. 1978. A predator's eye view of animal colors. *Evolutionary Biology* 11, 319–364.

Endler, J. A. 1980. Natural selection on color patterns in *Poecilia reticulata*. *Evolution* 34, 76–91.

Endler, J. A. 1987. Predation, light intensity and courtship behaviour in *Poecilia reticulata* (Pisces: Poeciliidae). *Animal Behaviour* 35, 1376–1385.

Evans, C. S. and Marler, P. 1984. Food calling and audience effects in male chickens, *Gallus gallus*: their relationships to food availability, courtship and social facilitation. *Animal Behaviour* 47, 1159–1170.

Fishelson, L. 1963. Observation on littoral fishes of Israel. 1. Behaviour of *Blennius pavo* Risso (Teleostei: Blenniidae). *Israel Journal of Zoology* 12, 67–80.

Fuller, R. and Berglund, A. 1996. Behavioral responses of a sex-role reversed pipefish to a gradient of perceived predation risk. *Behavioral Ecology* 7, 69–75.

Garant, D., Fontaine, P.-M., Good, S. P., Dodson, J. J., and Bernatchez, L. 2002. Influence of male parental identity on growth and survival of offspring in Atlantic salmon (*Salmo salar*). *Evolutionary Ecology Research* 4, 537–549.

Gonçalves, D., Simões, P. C., Chumbinho, A. C., *et al.* 2002. Fluctuating asymmetries and reproductive success in the peacock blenny. *Journal of Fish Biology* 60, 810–820.

Gonçalves, D., Fagundes, T., and Oliveira, R. F. 2003a. Field reproductive behaviour of sneaker males of the peacock blenny. *Journal of Fish Biology* 63, 528–532.

Gonçalves, D., Oliveira, R. F., Körner, K., and Schlupp, I. 2003b. Intersexual copying by sneaker males of the peacock blenny. *Animal Behaviour* 65, 355–361.

Gonçalves, D., Matos, M., Fagundes, T., and Oliveira, R. F. 2005. Bourgeois males of the peacock blenny, *Salaria pavo*, discriminate females from female-mimicking males. *Ethology* 111, 559–572.

Gonçalves, E. J., Almada, V. C., Oliveira, R. F., and Santos, A. J. 1996. Female mimicry as a mating tactic in males of the blenniid fish *Salaria pavo*. *Journal of the Marine Biological Association of the United Kingdom* 76, 529–538.

Gross, M. R. 1991. Evolution of alternative reproductive strategies: frequency-dependent selection in male bluegill sunfish. *Philosophical Transactions of the Royal Society of London B* 332, 59–66.

Gross, M. R. and Charnov, E. L. 1980. Alternative male life histories in bluegill sunfish. *Proceedings of the National Academy of Sciences of the United States of America* 77, 6937–6940.

Gyger, M., Karakashian, S. J., and Marler, P. 1986. Avian alarm calling: is there an audience effect? *Animal Behaviour* 34, 1570–1572.

Hammond, T. J. and Bailey, W. J. 2003. Eavesdropping and defensive auditory masking in an Australian bushcricket, *Caedicia* (Phaneropterinae: Tettigoniidae: Orthoptera). *Behaviour* 140, 79–95.

Hartley, I. R. and Davies, N. B. 1994. Limits to cooperative polyandry in birds. *Proceedings of the Royal Society of London B* 257, 67–73.

Hatchwell, B. J., Ross, D. J., Chaline, N., Fowlie, M. K., and Burke, T. 2002. Parentage in cooperatively breeding long-tailed tits. *Animal Behaviour* 64, 55–63.

Hector, A. C. K., Seyfarth, R. M., and Raleigh, M. 1989. Male parental care, female choice and the effect of an audience in vervet monkeys. *Animal Behaviour* 38, 262–271.

Henson, S. A. and Warner, R. R. 1997. Male and female alternative reproductive behaviors in fishes: a new approach using intersexual dynamics. *Annual Review of Ecology and Systematics* 28, 571–592.

Hogan-Warburg, A. L. 1966. Social behaviour of the ruff, *Philomachus pugnax* (L.). *Ardea* 54, 109–229.

Hogan-Warburg, A. L. 1993. Female choice and the evolution of mating strategies in the ruff *Philomachus pugnax* (L.). *Ardea* 80, 395–403.

Höglund, J. and Alatalo, R. 1995. *Leks*. Princeton, NJ: Princeton University Press.

Höglund, J., Montgomerie, R., and Widemo, F. 1993. Costs and consequences of variation in the size of ruff leks. *Behavioral Ecology and Sociobiology* 32, 31–39.

Höjesjö, J., Johnsson, J. I., Petersson, E., and Järvi, T. 1998. The importance of being familiar: individual recognition and social behavior in sea trout (*Salmo trutta*). *Behavioral Ecology* 9, 445–451.

Hugie, D. M. and Lank, D. B. 1997. The resident's dilemma: a female-choice model for the evolution of alternative male reproductive strategies in lekking male Ruffs (*Philomachus pugnax*). *Behavioral Ecology* 8, 218–225.

Hunt, J. and Simmons, L. W. 2002. Confidence of paternity and paternal care: covariation revealed through the experimental manipulation of the mating system in the beetle *Onthophagus taurus*. *Journal of Evolutionary Biology* 15, 784–795.

Johnstone, R. A. 2000. Conflicts of interest in signal evolution. In Y. Espmark, T. Amundsen, and G. Rosenqvist (eds.) *Animal Signals: Signalling and Signal Design in Animal Communication*, pp. 465–485. Trondheim, Norway: Tapir Academic Press.

Johnstone, R. A. 2001. Eavesdropping and animal conflict. *Proceedings of the National Academy of Sciences of the United States of America* 98, 9177–9180.

Johnstone, R. A. and Cant, M. A. 1999. Reproductive skew and the threat of eviction: a new perspective. *Proceedings of the Royal Society of London B* 266, 275–279.

Johnstone, R. A. and Grafen, A. 1992. The continuous Sir Philip Sidney game: a simple model of biological signaling. *Journal of Theoretical Biology* 156, 215–234.

Karavanich, C. and Atema, J. 1998. Individual recognition and memory in lobster dominance. *Animal Behaviour* 56, 1553–1560.

Kiflawi, M. and Gray, D. A. 2000. Size-dependent response to conspecific mating calls by male crickets. *Proceedings of the Royal Society of London B* 267, 2157–2161.

Klump, G. M. and Gerhardt, H. C. 1987. Use of non-arbitrary acoustic criteria in mate choice by female gray tree frogs. *Nature* 326, 286–288.

Kokko, H. 2003. Are reproductive skew models evolutionarily stable? *Proceedings of the Royal Society of London B* 270, 265–270.

Kokko, H., Johnstone, R. A., and Wright, J. 2002. The evolution of parental and alloparental care in cooperatively breeding groups: when should helpers pay to stay? *Behavioral Ecology* 13, 291–300.

Krebs, J. R. and Dawkins, R. 1984. Animal signals: mind-reading and manipulation. In J. R. Krebs and N. B. Davies (eds.) *Behavioural Ecology: An Evolutionary Approach*, 2nd edn, pp. 380–402. Oxford, UK: Blackwell Scientific.

Lank, D. B. and Dale, J. 2001. Visual signals for individual identification: the silent "song" of ruffs. *Auk* 118, 759–765.

Lank, D. B. and Smith, C. M. 1992. Females prefer larger leks: an experimental study with ruffs *Philomachus pugnax*. *Behavioral Ecology and Sociobiology* 30, 323–329.

Lank, D. B., Smith, C. M., Hanotte, O., Burke, T. A., and Cooke, F. 1995. Genetic polymorphism for alternative mating strategies in lekking male ruff, *Philomachus pugnax*. *Nature* 378, 59–62.

Le Comber, C., Faulkes, C. G., Formosinho, J., and Smith, C. 2003. Response of territorial males to the threat of sneaking in the three-spined stickleback, *Gasterosteus aculeatus*: a field study. *Journal of Zoology (London)* 261, 15–20.

Luyten, P. H. and Liley, N. R. 1985. Geographic variation in the sexual behaviour of the guppy, *Poecilia reticulata* (Peters). *Behaviour* 95, 164–179.

Magurran, A. E. and Seghers, B. H. 1990. Risk sensitive courtship in the guppy (*Poecilia reticulata*). *Behaviour* 112, 194–201.

Marler, P. and Evans, C. 1996. Bird calls: just emotional displays or something more? *Ibis* 138, 26–33.

Martin, E. and Taborsky, M. 1997. Alternative male mating tactics in a cichlid, *Pelvicachromis pulcher*: a comparison of reproductive effort and success. *Behavioral Ecology and Sociobiology* 41, 311–319.

Matos, R. J. and McGregor, P. K 2002. The effect of the sex of an audience on male–male displays of siamese fighting fish (*Betta splendens*). *Behaviour* **139**, 1211–1221.

Matos, R. J. and Schlupp, I. 2005. Performing in front of an audience: signalers and the social environment. In P. K. McGregor (ed.) *Animal Communication Networks*, pp. 63–83. Cambridge, UK: Cambridge University Press.

Matos, R. J., Peake, T. M., and McGregor, P. K. 2003. Timing of presentation of an audience: aggressive priming and audience effects in male displays of Siamese fighting fish (*Betta splendens*). *Behavioural Processes* **28**, 53–61.

Maynard Smith, J. 1991. Honest signalling: the Philip Sydney game. *Animal Behaviour* **42**, 1034–1035.

McGregor, P. K. 1993. Signalling in territorial systems: a context for individual identification, ranging and eavesdropping. *Philosophical Transactions of the Royal Society of London B* **340**, 237–244.

McGregor, P. K. and Dabelsteen, T. 1996. Communication networks. In D. E. Kroodsma and E. H. Miller (eds.) *Ecology and Evolution of Acoustic Communication in Birds*, pp. 409–425. Ithaca, NY: Cornell University Press.

McGregor, P. K. and Peake, T. M. 2000. Communication networks: social environments for receiving and signaling behaviour. *Acta Ethologica* **2**, 71–81.

McKibben, J. R. and Bass, A. H. 1998. Behavioral assessment of acoustic parameters relevant to signal recognition and preference in a vocal fish. *Journal of the Acoustic Society of America* **104**, 3520–3533.

Mougeot, F. and Bretagnolle, V. V. 2000. Predation as a cost of sexual communication in nocturnal seabirds: an experimental approach using acoustic signals. *Animal Behaviour* **60**, 647–656.

Neff, B. D. 2004. Increased performance of offspring sired by parasitic males in bluegill sunfish. *Behavioral Ecology* **15**, 327–331.

Oliveira, R. F., McGregor, P. K., and Latruffe, C. 1998. Know thine enemy: fighting fish gather information from observing conspecific interactions. *Proceedings of the Royal Society of London B* **265**, 1045–1049.

Olsson, M. 1994. Rival recognition affects male contest behavior in sand lizards (*Lacerta agilis*). *Behavioral Ecology and Sociobiology* **35**, 249–252.

Patzner, R. A., Seiwald, M., Adlgasser, M., and Kaurin, G. 1986. The reproduction of *Blennius pavo*. 5. Reproductive behaviour in the natural environment. *Zoologisches Anzeiger* **216**, 338–350.

Peake, T. M. 2005. Eavesdropping in communication networks. In P. K. McGregor (ed.) *Animal Communication*

Networks, pp. 13–37. Cambridge, UK: Cambridge University Press.

Pfeiffer, T., Rutte, C., Killingback, T., Taborsky, M., and Bonhoffer, S. 2005. Evolution of cooperation by generalized reciprocity. *Proceedings of the Royal Society of London B* **272**, 1115–1120.

Reeve, H. K., Emlen, S. T., and Keller, L. 1998. Reproductive sharing in animal societies: reproductive incentives or incomplete reproductive control by dominant breeders? *Behavioral Ecology* **9**, 267–276.

Ruchon, F., Laugier, T., and Quignard, J. P. 1995. Alternative male reproductive strategies in the peacock blenny. *Journal of Fish Biology* **47**, 826–840.

Ryan, M. J. and Rosenthal, G. G. 2001. Variation and selection in swordtails. In L. A. Dugatkin (ed.) *Model Systems in Behavioral Ecology*, pp. 133–148. Princeton, NJ: Princeton University Press.

Sato, T. 1994. Active accumulation of spawning substrate: a determinant of extreme polygyny in a shell-brooding cichlid fish. *Animal Behaviour* **48**, 669–678.

Sato, T., Hirose, M., Taborsky, M., and Kimura, S. 2004. Size-dependent male alternative reproductive tactics in the shell-brooding cichlid fish *Lamprologus callipterus* in Lake Tanganyika. *Ethology* **110**, 49–62.

Sayigh, L. S., Tyack, P. L., Wells, R. S., *et al.* 1999. Individual recognition in wild bottlenose dolphins: a field test using playback experiments. *Animal Behaviour* **57**, 41–50.

Shuster, S. M. and Wade, M. J. 2003. *Mating Systems and Strategies*. Princeton, NJ: Princeton University Press.

Sisneros, J. A. and Bass, A. H. 2003. Seasonal plasticity of peripheral auditory frequency sensitivity. *Journal of Neuroscience* **23**, 1049–1058.

Sisneros, J. A., Forlano, P. M., Deitcher, F. L., and Bass, A. H. 2004. Steroid-dependent auditory plasticity leads to adaptive coupling of sender and receiver. *Science* **305**, 404–407.

Smith, C. and Reichard, M. 2005. Females solicit sneakers to improve fertilization success in the bitterling fish (*Rhodeus sericeus*). *Proceedings of the Royal Society of London B* **272**, 1683–1688.

Taborsky, M. 1994. Sneakers, satellites and helpers: parasitic and cooperative behavior in fish reproduction. *Advances in the Study of Behavior* **23**, 1–100.

Taborsky, M. 1997. Bourgeois and parasitic tactics: do we need collective, functional terms for alternative reproductive behaviours? *Behavioral Ecology and Sociobiology* **41**, 361–362.

Taborsky, M. 1998. Sperm competition in fish: "Bourgeois" males and parasitic spawning. *Trends in Ecology and Evolution* 13, 222–227.

Taborsky, M. 1999. Conflict or cooperation: what determines optimal solutions to competition in fish reproduction? In V. C. Almada, R. F. Oliveira, and E. J. Gonçalves (eds.) *Behaviour and Conservation of Littoral Fishes*, pp. 301–349. Lisbon: Instituto Superior de Psicologia Aplicada.

Taborsky, M., Hudde, B., and Wirtz, P. 1987. Reproductive behaviour and ecology of *Symphodus* (*Crenilabrus*) *ocellatus*, a European wrasse with four types of male behaviour. *Behaviour* 102, 82–118.

Valera, F., Hoi, H., and Krištn, A. 2003. Male shrikes punish unfaithful females. *Behavioral Ecology* 14, 403–408.

van den Berghe, E. P., Wernerus, F., and Warner, R. R. 1989. Female choice and the mating cost of peripheral males. *Animal Behaviour* 38, 875–884.

van Rhijn, J. G. 1973. Behavioral dimorphism in male ruffs *Philomachus pugnax* (L.). *Behaviour* 47, 153–229.

van Rhijn, J. G. 1991. *The Ruff.* London: Poyser.

van Staaden, M. J., Rieser, M., Ott, S. R., Pabst, M. A., and Romer, H. 2003. Serial hearing organs in the atympanate grasshopper *Bullacris membracioides* (Orthoptera, Pneumoridae). *Journal of Comparative Neurology* 465, 579–592.

Wagner Jr., W. E. 1992. Deceptive or honest signalling of fighting ability? A test of alternative hypotheses for the function of changes in call dominant frequency by male cricket frogs. *Animal Behaviour* 44, 449–462.

Warner, R. R. 1987. Female choice of site versus mates in a coral reef fish, *Thalassoma bifasciatum. Animal Behaviour* 35, 1470–1478.

Warner, R. R. 1990. Male versus female influences on mating site determination in a coral reef fish. *Animal Behaviour* 39, 540–548.

Watters, J. V. 2005. Can the alternative male tactics "fighter" and "sneaker" be considered "coercer" and "co-operator" in coho salmon? *Animal Behaviour* 70, 1055–1062.

Welch, A. M., Semlitsch, R. D., and Gerhardt, H. C. 1998. Call duration as an indicator of genetic quality in male gray tree frogs. *Science* 280, 1928–1930.

Werner, N. Y., Balshine, S., Leach, B., and Lotem, A. 2003. Helping opportunities and space segregation in cooperatively breeding cichlids. *Behavioral Ecology* 14, 749–756.

Widemo, F. 1998. Alternative reproductive strategies in the ruff, *Philomachus pugnax*: a mixed ESS? *Animal Behaviour* 56, 329–336.

Widemo, F. and Owens, I. P. 1995. Lek size, male mating skew and the evolution of lekking. *Nature* 373, 148–151.

Widemo, F. and Owens, I. P. 1999. Size and stability of vertebrate leks. *Animal Behaviour* 58, 1217–1221.

Witte, K. and Ryan, M. J. 2002. Mate choice copying in the sailfin molly, *Poecilia latipinna*, in the wild. *Animal Behaviour* 63, 943–949.

Witte, K. and Ueding, K. 2003. Sailfin molly females (*Poecilia latipinna*) copy the rejection of a male. *Behavioral Ecology* 14, 389–395.

White, E. M., Gonçalves, D. M., Partridge, J. C., and Oliveira, R. F. 2004. Vision and visual variation in the peacock blenny. *Journal of Fish Biology* 65, 227–250.

Whitfield, D. P. 1987. Plumage variability, status signaling and individual recognition in avian flocks. *Trends in Ecology and Evolution* 2, 13–18.

CHAPTER SUMMARY

Many mating systems are characterized by male alternative life histories that utilize different mating tactics to reproduce. Bourgeois males attempt to monopolize mating access to females, and in fish, many of these males provide sole parental care to the developing young. Parasitic males use behavior patterns such as sneaking to steal fertilizations from bourgeois males. Modeling has shown that when bourgeois males provide higher genetic benefits – i.e., alleles leading to increased condition and higher fitness of their offspring – than parasitic males, females maximize both indirect and direct (parental care) benefits by mating exclusively with bourgeois males. However, when parasitic males have higher genetic benefits than bourgeois males, females must trade off genetic quality of their offspring with reduced parental care. Here I develop a model to examine such trade-offs and show that as the relative genetic benefits of parasitic versus bourgeois males increase or as the fitness benefit of parental care decreases, females maximize their fitness by having a greater proportion of their offspring sired by parasitic males. The optimal breeding situation, which maximizes individual fitness, differs for females, parasitic males, and bourgeois males and this should lead to sexual conflict. I test the model with data from bluegill sunfish (*Lepomis macrochirus*), where parasitic males may provide greater genetic benefits to females than bourgeois males. I show that high-quality females, as measured by three phenotypic measures, spawn in nests that have higher bourgeois male paternity and their offspring subsequently receive greater parental care. Assuming high-quality females are in better control of mating than low-quality females, these data suggest that the latter are in greater conflict with parasitic males.

17.1 INTRODUCTION

Given the enormous diversity of reproductive behavior observed in nature, there remains much to be learned about the complex social interactions and decisions made by individuals during mate choice and parental care (reviewed by Andersson 1994, Godin 1997, Henson and Warner 1997, Birkhead and Møller 1998). Many mating systems are complicated by having discrete life histories within the sexes that utilize alternative mating tactics (reviewed by Gross 1984, 1996, Taborsky 1998). These discrete life histories usually take the form of a precociously maturing male that adopts a parasitic mating tactic and a late-maturing male that adopts a monopolizing or "bourgeois" tactic and often provides sole parental care for the young (Taborsky 1994, 1997).

It has been proposed that alternative life histories within the sexes commonly evolve as a conditional strategy, whereby (usually) a male may develop into either tactic based on its condition or state (Dominey 1984, Gross 1996). Thus, life history is a plastic trait whereby a developing male can express either the parasitic or bourgeois phenotype. The conditional strategy predicts that the highest-quality males within a population – those with the highest condition or state – will adopt the tactic with the greater fitness benefit (Gross 1996). Because condition can have a genetic component through additive genetic variance (Rowe and Houle 1996, Blanckenhorn and Hosken 2003), there can be a predisposition (i.e., inheritance) of the life histories. This in turn should lead to one life history having higher fitness. Nevertheless, modeling shows that the alternative life histories can be evolutionarily stable (Repka and Gross 1995, Gross and Repka 1998).

When there is a difference in the genetic benefits provided by the alternative male life histories, females may

Alternative Reproductive Tactics, ed. Rui F. Oliveira, Michael Taborsky, and H. Jane Brockmann. Published by Cambridge University Press.
© Cambridge University Press 2008.

face a decision between choosing a mate for indirect benefits ("good genes") versus direct benefits such as parental care. Consider a mating system where bourgeois males provide sole parental care for the young and parasitic males steal fertilizations from bourgeois males. Cuckoldry by parasitic males will often lead to reduced care to the offspring from the bourgeois male (Trivers 1972, Westneat and Sherman 1993, Dixon et al. 1994, Neff 2003a). When bourgeois males provide greater genetic benefits than parasitic males, both bourgeois males and females should share a common interest in avoiding mating with parasitic males. In this case, a female will maximize both direct and indirect benefits by mating exclusively with bourgeois males. However, when parasitic males provide greater genetic benefits than bourgeois males, a trade-off arises that can lead to a conflict between the interests of females and the interests of the care-providing bourgeois males. Thus, mate choice and parental care can be complicated by intersexual conflict (Alonzo and Warner 2000a, b). Alonzo (Chapter 18) discusses the importance of sexual conflict for understanding the evolution and maintenance of alternative mating tactics. In this chapter, I focus on the effects of genetic benefits and parental care on mating dynamics.

Most previous models of female mate choice in the context of alternative mating tactics have assumed that there is no difference in genetic benefits provided by the alternatives (e.g., Henson and Warner 1997, Alonzo and Warner 2000b). Generally, these models confirm that when parasitic spawning reduces the probability of bourgeois males providing care, females should avoid parasitic males. However, sneak copulations by parasitic males can circumvent female choice, or females may even tolerate parasitic spawning when they occur in nests that are expected to receive a high degree of parental care for some other reason, such as perhaps a large brood size (e.g., Alonzo and Warner 2000a).

Two models have examined the effect of variation in genetic benefits among males employing alternative mating tactics on female mate choice. Based on the mating system of the side-blotched lizard Uta stansburiana, Alonzo and Sinervo (2001) show that it may be adaptive for females to adjust their mating preference based on the relative frequency of males of different life histories. The model assumes that the life histories are heritable and have equal fitness at equilibrium, but are negatively frequency dependent. Thus, the genetic benefits provided by a male are context-dependent; i.e., they depend on the frequency of his life history within the population. The model shows that females should assess this frequency as a basis of their mating preference. However, the model does not specifically address differences in direct benefits provided by the males such as parental care. Shellman-Reeve and Reeve (2000) develop a model based on skew theory that trades off paternal genetic benefits and care. Females form pair bonds with social mates but may solicit copulations from extra-pair mates. Paternal care is dichotomous with social mates either providing full care or not. The results show that infidelity will be higher when a female's potential extra-pair mate provides greater genetic benefits than her social mate and when the value of paternal care is relatively low.

In this chapter, I develop a model of mate choice based on the parasitic and bourgeois life histories that are characteristic of many mating systems. The model assumes that parasitic males provide greater genetic benefits than bourgeois males, but only bourgeois males provide direct benefits in the form of parental care. I examine the effects of genetic benefits and parental care on the optimal trade-off for females during mate choice. I also examine the optimal mating decisions of bourgeois and parasitic males and use these predictions to discuss sexual conflict. Finally, I present data from my research on bluegill sunfish Lepomis macrochirus that show that high-quality females may select nests with preferred rates of cuckoldry, but once in a nest, they may have less control over who actually sires their offspring.

17.2   GENETIC BENEFITS AND PARASITIC LIFE HISTORY

Only a few studies have examined the genetic quality of offspring sired by males from alternative life histories. In Atlantic salmon Salmo salar, Garant and colleagues (2002) show that offspring of parasitic males grow faster and may have higher survivorship during the endogenous feeding period than the offspring from the larger anadromous (bourgeois) males. Because growth rate is an important component of survivorship in fish, these data suggest that parasitic males may provide greater genetic benefits (also see Hutchings and Jones 1998, Garant et al. 2003). In bluegill sunfish, comparison of maternal half-siblings sired in vitro similarly showed that offspring of parasitic males grow faster and to a larger size than offspring of parental males while feeding endogenously on their yolk sac (Neff 2004a). The size advantage of parasitic offspring was estimated to significantly increase survivorship because it reduces susceptibility to predation by Hydra canadensis, a major predator of bluegill fry (Elliott et al. 1997). However, it is unclear if the genetic differences in growth persist during

he exogenous feeding stage or if environmental effects are
1ore important. For example, Gross (1982) found no dif-
erence in growth rate between (mature) cuckolders and
arentals based on back calculations from scale samples (but
:e discussion in Neff 2004a).

Conversely, in the side-blotched lizard, it has been
1own that genetic benefits are dependent on the frequency
f the morph (Sinervo and Lively 1996), and therefore
arasitic males provide greater genetic benefits only when
1ey are relatively rare. In the dung beetle *Onthophagus*
urus, data suggest that parasitic males provide fewer
:netic benefits because they appear to have lower fitness
1an the bourgeois morphs and females are less likely to
1vest in parasitic offspring (Hunt and Simmons 2001,
.otiaho *et al*. 2003; also see Sheldon 2000). Similarly in the
:orpionfly *Panorpa vulgaris*, bourgeois males, which pro-
1uce a nutrient gift for females, produced sons that had
1igher fitness, suggesting that these males provide greater
:netic benefits than the parasitic males, which do not
roduce a gift (Sauer *et al*. 1998). In most other systems, the
lternative life histories are presumed to have equal fitness
nd therefore no difference in genetic benefits has been
.ferred (e.g., Shuster and Wade 1991, Ryan *et al*. 1992,
.ank *et al*. 1995).

7.3   AN OPTIMIZATION MODEL FOR
FEMALE CHOICE

Iere I develop a model of female choice when the quality of
:nes (indirect benefit) trade off with the quality of care
lirect benefit). Because a female cannot maximize both
1direct and direct benefits, she must optimize them to
1aximize her fitness. The model considers two mating tac-
:s/life histories. The first is a "bourgeois" tactic in which
1ales defend breeding sites and provide sole parental care.
'he second is a "parasitic" tactic in which males steal fer-
1izations from bourgeois males but provide no care for the
oung. The model makes the following four assumptions:

l) A bourgeois male's parental investment increases with
his paternity (Trivers 1972, Westneat and Sherman
1993).

?) Offspring survivorship increases with increased parental
investment but with diminishing returns (Whittingham
et al. 1992).

1) Parasitic males provide greater genetic benefits that
confer higher survivorship to their offspring than
bourgeois males (e.g., Garant *et al*. 2002, Neff 2004a).

(4) The benefits from parental care and the benefits from
genetic quality are independent.

The model does not consider variation in quality within
a mating tactic. The effect of such variation on female mate
choice has been addressed elsewhere (e.g., Kirkpatrick
1985). Instead I model the mean quality of the two mating
tactics and focus on the trade-off between the direct benefits
from parental care and the indirect benefits from good
genes. I also do not specifically consider variation in offspring number, but instead focus on the proportion of
offspring fertilized by males from the two tactics. I use the
term "paternity" to refer to these proportions. In a given
brood, the paternity of the bourgeois male plus the paternity
of all genetically contributing parasitic males totals 100%.
A female is thus faced with optimizing the paternity of the
bourgeois male (or analogously optimizing the paternity of
the parasitic males) to maximize her fitness.

Along the x-axis, bourgeois male paternity is plotted
across the values of 0 to 1 (Figure 17.1). A paternity of 0
represents a nest with all parasitic offspring and a paternity of
1 represents a nest with all bourgeois offspring. The y-axis
plots both the fitness benefits from parental care (C_p) and
paternal genetic benefits (G_p). These benefits are modeled as
an increase in survivorship and have equivalent units. The
subscript p in the functions denotes the paternity of the
bourgeois male. The benefit from parental care increases with
paternity with diminishing returns (assumptions 1 and 2),
while the benefit from genes decreases with paternity
(assumption 3). This latter function is linear because the
genetic quality of a female's offspring depends only on the
proportion that are fertilized by males from each tactic.

The genetic benefits provided by bourgeois males is
defined by G_p at $p = 1$ because this represents a brood sired
entirely by the bourgeois male. Conversely, the genetic
benefits provided by parasitic males is defined by G_p at $p = 0$
because this represents a brood sired entirely by parasitic
males. Thus, the slope of G_p represents the genetic benefits
provided by parasitic males relative to bourgeois males – the
steeper the slope the greater the difference in genetic
benefits between the two mating tactics.

A female's fitness ($^f W_p$) is calculated from the product of
the two benefit curves

$$^f W_p = C_p \times G_p, \tag{17.1}$$

which is maximized when

$$\frac{\mathrm{d}^f W_p}{\mathrm{d}p} = 0, \tag{17.2}$$

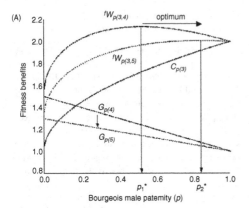

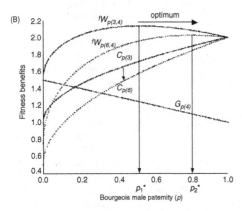

Figure 17.1 The trade-off between the fitness females gain from genetic benefits (G_p) and parental care (C_p) of their bourgeois male partners. The number in parentheses beside each curve denotes the equation in the text that the curve is based on. (A) As the difference in genetic benefits between the life histories decreases ($G_{p(4)} \rightarrow G_{p(5)}$), the optimum breeding situation for females (^fW_p) shifts to higher bourgeois male paternity ($p_1^* \rightarrow p_2^*$: $0.52 \rightarrow 0.84$). (B) As the importance of parental care to offspring survivorship increases ($C_{p(3)} \rightarrow C_{p(6)}$), the optimum breeding situation for females (^fW_p) shifts to higher bourgeois male paternity ($p_1^* \rightarrow p_2^*$: $0.52 \rightarrow 0.80$).

where d represents the derivative of ^fW_p with respect to the bourgeois male's paternity p (dp). When Eq. (17.2) has no solution for $0 \le p \le 1$, W_p is maximized at either $p = 0$ (when d^fW_p/d$p < 0$) or $p = 1$ (when d^fW_p/d$p > 0$).

For simplicity I have considered C_p and G_p as independent functions (assumption 4). A more complex model

could consider interactions between the benefits from genes and parental care. For example, a specific "good gene" might be more valuable (as measured by its effect on survivorship) when the offspring receives low rather than high levels of care.

To demonstrate the properties of the model, I initially define two specific functions for C_p and G_p:

$$C_p = 1 + \sqrt{p}; \qquad (17.3)$$

$$G_p = 1.5 - 0.5p. \qquad (17.4)$$

From these equations the optimal value of p that maximizes female fitness is 0.52; i.e., when 52% of the brood is fertilized by the bourgeois male and 48% is fertilized by parasitic males (Figure 17.1A). Next, to examine the effect of the relative importance of direct and indirect benefits, I adjusted the slope of the genetic benefits curve so that it was less steep:

$$G_p = 1.3 - 0.3p. \qquad (17.5)$$

In comparison to Eq. (17.4), this equation serves to reduce the indirect benefit of mating with a parasitic male. Consequently, the optimum p shifts to the right to a value of 0.84 (Figure 17.1A). In this case, females should be less willing to yield paternity to parasitic males. Incidentally, when G_p is flat (no difference in genetic benefits between the life histories) or positive (bourgeois males provide greater genetic benefits than parasitic males), the optimum p is always 1.

Similar results are obtained when the importance of care is increased by increasing the slope of the care curve so that it is steeper:

$$C_p = 0.5 + 1.5\sqrt{p}. \qquad (17.6)$$

Combining Eq. (17.6) with the original genetic benefit curve in Eq. (17.4) yields an optimum paternity of $p = 0.80$ (a shift to the right from the previous optimum of $p = 0.52$) (Figure 17.1B).

17.3.1 Sexual conflict over paternity

What is the optimum paternity from the perspective of each male mating tactic? First we must define the fitness of each male type. For a bourgeois male, fitness is defined by

$$^bW_p = p \times C_p \times G_{p=1}, \qquad (17.7)$$

where b denotes bourgeois male and $G_{p=1}$ denotes the fitness benefit from the genes of bourgeois males

When C_p is an increasing function (i.e., its derivative is always positive for $0 \leq p \leq 1$) Eq. (17.7) will be maximized at $p = 1$. When there is a single parasitic male spawning in a nest, his optimum fitness is defined by

$$W_p = (1 - p) \times C_p \times G_{p=0}, \qquad (17.8)$$

where s denotes parasitic male (s is used instead of p to avoid confusion with paternity) and $G_{p=0}$ denotes the fitness benefit from the genes of parasitic males.

Unlike the fitness function for bourgeois males, parasitic males typically have an intermediate optimum such that $0 < p < 1$. This is because although a parasitic male may seem to increase his fitness as he fertilizes a greater proportion of the brood (decreasing values of p), he pays a direct cost in terms of the reduced care that his offspring receive. Returning to Eqs. (17.3) and (17.4), for example, a parasitic male maximizes his fitness at $p = 0.11$; i.e., when the bourgeois male fertilizes 11% and he fertilizes 89% of the brood. Based on these same two equations, I previously showed that the optimum p from the perspective of the female is 0.52 and bourgeois males is 1. Thus, females, bourgeois males, and parasitic males have different optima, which will lead to sexual conflict (Figure 17.2).

How might this conflict be resolved? Suppose that we have a population where females have complete control of mating. In this case, 52% of the brood will be fertilized by the bourgeois male and 48% will be fertilized by a parasitic male. However, males should be selected to circumvent such female control. A bourgeois male might try to manipulate a female through, for example, monopolization, in an attempt to shift p towards his optimum of 100% paternity. Conversely, a parasitic male might use behavior patterns such as sneaking to access females or to exclude other parasitic males to shift p in the opposite direction towards his optimum of $p = 0.11$ (11% bourgeois male paternity and 89% parasitic male paternity). All else being equal, the marginal value of circumventing female choice for bourgeois and parasitic males can be calculated from the derivative of their respective fitness functions (Eqs. 17.7 and 17.8) evaluated at a given value of p. At the female's optimum, these derivatives are equal in magnitude for both male mating tactics. Thus, the equilibrium p will depend on, in part, the relative cost for each male to manipulate the breeding situation as well as the female's ability to resist such manipulation and exercise her own control. The cost for the female may differ between the two male types. For example, if bourgeois males are much

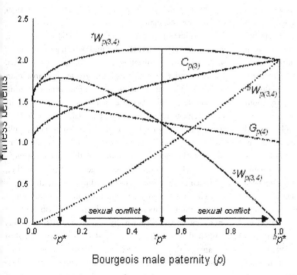

Bourgeois male paternity (p)

Figure 17.2 Fitness functions of females ($^f W_p$), bourgeois males ($^b W_p$), and parasitic males ($^s W_p$) given the relationships between bourgeois male paternity and genetic benefits (G_p) or parental care (C_p). The number in parentheses beside each curve denotes the equation in the text that the curve is based on. The optimum bourgeois male paternity for females is $^f p^* = 0.52$, for bourgeois males it is $^b p^* = 1$, and for parasitic males it is $^s p^* = 0.11$. These differences will lead to sexual conflict.

larger than females, a female may pay a high cost to resisting manipulation by the bourgeois male (e.g., when the manipulation is physical in nature).

The equilibrium p will also depend on the marginal value of resistance for the female. For example, in Figure 17.2, the female has more to lose by succumbing to manipulation by parasitic males than bourgeois males: her fitness curve is "steeper" to the left of her optimum as compared to the right (i.e., the unsigned derivative of her fitness function is greater to the left than the right of her optimum). Thus, in this example, all else being equal, there is stronger selection for females to resist, or counter, manipulation from parasitic males than manipulation from bourgeois males.

17.4   AN EMPIRICAL INVESTIGATION IN BLUEGILL SUNFISH

17.4.1   Parentage analysis

The natural history of bluegill is described in Box 17.1 In relation to the model, bluegill cuckolders (both sneakers and satellites) are the parasitic males and bluegill parentals are the bourgeois males. A bluegill colony was carefully selected at my study site in Lake Opinicon. Details of the colony and fish are published elsewhere (Neff 2001, Neff and Gross 2001). Briefly, a large enclosure was constructed to ensure that all spawning individuals could be collected. The colony occupied about 15% of the enclosure and all natural

---

Box 17.1   Bluegill sunfish (*Lepomis macrochirus*) natural history

Male bluegill are characterized by a discrete polymorphism in life histories termed "parental" and "cuckolder" (Gross 1982, 1991; also see Dominey 1980). In Lake Opinicon (Ontario, Canada), parentals mature at about age 7 years (Figure 17.3) and compete to construct nests in densely packed colonies. Nesting parentals court and spawn with multiple females (sequentially) and provide sole parental care for the developing eggs and fry in their nests (Gross 1982). By contrast, cuckolders do not build nests of their own or care for their offspring. Cuckolders mature precociously and steal fertilizations in the nests of parentals through two tactics: "sneakers" (age 2–3 years) hide behind plants and debris near the nest edge, but are visible after darting into the nest during female egg releases; "satellites" (age 4–5 years) are about the size of mature females (age 4–8 years) and by expressing female color and behavior are able to lead parentals into misidentifying them as a second female in the nest (Gross 1982, Neff and Gross 2001). Cuckolders are superior sperm competitors to parentals, fertilizing nearly 80% of the eggs released in a single dip – females release batches of about 30 eggs in distinctive actions called "dips" – when they are successful at intruding into a nest (Fu *et al.* 2001). Cuckolders appear to die before the age of mature parentals and do not become parentals themselves (Gross and Charnov 1980, Gross 1982).

During spawning, parentals readily detect and attempt to chase sneakers out of their nests but are relatively unsuccessful at detecting satellites. During the egg phase of care, parentals use the frequency of sneaker intrusions, but not the frequency of satellite intrusions, to assess their paternity and allocate care (Neff and Gross 2001, Neff 2003a, b). Specifically, parentals that are heavily cuckolded by sneakers are less willing to defend their nest from brood predators, fan their eggs less, and are more likely to cannibalize some of the brood. Once the eggs hatch, parentals can use an olfactory cue to assess cuckoldry by both sneakers and satellites (Neff and Sherman 2005). At this point parentals that are heavily cuckolded by either sneakers or satellites provide less care for the young. Thus, cuckoldry by sneakers has a greater impact on the overall care level received by the brood than cuckoldry by satellites.

Cuckolders may provide greater genetic benefits (i.e., good genes) than parentals. I found by using experiments of split, in vitro fertilization that cuckolder offspring were larger (standard length) and had greater eye area than parental offspring at the end of the endogenous feeding period (Neff 2004a). Because each female's eggs were split in half and fertilized by both a parental and a cuckolder, these differences could not be attributed to either yolk quality or other maternal effects. Similar results were found in the field, in which nests with a greater proportion of cuckolder offspring had larger fry at swim-up (i.e., at the end of the endogenous feeding period) than nests that contained mostly parental offspring. Growth rate is an important component of survivorship in fish, and the increased growth shown by cuckolder offspring was estimated to confer an estimated threefold survivorship advantage through reduced predation by *Hydra canadensis*, a major predator of bluegill fry (Elliott *et al.* 1997; also see Lister and Neff 2006). The long-term effects of this early differential in growth rate, however, are not yet known. It is possible that the increased growth of parasitic offspring has negative fitness consequences later in life.

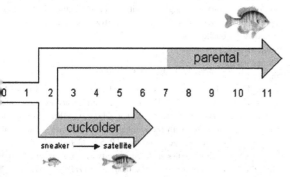

behaviors appeared to occur. At the end of the care period, microsatellite genetic markers were used to assign the paternity and maternity of each spawning individual (see Neff 2001). The parentage models identified the proportion of young fertilized by each putative parent. For the current analysis, this was done for 38 nests collected from the colony. Thus, I was able to track which nests each female spawned in and then calculate the proportion of the young in that nest sired by the tending parental and all sneakers and satellites. When females spawned in multiple nests, I averaged the paternity of each male type (i.e., parentals, sneakers, and satellites) across the nests weighted by the proportion of the female's total eggs spawned that were in each nest:

$$\overline{pat} = \sum \frac{mat_i \times brood\ size_i \times pat_i}{\sum(mat_i \times brood\ size_i)}, \qquad (17.9)$$

where pat on the left side of the equation represents the mean paternity of one of the types of males, and in the ith nest, mat_i is the maternity of the female, brood size$_i$ is the number of offspring, and pat_i is the paternity of either all sneakers, all satellites, or the nest-tending parental. Both summations are over all broods that the female spawned in. The number of offspring in a nest was estimated based on the dry weight of all the fry collected from the nest. Offspring were not counted because there can be tens of thousands in a nest, which impedes accurate counting.

I also applied a variation of the two-sex parentage model (Neff et al. 2000) to each female with each sneaker

male or each satellite male across the colony as a whole. This analysis included offspring from an additional nest of a bluegill–pumpkinseed hybrid parental male that spawned in the colony (see Neff 2001). Because of the high resolving power of the combined 11 loci to assign offspring to parent pairs, the two-sex parentage model is nearly equivalent to a straight exclusion approach (see Fiumera et al. 2002). In the analysis, each offspring was weighted by the brood size of the nest from which it was obtained. This was done because, although approximately equivalent numbers of offspring were sampled from each nest for the parentage analysis, the total number of fry in each nest varied across the colony.

17.4.2 Parental care

The level of parental care in each nest was quantified using two measures (see Neff and Gross 2001). First, brood defense was quantified by presenting a live brood predator (pumpkinseed sunfish, *Lepomis gibbosus*) in a clear bag at the edge of each parental's nest. A trial consisted of presenting the predator for 30 seconds, removing it for 30 seconds, and then presenting the predator for another 30 seconds. An index of the parental's willingness to defend his brood was later calculated from the equation:

$$brood\ defense = 1 \times LD + 2 \times OF + 3 \times Bi, \qquad (17.10)$$

where LD, OF, and Bi are the total number of lateral displays, opercular flares, and bites performed by the parental during the trial. The coefficients were selected to reflect the relative intensity of the parental's reaction and the potential for personal injury. Brood defense was tested twice: once during the egg stage (the day after spawning) and once during the fry stage (the day after eggs hatched). The two scores were summed to provide a single index of defense.

Second, fanning rate was calculated from three 5-minute observations taken on each of the 3 days that the eggs were present before hatching. The overall rate was calculated as the number of fanning motions observed per minute of observation based on all three periods. To construct a single index of parental care, brood defense and fanning rate were each standardized using z-scores and then summed. This combined index places an equal weighting on the two measures of parental care.

17.4.3   Female quality

A total of 45 females were collected from within the enclosure. One of these females was omitted from the current analysis because she was believed to be a bluegill–pumpkinseed hybrid. For the remaining 44 females, four phenotypic measures were assessed: standard length, Fulton's condition factor, active parasite load, and fluctuating asymmetry (FA). Fulton's condition factor was measured as body weight divided by the cube of standard length. Fulton's condition factor reflects the mobile lipid content and hence the energy reserves of a fish (Sutton *et al.* 2000, Neff and Cargnelli 2004). Active parasite load was based on the total weight (expressed relative to the host's weight) of five active parasites (i.e., parasites that feed on its host) obtained during complete dissections: *Dactylogyrus* sp., *Ergasilus caeruleus*, *Proteocephalus* sp., *Spinitectus* sp., and *Leptorhynchoides* sp. (Neff and Cargnelli 2004; also see Muzzall and Peebles 1998). Fluctuating asymmetry was calculated from 11 bilateral traits consisting of the number of pectoral fin rays, length of longest pectoral fin ray, number of pelvic fin rays, length of longest pelvic fin ray, number of teeth (left and right side of upper palette), number of gill rakers (four sets), dry weight of black opercular flap extension, and dry weight of otiliths. All traits were symmetrically distributed about a mode of (nearly) perfect symmetry. Each trait was first standardized (using z-scores) and then the standardized values were summed to provide an index of total FA (Leung *et al.* 2000).

An index of female quality was constructed by combining Fulton's condition factor, parasite load, and FA. This was accomplished by z-scoring each measure and then subtracting parasite load and fluctuating asymmetry from the condition factor. The z-score places an equal weighting on each measure in the overall index. The final index was again z-scored so that zero represented a female of average quality and lower (negative) values represented females of lower quality, while higher (positive) values represented females of higher quality. Female length, which highly correlates with age, was analyzed separately from the quality index.

17.4.4   Female mate choice

On average, females spawned in 4.8 nests (median = 4 range = 1 – 9). Female length and quality were not correlated ($r = 0.16$, $P = 0.29$, $n = 44$) and neither measure was correlated with the number of nests a female spawned in or the average brood size ($P > 0.43$ for each). Higher-quality females spawned in nests that had higher parental paternity ($r = 0.54$, $P < 0.001$, $n = 44$) (Figure 17.4A) and lower sneaker paternity ($r = -0.53$, $P < 0.001$, $n = 44$) (Figure 17.4B) and satellite paternity ($r = -0.43$, $P = 0.004$, $n = 44$) (Figure 17.4C). There was a positive trend between female quality and the average proportion of cuckoldry performed by satellites within each nest they spawned in ($r = 0.30$, $P = 0.050$, $n = 44$). However, the relationship may have been driven by a few extreme data points because a nonparametric analysis indicated no association (Spearman's $r = 0.16$, $P = 0.31$, $n = 44$). Female length was not correlated with any of the three paternity measures or the proportion of cuckoldry performed by satellites ($P > 0.24$ for each).

Across the entire colony, there was no relationship between female quality or length and the proportion of the females' offspring fertilized by parentals, sneakers, or satellites ($P > 0.12$ for each). Higher-quality females did not appear to have a greater proportion of their cuckolded offspring fertilized by satellite males ($r = 0.10$, $P = 0.54$, $n = 44$), but female length was negatively correlated with this proportion ($r = -0.30$, $P = 0.047$, $n = 44$). However this latter relationship is not significant when adjusted for multiple comparisons (corrected $a = 0.05/3 = 0.017$).

There was a positive relationship between female quality and combined index of parental care that their offspring received ($r = 0.34$, $P = 0.024$, $n = 44$; corrected $a = 0.05/2 = 0.025$) (Figure 17.5). There was no relationship between female length and parental care ($r = 0.23$, $P = 0.14$, $n = 44$). Finally, there was no apparent relationship between female quality and the average length or quality of the parental males they spawned with or between female length and the average parental male length ($P > 0.24$ for each). There was, however, a positive relationship between female length and the average quality of the parental males they spawned with ($r = 0.55$, $P < 0.001$, $n = 44$; corrected $a = 0.05/2 = 0.025$).

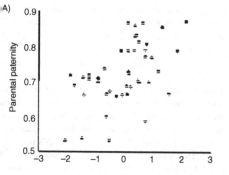

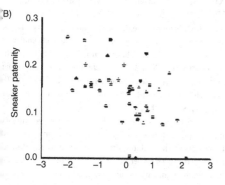

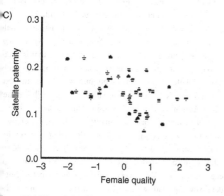

Figure 17.4 The relationship between bluegill (*Lepomis macrochirus*) female quality and the average (A) parental paternity, (B) sneaker paternity, or (C) satellite paternity in the nests they spawned. Paternity is presented as a proportion and female quality based on three phenotypic measures comprising parasite load, fluctuating asymmetry, and body condition (see text).

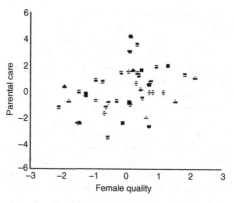

Female quality

Figure 17.5 The relationship between female quality and the mean parental care received by their offspring. Parental care is based on two standardized indices comprising fanning rate and brood defense, and female quality is based on three phenotypic measures comprising parasite load, fluctuating asymmetry, and body condition (see text).

17.5 DISCUSSION

When parasitic males provide greater genetic benefits than bourgeois males, the optimization model makes two general predictions. First, as the difference in genetic benefits between the two life histories increases, all else being equal, a female's optimum breeding situation will shift towards higher parasitic male paternity. This was demonstrated in the model by increasing the slope of the genetic benefits curve while holding the parental care benefit curve constant. Conversely, as parental care becomes increasingly important to survivorship of the young, a female's optimum breeding situation will shift towards higher bourgeois male paternity. Similar predictions have been made by a cost–benefit model developed by Shellman-Reeve and Reeve (2000) that treats paternal care as a dichotomous variable (i.e., a bourgeois male either provides care or does not). Second, the optimum paternity from the perspectives of females, parasitic males, and bourgeois males typically differs, leading to sexual conflict. The resolution of this conflict will depend on the relative costs and benefits to parasitic and bourgeois males for manipulation of females as well as the costs and benefits to females for resisting or counteracting the manipulation.

Bluegill sunfish provide an interesting preliminary test of the model. In bluegill, the parasitic males are the cuckolders and they use two age-dependent tactics to steal fertilizations from parentals (the bourgeois males). When young, cuckolders use a sneaking tactic, but they are visible to parentals as they dart into the nest. Parentals adjust their parental care behavior based on the number of sneakers that are near their nests during spawning; if there are too many sneakers they may even abandon the nest completely (Neff and Gross 2001, Neff 2003a). When older, cuckolders use a satellite tactic whereby they mimic females. During the egg phase of care, parentals actually increase their level of care when they spawn with more satellites, suggesting that parentals perceive satellites as second females in their nest and hence as an increase to their reproductive success (Neff and Gross 2001). Once the eggs hatch, the deception of satellites is revealed, likely through olfactory cues used for kin recognition (Neff and Sherman 2003, 2005; also see Brown and Brown 1996), and parentals adjust their care behavior based on the new information of paternity. Nevertheless, because cuckoldry by satellites goes undetected during the egg phase of care – the most energetically demanding portion of the care period (Coleman and Fischer 1991) – satellites have a lower impact on overall care relative to cuckoldry by sneakers.

Cuckoldry by satellites may further have a lower impact on parental care because of the effect that past investment has on the expected future fitness of parentals. Sargent and Gross (1993) developed a model that showed that larger past investment during parental care reduces an individual's future reproductive success and consequently increases the relative value of a current brood. Furthermore, because older young are often more likely to survive to maturity, the present value of the brood should also increase with its age (Sargent and Gross 1993). Thus, both increasing present value of a brood and decreasing future expectations select for greater parental investment with a brood's age. Their model thereby provided a resolution for the so-called "Concorde fallacy" (also see Dawkins and Brockmann 1980).

Coleman and colleagues (1985) tested this theory in bluegill by manipulating past investment in nesting parentals by experimentally reducing brood sizes either early or late in the care period. They showed that males that had made large past investment (late brood reduction) were more willing to invest in their brood as compared to males that had made only small past investment (early brood reduction). Thus, in addition to the direct positive effect that cuckoldry by satellites has during the egg phase of care presumably due to successful deception of parentals, it should have a lower impact on parental care than cuckoldry by sneakers because of its delayed detection and hence the

increased past investment parentals make under the pretense of higher paternity. Consequently, a female would maximize her fitness by having all her cuckolded young sired by satellites versus sneakers.

Here, I found that higher-quality females, as measured by Fulton's condition factor, parasite load, and symmetry, spawned in nests that had lower overall rates of cuckoldry, but proportionally more of the cuckoldry may have been performed by satellites. The offspring of these females subsequently received more parental care. This increased care appeared to relate directly to the parental male's paternity because there was no relationship between female quality and parental male length or quality, although there was a positive relationship between female length and parental male quality. This result suggests that high-quality females are avoiding nests with higher cuckoldry rates and particularly may be avoiding nests that are especially susceptible to cuckoldry by sneakers. Similar results have been reported for the Mediterranean wrasse *Symphodus ocellatus*, where females prefer nests with fewer sneaker males (Alonzo and Warner 2000a and references within). The bluegill results thus provide some support for the predictions of the model.

Across the entire bluegill colony, however, high-quality females did not appear to spawn a greater proportion of their eggs with satellites than lower-quality females, as might have been expected from the model. No doubt females, high quality or otherwise, are not in complete control of mating. For example, the cryptic behavior of sneaker males is one mechanism that these individuals use to circumvent female control (as well as parental male control). Instead, a high-quality female may be able to select a good breeding situation – for example, a nest that is likely to receive a desired level of cuckoldry or proportionately more cuckoldry by satellites – but once she enters the nest, she may have less control over which males actually sire her eggs. Assuming that high-quality females are in better control of mating than low-quality females, the increased parental care that the former females' offspring receive should outweigh their reduced (paternal) genetic quality. This could be tested directly by examining the fitness of eggs spawned by low- versus high-quality females. Such data would provide a valuable test of the model and of the state of sexual conflict in bluegill.

17.6   FUTURE RESEARCH

There are several avenues for future research. First, the difference in genetic benefits provided by the life histories needs further empirical investigation. There are only a few studies to my knowledge that have addressed this issue (e.g., Garant et al. 2002, Kotiaho et al. 2003, Neff 2004a; also see Welch et al. 1998, Barber and Arnott 2000, Sheldon et al. 2003). If there is in fact a difference in the genetic benefits provided by the life histories that leads to a difference in fitness of the life histories, then the underlying evolutionary mechanism cannot be a genetic polymorphism (i.e., alternative strategies). Instead the mechanism likely would be a conditional strategy. Genetic benefits also can include "compatible genes" such as those involved in the immune response (e.g., major histocompatibilty complex; for an example, see Roberts and Gosling 2003; for reviews see Edwards and Hedrick 1998, Tregenza and Wedell 2000, Mays and Hill 2004, Neff and Pitcher 2005). In bluegill it is likely that females also are seeking compatible genes for their offspring (Neff 2004b). Thus, female choice for genetic benefits may be more complex than just choice for a male from one life history or the other.

Second, the model makes explicit predictions about both intra- and intersexual conflict as it relates to the relative paternities of parasitic and bourgeois males. For example, in bluegill, conflict between females and sneakers should increase as sneaker paternity increases in a nest. Females should be dynamic in their mating behavior as information on paternity changes. When the parental's paternity is high, she should form coalitions with cuckolders, possibly by facilitating access to the nest by aiding in the deception of a satellite or even shielding a sneaker from the parental. When the parental's paternity is low, a female should instead form coalition with the parental, preventing additional cuckoldry, or even leave the nest to find another, more optimal nest to spawn in. Indeed, bluegill females spawned in an average of four to five nests, which may reflect such decision-making. Understanding sexual conflict between females, parasitic males, and bourgeois males should also help us to understand the evolution of the alternative life histories (see Chapter 18).

Third, the model could be expanded to include interactions between genetic benefits and parental care (G_p and C_p) that are likely to occur. For example, genetic benefits from a cuckolder may be more valuable (with respect to fitness) in a harsh environment – i.e., at lower levels of parental care (for an example of such an environmental effect see Welch 2004). Differences in egg number also could be modeled as opposed to only the proportion of a brood fertilized by each male type. When brood size increases with cuckoldry, bourgeois males may, in fact,

tolerate some level of cuckoldry, particularly when parental care is shareable (see Shellman-Reeve and Reeve 2000). Such tolerant behavior may be similar to "incentives," which have been modeled in reproductive skew theory (e.g., Reeve and Keller 2001). Within-tactic conflict could be considered: the model could be expanded to examine the effects of multiple parasitic males spawning in each nest. While the general predictions for females and bourgeois males would not change (i.e., their optima are not dependent on the number of parasitic males per se), the total cuckoldry within a nest likely would exceed the optimum of any one of the parasitic males.

Finally, calculating specific benefit curves for genetic benefits and parental care from empirical data would provide explicit predictions as to the optimal breeding situation for females and males. Comparing observed values to the predictions would provide insight into the evolutionary state of sexual conflict – who is winning the evolutionary arms race between the sexes (Arnqvist and Rowe 2002)? Such data could provide insight into current thinking about sexual conflict (Chapman et al. 2003).

References

Alonzo, S. H. and Sinervo, B. 2001. Mate choice games, context-dependent good genes, and genetic cycles in the side-blotched lizard, *Uta stansburiana*. *Behavioral Ecology and Sociobiology* 49, 176–186.

Alonzo, S. H. and Warner, R. R. 2000a. Dynamic games and field experiments examining intra- and intersexual conflict: explaining counterintuitive mating behavior in a Mediterranean wrasse, *Symphodus ocellatus*. *Behavioral Ecology* 11, 56–70.

Alonzo, S. H. and Warner, R. R. 2000b. Female choice, conflict between the sexes and the evolution of male alternative reproductive behaviors. *Evolutionary Ecology Research* 2, 149–170.

Andersson, M. 1994. *Sexual Selection*. Princeton, NJ: Princeton University Press.

Arnqvist, G. and Rowe, L. 2002. Antagonistic coevolution between the sexes in a group of insects. *Nature* 415, 787–789.

Barber, I. and Arnott, S. A. 2000. Spit-clutch IVF: a technique to examine indirect fitness consequences of mate preferences in sticklebacks. *Behaviour* 137, 1129–1140.

Birkhead, T. R. and Møller, A. P. (eds.) 1998. *Sperm Competition and Sexual Selection*. San Diego, CA: Academic Press.

Blanckenhorn, W. U. and Hosken, D. J. 2003. Heritability of three condition surrogates in the yellow dung fly. *Behavioral Ecology* 14, 612–618.

Brown, G. E. and Brown, J. A. 1996. Kin discrimination in salmonids. *Reviews in Fish Biology and Fisheries* 6, 201–219.

Chapman, T., Arnqvist, G., Bangham, J., and Rowe, L. 2003. Sexual conflict. *Trends in Ecology and Evolution* 18, 41–47.

Coleman, R. M. and Fischer, R. U. 1991. Brood size, male fanning effort and the energetics of a nonshareable parental investment in bluegill sunfish, *Lepomis macrochirus* (Teleostei, Centrarchidae). *Ethology* 87, 177–188.

Coleman, R. M., Gross, M. R., and Sargent, R. C. 1985. Parental investment decision rules: a test in bluegill sunfish. *Behavioral Ecology and Sociobiology* 18, 59–66.

Dawkins, R. and Brockmann, H. J. 1980. Do digger wasps commit the Concorde fallacy? *Animal Behaviour* 28, 892–896.

Dixon, A., Ross, D., Omalley, S. L. C., and Burke, T. 1994. Paternal investment inversely related to degree of extra-pair in the reed bunting. *Nature* 371, 698–700.

Dominey, W. J. 1980. Female mimicry in male bluegill sunfish: a genetic polymorphism? *Nature* 284, 546–548.

Dominey, W. J. 1984. Alternative mating tactics and evolutionarily stable strategies. *American Zoologist* 24, 385–396.

Edwards, S. V. and Hedrick, P. W. 1998. Evolution and ecology of MHC molecules: from genomics to sexual selection. *Trends in Ecology and Evolution* 13, 305–311.

Elliott, J. K., Elliott, J. M., and Leggett, W. C. 1997. Predation by hydra on larval fish: field and laboratory experiments with bluegill (*Lepomis macrochirus*). *Limnology and Oceanography* 42, 1416–1423.

Fiumera, A. C., Dewoody, J. A., Asmussen, M. A., and Avise, J. C. 2002. Estimating the proportion of offspring attributable to candidate adults. *Evolutionary Ecology* 16, 549–565.

Fu, P., Neff, B. D., and Gross, M. R. 2001. Tactic-specific success in sperm competition. *Proceedings of the Royal Society of London B* 268, 1105–1112.

Garant, D., Fontaine, P. M., Good, S. P., Dodson, J. J., and Bernatchez, L. 2002. The influence of male parental identity on growth and survival of offspring in Atlantic salmon (*Salmo salar*). *Evolutionary Ecology Research* 4, 537–549.

Garant, D., Dodson, J. J., and Bernatchez, L. 2003. Differential reproductive success and heritability of alternative reproductive tactics in wild Atlantic salmon (*Salmo salar* L.). *Evolution* 57, 1133–1141.

Godin, J. G. J. (ed.) 1997. *Behavioral Ecology of Teleost Fishes.* New York: Oxford University Press.

Gross, M. R. 1982. Sneakers, satellites and parentals: polymorphic mating strategies in North American sunfishes. *Zeitschrift für Tierpsychologie* 60, 1–26.

Gross, M. R. 1984. Sunfish, salmon, and the evolution of alternative reproductive strategies and tactics in fishes. In G. Potts and R. Wootton (eds.) *Fish Reproduction: Strategies and Tactics*, pp. 57–75. London: Academic Press.

Gross, M. R. 1991. Evolution of alternative reproductive strategies: frequency-dependent sexual selection in male bluegill sunfish. *Philosophical Transactions of the Royal Society of London B* 332, 59–66.

Gross, M. R. 1996. Alternative reproductive strategies and tactics: diversity within sexes. *Trends in Ecology and Evolution* 11, 92–98.

Gross, M. R. and Charnov, E. L. 1980. Alternative male life histories in bluegill sunfish. *Proceedings of the National Academy of Sciences of the United States of America* 77, 6937–6940.

Gross, M. R. and Repka, J. 1998. Game theory and inheritance in the conditional strategy. In L. A. Dugatkin and H. K. Reeve (eds.) *Game Theory and Animal Behavior*, pp. 168–187. New York: Oxford University Press.

Henson, S. A. and Warner, R. R. 1997. Male and female alternative reproductive behaviors in fishes: a new approach using intersexual dynamics. *Annual Review of Ecology and Systematics* 28, 571–592.

Hunt, J. and Simmons, L. W. 2001. Status-dependent selection in the dimorphic beetle *Onthophagus taurus*. *Proceedings of the Royal Society of London B* 268, 2409–2414.

Hutchings, J. A. and Jones, M. E. B. 1998. Life history variation and growth rate thresholds for maturity in Atlantic salmon, *Salmo salar*. *Canadian Journal of Fisheries and Aquatic Sciences* 55 (Suppl.), 22–47.

Kirkpatrick, M. 1985. Evolution of female choice and male parental investment in polygynous species: the demise of the sexy son. *American Naturalist* 125, 788–810.

Kotiaho, J. S., Simmons, L. W., Hunt, J., and Tomkins, J. L. 2003. Males influence maternal effects that promote sexual selection: a quantitative genetic experiment with dung beetles *Onthophagus taurus*. *American Naturalist* 161, 852–859.

Lank, D. B., Smith, C. M., Hanotte, O., Burke, T., and Cook, F. 1995. Genetic polymorphism for alternative mating behavior in lekking male ruff *Philomachus pugnax*. *Nature* 378, 59–62.

Leung, B., Forbes, M.R., and Houle, D. 2000. Fluctuating asymmetry as a bioindicator of stress: comparing efficacy of analyses involving multiple traits. *American Naturalist* 155, 101–115.

Lister, J.S. and Neff, B.D. 2006. Paternal genetic effects on foraging decision-making under the risk of predation. *Ethologist* 112, 963–970.

Mays, H.L. and Hill, G.E. 2004. Choosing mates: good genes versus genes that are a good fit. *Trends in Ecology and Evolution* 19, 554–559.

Muzzall, P.M. and Peebles, C.R. 1998. Parasites of bluegill, *Lepomis macrochirus*, from two lakes and a summary of their parasites from Michigan. *Journal of the Helminthological Society of Washington* 65, 201–204.

Neff, B.D. 2001. Genetic paternity analysis and breeding success in bluegill sunfish (*Lepomis macrochirus*). *Journal of Heredity* 92, 111–119.

Neff, B.D. 2003a. Decisions about parental care in response to perceived paternity. *Nature* 422, 716–719.

Neff, B.D. 2003b. Paternity and condition affect cannibalistic behavior in nest-tending bluegill sunfish. *Behavioral Ecology and Sociobiology* 54, 377–384.

Neff, B.D. 2004a. Increased performance of offspring sired by parasitic males in bluegill sunfish. *Behavioral Ecology* 15, 327–331.

Neff, B.D. 2004b. Stabilizing selection on genomic divergence in a wild fish population. *Proceedings of the National Academy of Sciences of the United States of America* 101, 2381–2385.

Neff, B.D. and Cargnelli, L.M. 2004. Relationships between condition factors, parasite load and paternity in bluegill sunfish, *Lepomis macrochirus*. *Environmental Biology of Fishes* 71, 297–304.

Neff, B.D. and Gross, M.R. 2001. Dynamic adjustment of parental care in response to perceived paternity. *Proceedings of the Royal Society of London B* 268, 1559–1565.

Neff, B.D. and Pitcher, T.E. 2005. Genetic quality and sexual selection: an integrated framework for good genes and compatible genes. *Molecular Ecology* 14, 19–38.

Neff, B.D. and Sherman, P.W. 2003. Nestling recognition via direct cues by parental male bluegill sunfish (*Lepomis macrochirus*). *Animal Cognition* 6, 87–92.

Neff, B.D. and Sherman, P.W. 2005. In vitro fertilization reveals offspring recognition via self-referencing in a fish with paternal care and cuckoldry. *Ethology* 111, 425–438.

Neff, B.D., Repka, J., and Gross, M.R. 2000. Parentage analysis with incomplete sampling of candidate parents and offspring. *Molecular Ecology* 9, 515–528.

Reeve, H.K. and Keller, L. 2001. Tests of reproductive-skew models in social insects. *Annual Review of Entomology* 46, 347–385.

Repka, J. and Gross, M.R. 1995. The evolutionarily stable strategy under individual condition and tactic frequency. *Journal of Theoretical Biology* 176, 27–31.

Roberts, S.C. and Gosling, L.M. 2003. Genetic similarity and quality interact in mate choice decisions by female mice. *Nature Genetics* 35, 103–106.

Rowe, L. and Houle, D. 1996. The lek paradox and the capture of genetic variance by condition dependent traits. *Proceedings of the Royal Society of London B* 263, 1415–1421.

Ryan, M.J., Pease, C.M., and Morris, M.R. 1992. A genetic polymorphism in the swordtail *Xiphophorus nigrensis*: testing the prediction of equal fitnesses. *American Naturalist* 139, 21–31.

Sargent, R.C. and Gross, M.R. 1993. Williams' principle: an explanation of parental care in teleost fishes. In T.J. Pitcher (ed.) *The Behavior of Teleost Fishes*, pp. 275–293. New York: Chapman and Hall.

Sauer, K.P., Lubjuhn, T., Sindern, J., et al. 1998. Mating system and sexual selection in the scorpionfly *Panorpa vulgaris* (Mecoptera: Panorpidae). *Naturwissenschaften* 85, 219–228.

Sheldon, B.C. 2000. Differential allocation: tests, mechanisms and implications. *Trends in Ecology and Evolution* 15, 397–402.

Sheldon, B.C., Arponen, H., Laurila, A., Crochet, P.A., and Merila, J. 2003. Sire coloration influences offspring survival under predation risk in the moorfrog. *Journal of Evolutionary Biology* 16, 1288–1295.

Shellman-Reeve, J.S. and Reeve, H.K. 2000. Extra-pair paternity as the result of reproductive transactions between paired mates. *Proceedings of the Royal Society of London B* 267, 2543–2546.

Shuster, S.M. and Wade, M.J. 1991. Equal mating success among male reproductive strategies in a marine isopod. *Nature* 350, 608–610.

Sinervo, B. and Lively, C.M. 1996. The rock–paper–scissors game and the evolution of alternative male strategies. *Nature* 380, 240–243.

Sutton, S.G., Bult, T.P., and Haedrich, R.L. 2000. Relationships among fat weight, body weight, water weight, and condition factors in wild Atlantic salmon parr. *Transactions of the American Fisheries Society* 129, 527–538.

Taborsky, M. 1994. Sneakers, satellites, and helpers: parasitic and cooperative behavior in fish reproduction. *Advances in the Study of Behavior* 23, 1–100.

Taborsky, M. 1997. Bourgeois and parasitic tactics: do I need collective, functional terms for alternative reproductive behaviors? *Behavioral Ecology and Sociobiology* **41**, 361–362.

Taborsky, M. 1998. Sperm competition in fish: "bourgeois" males and parasitic spawning. *Trends in Ecology and Evolution* **13**, 222–227.

Tregenza, T. and Wedell, N. 2000. Genetic compatibility, mate choice and patterns of parentage. *Molecular Ecology* **9**, 1013–1027.

Trivers, R. L. 1972. Parental investment and sexual selection. In B. Campbell (ed.) *Sexual Selection and the Descent of Man*, pp. 136–179. Chicago, IL: Aldine Press.

Welch, A. M. 2004. Genetic benefits of a female mating preference in gray tree frogs are context-dependent. *Evolution* **57**, 883–893.

Welch, A. M., Semlitsch, R. D., and Gerhardt, H. C. 1998. Call duration as an indicator of genetic quality in male gray tree frogs. *Science* **280**, 1928–1930.

Westneat, D. F. and Sherman, P. W. 1993. Parentage and the evolution of parental behavior. *Behavioral Ecology* **4**, 66–77.

Whittingham, L. A., Taylor, P. D., and Robertson, R. J. 1992. Confidence of paternity and male parental care. *American Naturalist* **139**, 1115–1125.

18 · Conflict between the sexes and alternative reproductive tactics within a sex

SUZANNE H. ALONZO

CHAPTER SUMMARY

Many examples of male alternative reproductive strategies have been identified and studied in detail and mechanisms that can allow the maintenance of alternatives have been identified. However, very little research has considered the role of intersexual interactions in the evolution of alternative reproductive tactics (ARTs). In this chapter, I first examine how alternative reproductive tactics within a sex can lead to conflict between the sexes as well as how conflict between the sexes may influence the evolution of alternative reproductive tactics within a sex. I then describe a few empirical examples of species with alternative reproductive tactics where interactions within and between the sexes have been studied. These examples illustrate that a more complete understanding of the evolution and expression of alternatives can be gained by thinking about interactions within and between the sexes concurrently. I also discuss female alternative reproductive tactics and describe a few empirical examples. I then suggest future empirical and theoretical directions needed for a co-evolutionary understanding of the evolution of alternative reproductive behavior patterns in both males and females.

8.1 INTRODUCTION

The chapters in this book are a testament to how much we now about the evolution, expression, and diversity of alternative reproductive tactics (ARTs). A number of classic examples of alternatives within a population are now well studied and understood (e.g., Dominey 1980, Gross and Charnov 1980, Gross 1982, Lank and Smith 1987, Shuster 1989, Gross 1991, Shuster and Wade 1991, Lank et al. 1995, Widemo and Owens 1995, Sinervo and Lively 1996, Shuster and Sassaman 1997, Widemo 1998, Sinervo et al. 2000). Many examples of male ARTs are believed to result

from competition among males for access to mates or reproductive resources. Yet, even when conflict between males leads to the evolution of ARTs (such as the coexistence of territorial and sneaker males), male fitness (and hence the evolution of male reproductive tactics) will be affected by female choice among males. This chapter focuses on how interactions between the sexes are affected by and affect the evolution of ARTs within a sex.

In general, the evolution of male and female reproductive behavior patterns is concurrently influenced by natural and sexual selection (Darwin 1871, Andersson 1994). Natural selection exists when there is individual variation in survival or fecundity (or fertility in males). In contrast, sexual selection is the result of differential mating success due to the complex interplay between competition within a sex and interactions between the sexes (Bateman 1948, Trivers 1972, Kirkpatrick 1982, Clutton-Brock and Parker 1992, Andersson 1994, Henson and Warner 1997, Alonzo and Warner 2000b, c). Both cooperation and conflict can be the outcome of interactions between the sexes. Although cooperation within and among the sexes does occur, conflict between the sexes is more common and arises when the fitness of one sex is not maximized by the behavior of the opposite sex (e.g., Trivers 1972, Parker 1979, Hammerstein and Parker 1987, Rowe et al. 1994, Arnqvist and Rowe 2002, Chapman et al. 2003, Eberhard and Cordero 2003, Houston et al. 2005). For example, conflict between the sexes often exists over parental care where individuals of both sexes would usually have greater fitness if individuals of the opposite sex invested more energy in care. It is also common for conflict to arise between the sexes over mating. For example, female choice among males and male mate guarding can generate conflict between the sexes. The inclusion of sexual conflict into our understanding of male and female reproductive tactics has greatly improved our ability

Alternative Reproductive Tactics, ed. Rui F. Oliveira, Michael Taborsky, and H. Jane Brockmann. Published by Cambridge University Press.
© Cambridge University Press 2008.

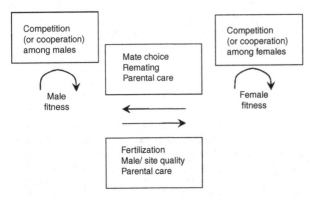

Figure 18.1 Both male and female fitness (and thus the behavior patterns that are expected to evolve) are affected simultaneously by interactions within and between the sexes. For example, the outcome of competition among males will determine the costs and benefits of female mate choice and remating frequency. However, female choice and mating strategies will concurrently affect the costs and benefits of competition among males and thus influence the evolution of interactions among males.

to understand the evolution of reproductive behavior patterns such as female mate choice and parental care (e.g., Hammerstein and Parker 1987, Brockmann and Grafen 1989, Davies 1989, Veiga 1990, Ahnesjo et al. 1992, Part et al. 1992, Oring et al. 1993, Rowe et al. 1994, Gwynne and Snedden 1995, Warner et al. 1995a, Gray 1996, Alonzo and Warner 1999, 2000b, Hardling 1999, Arnqvist and Rowe 2002, Smith et al. 2002, Wedell et al. 2002). However, the importance of interactions between the sexes to understanding alternative reproductive tactics within a sex is not generally recognized (but see Henson and Warner 1997, Hugie and Lank 1997, Alonzo and Warner 2000b, c, Alonzo and Sinervo 2001, Luttbeg 2004, Reichard et al. 2005), even though male and female reproductive fitness, and thus also presumably male and female reproductive strategies, will be influenced by interactions within and between the sexes simultaneously (Figure 18.1).

18.2 ALTERNATIVE REPRODUCTIVE TACTICS MAY LEAD TO CONFLICT BETWEEN THE SEXES

As described above, conflict between the sexes often arises over mating and parental care. One common pattern is the coexistence of territorial males and other nonterritorial tactics such as sneaking copulations (Taborsky 1994, 1997).

Although these alternatives are often believed to arise due to male competition, sneaking tactics may also circumvent female choice if females prefer territorial males (e.g., Cade 1980, Austad 1984, Gross 1984, Forsyth and Montgomerie 1987, Taborsky 1994, Lank et al. 1995). In this case, conflict between females and sneaker males will exist. In many species of fish with alternative reproductive tactics, nonterritorial males do not provide parental care (van den Berghe 1990, Taborsky 1994, 1997, Warner et al. 1995b). For example, in Symphodus tinca, parental care is facultative with territorial males providing care while nonterritorial males do not (van den Berghe 1990). Females prefer to spawn with territorial males. However, the amount of time females will search for a territory depends on individual experience and time in the reproductive season (Warner et al. 1995b). Because offspring survival and hence female fitness is higher in this species when males provide care, conflict between females and nonterritorial males will exist (Warner et al. 1995b). In other related species, differences between territorial and sneaker males in sperm production and fertilization rates have been found that may lead to conflict between the sexes if lowered fertilization rates lead to lower female fitness (Warner et al. 1995a, Warner 1997, Alonzo and Warner 2000a, Petersen et al. 2001).

In general, any differences between male alternatives that affect female fitness have the potential to generate conflict between the sexes. Similarly, any female discrimination between males based on tactic will lead to conflict between males and females over mating. When would we expect female choice among male tactics to evolve? If a female's expected fitness is the same when she mates with either male tactic, then no choice by females among male tactics would be expected. This could occur when negative frequency-dependent selection due to male competition leads to both tactics having equal expected male fitness at equilibrium or when condition-dependent tactics have no heritable differences. If males do not have any direct effect on female survival or future fecundity, then in these cases females would not be expected to choose between male tactics.

In contrast, if the male tactic has any differential effect on female fitness, such as on female survival or condition in the future, then we would expect to observe conflict between the sexes over the existence of male alternative reproductive tactics, and female choice among male alternative reproductive tactics would be expected to evolve. Sexual conflict may be common when male alternative behavior patterns depend on male condition, such as size, age, or energy reserves. These factors will often have a direct effect on female current reproductive success by affecting male parental care, sperm production, territorial defense, or provisioning of females by males. Furthermore, male condition may also be correlated with genes that affect the fitness of a female's offspring. Thus condition-dependent tactics may commonly lead to conflict between the sexes over mating, parental care, provisioning, or other male behavior patterns that differ among male alternative reproductive tactics. In general, we can expect that male alternative reproductive tactics will usually generate conflict between the sexes (and often female choice among male tactics) except in cases where the male tactic has no effect on the current or future reproductive success of the female.

8.3   INTERACTIONS BETWEEN THE SEXES CAN AFFECT ALTERNATIVE REPRODUCTIVE TACTICS

The previous section illustrates that the existence of male alternative reproductive behavior patterns can not only lead to conflict between the sexes over mating but can also drive the evolution of female mate choice whenever male alternatives differentially affect female fitness. Furthermore, if male tactics affect female fitness, they may drive the evolution of female reproductive strategies other than female mate choice.

The question then arises, how will female behavior and mate choice affect the evolution of male alternative reproductive behavior patterns? First of all, female choice among males may generate fitness patterns that lead to the evolution of male alternative reproductive tactics. For example, female choice for larger or older males may lead to condition-dependent male reproductive success, which has been shown to allow the evolution of discrete variation in reproductive behavior (Henson and Warner 1997, Alonzo and Warner 2000c). Similarly, if female choice among males is negatively frequency dependent (e.g., females choose novel males), female behavior could lead to the evolution of male alternative reproductive behavior patterns. Furthermore, variation among females in mate choice, for example as a function of female age, condition, or location, could allow male variation in reproductive tactics as well (Henson and Warner 1997, Alonzo and Warner 2000c, Coleman et al. 2004). As described above, the potential exists for female choice to generate patterns of fitness that allow the stable coexistence of male alternative reproductive tactics, but there has been very little research empirically or theoretically examining whether female choice actually generates conditions that allow the evolution of male alternative reproductive tactics.

Female choice has been shown to suppress the coexistence of male alternatives even in cases where male competition alone would lead to the evolution of alternatives (Henson and Warner 1997, Alonzo and Warner 2000c). Models that examine male alternatives and female choice simultaneously have shown that the inclusion of female choice alters the conditions for the coexistence of male alternatives (Table 18.1). When frequency- or condition-dependent success in competition among males exists, female choice among male alternatives is predicted to alter the frequency of male alternatives or even to suppress their coexistence (Figure 18.2). Even if female control over mating is weak, the stable frequency of the less preferred male tactic would still be expected to be lower than predicted by male competition alone (Henson and Warner 1997, Alonzo and Warner 2000c) (Figure 18.2). It is difficult to study whether female choice has blocked the evolution of alternatives since there are many reasons that a species may not exhibit discrete variation in reproductive tactics. It would be interesting to know, for example, if nonterritorial male alternative tactics are less common in species with internal fertilization or in species in which forced copulations are not possible. It would also be interesting to

Table 18.1. *A summary of how the consideration of female mate choice can affect the predicted expression and coexistence of male alternative reproductive tactics*

Scenario	Ignoring female choice	Considering choice
Static frequency dependence	Alternatives co-occur and are stable at the frequency where the negative frequency-dependent male fitness curves cross	Male alternatives only occur if females show no preference between male alternatives
Dynamic frequency dependence (females trade off survival and mating success)	Same as above	Male alternatives occur when female preference is time dependent, or if females show no preference between male alternatives
Dynamic energy dependence	Alternatives co-occur and males switch between behaviors when energy reserves are high	Female preference can suppress the occurrence of the energetically costly alternative, but not the low-energy alternative
Dynamic frequency dependence and size dependence	Alternatives co-occur and males switch between behaviors when large	Female preference can suppress the occurrence of the large-size alternative, but not the small-size alternative
Female condition dependence	Male alternatives not predicted	Female variation in choice acts as a mechanism for maintaining male alternatives

Source: From Alonzo and Warner (2000c), with permission.

(A) No female preference

(B) Females prefer male behavior B

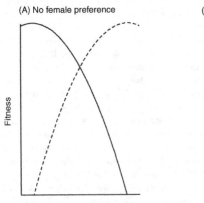
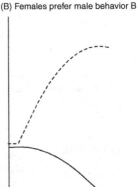

Fitness

Frequency of male behavior A

Figure 18.2 Female preference can influence predictions for male alternative reproductive patterns. For example, frequency-dependent fitness due to competition among males can maintain alternatives within a sex. Solid and dashed lines represent the fitness of two alternative behaviors A and B respectively as a function of the frequency of behavior A. (A) In the absence of female choice between male alternatives, the stable frequency of behavior A is predicted where the two lines cross. (B) If females prefer male behavior B, predictions can change dramatically and in this case is predicted to suppress the occurrence of male behavior A despite the existence of negative frequency dependence due to male competition. (Adapted from Henson and Warner 1997.)

xamine female mate choice in species that do not exhibit nale alternative tactics yet have the patterns of male com-•etition that would argue for the evolution of male alter-•atives.

8.4 EMPIRICAL EXAMPLES OF INTERACTIONS BETWEEN THE SEXES AND ALTERNATIVE REPRODUCTIVE TACTICS

Although few studies have addressed how interactions •etween the sexes influence alternatives within a sex, the esearch that exists is promising. Empirical examples exist n which interactions between the sexes are known to affect he fitness of tactics within a sex (e.g., Crespi 1988, van den Berghe et al. 1989, Davies 1992, Henson and Warner 997, Hugie and Lank 1997, Hunt and Simmons 2000, ones et al. 2001, Badyaev and Hill 2002, Garant et al. 002). As described above, it has also been shown that lternatives within a sex can lead to conflict between the exes (e.g., Davies 1992, Eadie and Fryxell 1992, Hattori nd Yamamura 1995, Henson and Warner 1997, Hunt and Simmons 2000, Sinervo and Zamudio 2001, Badyaev and Iill 2002, Candolin and Reynolds 2002b). Finally, species xist where alternatives occur in both sexes within a species nd interactions within and between the sexes affect the tness of alternatives within each sex (e.g., Reillo and Wise 988, Davies 1992, Hattori and Yamamura 1995, Hunt nd Simmons 2000, Sinervo et al. 2000, Sinervo and amudio 2001, Caillaud et al. 2002, Pienaar and Greeff 003). Rather than attempt an exhaustive review of sexual onflict and alternative reproductive behavior patterns, focus on describing four species where alternative repro-uctive behavior patterns exist and interactions between the exes are known to affect individual fitness.

8.4.1 Male alternative reproductive tactics lead to conflict between the sexes in the European bitterling (Rhodeus sericeus)

n this species of fish, some males defend territories that ontain a number of freshwater mussels where females pawn. Other males are not territorial and steal fertilizations n another male's territory (Reynolds et al. 1997). A female viposits her eggs into the gills of a mussel while males release perm into the inhalant siphon of the mussel to fertilize the emale's eggs. Sperm competition occurs between territorial

and nonterritorial males within the mussel (Candolin and Reynolds 2002a, b, Smith et al. 2002, 2003, Reichard et al. 2004a). Males increase their ejaculation rates (Candolin and Reynolds 2002a, Smith et al. 2003) and aggression rates toward rival males in the presence of sperm competition (Smith et al. 2003). The fitness of male alternative tactics is also density dependent in this species (Reichard et al. 2004b). The embryos develop in the gills of the mussel over a period of weeks (Smith et al. 2001). Females are attracted to territories based on territorial male traits (Candolin and Reynolds 2001, Smith et al. 2002, Reichard et al. 2005), but females inspect mussels directly to choose oviposition sites (Candolin and Reynolds 2001). Females prefer to spawn in species of mussels that lead to higher offspring survival (Smith et al. 2000, Mills and Reynolds 2002, Reichard et al. 2005). Females also prefer mussels with fewer embryos already in the mussel and offspring survival within mussels is negatively density dependent (Smith et al. 2000). In general, males are less likely to lead a female to mussels with a high number of embryos (Smith et al. 2003), but they will try to lead females to mussels with lower expected embryo survival if the risk of sperm competition from an intruding sneaker male is lower at those mussels (Smith et al. 2002, 2003) (Figure 18.3).

Sexual conflict occurs within and between the sexes in this species: territorial and nonterritorial males are always in conflict over sperm competition while females may be in conflict with other females over mussel use if high-quality mussels are limiting (Smith et al. 2000). In this example, if good mussel sites are limiting, conflict between the sexes will occur over spawning rate and oviposition site (Smith et al. 2002, Reichard et al. 2005). Thus, to understand patterns of fitness, reproductive behavior patterns, and the evolution of alternative reproductive behavior patterns in this species, it is necessary to consider interactions within and between the sexes. Female choice among territorial males may depend on the frequency of nonterritorial males and number of eggs in each mussel on a territory. Furthermore, the fitness of male alternatives will be affected by sperm competition and by female choice among males and among mussels within a territory. The exact role of intersexual selection in the evo-lution of male alternatives is not yet understood in this spe-cies. It would be very interesting to know if fertilization rates and female fitness are affected by nonterritorial males. In this species, understanding the evolution and maintenance of male behavior patterns requires considering how female behavior affects male fitness.

European bitterling, *Rhodeus serviceus*. (From A. C. Wheeler (1969) *The Fishes of the British Isles and North-west Europe*. East Lansing: Michigan State University Press.)

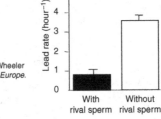

Figure 18.3 In the European bitterling, territorial males are less likely to lead females to spawning sites with sneaker (rival) male sperm. This may cause females to spawn in lower-quality mussels since male leading and female spawning are strongly associated. Mean lead rate per hour (±1 SE) is shown. (From Smith *et al.* 2002. with permission.)

18.4.2 In a Mediterranean wrasse, *Symphodus ocellatus*, conflict between the sexes may allow the stable coexistence of male alternative reproductive behavior patterns.

In this species of fish, multiple male alternative reproductive tactics exist (Taborsky *et al.* 1987, Alonzo *et al.* 2000). Large males defend nest sites where all of the spawning in the species occurs (Lejeune 1985). Females visit these sites and spawn with territorial males. These males provide parental care defending and aerating the eggs until they hatch 3–5 days later (Lejeune 1985, Taborsky *et al.* 1987). Small males in the population, called sneakers, steal fertilizations and do not court females, defend sites, or provide care (Soljan 1930, Lejeune 1985, Taborsky *et al.* 1987). Males of intermediate size, called satellites, court females and are allowed closer to the nest by the nesting male but also steal fertilizations (Taborsky *et al.* 1987). Multiple life-history pathways appear to exist (Alonzo *et al.* 2000). Territorial males tend to be older than sneakers and have higher early growth rates than sneakers. Males may be satellites either as 1- or 2-year-olds while sneakers tend to be 1 year old. Females appear to prefer mating with territorial males and, although often joined by sneaker males, will only spawn in the presence of a territorial male (author's personal observation). Nesting males compete with other nesting males for access to females and nest sites; sneakers compete for access to spawning females. Although satellites are tolerated by territorial males more than sneakers (Taborsky *et al.* 1987), they are still in conflict with nesting males over spawning rates and with females over mating. Females and nesting males are in conflict over male desertion of the nest which tends to occur at nests with low spawning rates (Taborsky *et al.* 1987, Alonzo and Warner 2000b). Thus, females exhibit a strong preference for nests with high mating rates but attempt to avoid spawning with sneakers (van den Berghe *et al.* 1989, Alonzo and Warner 1999, 2000b). However, nests with high mating rates attract sneakers and satellites, and thus sperm competition and sneaker spawning is higher at nests where females prefer to mate (Lejeune 1985, Alonzo and Warner 1999, 2000b). In this species, conflict occurs within and between the sexes and each of these interactions affects the fitness of individuals who follow different tactics such as female choice or male alternatives (Figure 18.4). Female behavior influences the fitness of male alternative behavior patterns, but it is conflict between females and territorial males over parental care that may allow sneaker males to exist since females avoid spawning with sneaker males if possible (Alonzo and Warner 1999, 2000b, Alonzo *et al.* 2000). This means that an understanding of male alternative reproductive tactics requires the consideration of multiple interactions within and between the sexes.

18.4.3 Male and female alternatives drive conflict within and between the sexes in the dunnock, *Prunella modularis*

Alternative reproductive tactics are facultative in both males and females of this species of bird (Davies 1986, 1992, Davies and Houston 1986, Davies and Hatchwell 1992, Sozou and Houston 1994, Davies and Hartley 1996, Davies *et al.* 1996, Langmore and Davies 1997, Langmore *et al.* 2002) (Figure 18.5). Individuals may mate monogamously

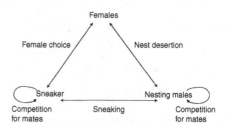

Ocellated wrasse, *Symphodus ocellatus*. (From Whitehead PJP, Bauchot ML, Hureau JC, Nielson J, Tortonese E. (1984) *Fishes of the North-Eastern Atlantic and the Mediterranean*. Paris: Unesco.)

Figure 18.4 Many conflicts within and between the sexes occur simultaneously in *Symphodus ocellatus*. Females are in conflict with sneakers over sneaking and with nesting males over mating as well as over desertion of the nest by the nesting male. Sneakers are in competition with other sneakers to fertilize eggs and in

conflict with nesting males over sneaking. Nesting males are in competition with other nesting males for access to nest sites and mates. All of these interactions simultaneously affect male and female fitness and behavior.

Figure 18.5 Dunnock, *Prunella modularis*. (From F. O. Morris [1891] *A History of British Birds*, 3rd edn. London: John C. Nimmo.)

polyandrously, polygynously, and polygynandrously where both male and female behavior determines the mating pattern. On average, female fitness is highest when mated to two males and lowest when sharing male care with multiple females and only one male (Hatchwell and Davies 1990, Davies and Hatchwell 1992). Similarly, male fitness is reduced by sharing paternity with other males and highest when mating alone with multiple females (Davies 1986, Davies and Houston 1986, Hatchwell and Davies 1990, 1992, Davies *et al.* 1992). When multiple males mate with a single female, they are either dominant (alpha males) or subordinate (beta males) (Davies 1992). Although alpha males attempt to guard females, female behavior allows males following the beta strategy to mate. The female elicits

copulations with males following the beta strategy and evades the mate-guarding tactics of the alpha male. This behavior increases female fitness because males following the beta strategy invest in parental care in proportion to their mating success with the female (Davies and Houston 1986, Davies 1992, Hartley and Davies 1994, Davies *et al.* 1996). Conflict also exists between females in this species. Females compete for male parental investment and the amount of conflict between females depends on the number of males and females in the mating group (Davies and Hatchwell 1992, Hartley and Davies 1994, Langmore and Davies 1997, Langmore *et al.* 2002). Conflict within and between the sexes occurs in this species and affects the behavior patterns and fitness of males and females. Focusing on male competition alone would not explain how beta males obtain mating success. Instead, in order to explain the observed patterns of male fitness as a function of tactic, it is necessary to understand that female fitness is increased by mating with multiple males (because males provide parental care).

18.4.4 Interactions within and between the sexes allow the coexistence of male and female alternative reproductive morphs in the side-blotched lizard, *Uta stansburiana*

In this species (Figure 18.6), male and female throat color morphs exist that are due to genetic differences (Sinervo and Lively 1996, Sinervo *et al.* 2000): three male color morphs coexist that differ in their territorial behavior, mating success, and aggression. Orange males are aggressive and defend larger territories with high mating rates. Blue

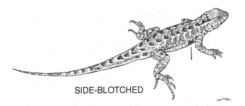

SIDE-BLOTCHED

Figure 18.6 Side-blotched lizard, *Uta stansturiana*. (From R. C. Stebbins [1985] *Western Reptiles and Amphibians*. Boston, MA: Houghton Mifflin.)

males are territorial but less aggressive and defend smaller territories. Yellow males sneak copulations on the territories of orange and blues males. However, orange males are more susceptible to parasitism by yellow males. Thus, orange males do well when competing with blue males, but do less well when yellow and other orange males are common. Similarly, yellow males do better when orange males are common and blue males do best when in competition with yellow males. Thus, male mating success is negatively frequency dependent and both male reproductive success and the frequency of each of the male morphs have been observed to cycle over time.

Although initial research focused solely on the existence of male alternative reproductive tactics (Sinervo and Lively 1996), variation in female throat color has now been found to be associated with discrete variation in female reproductive behavior (Sinervo *et al.* 2000, 2001, Sinervo and Zamudio 2001, Svensson *et al.* 2002, Comendant *et al.* 2003). Through careful consideration of female clutch size, egg size, and behavior, Sinervo *et al.* (2000) found that two discrete female morphs exist. Yellow females produce few small eggs; orange females produce many smaller eggs. When population density is high, the offspring of yellow females have an advantage because large eggs lead to greater offspring survival at high density. When population density is low, orange females have greater reproductive success because they produce more offspring. In this annual species, when the population density is high in one year, offspring survival is low causing the population to crash in the next year. However, since orange females produce many eggs and have a fitness advantage at low population density, the population increases again to a high density in the following year. Since yellow females do well in a high-density year while orange females do well in the following low-density year, the fitness and frequency of female morphs oscillate with population density in a 2-year cycle (Sinervo *et al.* 2000).

In this species, competition for mates occurs among males, and females compete for territories and through their offspring. Thus intrasexual competition affects fitness within each sex (Sinervo and Lively 1996, Sinervo 1998, 1999, Zamudio and Sinervo 2000, Sinervo and Zamudio 2001). However, females have been shown to differ in their mate choice patterns and a female's tactic affects the performance of her male offspring as well. Females also appear to have some control over which male fertilizes their eggs (Calsbeek and Sinervo 2002). Thus, patterns of female choice and maternal effects that differ between female morphs also affect the fitness of male morphs (Sinervo 1998, Alonzo and Sinervo 2001). The assumption that male competition alone determines male mating success fails to explain observed patterns of male fitness because female mate choice and maternal effects also influence the fitness of male tactics. In fact, models examining only male competition and assuming no effect of females on male mating success could not recreate the observed changes in the frequency of male morphs through time (Alonzo and Sinervo 2001). In contrast, models with incomplete female control over mating were able to predict the observed cycles (Alonzo and Sinervo 2001). In this species, both male and female alternative reproductive tactics exist where the fitness of each tactic depends on interactions within and between the sexes in a very complex way. It is not known to what extent competition within a sex and interactions between the sexes affect individual fitness and further studies are needed to determine the relative importance of intra- and intersexual interactions on the fitness and coexistence of male and female alternative reproductive patterns. However, this species clearly demonstrates that both males and females can exhibit alternative reproductive tactics within one species, that female competition can maintain discrete variation in female reproductive behavior, and that the fitness of male alternative reproductive tactics can depend simultaneously on male competition and female mate-choice behavior.

18.4.5 Future directions for empirical research

The excellent research on these four species illustrates that we can gain a better understanding of alternative reproductive tactics and the factors determining their fitness by thinking about the concurrent effect of interactions within a sex and conflict (or cooperation) between the sexes. However, few if any studies have expressly examined the relative importance of inter- and intrasexual interactions or onsidered how the co-evolution of male and female behavior patterns may alter the evolution of alternative reproductive tactics. The empirical patterns to date simply argue for the potential of this approach. How female choice patterns change as a result of changes in male competition or the frequency of male alternative tactics remains an interesting question. Although the effect of the frequency of male alternatives on male fitness is sometimes examined empirically, I am aware of no studies that have determined whether female fitness or behavior is dependent on the frequency of male tactics. It is possible that changes in interactions between males may have indirect effects on female fitness and thus have cascading effects on female choice and the fitness of male alternatives.

8.5 FEMALE ALTERNATIVE REPRODUCTIVE TACTICS

As some of the empirical examples described above illustrate, female alternative reproductive behavior patterns do exist. Although it is not known if alternative reproductive tactics are less common in females than in males, they are certainly less well studied than variation in male reproductive behavior. It is possible that female alternatives, like many female reproductive strategies, are simply less easily detected than their male counterparts. For example, the female alternative reproductive tactics in the side-blotched lizard were found only after years of research on the male alternative reproductive behavior patterns in this species, despite the fact that males exhibit discrete variation in throat color. Yet, the females of this species clearly exhibit discrete variation in reproductive tactics similar to the classic examples of ARTs in males. Although variation in female life-history patterns or reproductive allocation will be common, it is possible that this variation among females in reproductive behavior may more often be continuous than discrete. However, the same mechanisms hypothesized to allow the stable coexistence of alternatives in males could allow their coexistence in females. When will females experience negative frequency-dependent selection or condition-dependent fitness? Will they experience it as often as males, especially given that male competition is more commonly observed than direct competition among females? The kind of competition that leads to negative frequency-dependent selection in male fitness will probably be uncommon in females. However, if we define ARTs as discrete variation in reproductive tactics, I expect that female alternatives due to condition-dependent fitness will be relatively common, but recognizing discrete variation

in mate choice, egg size, provisioning behavior, or other typical female behavior patterns may be more difficult than recognizing the presence or absence of male courtship displays or territorial behavior. Although females may not exhibit the same kind of direct competition for mates and resources as often as we observe it in males, female competition for resources, mates, and sites is not uncommon. Furthermore, even when females do not compete directly, their offspring will compete with the offspring of other females which can indirectly generate frequency-dependent and condition-dependent reproductive success in females just as we observe in males due to male competition.

To determine whether alternative reproductive tactics are less common in females than in males, we need more empirical research that is expressly focused on describing and understanding individual variation in female reproduction. At present, we do not have sufficient research on female alternative tactics to know if female alternative reproductive behavior patterns are common in the wild. We also need theory that is focused on understanding the specific conditions that will allow the evolution of discrete variation in female reproductive tactics. In general, however, examples of classic alternative reproductive tactics in females are known and indicate that female alternative reproductive behavior patterns represent an interesting and promising area of research.

For example, in a variety of damselfly species (Odonata), female morphs are observed that differ in their color patterns and reproductive behavior (Forbes 1994 , Forbes et al. 1997, Cordero et al. 1998, Andres and Rivera 2001, Sherratt 2001, Van Gossum et al. 2001, Andres et al. 2002, Svensson et al. 2005). Usually, some females resemble males of the same species (androchromes) while the others (gynochromes) exhibit more classic female color patterns. It has been hypothesized that females may "mimic" male color patterns in order to avoid harassment by males (Forbes 1994, Forbes et al. 1997, Cordero et al. 1998, Sherratt 2001, Van Gossum et al. 2001, Andres et al. 2002). Consequently, these damselfly species may represent an example of conflict between the sexes over mating that leads to alternative reproductive patterns in females. In one species of damselfly, Ischnura elegans, three female morphs exist. Two color morphs are drab while the other morph is blue and resembles males. Although the three morphs appear to have similar survival, they differ in their mating frequencies in the wild with gynochromes mating more often than androchrome females (Cordero et al. 1998). Androchrome females may have lower mating opportunities but also experience reduced mating costs (Cordero *et al.* 1998). Harassment of females also appears to depend on the frequency of female color morphs and population density (Van Gossum *et al.* 2001, Andres *et al.* 2002, Svensson *et al.* 2005). It has also been argued that males may simply prefer the most common female morph leading to positive frequency-dependent harassment of females and thus negative frequency-dependent fitness of female color morphs (Van Gossum *et al.* 1999, 2001, Svensson *et al.* 2005). Further empirical research examining multiple variables simultaneously is needed to determine what mechanisms allow the coexistence of these alternative female morphs.

Another example of discrete variation in female reproductive behavior is intraspecific brood parasitism (Eadie and Fryxell 1992, Nee and May 1992, Lyon 1993, Johnston 1994, Eadie and Lyon 1998, Ahlund and Andersson 2001). However, unlike in Odonata, this pattern of variation tends to be age or context specific in expression (Sorenson 1991, Ahlund and Andersson 2001). For example, in the American coot *Fulica americana*, conspecific brood parasitism is common and generates conflict between females (Lyon 1993). Some females (called floaters) do not have their own nests and only lay eggs parasitically. Other females lay eggs parasitically before laying in their own nests (Lyon 1993, 1998, 2003a). Females lay smaller clutches of their own eggs when parasitized and parasitism causes deviations from the optimal clutch size (Lyon 1998, 2003b). Brood parasites tend to be older and have higher fecundity than females that do not lay eggs parasitically (Lyon 2003a). There is currently no evidence for conflict between the sexes over this strategy and males do not appear to mate with parasitizing females or father the parasitic eggs in their own nest (Lyon *et al.* 2002), although it may occur in other species.

More subtle examples of female alternatives also exist in the form of context-dependent female choice (e.g., Sih and Krupa 1992, Alonzo and Sinervo 2001, Qvarnstrom 2001) and state-dependent reproductive patterns (e.g., Stearns 1992, Ruppell and Heinze 1999, Roff 2002). For example, as observed in dunnocks, discrete context-dependent mating patterns (e.g., monogamy, polygyny, polyandry, and even polygynandry) coexist within a population. In other species, females exhibit age-dependent variation in reproductive behavior by helping at the nest when young and becoming independent breeders later in life (e.g., Komdeur 1994, 1996, Komdeur *et al.* 1997). Female copying of the mate choice decisions of other females often depends on female age and experience leading to a state-dependent expression of the reproductive tactic (e.g., Dugatkin and Godin 1992,

Goldschmidt *et al.* 1993, Brooks 1998, Westneat *et al.* 2000). Sex ratio allocation has also been observed to depend on female age and condition (e.g., Komdeur 1996, Svensson and Nilsson 1996, Komdeur *et al.* 1997, Kojola 1998, Kruuk *et al.* 1999, Pen *et al.* 1999, Bourke 2001, Albrecht and Johnson 2002, Pienaar and Greeff 2003). Finally, maternal effects often vary among and within individuals and can cause discrete variation in offspring phenotypes (Sinervo 1998, Gil *et al.* 1999, Hunt and Simmons 2000, Badyaev *et al.* 2002, Calsbeek and Sinervo 2002). For example, in the dung beetle *Onthophagus taurus*, alternative male phenotypes coexist: horned males are large, territorial, and aggressive; hornless males are small and sneak copulations (Moczek and Emlen 2000, Tomkins and Simmons 2000, Hunt and Simmons 2001). This appears to be a classic example of condition-dependent male alternative reproductive tactics (Hunt and Simmons 2001). The horned strategy has greater fitness when males are large, but males do better adopting the hornless sneaker male tactic when small. However, it has also been shown that differences in female provisioning of offspring leads to the variation in male size that generates the phenotypic polymorphism in males (Hunt and Simmons 2000, 2001). Hence, this classic example of male alternative reproductive tactics is actually driven by variation in female provisioning behavior that generates the necessary variation in male size for the male alternatives to be expressed.

These examples clearly illustrate that discrete variation in female reproductive tactics exists in a variety of forms. Female alternatives may not be the same as male alternative reproductive tactics, and they may not be driven predominantly by direct competition among females. However, these examples clearly represent female alternative reproductive tactics arising because of condition- or context-dependent reproductive success in females.

18.6   MOVING FORWARD IN OUR UNDERSTANDING OF ALTERNATIVE REPRODUCTIVE TACTICS

Considering co-evolution between males and females is one important way in which we can advance our understanding of male and female alternative reproductive tactics. Although it is now generally recognized that negative frequency-dependent selection and condition-dependent fitness can allow the stable coexistence of alternative reproductive tactics, we do not have general predictions regarding how co-evolution between the sexes will affect the evolution of variation within a sex. Neither female choice nor male behavior will be static even though they are treated this way in most models of alternative reproductive strategies. Instead, reproductive tactics in each sex are the outcome of interactions within and between the sexes. It is clear, however, that female choice has the potential to either suppress or allow the stable coexistence of male alternatives (Alonzo and Warner 2000c, Alonzo and Sinervo 2001) (Figure 18.2, Table 18.1). Furthermore, it is also possible that variation in reproductive tactics in one sex may allow the evolution of alternative tactics in the other sex (Alonzo and Warner 2000c, Alonzo and Sinervo 2001) (Figure 18.2, Table 18.1). More empirical examples are needed where male and female fitness and behavior are well studied within one species so that we can determine how female choice among tactics affects male alternatives and how the existence of male alternatives affects the evolution of female mate choice and other reproductive strategies. We also need co-evolutionary theory based on genetic models examining the evolution of male behavior and female choice concurrently. Finally, further empirical and theoretical research on female alternative reproductive tactics will lead to a greater understanding of female behavior and the evolution and expression of alternative reproductive tactics in general.

Acknowledgments

I thank Jamie Alonzo, Jane Brockmann, Ryan Calsbeek, Alexis Chaine, Tosha Comendant, Alasdair Houston, Bruce Lyon, Marc Mangel, John McNamara, Barry Sinervo, Erik Svensson, Michael Taborsky, Bob Warner, and countless others for the many discussions that have helped to develop the ideas contained here. I was supported by a National Science Foundation Grant IBN-0110506 and by Yale University during this research.

References

Ahlund, M. and Andersson, M. 2001. Brood parasitism: female ducks can double their reproduction. *Nature* 414, 600–601.

Ahnesjo, I., Vincent, A., Alatalo, R., Halliday, T., and Sutherland, W.J. 1992. The role of females in influencing mating patterns. *Behavioral Ecology* 4, 187–189.

Albrecht, D.J. and Johnson, L.S. 2002. Manipulation of offspring sex ratio by second-mated female house wrens. *Proceedings of the Royal Society of London B* 269, 461–465.

Alonzo, S.H. and Sinervo, B. 2001. Mate choice games, context-dependent good genes, and genetic cycles in the side-blotched lizard, *Uta stansburiana*. *Behavioral Ecology and Sociobiology* 49, 176–186.

Alonzo, S.H. and Warner, R.R. 1999. A trade-off generated by sexual conflict: Mediterranean wrasse males refuse present mates to increase future success. *Behavioral Ecology* 10, 105–111.

Alonzo, S.H. and Warner, R.R. 2000a. Allocation to mate guarding or increased sperm production in a Mediterranean wrasse. *American Naturalist* 156, 266–275.

Alonzo, S.H. and Warner, R.R. 2000b. Dynamic games and field experiments examining intra- and inter-sexual conflict: explaining counterintuitive mating behavior in a Mediterranean wrasse, *Symphodus ocellatus*. *Behavioral Ecology* 11, 56–70.

Alonzo, S.H. and Warner, R.R. 2000c. Female choice, conflict between the sexes and the evolution of male alternative reproductive behaviours. *Evolutionary Ecology Research* 2, 149–170.

Alonzo, S.H., Taborsky, M., and Wirtz, P. 2000. Male alternative reproductive behaviours in a Mediterranean wrasse, *Symphodus ocellatus*: evidence from otoliths for multiple life-history pathways. *Evolutionary Ecology Research* 2, 997–1007.

Andersson, M.B. 1994. *Sexual Selection*. Princeton, NJ: Princeton University Press.

Andres, J.A. and Rivera, A.C. 2001. Survival rates in a natural population of the damselfly *Ceriagrion tenellum*: effects of sex and female phenotype. *Ecological Entomology* 26, 341–346.

Andres, J.A., Sanchez-Guillen, R.A., and Rivera, A.C. 2002. Evolution of female colour polymorphism in damselflies: testing the hypotheses. *Animal Behaviour* 63, 677–685.

Arnqvist, G. and Rowe, L. 2002. Antagonstic coevolution between the sexes in a group of insects. *Nature* 415, 787–789.

Austad, S.N. 1984. A classification method of alternative reproductive behaviors and methods for field-testing ESS models. *American Zoologist* 24, 309–319.

Badyaev, A.V. and Hill, G.E. 2002. Paternal care as a conditional strategy: distinct reproductive tactics associated with elaboration of plumage ornamentation in the house finch. *Behavioral Ecology* 13, 591–597.

Badyaev, A.V., Hill, G.E., Beck, M.L., *et al.* 2002. Sex-biased hatching order and adaptive population divergence in a passerine bird. *Science* 295, 316–318.

Bateman, A.J. 1948. Intra-sexual selection in *Drosophila*. *Heredity* 2, 349–368.

Bourke, A. F. G. 2001. Reproductive skew and split sex ratios in social hymenoptera. *Evolution* 55, 2131–2136.

Brockmann, H. J. and Grafen, A. 1989. Mate conflict and male behavior in a solitary wasp, *Trypoxylon (Tryparigilum) politum* (Hymenoptera: Sphecidae). *Animal Behaviour* 37, 232–255.

Brooks, R. 1998. The importance of mate copying and cultural inheritance of mating preferences. *Trends in Ecology and Evolution* 13, 45–46.

Cade, W. 1980. Alternative male reproductive behaviors. *Florida Entomology* 63, 30–45.

Caillaud, M. C., Boutin, M., Braendle, C., and Simon, J. C. 2002. A sex-linked locus controls wing polymorphism in males of the pea aphid, *Acyrthosiphon pisum* (Harris). *Heredity* 89, 346–352.

Calsbeek, R. and Sinervo, B. 2002. Uncoupling direct and indirect components of female choice in the wild. *Proceedings of the National Academy of Sciences of the United States of America* 99, 14897–14902.

Candolin, U. and Reynolds, J. D. 2001. Sexual signaling in the European bitterling: females learn the truth by direct inspection of the resource. *Behavioral Ecology* 12, 407–411.

Candolin, U. and Reynolds, J. D. 2002a. Adjustments of ejaculation rates in response to risk of sperm competition in a fish, the bitterling (*Rhodeus sericeus*). *Proceedings of the Royal Society of London B* 269, 1549–1553.

Candolin, U. and Reynolds, J. D. 2002b. Why do males tolerate sneakers? Tests with the European bitterling, *Rhodeus sericeus*. *Behavioral Ecology and Sociobiology* 51, 146–152.

Chapman, T., Arnqvist, G., Bangham, J., and Rowe, L. 2003. Sexual conflict. *Trends in Ecology and Evolution* 18, 41–47.

Clutton-Brock, T. H. and Parker, G. A. 1992. Potential reproductive rates and the operation of sexual selection. *Quarterly Review of Biology* 67, 437–456.

Coleman, S. W., Patricelli, G. L., and Borgia, G. 2004. Variable female preferences drive complex male displays. *Nature* 428, 742–745.

Comendant, T., Sinervo, B., Svensson, E. I., and Wingfield, J. 2003. Social competition, corticosterone and survival in female lizard morphs. *Journal of Evolutionary Biology* 16, 948–955.

Cordero, A., Carbone, S. S., and Utzeri, C. 1998. Mating opportunities and mating costs are reduced in androchrome female damselflies, *Ischnura elegans* (Odonata). *Animal Behaviour* 55, 185–197.

Crespi, B. J. 1988. Alternative male mating tactics in a thrips: effects of sex ratio variation and body size. *American Midland Naturalist* 119, 83–92.

Darwin, C. 1871. *The Descent of Man, and Selection in Relation to Sex*. London: John Murray.

Davies, N. B. 1986. Reproductive success of dunnocks, *Prunella modularis*, in a variable mating system. 1. Factors influencing provisioning rate, nestling weight and fledgling success. *Journal of Animal Ecology* 55, 123–138.

Davies, N. B. 1989. Sexual conflict and the polygamy threshold. *Animal Behaviour* 38, 226–234.

Davies, N. B. 1992. *Dunnock Behaviour and Social Evolution*. Oxford, UK: Oxford University Press.

Davies, N. B. and Hartley, I. R. 1996. Food patchiness, territory overlap and social systems: an experiment with dunnocks *Prunella modularis*. *Journal of Animal Ecology* 65, 837–846.

Davies, N. B. and Hatchwell, B. J. 1992. The value of male parental care and its influence on reproductive allocation by males and female dunnocks. *Journal of Animal Ecology* 61, 259–272.

Davies, N. B. and Houston, A. I. 1986. Reproductive success of dunnocks, *Prunella modularis*, in a variable mating system. II. Conflicts of interest among breeding adults. *Journal of Animal Ecology* 55, 139–154.

Davies, N. B., Hatchwell, B. J., Robson, T., and Burke, T. 1992. Paternity and parental effort in dunnocks, *Prunella modularis*: how good are male chick feeding rules? *Animal Behaviour* 43, 729–745.

Davies, N. B., Hartley, I. R., Hatchwell, B. J., and Langmore, N. E. 1996. Female control of copulations to maximize male help: a comparison of polygynandrous alpine accentors, *Prunella collaris*, and dunnocks, *P. modularis*. *Animal Behaviour* 51, 27–47.

Dominey, W. J. 1980. Female mimicry in male bluegill sunfish: a genetic polymorphism? *Nature* 284, 546–548.

Dugatkin, L. A. and Godin, J.-G. J. 1992. Reversal of female mate choice by copying in the guppy (*Poecilia reticulata*). *Proceedings in the Royal Society of London B* 249, 179–184.

Eadie, J. M. and Fryxell, J. M. 1992. Density depedence, frequency depedendence, and alternative nesting strategies in goldeneyes. *American Naturalist* 140, 621–641.

Eadie, J. M. and Lyon, B. E. 1998. Cooperation, conflict, and creching behavior in goldeneye ducks. *American Naturalist* 152, 397–408.

Eberhard, W. G. and Cordero, C. 2003. Sexual conflict and female choice. *Trends in Ecology and Evolution* 18, 438–439

Forbes, M. 1994. Tests of hypotheses for female-limited polymorphism in the damselfly, *Enallagma boreal* (Selys). *Animal Behaviour* 47, 724–726.

Forbes, M. R., Schalk, G., Miller, J. G., and Richardson, J. M. L. 1997. Male–female morph interactions in the damselfly *Nehalennia irene* (Hagen). *Canadian Journal of Zoology* 75, 253–260.

Forsyth, A. and Montgomerie, R. D. 1987. Alternative reproductive tactics in the territorial damselfly *Calopteryx maculata*: sneaking by older males. *Behavioral Ecology and Sociobiology* 21, 73–81.

Garant, D., Fontaine, P. M., Good, S. P., Dodson, J. J., and Bernatchez, L. 2002. The influence of male parental identity on growth and survival of offspring in Atlantic salmon (*Salmo salar*). *Evolutionary Ecology Research* 4, 537–549.

Gil, D., Graves, J., Hazon, N., and Wells, A. 1999. Male attractiveness and differential testosterone investment in zebra finch eggs. *Science* 286, 126–128.

Goldschmidt, T., Bakker, T. C., and Bruin, E. F.-D. 1993. Selective copying in mate choice of female sticklebacks. *Animal Behaviour* 45, 541–547.

Gray, E. M. 1996. Female control of offspring paternity in a western population of red-winged blackbirds (*Agelaius phoeniceus*). *Behavioral Ecology and Sociobiology* 38, 267–268.

Gross, M. R. 1982. Sneakers, satellites and parentals: polymorphic mating strategies in North American sunfishes. *Zeitschrift für Tierpsychologie* 60, 1–26.

Gross, M. R. 1984. Sunfish, salmon, and the evolution of alternative reproductive strategies and tactics in fishes. In G. W. Potts and R. J. Wootton (eds.) *Fish Reproduction: Strategies and Tactics*, pp. 55–75. London: Academic Press.

Gross, M. R. 1991. Evolution of alternative reproductive strategies: frequency-dependent selection in male bluegill sunfish. *Philosophical Transactions of the Royal Society of London B* 332, 59–66.

Gross, M. R. and Charnov, E. L. 1980. Alternative male life histories in bluegill sunfish. *Proceedings of the National Academy of Sciences of the United States of America* 11, 6937–6940.

Gwynne, D. T. and Snedden, A. W. 1995. Paternity and female remating in *Requena verticalis* (Orthoptera: Tettigoniidae). *Ecological Entomology* 20, 191–194.

Hammerstein, P. and Parker, G. A. 1987. Sexual selection: games between the sexes. In J. W. Bradbury and M. B. Andersson (eds.) *Sexual Selection: Testing the Alternatives*, pp. 119–142. New York: John Wiley.

Hardling, R. 1999. Arms races, conflict costs and evolutionary dynamics. *Journal of Theoretical Biology* 196, 163–167.

Hartley, I. R. and Davies, N. B. 1994. Limits to cooperative polyandry in birds. *Proceedings of the Royal Society of London B* 257, 67–73.

Hatchwell, B. J. and Davies, N. B. 1990. Provisioning of nestlings by dunnocks, *Prunella modularis*, in pairs and trios: compensation reactions by males and females. *Behavioral Ecology and Sociobiology* 27, 199–210.

Hatchwell, B. J. and Davies, N. B. 1992. An experimental study of mating competition in monogamous and polyandrous dunnocks, *Prunella modularis*. 2. Influence of removal and replacement experiments on mating systems. *Animal Behaviour* 43, 611–622.

Hattori, A. and Yamamura, N. 1995. Co-existence of subadult males and females as alternative tactics of breeding post acquisition in a monogamous and protandrous anemonefish. *Evolutionary Ecology* 9, 292–303.

Henson, S. A. and Warner, R. R. 1997. Male and female alternative reproductive behaviors in fishes: a new approach using intersexual dynamics. *Annual Review of Ecology and Systematics* 28, 571–592.

Houston, A. I., Szekely, T., and McNamara, J. M. 2005. Conflict between parents over care. *Trends in Ecology and Evolution* 20, 33–38.

Hugie, D. M. and Lank, D. B. 1997. The resident's dilemma: a female choice model for the evolution of alternative mating strategies in lekking male ruffs (*Philomachus pugnax*). *Behavioral Ecology* 8, 218–225.

Hunt, J. and Simmons, L. W. 2000. Maternal and paternal effects on offspring phenotype in the dung beetle *Onthophagus taurus*. *Evolution* 54, 936–941.

Hunt, J. and Simmons, L. W. 2001. Status-dependent selection in the dimorphic beetle *Onthophagus taurus*. *Proceedings of the Royal Society of London B* 268, 2409–2414.

Johnston, C. E. 1994. The benefit to some minnows of spawning in the nests of other species. *Environmental Biology of Fishes* 40, 213–218.

Jones, A. G., Walker, D., Kvarnemo, C., Lindstroem, K., and Avise, J. C. 2001. How cuckoldry can decrease the opportunity for sexual selection: data and theory from a genetic parentage analysis of the sand goby, *Pomatoschistus minutus*. *Proceedings of the National Academy of Sciences of the United States of America* 98, 9151–9156.

Kirkpatrick, M. 1982. Sexual selection and the evolution of female choice. *Evolution* 36, 1–12.

Kojola, I. 1998. Sex ratio and maternal investment in ungulates. *Oikos* 83, 567–573.

Komdeur, J. 1994. The effect of kinship on helping in the cooperative breeding Seychelles warbler (*Acrocephalus sechellensis*). *Proceedings of the Royal Society of London B* **256**, 47–52.

Komdeur, J. 1996. Facultative sex ratio bias in the offspring of Seychelles warblers. *Proceedings of the Royal Society of London B* **263**, 661–666.

Komdeur, J., Daan, S., Tinbergen, J., and Mateman, C. 1997. Extreme adaptive modification in sex ratio of the Seychelles warbler's eggs. *Nature* **385**, 522–525.

Kruuk, L. E. B., Clutton-Brock, T. H., Albon, S. D., Pemberton, J. M., and Guinness, F. E. 1999. Population density affects sex ratio variation in red deer. *Nature* **399**, 459–461.

Langmore, N. E. and Davies, N. B. 1997. Female dunnocks use vocalizations to compete for males. *Animal Behaviour* **53**, 881–890.

Langmore, N. E., Cockrem, J. F., and Candy, E. J. 2002. Competition for male reproductive investment elevates testosterone levels in female dunnocks, *Prunella modularis*. *Proceedings of the Royal Society of London B* **269**, 2473–2478.

Lank, D. B. and Smith, C. M. 1987. Conditional lekking in ruff (*Philomachus pugnax*). *Behavioral Ecology and Sociobiology* **20**, 137–145.

Lank, D. B., Smith, C. M., Hanotte, O., Burke, T., and Cooke, F. 1995. Genetic polymorphism for alternative mating behaviour in lekking male ruff *Philomachus pugnax*. *Nature* **378**, 59–62.

Lejeune, P. 1985. Etude écoéthologique des comportements reproducteurs et sociaux des Labridae méditerranéens des genres *Symphodus* (Rafinesque 1810) et *Coris* (Lacepede 1802). *Cahiers d'Ethologie Appliquée* **5**, 1–208.

Luttbeg, B. 2004. Female mate assessment and choice behavior affect the frequency of alternative male mating tactics. *Behavioral Ecology* **15**, 239–247.

Lyon, B. E. 1993. Conspecific brood parasitism as a flexible female reproductive tactic in American coots. *Animal Behaviour* **46**, 911–928.

Lyon, B. E. 1998. Optimal clutch size and conspecific brood parasitism. *Nature* **392**, 380–383.

Lyon, B. E. 2003a. Ecological and social constraints on conspecific brood parasitism by nesting female American coots (*Fulica americana*). *Journal of Animal Ecology* **72**, 47–60.

Lyon, B. E. 2003b. Egg recognition and counting reduce costs of avian conspecific brood parasitism. *Nature* **422**, 495–499.

Lyon, B. E., Hochachka, W. M., and Eadie, J. M. 2002. Paternity–parasitism trade-offs: a model and test of host–parasite cooperation in an avian conspecific brood parasite. *Evolution* **56**, 1253–1266.

Mills, S. C. and Reynolds, J. D. 2002. Host species preferences by bitterling, *Rhodeus sericeus*, spawning in freshwater mussels and consequences for offspring survival. *Animal Behaviour* **63**, 1029–1036.

Moczek, A. P. and Emlen, D. J. 2000. Male horn dimorphism in the scarab beetle, *Onthophagus taurus*: do alternative reproductive tactics favour alternative phenotypes? *Animal Behaviour* **59**, 459–466.

Nee, S. and May, R. M. 1992. Population-level consequences of conspecific brood parasitism in birds and insects. *Journal of Theoretical Biology* **161**, 95–109.

Oring, L. W., Reed, J. M., and Alberico, J. A. R. 1993. Female control of paternity: more than meets the eye. *Trends in Ecology and Evolution* **8**, 259.

Parker, G. A. 1979. Sexual selection and sexual conflict. In M. S. Blum and N. A. Blum (eds.) *Sexual Selection and Reproductive Competition in Insects*, pp. 123–166. New York: Academic Press.

Part, T., Gustafsson, L., and Moreno, J. 1992. "Terminal investment" and sexual conflict in the collared flycatcher (*Ficedula albicollis*). *American Naturalist* **140**, 868–882.

Pen, I., Weissing, F. J., and Daan, S. 1999. Seasonal sex ratio trend in the European kestrel: an evolutionarily stable strategy analysis. *American Naturalist* **153**, 384–397.

Petersen, C. W., Warner, R. R., Shapiro, D. Y., and Marconato, A. 2001. Components of fertilization success in the bluehead wrasse. *Behavioral Ecology* **12**, 237–245.

Pienaar, J. and Greeff, J. M. 2003. Maternal control of offspring sex and male morphology in the *Otitesella* fig wasps. *Journal of Evolutionary Biology* **16**, 244–253.

Qvarnstrom, A. 2001. Context-dependent genetic benefits from mate choice. *Trends in Ecology and Evolution* **16**, 5–7.

Reichard, M., Jurajda, P., and Smith, C. 2004a. Male–male interference competition decreases spawning rate in the European bitterling (*Rhodeus sericeus*). *Behavioral Ecology and Sociobiology* **56**, 34–41.

Reichard, M., Smith, C., and Jordan, W. C. 2004b. Genetic evidence reveals density-dependent mediated success of alternative mating behaviours in the European bitterling (*Rhodeus sericeus*). *Molecular Ecology* **13**, 1569–1578.

Reichard, M., Bryja, J., Ondrackova, M., et al. 2005. Sexual selection for male dominance reduces opportunities for female mate choice in the European bitterling (*Rhodeus sericeus*). *Molecular Ecology* **14**, 1533–1542.

.eillo, P. R. and Wise, D. H. 1988. An experimental evaluation of selection on color morphs of the polymorphic spider *Enoplognatha ovata* (Araneae: Theridiidae). *Evolution* 42, 1172–1189.

.eynolds, J. D., Debuse, V. J., and Aldridge, D. C. 1997. Host specialization in an unusual symbiosis: European bitterlings spawning in freshwater mussels. *Oikos* 78, 539–545.

.off, D. A. 2002. *Life History Evolution.* Sunderland, MA: Sinauer Associates.

.owe, L., Arnqvist, G., Sih, A., and Krupa, J. J. 1994. Sexual conflict and the evolutionary ecology of mating patterns: water striders as a model system. *Trends in Ecology and Evolution* 9, 289–293.

.uppell, O. and Heinze, J. 1999. Alternative reproductive tactics in females: the case of size dimorphism in winged ant queens. *Insectes Sociaux* 46, 6–17.

.herratt, T. N. 2001. The evolution of female-limited polymorphisms in damselflies: a signal detection model. *Ecology Letters* 4, 22–29.

.huster, S. M. 1989. Male alternative reproductive strategies in marine isopod crustacean (*Paracerceis sculpta*): the use of genetic markers to measure differences in fertilization success among alpha, beta and gamma males. *Evolution* 43, 1683–1698.

.huster, S. M. and Sassaman, C. 1997. Genetic interaction between male mating strategy and sex ratio in a marine isopod. *Nature* 388, 373–377.

.huster, S. M. and Wade, M. J. 1991. Equal mating success among male reproductive strategies in a marine isopod. *Nature* 350, 608–610.

.ih, A. and Krupa, J. J. 1992. Predation risk, food deprivation and non-random mating by size in the stream water strider, *Aquarius remigis*. *Behavioral Ecology and Sociobiology* 31, 51–56.

.inervo, B. 1998. Adaptation of maternal effects in the wild: path analysis of natural variation and experimental tests of causation. In T. A. Mousseau, B. Sinervo, and J. Endler (eds.) *Adaptive Genetic Variation in the Wild*, pp. 41–64. New York: Oxford University Press.

.inervo, B. 1999. Mechanistic analysis of natural selection and a refinement of Lack's and Williams's principles. *American Naturalist* 154, S26–S42.

.inervo, B. and Lively, C. M. 1996. The rock–paper–scissors game and the evolution and alternative male strategies. *Nature* 380, 240–243.

.inervo, B. and Zamudio, K. R. 2001. The evolution of alternative reproductive strategies: fitness differential,

heritability, and genetic correlation between the sexes. *Journal of Heredity* 92, 198–205.

Sinervo, B., Svensson, E., and Comendant, T. 2000. Density cycles and an offspring quantity and quality game driven by natural selection. *Nature* 406, 985–988.

Sinervo, B., Bleay, C., and Adamopoulou, C. 2001. Social causes of correlational selection and the resolution of a heritable throat color polymorphism in a lizard. *Evolution* 55, 2040–2052.

Smith, C., Reynolds, J. D., Sutherland, W. J., and Jurajda, P. 2000. Adaptive host choice and avoidance of superparasitism in the spawning decisions of bitterling (*Rhodeus sericeus*). *Behavioral Ecology and Sociobiology* 48, 29–35.

Smith, C., Rippon, K., Douglas, A., and Jurajda, P. 2001. A proximate cue for oviposition site choice in the bitterling (*Rhodeus sericeus*). *Freshwater Biology* 46, 903–911.

Smith, C., Douglas, A., and Jurajda, P. 2002. Sexual conflict, sexual selection and sperm competition in the spawning decisions of bitterling, *Rhodeus sericeus*. *Behavioral Ecology and Sociobiology* 51, 433–439.

Smith, C., Reichard, M., and Jurajda, P. 2003. Assessment of sperm competition by European bitterling, *Rhodeus sericeus*. *Behavioral Ecology and Sociobiology* 53, 206–213.

Soljan, T. 1930. Die Fortpflanzung und das Wachstum von *Crenilabrus ocellatus* (Forskal), einem Lippfisch des Mittelmeeres. *Zeitschrift für wissenschaftliche Zoologie* 137, 156–174.

Sorenson, M. D. 1991. The functional significance of parasitic egg laying and typical nesting in redhead ducks: an analysis of individual behavior. *Animal Behaviour* 42, 771–796.

Sozou, P. D. and Houston, A. I. 1994. Parental effort in a mating system involving two males and two females. *Journal of Theoretical Biology* 171, 251–266.

Stearns, S. C. 1992. *The Evolution of Life Histories.* Oxford, UK: Oxford University Press.

Svensson, E. and Nilsson, J.-A. 1996. Mate quality affects offspring sex ratio in blue tits. *Proceedings of the Royal Society of London B* 263, 357–361.

Svensson, E. I., Abbott, J., and Hardling, R. 2005. Female polymorphism, frequency dependence, and rapid evolutionary dynamics in natural populations. *American Naturalist* 165, 567–576.

Svensson, E. I., Sinervo, B., and Comendant, T. 2002. Mechanistic and experimental analysis of condition and reproduction in a polymorphic lizard. *Journal of Evolutionary Biology* 15, 1034–1047.

Taborsky, M. 1994. Sneakers, satellites, and helpers: parasitic and cooperative behavior in fish reproduction. *Advances in the Study of Behavior* 23, 1–100.

Taborsky, M. 1997. Bourgeois and parasitic tactics: do we need collective, functional terms for alternative reproductive behaviours? *Behavioral Ecology and Sociobiology* 41, 361–362.

Taborsky, M., Hudde, B., and Wirtz, P. 1987. Reproductive behavior and ecology of *Symphodus ocellatus:* a European wrasse with four types of male behavior. *Behaviour* 102, 82–118.

Tomkins, J. L. and Simmons, L. W. 2000. Sperm competition games played by dimorphic male beetles: fertilization gains with equal mating access. *Proceedings of the Royal Society of London B* 267, 1547–1553.

Trivers, R. L. 1972. Parental investment and sexual selection. In B. Campbell (ed.) *Sexual Selection and the Descent of Man*, pp. 136–179. Chicago, IL: Aldine Press.

Van den Berghe, E. 1990. Variable parental care in a labrid fish: how care might evolve. *Ethology* 84, 319–333.

Van den Berghe, E. P., Wernerus, F., and Warner, R. R. 1989. Female choice and the mating cost of peripheral males. *Animal Behaviour* 38, 875–884.

Van Gossum, H., Stoks, R., and De Bruyn, L. 2001. Frequency-dependent male mate harrassment and intra-specific variation in its avoidance by females of the damselfly *Ischnura elegans. Behavioral Ecology and Sociobiology* 51, 69–75.

Van Gossum, H., Stoks, R., Matthysen, E., Valck, F. and De Bruyn, L. 1999. Male choice for female colour morphs in *Ischnura elegans* (Odonata, Coenagrionidae): testing the hypotheses. *Animal Behaviour* 57, 1229–1232.

Veiga, J. P. 1990. Sexual conflict in the house sparrow: interference between polygynously mated females versus asymmetric male investment. *Behavioral Ecology and Sociobiology* 27, 345–350.

Warner, R. R. 1997. Sperm allocation in coral reef fishes: strategies for coping with demands on sperm production. *BioScience* 47, 561–564.

Warner, R. R., Shapiro, D. Y., Marcanato, A., and Petersen, C. W. 1995a. Sexual conflict: males with highest mating success convey the lowest fertilization benefits to females. *Proceedings of the Royal Society of London B* 262, 135–139.

Warner, R. R., Wernerus, F., Lejeune, P., and Van den Berghe, E. 1995b. Dynamics of female choice for parental care in a fish species where care is facultative. *Behavioral Ecology* 6, 73–81.

Wedell, N., Gage, M. J. G., and Parker, G. A. 2002. Sperm competition, male prudence and sperm-limited females. *Trends in Ecology and Evolution* 17, 313–320.

Westneat, D. F., Walters, A., McCarthy, T. M., Hatch, M. I. and Hein, W. K. 2000. Alternative mechanisms of nonindependent mate choice. *Animal Behaviour* 59, 467–476.

Widemo, F. 1998. Alternative reproductive strategies in the ruff, *Philomachus pugnax:* a mixed ESS? *Animal Behaviour* 56, 329–336.

Widemo, F. and Owens, I. P. F. 1995. Lek size, male mating skew and the evolution of lekking. *Nature* 373, 148–151.

Zamudio, K. R. and Sinervo, B. 2000. Polygyny, mate-guarding, and posthumous fertilization as alternative male mating strategies. *Proceedings of the National Academy of Sciences of the United States of America* 97, 14427–14432.

19 · Cooperative breeding as an alternative reproductive tactic

WALTER D. KOENIG AND JANIS L. DICKINSON

CHAPTER SUMMARY

Cooperative breeding, in which more than a pair of individuals cooperates to produce young, is found in a small proportion of birds, mammals, and fishes. Cooperative breeding encompasses a variety of alternative tactics. Some of these, such as mate sharing by males and joint nesting by females, explicitly concern reproduction, while others, such as cooperative courtship and helping at the nest by nonbreeders, involve activities leading to indirect reproduction through kin. In addition to these behavior patterns, cooperative breeding systems less commonly exhibit traditional alternative reproductive tactics (ARTs), including sneaking by males and parasitic egg laying by females. In virtually all cases, cooperative breeding behavior is conditional, facultative, and potentially frequency dependent, at least within groups. However, few attempts have been made to understand the expression or diversity of reproductive tactics observed in cooperative breeding systems using the theoretical framework provided by ART theory. Viewing alternatives as ARTs may help to clarify the selective forces that promote helping at the nest and mate sharing and help to explain the infrequent occurrence of parasitic reproductive behavior patterns in cooperative breeders.

9.1 INTRODUCTION

Cooperative breeding is a phenomenon in which more than a pair of individuals shares the tasks of producing young in a single nest or litter. It is known to occur in about 3% of birds and mammals (Brown 1987, Arnold and Owens 1998, Russell 2004), as well as some fishes (Taborsky 1994, 2001), but may be considerably more frequent given that the mating systems of many species remain to be determined (Cockburn 2003). We will restrict our discussion to these vertebrate taxa. However, cooperative breeding can be considered a continuum culminating in eusociality (Sherman et al. 1995)

suggesting that the ideas we discuss might also be applicable to social bees and wasps (Strassman et al. 1994, Crespi and Yanega 1995, Sherman et al. 1995, Brockmann 1997).

Cooperative breeding is complex and includes a diversity of behavior, even when restricted to vertebrates. Most commonly, it involves one or more offspring that remain with their parents as "helpers at the nest." Such helpers typically feed young but do not participate in direct reproduction due to a combination of incest avoidance (Koenig et al. 1998, Koenig and Haydock 2004) and social dominance (Keller and Reeve 1994). Instead, they generally gain kin-selected, inclusive fitness benefits by helping to raise younger siblings (Dickinson 2004b, Dickinson and Hatchwell 2004). In these cases helping is clearly an alternative behavioral tactic, but since helpers are not actually reproducing, it is debatable whether one can consider this behavior an alternative *reproductive* tactic, unless behavior enhancing indirect fitness is included in this category. In other species, however, helpers exhibit behavior patterns that lead to successful breeding, sometimes coincident with helping, including parasitizing reproduction by dominants within their home group (when incest can be avoided) and engaging in extra-group mating attempts outside their home group (Double and Cockburn 2003, Cockburn 2004). In such cases helpers are clearly engaging in alternative reproductive tactics (ARTs), of which helping at the nest may be considered part of the overall strategy. Here we take the approach that it is worthwhile to examine helping at the nest in the context of traditional ART approaches, even when personal reproduction is not part of the tactic. Our rationale for this is, in part, because of our belief that the fitness consequences of ARTs, like other social behavior patterns, should be measured in terms of both direct and indirect benefits.

Beyond standard helping at the nest, there are a variety of mating systems that can be considered under the rubric of cooperative breeding. To the extent that these alternative

Alternative Reproductive Tactics, ed. Rui F. Oliveira, Michael Taborsky, and H. Jane Brockmann. Published by Cambridge University Press. © Cambridge University Press 2008.

mating systems co-occur with alternatives within the same population, they clearly fit the definition of an ART. These include mate sharing (or cobreeding) by males and joint nesting (or true communal nesting) by females (Faaborg and Patterson 1981, Hartley and Davies 1994, Vehrencamp and Quinn 2004). They also include cooperative courtship, a behavior found in a few avian taxa including manakins (family Pipridae) (McDonald 1989, McDonald and Potts 1994), and the wild turkey *Meleagris gallopavo* (Watts and Stokes 1971, Krakauer 2005). Although the latter phenomenon does not extend to raising young, these behavior patterns all unambiguously involve reproduction, can entail considerable intrasexual conflict, and are often highly variable within a population – some individuals may share mates or display cooperatively while others may not.

Despite the plethora of reproductive alternatives found in cooperative breeders, this phenomenon has rarely been discussed within the context of ART theory. Here we do this with the goal of focusing attention on questions that are generally overlooked with more traditional approaches.

19.2 "HELPING AT THE NEST" AS AN ART

The extremes of behavior among helpers at the nest are represented by the superb fairy-wren *Malurus cyaneus* in Australia and the acorn woodpecker *Melanerpes formicivorus* in California. In the superb fairy-wren virtually all males, regardless of status within their home groups, attempt to engage in extra-group matings with the result that a majority of successful fertilizations are extra-group in origin (Brooker et al. 1990, Mulder et al. 1994, Dunn and Cockburn 1999, Double and Cockburn 2003). In such a case, the very concept of a "helper" is ambiguous to the extent that virtually all males end up feeding young to which they are not related (Dunn et al. 1995). In contrast, in acorn woodpeckers, no extra-group mating occurs and incest avoidance generally constrains helpers from attempting to reproduce within their home group (Koenig et al. 1998, 1999, Koenig and Haydock 2004). Helpers in this case are nonbreeders, but their behavior generates indirect fitness benefits that are clearly an alternative to breeding independently (Dickinson 2004b, Dickinson and Hatchwell 2004).

19.2.1 Fitness consequences of helping by nonbreeders

By definition, nonbreeding helpers do not achieve fitness benefits through direct reproduction. Such helpers are most

frequently offspring of the breeders, and are often known or presumed to be refraining from reproduction because of incest avoidance (Koenig and Haydock 2004). Such species constitute the most common type of cooperative breeding system. Nonbreeding helpers are often primarily males, but in some cooperative species, they can be either males or females, and in a few, including the Seychelles warbler *Acrocephalus sechellensis* (Komdeur 1994, 1996), white-throated magpie-jay *Calocitta formosa* (Innes and Johnston 1996), and American crow *Corvus brachyrhynchos* (Caffrey 1992), helpers are mostly or exclusively female.

A good example of a species with nonbreeding helpers of both sexes is the Florida scrub-jay *Aphelocoma coerulescens*. This well-studied species lives in year-round territories and is socially and genetically monogamous (Quinn et al. 1999). About half of the groups consist of a breeding pair only, while the other half have one to eight nonbreeding helpers, both males and females that are generally offspring of the breeding pair (Woolfenden and Fitzpatrick 1984). Helpers have been demonstrated experimentally to increase the productivity of groups, primarily by decreasing predation on nestlings and increasing fledgling survival (Mumme 1992). Helpers thus gain indirect fitness benefits by aiding in the reproduction of relatives. They potentially achieve other fitness benefits as well, including an increased probability of inheriting part of their natal territory (Woolfenden and Fitzpatrick 1986).

It is clear that the fitness benefits of being a Florida scrub-jay breeder far outweigh those of being a helper. Emlen (1978), for example, estimated that the annual inclusive fitness achieved by a helper was 0.16 offspring equivalents (OEs) compared to a minimum of 0.51 offspring produced by independent pairs. Similar conclusions have been reached in other species with offspring that remain as nonbreeding helpers, even though details of the mating systems differ. For example, in the western bluebird *Sialia mexicana*, a species with occasional male helpers in which extra-pair matings are relatively common (Dickinson and Akre 1998), the estimated annual inclusive fitness of sons that help was only 0.41 OEs compared to 3.02 OEs for sons that breed (Dickinson 2004a). Helping at the nest can thus be considered an example of an alternative behavioral (if not reproductive) tactic in which the fitness pay-off of one alternative is clearly inferior to the other. Faced with such inequalities, it is unsurprising that helping in these species is generally considered a default strategy pursued by individuals that are unable to attain breeding status and that helpers are usually believed to be making the best of a bad job (Dickinson and Hatchwell 2004).

19.2.2   Conditionality and plasticity

Like many, if not most, alternative tactics, helping is a conditional strategy (Gross 1996). This does not imply the lack of a genetic component to the expression of cooperative breeding. Although such data are rare, there is evidence in pinyon jays *Gymnorhinus cyanocephalus* that helping tends to occur in certain lineages (Marzluff and Balda 1990), and a similar genetic basis most likely forms the foundation for helping in other cooperative breeders. However, there is no evidence from any cooperative breeding species that the tactics expressed represent underlying genetic differences among individuals (Austad 1984).

Helping is typically age related, with individuals helping when young and switching as they become older or as the opportunity arises. However, individuals do sometimes switch back and forth between the two strategies, depending on ecological factors and on their own individual success at breeding. For example, all helpers in long-tailed tits *Aegithalos caudatus* and many in Galápagos mockingbirds *Nesomimus parvulus*, western bluebirds, and white-fronted bee-eaters *Merops bullockoides* are failed breeders that return home after their own nests have failed to help at the nest of their parent or other close relative. Other Galápagos mockingbirds help after fledging a successful nest of their own, and some western bluebird sons do so while simultaneously breeding themselves (Curry and Grant 1990, Emlen 1990, Dickinson et al. 1996, Dickinson and Akre 1998, Hatchwell et al. 2001). Such examples are of interest because they clearly illustrate ways in which helping is an alternative behavioral tactic that can be employed whether or not helpers are engaged in personal reproduction.

The facultative nature of most helping behavior is clearly demonstrated by its plasticity, facilitated by a lack of clear morphological or even substantive physiological differences between helpers and breeders apart from a correlation between status and age. The plastic nature of helping and cooperative breeding even appears to hold across populations, a finding recently demonstrated in carrion crows *Corvus corone*, where at least some offspring transplanted from a noncooperatively breeding population in Switzerland to a cooperatively breeding population in Spain delayed dispersal and became helpers (Baglione et al. 2002).

Most work studying potential physiological differences between helpers and breeders has been endocrinological. In red-cockaded woodpeckers *Picoides borealis*, for example, where young males commonly help their parents, there are no differences in plasma testosterone (T) levels between male breeders and helpers, with T levels peaking during the copulation period (Khan et al. 2001). This could be indicative of the possibility that helpers are pursuing some other ART and potentially mating outside their home group. If so, however, they appear never to be successful, as pairs are genetically monogamous (Haig et al. 1994).

In the cases in which reproductive hormones have been found to be lower in helpers compared to breeders, differences can often be attributed to dominance rather than to reproductive incompetence or physiological differences between breeders and helpers. For example, in superb fairy-wrens, T levels are higher in dominants than subordinates but there are no intrinsic differences between breeders and helpers, as levels in helpers are similar to levels in breeders from groups without helpers (Peters et al. 2001). Rather than distinguishing helpers, such differences are more plausibly consistent with the "challenge hypothesis" (Wingfield et al. 1990), which suggests that the challenges dominant males experience as they interact aggressively with same-sex group members cause the observed elevated T levels. A similar idea was proposed by Reyer et al. (1986) to explain the differences in T levels found between helper pied kingfishers *Ceryl rudis* that are related to the breeders and thus, because of incest avoidance, not competing with dominants for matings, and those that are unrelated and thus potential reproductive rivals.

In summary, hormonal differences distinguishing helpers have been observed in some cases, but these are generally believed to be a consequence of behavioral differences rather than a cause of those differences. In short, helpers do not appear to differ from breeders endocrinologically in ways that present a substantive block to potential reproduction (Schoech et al. 2004).

19.2.3   Frequency dependence

As with many ARTs, helping at the nest is generally frequency dependent. Frequency dependence can be examined on at least three levels. First, the incidence of helping is in many cases density dependent with higher frequencies of helping in denser populations (Koenig and Mumme 1987, Emlen 1991) or, alternatively, in years when a particular population is larger (Woolfenden and Fitzpatrick 1984, Koenig and Mumme 1987). This suggests that the ecological factors resulting in helping behavior are sensitive to population density; that is, as population density increases, there are fewer breeding opportunities and a greater proportion of the population settles for helping. An unambiguous example is provided by the experiments in which groups of the cooperatively breeding Seychelles warbler were transplanted to an uninhabited island where, at low density and in the sudden absence of competition, all birds bred independently and helping did not occur (Komdeur 1992).

Second is frequency dependence of helping within a population. Unfortunately, since the incidence of non-breeding helpers is almost always density dependent, it is difficult to quantify the potential effect of frequency on the fitness of helpers independent of population density. For the most part, however, there is no reason to expect that the fitness benefits of helping are affected by its frequency in the population as a whole, except to the extent that helpers are potentially pursuing direct reproductive tactics leading to extra-pair fertilizations (see below). Instead, in most cases the occurrence of this behavior reflects variation in individual quality combined with spatial and temporal variation in the ecological conditions that affect the ability of individuals to pursue the superior behavioral alternative of independent reproduction (Brockmann 2001, Dickinson and Hatchwell 2004).

The third way that frequency dependence of helping can be examined is within social groups. In particular, the fitness benefits of helping are typically inversely related to its frequency within a group. For example, in Florida scrub-jays, annual fledgling production is greater among groups containing at least one helper compared to pairs, but there are no differences in success among groups with one helper compared to those with more than one helper (Woolfenden and Fitzpatrick 1984, 1990). Consequently, virtually all the fitness benefits to be gained by helping are accrued by the first helper. Additional helpers beyond the first usually confer little or no benefit to group survival and reproduction, although variations exist, such as in the stripe-backed wren *Campylorhynchus nuchalis*, where it takes two helpers for a group to outperform groups with no helpers, but additional helpers beyond two confer little or no apparent benefit (Rabenold 1990).

Frequency dependence within groups provides a sound basis for considering the phenomenon of cooperative breeding using a modeling approach similar to that applied to many ARTs. Consider first how fitness is related to frequency, or number, of helpers. At low frequency (e.g., a single helper), the indirect fitness benefits of helping are fairly high, although not nearly as great as the benefits of breeding (Figure 19.1A). However, with few, if any, additional indirect fitness benefits gained by additional helpers, their fitness decreases more or less inversely with their

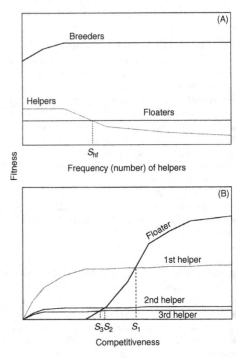

Figure 19.1 A simple model viewing helping-at-the-nest as an ART. (A) Fitness of breeders, helpers, and floaters as a function of helper frequency. The fitness of floaters is independent of helping frequency, while breeders gain fitness with up to one helper, additional helpers provide little or no fitness benefit. Helpers gain inclusive fitness benefits that decline as they are divided among an increased frequency (number) of helpers. (B) Fitness of helpers and floaters as a function of condition (competitive ability). Helping is assumed to be the best strategy for young, uncompetitive individuals, while the potential advantages to floating increase as individuals get older and more competitive. However, the major benefits of helping are achieved by the first helper, while additional helpers achieve little fitness regardless of their competitive ability. Consequently, the $(n+1)$th potential helper should switch to become a floater sooner and at a younger age than the nth helper.

frequency once it surpasses a single helper. At some point (S_{hf}), the fitness curve crosses that of leaving the group and becoming a floater. At that point, individuals faced with the alternative of staying to help or leaving the group should opt for the latter, even when no breeding opportunity is available. If a breeding opportunity is available, breeding is always the superior strategy.

Second, consider how condition, as it relates to ompetitiveness, might affect the propensity for an indi-idual to gain an independent breeding position rather than ecome the nth helper in their home group (Figure 19.1B). ighly competitive individuals, which are older or in good ondition, will be better off leaving the group and floating. ess competitive individuals, however, will achieve higher tness by remaining as a helper, with the switch point for ie sole helper in a group at S_1. Because of the frequency-ependent fitness benefits of helping, however, individuals iould only remain as successive helpers in a group if they 'e of even poorer competitive ability, switching at the ever-wer point S_n to become the nth helper in the group.

To the extent that this kind of model accurately reflects ie relationship between condition and the fitness pay-offs f breeding versus helping, it should be possible to predict ie tactic pursued by offspring based not just on the xternal spatial and temporal factors influencing availability f breeding opportunities, but also on the basis of individual uality, condition, or (suitably modified) other factors that ıay influence fitness of nondispersing offspring. Kokko and kman (2002), for example, explored a formal version of ıch a model tailored to the biology of the Siberian jay erisoreus infaustus, a species in which offspring stay at home ıt do not help. Based on the model, they were able to iccessfully fit observed group sizes in this species and ɔnfirm the importance to offspring of using the natal ter-tory as a "safe haven" while waiting for breeding oppor-ınities. In species with helpers, the benefits of staying ıme and helping may include the importance of family ɔup living and family-owned resources outside the 'eeding season, which would not show up in an analysis of 'eeding effects (Dickinson and McGowan 2005).

In general, however, applying this sort of model to 'edict individual behavior in cooperative breeders will be fficult, both because of the problems of measuring con-tion independent of age and because of the many ways that e fitness consequences of the decisions made by different dividuals can influence the fitness of others. Most obvi-ısly, the decision by an individual to remain as a second :lper influences the fitness of all other individuals, cluding that of the first helper. There are potentially other ss obvious effects as well. For example, helpers in a few ecies, such as acorn woodpeckers, form coalitions whose embers are more competitive in obtaining reproductive ıportunities than individuals are by themselves (Hannon al. 1985). This functionally increases the competitiveness ' multiple helpers and decreases the switch point for

helping beyond the first (S_2, S_3, etc.). Consequently, one would predict that groups are unlikely to contain multiple helpers unless the helpers are of such poor quality that they are unable to form competitive coalitions. Coalition for-mation is also found in other contexts, most notably among male African lions (Bygott et al. 1979), which are similarly more competitive for gaining control of female prides than singletons.

A third potential factor affecting the fitness pay-offs is the frequency of extra-pair paternity. In the cooperatively breeding western bluebird, extra-pair fertilizations are fairly common, with 19% of the offspring in 45% of the nests sired by males other than the social mate (Dickinson and Akre 1998). Young males sometimes breed and help sim-ultaneously, moving between their own nest and that of their parents, which is often next door. The fitness benefits of these alternatives are dependent, in part, on the incidence of extra-pair paternity in both the helper's own nest and the nest of his parents where he may potentially help. How these factors interact to influence the effort birds should devote to the behavioral alternatives of breeding versus helping was explored by Dickinson et al. (1996), who cal-culated that birds should, under most circumstances, only help if they have less than complete confidence of paternity at their own nests (Figure 19.2). Such factors are also likely to play a role in the behavior of other species in which extra-group parentage is common, such as the Australian fairy-wrens (Dunn et al. 1995, Dunn and Cockburn 1996, Double and Cockburn 2003).

19.2.4   Within-group reproduction by helpers

In many cooperative breeders, individuals that would otherwise be nonbreeding helpers can "inherit" and become breeders in their home group. In most, although not all cases, inheritance is constrained by incest avoidance and thus only occurs after the breeder of the opposite sex, usually the parent, has died and been replaced by a new, unrelated individual (Koenig and Haydock 2004). Examples include superb fairy-wrens (Dunn et al. 1995), stripe-backed wrens (Rabenold et al. 1990, Piper and Slater 1993), acorn woodpeckers (Koenig et al. 1998, Koenig et al. 1999), dwarf mongooses Helogale parvula (Rood 1990), meerkats Suricata suricatta (O'Riain et al. 2000), and Damaraland mole-rats Cryptomys damarensis (Bennett et al. 1996, Cooney and Bennett 2000). In some cases, inheritance of breeding status is one way in which same-sex individuals form breeding coalitions that compete for and

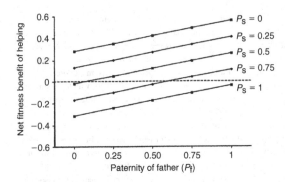

Figure 19.2 Estimates of the fitness benefits potentially gained by a male western bluebird that has the choice of helping at his own nest or helping at his father's nest as a function of the proportion of young the bird sires in his own nest (P_s) and the proportion of young his father sired at his nest (*y*-axis, P_f). Note that if the son subsequently share breeding opportunities. In acorn sires no offspring in his own nest then he should always help at his father's nest, regardless of the latter's share of paternity, while if the son can expect to sire all offspring in his own nest he should not help even if his father achieves 100% paternity. (From Dickinson *et al.* 1996, with permission of Oxford University Press

subsequently share breeding opportunities. In acorn woodpeckers, for example, offspring of either sex may become a cobreeder within a group through inheritance, cooperative dispersal with same-sex siblings, or by joining a same-sex relative (usually a sibling) that previously dispersed to another group (Koenig *et al.* 1998).

Fishes exhibit an even greater diversity of tactics leading to within-group reproduction (Taborsky 1994, 2001). Besides territorial inheritance when dominant breeders are removed (Balshine-Earn *et al.* 1998), small males of some species exhibit traditional "satellite" or "sneaker" behavior, living adjacent to a dominant "bourgeois" male and parasitizing his reproductive effort without being explicitly tolerated. In others, satellite males are accepted, at least to some extent, by the dominant male (Taborsky *et al.* 1987, Martin and Taborsky 1997, Oliveira *et al.* 2002). In most of these cases, kinship appears to be low between the dominant and satellite males, and thus when the latter are tolerated, it is presumably because of the reciprocal fitness benefits they offer to dominants. A well-studied example is that of the cichlid *Neolamprologus brichardi* in which young often remain on their natal territory long after their parents have died and eventually become reproductively mature helpers sharing the territory with a dominant, unrelated pair of breeders (Taborsky 1984, 1985). Such helpers act furtively and sneak fertilizations from the dominant male, successfully gaining direct reproduction and protection against predation in apparent exchange for increasing the productivity of the group (Taborsky 2001).

Although the origins of these ARTs differ among taxa, they are all mechanisms by which helpers share reproduction within a social unit. Such cobreeding (also known as joint nesting when it involves females nesting together) is discussed in greater detail in Section 19.3.

19.2.5 Extra-pair and extra-group reproduction by males

In a few species, nonbreeding helpers within their home group also pursue strategies potentially leading to extra-pair matings with individuals outside their home group. Helpers in these species are unambiguously pursuing the traditional ART of attempting to circumvent reproductive monopolization by dominants through sneaking, while simultaneously engaging in the less common ART of helping at the nest where they gain indirect fitness benefits.

In the plural-breeding Mexican jay *Aphelocoma ultramarina*, groups consist of multiple pairs of birds that build nests independently along with offspring from prior breeding attempts that remain as potential helpers within the group (Brown and Brown 1990). With both retention of young and dispersal among social groups, kinship relationships within groups are complex and ever changing. Pairs build nests independently, but unpaired males pursue paired females within groups and account for a high proportion (40%) of overall parentage through extra-pair matings (Li and Brown 2000). These unpaired males are more likely to care for nestlings if they have sired young in

he brood, suggesting that females engage in extra-pair
matings primarily to gain additional paternal care (Li and
Brown 2002).

A second example is that of the Australian fairy-wrens
genus *Malurus*). In the particularly well-studied superb
fairy-wren, for example, social groups are variable but frequently consist of a social pair along with one or more
subordinate male offspring. However, in contrast to other
helper systems, extra-group matings are not just common,
but constitute the majority of fertilizations, with over 65%
of the offspring sired by males outside the social unit
(Brooker *et al.* 1990, Mulder *et al.* 1994). Given this situation, all birds in the population, including dominant males,
dominant females, and subordinate ("helper") males, can be
considered to be at least potentially pursuing the ART of
mating outside their social group. Extra-pair fertilizations
appear to take place primarily before dawn when females
foray into other territories seeking out older, high-quality
males that molt earlier and remain in breeding plumage
longer (Dunn and Cockburn 1999, Green *et al.* 2000).
Meanwhile, younger, unrelated males that are not likely
to achieve extra-pair matings appear to provide paternal
care in return for some reproductive access to their social
mates (Dunn and Cockburn 1996, Cockburn *et al.* 2003).
However, males helping at nests of dominant males do gain
extra-group matings, apparently by acting as satellites and
copulating with females that visit the dominant male during
predawn forays (Double and Cockburn 2000, 2003).

This extraordinary system appears to entail at least four
partly overlapping ARTs that are primarily condition
dependent, but also influenced, in part, by group compos-
ion (Box 19.1). First is the "stud" ART of being highly
attractive to extra-group females and achieving many extra-
group matings. This tactic is only available to older, high-
quality males that are able to molt early (Peters 2000, Peters
et al. 2000). Second is the "steady" territorial tactic of
achieving within-pair matings with a social mate. In contrast
to most other species, this appears to be a "best-of-a-bad-
job" alternative pursued by males that are younger and
otherwise unattractive to foraging females. Third is parasitizing stud males, a tactic that can be successfully pursued
by subordinate helpers fortunate enough to be in a social
group with an old, highly attractive dominant male. Fourth
is the tactic of gaining kin-selected indirect fitness benefits
by helping raise related offspring, an ART pursued by
younger males that are living in a social unit with their
mothers and otherwise unable to achieve extra-group matings. Determining the fitness costs and benefits of these

Box 19.1   Superb fairy-wrens

Male superb fairy-wrens, a spectacularly plumaged,
small (<14 cm in length) species common in eastern
Australia (Figure 19.3), exhibit one of the most complex
sets of alternative reproductive tactics of any known
male vertebrate (Dunn and Cockburn 1996, 1999,
Cockburn *et al.* 2003). Particularly attractive males are
able to hold territories while simultaneously obtaining
considerable fitness by means of extra-group matings.
Other males, unsuccessful at such extracurricular
courtship, maintain territories and gain direct fitness by
mating with their social mates. Unattractive territorial
males and young males still living with their mother may
end up as helpers raising what in the former case are
unrelated offspring. However, sometimes such males
may act as satellites mating parasitically with females
drawn to their group due of the presence of a particularly attractive dominant male (Table 19.1).

Figure 19.3 Superb fairy-wren, *Malurus cyancus*. (Photograph
© 2005 Ros Runciman and Yeranda Images, used with
permission.)

options, together with the points at which birds would be
expected to switch from one tactic to another, is clearly a
challenge.

19.2.6   Intraspecific brood parasitism
by females

In some cooperative breeding systems, such as those found
in rails (Rallidae), females engage in ARTs as diverse as
those used by male superb fairy-wrens. Three species
have been studied in detail (common moorhen *Gallinula*

Table 19.1. *Summary of male ARTs in superb fairy-wrens*

	Stud	Steady	Parasite	Helper
Characteristics	Attractive territorial dominants	Nonattractive territorial dominants	Territorial subordinates	Offspring
Offspring of breeder female?	No	No	No	Yes
Obtains within-group matings?	Yes	Yes	Sometimes	No
Obtains extra-group matings?	Yes	No	Yes	No
Acts as satellite?	No	No	Yes	No
Acts as helper?	No	No	Yes	Yes
Primary fitness gains	Direct	Direct	Indirect	Indirect

Box 19.2   The common moorhen

Female common moorhens, a rail found throughout much of the Americas as well as Europe, Asia, Africa, and Australasia, are the analog of male superb fairy-wrens in terms of exhibiting a wide range of alternative reproductive tactics (McRae 1996, 1998) (Figure 19.4). Some females remain in their natal group as subordinates and act as nonbreeding helpers, others become nonterritorial floaters that lay eggs parasitically in other females' nests, while still others remain in their natal group as subordinates nesting jointly with their mother and breeding incestuously with their father (Table 19.2). Dominant, territorial females lay eggs in their own group and in addition sometimes lay eggs parasitically in the nests of other females.

Figure 19.4 Common moorhen, *Gallinula chloropus*.
(Photograph © 2005 Richard Ditch, used with permission.)

chloropus, pukeko *Porphyrio porphyrio*, and Tasmanian native hen *Gallinula mortierii*), the most complex of which appears to be the common moorhen (Box 19.2). In addition to joint nesting (see Section 19.3), over 25% of breeding females of this species parasitize a neighbor's nest. The majority of these females lay eggs either before initiating their own nest or following predation. However, at least some females are nonterritorial floaters that attempt to circumvent the reproductive monopolies of territorial females by parasitism (McRae 1998). Once they have started laying eggs in their own nests, parasitized females generally accept foreign eggs (McRae 1995). Although the reasons for host tolerance are not entirely clear, quasi-parasitism, in which the parasitic female mates with the host's mate, is not involved. However, hosts are often genetic relatives and thus kin selection may play a role (McRae and Burke 1996).

Common moorhens also lay eggs communally and young birds often remain in their natal group, sometimes as nonbreeding helpers, but more commonly and extraordinarily as incestuous cobreeders committing father–daughter incest and laying eggs jointly with their mother (McRae 1996). Thus, female ARTs in this system include (1) being a nonbreeding helper, presumably gaining indirect fitness benefits by helping to raise younger siblings; (2) becoming a floater and laying eggs parasitically in the nests of other females; (3) mating incestuously in the natal group and laying eggs jointly with the mother; (4) becoming a dominant territorial female in laying eggs in a group; and (5) laying eggs parasitically in another female's nest while being a breeder in a territorial group.

Of these tactics, helping by nonbreeders (1), joint nesting (3, although not necessarily incestuously), and being a dominant breeder (4) have also been reported in pukeko (Craig and Jamieson 1990, Jamieson *et al.* 1994).

Table 19.2. *Summary of female ARTs in common moorhens*

	Dominant	Incestuous	Floaters	Helper
Characteristics	Territorial	Subordinate offspring	Nonterritorial	Offspring
Lays eggs in home nest?	Yes	Yes	–	No
Lays parasitic eggs in other nests?	Yes	No	Yes	No
Acts as helper?	No	No	No	Yes
Primary fitness gains	Direct	Direct/indirect	Direct	Indirect

Lambert *et al.* 1994, Jamieson 1997) and Tasmanian native hens (Goldizen *et al.* 1998, 2000). Intraspecific brood parasitism in these species has not been confirmed, but it would not be surprising if it were eventually found. The perspective of ARTs can help define the data required to sort out the conditional and ecological factors that influence the range and choice of tactics observed within these populations.

19.2.7 Additional considerations

Helping at the nest is a diverse phenomenon encompassing a variety of highly plastic, potentially condition-dependent tactics yielding either indirect fitness benefits, direct reproductive success, or both. In most cases where helpers gain only indirect fitness benefits but do not breed or achieve direct reproduction within their social groups, they are cooperating with the dominant resource holder. In cases where helpers parasitize the reproductive effort of dominant males (such as in the fish *Neolamprologus pulcher*), mate with paired females (as in the Mexican jay), or parasitize the attractiveness of dominant stud males (as in the superb fairy-wren), the satellites or helpers are sneaking in order to evade the monopoly on reproduction that dominant resource-holding males would otherwise achieve.

In these latter cases, helpers use unambiguous subordinate ARTs to achieve some degree of reproduction, though their fitness outcome is lower than that of older, resource-holding, dominant males. However, these cases grade into those where subordinates appear to cooperate with dominants to achieve some measure of direct reproduction. This continuum is most clearly shown by cooperatively breeding fishes, where subordinate, satellite males vary from those that parasitize dominant bourgeois males that exclude the satellites from their territories whenever possible to those that apparently cooperate with and are tolerated by bourgeois males despite achieving some measure of paternity

(Taborsky 2001). These cases are discussed further below, where we focus explicitly on mate sharing and reproductive sharing in cooperative breeders.

19.3 COOPERATIVE POLYGAMY AND MATE SHARING AS ARTs

As discussed above, helpers in some cooperative breeding systems inherit and achieve direct reproduction within their home group, often sharing breeding status and mates with older breeders of the same sex. In others, however, individuals form same-sex coalitions that compete for reproduction with other such coalitions to obtain breeding positions and then, once they succeed, compete among themselves for reproduction. Coalitions may only compete for matings, as in several species of manakins (genus *Chiroxiphia*) (Foster 1981, McDonald and Potts 1994) and wild turkeys (Watts and Stokes 1971, Krakauer 2005). More commonly, coalitions eventually share reproduction once a breeding opportunity is achieved. Such coalitions may be composed of related individuals, as in acorn woodpeckers (Koenig *et al.* 1984) and Tasmanian native hen (Maynard Smith and Ridpath 1972), unrelated individuals, as in dunnocks *Prunella modularis* (Davies 1992), the alpine accentor *Prunella collaris* (Davies *et al.* 1995, Nakamura 1998a, b), Arabian babblers *Turdoides squamiceps* (Lundy *et al.* 1998), and white-winged choughs *Corcorax melanorhamphos* (Heinsohn *et al.* 2000); or a mixture of related or unrelated males, as in African lions (Packer *et al.* 1991, 2001). With few exceptions, these systems are facultative and result in variation in actual reproductive access within groups. Diversity among female mate-sharing systems is as great as among male systems.

The evolution of mate sharing is not well understood. However, most systems in which mate sharing is typically among relatives also have nonbreeding helpers, suggesting that factors similar to, and perhaps more intense than, the kinds of constraints promoting helping at the nest lead ultimately to mate sharing and cooperative polygamy (Dickinson and Hatchwell 2004). In contrast, species in which cobreeders are typically unrelated generally do not have helpers, thus suggesting an alternative route to the evolution of such systems, most likely sexual conflict (Davies 1992).

19.3.1   Male mate-sharing systems

Mate sharing (cobreeding) by males, or cooperative poly-andry, is much less common than helping at the nest (Brown 1987). However, several variations of cooperative polyandry are found in vertebrates. In acorn woodpeckers, for example, breeding coalitions of males are formed either by inheritance of breeding status by helpers following the death and replacement of related female breeders in a group or by brothers forming coalitions that compete for repro-ductive vacancies together (Hannon et al. 1985). In both cases, cobreeders subsequently compete with each other for paternity (Haydock and Koenig 2002, 2003). Currently, it is unclear to what extent reproduction is monopolized by only one of the cobreeder males. Although one male usually fathers most or all the young in a particular brood, paternity of offspring within broods appears to be an all-or-nothing affair, and it is possible that cobreeding males have an equal probability of fathering any particular offspring. Similar coalitions of brothers or other kin are found in several other species including the Tasmanian native hen (Maynard Smith and Ridpath 1972), human villages of the Ladakh in Himalayan Tibet (Crook and Osmaston 1994), and the cichlid Neolamprologus multifasciatus (Taborsky 2001).

Determining the precise fitness consequences of mate sharing in these systems requires extensive molecular ana-lyses, which are mostly still in their early stages. However, analyses thus far indicate that, as with nonbreeding helpers, there may be fitness benefits for relatively small cobreeding coalitions but not for those as large as are found in the population. For example, there are clear fitness advantages to cobreeding male acorn woodpecker duos over singletons due to both increased survivorship and higher reproductive success (Koenig and Mumme 1987). However, there appear to be no per capita fitness benefits to breeding coalitions of more than two males, which are nonetheless fairly common, with nearly 20% of groups containing three to as many as seven cobreeder males.

A second, very different form of cooperative polyandry is found in the dunnock (Davies 1992). The two sexes defend territories independently in this species. Females frequently defend a territory that overlaps the territories of more than one male, which may subsequently coalesce into a cobreeding coalition. In addition, males are sometimes unable to maintain exclusive control over their territory, which is then invaded by a second or even third male to form a polyandrous or polygynandrous breeding group. Mate-sharing males are unrelated and provide paternal care to nests in relation to their degree of copulatory access to a particular female.

This latter case illustrates a relatively frequent situa-tion, also found in the alpine accentor (Hartley et al. 1995), Smith's longspur Calcarius pictus (Briskie et al. 1998), Galápagos hawk Buteo galapagoensis (Faaborg et al. 1995), Eclectus parrot Eclectus roratus (Heinsohn and Legge 2003), among others (Stacey 1982), in which the benefit to females mating with more than one male appears to be the potential for inducing additional males to provide paternal care to their offspring. To the extent this is true, they provide excellent examples of the importance of sexual conflict in ARTs (Henson and Warner 1997, Alonzo and Warner 2000). Specifically, they are examples in which one sex, in this case females, is apparently able to manipulate the behavior of the other (males), and males, in most cases, only feed young or otherwise provide paternal care to offspring that have been parented by known relatives (as in the vast majority of nonbreeding helpers at the nest) or to young that they had some chance of siring.

Closely related to mate sharing is the phenomenon of cooperative courtship, in which two to several males per-form a coordinated song and dance to attract females. Cooperative courtship occurs in wild turkeys in North America (Watts and Stokes 1971, Krakauer 2005) and several species of manakins (genus Chiroxiphia) in Central and South America (Foster 1981, McDonald and Potts 1994) and is also known from various families of fishes including suckers (family Catostomidae), in which spawn-ing occurs in trios of two male partners pressing against the flanks on either side of the female (Page and Johnston 1990, Taborsky 1994). Females in the avian species strongly prefer to mate with sets of males engaged in the elaborate courtship displays afforded by two or sometimes even more males displaying together, thus forcing males into coali-tions. The fitness consequences of cooperative courtship have yet to be studied in fishes, but in the bird examples the dominant male within a coalition appears to gain the vast majority of copulations in at least two species – the wild turkey (Krakauer 2005) and long-tailed manakin

Chiroxiphia linearis (McDonald and Potts 1994). In contrast, subordinate males appear to gain fitness benefits via completely different means in these two species: in wild turkeys, subordinates are closely related to dominants and achieve fitness indirectly through kinship, whereas in the long-tailed manakin, males are unrelated and instead benefit by potentially inheriting the dominant's status following the latter's death.

9.3.2 Female mate-sharing systems

Even less common than male mate sharing are systems in which females nest cooperatively in the same social unit. However, the diversity of such systems is at least as great as that of male mate sharing and can include both joint-nesting species, in which females to some extent produce and raise young together, and plural-breeding species, in which females breed separately but in the same social unit. Both phenomena are invariably facultative. Joint nesting may either lead to cooperative polygyny, in which more than one female shares a male, or cooperative polygynandry, in which multiple females share multiple males.

Among birds, joint nesting is known to occur regularly in ratites (the ostrich *Struthio camelus*, greater rhea *Rhea americana*, and emu *Dromaius novaehollandiae*), the magpie goose *Anseranas semipalmata*, several species of swamphens (see Section 19.6), the acorn woodpecker, all four species of anis (subfamily Crotophaginae), the Taiwan yuhina *Yuhina brunneiceps*, and the Seychelles warbler (Richardson *et al.* 2002), as well as rarely in several other cooperatively breeding species (Vehrencamp 2000, Vehrencamp and Quinn 2004).

In the ostrich, males defend large territories and form a pair bond with a single "major" hen that lays eggs in the male's nest and subsequently shares incubation and fledging care with the male (Hurxthal 1979, Bertram 1992). In addition, however, major hens may lay eggs parasitically in the nests of other females (Kimwele and Graves 2003), while groups of "minor" hens that are not associated with a particular male visit multiple nests where they mate and are allowed to lay eggs without contributing to subsequent parental care (Hurxthal 1979, Bertram 1992). Major hens are to some extent able to discriminate among eggs in their nest and favor their own over those laid by other females, but patterns of parentage are complex, with the territorial male siring about 70% of incubated eggs laid by the major hen and about one-third of the eggs laid by minor hens (Kimwele and Graves 2003). Thus, female ARTs in this

system include (1) being a major hen, bonding with a male, and laying eggs in his nest; (2) being a major hen and laying eggs parasitically in another male's nest; (3) being a lone minor female, mating and laying eggs quasi-parasitically in the nests of several males; and (4) forming a coalition with a group of minor females that mate and lay eggs in the nests of several males. Molecular studies quantifying the fitness consequences of these various tactics have only begun to be performed (Kimwele and Graves 2003).

Female common moorhens, discussed in Box 19.2, also pursue a comparable set of ARTs, including intraspecific brood parasitism and joint nesting, the latter usually between a mother and her daughter (McRae 1996). Nest parasitism has also been detected in several other joint-nesting species, including the groove-billed ani *Crotophaga sulcirostris* (Vehrencamp *et al.* 1986) and smooth-billed ani *Crotophaga ani* (Loflin 1983). In others, however, nest parasitism is rare, if it occurs at all. In acorn woodpeckers, for example, joint nesting between close relatives, either sisters or a mother and daughter, occurs in about 20% of nests (Koenig and Mumme 1987). All mating in this species appears to be within the group, and no cases of intraspecific brood parasitism have been detected (Dickinson *et al.* 1995, Haydock *et al.* 2001).

Joint nesting in acorn woodpeckers usually involves only two females, but even so, competition is considerable and females regularly destroy eggs laid by their cobreeders prior to laying their own (Mumme *et al.* 1983, Koenig *et al.* 1995), a process that results in equally divided maternity between incubated eggs (Haydock and Koenig 2002). Estimates of lifetime fitness nonetheless do not generate a net benefit for joint nesting over solitary breeding, and thus the existence of joint nesting appears to be a result of other correlated factors such as the increased competitive ability of coalitions in acquiring and defending high-quality breeding territories (Koenig and Mumme 1987).

Among social mammals, plural breeding by females occurs to at least some extent in a variety of species (Lewis and Pusey 1997, Russell 2004), depending on the degree to which there is a clear dominance hierarchy with the dominant female able to suppress reproduction by subordinates (Creel and Waser 1997). In meerkats, for example, groups contain a single dominant female, usually the oldest and heaviest in the group, who mothers about 80% of the litters. Subordinate females, who are reproductively capable and produce the remaining litters, are able to do so only when the dominant's capacity for control is reduced, such as immediately after a new female succeeds to the dominant position and when subordinates are older and particularly competitive (Clutton-Brock *et al.* 2001). In contrast, in African lions, related females form prides within which all individuals breed, communally caring for their young in crèches during lactation (Lewis and Pusey 1997). Reproductive suppression usually does not occur, except when group size and competition for food is particularly high (Packer *et al.* 2001).

As usual, analogous examples of communal brood care occur in fishes, where several species of cichlids have been observed communally caring for eggs or young. In *Neolamprologus multifasciatus*, for example, two or even three female group members may contribute to the communal brood offspring (Taborsky 2001) (Box 19.3). Joint-nesting

---

Box 19.3   Cichlid fishes

African cichlid fishes exhibit virtually the entire range of alternative tactics found in cooperatively breeding birds and mammals, including both helping at the nest and cobreeding (Taborsky 1994, 2001). In *Neolamprologus brichardi*, for example, male offspring often remain in their natal groups as nonbreeding helpers but may become sneakers attempting to parasitize the dominant male after the death and replacement of their parents by an unrelated pair. Such individuals gain direct reproduction and protection against predation in apparent exchange for increasing the productivity of the group. In *Neolamprologus multifasciatus* (Figure 19.5), a small species that nests in snail shells, both males and females form cobreeding coalitions of siblings, suggesting the importance of kin-selected benefits.

Figure 19.5 Cichlid fish *Neolamprologus multifasciatus*.
(Photograph © 2005 Sam Borstein, used with permission.)

---

females are apparently relatives, so kin selection may be important along with group benefits such as increased survival and sharing of the costs of nest building. In general, however, cases of communal brood care in fishes appear to be rare, presumably because the benefits of leaving care completely to alloparents is likely to outweigh the costs of increased predation on one's young from desertion and not helping (Taborsky 1994).

19.4   EXPLAINING THE DIVERSITY

19.4.1   Reproductive skew

Currently the most common means of approaching the diversity of mate-sharing and joint-nesting systems is reproductive skew theory, a predictive framework based on inclusive fitness theory specifically geared to understanding how reproduction is partitioned among potential breeders. Although models vary, optimal, or concessions, models assume that there is a dominant breeder that controls the distribution of reproduction among same-sex individuals within groups and goes on to determine the share of the group's reproductive output that must be achieved by the subordinate in order to make it worthwhile for him (or her) to remain in the group rather than breed on his (or her) own. Factors explicitly included as potentially influencing this decision include the relative reproductive advantages of groups, the relatedness of cobreeders, and the difficulty of obtaining an independent reproductive position elsewhere in the population (Vehrencamp 1983a, b, Keller and Reeve 1994, Johnstone 2000, Magrath *et al.* 2004).

The idea of optimal skew models is that, depending on the particular set of ecological conditions, it is in the best interests of the dominant to concede some reproduction to subordinates in order to keep the latter from leaving the group. Both dominants and subordinates are assumed to interact with each other to result in fitness pay-offs for both individuals that are greater together than if the subordinate leaves the group or the dominant evicts him (or her). Thus the basis for these models is transactional in that they postulate that dominants limit their own share of reproduction (or concede some measure of reproduction to subordinates) in return for stable cooperation by the subordinate (Keller and Reeve 1994).

Whether this kind of model represents the social relationships among cobreeders within cooperative breeding groups and thus accurately predicts reproductive partitioning is currently an issue of considerable interest and debate (Johnstone 2000, Magrath *et al.* 2004). However, despite some evidence supporting the ability of optimal skew theory to predict patterns of reproduction in several vertebrates, including pukeko (Jamieson 1997) and African lions (Packer *et al.* 1990), examples from birds, mammals, and fishes suggest that many such systems do not meet the assumptions nor match the predictions of optimal skew theory (Clutton-Brock *et al.* 2001, Haydock and Koenig 2003, Skubic *et al.* 2004).

In the polygynandrous acorn woodpecker (Haydock and Koenig 2002, 2003) joint-nesting females divide maternity among broods more equally than would be expected by chance, whereas paternity within broods of mate-sharing males is highly skewed toward one male. Given demographic differences between the sexes, this contrast is generally consistent with optimal skew theory. However, other features of this system are not. Joint-nesting females, for example, are close relatives and endure sufficient constraints to independent reproduction that they are predicted to exhibit considerable reproductive skew. This is not the case, apparently, because egg destruction by joint-nesting females eliminates the possibility of one of the females maintaining control over clutch composition. As for males, although a single male dominates paternity within each brood, the identity of the male often switches from one brood to the next, and there appears to be no character that allows a priori prediction of which male will dominate paternity within a group. These observations suggest that dominants may have, at best, limited control over reproductive partitioning within groups and that other factors, most notably sexual conflict, are probably playing an important role in determining the patterns of reproductive skew found in this species.

A well-investigated mammalian example is that of meerkats (Clutton-Brock *et al.* 2001). As discussed in section 19.3.2, subordinate females are sometimes able to reproduce. Subordinate reproduction does not appear to be affected either by the subordinate's relatedness to the dominant or by factors such as group size that in turn correlate with the benefits of group living. Furthermore, the probability of subordinate females leaving the group is not influenced by whether they breed or not. These results generally contradict predictions of optimal skew theory, again supporting the hypothesis that dominant females have limited control over reproductive partitioning and that subordinates breed whenever dominants, for whatever reason, are unable to stop them.

19.4.2 ARTs as an alternative theoretical framework

Given the limited success of optimal skew theory in predicting patterns of reproductive partitioning in vertebrates, it is clearly time to consider alternative approaches. ART theory is particularly promising in that it explicitly focuses on differences in competitive ability and readily accommodates the particular factors that may be important in a specific system. Consider the meerkats. In contrast to the shortcomings of optimal skew theory, a traditional ART approach might examine the fitness benefits of subordinate reproduction versus helping as a function of the difference in competitive abilities between dominant and subordinate breeders, including factors such as condition and age (Figure 19.6). Although the fitness benefits of helping and breeding will both increase as the condition of the subordinate increases, benefits of breeding increase faster than benefits of helping. With appropriate data including all relevant variables, it should be possible to predict where the switch point between the alternative tactics (S_{hb}) should occur.

An example of such an analysis is that of the cichlid fish *Neolamprologus pulcher*, in which large helpers may either postpone reproduction and invest in cooperative alloparental care or attempt to parasitize matings by the territorial dominant male (Taborsky 1984, 1985). In contrast to birds and mammals, indeterminate growth in fishes means that the costs and benefits of parasitism change as the helper

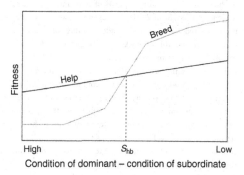

Figure 19.6 A simple graphical model viewing the alternatives of helping versus breeding in a subordinate female meerkat as a function of the difference in condition between the dominant and subordinate. The fitness benefits of breeding increase faster than those of helping, leading to a switch point (S_{hb}) determining at what point subordinates should attempt to circumvent the dominant's reproductive monopoly and reproduce.

grows, and it is therefore critical to consider the alternatives available to a helper compared to its future fitness expectations at each step of its life history. In an attempt to incorporate such life-history and behavioral variables potentially affecting the optimal reproductive strategy of subordinates, Skubic *et al.* (2004) used a dynamic programming model to predict the conditions when helpers should parasitize male dominants. Their results identify several factors that should be important to this decision, including the expulsion risk incurred by parasitism, the relatedness of the subordinate to the dominant, and the fraction of the total reproductive output of the group the subordinate can expect to parasitize.

Although it is difficult to compare the results of their model to what would be predicted from using optimal skew theory, the dynamic programming approach, which views parasitism and helping as ARTs and seeks to predict when switching should occur relative to body size, affords more flexibility by allowing an unlimited number of factors to be considered in determining the expected outcome. This avoids the most limiting features of skew theory, including its focus on a rigid set of ecological factors and, at least as traditionally envisioned, neglect of issues such as mate choice and intersexual conflict.

Sexual conflict is a particularly critical omission in the reproductive skew literature since it is known to play an important role both in the determination of the fitness consequences of ARTs (Henson and Warner 1997) and in the evolution of mating systems of cooperative breeders in general (Cockburn 2004). For example, sexual conflict is considered to be key to understanding the variable mating system of the dunnock as a consequence of differing mating optima for males and females (Davies 1991, 1992). For males, the increased production of young in cooperatively polyandrous groups does not compensate for shared paternity, and thus monogamy yields higher fitness than cooperative polyandry. This is in conflict with the interests of females for whom additional paternal care means higher reproductive success for polyandrous trios. Meanwhile, the combined output of two females exceeds that of one, and thus polygyny is more profitable for males than monogamy. As females experience a fitness loss by sharing a male, their interests are again in conflict with those of males. These inequalities are influenced by environmental quality, leading to complex predictions of the mating mode that are nonetheless readily accommodated by a traditional ART approach (Figure 19.7). Such conflicts are not considered by traditional optimal skew models.

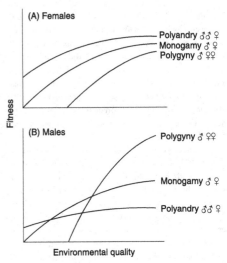

Figure 19.7 A graphical illustration of sexual conflict in the dunnock. (A) Female fitness is greatest when she gains the paternal care of two males (polyandry), less when she has the undivided care of a single male (monogamy), and least when she shares paternal care of a single male with a second female (polygyny). The differences among these options are present regardless of the quality of the environment, but the importance of paternal care is greater in poor environments. (B) Male fitness vis-à-vis the mating mode is the opposite of females in good environments, with males benefiting most by polygyny (which is twice the fitness of a polygynous female) and least by polyandry (which is one-half the female polyandry curve). However, additional male paternal care becomes important in poor environments, leading to potential switch points where first monogamy and then polyandry is most beneficial to males. (From Davies 1991, with permission of Blackwell Scientific Publications.)

19.5   CONCLUSION

Cooperative breeding is a diverse phenomenon and understanding it is likely to be most successful if a variety of theoretical approaches are employed. Both helping at the nest and mate sharing, the two most characteristic patterns of cooperative breeders, are almost always facultative and highly plastic, and thus analyzing this behavior as ARTs can offer a new perspective into their diversity and occurrence. An ART approach is particularly useful in resolving several unanswered questions regarding cooperative breeding, including the tactics pursued by potential helpers in groups where the frequency of helping is already above he level at which helpers provide additional fitness benefits, he behavior of cobreeders in species where sexual conflict is mportant, and the factors influencing choices in species where multiple reproductive tactics are available, including elping at the nest, parasitizing the reproductive efforts of thers, breeding singly, and sharing breeding status. iewing these alternatives as ARTs, rather than using pproaches that limit the factors that are presumed to affect ehavioral choices, may provide new insights or, at the ery least, help clarify the selective forces promoting the iversity of behavior exhibited by cooperative breeders.

Acknowledgments

We thank M. Taborsky and an anonymous reviewer for omments on the manuscript. Our work on cooperative reeding has been supported by the National Science 'oundation.

References

lonzo, S. H. and Warner, R. R. 2000. Female choice, conflict between the sexes and the evolution of male alternative reproductive tactics. *Evolutionary Ecology Research* 2, 149–170.

rnold, K. E. and Owens, I. P. F. 1998. Cooperative breeding in birds: a comparative test of the life history hypothesis. *Proceedings of the Royal Society of London B* 265, 739–745.

ustad, S. N. 1984. A classification of alternative reproductive behaviors and methods for field-testing ESS models. *American Zoologist* 24, 309–319.

aglione, V., Canestrari, D., Marcos, J. M., Griesser, M., and Ekman, J. 2002. History, environment and social behaviour: experimentally induced cooperative breeding in the carrion crow. *Proceedings of the Royal Society of London B* 269, 1247–1251.

alshine-Earn, S., Neat, F. C., Reid, H., and Taborsky, M. 1998. Paying to stay or paying to breed? Field evidence for direct benefits of helping behavior in a cooperatively breeding fish. *Behavioral Ecology* 9, 432–438.

ennett, N. C., Faulkes, C. G., and Molteno, A. J. 1996. Reproductive suppression in subordinate, non-breeding female Damaraland mole-rats: two components to a lifetime of socially induced infertility. *Proceedings of the Royal Society of London B* 263, 1599–1603.

ertram, B. C. R. 1992. *The Ostrich Communal Nesting System*. Princeton, NJ: Princeton University Press.

riskie, J. V., Montgomerie, R., Pöldmaa, T., and Boag, P. T. 1998. Paternity and paternal care in the polygynandrous

Smith's longspur. *Behavioral Ecology and Sociobiology* 43, 181–190.

Brockmann, H. J. 1997. Cooperative breeding in wasps and vertebrates: the role of ecological constraints. In J. C. Choe and B. J. Crespi (eds.) *The Evolution of Social Behavior in Insects and Arachnids*, pp. 347–371. Cambridge, UK: Cambridge University Press.

Brockmann, H. J. 2001. The evolution of alternative strategies and tactics. *Advances in the Study of Behavior* 30, 1–51.

Brooker, M. G., Rowley, I., Adams, M., and Baverstock, P. R. 1990. Promiscuity: an inbreeding avoidance mechanism in a socially monogamous species? *Behavioral Ecology and Sociobiology* 26, 191–200.

Brown, J. L. 1987. *Helping and Communal Breeding in Birds*. Princeton, NJ: Princeton University Press.

Brown, J. L. and Brown, E. R. 1990. Mexican jays: uncooperative breeding. In P. B. Stacey and W. D. Koenig (eds.) *Cooperative Breeding in Birds: Long-Term Studies of Ecology and Behavior*, pp. 267–288. Cambridge, UK: Cambridge University Press.

Bygott, J. D., Bertram, B. C. R., and Hanby, J. P. 1979. Male lions in large coalitions gain reproductive advantages. *Nature* 282, 839–841.

Caffrey, C. 1992. Female-biased delayed dispersal and helping in American crows. *Auk* 109, 609–619.

Clutton-Brock, T. H., Brotherton, P. N. M., Russell, A. F., *et al.* 2001. Cooperation, control, and concession in meerkat groups. *Science* 291, 478–481.

Cockburn, A. 2003. Cooperative breeding in oscine passerines: does sociality inhibit speciation? *Proceedings of the Royal Society of London B* 270, 2207–2214.

Cockburn, A. 2004. Mating systems and sexual conflict. In W. D. Koenig and J. L. Dickinson (eds.) *Ecology and Evolution of Cooperative Breeding in Birds*, pp. 81–101. Cambridge, UK: Cambridge University Press.

Cockburn, A., Osmond, H. L., Mulder, R. A., Green, D. J., and Double, M. C. 2003. Divorce, dispersal, density-dependence and incest avoidance in the cooperatively breeding superb fairy-wren *Malurus cyaneus*. *Journal of Animal Ecology* 72, 189–202.

Cooney, R. and Bennett, N. C. 2000. Inbreeding avoidance and reproductive skew in a cooperative mammal. *Proceedings of the Royal Society of London B* 267, 801–806.

Craig, J. L. and Jamieson, I. G. 1990. Pukeko: different approaches and some different answers. In P. B. Stacey and W. D. Koenig (eds.) *Cooperative Breeding in Birds: Long-Term Studies of Ecology and Behavior*, pp. 387–412. Cambridge, UK: Cambridge University Press.

Creel, S. R. and Waser, P. M. 1997. Variation in reproductive suppression among dwarf mongooses: interplay between mechanisms and evaluation. In N. G. Solomon and J. A. French (eds.) *Cooperative Breeding in Mammals*, pp. 150–170. Cambridge, UK: Cambridge University Press.

Crespi, B. J. and Yanega, D. 1995. The definition of eusociality. *Behavioral Ecology* 6, 109–115.

Crook, J. H. and Osmaston, H. 1994. *Himalayan Buddhist Villages: Environment, Resources, Society and Religious Life in Zangskar, Ladakh.* Bristol, UK: University of Bristol Press.

Curry, R. L. and Grant, P. R. 1990. Galápagos mockingbirds: territorial cooperative breeding in a climatically variable environment. In P. B. Stacey and W. D. Koenig (eds.) *Cooperative Breeding in Birds: Long-Term Studies of Ecology and Behavior*, pp. 289–331. Cambridge, UK: Cambridge University Press.

Davies, N. B. 1991. Mating systems. In J. R. Krebs and N. B. Davies (eds.) *Behavioural Ecology: An Evolutionary Approach*, 3rd edn, pp. 263–294. Oxford, UK: Blackwell Scientific.

Davies, N. B. 1992. *Dunnock Behaviour and Social Evolution.* Oxford, UK: Oxford University Press.

Davies, N. B., Hartley, I. R., Hatchwell, B. J., *et al.* 1995. The polygynandrous mating system of the alpine accentor, *Prunella collaris.* 1. Ecological causes and reproductive conflicts. *Animal Behaviour* 49, 769–788.

Dickinson, J. L. 2004a. Facultative sex ratio adjustment by western bluebird mothers with stay-at-home helpers-at-the-nest. *Animal Behaviour* 68, 373–380.

Dickinson, J. L. 2004b. A test of the importance of direct and indirect fitness benefits for helping decisions in western bluebirds. *Behavioral Ecology* 15, 233–238.

Dickinson, J. L. and Akre, J. J. 1998. Extrapair paternity, inclusive fitness, and within-group benefits of helping in western bluebirds. *Molecular Ecology* 7, 95–105.

Dickinson, J. L. and Hatchwell, B. J. 2004. Fitness consequences of helping. In W. D. Koenig and J. L. Dickinson (eds.) *Ecology and Evolution of Cooperative Breeding in Birds*, pp. 48–66. Cambridge, UK: Cambridge University Press.

Dickinson, J. L. and McGowan, A. 2005. Winter resource wealth drives delayed dispersal and family-group living in western bluebirds. *Proceedings of the Royal Society of London B* 272, 2423–2428.

Dickinson, J. L., Haydock, J., Koenig, W. D., Stanback, M. T., and Pitelka, F. A. 1995. Genetic monogamy in single-male groups of acorn woodpeckers, *Melanerpes formicivorus. Molecular Ecology* 4, 765–769.

Dickinson, J. L., Koenig, W. D., and Pitelka, F. A. 1996. Fitness consequences of helping behavior in the western bluebird. *Behavioral Ecology* 7, 168–177.

Double, M. C. and Cockburn, A. 2000. Pre-dawn infidelity: females control extra-pair mating in superb fairy-wrens. *Proceedings of the Royal Society of London B* 267, 465–470.

Double, M. C. and Cockburn, A. 2003. Subordinate superb fairy-wrens (*Malurus cyaneus*) parasitize the reproductive success of attractive dominant males. *Proceedings of the Royal Society of London B* 270, 379–384.

Dunn, P. O. and Cockburn, A. 1996. Evolution of male parental care in a bird with almost complete cuckoldry. *Evolution* 50, 2542–2548.

Dunn, P. O. and Cockburn, A. 1999. Extrapair mate choice and honest signaling in cooperatively breeding superb fairy-wrens. *Evolution* 53, 938–946.

Dunn, P. O., Cockburn, A., and Mulder, R. A. 1995. Fairy-wren helpers often care for young to which they are unrelated. *Proceedings of the Royal Society of London B* 259, 339–343.

Emlen, S. T. 1978. The evolution of cooperative breeding in birds. In J. R. Krebs and N. B. Davies (eds.) *Behavioral Ecology: An Evolutionary Approach*, pp. 245–281. Sunderland, MA: Sinauer Associates.

Emlen, S. T. 1990. White-fronted bee-eaters: helping in a colonially nesting species. In P. B. Stacey and W. D. Koenig (eds.) *Cooperative Breeding in Birds: Long-Term Studies of Ecology and Behavior*, pp. 489–526. Cambridge, UK: Cambridge University Press.

Emlen, S. T. 1991. Evolution of cooperative breeding in birds and mammals. In J. R. Krebs and N. B. Davies (eds.) *Behavioural Ecology: An Evolutionary Approach*, 3rd edition, pp. 301–337. Oxford, UK: Blackwell Scientific.

Faaborg, J. and Patterson, C. B. 1981. The characteristics and occurrence of cooperative polyandry. *Ibis* 123, 477–484.

Faaborg, J., Parker, P. G., Delay, L., *et al.* 1995. Confirmation of cooperative polyandry in the Galápagos hawk (*Buteo galapagoensis*). *Behavioral Ecology and Sociobiology* 36, 83–90.

Foster, M. S. 1981. Cooperative behavior and social organization of the swallow-tailed manakin (*Chiroxiphia caudata*). *Behavioral Ecology and Sociobiology* 9, 167–177.

Goldizen, A. W., Putland, D. A., and Goldizen, A. R. 1998. Variable mating patterns in Tasmanian hens (*Gallinula mortierii*): correlates of reproductive success. *Journal of Animal Ecology* 67, 307–317.

Goldizen, A. W., Buchan, J. C., Putland, D. A., Goldizen, A. R., and Krebs, E. A. 2000. Patterns of mate-sharing in a population of Tasmanian native hens *Gallinula mortierii*. *Ibis* 142, 40–47.

Green, D. J., Osmond, H. L., Double, M. C., and Cockburn, A. 2000. Display rate by male fairy-wrens (*Malurus cyaneus*) during the fertile period of females has little influence on extra-pair mate choice. *Behavioral Ecology and Sociobiology* 48, 438–446.

Gross, M. R. 1996. Alternative reproductive strategies and tactics: diversity within sexes. *Trends in Ecology and Evolution* 11, 92–98.

Haig, S. M., Walters, J. R., and Plissner, J. H. 1994. Genetic evidence for monogamy in the cooperatively breeding red-cockaded woodpecker. *Behavioral Ecology and Sociobiology* 34, 295–303.

Hannon, S. J., Mumme, R. L., Koenig, W. D., and Pitelka, F. A. 1985. Replacement of breeders and within-group conflict in the cooperatively breeding acorn woodpecker. *Behavioral Ecology and Sociobiology* 17, 303–312.

Hartley, I. R. and Davies, N. B. 1994. Limits to cooperative polyandry in birds. *Proceedings of the Royal Society of London B* 257, 67–73.

Hartley, I. R., Davies, N. B., Hatchwell, B. J., *et al.* 1995. The polygynandrous mating system of the alpine accentor, *Prunella collaris*. 2. Multiple paternity and parental effort. *Animal Behaviour* 49, 789–803.

Hatchwell, B. J., Anderson, C., Ross, D. J., Fowlie, M. K., and Blackwell, P. G. 2001. Social organization of cooperatively breeding long-tailed tits: kinship and spatial dynamics. *Journal of Animal Ecology* 70, 820–830.

Haydock, J. and Koenig, W. D. 2002. Reproductive skew in the polygynandrous acorn woodpecker. *Proceedings of the National Academy of Sciences of the United States of America* 99, 7178–7183.

Haydock, J. and Koenig, W. D. 2003. Patterns of reproductive skew in the polygynandrous acorn woodpecker. *American Naturalist* 162, 277–289.

Haydock, J., Koenig, W. D., and Stanback, M. T. 2001. Shared parentage and incest avoidance in the cooperatively breeding acorn woodpecker. *Molecular Ecology* 10, 1515–1525.

Heinsohn, R. G. and Legge, S. 2003. Breeding biology of the reverse-dichromatic, cooperative parrot, *Eclectus roratus*. *Journal of Zoology (London)* 259, 197–208.

Heinsohn, R. G., Dunn, P. O., Legge, S., and Double, M. C. 2000. Coalitions of relatives and reproductive skew in cooperatively breeding white-winged choughs. *Proceedings of the Royal Society of London B* 267, 243–249.

Henson, S. A. and Warner, R. R. 1997. Male and female alternative reproductive behaviors in fishes: a new approach using intersexual dynamics. *Annual Review of Ecology and Systematics* 28, 571–592.

Hurxthal, L. M. 1979. *Breeding Behaviour of the Ostrich, Struthio camelus massaicus, in Nairobi National Park*. Nairobi: University of Nairobi.

Innes, K. E. and Johnston, R. E. 1996. Cooperative breeding in the white-throated magpie-jay: how do auxiliaries influence nesting success? *Animal Behaviour* 51, 519–533.

Jamieson, I. G. 1997. Testing reproductive skew models in a communally breeding bird, the pukeko, *Porphyrio porphyrio*. *Proceedings of the Royal Society of London B* 264, 335–340.

Jamieson, I. G., Quinn, J. S., Rose, P. A., and White, B. N. 1994. Shared paternity among non-relatives is a result of an egalitarian mating system in a communally breeding bird, the pukeko. *Proceedings of the Royal Society of London B* 257, 271–277.

Johnstone, R. A. 2000. Models of reproductive skew: a review and synthesis. *Ethology* 106, 5–26.

Keller, L. and Reeve, H. K. 1994. Partitioning of reproduction in animal societies. *Trends in Ecology and Evolution* 9, 98–102.

Khan, M. Z., McNabb, F. M. A., Walters, J. R., and Sharp, P. J. 2001. Patterns of testosterone and prolactin concentrations and reproductive behavior of helpers and breeders in the cooperatively breeding red-cockaded woodpecker (*Picoides borealis*). *Hormones and Behavior* 40, 1–13.

Kimwele, C. N. and Graves, J. A. 2003. A molecular genetic analysis of the communal nesting of the ostrich (*Struthio camelus*). *Molecular Ecology* 12, 229–236.

Koenig, W. D. and Haydock, J. 2004. Incest and incest avoidance. In W. D. Koenig and J. L. Dickinson (eds.) *Ecology and Evolution of Cooperative Breeding in Birds*, pp. 142–156. Cambridge, UK: Cambridge University Press.

Koenig, W. D. and Mumme, R. L. 1987. *Population Ecology of the Cooperatively Breeding Acorn Woodpecker*. Princeton, NJ: Princeton University Press.

Koenig, W. D., Mumme, R. L., and Pitelka, F. A. 1984. The breeding system of the acorn woodpecker in central coastal California. *Zeitschrift für Tierpsychologie* 65, 289–308.

Koenig, W. D., Mumme, R. L., Stanback, M. T., and Pitelka, F. A. 1995. Patterns and consequences of egg destruction among joint-nesting acorn woodpeckers. *Animal Behaviour* 50, 607–621.

Koenig, W. D., Haydock, J., and Stanback, M. T. 1998. Reproductive roles in the cooperatively breeding acorn woodpecker: incest avoidance versus reproductive competition. *American Naturalist* 151, 243–255.

Koenig, W. D., Stanback, M. T., and Haydock, J. 1999. Demographic consequences of incest avoidance in the cooperatively breeding acorn woodpecker. *Animal Behaviour* 57, 1287–1293.

Kokko, H. and Ekman, J. 2002. Delayed dispersal as a route to breeding: territorial inheritance, safe havens, and ecological constraints. *American Naturalist* 160, 468–484.

Komdeur, J. 1992. Importance of habitat saturation and territory quality for evolution of cooperative breeding in the Seychelles warbler. *Nature* 358, 493–495.

Komdeur, J. 1994. Experimental evidence for helping and hindering by previous offspring in the cooperative-breeding Seychelles warbler *Acrocephalus sechellensis*. *Behavioral Ecology and Sociobiology* 34, 175–186.

Komdeur, J. 1996. Influence of helping and breeding experience on reproductive performance in the Seychelles warbler: a translocation experiment. *Behavioral Ecology* 7, 326–333.

Krakauer, A. 2005. Kin selection and cooperative courtship in wild turkeys. *Nature* 434, 69–72.

Lambert, D. M., Millar, C. D., Jack, K., Anderson, S., and Craig, J. L. 1994. Single- and multilocus DNA fingerprinting on communally breeding pukeko: do copulations or dominance ensure reproductive success? *Proceedings of the National Academy of Sciences of the United States of America*, 91, 9641–9645.

Lewis, S. E. and Pusey, A. E. 1997. Factors influencing the occurrence of communal care in plural breeding mammals. In N. G. Solomon and J. A. French (eds.) *Cooperative Breeding in Mammals*, pp. 335–363. Cambridge, UK: Cambridge University Press.

Li, S.-H. and Brown, J. L. 2000. High frequency of extrapair fertilization in a plural breeding bird, the Mexican jay, revealed by DNA microsatellites. *Animal Behaviour* 60, 867–877.

Li, S.-H. and Brown, J. L. 2002. Reduction of maternal care: a new benefit of multiple mating? *Behavioral Ecology* 13, 87–93.

Loflin, R. K. 1983. Communal behaviors of the smooth-billed ani (*Crotophaga ani*). Ph.D. thesis, University of Miami, Coral Gables, FL.

Lundy, K. J., Parker, P. G., and Zahavi, A. 1998. Reproduction by subordinates in cooperatively breeding Arabian babblers is uncommon but predictable. *Behavioral Ecology and Sociobiology* 43, 173–180.

Magrath, R. D., Johnstone, R. A., and Heinsohn, R. G. 2004. Reproductive skew. In W. D. Koenig and J. L. Dickinson (eds.) *Ecology and Evolution of Cooperative Breeding in Birds*, pp. 157–176. Cambridge, UK: Cambridge University Press.

Martin, E. and Taborsky, M. 1997. Alternative male mating tactics in a cichlid, *Pelvicacharomis pulcher*: a comparison of reproductive effort and success. *Behavioral Ecology and Sociobiology* 41, 311–319.

Marzluff, J. M. and Balda, R. P. 1990. Pinyon jays: making the best of a bad situation by helping. In P. B. Stacey and W. D. Koenig (eds.) *Cooperative Breeding in Birds: Long-Term Studies of Ecology and Behavior*, pp. 199–237. Cambridge, UK: Cambridge University Press.

Maynard Smith, J. and Ridpath, M. G. 1972. Wife sharing in the Tasmanian native hen, *Tribonyx mortierii*: a case of kin selection? *American Naturalist* 106, 447–452.

McDonald, D. B. 1989. Correlates of male mating success in a lekking bird with male–male cooperation. *Animal Behaviour* 37, 1007–1022.

McDonald, D. B. and Potts, W. K. 1994. Cooperative display and relatedness among males in a lek-mating bird. *Science* 266, 1030–1032.

McRae, S. B. 1995. Temporal variation in responses to intraspecific brood parasitism in the moorhen. *Animal Behaviour* 49, 1073–1088.

McRae, S. B. 1996. Family values: costs and benefits of communal nesting in the moorhen. *Animal Behaviour* 52, 225–245.

McRae, S. B. 1998. Relative reproductive success of female moorhens using conditional strategies of brood parasitism and parental care. *Behavioral Ecology* 9, 93–100.

McRae, S. B. and Burke, T. 1996. Intraspecific brood parasitism in the moorhen: parentage and parasite–host relationships determined by DNA fingerprinting. *Behavioral Ecology and Sociobiology* 38, 115–129.

Mulder, R. A., Dunn, P. O., Cockburn, A., Lazenby-Cohen, K. A., and Howell, M. J. 1994. Helpers liberate female fairy-wrens from constraints on extra-pair mate choice. *Proceedings of the Royal Society of London B* 255, 223–229.

Mumme, R. L. 1992. Do helpers increase reproductive success? An experimental analysis in the Florida scrub jay. *Behavioral Ecology and Sociobiology* 31, 319–328.

Mumme, R. L., Koenig, W. D., and Pitelka, F. A. 1983. Reproductive competition in the communal acorn woodpecker: sisters destroy each other's eggs. *Nature* **305**, 583–584.

akamura, M. 1998a. Multiple mating and cooperative breeding in polygynandrous alpine accentors. 1. Competition among females. *Animal Behaviour*, **55**, 259–275.

akamura, M. 1998b. Multiple mating and cooperative breeding in polygynandrous alpine accentors. 2. Male mating tactics. *Animal Behaviour* **55**, 277–289.

liveira, R. F., Carvalho, N., Miranda, J., et al. 2002. The relationship between the presence of satellite males and nest-holders' mating success in the Azorean rock-pool blenny *Parablennius sanguinolentus parvicornis*. *Ethology* **108**, 223–235.

'Riain, M. J., Bennett, N. C., Brotherton, P. N. M., McIlrath, G. M., and Clutton-Brock, T. H. 2000. Reproductive suppression and inbreeding avoidance in wild populations of co-operatively breeding meerkats (*Suricata suricatta*). *Behavioral Ecology and Sociobiology* **48**, 471–477.

cker, C., Scheel, D., and Pusey, A. E. 1990. Why lions form groups: food is not enough. *American Naturalist* **136**, 1–19.

cker, C., Gilbert, D. A., Pusey, A. E., and O'Brian, S. J. 1991. A molecular genetic analysis of kinship and cooperation in African lions. *Nature* **351**, 562–565.

cker, C., Pusey, A. E., and Eberly, L. E. 2001. Egalitarianism in female African lions. *Science* **293**, 690–693.

ge, L. M. and Johnston, C. E. 1990. Spawning in the creek chubsucker, *Erimyzon oblongus*, with a review of spawning behavior in suckers (Catostomidae). *Environmental Biology of Fishes* **27**, 265–272.

ters, A. 2000. Testosterone treatment is immunosuppressive in superb fairy-wrens, yet free-living males with high testosterone are more immunocompetent. *Proceedings of the Royal Society of London B* **267**, 883–889.

ters, A., Astheimer, L. B., Boland, C. R. J., and Cockburn, A. 2000. Testosterone is involved in acquisition and maintenance of sexually selected male plumage in superb fairy-wrens, *Malurus cyaneus*. *Behavioral Ecology and Sociobiology* **47**, 438–445.

ters, A., Astheimer, L. B., and Cockburn, A. 2001. The annual testosterone profile in cooperatively breeding superb fairy-wrens, *Malurus cyaneus*, reflects their extreme infidelity. *Behavioral Ecology and Sociobiology* **50**, 519–527.

per, W. H. and Slater, G. 1993. Polyandry and incest avoidance in the cooperative stripe-backed wren of Venezuela. *Behaviour* **124**, 227–247.

Quinn, J. S., Woolfenden, G. E., Fitzpatrick, J. W., and White, B. N. 1999. Multi-locus DNA fingerprinting supports genetic monogamy in Florida scrub-jays. *Behavioral Ecology and Sociobiology* **45**, 1–10.

Rabenold, K. N. 1990. *Campylorhynchus* wrens: the ecology of delayed dispersal and cooperation in the Venezuelan savanna. In P. B. Stacey and W. D. Koenig (eds.) *Cooperative Breeding in Birds: Long-Term Studies of Ecology and Behavior*, pp. 159–196. Cambridge, UK: Cambridge University Press.

Rabenold, P. P., Rabenold, K. N., Piper, W. H., Haydock, J., and Zack, S. N. 1990. Shared paternity revealed by genetic analysis in cooperatively breeding tropical wrens. *Nature* **348**, 538–540.

Reyer, H.-U., Dittami, J. P., and Hall, M. R. 1986. Avian helpers at the nest: are they psychologically castrated? *Ethology* **71**, 216–228.

Richardson, D. S., Burke, T., and Komdeur, J. 2002. Direct benefits and the evolution of female-biased cooperative breeding in Seychelles warblers. *Evolution* **56**, 2313–2321.

Rood, J. P. 1990. Group size, survival, reproduction, and routes to breeding in dwarf mongooses. *Animal Behaviour* **39**, 566–572.

Russell, A. F. 2004. Mammals: comparisons and contrasts. In W. D. Koenig and J. L. Dickinson (eds.) *Ecology and Evolution of Cooperative Breeding in Birds*, pp. 210–227. Cambridge, UK: Cambridge University Press.

Schoech, S. J., Reynolds, S. J., and Boughton, R. K. 2004. Endocrinology. In W. D. Koenig and J. L. Dickinson (eds.) *Ecology and Evolution of Cooperative Breeding in Birds*, pp. 128–141. Cambridge, UK: Cambridge University Press.

Sherman, P. W., Lacey, E. A., Reeve, H. K., and Keller, L. 1995. The eusociality continuum. *Behavioral Ecology* **6**, 102–108.

Skubic, E., Taborsky, M., McNamara, J. M., and Houston, A. I. 2004. When to parasitize? A dynamic optimization model of reproductive strategies in a cooperative breeder. *Journal of Theoretical Biology* **227**, 487–501.

Stacey, P. B. 1982. Female promiscuity and male reproductive success in social birds and mammals. *American Naturalist* **120**, 51–64.

Strassman, J. E., Hughes, C. R., Turillazzi, S., Solis, C. R., and Queller, D. C. 1994. Genetic relatedness and incipient eusociality in stenogastrine wasps. *Animal Behaviour* **48**, 813–821.

Taborsky, M. 1984. Broodcare helpers in the cichlid fish *Lamprologus brichardi*: their costs and benefits. *Animal Behaviour* **32**, 1236–1252.

Taborsky, M. 1985. Breeder–helper conflict in a cichlid fish with broodcare helpers: an experimental analysis. *Behaviour* **95**, 45–75.

Taborsky, M. 1994. Sneakers, satellites, and helpers: parasitic and cooperative behavior in fish reproduction. *Advances in the Study of Behavior* **23**, 1–100.

Taborsky, M. 2001. The evolution of bourgeois, parasitic, and cooperative reproductive behaviors in fishes. *Journal of Heredity* **92**, 100–110.

Taborsky, M., Hudde, B., and Wirtz, P. 1987. Reproductive behavior and ecology of *Symphodus ocellatus*, a European wrasse with four types of male behaviour. *Behaviour* **102**, 82–118.

Vehrencamp, S. L. 1983a. A model for the evolution of despotic versus egalitarian societies. *Animal Behaviour* **31**, 667–682.

Vehrencamp, S. L. 1983b. Optimal degree of skew in cooperative societies. *American Zoologist* **23**, 327–335.

Vehrencamp, S. L. 2000. Evolutionary routes to joint-female nesting in birds. *Behavioral Ecology* **11**, 334–344.

Vehrencamp, S. L. and Quinn, J. S. 2004. Joint laying systems. In W. D. Koenig and J. L. Dickinson (eds.) *Ecology and Evolution of Cooperative Breeding in Birds*, pp. 177–196. Cambridge, UK: Cambridge University Press.

Vehrencamp, S. L., Bowen, B. S., and Koford, R. R. 1986. Breeding roles and pairing patterns within communal groups of groove-billed anis. *Animal Behaviour* **34**, 347–366.

Watts, C. R. and Stokes, A. W. 1971. The social order of turkeys. *Scientific American* **224**, 112–118.

Wingfield, J. C., Hegner, R. E., Dufty Jr., A. M., and Ball, G. F. 1990. The "challenge hypothesis": theoretical implications for patterns of testosterone secretion, mating systems, and breeding strategies. *American Naturalist* **136**, 829–845.

Woolfenden, G. E. and Fitzpatrick, J. W. 1984. *The Florida Scrub Jay: Demography of a Cooperative-Breeding Bird*. Princeton, NJ: Princeton University Press.

Woolfenden, G. E. and Fitzpatrick, J. W. 1986. Sexual asymmetries in the life history of the Florida scrub jay. In D. Rubenstein and R. W. Wrangham (eds.) *Ecological Aspects of Social Evolution: Birds and Mammals*, pp. 87–107. Princeton, NJ: Princeton University Press.

Woolfenden, G. E. and Fitzpatrick, J. W. 1990. Florida scrub jays: a synopsis after 18 years of study. In P. B. Stacey and W. D. Koenig (eds.) *Cooperative Breeding in Birds: Long-Term Studies of Ecology and Behavior*, pp. 241–266. Cambridge, UK: Cambridge University Press.

20 · Integrating mechanisms and function: prospects for future research

H. JANE BROCKMANN, RUI F. OLIVEIRA, AND MICHAEL TABORSKY

CHAPTER SUMMARY

In this chapter we pull together the common threads of the other chapters of this book. In doing this we identify a number of issues that need further research. Rather than repeating what has been said before, we identify the features that stand out because they are unexplained, previously unrecognized or just neglected. We argue that to understand alternative reproductive tactics (ARTs) we must use an approach that integrates the study of mechanisms and evolution.

20.1 WHAT IS NEXT IN THE STUDY OF ARTs?

Continuous variation in reproductive characters (behavior, morphology, physiology) is found in all species but the real puzzle comes in understanding the special cases in which variation is discontinuous and thus constitutes consistent, discretely different ways of achieving reproduction for animals within one population. If one phenotype were just a little less successful than the other, then we would expect it to be eliminated from the population over time by natural selection. It is for this reason that the maintenance of ARTs is an evolutionary puzzle. ARTs are also a puzzle to geneticists, physiologists, and developmental biologists who must explain how one genetic and developmental program can result in two different phenotypic outcomes. Our chief challenge is to draw together the genetic, developmental, behavioral, and physiological views of ARTs to understand the evolution of the mechanisms that we see as alternative phenotypes.

20.1.1 Categories of ARTs

Discontinuities in behavioral, morphological, or physiological traits can be difficult to detect (Eberhard and Gutiérrez 1991, Emlen 1996, Kotiaho and Tomkins 2001, Rowland and Qualls 2005, Rowland *et al.* 2005, Tomkins *et al.* 2005), but many clear examples are illustrated in the chapters of this book. In some cases authors describe continuous variation by the extremes and this has made it difficult to be sure whether particular cases are true ARTs or not. For example, singing to attract mates in male crickets may be highly variable with some individuals singing much of the night whereas others utter only a chirp or two, a continuous pattern in which the two ends of the continuum may be described as singing and nonsinging male behavior. Certainly the development, mechanisms, and maintenance of such variation is intriguing, but the processes involved are likely to be different from those acting on two (or more) discretely different kinds of males, singers and nonsingers described by a bimodal distribution (e.g., threshold mechanisms, disruptive selection). Often, suites of behavioral, morphological, and physiological traits are correlated with alternative phenotypes and it may be that some of these traits are discontinuous whereas others are continuous, but as long as the reproductive functions are discrete, then they are ARTs.

The study of ARTs has been hampered by typological thinking about mating systems that ignores significant and consistent variation. Parasitic tactics were often considered to be mistakes or desperate maneuvers by animals with no hope of achieving success. The result is that one of the most important unresolved issues for the study of ARTs is that many are poorly or incompletely described. Good descriptions are crucial to our ability to study the phenomenon. For example, some of the best-studied ARTs turn out to have three phenotypes (Chapters 9, 10, and 12), Differences between these phenotypes in social interactions (such as territoriality or aggressiveness), mating, and life-history patterns can result in cyclical dynamics such as rock–paper–scissors (Sinervo and Calsbeek 2006).

Alternative Reproductive Tactics, ed. Rui F. Oliveira, Michael Taborsky, and H. Jane Brockmann. Published by Cambridge University Press. Cambridge University Press 2008.

ARTs have been categorized in a number of different ways. As discussed in Chapter 1, "tactics" and "strategies" are not easily distinguished and imply a dichotomy between genetic and nongenetic control that is not useful (e.g., nature vs. nurture; see Section 20.3), so we do not support this distinction and refer instead to all cases as tactics. Tactics are governed by evolved decision-making rules, whatever the underlying mechanism. Many authors also distinguish ARTs that are associated with a genetic polymorphism from those that are conditional (i.e., influenced by individual status, condition, age, or environmental conditions). We now know that environmental conditions often affect the expression of genetic polymorphisms (Chapters 10 and 12), that thresholds of genetically based tactic expression may vary even within populations (Chapters 5, 8, and 10), and that genetic differences may affect the expression of condition-dependent tactics (Chapter 5). Therefore, the distinction between genetic polymorphism and conditional tactic does not seem useful. Rather, we consider gene–environment interactions to be of paramount importance in understanding reproductive tactics. Well-adapted animals are expected to switch from one tactic to another so that fitness is maximized, i.e., they should always be making the *best* of their situation. With ARTs individuals allocate resources to either one or the other (mutually exclusive) way of achieving the same functional end (reproductive tactic) using evolved decision-making rules (i.e., the tactics are adaptations). We suggest that ARTs should be categorized in the same way as other alternative allocation phenotypes such as sex allocation (Figure 1.1; Chapter 2) (Henson and Warner 1997, Taborsky 1998, Brockmann 2001) and alternative life histories (West-Eberhard 2003). When this is done, theory developed for other alternative allocation problems (e.g., sex allocation, alternative life histories) can be readily applied to ARTs.

In a remarkable convergence of views, those studying ARTs primarily from an evolutionary perspective (e.g., Taborsky 1998, Brockmann 2001; Chapter 1) and those studying the underlying mechanisms of ARTs (Chapters 6 and 7; Moore *et al.* 1998) have arrived at the same system for classifying ARTs (Figure 1.2). Both approaches view ARTs as either fixed or plastic during the life of an individual and among the plastic tactics, as either irreversible switches that occur at a particular age, condition, or under particular environmental situations, or fully reversible through the adult life of an individual. Fixed ARTs are thought to be due to organizational effects whereas plastic ARTs involve activational processes during the adult life of the individual. In many cases fixed ARTs are known to

involve a switch mechanism (Chapters 5, 6, 7, 8, and 10) during the development of the individual (e.g., horned beetles: Chapter 5), so the differences between the two processes may be a matter of developmental timing rather than a fundamental differences in underlying processes. Nonetheless, identifying these patterns has allowed both evolutionary and mechanistic studies to begin to identify the factors controlling the expression and evolution of ARTs.

The ARTs literature has emphasized males but cases of female ARTs are scattered among most taxonomic groups. Female ARTs include brood parasitism (Chapters 8 and 12); tactics to avoid male coercion (Chapters 8 and 18); reproductive dominance and suppression (Chapters 8, 14, and 15); alternative colony-founding tactics (Chapter 8) and cobreeding (Chapter 19); monandry and polyandry (Chapters 12 and 19); differences in fecundity or investment in eggs (Chapters 2 and 12); and consistent differences among females in preferences for males (Chapters 13, 17, and 18). As with males, female ARTs can be expected to evolve under conditions of intense intrasexual or social competition and high reproductive skew. Since females are generally thought to be under less intense sexual selection than males, it is not surprising that female ARTs are less common. Of course, it is also possible that female ARTs, like female sexual behaviors generally, are simply less easily observed than male ARTs. Certainly, condition-dependent and frequency-dependent fitness effects can occur in females so it is important to search for female ARTs more explicitly if we hope to understand the evolution of reproductive behavior.

20.1.2   Crossing fitness curves and beyond: the importance of modeling

ARTs are found across a wide range of taxonomic groups and across different stages of the reproductive process. ARTs occur most often in the context of intrasexual selection when (a) there are reproductive opportunities for those individuals that opt out of costly or high-risk male-male interactions to seek mating opportunities under less competitive circumstances (e.g., fighting vs. dispersing males); or when (b) parasitizing of costly male investment (e.g., bourgeois vs. parasitic males) can result in fitness gains, and (c) patterns arising in response to intersexual selection are much less common but include female choice for direct benefits (e.g., useless nuptial gifts) and mate conflict (e.g., noncourtship and forced copulation). ART ay also evolve when mating opportunities arise because emales are found in different habitats that require different male adaptations. For example, unmated female fig wasps Chapter 8) can be found both inside and outside the natal g; males that remain inside must fight for access to females o they develop enlarged heads and mandibles and have educed wings whereas males that leave the natal fig must y and they do not have to fight for females so they have ormal mandibles and wings. In all cases the model of crossing fitness curves has been used to understand the maintenance of alternative tactics (figures in various chapters). This simplistic model says little more than that individuals switch from one tactic to the other under the conditions that maximize their fitness or that population-ide frequencies of alternative tactics converge on the ESS. ven this simple model, however, is rarely tested experi-entally. It should be possible, for example, to change the proportions of the tactics in a population and predict a hange in their fitness. It should also be possible to change he pay-offs to the tactics and thereby predict a change in neir frequencies. Such experimental approaches are equired to determine whether our view of the evolution of RTs is supported.

Better models are needed for understanding ARTs. Current theory is generally based on fixed rather than dynamic fitness functions. New models need to incorporate differences in individual condition and status and the mechanisms and decision rules by which tactics are witched, as well as frequency and density dependence (and neir interactions) (Chapter 4). Such models will require dynamic game modeling as well as measures of physiological condition and the benefits that accrue to alternative tactics. he dynamics of between-sex interactions are as important understand as the dynamics of within-sex interactions. or this reason both male and female tactics and male–male conflict need to be incorporated into modeling forts (Chapter 18). Models predict that females should adjust their mating decisions to males of different tactics epending on their condition and the relative fitness gains sociated with direct vs. indirect benefits. Multilevel games ill be important when female choice affects the pay-offs to ales choosing among ARTs and when the frequency of ale tactics affects the pay-offs to females in their choice mong reproductive tactics. Furthermore, the success * alternative phenotypes often depends on spatial and mporal dynamics. In some well-studied cases (e.g., hapters 12 and 13), fitness is affected by the number of, stance from, and morph of neighbors, as well as habitat

characteristics and seasonal patterns. Recent modeling efforts (Formica *et al.* 2004, Hamilton *et al.* 2006, Koseki and Fleming 2006, Vercken *et al.* 2007) show that the frequency and coexistence of ARTs are shaped by such spatial and temporal dynamics. Modeling of complex systems like ARTs is crucial because it reveals hidden assumptions and often results in counterintuitive predictions.

20.1.3 Equality of fitness?

As Darwin (1871) observed, when two differing male forms are found in the same population at the same time, both must have "certain special, but nearly equal advantages from their differently shaped organs." This general expectation for equality of fitness and evolutionary stability for alternative phenotypes has been supported in a number of cases where mixed evolutionarily stable states (ESSs) and frequency dependence are involved (e.g., Chapters 9, 10, and 12). However, many studies have demonstrated that alternative phenotypes are maintained in populations without equal fitness. It is generally agreed that the majority of these ARTs are conditional on environmental, social, or individual status (e.g., Chapters 8, 11, 14, and 15). When each individual follows a conditional tactic that maximizes its fitness, equal success among the tactics is unlikely (Chapter 2; Repka and Gross 1995). There are studies, however, in which differences in success between morphs are almost certainly attributable to the fact that it is very difficult to measure the reproductive success of both tactics with equal reliability. In general the sedentary, monopolizing, and higher-variance or bourgeois tactic has been found to have higher fitness than the dispersing, sneaker, lower-variance or parasitic tactic, which is also the tactic whose fitness would be most difficult to measure. In particular, the success of all males must be counted including those that disperse, those that die before reproducing, and those that never inseminate any females (Chapter 9). This almost certainly means that some studies have prematurely claimed unequal fitness between tactics (Shuster and Wade 2003).

Many of the assumptions we have made in the past about costs and benefits of alternative tactics are being challenged. Most studies have assumed that females would not choose to mate with parasitic males if given the option and indeed there is experimental evidence for this in some species (Alonzo and Warner 2000, Gonçalves *et al.* 2005). However, in several recent studies, females have been shown to mate polyandrously both with the faster-growing parasitic males, thereby increasing their indirect benefits, and with parental males, thereby increasing their direct benefits (Chapter 17; Neff 2004). In still other species females apparently mate preferentially with parasitic males to increase genetic benefits but also with aggressive, monopolizer males to reduce the costs of male coercion (Watters 2005). Such female preferences may help to explain the frequency of "cuckoldry" in many ARTs systems and the apparent tolerance that some bourgeois males show toward satellite or parasitic males.

20.1.4 Competition and cooperation in ARTs

Although most male ARTs have been viewed as highly competitive, there is increasing evidence that cooperation can be an adaptive competitive tactic (Taborsky 2001). If females copy the mate choice decisions of other females or if females actively prefer multiple males (for example, to ensure fertilization of all the females' eggs), then a bourgeois male may gain from the presence of satellite males (Chapters 10 and 13). Cooperative behavior offered by reproductive parasites to bourgeois competitors may reduce the costs of competition for both parties, or provide incentives to the bourgeois male to be more tolerant of the parasite's presence. In some species males can increase their success by allowing particular males (e.g. dull yearlings or males of the same morph) to occupy adjacent territories (Chapters 12 and 13) or by forming alliances with other males (Chapters 14, 15, and 19). The pay-offs to competitive interactions are likely to be dynamic and the degree of reproductive skew variable, ranging from competitive to neutral to more beneficial for one party or the other. The spatial and temporal dynamics of reproductive pay-offs will be influenced by individual status, social conditions, and frequency dependence, which means that extensive modeling of these systems will be required.

20.1.5 Evolution of tactic frequencies and the importance of frequency dependence

Models with crossing fitness curves predict the stable frequency of alternative morphs in the population. Considerable theory exists for the evolution of sex ratios (stable frequencies of two sexes in a population: Charnov 1982, Hardy 2002), but there is little information on the population-wide frequency of other alternative phenotypes (Chapter 2). When alternative tactics are frequency dependent then they should evolve like sex ratios. Factors such as the relative costs of producing the two morphs

(cost ratios) and the relative reproductive values of the two morphs should affect equilibrium morph ratios as they do sex ratios. To add another level of complexity, combinations of conditional and pure or "mixed strategies" (e.g., gynodioecy) are well-known sex-allocation patterns (Chapter 2) and models suggest that such combinations should also occur in ARTs (Hazel et al. 2004, Plaistow et al. 2004). Trimorphic ARTs are also frequency dependent, often with rock–paper–scissors intransitivities that result in cyclical patterns of morph frequencies (Sinervo and Calsbeek 2006). However, not all ARTs are frequency dependent; in some cases ARTs evolve in niches in which some fitness can be achieved. In these cases morph ratios should reflect the relative frequencies of the two niches in the environment (Chapter 2).

Both frequency dependence and density dependence can maintain alternative phenotypes in a population. Frequency and density dependence often interact in complex ways (Chapter 19). For example, in the damselfly Ischnura ramburi, one female morph mimics males thus avoiding costly matings, but since males learn to recognize male mimics, the system shows negative frequency dependence (the rarer morph is more successful; Chapter 8). However the mating cost to females (male mimics) is incurred only at high densities, so the intensity of frequency-dependent selection in this system changes with density (Brockmann 2001, Sirot et al. 2003). This means that population-level differences in demographic structure (e.g., density and operational sex ratio) can affect the social context for mating and influence the direction of sexual selection at different temporal and spatial scales (Chapter 11).

Although generally regarded as important for the maintenance of ARTs (and for maintaining other polymorphisms), few studies have evaluated frequency dependence experimentally (Conover and van Voorhee 1990, Hori 1993, Basolo 1994, Roff 1996, Giraldeau and Livoreil 1998, Olendorf et al. 2006). If frequencies of morph are manipulated then changes should occur in the reproductive success of the manipulated and alternative morphs and predicted population-wide adjustments in frequencies should follow.

20.1.6 Information is important

Those studying ARTs often stress the importance of knowing the environmental information available to the individual and the value of that information in affecting the expression of alternative tactics. For example, when little

formation is available about the association between
ehavior and fitness, then mixed strategies are favored
Brockmann 2001). Environmental predictability is often
mentioned as a selective pressure favoring alternative tactics
ut few studies have examined the ability of individuals to
etect and evaluate this variable and to act on the basis of
uch information. Clearly, the amount and nature of the
formation available to individuals and the benefits and
osts associated with gathering and responding to this
formation will influence the evolution of ARTs (Leimar *et*
l. 2006).

ARTs are also shaped by the communication networks in
hich they occur. The presence of eavesdroppers such as
ocial competitors and predators will affect the evolution of
gnals associated with reproductive tactics (Chapter 16).
or example, bourgeois males reduce their courtship sig-
aling behavior when the reproductive success of both
males and bourgeois males declines with increasing num-
ers of parasitic males (Alonzo and Warner 1999). Selection
ill favor females that either thwart or encourage eaves-
ropping by parasitic males depending on whether females
se or gain fitness from mating with these individuals.

Some ARTs depend on misinformation. For example,
ale damselflies apparently cannot distinguish male mimics
om males (Sherratt 2001) and male peacock blennies
nnot distinguish female mimics from females (Gonçalves
al. 2005). Such mimicry systems depend on morph-
ecific costs and benefits as well as on an underlying
echanism for mate recognition. If, for example, mating is
stly to females and males learn to recognize females then
re morphs will always have an advantage, a frequency-
ependent effect that will maintain variation in female
aits (Fincke *et al.* 2005). A quantitative, decision-theory
pproach (Dall *et al.* 2005) to analyzing the information
vailable to individuals is needed to understand both the
volution of and mechanisms underlying ARTs.

0.1.7 Understanding trade-offs

he evolution of alternative tactics is grounded in the
otion of trade-offs: allocating resources in one direction will
duce alternative possibilities. For example, dispersing
ales reduce the costs of fighting or maintaining territories
ut they do not attract females, they are exposed to increased
vels of sperm competition, and they may have a shorter
espan. Animals that need wings for flight must expend
sources on developing flight muscles, resources that then
nnot be used for other activities such as producing eggs.

This intuitive concept of trade-offs as direct competition
for limited resources among different body functions has
been examined in studies of wing-dimorphic crickets. The
greater ovarian growth and reduced lipid biosynthesis of the
wingless morph relative to the winged morph was found to
result from the differential allocation of internal reserves to
the two traits (Zera and Harshman 2001). Knowledge of
the underlying basis for trade-offs will help evolutionary
biologists understand the likely evolutionary trajectory for
ARTs and assist physiologists in understanding the developmental expression of ARTs (Emlen 2001, Emlen *et al.*
2005, Simmons and Emlen 2006).

20.1.8 Origin and evolution of ARTs

ARTs are found more commonly in some groups than in
others (e.g., Chapters 8 and 11). To some degree this bias
results from differences between taxonomic groups in their
depth of study, but clearly this is not the only explanation.
ARTs occur most commonly in groups with intense
sexual selection; where there are large investments by males
or females; when males have the ability to monopolize
some mating opportunities; and when opportunities exist
to partially fertilize clutches of polyandrous females
(Chapters 2, 8, and 11). ARTs can magnify intrasexual
variance in reproductive success and may lead to more
intense sexual selection on some male tactics than on others.

ARTs seem particularly common among fishes
(Chapter 10). The factors that may influence the frequent
evolution of ARTs in fishes include the presence of indeterminate growth, which results in large size differences
among males; external fertilization, which selects for
large quantities of sperm and the opportunity for sperm
competition; and the frequency of parental care, a high-
investment tactic that can be exploited by other males. One
tentative conclusion about the phylogeny of male ARTs
(as described for three groups of fishes) is that although
ARTs arise repeatedly during the course of evolution, they
are found only at the tips of branches (Chapter 3). This
means that ARTs rarely become permanent features of
deeper clades. This is not surprising since the frequency of
ARTs is so responsive to environmental and social conditions, which means that one of the morphs can easily go
extinct (Chapter 3). However, once ARTs have evolved and
there is a developmental uncoupling of dimorphic structures and behavior, then the alternative tactics may evolve at
least partially independently (Chapter 5; West-Eberhard
2003).

20.1.9   Integrating mechanisms with evolution

One important remaining challenge for the study of ARTs is to understand how the mechanisms that cause variants to arise are favored by selection. When we say that ARTs have evolved in a lineage, we mean that a developmental process has caused a change in the phenotype of some individuals in that lineage. That process is likely affected by genetic, physiological, and environmental factors. There are two established mechanisms that will produce dimorphism, a threshold mechanism and a change in the scaling of the power relationships among traits (Chapters 5, 6, 9, and 10). When differences in phenotype are due to heritable differences in the underlying mechanism and when these differences result in fitness differences then selection on the mechanisms will occur. When disruptive selection is combined with heritable developmental threshold mechanisms, then genetic correlations between tactics will be minimized by partially uncoupling gene expression of the alternative tactics permitting them to evolve along independent trajectories. This mechanistic model for the evolution of ARTs needs to be verified and many questions remain. What genes are involved in these threshold mechanisms and are there patterns or homologies for the genes affecting ARTs? How does the genetic architecture of ARTs change as tactics evolve and are there common patterns or underlying constraints in this process? Are there constraints on the underlying mechanisms that affect the kinds of options that are available as ARTs or that make some ARTs more common than others (e.g., the frequency of sexual mimicry would suggest that this is the case)? A fully integrative approach is needed to answer these questions, an approach that unites mechanistic studies with evolutionary modeling and empirical tests.

20.2   WHAT DOES THE STUDY OF ARTs TELL US ABOUT OTHER BIOLOGICAL PROBLEMS?

ARTs provide a valuable tool for understanding biology, from genes through development and physiology to behavior and morphology. The reviews in this book have demonstrated that ARTs are common among organisms and that alternative phenotypic pathways have evolved multiple times (Chapters 3 and 5; Emlen et al. 2005). This means that ARTs provide a rich pallet from which to investigate a number of important biological questions. ARTs are also part of a much larger category of alternative allocation phenotypes that covers a wide diversity of animal adaptations (Chapter 2). Just as the study of sex allocation or game theory can improve our understanding of ARTs, the study of ARTs can provide insights into the mechanisms, evolution and maintenance of variation in populations.

20.2.1   The control, origin, and evolution of complex phenotypes (suites of characters)

From a mechanistic perspective, one of the most remarkable (and valuable) features of ARTs is that there is a dissociation between the control and expression of ARTs (Chapter 7, Figure 7.1). Most animals show clear differences in morphology, physiology, and behavior between males and females. But in species with alternative tactics one sex may look and act like the opposite sex. For example in the damselfly Ischnura ramburi, females of one morph (andromorphs) are physically and behaviorally similar to males and they are often treated like males (Chapter 8; Robertson 1985, Sirot et al. 2003). In many fish species such as bluegill sunfish and peacock blennies, parasitic males mimic females in appearance and behavior and thereby escape bourgeois male aggression (Chapter 10; Dominey 1980, Gonçalves et al. 1996, 2005; see Taborsky 1994 for review). This decoupling of the expression of behavioral and morphological male traits (i.e., secondary sexual characters) from gametogenesis offers unique opportunities to study the proximate mechanisms of reproduction (Moore 1991, Moore et al. 1998, Oliveira et al. 2005, Oliveira 2006). Causal mechanisms underlying individual variation in reproduction can be studied much better in species with ARTs, since within-sex variation in reproductive traits is not confounded by the effects of gender.

The expression of alternative reproductive tactics is often a function of behavioral plasticity (Gross 1996) which is primarily triggered by structural reorganization (for slow and long-lasting changes) and biochemical switching (involving neuromodulators such as catecholamines, which allows faster and reversible changes: Zupanc and Lamprecht 2000). Both mechanisms usually depend on organizational and activational effects of hormones (Arnold and Breedlove 1985, Moore 1991, Moore et al. 1998, Oliveira 2006). In vertebrates sex steroids, glucocorticoids, and neuropeptides are major candidates for the differentiation and maintenance of ARTs (Chapter 7, Brantley et al. 1993, Nelson 2005). Because similar regulatory mechanisms are involved in the development and

ifferentiation of sex (Crews 1998, Crews *et al.* 1998, Devlin
nd Nagahama 2002, Godwin and Crews 2002, Oliveira
005, Vigers *et al.* 2005), the study of proximate mechan-
ms underlying ARTs has important general implications.
n both cases developmental processes need to be con-
idered to understand how factors taking effect early in
ntogeny can shape consistent individual variation within
nd between sexes (see Crews and Groothuis 2005).
Jnderstanding the proximate mechanisms underlying the
egulation of ARTs might also help to explain other prob-
ems such as alternative life-history patterns (Turner and
irosse 1980, Meyer 1987, Lu and Bernatchez 1999, Jonsson
nd Jonsson 2001, Kurdziel and Knowles 2002) or the mode
f action and importance of threshold mechanisms (Chapter
, West-Eberhard 1989, Roff 1996, Zera and Denno 1997,
Hartfelder and Emlen 2005).

0.2.2   The proximate causes of ARTs and functional genomics

RTs offer a unique opportunity to study the physiological
nechanisms and the genetic architecture underlying the
ecoupling of traits that are usually present in concordance,
nce in the parasitic tactic a mosaic of both male and female
aits may be present instead of a constellation of gender-
pecific traits (e.g., species with parasitic males that mimic
males in which the maturation of the male gonad is
ecoupled from the expression of male secondary sex
haracters and male sexual behavior). Therefore, ARTs
rovide insight into the proximate mechanisms linking male
nd female phenotypes under conditions that are not
athological.

Neuroendocrinological studies have shown that dif-
erent neuropeptides (e.g., GnRH, AVT) and steroids are
perating either independently or in concert to coordinate
e expression of a suite of characters, characteristic of a
ven tactic (see Chapters 6 and 7). These studies have also
emonstrated that the decoupling of different male traits
parasitic males may be achieved either by differences in
ormone levels or by varying the local microenvironments
the different target tissues, due to differential expres-
on of receptors or to differential levels of activity of
atabolic enzymes that modulate the availability of the
tive hormone to specific targets (e.g., 11-β-hydroxilase
d 11-β-HSD that metabolize T into KT). These are key
eps in the expression of male secondary sex characters,
ermatogenesis, and the modulation of the expression
reproductive behavior in male teleosts (Oliveira 2006).

For example, in the protogynous wrasse *Halichoeres
trimaculatus*, which has ARTs, the relative levels of brain
steroid receptors vary between alternative reproductive
phenotypes, with levels of androgen receptor transcripts
being significantly higher in the brain of terminal-phase
males than in initial-phase males, whereas no other significant differences in gene expression were observed
either for androgen or for estrogen receptors in the gonads
or for estrogen receptors both in the brain and in the
gonads (Kim *et al.* 2002). Thus, by varying the expression
of androgen receptors in specific tissues (brain vs. gonad),
terminal-phase males can both increase their sensitivity to
circulating androgen levels in specific targets (the brain),
and at the same time decouple the effects of androgens
in different target tissues by varying androgen receptor
densities, so that unwanted effects of androgens can be
avoided. This mechanism hypothetically makes it possible
to activate the expression of an androgen-dependent
reproductive behavior in bourgeois males without having
the associated costs of increasing spermatogenesis or the
expression of a sex character, since the androgen action can
be independently modulated at each compartment (brain
vs. gonad vs. morphological secondary sex character).
Studies focusing on target tissues are thus a major avenue
for future research in this area.

One emerging approach when studying target tissues has
been the use of functional genomics tools that are now
becoming more accessible. DNA microarrays allow the
monitoring of large sets of genes (thousands) in key tissues
(brain, gonads, somatic ornaments), hence making it possible to identify genes and regulatory networks which are
consistently up- or downregulated between alternative sex
types. Genes that are differentially expressed in alternative
morphs are potential candidates to be involved in the
expression of the alternative tactics (Hofmann 2003). Since
ARTs involve differences in the expression of reproductive
behavior between alternative phenotypes, this approach has
concentrated on comparing the brain gene expression profiles between morphs. In the Atlantic salmon *Salmo salar*,
gene expression profiles were compared between sneaker
males and immature juveniles (of the same age) that represent alternative life histories (Aubin-Horth *et al.* 2005).
The immature males will later migrate and then return to
the breeding grounds where they will reproduce as bourgeois males. Fifteen percent of the genes included in the
array (ca. 3000) were expressed differentially between the
sneaker and the juvenile immature males (Aubin-Horth
et al. 2005). In sneaker males most of the upregulated genes

are involved in reproduction and related processes (e.g., gonadotrophins, growth hormone, prolactin, and POMC genes) whereas in immature males upregulated genes are associated with somatic growth (e.g., genes involved in transcription regulation and protein synthesis, folding, and maturation). This reflects at the cellular level the classic life-history trade-off between reproduction and growth, illustrated by these two alternative phenotypes. Another set of genes upregulated in sneakers are involved in neural plasticity (e.g., genes coding for synaptic function and for cell-adhesion glycoproteins that have been implicated in memory formation) and in neural signaling (e.g., genes coding for nitric oxide synthesis, a neurotransmitter involved in the regulation of neuropeptide action). This difference has been interpreted as suggesting that the expression of the sneaker tactic is more demanding at the cognitive level (Aubin-Horth et al. 2005). Therefore, a functional genomics approach may not only allow the confirmation of expected differences in the profiles of gene expression of specific target tissues between alternative phenotypes, but it may also reveal differences in gene expression between morphs in otherwise unsuspected biological processes (e.g., neural plasticity).

20.2.3 Behavioral syndromes

Ethologists have long realized that there is often consistent behavioral variation between individuals of a species (Bagg 1916, Lorenz 1935, Tinbergen 1951, van Oortmerssen 1971, Huntingford 1976, Benus et al. 1987, Clark and Ehlinger 1987, Riechert and Hedrick 1993, Verbeek et al. 1994). Interestingly, a broader systematic study of the mechanisms underlying such individual variation has begun only recently, but with all the more vigor (e.g., van Oortmerssen and Bakker 1981, Ehlinger and Wilson 1988, Benus et al. 1991, Hessing et al. 1994, Dingemanse et al. 2002, van Oers et al. 2004, 2005, Both et al. 2005, Kralj-Fiser et al. 2007; reviews in Wilson 1998, Koolhaas et al. 1999, Gosling 2001, Sih et al. 2004a, b, Groothuis and Carere 2005). The consistent tendency of individuals to behave in a certain way, either in a particular behavioral context (e.g., in resource competition) or across contexts (e.g., in exploration, predator avoidance, and dominance interactions) has been referred to as an animal's coping style, behavioral type, profile, or tendency, or – in analogy to a term used in human psychology – "personality" (Gosling and John 1999, Kolhaas et al. 1999, Carere and Eens 2005, Groothuis and Carere 2005). Suites of correlated behaviors

have been called behavioral syndromes (Sih 2004a, b), which refers to a property of a population of individuals denoting a correlation between rank-order differences of individuals through time or across situations (Bell 2007).

The study of ARTs and behavioral syndromes has much in common. Variation in behavior, regardless of whether it is discontinuous or continuous, is strongly influenced by developmental processes (Caro and Bateson 1986, Meaney 2001, Stamps 2003); it is heritable (Benus et al. 1991, Sinervo and Zamudio 2001, Drent et al. 2003); it depends on the abiotic, biotic, and social environments (Emlen 1997 Benus and Henkelmann 1998, Groothuis and Carere 2005) and it has profound fitness consequences (Shuster 1989 Ryan et al. 1992, Dingemanse and Reale 2005), which affect populations and thereby have important ecological and evolutionary implications (Gross 1991, Bolnick et al. 2003 Sih et al. 2004a). The most significant difference between alternative allocation phenotypes (AAP) such as ARTs and behavioral syndromes (BS) seems to be the form of the trait distributions. AAP are characterized by discontinuous or bimodal/multimodal phenotype distributions (Chapters 1 and 2), whereas BS usually show a continuous, unimodal distribution (Wilson et al. 1994). In both cases, it is of paramount interest to understand the ultimate and proximate mechanisms causing and maintaining behavioral variation within populations.

Students of ARTs can learn from BS research that individual differences in behavior can represent limited plasticity. If behavioral tendencies "carry over" between different contexts due to the make-up of an organism "optimal" responses to particular problems might not be expected (Sih et al. 2004b). Imagine a benefit exists in a population for high aggression levels in competition for food resources. Individuals thereby selected to be very aggressive may not be able to overcome such tendencies in the reproductive context. In other words, they may be predisposed to perform a bourgeois reproductive tactic even if a parasitic tactic may provide higher rewards. The important message here is that correlations among traits can act as evolutionary constraints (Duckworth 2006). We may not find the expected optimal behavior because correlated responses to selection on nontarget traits can depend on genetic correlations (Lande and Arnold 1983, Roff and Fairbairn 1993). Traits may evolve together as packages. I would be worth studying also in the context of ARTs, which behaviors are correlated across which contexts, how stable these correlations are, and which evolutionary processes and physiological mechanisms might be responsible for the

xistence of such correlations. This is an addition to the
2 pertinent questions asked in Chapter 1 to aim at a
omprehensive understanding of the evolution of ARTs.

In return, research on BS may benefit from the knowledge
f principles developed in the long history of AAP studies
Brockmann 2001). The theory developed to explain sex
llocation, alternative life histories, and ARTs may yield
dequate approaches and methods to resolve questions of
ae coexistence of alternative coping styles (Chapter 2).
Conceptual understanding and empirical results have accu-
ulated in the research of ARTs (Chapter 1) that may
gnificantly further the comprehension of BS, at least by
rning the focus towards the most urgent questions. The
aajority of the 12 questions asked in Chapter 1 are also
elevant for the study of BS. For example, it seems presently
nclear to what extent behavioral types such as shyness or
oldness are flexible (or reversible) over a lifetime (Wilson
al. 1993, Coleman and Wilson 1998, Frost et al. 2007). Are
reshold mechanisms and developmental switches involved
the generation of diverging behavioral profiles (Groothuis
d Carere 2005), and if so, to what degree are these switches
d threshold mechanisms subject to natural selection (Roff
998)? The integrative approach to the study of ARTs
Chapters 2, 5, 7, 8, 9, 10, 12, and 15) may be particularly
luable for BS research, because both physiological and
volutionary mechanisms must be considered for a compre-
ensive understanding of individual variation in behavior
Coolhaas et al. 1999, Oliveira et al. 2005).

The study of behavioral syndromes still suffers from a
ck of quantitative information in most natural systems
ih et al. 2004b). A crucial question in BS research is the
ature of phenotype distributions. Is the distribution of shy
d bold, proactive and reactive, sedentary and roaming
havioral types indeed continuous and unimodal as is
ually assumed or is there sometimes evidence for
nderlying disruptive selection processes? Surprisingly,
is question is hitherto largely neglected in the study of BS
ih et al. 2004a, b, Bell 2007). If disruptive selection is
volved, concepts developed in the study of AAP could be
plied. If instead BS are characterized by uniform distri-
tions, the persistence of systematic individual differences
behavior despite apparent absence of disruptive selection
eeds to be explained. What is the importance of density-
d frequency-dependent selection under these conditions
all et al. 2004, Wilson et al. 1994)? Why is behavioral
asticity hampered (DeWitt et al. 1998, West-Eberhard
03)? To what extent is state dependence involved in the
pression of behavioral phenotypes (Dall et al. 2004)? This

is where the study of ARTs and BS can effectively
complement one another.

20.2.4   The role of ARTs in speciation

Intraspecific alternative adaptations predispose populations
to speciation because they represent fitness trade-offs under
natural selection (West-Eberhard 2003). When divergence
in the form of alternative phenotypes has developed, particular variants may be fixed in certain subpopulations due
to assortative mating, environmentally mediated change in
expression, or frequency-dependent selection. Examples
include (1) socially parasitic inquiline ants that reproduce by
laying eggs in the colonies of other ant species; there
is evidence that these ants have evolved by sympatric speciation due to parallel size-related alternatives in the
two sexes (Buschinger 1986; see Buschinger 1990, Bourke
and Franks 1991, West-Eberhard 2003 for review). (2) In
pacific sockeye salmon Oncorhynchus nerka, some individuals do not migrate to sea but stay in the rivers where they
were born to reproduce earlier than the anadromous conspecifics (Thorpe 1989). In the male sex, this usually
involves parasitic sneaking behavior. In some populations,
the marked size difference between stationary and migratory individuals of both sexes apparently leads to assortative
mating by size (Foote 1988, Foote and Larkin 1988),
which creates genetic divergence between the anadromous
sockeye and nonmigratory kokanee forms (see Foote
and Larkin 1988 for references). The ultimate cause for
assortative mating in systems with extensive, discontinuous
size variation may be the benefit of mating among mates
with a similar, precocious life-history type. (3) Lizards
have been suggested to nicely demonstrate the importance
of secondary sexual signals for the evolution of intrapopu-
lation divergence and sympatric speciation (Lande 1982,
West-Eberhard 1983). In lizards with intraspecific and
intrasexual color polymorphisms and alternative mating
behaviors, frequency-dependent selection in combination
with assortative mating between like-types and reduced
hybrid fitness may further genetic diversification and speciation (Hochberg et al. 2003, Sinervo and Calsbeek 2006).
When assortative mating is not linked to resource competition, genetic drift may break the linkage equilibrium
between the trait responsible for mate selection and the
respective ecological traits (Dieckmann and Doebeli 1999),
hence leading to reproductive isolation between ecologically diverging subpopulations. (4) Color polymorphisms
and negatively frequency-dependent selection that might

be associated also with alternative mating behaviors may cause speciation in the rapidly radiating Lake Victoria cichlids (Seehausen and Schluter 2004). Male–male competition and aggression focused on like-types generates negative assortment of nuptial color patterns among habitats. Whether and to what extent alternative mating behaviors may be involved in this diversification process is yet unclear.

In species with male ARTs, reproductive parasites or "sneakers" may participate in fertilization attempts of other species (Crapon de Caprona 1986, Taborsky 1994, Jansson and Ost 1997, Wirtz 1999) or forcefully copulate with heterospecific females (Seymour 1990, Russell *et al.* 2006). This causes hybridization, a speciation mechanism that is often underrated (Mallet 2007). It may be particularly important in lineages with rapid adaptive radiation such as the cichlids of the Great African Lakes (Salzburger *et al.* 2002, Seehausen 2004). Interspecific fertilization by reproductive parasites is probably a frequent phenomenon because it is inherently "cheap," i.e., reproductive parasites only contribute sperm and do not invest in secondary sexual characters, courtship, nest building, and brood care (Taborsky 1994). This is one likely cause of unidirectional hybridization, which is apparently more frequent than reciprocal hybridization (Wirtz 1999). Phylogenetic analyses of ARTs in fishes suggest that they are distributed near the tips of the phylogenetic trees (Chapter 3), which might indicate a functional link between speciation and the evolution of alternative mating tactics.

ARTs may result from divergent reproductive niches due to habitat differences or when same-sex competitors show bimodal or multimodal trait distributions caused by natural selection (Chapter 2; Denno 1994, Skúlason and Smith 1995, Pigeon *et al.* 1997, Danforth and Desjardins 1999, Jonsson and Jonsson 2001). Trophic morph divergence, for example, may strongly affect reproductive options and thereby relate to the mating tactics used, such as in Arctic charr (Jonsson and Jonsson 2001, Snorrason and Skúlason 2004). This may drive speciation by disruptive selection under conditions of at least minimum ecological contact between the diverging lines (Smith and Skúlason 1996), whereby reproductive isolation may evolve surprisingly quickly (Hendry *et al.* 2000). When gene flow becomes severely restricted, further morph specialization may ensue, which finally gives rise to new species (Snorrason and Skúlason 2004). This is a promising area for future research into evolutionary mechanisms underlying speciation on the basis of intraspecific morph divergence.

20.3   BROADER IMPLICATIONS OF ARTs

The study of ARTs can contribute much to our understanding of fundamental issues in biology (as discussed above), but it can also contribute to our understanding of applied problems. In this section we address a few of these topics.

20.3.1   ARTs in conservation

Conservation is a growing field of knowledge in which the maintenance of biodiversity is a key goal. In conservation the species is the commonly used unit of biodiversity and intraspecific variations are usually overlooked (but see Bolnick *et al.* 2003). However, intraspecific variation represents an important component of ecologically functional diversity within a species and this translates into adaptive genetic variation of the population and hence its evolutionary potential. This means that polyphenisms including ARTs are a valuable part of biological diversity and should be considered in conservation actions. For example, the occurrence of sneaker males in the peacock blenny *Salaria pavo* is limited to lagoon populations in southern France and in southern Portugal (Ruchon *et al.* 1995, Gonçalves *et al.* 1996). Apparently, the scarcity of nest sites in the nearshore lagoon environments poses a constraint on male reproductive rate in this crevice-nesting species and the operational sex ratios become female biased (J. Saraiva and R. F. Oliveira, unpublished data). This leads to a sex-role reversal in courtship behavior, with females taking the leading role in courtship in the lagoon population (Almada *et al.* 1995). ARTs are also present among males in these populations, with younger and smaller males mimicking female courtship behavior in order to gain access to nests during spawning episodes (Gonçalves *et al.* 1996). Hence although this species is common in rocky shores in the Mediterranean Sea and in adjacent areas from northern Morocco to the Bay of Biscay (Zander 1986), the lagoon populations deserve a special conservation status for the intraspecific variation they exhibit.

The occurrence of intraspecific diversity should be preserved not only because of their intrinsic interest as unique biological phenomena, but also because such local adaptation adds to the species genetic assets and flexibility and thus can contribute to the evolutionary potential and long-term survival of the species (Buchholz and Clemmons 1997). This is illustrated by our discussion of the potential le of ARTs in speciation. But species with ARTs may also maintain greater intraspecific genetic diversity than populations without such variation. For example, alternative dispersal tactics, which are often associated with ARTs (chapters 2 and 8), will influence genetic variation and population viability as well as the effectiveness of release programs (Thomas *et al.* 2000). One problem in the conservation of highly sexually selected species, which are some of our most spectacular species, is that genetic diversity and the effective size of the breeding population is constrained by the presence of a few highly preferred males (Parker and Waite 1997). However, when ARTs are present, genetic variance is more likely to be maintained and inbreeding problems reduced.

The presence of ARTs should be considered whenever human intervention is planned. For example, an effort to increase the number of nest sites for rare cavity-nesting birds by placing nest-boxes in the environment at high densities had an unanticipated result, an increase in intraspecific brood parasitism (Eadie *et al.* 1998). This facultative, female reproductive tactic turned out to be density dependent and an increase in nesting sites ended up decreasing population growth. ARTs should also be considered in resource exploitation, since harvesting may lead to the selective removal of one of the reproductive morphs. For example, because bourgeois males are larger than parasitic males, and because harvest activities are usually directed towards the largest individuals in the population, bourgeois males are more likely to be removed. The effects of selective removal of specific morphs on population persistence and genetic diversity are still poorly understood. However, it can be predicted that the removal of larger individuals may influence life-history decisions leading to a reduction in size at sexual maturity and concomitantly to a reduction in female fecundity (assuming size-dependent fecundity) (Vincent and Sadovy 1998). Also, the differential removal of males in a population may lead to sperm limitation and consequently to a reduction in female fecundity with an impact on population persistence (e.g., male biased culture in ungulates: Ginsberg and Milner-Gulland 1994). Knowledge of ARTs has already influenced management decisions. In the United States, sunfishes (*Lepomis* sp.) are an important resource for sport fishing and some species of sunfish exhibit ARTs consisting of three male types: nest-guarding parental males, female-mimicking satellite males, and sneaker males that dart into the nests of parental males during spawning to release sperm (Gross 1982). The fishery disproportionately removes parental

males both because they are larger and because they are site-attached to their nests making them easier to target. As a consequence the unguarded eggs and larvae of captured parental males are cannibalized by other individuals, thus reducing the survival of the young. To control the impact of this selective removal, fisheries management policies were revised and the frequency of alternative males in the population is being controlled (M. Gross, personal communication, in Vincent and Sadovy 1998). Similarly, increased fishing pressure on the large, anadromous, hooknose male salmon may end up increasing the proportion of the less desired, small jacks in the population (Gross 1991b).

20.3.2   ARTs in pest management

The occurrence of alternative reproductive phenotypes can play a key role in the effectiveness of pest management strategies, since variants can be maintained or selectively targeted by specific pest management practices. Red imported fire ants *Solenopsis invicta* introduced into the United States from South America in the 1930s and 1940s rapidly became a pest in southern states because of their negative economic impact. They produce large nest mounds that may damage agricultural equipment or even promote the collapse of road sections by removing the soil under the asphalt; they inflict significant damage to agricultural crops (e.g., soybeans, eggplant, corn, etc.) and livestock; and they have a painful sting that may be dangerous to sensitized people that develop an allergic reaction to their venom. Different pest management methods have been developed to control the populations of the red imported fire ant, including pesticides and biological control agents. Recently a microsporidian pathogen (*Thelohania solenopsae*) with a high prevalence rate that may reduce or even kill colonies has been detected in the US populations of fire ants, thus having a potential role as a biological control agent (Williams *et al.* 2003). Since *T. solenopsae* can be transmitted by the introduction of infected broods into a colony, the degree of inter-colony brood transfer is a key factor for its spread. In fire ants two types of social organization are commonly present: monogyny, in which a single egg-laying queen is present per colony, and polygyny, in which multiple queens are present per colony (Ross and Keller 1995). These two ethotypes also differ in their social behavior: polygyne colonies exchange workers, food, brood and mated females, whereas monogyne colonies are very territorial (Tschinkel 1998). Therefore, the social structure of the colonies is expected to moderate the infection by *T. solenopsae*. As predicted, polygyne colonies have a much higher prevalence of this infection in the field (Oi *et al.* 2004, Fuxa *et al.* 2005) and a longer persistence and faster spread of infections started in the laboratory than monogyne colonies (Oi 2006, Preston *et al.* 2007). Therefore, by affecting the dynamics of the microsporidium infection, the multiple social forms of these fire ant populations play a key role when considering the potential of this pathogen as a biological agent in pest management.

20.3.3   ARTs in medicine

Medical research is mainly focused on explaining how the body systems work and why some people are more vulnerable to a particular disease than others. The proximate mechanisms of disease vulnerability are commonly seen as resulting from evolutionary "defects" and random processes. More recently, the role of evolutionary constraints and host–parasite co-evolution have also been implicated in evolutionary approaches to the study of disease (Nesse and Williams 1998). The study of ARTs provides a different conceptual framework that promotes the view that genetic variants are likely to represent evolved alternative adaptations for different environments instead of the classic view as malfunctioning phenotypes. This ARTs view of adaptive variation is relevant to understanding the evolution of host–parasite interactions, the evolution of and spread of disease organisms, and antibiotic resistance.

20.3.4   ARTs in evolutionary psychology

The study of ARTs may help to explain the evolution of apparently nonadaptive human characters such as homosexuality. Same-sex sexual behavior is present in a large number of species from different vertebrate taxa, and it is commonly associated with ARTs (Bagemihl 1999). In humans sexual orientation shows marked sex differences: male homosexuality presents a somewhat bimodal distribution, whereas female homosexuality displays a more continuous distribution, from strictly heterosexual to strictly homosexual individuals (LeVay 1996). Therefore, the distribution of male homosexuality in humans resembles the discrete distribution of ARTs. However, despite displaying a similar pattern to same-sex sexuality in animals (e.g., female mimicry in males in order to get access to breeding females), human homosexuality does not share the same functional explanation. On the contrary, male

homosexual behavior is an evolutionary paradox since it is associated with decreased direct reproduction and other hypotheses, such as kin selection, also fail to account for its maintenance in the population (Bobrow and Bailey 2001). This paradox is further stressed by the prevalence of homosexuality in different cultures and by the fact that at least part of the variation in sexual orientation has a genetic basis (see Bailey *et al.* 2000 and references therein).

Thus, by providing the theoretical basis and the methodologies for the study of discrete, within-sex variation in sexual behavior, the study of ARTs may also contribute to the rigorous study of the complexities of human social behavior.

20.3.5   ARTs in education and the public understanding of science

Finally, we would like to finish the last chapter of this book by drawing attention to the fact that ARTs are an excellent topic for popular science films and books since they provide strong narratives that may increase the impact of science communication (Dingwall and Aldridge 2006). In this respect it is worth mentioning that one of the first scientific studies of ARTs was published by Desmond Morris (Morris 1952), one of the most active popular-science writers today. Moreover, since the boom in popular-science publishing during the 1990s, ARTs are being portrayed in many wildlife films and documentaries and featured in popular books (e.g., Judson 2002, Crump 2005). These popular accounts of ARTs communicate to the public both fascinating biology and important biological concepts such as the role of evolution in maintaining variation in populations.

We hope that in the near future ARTs can be further used in conveying to the general public one of the basic advances in biology in the last few decades that two levels of explanation are needed for the full understanding of any biological trait: (1) an evolutionary explanation regarding its function; and (2) a proximate explanation for how it works. These two levels should not be taught separately since they complement one another; the teaching of this emerging corollary of modern biology to the general public will be a challenge that represents "integration" at still a higher level.

References

Almada, V. C., Gonçalves, E. J., Oliveira, R. F., and
   Santos, A. J. 1995. Courting females: ecological constraint

affect sex roles in a natural population of the blenniid fish *Salaria pavo*. *Animal Behaviour* 49, 1125–1127.

lonzo, S. H. and Warner, R. R. 1999. A trade-off generated by sexual conflict: Mediterranean wrasse males refuse present mates to increase future success. *Behavioral Ecology* 10, 105–111.

lonzo, S. H. and Warner, R. R. 2000. Dynamic games and field experiments examining intra-and intersexual conflict: explaining counterintuitive mating behavior in a Mediterranean wrasse, *Symphodus ocellatus*. *Behavioral Ecology* 11, 56–70.

rnold, A. B. and Breedlove, S. M. 1985. Organizational and activational effects of sex steroids on brain and behavior: a reanalysis. *Hormones and Behavior* 19, 469–498.

ubin-Horth, N., Landry, C., Letcher, B., and Hofmann, H. 2005. Alternative life histories shape brain gene expression profiles in males of the same population. *Proceedings of the Royal Society of London B* 272, 1655–1662.

agemihl, B. 1999. *Biological Exuberance: Animal Homosexuality and Natural Diversity*. New York: St. Martin's Press.

agg, H. J. 1916. Individual differences and family resemblances in animal behavior. *American Naturalist* 50, 222–236.

ailey, J. M., Dunne, M. P., and Martin, N. G. 2000. Genetic and environmental influences on sexual orientation and its correlates in an Australian twin sample. *Journal of Personality and Social Psychology* 78, 524–536.

asolo, A. L. 1994. The dynamics fisherian sex-ratio evolution: theoretical and experimental investigations. *American Naturalist* 144, 473–490.

ell, A. M. 2007. Future directions in behavioural syndromes research. *Proceedings of the Royal Society B-Biological Sciences* 274, 755–761.

enus, R. F. and Henkelmann, C. 1998. Litter composition influences the development of aggression and behavioural strategy in male *Mus domesticus*. *Behaviour* 135, 1229–1249.

enus, R. F., Koolhaas, J. M., and van Oortmerssen, G. A. 1987. Individual differences in behavioural reaction to a changing environment in mice and rats. *Behaviour* 100, 105–122.

enus, R. F., Bohus, B., Koolhaas, J. M., and van Oortmerssen, G. A. 1991. Heritable variation for aggression as a reflection of individual coping strategies. *Experientia* 47, 1008–1019.

obrow, D. and Bailey, J. M. 2001. Is male homosexuality maintained via kin selection? *Evolution and Human Behavior* 22, 361–368.

Bolnick, D. I., Svanbäck, R., Fordyce, J. A., et al. 2003. The ecology of individuals: incidence and implications of individual specialization. *American Naturalist* 161, 1–28.

Both, C., Dingemanse, N. J., Drent, P. J., and Tinbergen, J. M. 2005. Pairs of extreme avian personalities have highest reproductive success. *Journal of Animal Ecology* 74, 667–674.

Bourke, A. F. G. and Franks, N. R. 1991. Alternative adaptations, sympatric speciation and the evolution of parasitic, inquiline ants. *Biological Journal of the Linnean Society* 43, 157–178.

Brantley, R. K., Wingfield, J. C., and Bass, A. H. 1993. Sex steroid levels in *Porichthys notatus*, a fish with alternative reproductive tactics, and a review of the hormonal bases for male dimorphism among teleost fishes. *Hormones and Behavior* 27, 332–347.

Brockmann, H. J. 2001. The evolution of alternative strategies and tactics. *Advances in the Study of Behavior* 30, 1–51

Buchholz, R. and Clemmons, J. 1997. Behavioral variation: a valuable but neglected biodiversity. In J. Clemmons and R. Buchholz (eds.) *Behavioral Approaches to Conservation in the Wild*, pp. 181–211. Cambridge, UK: Cambridge University Press.

Buschinger, A. 1986. Evolution of social parasitism in ants. *Trends in Ecology and Evolution* 1, 155–160.

Buschinger, A. 1990. Sympatric speciation and radiative evolution of socially parasitic ants: heretic hypotheses and their factual background. *Zeitschrift für zoologische Systematik und Evolutionsforschung* 28, 241–260.

Carere, C. and Eens, M. 2005. Unravelling animal personalities: how and why individuals consistently differ. *Behaviour* 142, 1149–1157.

Caro, T. M. and Bateson, P. 1986. Organization and ontogeny of alternative tactics. *Animal Behaviour* 34, 1483–1499.

Charnov, E. L. 1982. *The Theory of Sex Allocation*. Princeton, NJ: Princeton University Press.

Clark, A. B. and Ehlinger, T. J. 1987. Pattern and adaptation in individual behavioral differences. In: *Perspectives in Ethology, Vol. 7* (Ed. by P. P. G. Bateson & P. H. Klopfer), pp. 1–47. New York, Plenum.

Coleman, K. and Wilson, D. S. 1998. Shyness and boldness in pumpkinseed sunfish: individual differences are context-specific. *Animal Behaviour* 56, 927–936.

Conover, D. and van Voorhees, D. 1990. Evolution of a balanced sex ratio by frequency dependent selection in a fish. *Science* 250, 1556–1558.

Crapon de Caprona, M. D. 1986. Are preferences and tolerances in cichlid mate choice important for speciation? *Journal of Fish Biology* 29, 151–158.

Crews, D. 1998. On the organization of individual differences in sexual behavior. *American Zoologist* 38, 118–132.

Crews, D. and Groothuis, T. 2005. Tinbergen's fourth question, ontogeny: sexual and individual differentiation. *Animal Biology* 55, 343–370.

Crews, D., Sakata, J., and Rhen, T. 1998. Developmental effects on intersexual and intrasexual variation in growth and reproduction in a lizard with temperature-dependent sex determination. *Journal of Comparative Physiology C* 119, 229–241.

Crump, M. 2005. *Headless Males Make Great Lovers*. Chicago, IL: University of Chicago Press.

Dall, S. R. X., Houston, A. I., and McNamara, J. M. 2004. The behavioural ecology of personality: consistent individual differences from an adaptive perspective. *Ecology Letters* 7, 734–739.

Dall, S. R. X., Giraldeau, L.-A., Olsson, O., McNamara, J. M., and Stephens, D. W. 2005. Information and its use by animals in evolutionary ecology. *Trends in Ecology and Evolution* 20, 187–193.

Danforth, B. N. and Desjardins, C. A. 1999. Male dimorphism in *Perdita portalis* (Hymenoptera, Andrenidae) has arisen from preexisting allometric patterns. *Insectes Sociaux* 46, 18–28.

Darwin, C. 1871. *The Descent of Man, and Selection in Relation to Sex*. London: John Murray.

Denno, R. F. 1994. The evolution of dispersal polymorphism in insects: the influence of habitats, host plants and mates. *Researches on Population Ecology* 36, 127–135.

Devlin, R. H. and Nagahama, Y. 2002. Sex determination and sex differentiation in fish: an overview of genetic, physiological, and environmental influences. *Aquaculture* 208, 191–364.

DeWitt, T. J., Sih, A., and Wilson, D. S. 1998. Costs and limits of phenotypic plasticity. *Trends in Ecology and Evolution* 13, 77–81.

Dieckmann, U. and Doebeli, M. 1999. On the origin of species by sympatric speciation. *Nature* 400, 354–357.

Dingemanse, N. J. and Reale, D. 2005. Natural selection and animal personality. *Behaviour* 142, 1159–1184.

Dingemanse, N. J., Both, C., Drent, P. J., van Oers, K., and van Noordwijk, A. J. 2002. Repeatability and heritability of exploratory behaviour in great tits from the wild. *Animal Behaviour* 64, 929–938.

Dingwall, R. and Aldridge, M. 2006. Television wildlife programming as a source of popular scientific information: a case study of evolution. *Public Understanding of Science* 15, 131–152.

Dominey, W. J. 1980. Female mimicry in male bluegill sunfish: a genetic polymorphism? *Nature* 284, 546–548.

Drent, P. J., van Oers, K., and van Noordwijk, A. J. 2003. Realized heritability of personalities in the great tit (*Parus major*). *Proceedings of the Royal Society of London B* 270, 45–51.

Duckworth, R. A. 2006. Behavioral correlations across breeding contexts provide a mechanism for a cost of aggression. *Behavioral Ecology* 17, 1011–1019.

Eadie, J., Sherman, P., and Semel, B. 1998. Conspecific brood parasitism, population dynamics, and the conservation of cavity-nesting birds. In T. Caro (ed.) *Behavioral Ecology and Conservation Biology*, pp. 306–340. Oxford, UK: Oxford University Press.

Eberhard, W. G. and Gutiérrez, E. E. 1991. Male dimorphisms in beetles and earwigs and the question of developmental constraints. *Evolution* 45, 18–28.

Ehlinger, T. J. and Wilson, D. S. 1988. Complex foraging polymorphism in bluegill sunfish. *Proceedings of the National Academy of Sciences of the United States of America* 85, 1878–1882.

Emlen, D. J. 1996. Artificial selection on horn length–body size allometry in the horned beetle *Onthophagus acuminatus*. *Evolution* 50, 1219–1230.

Emlen, D. J. 1997. Diet alters male horn allometry in the beetle *Onthophagus acuminatus* (Coleoptera: Scarabaeidae). *Proceedings of the Royal Society of London B* 264, 567–574.

Emlen, D. J. 2001. Costs and the diversification of exaggerated animal structures. *Science* 291, 1534–1536.

Emlen, D. J., Marangelo, J., Ball, B., and Cunningham, C. W. 2005. Diversity in the weapons of sexual selection: horn evolution in the beetle genus *Onthophagus* (Coleoptera: Scarabaeidae). *Evolution* 59, 1060–1084.

Fincke, O. M., Jodicke, R., Paulson, D. R., and Schultz, T. D. 2005. The evolution and frequency of female color morphs in Holarctic Odonata: why are male-like females typically the minority? *International Journal of Odontology* 8, 183–212.

Foote, C. J. 1988. Male mate choice dependent on male size in salmon. *Behaviour* 106, 63–80.

Foote, C. J. and Larkin, P. A. 1988. The role of male choice in the assortative mating of anadromous and non-anadromous sockeye salmon (*Oncorhynchus nerka*). *Behaviour* 106, 43–62.

ormica, V. A., Gonser, R. A., Ramsay, S., and Tuttle, E. M. 2004. Spatial dynamics of alternative reproductive strategies: the role of neighbors. *Ecology* **85**, 1125–1136.

rost, A. J., Winrow-Giffen, A., Ashley, P. J., and Sneddon, L. U. 2007. Plasticity in animal personality traits: does prior experience alter the degree of boldness? *Proceedings of the Royal Society of London B* **274**, 333–339.

uxa, J. R., Milks, M. L., Sokolova, Y. Y., and Richter, A. R. 2005. Interaction of an entomopathogen with an insect social form: an epizootic of *Thelohania solenopsae* (Microsporidia) in a population of the red imported fire ant, *Solenopsis invicta*. *Journal of Invertebrate Pathology* **88**, 79–82.

arant, D., Fontaine, P.-M., Good, S. P., Dodson, J. J., and Bernatchez, L. 2002. Influence of male parental identity on growth and survival of offspring in Atlantic salmon (*Salmo salar*). *Evolutionary Ecology Research* **4**, 537–549.

insberg, J. R. and Milner-Gulland, E. J. 1994. Sex-biased harvesting and population dynamics in ungulates: implications for conservation and sustainable use. *Conservation Biology* **8**, 157–166.

iraldeau, L.-A. and Livoreil, B. 1998. Game theory and social foraging. In L. Dugatkin and H. K. Reeve (eds.) *Game Theory and Animal Behavior*, pp. 16–37. Oxford, UK: Oxford University Press.

odwin, J. and Crews, D. 2002. Hormones, brain and behavior in reptiles. In D. W. Pfaff, A. P. Arnold, A. M. Etgen, S. E. Farbach, and R. T. Rubin (eds.) *Hormones, Brain and Behavior*, vol. 2, pp. 649–798. New York: Academic Press.

onçalves, D., Matos, M., Fagundes, T., and Oliveira, R. F. 2005. Bourgeois males of the peacock blenny, *Salaria pavo*, discriminate females from female-mimicking males? *Ethology* **111**, 559–572.

onçalves, E. J., Almada, V. C., Oliveira, R. F., and Santos, A. J. 1996. Female mimicry as a mating tactic in males of the blenniid fish *Salaria pavo*. *Journal of the Marine Biological Association of the United Kingdom* **76**, 529–538.

osling, S. D. 2001. From mice to men: what can we learn about personality from animal research? *Psychological Bulletin* **127**, 45–86.

osling, S. D. and John, O. P. 1999. Personality dimensions in nonhuman animals: a cross-species review. *Current Directions in Psychological Science* **8**, 69–75.

roothuis, T. G. G. and Carere, C. 2005. Avian personalities: characterization and epigenesis. *Neuroscience and Biobehavioral Reviews* **29**, 137–150.

Gross, M. R. 1982. Sneakers, satellites and parentals: polymorphic mating strategies in North American sunfishes. *Zeitschrift für Tierpsychologie* **60**, 1–26.

Gross, M. R. 1991a. Evolution of alternative reproductive strategies: frequency-dependent sexual selection in male bluegill sunfish. *Philosophical Transactions of the Royal Society of London B* **332**, 59–66.

Gross, M. R. 1991b. Salmon breeding behavior and life history evolution in changing environments. *Ecology* **72**, 1180–1186.

Gross, M. R. 1996. Alternative reproductive strategies and tactics: diversity within sexes. *Trends in Ecology and Evolution* **11**, 92–98.

Hamilton, I. M., Haesler, M. P., and Taborsky, M. 2006. Predators, reproductive parasites, and the persistence of poor males on leks. *Behavioral Ecology* **17**, 97–107.

Hardy, I. C. W. (ed.) 2002. *Sex Ratios: Concepts and Research Methods*. Cambridge, UK: Cambridge University Press.

Hartfelder, K. and Emlen, D. J. 2005. Endocrine control of insect polyphenism. In L. I. Gilbert, K. Iatrou, and S. S. Gill (eds.) *Comprehensive Molecular Insect Science*, vol. 3, *Endocrinology*, pp. 651–703. Boston, MA: Elsevier.

Hazel, W., Smock, R., and Lively, C. M. 2004. The ecological genetics of conditional strategies. *American Naturalist* **163**, 888–900.

Hendry, A. P., Wenburg, J. K., Bentzen, P., Volk, E. C., and Quinn, T. P. 2000. Rapid evolution of reproductive isolation in the wild: evidence from introduced salmon. *Science* **290**, 516–518.

Henson, S. A. and Warner, R. R. 1997. Male and female alternative reproductive behaviors in fishes: a new approach using intersexual dynamics. *Annual Review of Ecology and Systematics* **28**, 571–592.

Hessing, M. J. C., Hagelso, A. M., Schouten, W. G. P., Wiepkema, P. R., and van Beek, J. A. M. 1994. Individual behavioral and physiological strategies in pigs. *Physiology and Behavior* **55**, 39–46.

Hochberg, M. E., Sinervo, B., and Brown, S. P. 2003. Socially mediated speciation. *Evolution* **57**, 154–158.

Hofmann, H. A. 2003. Functional genomics of neural and behavioral plasticity. *Journal of Neurobiology* **54**, 272–282.

Hori, M. 1993. Frequency-dependent natural selection in the handedness of scale-eating cichlid fish. *Science* **260**, 216–219.

Huntingford, F. A. 1976. Relationship between anti-predator behaviour and aggression among conspecifics in threespined stickleback, *Gasterosteus aculeatus*. *Animal Behaviour* **24**, 245–260.

Jansson, H. and Ost, T. 1997. Hybridization between Atlantic salmon (*Salmo salar*) and brown trout (*S. trutta*) in a restored section of the River Dalalven, Sweden. *Canadian Journal of Fisheries and Aquatic Sciences* 54, 2033–2039.

Jonsson, B. and Jonsson, N. 2001. Polymorphism and speciation in Arctic charr. *Journal of Fish Biology* 58, 605–638.

Judson, O. 2002. *Dr. Tatiana's Sex Advice to All Creation*. New York: Henry Holt.

Kim, S. J., Ogasawara, K., Park, J. G., Takemura, A., and Nakamura, M. 2002. Sequence and expression of androgen receptor and estrogen receptor gene in the sex types of protogynous wrasse, *Heliochoeres trimaculatus*. *General and Comparative Endocrinology* 127, 165–173.

Koolhaas, J. M., Korte, S. M., De Boer, S. F., et al. 1999. Coping styles in animals: current status in behavior and stress-physiology. *Neuroscience and Biobehavioral Reviews* 23, 925–935.

Koseki, Y. and Fleming, I. A. 2006. Spatio-temporal dynamics of alternative male phenotypes in coho salmon populations in response to ocean environment. *Journal of Animal Ecology* 75, 445–455.

Kotiaho, J. S. and Tomkins, J. L. 2001. The discrimination of alternative male morphologies. *Behavioral Ecology* 12, 553–557.

Kralj-Fiser, S., Scheiber, I. B. R., Blejec, A., Moestl, E., and Kotrschal, K. 2007. Individualities in a flock of free-roaming greylag geese: behavioral and physiological consistency over time and across situations. *Hormones and Behavior* 51, 239–248.

Kurdziel, J. P. and Knowles, L. L. 2002. The mechanisms of morph determination in the amphipod *Jassa*: implications for the evolution of alternative male phenotypes. *Proceedings of the Royal Society of London B* 269, 1749–1754.

Lande, R. 1982. Rapid origin of sexual isolation and character divergence in a cline. *Evolution* 36, 213–223.

Lande, R. and Arnold, S. J. 1983. The measurement of selection on correlated characters. *Evolution* 37, 1210–1226.

Leimar, O., Hammerstein, P., and Van Dooren, J. M. 2006. A new perspective on developmental plasticity and the principles of adaptive morph determination. *American Naturalist* 167, 367–376.

LeVay, S. 1996. *Queer Science: The Use and Abuse of Research into Homosexuality*. Cambridge, MA: MIT Press.

Lorenz, K. 1935. Der Kumpan in der Umwelt des Vogels. *Journal für Ornithologie* 83, 137–215; 289–413.

Lu, G. Q. and Bernatchez, L. 1999. Correlated trophic specialization and genetic divergence in sympatric lake whitefish ecotypes (*Coregonus clupeaformis*): support for the ecological speciation hypothesis. *Evolution* 53, 1491–1505.

Mallet, J. 2007. Hybrid speciation. *Nature* 446, 279–283.

Meaney, M. J. 2001. Maternal care, gene expression, and the transmission of individual differences in stress reactivity across generations. *Annual Review of Neuroscience* 24, 1161–1192.

Meyer, A. 1987. Phenotypic plasticity and heterochrony in *Cichlasoma managuense* (Pisces, Cichlidae) and their implications for speciation in cichlid fishes. *Evolution* 41, 1357–1369.

Moore, M. C. 1991. Application of organization–activation theory to alternative male reproductive strategies: a review. *Hormones and Behavior* 25, 154–179.

Moore, M. C., Hews, D. K., and Knapp, R. 1998. Hormonal control and evolution of alternative male phenotypes: generalizations of models for sexual differentiation. *American Zoologist* 38, 133–151.

Morris, D. 1952. Homosexuality in the ten-spined stickleback (*Pygosteus pungitus* L.). *Behaviour* 4, 233–261.

Neff, B. D. 2004. Increased performance of offspring sired by parasitic males in bluegill sunfish. *Behavioral Ecology* 15, 327–331.

Nelson, R. J. 2005. *An Introduction to Behavioral Endocrinology*, 3rd edn. Sunderland, MA: Sinauer Associates.

Nesse, R. M. and Williams, G. C. 1998. Evolution and the origins of disease. *Scientific American* 279, 86–93.

Oi, D. H. 2006. Effect of mono- and polygyne social forms on transmission and spread of a microsporidium in fire ant populations. *Journal of Invertebrate Pathology* 92, 146–151.

Oi, D. H., Valles, S. M., and Pereira, R. M. 2004. Prevalence of *Thelohania solenopsae* (Microsporidae: Thelohaniidae) infection in monogyne and polygyne red imported fire ant (Hymenoptera: Formicidae). *Environmental Entomology* 33, 340–345.

Olendorf, R., Rodd, F. H., Punzalan, D., et al. 2006. Frequency-dependent survival in natural guppy populations. *Nature* 441, 633–636.

Oliveira, R. F. 2006. Neuroendocrine mechanisms of alternative reproductive tactics in fish. In K. A. Sloman, R. W. Wilson, and S. Balshine (eds.) *Fish Physiology*, vol. 24, *Behavior and Physiology of Fish*, pp. 297–357. New York: Elsevier.

Oliveira, R. F., Ros, A. F. H., and Gonçalves, D. M. 2005. Intra-sexual variation in male reproduction in teleost fish: a comparative approach. *Hormones and Behavior* 48, 430–439.

arker, P. G. and Waite, T. 1997. Mating systems, effective population size, and conservation of natural populations. In J. Clemmons and R. Buchholz (eds.) *Behavioral Approaches to Conservation in the Wild*, pp. 243–261. Cambridge, UK: Cambridge University Press.

igeon, D., Chouinard, A., and Bernatchez, L. 1997. Multiple modes of speciation involved in the parallel evolution of sympatric morphotypes of lake whitefish (*Coregonus clupeaformis*, Salmonidae). *Evolution* 51, 196–205.

laistow, S., Johnstone, R. A., Colegrave, N., and Spencer, M. 2004. Evolution of alternative mating tactics: conditional versus mixed strategies. *Behavioral Ecology* 15, 534–542.

reston, C. A., Fritz, G. N., and Vander Meer, R. K. 2007. Prevalence of *Thelohania solenopsae* infected *Solenopsis invicta* newly mated queens within areas of differing social form distributions. *Journal of Invertebrate Pathology* 94, 119–124.

epka, J. and Gross, M. R. 1995. The evolutionarily stable strategy under individual condition and tactic frequency. *Journal of Theoretical Biology* 176, 27–31.

iechert, S. E. and Hedrick, A. V. 1993. A test for correlations among fitness-linked behavioral traits in the spider *Agelenopsis aperta* (Aranae, Agelenidae). *Animal Behaviour* 46, 669–675.

obertson, H. 1985. Female dimorphism and mating behaviour in a damselfly, *Ischnura ramburi*: females mimicking males. *Animal Behaviour* 33, 805–809.

off, D. A. 1994. Evidence that the magnitude of the trade-off in a dichotomous trait is frequency dependent. *Evolution* 48, 1650–1656.

off, D. A. 1996. The evolution of threshold traits in animals. *Quarterly Review of Biology* 71, 3–35.

off, D. A. 1998. The maintenance of phenotypic and genetic variation in threshold traits by frequency-dependent selection. *Journal of Evolutionary Biology* 11, 513–529.

off, D. A. and Fairbairn, D. J. 1993. The evolution of alternate morphologies: fitness and wing morphology in male sand crickets. *Evolution* 47, 1572–1584.

oss, K. G. and Keller, L. 1995. Ecology and evolution of social organization: insights from fire ants and other highly eusocial insects. *Annual Review of Ecology and Systematics* 26, 631–656.

owland, J. M. and Qualls, C. R. 2005. Likelihood models for discriminating alternative phenotypes in morphologically dimorphic species. *Evolutionary Ecology Research* 7, 421–434.

Rowland, J. M., Qualls, C. R., and Beaudoin-Ollivier, L. 2005. Discrimination of alternative male phenotypes in *Scapanes australis* (Boisduval) (Coleoptera: Scarabaeidae: Dynastinae). *Australian Journal of Entomology* 44, 22–28.

Ruchon, F., Laugier, T., and Quignard, J. P. 1995. Alternative male reproductive strategies in the peacock blenny. *Journal of Fish Biology* 47, 826–840.

Russell, S. T., Ramnarine, I. W., Mahabir, R., and Magurran, A. E. 2006. Genetic detection of sperm from forced copulations between sympatric populations of *Poecilia reticulata* and *Poecilia picta*. *Biological Journal of the Linnean Society* 88, 397–402.

Ryan, M. J., Pease, C. M., and Morris, M. R. 1992. A genetic polymorphism in the swordtail, *Xiphophorus nigrensis*: testing the prediction of equal fitnesses. *American Naturalist* 139, 21–31.

Salzburger, W., Baric, S., and Sturmbauer, C. 2002. Speciation via introgressive hybridization in East African cichlids? *Molecular Ecology* 11, 619–625.

Seehausen, O. 2004. Hybridization and adaptive radiation. *Trends in Ecology and Evolution* 19, 198–207.

Seehausen, O. and Schluter, D. 2004. Male–male competition and nuptial-colour displacement as a diversifying force in Lake Victoria cichlid fishes. *Proceedings of the Royal Society of London B* 271, 1345–1353.

Seymour, N. R. 1990. Forced copulation in sympatric American black ducks and mallards in Nova Scotia. *Canadian Journal of Zoology/Revue Canadienne de Zoologie* 68, 1691–1696.

Shuster, S. M. 1989. Male alternative reproductive strategies in a marine isopod crustacean (*Paracerceis sculpta*): the use of genetic markers to measure differences in fertilization success among a-, B-, and g-males. *Evolution* 43, 1683–1698.

Shuster, S. M. and Wade, M. J. 2003. *Mating Systems and Strategies*. Princeton, NJ: Princeton University Press.

Sih, A., Bell, A., and Johnson, J. C. 2004a. Behavioral syndromes: an ecological and evolutionary overview. *Trends in Ecology and Evolution* 19, 372–378.

Sih, A., Bell, A. M., Johnson, J. C., and Ziemba, R. E. 2004b. Behavioral syndromes: an integrative overview. *Quarterly Review of Biology* 79, 241–277.

Simmons, L. W. and Emlen, D. J. 2006. Evolutionary trade-off between weapons and testes. *Proceedings of the National Academy of Sciences of the United States of America* 103, 16346–16351.

Sinervo, B. and Calsbeek, R. 2006. The developmental, physiological, neural, and genetical causes and consequences of frequency-dependent selection in the wild. *Annual Review of Ecology Evolution and Systematics* 37, 581–610.

Sinervo, B. and Zamudio, K. 2001. The evolution of alternative reproductive strategies: fitness differential, heritability, and genetic correlation between the sexes. *Journal of Heredity* 92, 198–205.

Sirot, L. K., Brockmann, H. J., Marinis, C., and Muschett, G. 2003. Maintenance of a female-limited polymorphism in *Ischnura ramburi* (Zygoptera: Coenagrionidae). *Animal Behaviour* 66, 763–775.

Skúlason, S. and Smith, T. B. 1995. Resource polymorphisms in vertebrates. *Trends in Ecology and Evolution* 10, 366–370.

Smith, T. B. and Skúlason, S. 1996. Evolutionary significance of resource polymorphisms in fish, amphibians and birds. *Annual Review of Ecology and Systematics* 27, 111–134.

Snorrason, S. S. and Skúlason, S. 2004. Adaptive speciation in northern freshwaterfishes: patterns and processes. In U. Dieckmann, H. Metz, M. Doebeli, and D. Tautz (eds.) *Adaptive Speciation*, pp. 210–228. Cambridge, UK: Cambridge University Press.

Stamps, J. 2003. Behavioural processes affecting development: Tinbergen's fourth question comes of age. *Animal Behaviour* 66, 1–13.

Taborsky, M. 1994. Sneakers, satellites, and helpers: parasitic and cooperative behavior in fish reproduction. *Advances in the Study of Behavior* 23, 1–100.

Taborsky, M. 1998. Sperm competition in fish: "bourgeois" males and parasitic spawning. *Trends in Ecology and Evolution* 13, 222–227.

Taborsky, M. 2001. The evolution of parasitic and cooperative reproductive behaviors in fishes. *Journal of Heredity* 92, 100–110.

Thomas, C. D., Bagguette, M., and Lewis, O. T. 2000. *Butterfly movement and conservation in patchy landscapes*. In L. M. Gosling and W. J. Sutherland (eds.) *Behaviour and Conservation*, pp. 85–104. Cambridge, UK: Cambridge University Press.

Thorpe, J. E. 1989. Developmental variation in salmonid populations. *Journal of Fish Biology* 35 (Suppl. A), 295–303.

Tinbergen, N. 1951. *The Study of Instinct*. Oxford, UK: Clarendon Press.

Tomkins, J. L., Kotiaho, J. S., and LeBas, N. R. 2005. Matters of scale: positive allometry and the evolution of male dimorphisms. *American Naturalist* 165, 389–402

Tschinkel, W. R. 1998. The reproductive biology of fire ant societies. *BioScience* 48, 593–605.

Turner, B. J. and Grosse, D. J. 1980. Trophic differentiation in *Ilyodon*, a genus of stream-dwelling goodeid fishes: speciation versus ecological polymorphism. *Evolution* 34, 259–270.

van Oers, K., Drent, P. J., de Goede, P., and van Noordwijk, A. J. 2004. Realized heritability and repeatability of risk-taking behaviour in relation to avian personalities. *Proceedings of the Royal Society of London B* 271, 65–73.

van Oers, K., de Jong, G., van Noordwijk, A. J., Kempenaers, B., and Drent, P. J. 2005. Contribution of genetics to the study of animal personalities: a review of case studies. *Behaviour* 142, 1185–1206.

van Oortmerssen, G. A. 1971. Biological significance, genetics and evolutionary origin of variability in behaviour within and between inbred strains of mice (*Mus musculus*): behaviour genetic study. *Behaviour* 38, 1–92.

van Oortmerssen, G. A. and Bakker, T. C. M. 1981. Artificial selection for short and long attack latencies in wild *Mus musculus domesticus*. *Behavior Genetics* 11, 115–126.

Verbeek, M. E. M., Drent, P. J. and Wiepkema, P. R. 1994. Consistent individual differences in early exploratory behaviour of male great tits. *Animal Behaviour* 48, 1113–1121.

Vercken, E., Massot, M., Sinervo, B. and Clobert, J. 2007. Colour variation and alternative reproductive strategies in females of the common lizard *Lacerta vivipara*. *Journal of Evolutionary Biology* 20, 221–232.

Vigers, R. S., Silversides, D. W., and Tremblay, J. J. 2005. New insights into the regulation of mammalian sex determination and male sex differentiation. *Vitamins and Hormones: Advances in Research and Applications* 70, 387–413.

Vincent, A. and Sadovy, Y. 1998. Reproductive ecology in the conservation and management of fishes. In T. Caro (ed.) *Behavioral Ecology and Conservation Biology*, pp. 209–245. Oxford, UK: Oxford University Press.

Watters, J. V. 2005. Can the alternative male tactics "fighter" and "sneaker" be considered "coercer" and "cooperator" in coho salmon? *Animal Behaviour* 70, 1055–1062

West-Eberhard, M. J. 1983. Sexual selection, social competition, and speciation. *Quarterly Review of Biology* 58, 155–183.

West-Eberhard, M. J. 1989. Phenotypic plasticity and the origins of diversity. *Annual Review of Ecology and Systematics* 20, 249–278.

West-Eberhard, M. J. 2003. *Developmental Plasticity and Evolution*. Oxford, UK: Oxford University Press.

Williams, D. F., Oi, D. H., Porter, S. D., Pereira, R. M., and Briano, J. A. 2003. Biological control of imported fire ants (Hymenoptera: Formicidae). *American Entomologist* **49**, 150–163.

Wilson, D. S. 1998. Adaptive individual differences within single populations. *Philosophical Transactions of the Royal Society of London B* **353**, 199–205.

Wilson, D. S., Coleman, K., Clark, A. B., and Biederman, L. 1993. Shy–bold continuum in pumpkinseed sunfish (*Lepomis gibbosus*): an ecological study of a psychological trait. *Journal of Comparative Psychology* **107**, 250–260.

Wilson, D. S., Clark, A. B., Coleman, K., and Dearstine, T. 1994. Shyness and boldness in humans and other animals. *Trends in Ecology and Evolution* **9**, 442–446.

Wirtz, P. 1999. Mother species–father species: unidirectional hybridization in animals with female choice. *Animal Behaviour* **58**, 1–12.

Zander, C. D. 1986. Blenniidae. In P. J. P. Whitehead, M. L. Bauchot, J.-C. Hureau, J. Nielsen, and E. Tortonese (eds.) *Fishes of the North-Eastern Atlantic and the Mediterranean*, vol. 3, pp. 1096–1112. Paris: UNESCO.

Zera, A. J. and Denno, R. F. 1997. Physiology and ecology of dispersal polymorphism in insects. *Annual Review of Entomology* **42**, 207–230.

Zera, A. J. and Harshman, L. G. 2001. The physiology of life history trade-offs in animals. *Annual Review of Ecology and Systematics* **32**, 95–126.

Zupanc, G. K. H. and Lamprecht, J. 2000. Towards a cellular understanding of motivation: structural reorganization and biochemical switching as key mechanisms of behavioral plasticity. *Ethology* **106**, 467–477.

Index of species

Subject index

Printed in the United States
by Baker & Taylor Publisher Services